The Toledo Museum of Art

# ROMAN MOLD-BLOWN GLASS

THE TOLEDO MUSEUM OF ART

# Roman Mold-blown Glass

THE FIRST THROUGH SIXTH CENTURIES

E. Marianne Stern

« L'ERMA » di BRETSCHNEIDER
IN ASSOCIATION WITH THE TOLEDO MUSEUM OF ART

Research and manuscript preparation support
for this catalogue was provided
by The Andrew W. Mellon Foundation.
The catalogue has been published with the assistance of
The Getty Grant Program; the National Endowment for the Arts,
a federal agency; the Margaret H. Levis Publications Fund;
and Sotheby's.

First Edition


Monroe Street at Scottwood Avenue
P. O. Box 1013 Toledo, Ohio 43697, telephone 419-255-8000



Published by "L'Erma" di Bretschneider, Rome, Italy

Project Supervisor and Editor: Kurt T. Luckner
Managing Editor: Sandra E. Knudsen
Photography Assistants: Elizabeth M. Barry and Timothy A. Motz
Photography: Larry Rosenberg, Sylvania, Ohio; Photography, Inc., Toledo,
Ohio; and Tim Thayer, Oak Park, Michigan
Drawings: Lorene Sterner, Ann Arbor, Michigan,
and Timothy A. Motz, Toledo, Ohio

ISBN 88-7062-916-3

Black-and-white photographs and color transparencies of all the objects in this
catalogue are available from the Registrar, The Toledo Museum of Art.

# CONTENTS

# II



# III

# IV

# Foreword

The Toledo Museum of Art has one of the largest, most extensive, and most varied collections of Roman glass vessels and objects from the eastern Mediterranean currently housed in any museum. The importance of the Museum's collection lies in its holdings of table and storage wares of Syro-Palestinian origin, in terms of sheer numbers of intact examples, diversity of forms, longevity of these forms, and the numerous manufacturing techniques and decorative types represented. The collection includes groups of closely related objects that approach the encyclopedic in their completeness. For example, in this catalogue the Hexagonal Bottles with High Relief (Cat. Nos. 15-47) and the Date-shaped Bottles (Cat. Nos. 84-107) are prime instances where the ability to analyze a large group in one place has produced important new information.

This catalogue covers 193 objects of full-size mold-blown glass made between the early first century A.D. and the early seventh century. Author E. Marianne Stern's exceptional research has led her to propose new interpretations of ancient glass technology and dating. Her treatment of the material is exhaustive, covering not only glass-blowing, mold-making, and mold-blowing techniques but also aspects of decoration. The rigorous descriptions of the objects continues Toledo's commitment to setting an example for a new standard in ancient glass studies. Furthermore, she provides the reader with new insights into these vessels: what they contained, who bought them, and the critical distinction between where they were made and where they were found.

Most general art museums exhibit at least a few examples of Roman glass. But these are frustrating to describe, because we know simultaneously so much and so little about them. On the one hand, aesthetic admiration for such accidents of time as iridescence (imitated with great skill by glass artists of the early part of the twentieth century including Émile Gallé and Louis Comfort Tiffany; see also *Glossaries*) still obscures understanding of what Roman glass actually looked like. On the other hand, it is very exciting that we can pinpoint with great exactness when it was first discovered that glass can be inflated: about 90 B.C.; and where: along the eastern Mediterranean coast in or near Sidon, an independent city state allied to the expanding Roman empire. Although the new technique may have been guarded as a workshop secret for a while, the consequences of the invention cannot be overstated, because it meant that glass could be quickly and cheaply formed, anywhere, into tablewares and storage containers. Within a century vessels and objects made of glass ceased to be expensive luxuries and had come within the grasp of even the average household. The famous Roman natural historian Pliny the Elder (A.D. 23-79) witnessed the transition and marveled at it.[1]

In art museums objects of daily life — glass, pottery, fabric, metal — surviving from the centuries when ancient Rome ruled the known world are often exhibited only sketchily. At the Toledo Museum, however, the collection of glass was begun by the Museum's founder, glass industrialist Edward Drummond Libbey (1854-1925). He saw that a collection representing the history and art of glass would be important to the citizens of a community where glass was long the major industry. Determined to form a comprehensive collection tracing the entire history of glass, Mr. Libbey was able in a remarkably short period to acquire a series of notable existing collections. Among these was Mr. Libbey's greatest coup, the acquisition of the comprehensive collection of ancient and Islamic glass formed by Thomas E. H. Curtis, which in one astute move elevated Toledo's glass collection to international importance.[2]

The original Libbey gifts have been selectively enhanced by accessions to fill gaps or add luster to the existing material.

[1] *Natural History*, 5.76; 36.190-194. See below, *Glossaries: glass making* and Chapter I: *The First Century: The Role of Sidon.*

[2] On Thomas E. H. Curtis and his collection of glass: Grose 1989, 13, 17-20.

Toledo's ancient glass has been cared for by successive curators, notably Nell Jaffe Baer, Assistant to Directors Blake-More Godwin and Otto Wittmann 1919-58, and Rudolph Riefstahl, Curator of Decorative Arts (Medieval and Renaissance) 1958-71. In 1976-77 David F. Grose was the Museum's first Curator of Glass.[3] Since his employment as Curator of Ancient Art in 1969, Kurt T. Luckner has reinstalled the entire collection as well as acquiring significant examples. Thanks to Mr. Libbey's original vision and to continuing community interest in glass, the Museum's entire collection of almost seven thousand objects is on display. About two thousand of the finest works trace the development of glass from the Bronze Age to the present in the Art in Glass Gallery, inaugurated in 1970 as the gift of Mr. and Mrs. Harold Boeschenstein. And in 1976 matching funds from the National Endowment for the Arts and Libbey-Owens-Ford Inc. in memory of its past president made possible the John D. Biggers Glass Study Room, which exhibits the balance of the collection.

The intact examples of glass wares in museums around the world come mostly from votive or funerary deposits, but the vast bulk of the surviving evidence is fragmentary. Glass was broken, swept up, and discarded in Roman villas, temples, forts, and houses from Britain to Egypt and from Ukraine to Tunisia. Recently a handful of glass-working sites has been investigated, where broken cullet has been found layered knee-deep.[4] The Toledo collection, however, is notable because most of its ancient glass is complete or intact. We hope this and future catalogues of Toledo's well-preserved vessels will enable scholars to better interpret fragments of similar vessels elsewhere.

A catalogue of this international scope and cross-disciplinary focus is the product of the collaborative efforts of many people of different skills and interests in and outside the Museum. Many individuals are recognized in the Acknowledgments, but the extraordinary efforts of a few deserve special notice here. Kurt T. Luckner, Curator of Ancient Art, has supervised the ancient glass collections from the moment of his arrival in 1969, and the ancient glass catalogues would never have been possible without his persistence. In 1977 David F. Grose first proposed the concept of publishing all of the Museum's glass from antiquity to the present as a long-term research and publication project. The 1982 meeting of the International Association for the History of Glass in New York City; Corning, New York; and Toledo inspired the Museum to reevaluate the installation and labeling of its entire glass collection. E. Marianne Stern was hired in 1986 to write this catalogue and during her tenure as Guest Curator of Roman Glass until October 1992 created an exciting ferment of activity that brought together the experiments, knowledge, and energy of specialists from archaeologists to philologists to ornithologists to glass blowers, all focused on the objects in the Roman glass collection. Sandra E. Knudsen, Coordinator of Publications, took the editorial lead in shaping the catalogue manuscript, whose substance and detail owe much to her scholarly attention. Janet D. Jones and Timothy A. Motz devoted months to the painstaking research vital to preparing a finished manuscript. Donna Niehous compiled the index.

The draft manuscript of this book won the 1991 « L'Erma » di Bretschneider publication award. Held only every two, years, this prestigious contest attracts manuscripts from around the world in the fields of prehistoric, classical, oriental, and medieval archaeology and was juried by a panel of twelve distinguished archaeologists and art historians. The award is two-fold: a cash prize and a guarantee that the manuscript would be published at the expense of « L'Erma » di Bretschneider. The Museum was thus able to reduce its publication subvention in order to lower the retail price.

This catalogue could never have been achieved without the considerable financial assistance provided by the Andrew W. Mellon Foundation to support research on the collections and manuscript preparation. Foremost among the Museum's desires was systematic publication of the glass collection; the Mellon grant was therefore applied toward the preparation of several catalogues in different fields of glass studies. Only this farsighted grant enabled us to initiate and sustain a project of this magnitude and complexity. The result here is the most comprehensive work on Roman mold-blown glass ever written. It is a landmark accomplishment that will become an invaluable resource for reference and for future research in the field. Assistance with the color plates came from the copublisher, « L'Erma » di Bretschneider, which paid for twelve of the new transparencies. Publication assistance also came from the National Endowment for the Arts, a federal agency, Sotheby's, the Getty Grant Program, and the Margaret H. Levis Publications Fund.

David W. Steadman
Director

[3] Dr. Grose has continued his association with the Toledo Museum, publishing *The Toledo Museum of Art. Early Ancient Glass: Core-Formed, Rod-Formed, and Cast Vessels and Objects from the Late Bronze Age to the Early Roman Empire, 1600 B.C.-A.D. 50* (New York 1989), the comprehensive catalogue of 713 vessels, fragments, and objects in the pre-blown ancient glass collection, and assisting in important ways as a reviewer of this volume and as an advisor on the completion of future volumes on the collection of free-blown and expanded mold-blown Roman and early Byzantine glass.

[4] Most notably Jalame, in western Galilee, Israel, a site whose glass furnace has been precisely dated in the late fourth century: Weinberg 1988 and Stern 1992a.

# Acknowledgments

In late 1985 Kurt T. Luckner, Curator of Ancient Art, approached me about writing a catalogue of The Toledo Museum of Art's collection of Roman and Byzantine blown glass. After our initial discussions about the project, he telephoned me in Amsterdam just hours after I had returned from India where I had been investigating the role of glass in the early Roman India trade. The Museum had decided to use a substantial grant from the Andrew W. Mellon Foundation to support the preparation of publications focusing on the Museum's glass collections. I joined the staff of The Toledo Museum of Art as Guest Curator of Roman Glass in December 1986. My thanks go to Kurt T. Luckner and Roger Mandle, former Director, who offered me this position.

The completed manuscript was submitted to the Museum's Publications Office in August 1992. After my departure from the Museum, the Publications Office compiled the bibliography, incorporated outside readers' comments, transferred my charts to the computer, obtained most of the comparison photographs, drafted captions for the figures, and added a map. The Publications Office also composed Chapter IV. The entries for Chapter IV are based on notes I made between 1988 and 1990, with the exception of the Remarks for Cat. Nos. 189-192, which are the work of the Museum. While readying the manuscript for typesetting, a penultimate draft of Chapter III and Cat. Nos. 138-168 was conflated with the final manuscript. I had no opportunity before final printing to review this chapter in depth. Chapter IV and Cat. Nos. 189-192 were also left unchanged.

Major projects such as this are always a collaboration among many people, so I wish to acknowledge and thank those who helped me: Janet D. Jones, Andrew Kosak, Karen E. Lichtenbaum, and Timothy A. Motz helped with matters as diverse as complex library research and entering data into the project's computer. Ann F. Mather, Cathy Wirzylo, Alix Bowen, Becky Oles, Kathy Schwartz, John Henry Fullen, Erik W. Longton, and Timothy A. Motz typed and formatted the manuscript through several changes of software; Barbara Broer and Ann O'Leary organized project volunteers, several of whom deserve special recognition: Helen Roberts, Heide Klein, Joan Hassel, Carol Booth, Carol Kheiry, and Gretchen Vandenberghe for assistance with an astounding range of tasks such as filing, sorting, photocopying, binding photocopies, and generating index cards for a variety of research topics.

I wish to express my deep appreciation to those members of the staff of The Toledo Museum of Art who initially approved and oversaw the project to its completion: Roger Mandle, former Director; David W. Steadman, Director; Roger M. Berkowitz, Deputy Director; Robert F. Phillips, Chairman of the Curatorial Department; and Kurt T. Luckner, Curator of Ancient Art, who served as Project Director and Editor; to Sandra E. Knudsen, Coordinator of Publications, who supervised all administrative and editorial aspects as well as photography and drawings. Jeffrey Boyer, Technical Assistant, was always at hand to bring whatever objects were needed and to assist in cleaning and restoration. Elizabeth M. Barry, Publications Assistant, oversaw details of new photography. Darlene Lindner, Administrative Curatorial Department Assistant, and her efficient corps of clerk typists. Patricia J. Whitesides, Registrar; Steve Nowak, Associate Registrar; and Lee Mooney, Registrarial Secretary, were helpful providing access to the ancient glass during the Museum's lengthy renovation project. Anne O. Morris, Librarian; Joan Sepessy and Judith Friebert, Assistant Librarians; and Margaret Buhl, Library Assistant, answered queries. George Hartman, Superintendent, Facilities Services, Harlan Horner, and Phil Johnston, Facility Services, who saw that my office was always clean and comfortable. John S. Stanley, Assistant Director for Operations, advised in the choice of a computer for the project. Sharon Scott, Cataloguer, explored usages of the AskSam database program. Vivian Fitzgerald and Marilyn Mavis; William Hutton, Senior Curator, Tom Loeffler, Technical Assistant, and Steven Frushour, Associate Coordinator of Exhibitions, for their interest in the project. Steve Flower, former Comptroller, and Carol Bintz, Comptroller, oversaw the financial aspects of the project. Joan Leiter and Nancy Hasselkus provided notary services. Nancy Dermeyer, Carol Capelletti, and Pat Lark attended to numerous shipping and receiving requests. Mykk Gauger, Administrative Assistant to the Director, Robert J. Oates, Manager of Human Resources, and Diann Tohle, Assistant Manager of Human Resources, provided advice and assistance whenever it was sought.

Three colleagues merit special mention. Kurt T. Luckner invited me to undertake this project and with a mixture of bubbling enthusiasm and ever new challenges he directed its development. As the Toledo Museum's Curator of Ancient Art, he acquired some of the finest Roman glass in the collection. Sandra E. Knudsen brought professional insights and editorial expertise to the project. She not only carried administrative and editorial responsibilities, she also attended to such diverse matters as obtaining visas, an alien registration card, and housing, and introducing me to the American way of life. Timothy A. Motz began work as a library research assistant, patiently fielding innumerable requests for publications from the University of Michigan's Graduate Library in Ann Arbor. He became the publications assistant for the last two years of manuscript preparation and checked all bibliographic references, researched special topics, compiled the bibliography, ordered and prepared comparison illustrations, and designed the charts. He also supervised the work of volunteer Shantanu Agrawal, who compiled the maps, and development of the database for the index with Donna Niehous.

Carl J. Schulz and Raymond M. Sess, of Photography, Inc. Toledo, made most of the photographs. The color plates and some black-and white photographs were the work of Tim Thayer, Oak Park, Michigan. Some of the black-and-white comparison photographs benefited from reprinting or retouching by Timothy A. Motz.

Lorene Sterner, Ann Arbor, Michigan, served as technical draftsperson and created the drawings for Cat. Nos. 1-137 and 169-193. Timothy A. Motz made the drawings for Cat. Nos. 138-168 and assembled the complex charts of mold generation relationships. Maureen Blake, Lois Dorfman, Timothy A. Motz, and I made photographs in the Toledo Museum's Glass Studio that are included in the introductory essays and glossaries.

An important aspect of this catalogue is the analysis of ancient glass-blowing techniques. The Toledo Museum of Art's Glass Studio provided the unique opportunity to experiment in re-creating these techniques. My thanks go to Rose M. Glennon, former Chair, Museum Education, who initially enrolled me in the Museum's glass-blowing program; to Claire Schaefer, Acting Chair, Museum Education, Creative Arts Education, whose constant interest and support encouraged me to organize a Toledo Seminar in Experimental Archaeology (April 1992); and to Kathleen M. McCarthy and Leonard Marty, Supervisors of the Museum's Glass Studio, for instruction. Kathleen McCarthy's outstanding qualities as a teacher inspired me first to persevere in learning and then to pursue new adventures in this physically demanding and absorbing art. Her lively interest in the progress of the catalogue was a continuing source of inspiration. We discussed endless possibilities of ancient glass-making processes while she helped me with experiments. She appears in many of the photographs which demonstrate glassblowing techniques. Through the years, Jeffrey Boyer shared my glass-blowing bench; without his assistance, coping with the weight and mass of the hot glass would have exceeded my own weight and mass.

Outside Toledo, many specialists made important contributions. Gladys Davidson Weinberg, Columbia, Missouri, was the first to recommend me to the Museum for this project and was a constant source of encouragement and advice. She read many parts of the manuscript in various stages of completion and offered invaluable suggestions. Many friends and colleagues in the fields of ancient and later glass provided advice, information, off prints, and photographs: Heidi Amrein, Lausanne; Susan H. Auth, Newark; Dan P. Barag, Jerusalem; Simonetta Biaggio-Simona, Locarno; Dela von Boeselager, Cologne; Robert H. Brill, Corning; Pilar Caldera de Castro, Merida; Carina Calvi, Padova; Üzlifat Canav-Özgümüs, Istanbul; Barbara Czurda, Vienna; Emel Erten Yagci, Ankara; Wendy Evans, London; Daniele Foy, Aix-en-Provence; Sylvia Fünfschilling, Augst; David F. Grose, Amherst; Donald B. Harden, London; Anne Hochuli-Gysel, Cestas; Clasina Isings, Utrecht; Gusta Jacobson-Lehrer, Tel Aviv; Rosemarie Lierke, Frankfurt am Main; Sophia M. E. van Lith, Amsterdam; Maria Manzari Cohen, New York; Gioia Meconcelli Notarianni, Bologna; Martine Newby, London; Roy G. Newton, Sheffield; Mary R. De Maine, Minneapolis; Lucia Pirzio Biroli Stefanelli, Rome; Jennifer Price, Durham; Helmut Ricke, Düsseldorf; Ursula Riemenschneider, Altenessen; Pieter C. Ritsema van Eck, Amsterdam; Andrea Rottloff, Gersthofen; Beat Rütti, Augst; Axel von Saldern, formerly Hamburg; Lucia A. Scatozza Höricht, Rome; Nina P. Sorokina, Moscow; Marian Wenzel, Worcester; David F. Whitehouse, Corning; Kenneth M. Wilson, Dearborn and Punta Gorda.

Other museum curators, archaeologists, archivists, philologists, and librarians who provided useful information and encouragement are: Maxwell Anderson, Atlanta; Dietrich Boschung, Munich; Simonetta Bonomi, Adria; Hermann and Andrea Büsing, Bochum; Brian F. Cook, London; Kimbeth Coventry, Chicago; J. Derksen, Amsterdam; Gideon Foerster, Jerusalem; Christoff L. Frommel, Rome; Antoinette Gerhartl, Nijmegen; Norma and Bernard Goldman, Detroit; Anne E. Havinga, formerly Northampton; John W. Hayes, formerly Toronto; Dietrich Hertel, Cologne; Norma P. H. Jenkins, Corning; John W. Keefe, New Orleans; Richard M. Krill, Toledo; Hans-Peter Laubscher, Munich; Joan D. Mertens, New York; James D. Muhly, Philadelphia; Constantina Oldknow, formerly Los Angeles; Elsie H. and William H. Peck, Detroit; Carlos A. Picón, New York; Ellen D. Reeder, Baltimore; Susan Roberts-Manganelli, Stanford; Connie Rodriguez, New Orleans; Susan Rotroff, New York; Jan Willem Salomonson, Bilthoven; Ingeborg Scheibler, Munich; Marion Schröder, Rome; Alan Shestack, Boston; Erika Simon, Würzburg; Marilyn L. Simon, Flushing, New York; Penelope Slough, Detroit; Richard Stern, Amsterdam; Charles I. Terbille, Toledo; Klaus Vierneisel, Munich; Saul Weinberg,

Columbia; Annet van Wiechen, Hazerswoude; Elizabeth Lyding Will, Amherst; G. Zampieri, Padova.

Many specialists have assisted with information on the subjects of their expertise. On birds and plants: Elke and Hans-Joachim Boehr, Wiesbaden, and Sam Segal, Amsterdam; with translations from Arabic, Basil Collins, Toledo; Russian, Galya Levitina, Amsterdam; and Latin, David O. Ross, Ann Arbor; with information about inscriptions: Henry Macadam, Princeton; G. Neumann, Würzburg; Joyce Reynolds, Cambridge; C. J. Ruygh, Amsterdam; Antony E. Raubitschek, Palo Alto; and K. A. Worp, Amsterdam; with advice and bibliographical references regarding Judaica: Leslie Balkany, Toledo; Louis H. Feldman, New York; Joseph Gutmann, Detroit; Frank Peters, Toledo; Leonard V. Rutgers, Philadelphia; and Rabbi Alan M. Sokobin, Toledo; for information about and demonstrations of glassblowing techniques: Frederick Birkhill, Pinckney, Michigan; Henry Halem, Kent, Ohio; Paul Krebs, Toledo; John Littleton, Bakersville, North Carolina; William Morris, Stanwood, Washington; Richard Ritter, Bakersville, North Carolina; and Michael Zelenka, Toledo.

My thanks are due to the staff of the University Graduate Library at the University of Michigan, Ann Arbor for assistance in obtaining publications through Interlibrary Loan; and to Elaine Gazda, Director, the Kelsey Museum of Archaeology, Ann Arbor, and the staff of the Kelsey Museum for facilitating my research in many ways.

Special thanks are due to Seaside Software, Inc., Perry, Florida, for providing not only an evaluation package of their text-oriented database program AskSam, but also generous technical support. Tom Shay solved initial problems with setting up the project database and, at a later stage, helped transfer the data to Wordperfect.

For their generous hospitality I thank Phita Stern, Amsterdam; Renée Kotting-Menko, Amsterdam; Renate Wallace, London; John H. and Phyllis Martin, Corning; Frank and Judy Platt, New York; Harriet S. Waterman, Toronto; Marilyn Symmes, New York; Georgine von den Steinen, Basel, Doreen Canaday Spitzer, Princeton.

Friends and colleagues read and commented on parts of the manuscript. They offered many valuable suggestions which have found their way into the text: Dan P. Barag, Jerusalem; David F. Grose, Amherst Massachusetts; Sandra E. Knudsen, Toledo; Rosemarie Lierke, Frankfurt am Main; Kurt T. Luckner, Toledo; Timothy A. Motz, Toledo; Roy G. Newton, Sheffield; Erika Simon, Würzburg; and Gladys D. Weinberg, Columbia, Missouri.

It is my special pleasure to thank the publishing house of « L'Erma » di Bretschneider, Rome, and its jury of scholars for conferring the publisher's award on the manuscript of *Roman Mold-blown Glass* in 1992.

To all these individuals and to others whose names may have been inadvertently omitted, I wish to express my sincere thanks and appreciation.

E. Marianne Stern

Toledo, Ohio
August 1994

# Introduction to the Catalogue

This catalogue publishes the examples of Roman and Byzantine mold-blown glass in the collection of The Toledo Museum of Art and to discuss them in their historical, technological, and artistic context. The 193 entries describe vessels that were made between the early first century and the beginning of the seventh century by standard (or full-size) mold blowing. This technique came into use in the wake of the discovery in the first century B.C. that glass can be inflated. How early in the first century B.C. glass blowing first occurred is still debated, and further investigations may well show that this could have been earlier than most scholars now think. All the vessels described in this volume received their final shape and mold-blown decoration in full-sized molds that had the same interior dimensions as the glass vessels made in them. Not included here are mold-blown utilitarian vessels (mostly prismatic bottles) for storage and transport that show a mold-blown design only on their undersides, nor vessels with mold-blown decoration that was subsequently expanded by free blowing. These will be covered in future catalogues of the Toledo Museum's collection.[1]

## SCOPE OF STUDY

My proposed format for this project was to write annotated entries of the approximately 1150 objects in the Museum's collection, without including an overview of the history of glass in this period since many important classes of Roman glass are not represented in the Museum's collection. However, my approach changed during the course of the project, first because I took up glass blowing in The Toledo Museum of Art's Glass Studio in order to better understand the nature of the medium. This made it possible to address certain technological and manufacturing questions, and to incorporate the results. Second, because David Grose's volume *Early Ancient Glass* (1989) with its chapters synthesizing the history of early glass was such a success, it was decided to enlarge this catalogue by adding analogous discussions that dealt with broad currents in Roman glass working. It soon became clear, based on budget restrictions, that such an approach to the entire Roman glass collection would not be possible. Therefore, the decision was made to publish this volume devoted solely to the Roman full-size mold-blown glass. After I left Museum's employ, the Museum decided to add to this volume its holdings of Byzantine mold-blown glass. Chapter IV was composed by the Museum. The Museum's collections of Roman and early Byzantine free-blown and expanded mold-blown glass will be published at a later date.

This catalogue is divided into four chapters, arranged in a loose chronological order; the time span of the head-shaped vessels discussed in the third chapter overlaps that of the first two chapters. The first chapter is devoted to the mold-blown vessels of the first century, from the invention of the multipart glass-blower's mold through the high point of mold blowing between about A.D. 25 and 75 to the decline of mold blowing around or shortly before 100. Chapter II presents the mold-blown wares of the second through the fourth centuries. The twenty-three vessels in this chapter accurately reflect the decreased interest in this technique throughout the Roman empire during this period. The third chapter is devoted to a discussion of head-shaped vessels, from the earliest datable example perhaps made as early as about A.D. 14 to the latest ones in the collection, dating to the second half of the fourth century. Chapter IV provides a concise survey of the basic research around the Toledo Museum's small collection of late Roman and early Byzantine full-size mold-blown vessels from Palestine and Syria. Objects of dubious authenticity have been included at the end of the respective sections where they would belong if authentic.

An introductory essay begins each chapter, placing Toledo's pieces within their historical, archaeological, and cultural

[1] TMA, *Roman and Early Byzantine Glass* (forthcoming).

contexts. These essays vary in scope according to the holdings of the collection and the issues addressed. The essay for Chapter I provides a technological, historical, and artistic survey of the development of mold blowing on the Syro-Palestinian coast in the first century from its experimental beginnings to its decline. The discussion of technology is based on physical examination of surviving glass objects, my own experiments in glass blowing, and the insights of Toledo-area glass blowers. It also draws heavily on published and unpublished observations of colleagues who have patiently answered my numerous inquiries about the undersides of vessels in their collections. The interpretation of iconographical motifs is included as a basis for an attempt to identify the cultural background of the clienteles for whom the vessels were created. The essay for Chapter II is brief because mold blowing was in decline during the second to fourth centuries, except for head-shaped vessels which are discussed in their own chapter. The essay for Chapter III presents a survey of excavated head-shaped vessels whose dates have been established by archaeological context. The essay for Chapter IV gives background information on the Toledo Museum's groups of late Roman and early Byzantine jugs and jars. Wherever information relating directly to vessels discussed in the catalogue is presented, subheads within the essays correspond to subheads in the catalogue.

Catalogue entries following each essay are organized in sections based on shape and Types. Color plates and black-and-white photographs as well as profile and roll-out drawings illustrate the catalogue. Charts, form tables, graphs, figure drawings, and a map supplement the text. Photographs of related objects from other collections accompany the essays and selected entries. Glossaries, concordances, and indices facilitate reference to specific topics and objects discussed in this volume.

Many Roman mold-blown vessel Types have never been the object of a major typological or iconological study. Detailed information relating to Types and mold Series (see *Glossaries: Descriptive Terms for Classification*) represented in the Toledo collection is found only in the entries, not in the chapter essays, which are reserved for background and information of a more general character that is important for a Type or Series as a whole. For example, in the discussion of Hexagonal Bottles with High Relief (Cat. Nos. 15-47), information on the development of the various Types and their relationships to each other can be found in the chapter essay, while information about the development and deterioration of the molds used to make the specific bottles in the Toledo collection is in the entries. As a rule, the first entry introducing a new Type or a new Series contains the typological, technical, iconographical, and iconological information as well as comparanda. The one exception is the section on date-shaped flasks (Cat. Nos. 84-107), where all the information relating to this Type has been placed in the chapter instead of in the first entry.

## ORGANIZATION OF ENTRIES

The objects in Chapter I are organized by function. Tablewares are presented first, then containers. Within these major divisions, the vessels are arranged by shape and by Type. Type is here defined as subject and design of mold decoration. Where possible, Types are subdivided into Series of molds derived from one or more archetype. On these terms, see *Glossaries: Descriptive Terms for Classification*. In the catalogue entries, individual objects are arranged in chronological order according to the proposed scheme by which the molds were generated. In Chapters I through IV objects are grouped by shape and chronology.

Very few vessels in the Toledo collection have been identified as products of specific workshops. Workshop attributions are discussed in the chapter essays, but in the catalogue entries the vessels are treated by Type within each major group of shapes, regardless of workshop attribution. Each sequence of groups begins with the most common Types and concludes with those less frequently encountered. This arrangement facilitates consulting the catalogue as a reference tool, while at the same time clarifying the relationships among Types, Series, and workshops.

It has previously been assumed that vessels belonging to one Type were made in one workshop. However, research on the Toledo collection has now proven that this assumption may be misleading. Technical idiosyncrasies, such as the variety of ways of constructing multipart molds, are sometimes, but not always, characteristics developed by individual workshops. Vessels belonging to one Type, and even to one Series, were sometimes made in different Mold Construction Types (MCTs). In such cases it is justifiable to conclude that more than one workshop was involved in their production, sometimes even that they were blown in different manufacturing centers. The fact that it is easy to make a new mold from any existing vessel, regardless of the technical construction of the mold (MCT), made this circumstance possible.

The relationship between workshops and MCTs varies. Some glass blowers, such as Ennion, used more than one MCT, depending on the shape to be made. On the other hand, a single workshop, the Workshop of the Floating Handles, made at least ten different subject Types but used only one MCT. Future research may identify other groups of vessels with different subject Types but a common MCT as products of one workshop.

## DESCRIPTION OF ENTRIES

Individual catalogue entries follow the format devised for David F. Grose, *Early Ancient Glass*, with modifications as needed. They include detailed information on each object, with discussions of technique, color, shape, and decoration. Full discussions of typological, technical, iconographical, and iconological data as well as comparanda accompany the lead entries of all groups and also entries to Types and Series represented by only one object. The descriptions follow a format developed specifically for this catalogue in order to convey accurately the complicated nature of ancient Mold Construction Types (MCTs) and the alterations to the surface decoration occurring when a single Type or Series was popular over an extended period. While some readers may object to the length of certain entries or to repetition in others, the alternative would have been to omit information which may prove important for future studies or to use a cumbersome cross-reference system.

Each entry begins with a short identification section listing title, origin, find spot, date of the object, and measurements:

Title. The generic name of a vessel is stated first, usually followed by the subject type of the decoration. Occasionally, the mold Series is identified in the title.

Origin. The general region where an object is thought to have been made is usually stated in adjectival form, such as "Syro-Palestinian," followed, where possible, by a more specific reference to its place of manufacture, for example, "probably made in Sidon." In order to avoid confusion with modern art historical associations with the term "Italian," objects made in this area are classified as "made in Italy." For the geographic designations used in this volume, see *Glossaries: Geographic Designations*. In many instances such attributions are a matter of informed opinion based on the impressions formulated from distribution patterns of vessels from excavations similar to those in Toledo's collection. In order to cite the distribution areas of a specific Type, parallels are quoted according to their find spots, usually from East to West if the distribution pattern precludes a western origin. The validity of this approach depends, of course, on the amount and reliability of information available.

Find spot. When a vessel is known or said to have been found, excavated, or acquired at a particular site or country, that information follows origin. The find spot of an object discovered in a controlled excavation is beyond doubt, but find spots attached to objects by vendors can range from innocent misconception to deliberate deception. In the case of vessels from old collections, it may be uncertain whether a provenance refers to the actual find spot of an object or to the place where it was acquired by its first recorded owner.

Date. The inclusive dates for the manufacture of a vessel are stated in quarter centuries to full centuries. All dates in this volume are Common Era (C.E. or A.D.) unless otherwise indicated. In so far as was possible a numerical date, by years, or more often, by quarter centuries, has been assigned to each object. Because it is not always possible to assign a vessel to a specific period, a scale of qualifications is used to indicate how reliable the dating is. A date given without any qualification indicates that there is sufficient evidence for a reliable date. If a date is preceded by the word "probably," the evidence is circumstantial. If a date is preceded by "perhaps," there is some evidence to indicate this date. If a date is followed by a question mark, it is the author's opinion. In all cases dating is restricted to as narrow a time span as possible. The date range assigned to a specific object may be shorter than the period during which its Type or Series was in production. Whenever possible, the chronological classification is based on parallels found in independently dated contexts. Objects discovered in tombs can present a confusing problem. In the Near East, many tombs were used for several generations; consequently, it is frequently impossible to determine which objects belong to which burial. In the West, where single burials were usually the rule, the question of dating can be approached in a more straightforward manner.

For the important group of Hexagonal Bottles with High Relief, the author has posited a new hypothesis in an attempt to refine the dates of manufacture. This hypothesis is based on calculating the minimum number of generations separating a specific object from the archetype of the first generation of molds. For some generations, independently dated examples from excavations provide the necessary "absolute" dates. For others, specifically vessels made in the Workshop of the Floating Handles, the estimated production period of the workshop is the basis for dating. And for objects where no datable parallels could be identified, dating may be based on aspects of technique like finishing of the rim, MCT, Type, and color. Occasionally, other details have been used in order to establish a more precise date for an object representing a Type used during an extended period. Because dating on purely stylistic grounds is only reliable in so far as an object can be related to a sequence of well-dated material (as is the case, for instance, with coins or Arretine pottery), dates put forward on these grounds are hypothetical, at times even speculative. Therefore dating on stylistic grounds has been avoided whenever more reliable criteria exist.

Measurements. All measurements are given in centimeters. The standard dimensions, in their normal order of sequence, are as follows:

**H** = the maximum height from the top of the rim to the base of a vessel when it is held upright, perpendicular to a level surface. Pres. (= preserved), is added in parenthesis if the object is not complete. If objects are asymmetrical or have a very irregular rim, two measurements separated by a dash are given, first the larger one, then the smaller one, as for example in Cat. No. 7. The first is the largest measurement at the rim, the second is the smallest.

**H w handle(s)** = one of two measurements: 1) if the vessel has a basket handle, the measurement is to the highest point even if the handle is lopsided; 2) if there is no basket handle, the measurement is to the topmost point of the (highest) side handle.

**D** = maximum external diameter, taken from mold seam to mold seam. If the external diameter measured between seams is much larger or smaller, it follows the first measurement, separated by an x, as in Cat. No. 14.

**D rim** = maximum diameter of the rim taken on the exterior. If the rim is asymmetrical or highly irregular, two measurements separated by a dash are given, first the larger one, then the smaller. If a rim or mouth was not meant to be circular, the largest and smallest measurement are given, separated by an x, as in Cat. No. 131.

**D base** = maximum diameter of the base, the resting surface. If the base is not circular, two dimensions are given, the first from seam to seam or along the seam, the second by the width between seams, separated by an x, as in Cat. No. 42.

**Th** = thickness at the rim if not otherwise indicated.

Pontil marks are measured to indicate the maximum diameter.

The identification section is followed by a series of paragraphs that include detailed descriptions presented in the following sequence: color and fabric, manufacturing technique and MCT, form, and decoration.

Color. Terms used to describe the colors of objects attempt to identify the original color intended by the glassmaker, insofar as that can be ascertained by visual observation. Scholars have no uniform method of describing glass colors. In this catalogue the names of the colors used to describe objects are those of the *Rock-Color Chart* distributed by the Geological Society of America, Boulder, Colorado. Hue, value, and chroma are added, in accordance with the Munsell system, by a numerical designation in parentheses. All colors were determined under the same circumstances, by daylight. Because the degree of transparency affects the perception of color, the glass is called transparent, translucent, or opaque depending on the amount of light transmitted. Black glass was not made by ancient glassmakers; glass appearing black is always a translucent dark color such as purple, green, blue, or brown. Color designations of comparanda employ the terminology of the sources quoted.

Fabric. Terms used to describe fabric are intended to identify the quality of the glass as it appeared when the glass worker had finished making the object. All observations are presented as seen under a 10x magnifying glass. The chemical composition of the glass has not been determined for any of the objects discussed in this volume.

Manufacturing Technique and MCT. Technique of manufacture and Mold Construction Type (MCT) are important criteria for determining the attribution of mold-blown glass vessels to a particular origin, Series, or date. For Mold Construction Types, see *Glossaries: Mold Construction Types.* MCT plays a key role in classifying mold-blown glass and in arranging entries belonging to one Type or Series. Nine different types of mold construction have been identified (MCTs I-IX, Fig. 15). Here vessels belonging to one Type or Series are classified according to a shared MCT. Observations on unusually crisp or blurred relief — which has significance for the age of the mold — are also made in this section.

Form. Form refers to the shape of the vessel and its individual parts; see *Glossaries: Descriptive Terms for Vessel Shapes.* Descriptions of form proceed from the top of the vessel to the bottom. Terms frequently used to describe the various parts of a vessel are listed in the *Glossaries: Descriptive Terms for Vessel Parts.*

Decoration. Descriptions of decoration proceed from the top of the vessel to the bottom. In the case of mold-blown glass, decoration is synonymous with Type.

Classification. Classification refers to the chronological and artistic place of a vessel within the sequence of a given Type or Series if this has been determined. The names of Type and Series are followed by a numerical designation indicating Mold and Generation. For an explanation of the terms Type, Series, Mold, and Generation and the system used for their classification, see the *Glossaries: Descriptive Terms for Classification.* Throughout this catalogue these terms are capitalized when used for the meaning defined in this section of the Glossary.

Condition. Vessels with no losses are described as "intact." Those that are either broken and repaired or virtually whole are described as "complete," but all losses are recorded. The effects of weathering are described, as are surfaces that are exceptionally well preserved. Modern cleaning, surface consolidation, repairs and restorations are noted.

Collections. The names and locations of public and private collections that formerly owned vessels now in the collection of The Toledo Museum of Art are given, in chronological order, with the inclusive dates. All but a few objects published in this catalogue were purchased for the Museum in 1919 by its founder, Edward Drummond Libbey, who in that year acquired the sizable collection of ancient and Islamic glass formed by Thomas E. H. Curtis of New York City and Plainfield, New Jersey. Between 1900 and 1915 Curtis acquired virtually all his glass collection of over five thousand objects from antiquities dealers in New York City.[2] The Ancient Arts Gallery, owned by Azeez Khayat, was Curtis's principal dealer. Khayat's own sources were in what was in antiquity northern Palestine and the southern coast of Phoenicia. Khayat conducted his own excavations, first in and around his birthplace, Tyre, and later in Haifa and its vicinity where he bought a house, and near his summer home in Nazareth. He also encouraged others to excavate and search for antiquities, which he then acquired directly from them.[3]

Exhibitions. The title, sponsoring institution, and dates of special exhibitions in which a Toledo object appeared are listed in chronological order. Catalogues of those exhibitions are listed under Publications only.

Publications. References are restricted to prior publication(s) of the object described in the entry. Studies that merely cross-reference or illustrate an object at Toledo are omitted unless they mention its historical or archaeological context.

Remarks. This section describes each object in its technological, historical, iconological, artistic, and cultural context as well as comparanda. If a Type or Series is represented by more than one object, this information is presented in the first entry of the group. Endnotes accompany the Remarks.

## ILLUSTRATIONS

Photographs. All objects in this volume are illustrated in

[2] On Thomas E. H. Curtis, see Grose 1989, 19-21.

[3] Bergmann 1974, 238-244.

black-and-white photographs. About one-third also appear in the color plates. Types with different decoration on various sides are shown in as many views as needed to illustrate all important aspects of the decoration. Scale, usually 1:1 or 1:2, is noted in the caption for all catalogue photographs. Comparison illustrations and color plates are variously scaled.

PROFILE DRAWINGS AND ROLL-OUTS. All objects in this volume are illustrated in profile drawings. Roll-out drawings are added for those objects which have individual design elements rather than a repeated pattern. All profile and roll-out drawings are printed to a scale of 1:1 unless otherwise noted.

COMPARISON PHOTOGRAPHS. The photographs of objects from archaeological excavations or from other collections that illustrate the introductory essays or entries were selected because they document important objects not found in Toledo's collection. Comparison photographs are not printed to scale, but basic dimensions are given if provided. Photographs of glass-working tools and glass-blowing techniques appear in the introductory essays and *Glossaries: Technical Terms*.

CHARTS, FORM TABLES, GRAPHS, AND MAP. The charts and form tables in the introductory essays illustrate different Types of mold construction (MCTs) (Fig. 15) and the historical development and evolution of specific Types and Series (Figs. 46, 50, and 54). In the discussion of *Advances in Roman Glass Technology* graphs (Figs. 16 and 17) illustrate the physical properties of ancient glass. The map following the index locates major ancient cities, archaeological sites, and regions of the Roman world mentioned in this catalogue.

# Glossaries

## TECHNICAL TERMS

Technical terms frequently used in this catalogue are listed below. The definitions have been adapted from several sources.[1] Additional terms can be found in the glossaries in Goldstein 1979, Weinberg 1988, and Grose 1989. Alternative terms are mentioned in regular type; cross references are indicated to words in italics. Some aspects of ancient glassmaking technology are discussed in more detail in the text and the short definitions here refer to those passages.

**annealing** Gradually cooling a finished glass object to relieve stresses resulting from different rates of cooling of its various parts such as the body, handle, and foot of a vessel, or its surface or skin and its interior. The thicker the glass, the longer the annealing time required. Although most Roman glass is thin-walled and probably did not require extended annealing, annealing appears to have been a problem in some workshops.[2] Some ancient glass blowers may have simply covered finished pieces with hot ashes in order to slow down the cooling process,[3] while others may have put their pieces in a specific area of the glass-working furnace[4] or, possibly, in a separate oven built specifically for annealing,[5] as is today the most common practice.

[1] The concept and organization of these glossaries are based on those which first appeared in Grose 1989. In the interest of standardizing these sections of The Toledo Museum of Art's catalogues, I have followed the wording of many areas of those glossaries when appropriate. Also helpful are the following publications: Scholes and Greene 1975; Newton and Davison 1989. My thanks are owed to Roy Newton and Rosemarie Lierke for reading and commenting on this section. Although I have incorporated many of their suggestions, the opinions expressed are my own.

My sincere thanks are owed to my colleague glass blowers for posing for workshop photographs: Frederick Birkhill (see Figs. 21, 22, and 25); Kathleen M. McCarthy (see Figs. 6, 7, 13, 14, 30, 32, and 33); Leonard Marty (see Figs. 9 and 22); and Jeffrey Boyer (see Fig. 41).

[2] For example, at a fourth-century glass-working furnace at Jalame, Israel: Weinberg 1988, 33.

[3] Lierke 1991, 312 and 315.

[4] For example, at Jalame: Weinberg 1988, 33.

[5] For example, at a first-century workshop in Avenches, Switzerland, active between 40 and 70: Morel et al. 1992, 56, fig. 7.

**batch** The mixture of raw materials prepared for heating to make glass (see *glass making*).

**bit** A shaped, hot gather of glass on a metal rod or pontil; generally used for *trailing* a thread or coil, or applying a handle or decoration to a vessel held on a blowpipe or pontil (Fig. 1).

**block, blocking mold** A hollowed block of fruit wood used wet to shape glass at various stages of blowing. Although common in glass studios today, this tool was probably not known in antiquity.[6]

**blowing spirals** (also called striations or striae) Streaks in imperfectly melted or imperfectly homogenized glass caused by the rotary motion of the blowpipe (see *cord* and ills. at Cat. No. 139).

**blowpipe, blowing-iron** A metal tube used for blowing glass (see Figs. 6, 13, and 27 and *Advances in Roman Glass Technology: The Invention of the Blowpipe*).

**bubble** An air pocket in glass (see also *bubbles*), especially an air pocket formed by deliberately blowing air into a hot gob of glass through a pipe (see Figs. 6, 13, and 27).

**bubbles** Air pockets accidentally trapped in glass during *glass making*, glass melting, or *glass working*. Very small bubbles are called "pinprick bubbles" or "seeds," larger ones "blisters." The distinction is arbitrary. In this catalogue a 10X magnifying glass was used to determine the size of bubbles. Ancient glass usually has more bubbles than modern glass, perhaps because the glass was too viscous for gaseous reaction products to escape or because the crucibles were made of clay (see *crucible*). Bubbles (oval, elongated, linear) tend to distort in the direction of the flow of the glass during manipulation while it is hot, and this can be useful for determining the method of manufacture (Fig. 2).

**casting** Filling a mold with amorphous glass. Most modern casting is done by pouring molten hot glass into a preheated mold and then annealing it. Before the introduction of hot glass in glass working (see *Advances in Roman Glass Technology: The Introduction of Hot Glass*), casting processes were variations of what is now commonly

[6] Stern 1992a, 493.

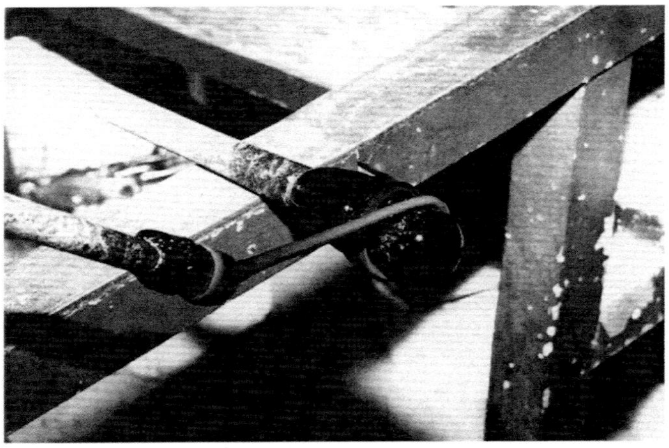

*Fig. 1.* **Bit**: *Trailing a bit onto a paraison for a decorative coil by rotating the paraison on the blowpipe. The coil can then be marvered or smoothed into the surface of the paraison or left in relief. Photo: Author.*

called chip casting.[7] The larger the glass chips, the more transparent the finished object tends to be. The mold can be filled while both mold and chips are cold. Additional chips can be added while the mold is heated; but if the sprue, or pour-channel, through which the chips are poured is large enough, it can be filled before heating, thus making it unnecessary to add chips when the mold is hot. After all the glass has melted, at a temperature of about 850°-965° C, the mold is slowly cooled and finally removed.

[7] Stern 1994, 48-53.

**chill marks** Fine wrinkles or ripples on the surface of a glass object resulting from its sudden contact when hot with a cold surface, such as a marver or metal mold (e.g. Cat. Nos. 43 and 44, Fig. 3). If the glass is reheated at a sufficiently high temperature, chill marks can be smoothed away.

**chunk gathering** (as opposed to *hot gathering*) Picking up a chunk of preheated raw glass or cullet on the end of a white-hot blowpipe or pontil. In order to work the glass, the chunk must be heated further until the interior is soft (see *Advances in Roman Glass Technology: Chunk Gathering* and Figs. 18-21).

**clamping tool, clamp** A tool, of unknown shape and construction in ancient times, used before the pontil technique was developed to finish the rim of a vessel by holding it in the fire after it was cracked off the blowpipe. The mold or the vessel was held with some kind of clasping instrument. Deformations and pinching marks from clamping tools can be seen on many ancient glass vessels. Vertical crimp marks at the base of the neck can be seen on Cat. Nos. 22, 26, 28-33, 40, 42, 49, 78, and 108. A different tool did not crimp the glass but squashed the neck so that it was no longer circular at the base: e.g. Cat. Nos. 161 and 162. Future research on such tool marks may provide insight into the construction

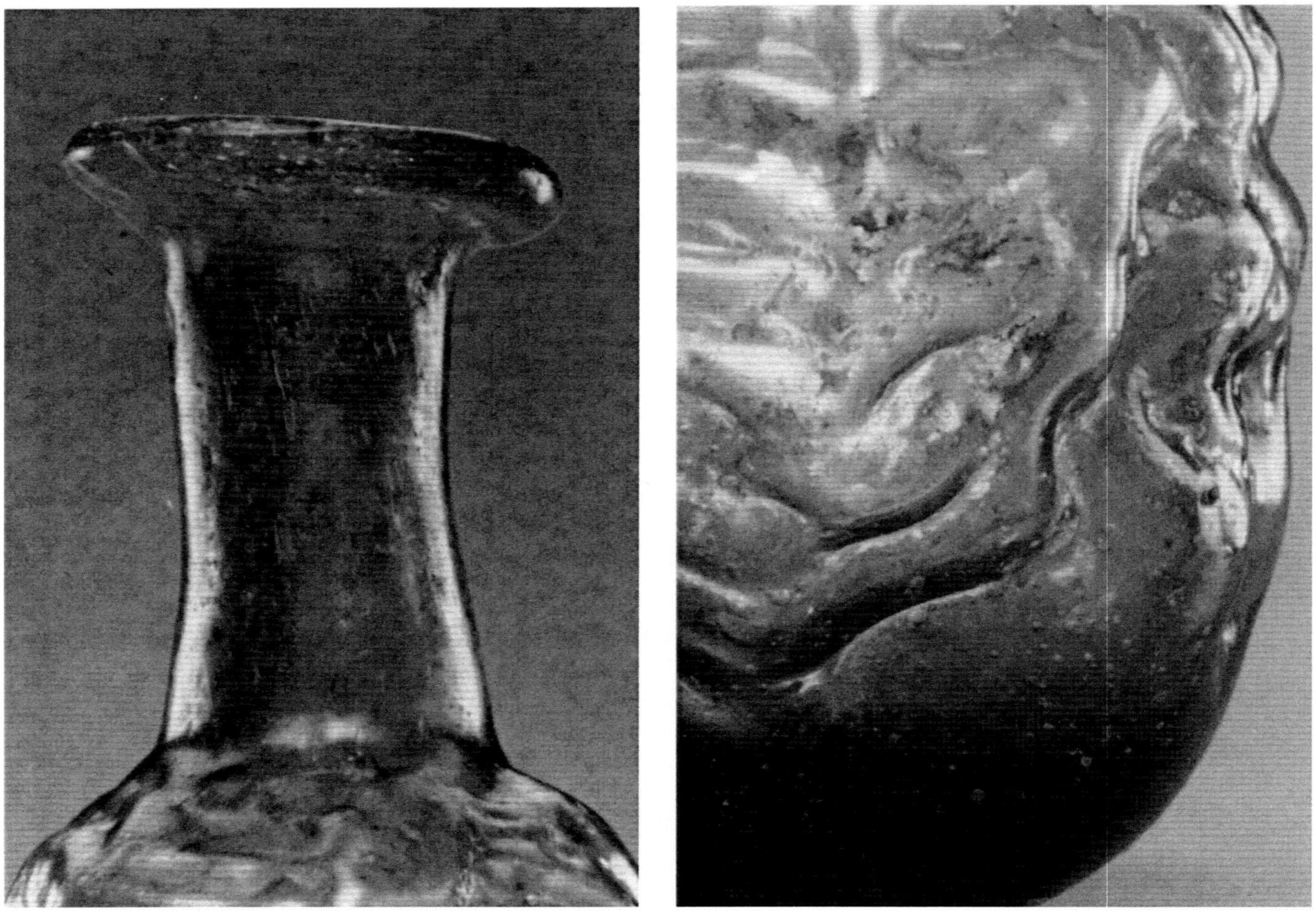

*Fig. 2.* **Bubbles**: *Enlarged details of neck (left) and base (right) of head-shaped vessel Cat. No. 144, showing air pockets trapped in the glass.*

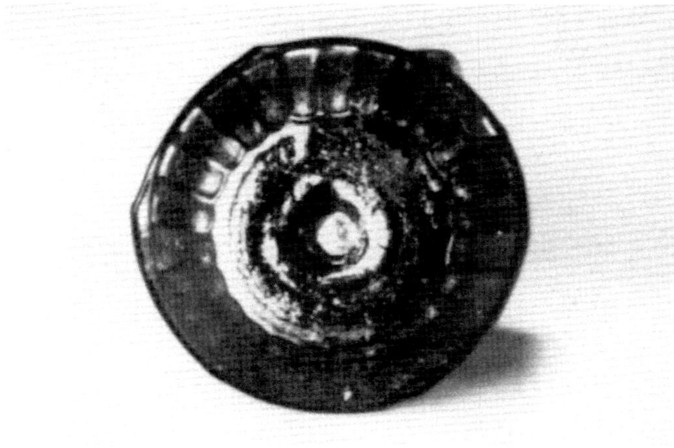

*Fig. 3.* **Chill marks**: *The underside of the base of Cat. No. 43, showing concentric chill marks caused by the contact of hot glass with a mold. See also detail photograph of the underside of the base at Cat. No. 44. Photo: Timothy A. Motz.*

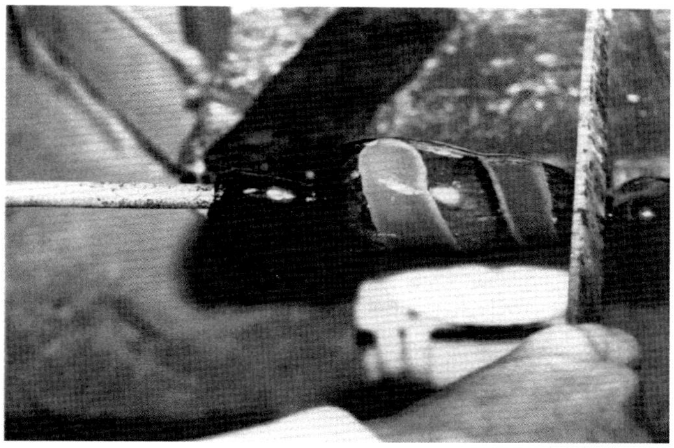

*Fig. 4.* **Cracking off**: *Filing between the blowpipe (on the right) and the vessel on the pontil (left) creates a stress line; if the vessel is tapped, a crack develops at this point and the vessel is removed from the blowpipe. (The blowpipe rests on the arms of the glass-blower's bench. On the bench lies a moist wooden paddle; a white pail filled with water is supported by a concrete block behind the bench. The setup appears more clearly in Figs. 34 and 35). Photo: Lois Dorfman.*

of ancient clamping devices. For use of a modern clamp, see Fig. 37.

**coil** A *trail* of hot glass wound around an object (e.g. Fig. 1 and Cat. No. 164).

**colorant** Usually a metallic oxide added deliberately to the batch to color the glass. Frequently used colorants in antiquity were the oxides of copper, iron, cobalt, and manganese. These produce different colors depending on the amount of oxygen in the atmosphere of the furnace. Such deliberately colored glasses are called *specialized glasses*. Glass that has not been *decolorized* or colored by deliberately introduced ingredients is called "natural bluish green." The natural bluish green color results from iron oxide, a common impurity in many sands. Depending on the amount of oxygen in the kiln, iron impurities account for a range of colors from blue to olive to green to amber (see Color Plates 19 and 29, pages 57 and 63).[8] See also *opacifier.*

**cord** A streak in glass caused by local differences in the refractive index of glass, resulting from local variations in the composition of the glass or solution of refractory material from the furnace or crucible. During blowing, these streaks are elongated in the direction of the flow of the glass and appear as *blowing spirals.*

**cracking off** Removing a vessel from the blowpipe (Fig. 4, see also Figs. 34 and 35).

**crucible** A container or pot placed in the furnace for making or remelting glass. Throughout antiquity and the Middle Ages crucibles were small, made of fired clay.[9] The use of clay crucibles may explain why the glass made in these periods always has some bubbles, because in modern glass-working practice, the "evolution of seeds from clay refractories is particularly troublesome when small quantities of glass are melted in crucibles. It is scarcely possible to obtain well-*fined* glass in such experimental melts unless a platinum crucible is employed."[10] See *glass making.*

**cullet** Chunks of glass or broken objects or waste from glass working such as *knock-offs, moils*, etc. (e.g. Fig. 35). When added to a *batch*, cullet promotes rapid melting and acts as a nucleus around which new glass forms. When the amount of cullet in a batch is high, the glass sometimes seems stiffer or higher in viscosity.[11]

**cut-out** A projecting roll or fold of glass on the exterior of a vessel (see detail of Cat. No. 120).

**decolorizer** An ingredient added to the *batch* to counteract the effect of impurities of iron oxide which give glass a yellowish or bluish-green color. In antiquity the most common decolorizing agents were antimony and *manganese*. These may not have been added deliberately as separate chemicals but at first accidentally (and usefully) introduced with a raw material which differed from that from other sources.[12] Decolorized glass is *specialized glass* (see Cat. Nos. 82, 163, 189, and 190).

**fabric** The chemical composition of the glass. In this catalogue, the term is also used to indicate the glass material itself.

**fining, plaining** The freeing of batch glass (see *batch*) from bubbles. In modern glass making this is accomplished by adding a fining

[8] On coloring caused by reducing and oxidizing atmospheres: Brill 1988, 269-276; C. Sellner, H. J. Oel, and B. Camara, "Untersuchung alter Gläser (Waldglas) auf Zusammenhang von Zusammensetzung, Farbe und Schmelzatmosphäre mit der Elektronenspektroskopie und der Elektronenspinresonanz (ESR)," *GlastechBer* 52 (1979) 255-264. Techniques for regulating the amount of oxygen in a kiln had already been perfected by Greek potters in the seventh century B.C.

[9] On household wares such as everyday cooking pots used as crucibles: Morel et al. 1992, 14-15, figs. 23-25 (crucibles excavated in a first-century workshop at Avenches, Switzerland); Danièle Foy et al., "Creusets de verrier de l'Antiquité tardive et du Haut Moyen-Age découverts dans le sud-est de la France," *AnnAIHV* 11 (Basel 1988) 199-216, esp. 210ff.; Stern 1992a.

[10] Scholes and Greene 1975, 445-446.

[11] Opinions differ about the degree of stiffness and about its cause. It may be due to loss of gases (water, carbon dioxide, oxygen) from previous exposure of the cullet to high temperatures, or it may result from volatilization of the alkalis (Roy Newton, pers. com.).

[12] R. G. Newton, "W. E. S. Turner: Recollections and Developments [the Eighth Turner Memorial Lecture]," *Journal of the Society of Glass Technology* 26 (1985) 98, notes the ability of some plants to concentrate in their tissues certain of the trace elements, such as manganese, present in the soil in which they grow.

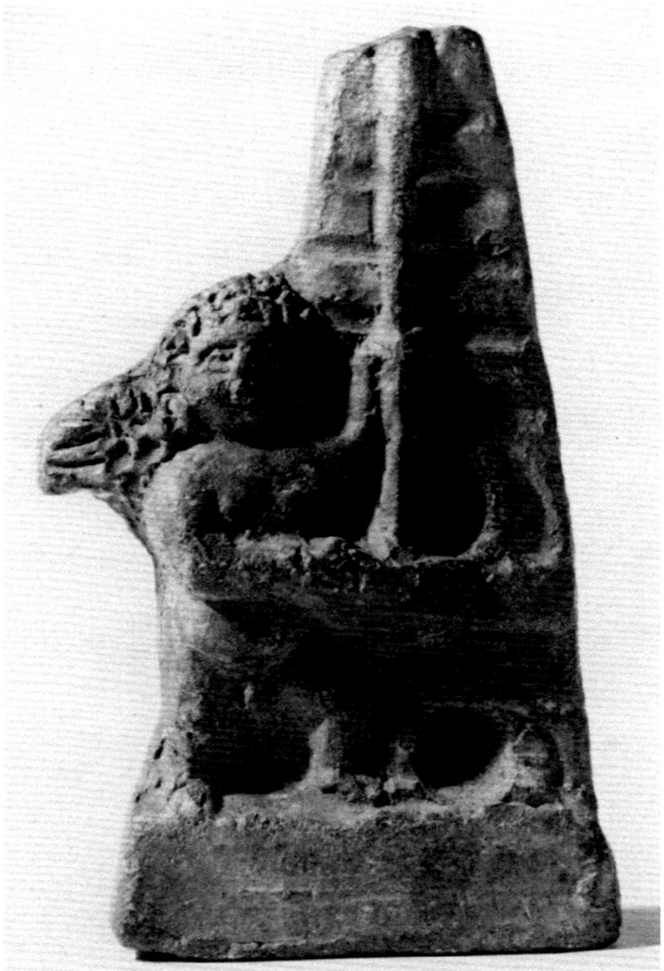

*Fig. 5.* **Roman glass-working furnace**: *Egyptian terra-cotta group showing a glass furnace, first or second century* A.D. *London, British Museum, no. GR 1986.10-3.2. Photo: Copyright British Museum.*

agent, such as sodium sulphate, arsenic, antimony, sodium chloride, or potassium chloride, to the melted glass. The reaction of the fining agent with elements in the glass increases the size of the bubbles and allows them to rise and escape.

**flash** Glass escaped through a crack in a mold or protruding as a *mold seam*.

**free blowing, offhand blowing** Forming objects without the aid of molds by manipulating with tools and using the natural forces of expansion, gravity, and centrifugal force.

**frit** See *glass making*.

**frosting** A network of tiny cracks on the surface of an object.

**furnace** Throughout antiquity *glass making* and *glass working* required different technologies and different types of furnaces.[13] This distinction is reflected in the Greek and Latin terminology for the glass industry.[14] The ancient Roman glass industry knew specialized furnaces for three different functions, one for each of the two steps of glass making and one for glass working. A glass-working furnace is depicted on the discuses of two clay oil lamps of the second half of the first century (Fig. 27, left and right),[15] and on an Egyptian terra-cotta group datable to the first or second century (Fig. 5).[16] The tall, truncated conical form of the furnace shown on the terra-cotta differs from the low, domed structures on the lamps, but it has the two small openings that are essential for a glass-working furnace, the lower one for stoking the fire, the upper one for working the glass. On the terra-cotta group, the edges of these openings are raised, suggesting that they might represent terra-cotta edges similar to those of first-century glass-working furnaces excavated at Avenches, western Switzerland,[17] and Saintes, southern France,[18] where they have been interpreted as the edge of the opening through which the glassblower worked. Important improvements in furnace technology appear to have taken place in Italy in the first century (see *glass making*). Remains of ancient Roman glass working furnaces have been noted in many countries; excavated examples are known from sites in Israel,[19] Ukraine,[20] Tunisia,[21] Croatia,[22] Switzerland,[23] France,[24] Germany,[25] Netherlands,[26] and Britain.[27]

13 Baldoni 1987; A. Gasparetto, "A proposito dell'officina vetraria torcellana," *JGS* 9 (1967) 50-75. The present studio glass practice of batching in and working from one small tank furnace is a very recent development, based on a furnace designed by Dominick Labino, Waterville, Ohio, in 1962.

14 On Greek and Latin terms for glass makers and glass workers: Stern 1989a, 121-122.

15 Two Roman lamps:

a) found near Ferrara (Italy), second half of the first century, Ferrara, Museo Archeologico Nazionale, no. 52196, L 11.5 cm, D 8.3 cm (here Fig. 27, left);
b) found at Asseria (Croatia), second half of the first century, Spalato, Croatia, Museum, no. 1094/30 (here Fig. 27, right).

On both lamps, see Baldoni 1987, 22-29; Grose 1989, 33, fig. 10.

16 London, BM, no. GR 1986.10-3.2, formerly Gustave Mustaki Collection, Price 1988a, 317-319, fig. 3, pl. LIVa.

17 Morel et al. 1992, 6-7, figs. 9-10.

18 A. Hochuli-Gysel, pers. com.

19 Avi-Yonah 1975-78, I, 18, s.v. *Accho*, mentions the excavation of a Hellenistic-Roman glass furnace at Acre. Weinberg 1988 provides the most comprehensive discussion to date of a Roman glass-working furnace, active during the fourth century at Jalame, Israel. Stern 1992a argues that the archaeological evidence indicates that glass workers at Jalame blew chunks of glass rather than gathering molten hot glass.

20 Sorokina 1977b, 96, pl. I, from Komarowo. Kotigoroshko 1989, 187-190, figs. 4-5, discusses the remains of a first-century B.C.-first-century A.D. glass furnace, clay crucibles, and tools excavated at Malaya Kopanya, in Carpathia; see also *Advances in Roman Glass Technology: The Invention of the Blowpipe*.

21 M. H. Fantar, "Le Verre à Carthage Punique," *BullAIHV* 6 (1971-1972), 17-18, mentions Roman furnaces excavated at Carthage and Kerkouane.

22 Clairmont 1975, 59ff., describes remains of a Roman glass factory at Salona.

23 Morel et al. 1992, 2-9, and Heidi Amrein, "Un atelier de verrier gallo-romain à Avenches," presented at the 12$^{e}$ Congrès de l'Association Internationale pour l'Histoire du Verre (Vienna August 26-1, 1991) (but not published in *AnnAIHV* 12 [1991, publ. 1993]), discuss the remains of four first-century glass-working furnaces at Avenches (ancient Aventicum). At Augst, remains of glass-working furnaces were excavated, dating to the late first century (Augst, Oberstadt) and second to third centuries (Augst, Unterstadt): Rütti 1991, esp. vol. 1, 150-168; there is no evidence of glass *making* at Augst.

24 Foy et al. 1989, 55-69, provides an overview of late Roman and early medieval furnaces excavated in France.

25 A. B. Follmann-Schulz, "Fours de verriers romains dans la province de Germanie inferieure," *AFAV* 4 (1989, published 1991) 35-40, provides an overview of Roman furnaces excavated in the Roman province of Germania Inferior; W. Gaitzsch, "Four de verriers romains en forêt de Hambach," *AFAV* 4 (1989, published 1991) 41-46, discusses remains of Roman furnaces excavated in the forest of Hambach, halfway between Cologne and Aachen.

26 A. E. van Giffen, C. W. Vollgraf, and G. van Hoorn, *Opgravingen op het Domplein te Utrecht* (Haarlem 1934) I, 63; II, pl. XXII.

27 Jennifer Price and H. E. M. Cool, "The Evidence for the Production

**gather, gathering** Glass picked up on a blowpipe or pontil. The gather can be either a preheated chunk of glass (*chunk gathering*), or a gob of molten hot glass taken from a crucible or tank in a glass-working furnace (*hot gathering*).

**Generation** A term used in the classification of mold-blown vessels belonging to one Series (see below: *Descriptive Terms for Classification*).

**glass** A product of fusion which upon cooling becomes increasingly viscous until it solidifies without crystallizing. Glass is not a specific chemical compound. There are innumerable recipes for the composition of glass. Ancient Roman glass is a mixture of silica (sand, crushed quartz, or crushed flint), alkali (in the form of plant ashes, i.e. soda or potash, or natron, for example from the Wadi Natrun, about 100 km southwest of Alexandria), and lime. Fused silica is the ideally durable glass, but the melting of pure silica requires extremely high temperatures which were not obtainable until modern times. Adding alkali, which acts as a flux, lowers the melting temperature of the glass batch but has the disadvantage of increasing its solubility in water. Lime counteracts this deleterious effect and acts as a stabilizer, making the glass durable. In antiquity, the lime needed for durability was not recognized as a separate component. There were two basic recipes, each of which made use of two ingredients to introduce the three components.[28] In Mesopotamia, glass was made from crushed quartzite pebbles, which provided silica, and plant ash, which provided soda and lime. In Roman glass from the Mediterranean and Europe, the alkali was a relatively pure form of soda called natron.[29] The lime came from impurities in the sand, the source of the silica.[30] The importance of lime as a stabilizer was not recognized in antiquity; this explains the renown of particular sands for glass making such as those from the Belus River in ancient Phoenicia,[31] and from the Liternus River in Campania, Italy.[32] Pliny noted that the sands from the Belus river were not fit for glass making until they had been washed by the sea. At that time the sands would have been mixed with debris from shells, washed in by the waves of the Mediterranean, which would have provided the necessary lime. Pliny recognized the importance of contact with the sea but was clearly not aware that it was lime which made the difference:

> The river [Belus] is muddy and flows in a deep channel, revealing its sands only when the tide ebbs. For it is not until they have been tossed by the waves and cleansed of impurities that they glisten. Moreover, it is only at that moment, when they are thought to be affected by the sharp, astringent properties of the brine, that they become fit for use [in glass making].[33]

(See also *glass making* and *Advances in Roman Glass Technology*: *The Properties of Glass*).

**glass blowing** Inflating glass with air blown through a pipe to form a hollow object (Fig. 6). Glass can be free blown, or it can be blown into various types of molds. See also *Advances in Roman Glass Technology*: *Inflation* and *The Invention of the Blowpipe* and Figs. 21-23.

**glass making** The process of making *glass* from basic raw materials. Throughout antiquity and the Middle Ages, glass making was a two-stage process requiring a different *furnace* for each stage.[34] First, a solid-state reaction between the silica and the alkali (soda or potash) was required, but this reaction should not allow the mass to melt, because the released gases would be trapped in the molten glass. This process is also called fritting. After the reacted frit was cooled, ground up finely, and mixed with minor ingredients such as colorants, opacifiers, and others, the second phase consisted of melting the mixture. Pliny (*Natural History* 36.194) describes improvements in making glass from a batch that were new in his time (A.D. 23/24-79) and appear to have been invented in Italy:

> Wherever it [white sand from the Volturnus River between Cumae and Liternum, in Campania] is softest, it is taken to be ground in a mortar or mill. Then it is mixed with three parts of soda, either by weight or by measure, and after being fused[35] is taken in its molten state[36] to other furnaces. There it forms a lump known in Greek as sand-soda [*hammonitrum*]. This is again melted and forms pure glass, and is indeed a lump of clear colorless glass.[37]

of Glass in Roman Britain," *AFAV* 4 (1989, publ. 1991) 23-30, provides an overview of Roman furnaces excavated in Roman Britain; J. D. Shepherd and M. Heyworth "Le travail du verres dans Londres romains (Londinium): un état de la question," *AFAV* 4 (Rouen 1989, publ. 1991) 13-22, discusses remains of furnaces excavated in Roman London.

28 Newton and Davison 1989, 54; Brill 1988, 264-269.

29 Recent evidence from a glass-working site in the Severn estuary suggests some locally made glass in Britain of Roman date may have been potash-lime-silica glass: J. R. L. Allen and M. G. Fulford, "Romano-British Settlement and Industry," *AntJ* 67 (1987) 275-280.

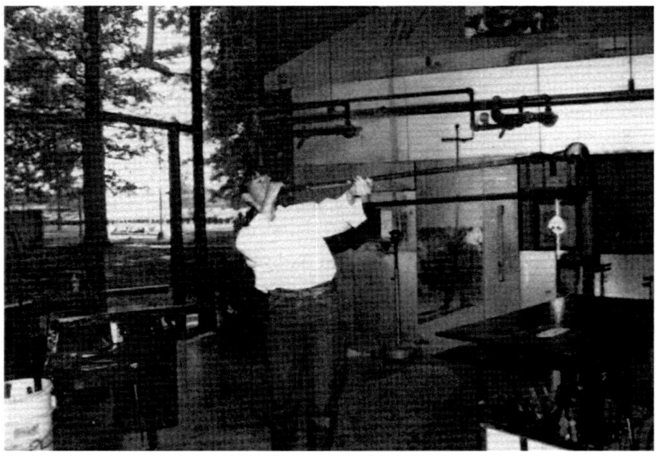

*Fig. 6.* **Glass blowing with a blowpipe**: *The glass-blower's posture lifting the paraison may be compared to that of the ancient glass blowers seen on Roman lamp disks (see Fig. 27). On the right (glowing hot) is the glory hole; on the left (with closed doors to keep the heat inside) is the furnace with the glass tank in the background, below the glass-blower's left hand. "Pigs" or stands for resting the blowpipe can be seen in front of the furnace and glory hole. Photo: Lois Dorfman.*

30 Pliny, *Natural History*, 36.192 mentions the occasional addition of shells (*conchae*) to the batch, which would have been a source of lime, but when he describes the "new method of making glass" (i.e. in his own time, A.D. 23/24-79) he names only two ingredients: sand and soda.

31 Strabo, *Geography*, 16.758; Pliny, *Natural History*, 36.190; Flavius Josephus, *Jewish War*, 2.10, 2; Tacitus, *Histories*, 5.7.

32 Pliny, *Natural History*, 36.194.

33 ... *limosus, vado profundus, non nisi refuso mari harenas fatetur; fluctibus enim volutatae nitescunt detritis sordibus. Tunc et marino creduntur adstringi morsu, non prius utiles*, translation by D. E. Eicholz, *The Loeb Classical Library*, no. 419 (London 1962, reprinted 1971, 1989) 151. "The sand from the mouth of the Belus River contained 9% lime," Turner 1956, 276T-300T; Brill 1988, 266-269.

34 Roy Newton, "Recent Views on Ancient Glasses," *Glass Technology* 21 (1980) 174-178; Newton and Davison 1989, 61-62.

35 Perhaps Pliny refers to the fritting process.

36 It would have been physically impossible to transport the glass while it was molten hot.

37 *(Harena) qua mollissima est, pila molave teritur. Dein miscetur III partibus nitri pondere vel mensura ac liquata in alias fornaces transfunditur. Ibi fit massa quae vocatur hammonitrum atque haec recoquitur et fit vitrum purum ac massa vitri candidi.* Pliny, *Natural History*, 36.194, translation by D. E. Eicholz, *The Loeb Classical Library*, no. 419 (London 1962, reprinted 1971, 1989).

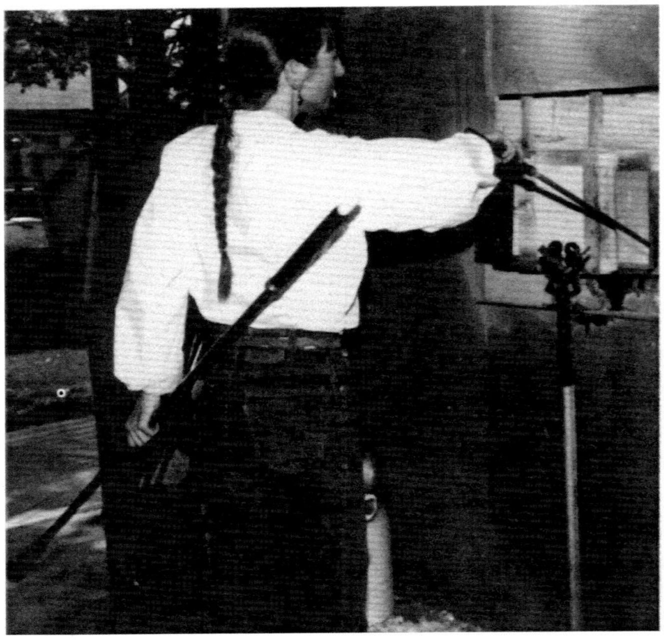

*Fig. 7.* **Hot gathering**: *Gathering a gob of hot glass from the furnace for a pontil. The blowpipe with the vessel is held in the left hand. A stand for supporting a blowpipe can be seen in front of the furnace. The mirror to the right of the stand is used today to allow the glass blower to see more accurately the shape and condition of glass at the end of a blowpipe. Photo: Lois Dorfman.*

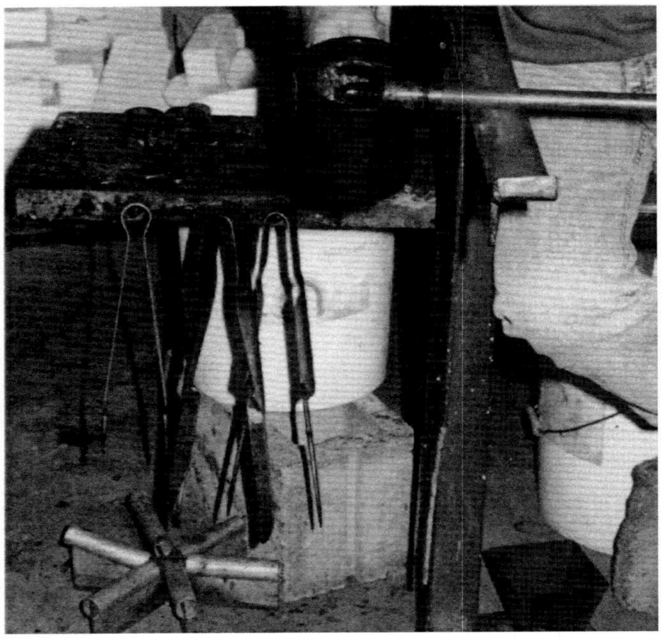

*Fig. 9.* **Modern glass-working tools**: *Detail of a glass blower sitting at a glass-blower's bench (not used in antiquity) shaping a gob of glass in a wood block-mold (not used in antiquity). On the bench lies a pair of modern diamond shears. Suspended from the bench are two pairs of pointed jacks and one pair of flat blade jacks. Behind the bench are two pails of water for keeping the wood tools wet. For examples of ancient glass-working tools, see Fig. 8. Photo: Author.*

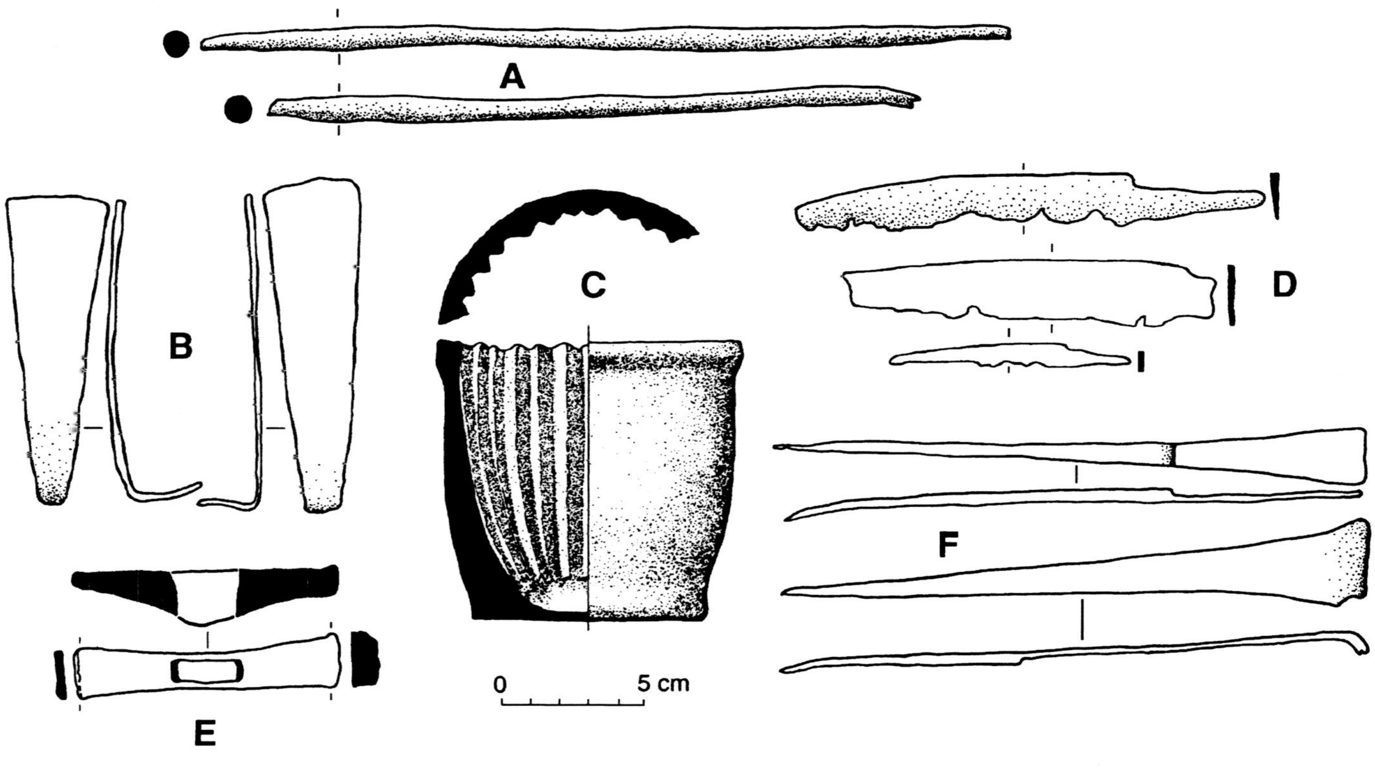


*Fig. 8.* **Ancient glass-working tools**: *Metal tools dated to the third or fourth century* A.D.*, excavated in a glass workshop at Komarowo, Ukraine. A: iron rods or tubes, D 0.8 cm (pontil rods or blowpipes?); B: iron tools; C: terra-cotta mold; D: iron blades (knives); E: iron hammer head; F: iron jacks. The jacks are similar in shape to modern glass-working jacks (see Fig. 9). Drawings: Nina Sorokina; redrawn by Timothy A. Motz.*

The exact nature of the improvements described by Pliny has not been researched, but the pureness of the glass suggests they may be related to a new furnace technology that allowed for higher temperatures during melting. Pliny describes the old method of making glass, *antiqua ratio vitri*, shortly before, in *Natural History* 36.194. Words that are modern conjectures by Pliny's editors for ancient readings in the manuscript are italicized in the following translation. Square brackets indicate words inserted by the translator.

A fire of light, dry wood is used for preparing the melt,[38] to which are added *copper* and soda, preferably *Egyptian soda*. [Glass], like copper, is melted in a series of furnaces, and dull black lumps are formed. [Molten glass] is everywhere so sharp that, before there is the least sensation, it cuts to the bone any part of the body on which it *splutters*. After being reduced to lumps, the glass is again fused[39] in the workshop and is tinted.[40]

Ancient glass melted at approximately 1000°-1100° C. The resulting raw glass was an article of trade; it was sold in *ingots* or chunks and transported to the workshops where *glass working* made it into objects and vessels. See also *Advances in Roman Glass Technology: The Properties of Glass.*

**glass melting** The preparation of molten glass by remelting fragments of cullet and/or raw glass for *glass working*. Glass melting requires only one heating operation. No knowledge of *glass making* is needed.

**glass working** The forming of glass into objects. This can be done when glass is heated (hot working) or while it is cold (cold working). Hot working procedures used by the Romans included various casting techniques, fusing, mold-forming and sagging, core-forming and spinning, and blowing. Cold working procedures included cutting, drilling, engraving, grinding, and polishing. In the Roman period, most hot working was done in a *furnace* built specially for that purpose. However, a built furnace, closed at the top, was not a precondition for working with hot glass.[41]

**glory hole** An opening in a modern glass-working furnace or a separate source of heat without glass. Used today for reheating during glass working (see Figs. 6, 19, and 31).

**gob** A mass of hot glass on the end of a pipe or pontil which has not been inflated.

**hot gathering** (as opposed to *chunk gathering*) Taking molten hot glass from the furnace on the end of a blowpipe or pontil (Fig. 7). In order to work the glass, the outer surface or skin of the glass must first be cooled. See also *Advances in Roman Glass Technology: The Introduction of Hot Glass.*

**hot glass** (molten hot glass) During hot working, glass is always hot. In this catalogue the term hot glass is used specifically to refer to molten glass that is fluid enough to be gathered at the end of a blowpipe or to be poured into a mold, i.e. in a viscosity range of about 4 to 3.2. See also *Advances in Roman Glass Technology: The Introduction of Hot Glass* and Figs. 16 and 17.

**impurity** See *stone.*

**inflation** See *glass blowing.*

**ingot** A mass of raw glass of a convenient shape for storage or transportation to be processed later.[42] Glass ingots were usually shaped like a circular or segmental cake or square, depending on the shape of the crucible in which the glass was made. The original shape of ingots of colored glass, fragments of which were excavated in a first-century glass-working factory at Avenches, Switzerland, could not be reconstructed.[43]

**iridescence** One form of *weathering* that has been observed on ancient glasses. It is a play of bright and varied colors produced by an optical effect called "thin film iridescence," similar to the effect seen on soap bubbles or oil-covered water. On glass, the layers of weathering film are extremely thin. In modern times iridescence is deliberately created by exposure to metallic fumes. (R. Brill, pers. com.).

**jacks** (also called tongs, pucellas) A metal tool used for grooving and shaping vessels. Two pairs of metal jacks were found in third- to fourth-century glass workshops at Merida, Spain,[44] and at Komarowo, western Ukraine[45] (Fig. 8, no. F). Modern jack blades can be flat, round, pointed, etc. (Fig. 9, see also Fig. 23).

**kiln** See *furnace.*

**knocking-off** Removing a completed vessel from a pontil (Fig. 10, see also Figs. 11 and 39).

**knock-off** Glass remaining on a *blowpipe* or *pontil* after the vessel is removed (Fig. 11, see also Fig. 39). Fig. 35 shows a knock-off remaining on a blowpipe. After separation from the pipe, it becomes *cullet.*

**manganese** A metallic oxide used increasingly from the first century as a *decolorizing* agent. When added in larger amounts, it produces a reddish color varying from purple to reddish brown (see Cat. Nos. 116, 117, 139, and 160 in Color Plates 18, 21, and 27).

**marver** A flat surface of stone or metal on which a *gather* or a *paraison* is rolled to smooth and shape (Fig. 12). A slab of polished marble with burn marks that may have served as a marver was excavated at a first-century glass-working site in Avenches, Switzerland.[46]

**MCT** The acronym for Mold Construction Type (see below, *Mold Construction Types* and Fig. 15).

**moil, moile** (also called overblow) The unwanted part of the *paraison* which is removed from the blowpipe with the unfinished vessel. After the vessel is annealed, the moil is cracked off along a scoring line.[47] Moils become *cullet* (Fig. 35). Moils have been excavated at Mérida, Spain, and in great numbers at Jalame, Israel.[48]

[38] Literally, "cooking," *coquitur.*

[39] Fused must here mean remelted.

[40] *Levibus autem aridisque lignis coquitur addito* Cyprio *ac nitro, maxime* Aegyptio. *Continuis fornacibus ut aes liquatur, massaeque fiunt colore pingui nigricantes. Acies tanta est quacumque ut citra sensum ullum ad ossa consecet quidquid* adflaverit *corporis. Ex massis rursus funditur in officinis tinguiturque.* Pliny, *Natural History*, 36.194, translation by D. E. Eicholz, *The Loeb Classical Library*, no. 419 (London 1962, reprinted 1971, 1989) 153. (Conjectures in the Latin text are printed here in roman type).

[41] Rosemarie Lierke, pers. com., notes that the absence of excavated furnaces from pre-Roman times is "almost proof" that they were not used. Tine Gam, "Perlemager af fag," *Skalk* 1 (1990) 12-15, describes medieval bead-making techniques at a simple open hearth in which it was easy to reach a temperature of 1000-1100° C in the hottest part by using a bellows. See also Stern 1994, 24-25.

[42] On trade in glass ingots: Charleston 1963, 54-67.

[43] Morel et al. 1992, 14, fig. 12. See also the Bronze Age ingots recovered from the shipwreck off Kaş, Turkey; Grose 1989, 54-55; George F. Bass, "Oldest Known Shipwreck Reveals Bronze Age Splendors," *National Geographic* 172 (1987) 693-733.

[44] Price 1973, 82, fig. 5.

[45] Sorokina 1977b, 96, pl. I, no. 3.

[46] Morel et al. 1992, 14-15.

[47] On the technique: Weinberg 1988, 89, fig. 4-45.

[48] Mérida: Price 1973, 80-82, fig. 4. Jalame: Weinberg 1988, 98-101, cat. nos. 497-516.

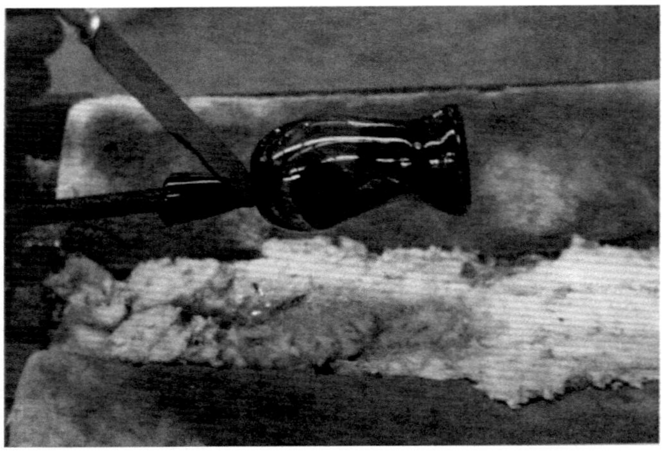

*Fig. 10.* **Knocking-off**: *Cooling the glass between the pontil and the vessel with an unheated piece of metal (a file). The sudden localized cooling in a narrow section of hot glass creates a stress line; a tap against the blowpipe will cause the vessel to break off. Photo: Lois Dorfman.*

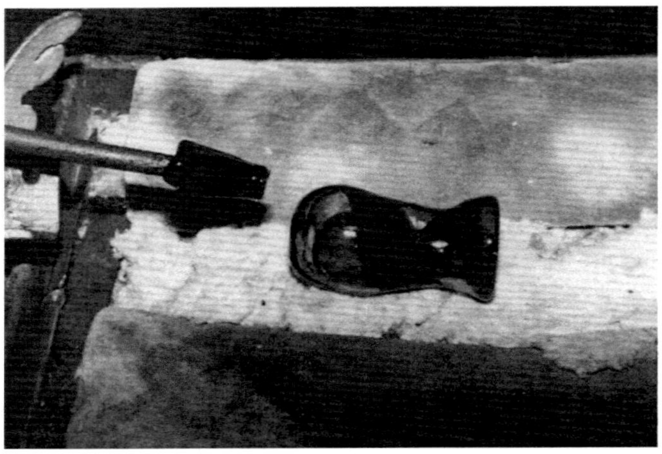

*Fig. 11.* **Knock-off**: *The portion of glass still adhering to the pontil after the vessel is removed. When the knock-off cools, it pops away from the pontil and becomes cullet. Photo: Lois Dorfman.*

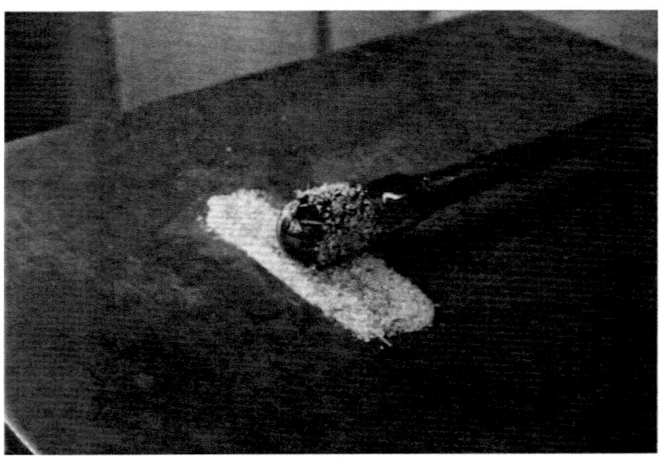

*Fig. 12.* **Marver**: *Rolling a gob of glass through glass chips on the marver (here a steel table). The gob can then be rolled on a free part of the marver to press the chips into the glass. The gob can then be further heated and worked. Photo: Author.*

**mold** A form of one or more sections in which an object receives its final shape (mold blowing), surface decoration (by expanding the paraison or by optical blowing [unknown in antiquity]), or both. All the vessels discussed in this catalogue received their final shape and decoration simultaneously in full-size molds that had the same interior dimensions as the exterior of the vessels made in them. Ancient Roman molds were made from a variety of materials (see Fig. 8 and *Mold Blowing: Mold Materials*). Parts of multipart molds are called "parts" or "sections;" for different types of mold construction, see below: *Mold Construction Types* and Fig. 15).

**Mold** is also a term used in the classification of vessels belonging to one Series (see below: *Descriptive Terms for Classification*).

**mold blowing** Blowing into a mold to give a vessel its final shape. The designs on the walls of the finished vessel appear in negative on the interior (see *Mold Blowing*, especially Figs. 40 and 41).

**mold seam(s)** Mark(s) on a mold-blown object where the force of the glass-blower's breath pushed the glass outward between the edges of the mold sections. Mold seams are the most important criterion used to determine the technical construction of a mold (MCT). Depending on the temperature of the glass when it was blown and how well the mold sections were aligned, mold seams can protrude strongly or be very faint.

**molten hot glass** See *hot glass.*

**offhand blowing** See *free blowing.*

**opacifier** An agent introduced deliberately to make glass opaque. In Roman glass making, antimony was the most common opacifier, but lead-tin oxide was also used.[49] Depending upon the firing conditions and the amount added, antimony can react as a *decolorizer* or as an opacifier producing opaque white or yellow. The opacity or translucency of most early ancient glasses is the result of low melting temperatures which allowed microscopic air bubbles to remain in the glass as it cooled.[50]

**overblow** Glass that bulges out over the edge of a mold when the paraison is expanded (e.g. Cat. Nos. 120-122, 147). See also *moil.*

**paraison, parison** An inflated gob or gather of glass (Fig. 13).

**pitting** Tiny cavities in the surface of the glass due to weathering.

**polishing** (1) Fire-polishing: smoothing and rounding glass by reheating. (2) Cold-polishing: grinding and smoothing by the use of a wheel and a polishing agent. A hand-held stone could also be used for polishing.

**pontil, pontil rod**, or *punty* A solid metal rod, to which vessels may be transferred from the blowpipe for further working (see *pontil technique*). Also used for applying *bits* for handles and *trails* (Fig. 14, see also Figs. 1, 4, 7, 8 (perhaps), 10, 11, 33, 35, and 36).

**pontil mark** A scar left by a pontil on the underside of a vessel (see also *pontil technique*) (see detail of underside of base of Cat. No. 120). In Roman glass working the scar was rarely ground off and polished. Ring-shaped pontil scars made by using the blowpipe itself instead of a pontil are also common (see Fig. 39). Many

[49] Mavis Bimson and A. E. Werner, "Two Problems in Ancient Glass: Opacifiers and Egyptian Core Materials," *AnnAIHV* 4 (1967, published 1969) 262-264; Robert H. Brill, "The Scientific Investigation of Ancient Glass," in *Studies in Glass History and Design. Papers Read to Committee B Sessions of the VIIIth International Congress on Glass* (London 1968, published 1969) 56-57.

[50] Newton and Davison 1989, 59.

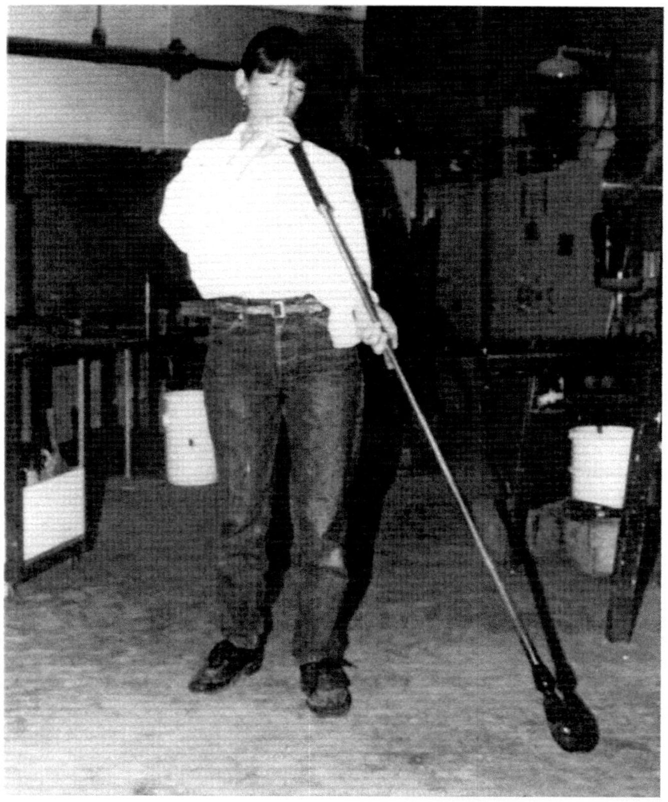

*Fig. 13.* **Paraison**: *A paraison on the end of the blowpipe is being inflated. Photo: Lois Dorfman.*

Roman vessels lack pontil marks; they were held with a *clamping instrument* of unknown nature.

**pontil technique** The technique of transferring a vessel to a *pontil* for further working after it has been cracked off the blowpipe. The technique did not become common before the end of the first century. See *Advances in Roman Glass Technology: The Pontil Technique* and Figs. 33-36, compare Figs. 37-39.

**projecting roll** A decoration formed either by an *overblow* (e.g. Cat. Nos. 120 and 147) or by pushing the glass outward with a blunt pointed tool (e.g. Cat. No. 128).

**punty** See *pontil.* Also the glass on the pontil that will form the bond between vessel and pontil.

**raw glass** See *glass making.* In ancient Greek texts the word for raw glass was *bolos*, "chunk." [51] The Greek word *bolos*, written in Hebrew letters, is used in the Babylonian Talmud for raw glass transported to factories in the form of chunks. [52]

**section** Part of a multipart glass-blower's mold.

[51] See *Advances in Roman Glass Technology: Chunk Gathering.* Robert H. Brill, "Some Thoughts on the Origin of the Chinese Word, 'Boli'," *Glass and Enamel* (Shanghai) 18, no. 5 (1991) 61-64, suggests that the Chinese word for glass, *bo-li*, may have been derived from the Greek *bolos.*

[52] Babylonian Talmud, Shabbath 154b, where Rabbi Huna (third century) explains a passage of the Mishnah relating to rules for observing the Sabbath: Weinberg 1988, 25, n. 2: "If a man's animal is loaded with glass and the Sabbath has commenced without his realizing it, he is instructed as follows (free translation): "If his animal is loaded with glass vessels, he may bring pillows and blankets and lay them underneath, then loosen the cords and the sack will fall down. But if his animal is loaded with [glass] chunks (boli), he must just loosen the cords and let the sacks fall down, even if the glass breaks."

**Series** All vessels blown in a mold derived from one archetype (see below: *Descriptive Terms for Classification*).

**shears** A metal tool used for cutting (see Fig. 38). In Roman times shears were constructed like tongs or *jacks.*

**specialized glass** Colored or colorless glass into which *colorants* or *decolorizers* were deliberately introduced. [53] These glasses were made in only a few centers and exported throughout the Roman world.

**stone** Any non-glassy material embedded in a piece of glass. The most common sources of stones in ancient glass were unfused batch materials, bits of refractory material from the furnace or pot, and accidental inclusions of dirt from the environment (see Cat. No. 131).

**thread** See *trail.*

**tongs** See *jacks.*

**trail, trailing** A coil of glass applied hot to a vessel, either to form a handle or for decoration; a very thin trail is called a thread (see Cat. No. 123); thick trails may be coils (see Fig. 1 and Cat. No. 126).

**tuyère** Mouthpiece of a bellows, usually made of clay.

[53] On specialized glasses: Charleston 1963, 54-68.

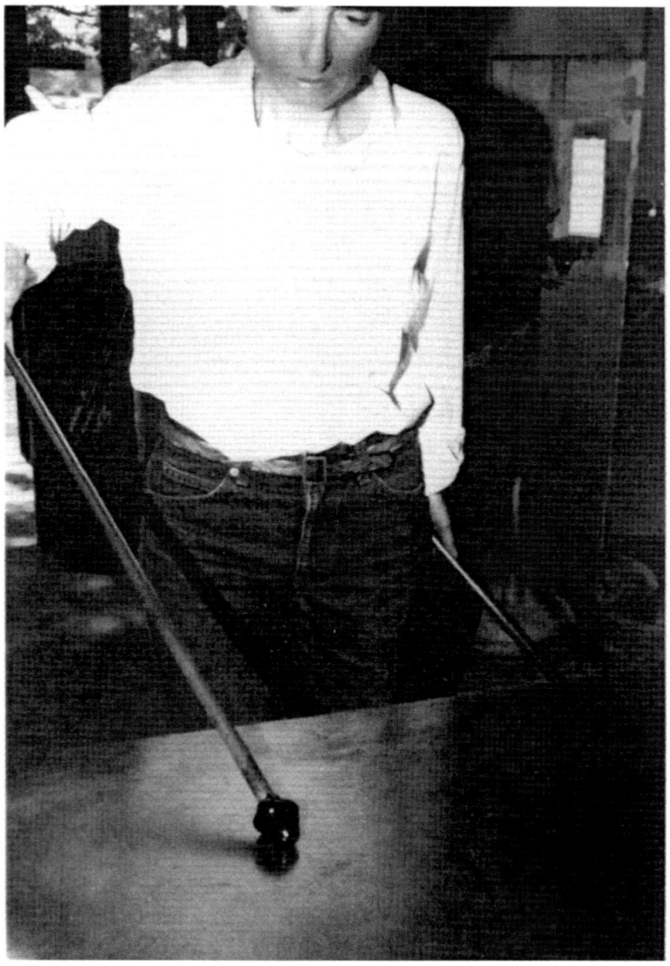

*Fig. 14.* **Pontil**: *Shaping a gob (i.e. preparing a pontil) on a pontil rod by rolling the gob on a marver. The glass blower holds a blowpipe with the vessel to which she will attach the pontil in her left hand. Photo: Lois Dorfman.*

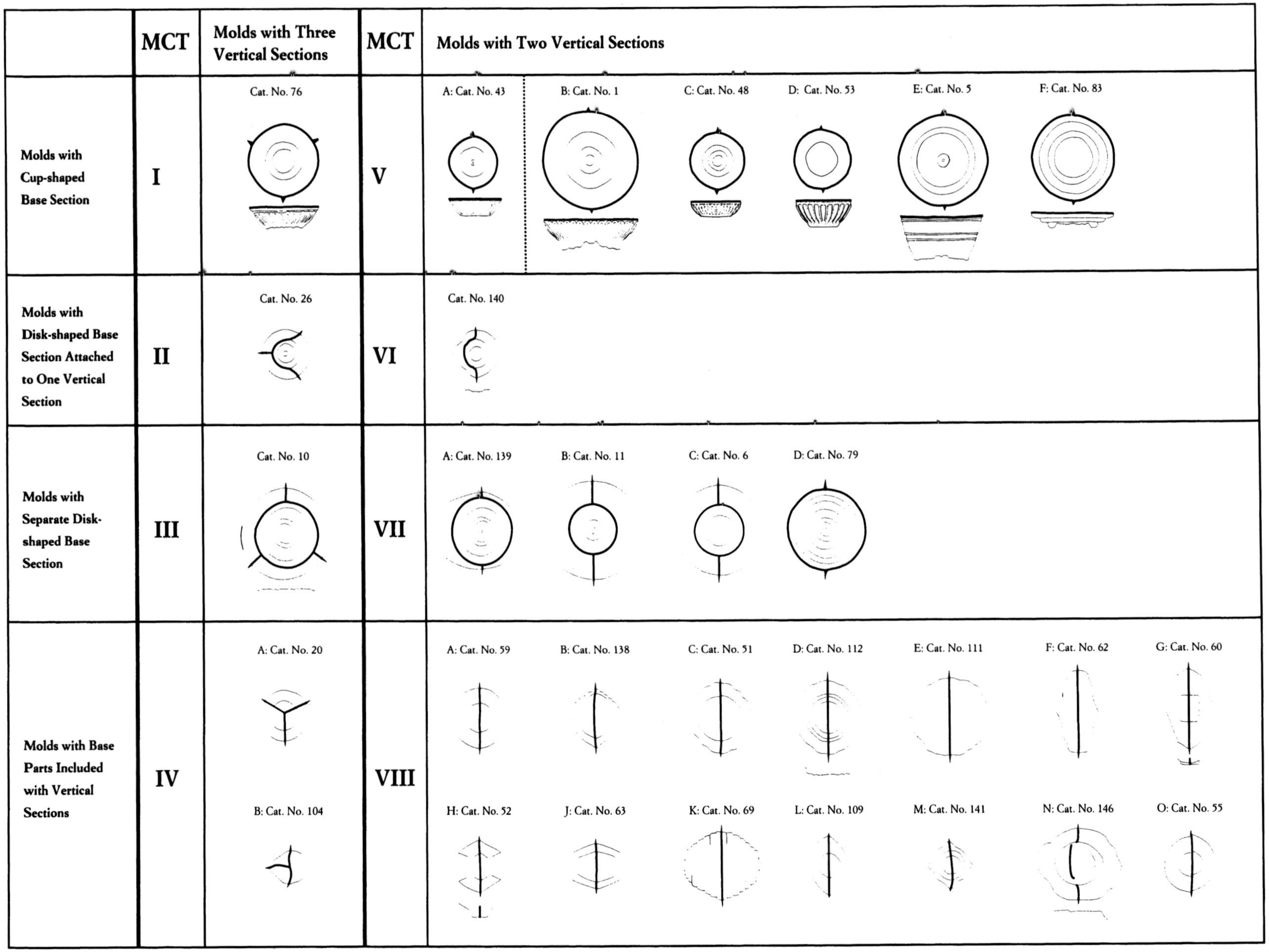


*Fig. 15. First-century Mold Construction Types (MCTs I-VIII). Bold lines indicate mold seams. Scale 1:4.5. The drawings represent actual vessels in the Toledo Museum collection. MCT V A is set apart because, although it shows Cat. No. 43, this particular MCT is a generic type, somewhat different from the other examples of MCT V (see discussion). Chart designed by the author.*

**Type** Synonymous with subject and design, can refer to mold-blown vessels and to the molds in which they were blown (see below: *Descriptive Terms for Classification*).

**viscosity, resistance to flow** Viscosity is the most important property of glass. All processes of glass working (blowing, trailing, marvering, pinching, etc.) depend upon it. The lower the viscosity, the softer and less cohesive the glass. In its molten state glass loses cohesion (see *hot glass*). A high viscosity makes the glassy state possible. Viscosity is measured in poise. Depending on the temperature and composition of glass, viscosity changes. Different glass-working operations require different viscosities (see *Advances in Roman Glass Technology: The Properties of Glass* and Figs. 16 and 17).

**waste** Any refuse from glass working, including *cullet, knock-offs, moils,* and deformed or broken vessels.

**weathering** Changes to the surface of glass produced by interaction with its environment.[54] These include iridescence, white film, dulling, white to brown enamel-like weathering, black layer, and — the most severe form — pitting and corrosion.

## MOLD CONSTRUCTION TYPES (MCTs I-IX)

Nine Mold Construction Types (MCTs) have been identified from vessels in the Toledo Museum collection (Fig. 15). There were four ways to shape the base of a vessel: a cup-shaped base section (MCTs I and V), a disk-shaped base section permanently attached to one of the vertical sections (MCTs II and VI), a separate disk-shaped base section (MCTs III and VII), and a base divided among as many parts as there were vertical mold sections, each of which included part of the base (MCTs IV and VIII). MCT IX was open at the bottom. In the Toledo Museum collection examples of each type of base exist with either two or three vertical sections. Thus MCTs I and V are basically the same, the only difference being that MCT I has three vertical sections and MCT V has two. In the same way, MCTs II and VI, MCTs III and VII, and MCTs IV and VIII have the same type of base construction, differing only in the number of vertical sections. MCT IX is represented in Toledo only by vessels with two vertical sections.

Not surprisingly, shape of vessel and construction of mold appear to have been closely related. Four types of MCTs were used for tablewares and larger vessels (MCTs I, III, V, and VII). Three types occur exclusively for small Hexagonal Bottles with High Relief (MCTs II, IV A, and V A). Three types were used for other small containers (MCTs I, V, and VIII). Four types have been noted for head-shaped vessels (MCTS VI, VII, VIII, and IX).

MCTs I-VIII were all invented during the first century. MCT IX was developed after the end of the first century.

A nineteenth-century mold type is found in two imitations of head flasks (Cat. Nos. 166 and 167).

MCT I Vessels made in MCT I molds have no seam on the underside of the base or on the lower part of the wall forming the bottom section of the vessel. The mold had three vertical sections and a cup-shaped base section. The greatest circumference of the mold was circular. Some MCT I molds may have been made of metal or plaster, but most were probably made of clay. MCT I molds were used for a few small closed containers (e.g. Cat. No. 76), but more commonly for tablewares (e.g. Ennion's jugs).[55] Date: first half of the first century, probably first quarter. Origin: Syro-Palestinian, probably Sidon.

MCT II Vessels made in MCT II molds have no seam on the underside of the base of the vessel, but rather two short diagonal seams slanting upward to two of the three vertical seams. The mold had three vertical sections and a disk-shaped base section permanently attached to one of the vertical sections. The greatest circumference of the mold was circular. The mold was made of clay. The use of an MCT II mold has been noted only for Hexagonal Bottles with High Relief; it is common in Series A3 of the Vessels Type, where the underside of the base is decorated with concentric circles (see photographs of the undersides of the bases of Cat. Nos. 26 and 27).[56] Date: first half of the first century, probably beginning slightly later than MCT I. Origin: Syro-Palestinian, probably Sidon. For discussion, see Cat. No. 26.

MCT III Vessels made in MCT III molds have no seam on the underside of the base of the vessel or around its circumference; three vertical seams end at the edge of the base. The mold had three vertical sections and a separate disk-shaped base section.[57] The greatest circumference of the mold was circular. It is not clear of what material the mold was made. The use of an MCT III mold has been noted for tablewares only; it is common for beakers decorated with knot-shaped bosses, where the underside of the base is often decorated with concentric circles (e.g. Cat. No. 10). MCT III molds were probably invented by Ennion.[58] Date: First through third (and fourth?) quarter of the first century. Origin: Syro-Palestinian, probably Sidon.

MCT IV molds had only three vertical sections. Two variations occur: vessels blown in MCT IV A molds have a flat base allowing the vessel to stand, while vessels blown in an MCT IV B mold lack a standing surface. Both MCTs appear to have been used for small containers only.

MCT IV A Bottles made in MCT IV A molds have a Y-shaped seam on the underside of the base. Each of the three vertical sections of the mold included a triangular section forming about one third of the base. The greatest circumference of the mold was circular. It was made of clay. MCT IV A was the most common mold used for Hexagonal Bottles with High Relief (e.g. Cat. No. 15); it does not appear to have been used for other shapes. Date: first half of the first century, probably first quarter. Origin: Syro-Palestinian, probably Sidon.

MCT IV B Bottles made in MCT IV B molds have one continuous seam around the walls and bottom, dividing the vessel into two halves, one of which is divided into two approximately equal parts by a shorter mold seam from neck to bottom only.

54 Newton and Davison 1989, 135-164.

55 Israeli 1983, 67-68.

56 Judging from the drawing published by Eisen 1927, I, 251, fig. 122, a cylindrical bottle belonging to the Plants Type (see Cat. No. 48) may have been blown in an MCT II mold.

57 The absence of a mold seam between the disk-shaped base section and the lower edge of the walls can be explained by the fact that during blowing into a mold vertically from above, the pressure of the glass is much stronger outward and upward than downward.

58 An amphora bearing Ennion's signature in St. Petersburg, Hermitage, no. P 1852.54, from Panticapaeum, was made in an MCT III mold, Kunina 1973, 101-106, esp. 103, figs. 1-6.

The mold had three sections, one large section for about one-half of the circumference and two smaller sections. It was made of clay or plaster. MCT IV B was used for date-shaped flasks (e.g. Cat. No. 104). Date: middle and second half of the first century. Origin: Syro-Palestinian, perhaps Sidon.

MCT V Vessels made in MCT V molds have no seam on the underside of the base or on the lower part of the wall forming the bottom portion of the vessel. The mold had two vertical sections and a cup-shaped base section. The greatest circumference of the mold was circular. MCT V A stands somewhat apart from the other MCT V molds because evidence from vessels made in this mold type suggests the molds were made of metal. This MCT is always indicated with the suffix "A" (MCT V A). In Fig. 15 it is therefore set apart from the other MCT V molds. Other examples of MCT V may have been made of clay or plaster. Date: first half of the first century, probably first quarter. Origin: Syro-Palestinian, probably Sidon.

MCT V A Bottles made in an MCT V A mold have a curious umbilicus in the center of the underside of the base. This mold was common only for Hexagonal Bottles of the Mixed Symbols Type (e.g. Cat. Nos. 43 and 44, see Fig. 3).

MCTs V B and V E were used for tablewares (e.g. Cat. Nos. 1 and 5).

MCTs V C and V D were used for small bottles and jugs (e.g. Cat. Nos. 48 and 53; compare 72).

MCT V F was used for pyxides (e.g. Cat. No. 83).

MCT VI Vessels made in MCT VI molds have no seam on the underside of the base but rather two short diagonal seams slanting upward to the two vertical seams. The mold had two vertical sections and a disk-shaped base section permanently attached to one of the vertical sections. The greatest circumference of the mold fitted within a circle. The mold was probably made of clay. The use of an MCT VI mold has been noted for one head flask (Cat. No. 140). Date: second (?) to third quarter of the first century. Origin: Syro-Palestinian, probably Sidon.

MCT VII Vessels made in MCT VII molds have no seam on the underside of the base or around its circumference; two vertical seams end at the edge of the base. The mold had two vertical sections and a separate disk-shaped base section.[59] The greatest circumference of the mold was either circular or fitted within a circle. The material used for the mold(s) has not been determined. The use of an MCT VII mold has been noted for tablewares, pyxides, and head flasks. The undersides of these vessels are usually decorated, most frequently with concentric circles (e.g. Cat. Nos. 79 and 139), but other patterns also occur (e.g. Cat. Nos. 142 and 148). A few, presumably late-first-century vessels have a plain undecorated base (e.g. Cat. No. 6). Date: probably invented in the third quarter of first century and still (or again) in use at the end of the second or early third century (e.g. Cat. Nos. 120-128). Origin not established; found in the East and in the West.

MCT VIII Vessels made in MCT VIII molds have a straight seam on the underside of the base, or across the bottom of the vessel if it did not have a standing surface. The mold had two vertical sections, each forming part of the base. The greatest circumference of the mold varied greatly in outline, but it was always symmetrical. Most MCT VIII molds were probably made of clay. This mold has not been noted in tablewares made in the eastern Mediterranean. It was the most common construction for all shapes of small containers except Hexagonal Bottles with High Relief. The earliest documented use is in the Workshop of the Floating Handles. Date: early first century to (end of?) the second century. Origin: Syro-Palestinian, probably Sidon; found in the East and in the West.

MCT IX Vessels made in MCT IX molds have no seam on the underside of the base, which can be flat, concave, or convex. A slight bulge in the glass between the lower edge of the wall and the base shows that the glass had room to expand outward. The surface of the underside of the vessel often has small pits and inclusions of dust picked up from the floor of the workshop. The mold was open at the bottom and had only two vertical sections.) Most molds were probably made of clay or plaster. This type of mold was commonly used for head-shaped vessels that can be dated to the third and fourth centuries (e.g. Cat. No. 153). MCT IX was developed after the end of the first century.[60]

## DESCRIPTIVE TERMS FOR CLASSIFICATION

In the description and classification of Hexagonal Bottles with High Relief, this catalogue uses the following terms: Type, Series, Mold, and Generation.

**Type** The term Type, capitalized, is introduced here to indicate vessels showing related motifs in the main zone of decoration. In previous glass studies these have often been called "Series."[61] However, Series has become the accepted term in terra-cotta studies to refer to vessels descended from a single archetype, and this term has been reserved for the same purpose here. Thus, the Vessels Type includes all bottles decorated with motifs of vessels; while the Fruit Type refers to those decorated with fruit.

**Series** One Type can include various Series. The term Series, capitalized, is used here to indicate vessels that descend from a single archetype.[62] Within each Type the Series are indicated by an uppercased letter (A, B, etc.).

**Mold** Each technical variation in molds belonging to the same Series is indicated by an Arabic numeral (1, 2, etc.). Thus the Arabic numerals refer to any technical variation, whether a different MCT or merely a different section division within one MCT. For example, Molds A1 and A2 of the Vessels Type were both MCT IV A molds belonging to Series A of the Vessels Type. However, Mold A1 (e.g. Cat. No. 15) had seams between panels 2 and 3, 4 and 5, 6 and 1, while Mold A2 (e.g. Cat. No. 25) had seams between panels 1 and 2, 3 and 4, 5 and 6.

**Generation** When a mold became worn out, it was usually replaced. Bottles blown into a new mold constitute a younger generation. The decorative motifs of the new mold could be freshened or modified: in this case the new Generation is indicated by a lowercased letter (a, b, etc.) following the numeral referring to the classification of the mold. If no modifications are visible in the new Generation, this is indicated by adding "-y" (for "younger

[59] On the absence of a mold seam between the disk-shaped base section and the lower edge of the walls, see discussion of MCT III.

[60] Brown 1968, 405, suggests that some first-century mold-blown beakers with tooled bases were blown in open molds, but these could just as well have been blown in molds that included a section for the base (see Cat. No. 8).

[61] Eisen 1927, I, 233ff.; Lightfoot 1985, 124.

[62] R. V. Nicholls, "Type, Group, and Series: a reconsideration of some coroplastic fundamentals," *BSA* 47 (1952) 217-226.

generation") or "-yy" (for "next younger generation") following the lowercased letter indicating the generation. Thus, bottles belonging to Generation A1a were presumably blown in the first mold made for Series A of a given Type and are the first to be discussed. Generation A1b indicates the first modification observed in a bottle descending from the original archetype. Generation A1b-y indicates a bottle descended from A1b but blown in a younger generation mold that was not noticeably modified.

## DESCRIPTIVE TERMS FOR VESSEL SHAPES

A universally accepted terminology of ancient glass shapes does not exist in spite of various past attempts to devise one.[63] In this catalogue, Greek and Latin names are used only when identification of a shape is beyond doubt and the term has come into general usage. Where possible, an English name is preferred. The following shape names are used:

**amphora** Originally, a two-handled storage or transport vessel used by Greeks and Romans for both liquids and solids. The shape was imitated in reduced size in glass. Glass amphoras and amphoriskoi (miniature amphoras) were used predominantly for liquids and semi-liquids (e.g. Cat. No. 59).

**beaker** A straight-sided cup that could have been used for drinking, playing with dice, or other functions (e.g. Cat. No. 8).

**bottle** A storage container for liquids with a narrow mouth that could be closed with a stopper,[64] for example, Cat. No. 15. In most cases bottles lack handles. If a vessel shape occurs predominantly without handles, it is here also called a bottle on the rare occurrences when the shape is found with one or more handles, for example, Cat. Nos. 50-70. Two-handled bottles imitating the shape of a Greek or Roman amphora are called amphoras.

[63] In 1979 the ICOM (International Committee of Museums) committee for glass appointed a Working Group for the Terminology of Ancient Glass, consisting of Clasina Isings, Renate Pirling, Jennifer Price, Nina P. Sorokina, and E. Marianne Stern. In 1985 the Working Group decided to abandon the task because it had been impossible to convene a meeting of all of the participants.

[64] Various materials are known to have been used in Roman times as stoppers for glass vessels. Glass stoppers: Cologne, RGM, lathe-turned of sea-green glass, dated first century, from Cologne Marienburg, Fritz Fremersdorf, *Römisches Buntglas aus Köln (Denkmäler des römischen Köln* 3, Cologne 1958), 55, pl. 124h; Athenian Agora, no. G 398, cast glass, and not trimmed, from a first-century B.C. context (G. D. Weinberg, pers. com.). Glass stoppers for tubular glass kohl jars dating from the mid-third to mid-fourth centuries are known: formerly Constable Maxwell Collection, *Maxwell* 1979, 68, no. 109, ill.; and formerly R. W. Smith Collection, acquired in Lebanon, Smith 1957, 155-157, no. 321, ill., both with a bird-like element as finial. A lead stopper for a glass aryballos was excavated in a first-century tomb at Monasteriaki Kephala, Crete: Carington Smith 1982, 282, no. 77, pl. 40a, b. A wooden stopper was found at Cologne in a sarcophagus dated to the first half of the third century; La Baume 1960, 82, fig. 26. A stopper made of grass was found in a first-century glass unguentarium from the Egyptian Fayum: Edgar 1905, 55, no. 32661, pl. VIII. An upside down pine cone used as a stopper was found in a Roman glass bottle in the Athenian Agora (G. D. Weinberg, pers. com.). Dried figs were used as stoppers for ceramic vessels in the Bronze Age shipwreck at Ulu Burun: G. Bass, "A Bronze Age Shipwreck at Ulu Burun (Kaş): 1984 Campaign," *AJA* 90 (1986) 285. The Tosephta, Shabbath 8 (9), 12 mentions the use of papyrus as a stopper: "who takes a sheet in order to wind it over the mouth of a small flask of spikenard," cited in *Function and Design in the Talmudic Period*, exh. cat., Haaretz Museum 1978 (Tel Aviv 1979, 2nd ed.) 53*. On ancient stoppers, see also Virginia R. Anderson-Stojanovic, "The Chronology and Function of Ceramic Unguentaria," *AJA* 91 (1987), 114, nn. 50-51.

**cup** A deep or shallow bowl used for drinking (e.g. Cat. No. 1).

**dropper flask** See *sprinkler.*

**flask** A container of bottle shape that was not meant to be used for storage and therefore usually was not closed with a stopper, e.g. a flask with a funnel-shaped neck used as a decanter (see Cat. No. 145).

**guttus** or **gutturnium** See *sprinkler.*

**jar** A wide-mouthed container that could be closed with a lid, usually lacking a handle.

**jug** A container with one handle for pouring liquids; the mouth can be circular, spouted, or trefoil.

**pyxis** A circular or multisided jar or box, usually small and with a lid, used for a variety of purposes (e.g. Cat. No. 79).

**sprinkler** (also called a dropper flask, guttus, or, more rarely, gutturnium) A vessel with an inner diaphragm at the junction of the neck and body. The Latin names guttus and gutturnium, used by ancient authors for vessels from which wine, oil (for drinking, cleansing, and for filling lamps), or water could be poured drop by drop, are technically appropriate for glass sprinklers, but they are not used in this catalogue because the actual shape(s) of a Roman guttus are not known, and no mention is made in antiquity of a guttus of glass.[65] Vessels of several different shapes could be used as sprinklers (see Chapter II: *The Second to Fourth Century: Shapes*).

## DESCRIPTIVE TERMS FOR VESSEL PARTS

Only terms which are unusual or which might cause confusion are defined or listed here. Other descriptive terms found in the entries are considered self-explanatory.

**base** The underside or resting surface of a vessel. It may be flat, convex, or slightly concave, depending on the shape of the mold and manipulation of the vessel after the mold was pulled off.

**body** The portion between rim (and neck, if present) and base. The upper part of the body is called the shoulder; the lower part, the bottom. Variations of body shapes include:

- Globular or spherical
- Ovoid
- Piriform
- Conical
- Cylindrical
- Bulbous (not regular enough to be called one of the above)
- Hexagonal

**bottom** The lowest part of the body, just above the base.

**mouth** Used only when it is a distinct feature, as on bottles or jugs. Standard variations of mouth shapes include:

- Circular
- Trefoil
- Spouted

[65] Hilgers 1969, 58-60 and 191-192.

**rim** The edge and directly adjacent upper part of a vessel. Types and finishing techniques of rims include:

- **Folded rim**, solid or hollow with air space in the fold. The thin glass lip could be made to collapse inward (folded inward) or outward (folded outward) by using natural forces of centrifugal force, gravity, and heat. Most folded rims were not tooled but resulted, sometimes intentionally, sometimes unintentionally, from manipulating and turning during fire-polishing of the rim.
- **Ground rim**, ground (and sometimes polished) after annealing.
- **Rounded rim**, fire-polished before annealing.
- **Unworked rim**, sharp, unfinished edge.

**description**

- **Flute**, vertical ornamental pattern composed of juxtaposed concave elements (e.g. the bottom of Cat. No. 40).
- **Tongue**, vertical ornamental pattern composed of juxtaposed convex elements (e.g. the bottom of Cat. No. 45).

## GEOGRAPHIC DESIGNATIONS

No uniformly accepted rules exist for the transliteration and translation of geographic names in the Near East, the region associated with most of the objects in this catalogue. Depending on whether the Arab, Hebrew, or Greek version of a name is used and whether the name is translated or transliterated into English, French, or another language, names referring to the same location can appear in very different spellings, for example Akko, Akka, Accho, Acre (called Ptolemais in the Hellenistic period); Beisan or Beth Shean (ancient Nysa, Skythopolis). To facilitate use, this catalogue uses place names as they are spelled in the source or publication cited, rather than attempting to adopt a uniform spelling. Alternative spellings occurring in major publications are noted in the Index. Locations of excavation sites usually are given with the name of the country in which they are now found. However, the political situation in several countries is currently so fluid and has changed so frequently in the past forty years that no political meaning should be attached to these designations.

Attributions such as "Syria" and "Beirut" found in old publications and documents do not translate to what is today the territory of the modern state of Syria or the city of Beirut. Many objects said to have been found generically in "Syria" were unearthed in villages along the coast of what are now northern Israel and southern Lebanon.[66] "Beirut" referred to the Vilayet of Beirut, an administrative province of the Ottoman empire. Because boundaries of modern countries do not reflect geographical units in antiquity and the names and borders of administrative units of the Roman Empire changed frequently in the Near East, the following terms are used to designate larger geographical areas for the origin and ancient distribution pattern of objects. This nomenclature is used in a strictly archaeological sense:

**Eastern Mediterranean** The combined territories of the modern countries and regions bordering the eastern Mediterranean Sea and their inland areas: Turkey, Syria, Lebanon, Israel, Jordan, Cyprus, and Egypt.

**Near East** The combined territories of western Syria, Lebanon, Israel, Jordan, and Cyprus.

**Palestine or Palestinian** Roughly the area of Lebanon south of Tyre; Israel, including the Gaza Strip; and the western half of Jordan. The northern part of the coastal area of this region overlaps with the area called Phoenicia. The area east of the River Jordan is here usually referred to as Transjordan or Inland Palestine. Inland Palestine can also refer to areas in Israel that do not border the Mediterranean, such as Galilee and Judea.

**Phoenicia** or **Phoenician** Roughly the area along the eastern Mediterranean coast from Akko (Acre) in Israel to Lattakya (ancient Antiochia) in Syria. In order to avoid confusion with the ethnic, historic, and cultural associations usually associated with the adjective Phoenician, "Phoenicia or Phoenician" is used in the classification of origin in this catalogue only as an indication of the geographic area described above. Pliny calls this area Phoenice or Coele ("Hollow") Syria, "that part of Syria which is known as Phoenicia (Phoenice) and borders on Judea contains... amid the lower slopes of Mount Carmel... near the colony of Ptolemais" (also known as Akko, Acre) (*Natural History* 36.190; cf. *Natural History* 5.75).

**Syria or Syrian** Roughly the territory of the Arab Republic of Syria, excluding the coastal area south of Lattakya which is here included in Phoenicia.

**Syro-Palestinian** The combined areas designated Palestinian, Phoenician, and Syrian (see above).

[66] For example, the preface to the catalogue of the Durighello Collection, auctioned in Paris in 1911, records that the glasses were found "in the villages of Chefah-Amer, Keppreh, Bassah, El-Zib, and in almost all localities between Acre (Akko) and Tyre, also further inland, as far as Lake Tiberias." The villages mentioned are in northern Palestine and southern Phoenicia, western Galilee, and Galilee. Bergman 1974, 243, states that the well-known New York antiquities dealer Azeez Khayat "spent every winter and spring excavating in Syria," but specifies the actual locations of his excavations as having been in and near Tyre and Haifa on the Syro-Palestinian coast and in the vicinity of Nazareth in Galilee.

## ABBREVIATIONS

| | |
|---|---|
| acc. no. | TMA accession number. At The Toledo Museum of Art the four digits to the left of the period represent the year in which the object was accessioned. In earlier publications the first two digits performed the same function. For example, the current acc. no. 1923.45 is the same as the former acc. no. 23.45. The digits following the period indicate the sequence of accession within the year; e.g. 1923. 45 indicates the forty-fifth work accessioned in 1923. |
| acq. | acquired, acquisition |
| anc. | ancient |
| App. | Appendix |
| Arch. | Archaeological |
| BM | British Museum, London |
| C | Centigrade or Celsius |
| ca. | circa (about, used for measurements only) |
| Carnegie | The Carnegie Museum of Art, Pittsburgh, Pa. |
| Cat. No(s). | catalogue number(s), cross reference for this volume. |
| cf. | *confer*, compare, see (the abbreviated form refers to other publications; see, written in full, refers to items in this volume) |
| cm | centimeter(s) |
| CMG | The Corning Museum of Glass, Corning, N.Y. |
| col. | column |
| color pl(s). | color plate(s) |
| comp. | compiled by |
| D | diameter |
| ed. | edition or edited by |
| esp. | especially |
| est. | estimated |
| exh. cat. | exhibition catalogue |
| ext. | exterior surface |
| f. | following page |
| ff. | following pages |
| fig(s). | figure(s) in cited references |
| Fig(s). | figure(s) in this volume |
| Gr. | Greek |
| H | height |
| idem | the same |
| ill. | illustration or illustrated |
| int. | interior surface |
| km | kilometer(s) |
| L | length |
| Lat. | Latin |
| lit. | literature |
| loc. cit. | *loco citato*, in the place already quoted |
| MCT | Mold Construction Type (see *Mold Construction Types* and Fig. 15) |
| MFA | Museum of Fine Arts, Boston |
| MKG | Museum für Kunst und Gewerbe, Hamburg |
| MMA | The Metropolitan Museum of Art, New York |
| mod. | modern |
| Mus. | Museum, Musée, Museo, etc. |
| n(n). | note(s) or footnote(s) (the abbreviated form refers to other publications; note, written in full, refers to notes in this volume) |
| Nat. | National, Nationale, Nazionale, etc. |
| NCG | Ny Carlberg Glyptotek, Copenhagen |
| n.d. | no date of publication |
| neg. no(s). | negative number(s) |
| Newark | The Newark Museum, Newark, N.J. |
| no(s). | number(s), used for reference to accession or inventory numbers in other museums |
| n.p. | no place of publication |
| N. S. | new series |
| opp. | opposite |
| pace | with deference to; expresses polite disagreement (Latin *pace*, ablative of *pax*, peace). |
| passim | throughout, scattered |
| pers. com. | personal communication |
| pl(s). | plate(s) |
| pres. | preserved |
| r. | ruled, reigned |
| rec. | reconstructed, reconstruction (used of measurements) |
| repr. | reprinted |
| RGM | Römisch-Germanisches Museum, Cologne |
| RGZM | Römisch-Germanisches Zentral Museum, Mainz |
| RM | The Rockefeller Museum, Jerusalem, formerly The Palestine Archaeological Museum |
| ROM | Royal Ontario Museum, Toronto, Canada |
| suppl. | supplement |
| s.v. | *sub voce*, at the word, entry |
| Tel Aviv | Eretz Israel Museum, Tel Aviv, formerly Haaretz Museum |
| TMA | The Toledo Museum of Art |
| V & A | Victoria and Albert Museum, London |
| vel sim. | *vel similis*, or similar |
| vol. | volume |
| Yale | Yale University Art Gallery, New Haven, Conn. |

# Advances in Roman Glass Technology

There is no record of or evidence for systematic investigation in antiquity into the properties of glass. The discoveries and inventions leading to glass blowing as it was practiced in the Roman world took place over an extended period of time, from the original discovery of the material glass, in the second half of the third millennium B.C.,[1] to the end of the first century A.D., when molten hot glass[2] and the use of a pontil had come into common use. These discoveries were due to a series of serendipitous workshop accidents that made glass workers aware of new properties of glass and prompted the invention of new tools and implements that could take full advantage of these properties. These new tools in turn led to important new discoveries concerning properties of glass and the development of new glass-working techniques.

## THE PROPERTIES OF GLASS

The most important physical property of glass to be taken into account during all glass-forming operations is its viscosity (Fig. 16). Viscosity is the opposite of fluidity. In its most viscous state glass is solid. The viscosity of glass depends on its temperature and composition. As temperature increases, viscosity decreases and glass becomes fluid. In practical terms of glass working, the duration of exposure to a given temperature can be even more important than the actual temperature.[3] The exact temperatures needed to achieve a specific degree of viscosity depend on the chemical composition of the glass. Most ancient soda-lime-silica glasses would have needed temperatures comparable to those calculated for the glass used at a fourth-century factory in Jalame because

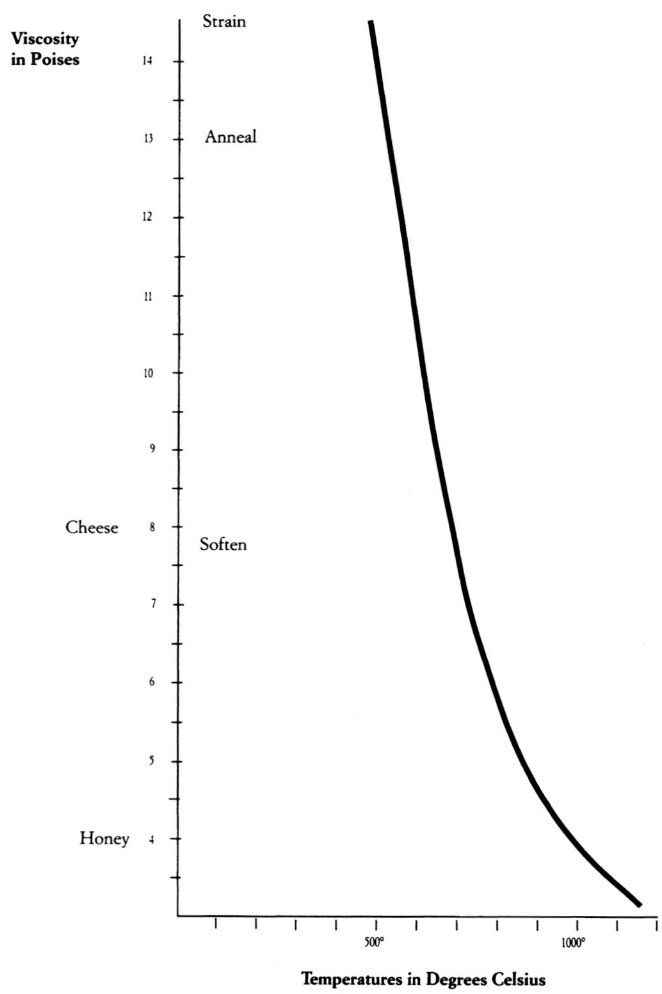


*Fig. 16. Viscosity and temperatures of ancient soda-lime-silica glass. Adapted from Brill 1988, 280, Table 9-12.*

[1] On the beginnings of glass making and glass working, see most recently Stern 1994, 87-88. The earliest chunks of glass were excavated in Mesopotamia at Tell Asmar (ancient Eshnuna) and Abu Shahrein (ancient Eridu) in stratigraphic contexts dating to the 23rd and 21st centuries B.C.: Dan P. Barag, "The Origin of Glass," *IX Congrès International du Verre, Versailles, 27 septembre-2 octobre 1971. Communications artistiques et historiques* (1972) 183-190. On the translucent blue chunk from Eridu, 21st century B.C., London, BM, no. 115474: Barag 1985a, 111, no. 179, pl. 20, color pl. A. See summary in Grose 1989, 45, n. 2.

[2] The terms "hot glass" and "molten hot glass" are used here to refer to glass that is fluid enough to be gathered at the end of a blowpipe or to be poured, i.e. in a viscosity range of about 4-3.2 (see Figs. 16 and 17).

[3] Robert H. Brill "A Note on the Scientist's Definition of Glass," *JGS* 4 (1962) 134-136 for a scientific explanation of this phenomenon.

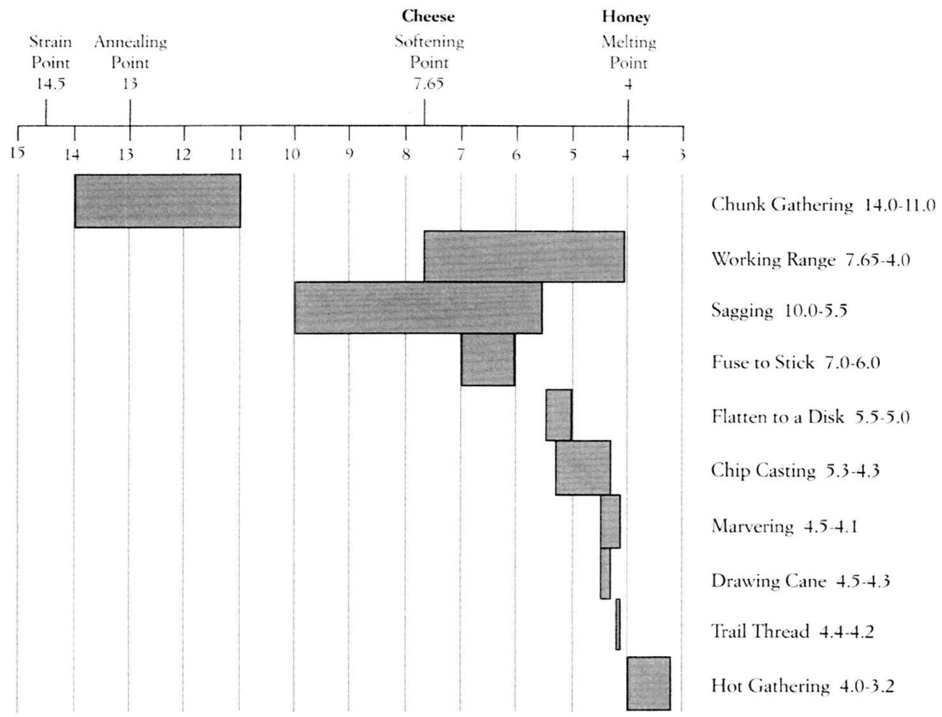


*Fig. 17A. Viscosities required for various glass-working operations using ancient soda-lime-silica glass. Chart designed by the author.*

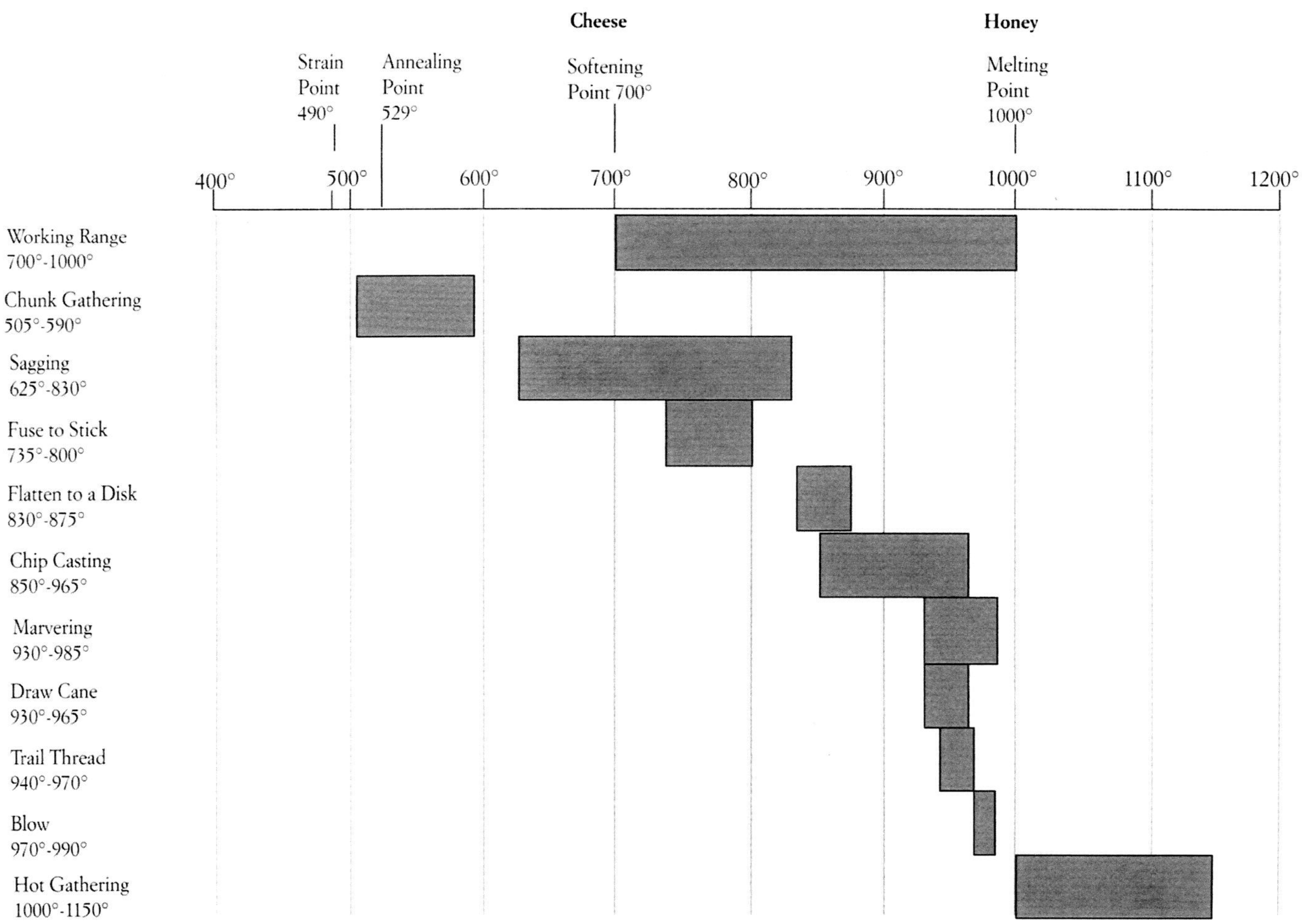


***Fig. 17B. Temperatures required for various glass-working operations using ancient soda-lime-silica glass. The working range is the difference between the gathering temperature and the softening temperature. Chart designed by the author.***

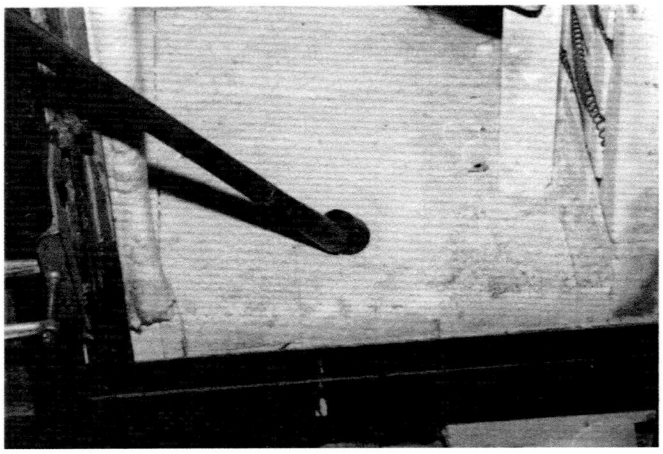

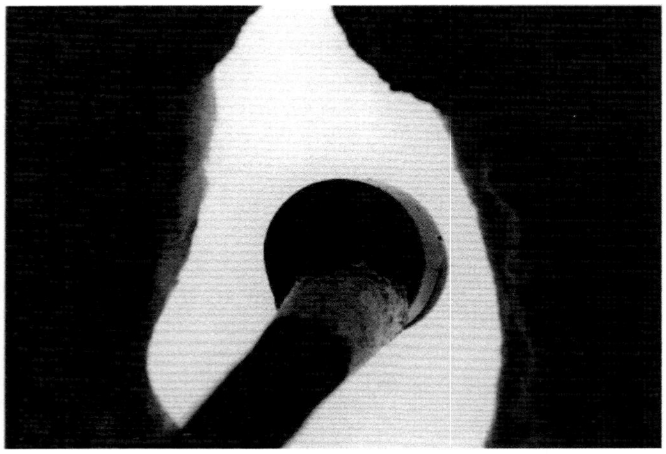

*Fig. 18. Picking up a chunk of colored glass on a metal blowpipe. Note that the glowing tip of the pipe normally visible to an observer does not show in the photograph. Photo: Author.*

*Fig. 19. Heating in the glory hole a chunk of glass picked up on a metal blowpipe. Photo: Lois Dorfman.*

their compositions were not very different.[4] The graph shows the correspondence between temperature and viscosity for ancient soda-lime-silica glass.

Today, the official working range of glass is expressed as between its softening point at 7.65 poise and its melting point at 4.0 poise (Fig. 17). Depending on the chemical composition of the glass, the working range can be long or short in time.[5] At a viscosity of 7.65 poise, glass has the consistency of a semi-hard cheese, and it is possible to poke a hole in it with a pointed tool; for ancient soda-lime-silica glasses the temperature needed to attain this viscosity would have been about 690° C. At a viscosity of 4.0 poise glass melts and acts like honey; for ancient glasses this happened at a temperature of about 1000° C.

Different glass-forming operations require different degrees of viscosity and thus different temperatures. In order to control viscosity a glass worker needs not only to reheat the piece on which he is working, he must also frequently cool it. Ancient glass workers must have estimated the temperature and viscosity of glass in much the same way as glass blowers do today, by observing its color and movement. The brighter and lighter the color, the hotter the glass.[6]

## CHUNK GATHERING

Chunk gathering is described in a poem written on a third-century papyrus found in Egypt.[7] The Greek text gives a vivid description of a glass blower making a vessel:

> First he heated the very point of the blowpipe, then snatched from nearby a chunk of bright glass and placed it skillfully within the

[4] Brill 1988, 278-280.

[5] Some operations common in the Hellenistic period, such as sagging and fusing, can be done before glass reaches its softening point.

[6] Marking the color of glass as an optical means of estimating its temperature is a recurrent theme in early Mesopotamian glass-making recipes: Oppenheim et al. 1970, 25, 39, 48.

[7] R. A. Coles, *The Oxyrhynchus Papyri*, 50 (London 1983) 57-59, no. 3536. Compare the use of *bolos*, "chunk," in *Anthologia Palatina* 16, no. 323 (a poem attributed to Mesomedes of Crete, second century). For a discussion of these texts: Stern 1994, 28.

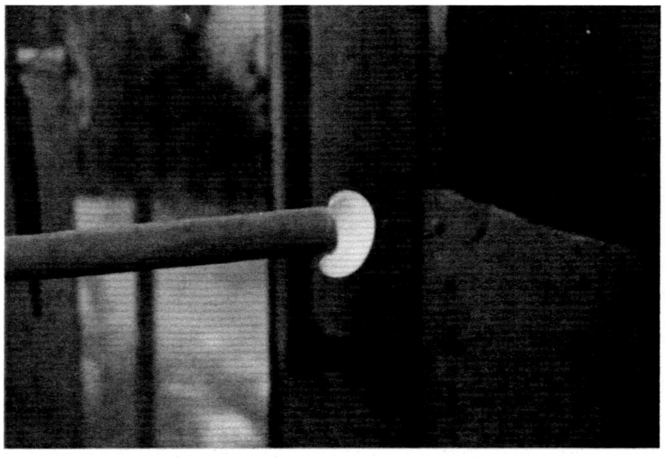

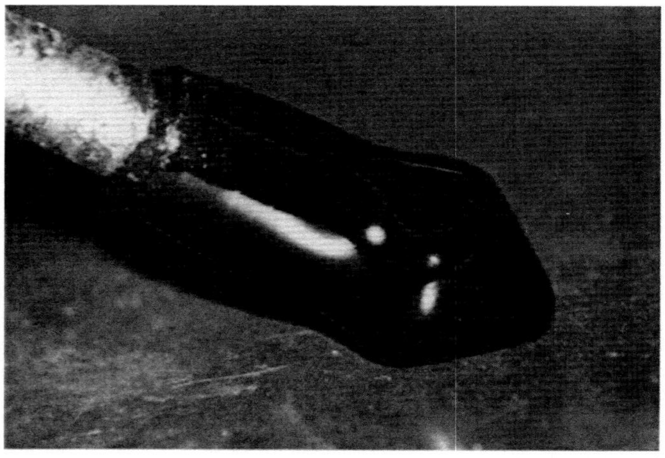

*Fig. 20. Left: Shaping the glass on a ceramic blowpipe by rotating the pipe. Right: Shaping the glass on a metal pipe by marvering. Photos: Lois Dorfman.*

hollow furnace. And the crystal as it tasted the heat of the fire was softened by the strokes of Hephaistos like... He blew in from his mouth a quick breath... like a man essaying the most delightful art of the flute. The glass received the force of his breath and became swollen out around itself like a sphere before it. It would receive another onslaught of the divine breath, for often, swinging it like an ox-herd his crook he would breathe into...

The fragment begins with a description of chunk gathering, called *harpazein bolon*, "to snatch a chunk." The ancient glass blower heated the end of his pipe so that the chunk which was being preheated nearby would stick to it.[8] Today picking up a chunk on the end of a blowpipe is common practice for glass blowers who wish to use colored glass but have no separate pot for melting color (Fig. 18; compare Figs. 29 and 30). The chunks have to be preheated to a temperature in the annealing range because glass will undergo thermal shock and fracture if it is cold on contact with a hot material. After the chunk has been picked up, it is heated further in a glory hole or working port until the glass is soft enough for further manipulation (Fig. 19, compare Fig. 20, left and right). Once a chunk has been heated to the correct temperature, every glass-forming operation can be performed: it can be shaped with a tool, drawn out thin to make canes and threads, and blown (see Fig. 24).

Compared with hot gathering, chunk gathering has several advantages. First, the highest temperature needed for any of the glass-working techniques performed before the invention of blowing would have been that for trailing threads. Based on data assembled for the glass used in a fourth-century factory at Jalame,[9] this could have been done at a temperature as low as 940° C to 950° C. Most other operations required lower temperatures (see Fig. 17). Second, to work with molten hot glass the pot(s) containing glass must be kept at a temperature of 1000° C to 1100° C throughout the duration of the work. Without a supply of molten hot glass temperatures can vary according to the operation performed, which allows for much more flexibility in temperature. Third, hot gathering reguires the use of a closed furnace, while chunks could have been picked up and heated above a live fire, even without a furnace.[10] Chunk gathering would have needed considerably less fuel and less manpower to take care of firing. A furnace for chunk gathering would have combined the functions performed in a modern glass studio by color oven and glory hole. Fourth, when a pot of glass is melted, much glass is lost because it sticks to the wall of the crucible and cannot be used. No glass is lost when preheated chunks are picked up.

The advantages of chunk gathering are as valid today as in antiquity. But they were of particular relevance in the earliest history of glass working when the craft of glass making was not yet widespread and glass workers usually imported potentially expensive colored glass from faraway glass-making centers.[11] A furnace for gathering chunks could have operated successfully at temperatures that were about 100° to 150° C lower than a furnace for gathering hot glass which would have required temperatures between 1000° C and 1100° C. The higher temperatures necessary for working with molten glass and the need to maintain those temperatures over a long period of time require a technological sophistication that would have been difficult to achieve with the technology available in early ancient times.

After the introduction of molten hot glass, chunk gathering probably remained normal practice in the Roman period for gathering small amounts of colored glass for decorative blobs, threads, and handles. During the first century, some glass blowers may have continued to gather chunks rather than hot glass depending on the type of furnace they were using and the kinds of objects they were producing. Most early mold-blown vessels are small and made of expensive colored glass that had to be imported. Experiments in The Toledo Museum of Art Glass Studio show that these could have been made from one or two small chunks of glass or cullet from broken objects. Literary and archaeological evidence indicates that in the Near East chunk gathering was still practiced for the blowing of vessels in the third and fourth centuries. The papyrus from Egypt quoted above with a description of a glass blower working from chunks has been dated to the third century. Archaeological evidence from a fourth-century glass-working factory at Jalame suggests that glass blowers there were still gathering chunks rather than hot glass.[12]

## INFLATION

The most significant discovery in antiquity about the properties of glass was that it can be inflated. Today, glass blowers use five basic methods to expand glass. Two are based on attaching an inflation tool directly to the glass, while the other three make use of hand-held tools to expand a piece of glass that is attached to a second tool. The glass is attached directly to and held by the blowing tool when a blowpipe (Figs. 21 and 25; see Figs. 6 and 32) is used or when a tube of glass is inflated, in which case the tube itself functions as the inflation tool. The three tools that do not form a unit with the glass being expanded are a short metal pipe with a narrow opening on either side known as a puffer (Fig. 22), an air hose (in place of which a bellows [see Fig. 26] would have been used in antiquity), and a wooden steam stick, a moist piece of wood that causes hot glass to expand by steam evaporating from it when poked into a hot mass of glass (Fig. 23). Any of the last four expansion methods might have been the origin of the discovery that glass can be inflated.

The archaeological evidence presently available indicates that in one workshop at least the blowing process began with

[8] The chunks do not stick to the metal pipe itself but to a thin coating of glass left over from previous use. When the pipe is heated, this residual glass melts and acts as an adhesive bond to pick up the preheated chunks of glass.

[9] Brill 1988, 280.

[10] Stern 1994, 24.

[11] In some cases imported raw glass was colored locally, e.g. in bead making. Pliny, *Natural History* 36.193 seems to imply that in Sidon the coloring was done in the workshops where the glass was formed into objects, see Chapter I: *Early Syro-Palestinian Mold Blowing: The Role of Sidon.*

[12] Stern 1992a, 491-494.

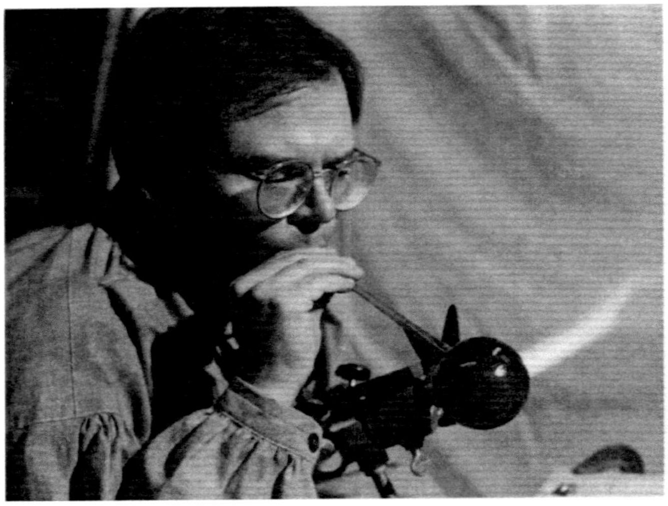

*Fig. 21. Inflating a chunk of glass attached to a tube of glass functioning as a blowpipe (lamp work). The chunk has been heated to a working temperature by the gas-oxygen torch seen in the background. Photo: Frederick and Jean Birkhill.*

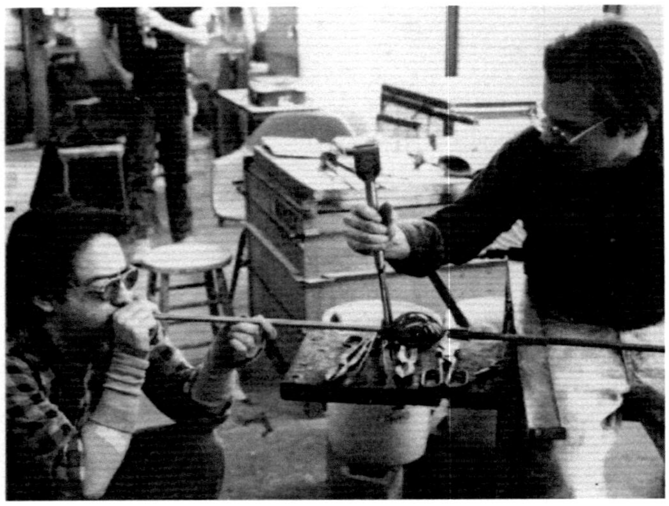

*Fig. 22. The glass blower seated at the bench jacks the mouth of a vessel while a second glass blower expands the vessel by blowing through a straight puffer. Photo: Author.*

the inflation of glass tubes. Quantities of glass-working waste, including inflated tubes, have been excavated in Jerusalem in a deposit dated about 50 B.C. or slightly earlier.[13] The technique of locally heating and inflating a glass tube suggests that the original discovery took place during the production of beads,[14] for although there were no beads among the Jerusalem waste, the tubes themselves are very similar in size and shape to tubes normally found in Hellenistic workshops producing beads.[15] The term "tube blowing"[16] is used here to distinguish the method used in antiquity from "lamp work," the technique in use in recent times,[17] because the ancient method of heating the tubes has not yet been investigated.

The importance of the fact that this discovery took place during work with tubes cannot be over emphasized. Tube blowing is the only method of inflating (apart from the use of a blowpipe) in which the tool for inflation is attached directly to the glass. All other methods are based on expanding glass attached to a second tool. In tube blowing, the tube acts as a precursory form of blowpipe, thus presenting the concept of expanding glass that was attached directly to the tool used for inflating.

Some small blown bottles and unguentaria may have been made by tube blowing, but it is not possible to prove this because of the difficulty in distinguishing between bottles

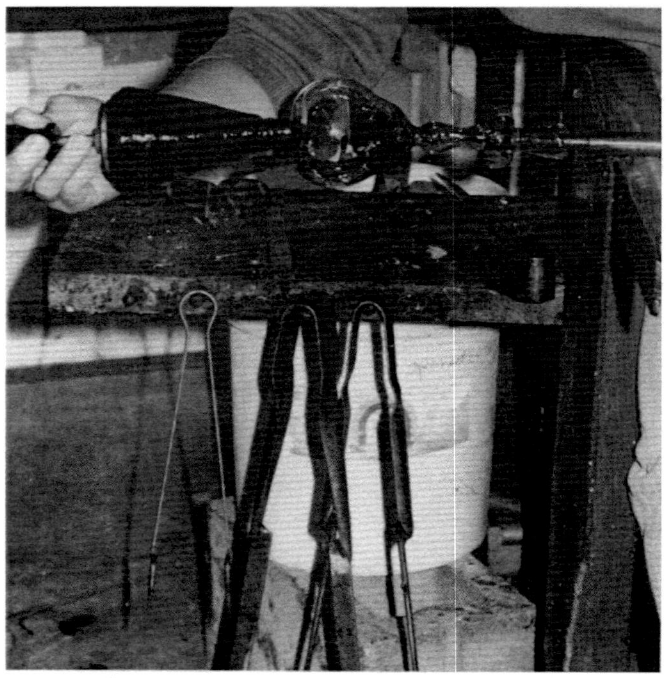

*Fig. 23. The glass blower presses a moist steam-stick, a tapered wooden tool, into a piece of hot glass to create steam, which causes the glass to inflate. Photo: Author.*

13 Avigad 1972, 198-200; Avigad 1980, 186-192; Israeli 1991, 46-55, pls. 13-14 (with lit.).

14 This was first suggested by Stern 1977a, 159, n. 6. So also: Israeli 1991, 53.

15 For example, glass tubes found in Rhodes, Gladys Davidson Weinberg, "Glass Manufacture in Hellenistic Rhodes," *ArchDelt* 24 (1969) 143-151, pls. 76-88.

16 The term tube blowing was first introduced by G. Eisen, who suggested that tube blowing preceded blowing with a pipe: "The Origin of Glass Blowing," *AJA* 20 (1916) 134-143; Eisen 1927, I, 214.

17 Lamp work, also called flame work, is based on the use of a single, concentrated flame produced by a lamp which heats the glass locally. In antiquity, the flame of an oil lamp probably could have been intensified by a puffer or bellows to produce a pointed flame of sufficient heat: R. Lierke, "Glasformung mit der Grubenlampe? Zum bergmännischen Geleucht im 16. und 17. Jahrhundert," *Der Anschnitt* 44 (1992) 92-97; Rosemarie Lierke, "Auf der Suche nach Lampenarbeit in der Geschichte," *AnnAIHV* 12 (1991, publ. 1993) 33-43. It has not yet been determined whether other methods could have been used. Whatever the source of heat, the tubes were expanded by mouth blowing. Today, the technique of shaping tubes by mouth blowing is used primarily for making laboratory instruments, but all shapes of vessels can be made this way. On lamp work: H. Ricke, "Lampengeblasenes Glas des Historismus," *JGS* 20 (1978) 45-99; Frederick Birkhill, "A Concise History of Lampworking," *Glass Art Society Journal* (1984-85) 146-149; Rosemarie Lierke, "Early History of Lampwork — Some Facts, Findings, and Theories," *GlastechBer* 63 (1990) 363-369.

made by tube blowing and bottles made on a blowpipe.[18] There are many ways to make a tube, several of which were used in antiquity.[19] The only conclusive evidence for tube blowing would be traces of folding in the case of folded tubes or of a thickened area in the center of the underside of a vessel where the tube was closed. However, not every thickened drop in the base of a vessel is the result of closing a tube. Depending on how glass is marvered, a bottle blown on a blowpipe can have a small thickened drop in the center of the base (Fig. 24).

## THE INVENTION OF THE BLOWPIPE

The discovery that glass can be inflated was destined not only to change glass-working techniques, it also prompted the invention of the blowpipe. After the chance discovery that glass can be inflated, the development of the blowpipe must have been pursued deliberately. The idea of lengthening a glass tube with an additional tube or pipe would have occurred naturally to an inventive glass worker exploring the possibilities of tube blowing and trying to avoid the heat of the fire. This might have been done first with a length of reed, because the glass tube with the added length of the reed would have been long enough to provide protection from the heat of the fire.

The primary challenge would have been to devise a tube-shaped tool that could be attached directly to a preheated chunk of glass.[20] Today we are so familiar with iron and steel blowpipes that it is difficult to imagine why an ancient glass blower might use a pipe or tube made of anything else, especially since iron was a common material in the first century B.C., the period when inflation techniques were being developed. However, iron tubes are difficult to make and were not widely used for any other purpose in this period. Thus iron tubes would not have been readily available to a glass blower wishing to experiment with any form of blowpipe. Rather than assuming that the first glass blowers went through the difficulty of making an iron pipe before they knew it could be used successfully to inflate glass, it seems logical that they would experiment with a material that was more readily available.

A conceivable interim stage in the development of the blowpipe might have been the use of separate glass tubes of different lengths and widths for picking up a chunk of glass and expanding it. Today this is the way glass blowers working at a lamp usually pick up colored glass (Fig. 25, see also Fig. 21). However, only very small vessels can be made this way.

The most obvious material for making experimental blowpipes would have been clay. Clay was available in or near every glass workshop. Glass workers made and used crucibles, containers, and other ceramic tools long before the invention of glass blowing. They were even familiar with ceramic tubes, because these were used as bellows or mouth blowers. Tubular ceramic mouth blowers were known in

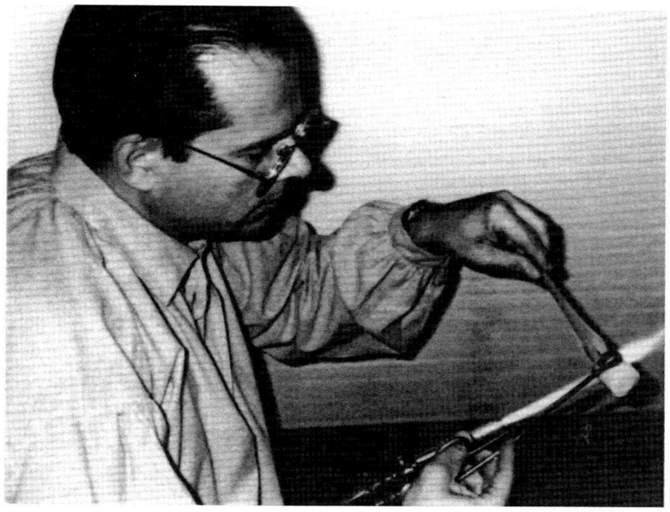

*Fig. 25. Holding a chunk of colored glass with tongs to attach it to a glass glass-blowing tube. Photo: Frederick and Jean Birkhill.*

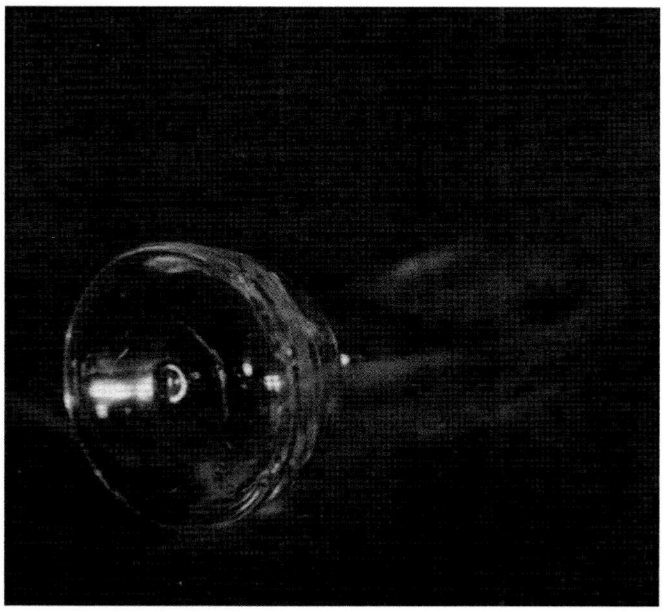

*Fig. 24. Base of a bottle blown from chunks picked up on a blowpipe. Tube blowing results in a vessel with a thickened drop in the base (where the end of the tube was closed), but, as this example demonstrates, vessels blown on a blowpipe can sometimes also have a small thickened drop in the center of the base; it depends on how the vessel was marvered. Photo: Maureen Blake.*

[18] On the difficulty of distinguishing tube-blown vessels from those made with a blowpipe: M. Spaer, "Some Egg-shaped Glass Vessels and Reflections on Early Glassblowing," *IMJ* 6 (1987) 51-58.

[19] The tubes found in Jerusalem were made by folding lengthwise: Israeli 1991, 48, pl. 12. Other tubes made before the invention of the blowpipe show no signs of folding, for example, tubes from Amarna in The Toledo Museum of Art: Grose 1989, 52-53, fig. 27. Tubes that were not made by folding could have been made by a variety of methods such as the *ladha* process, used today in South India: E. Marianne Stern, "The Secret of Papanaidupet," *GlastechBer* 60 (1987) 346-351, or simply by poking a hole in a bit of glass and drawing it out.

[20] The only suitable metal would have been iron, because it is the only metal known in antiquity that has a higher melting point than glass. Bronze, tin, lead, and other metals melt at lower temperatures than glass.

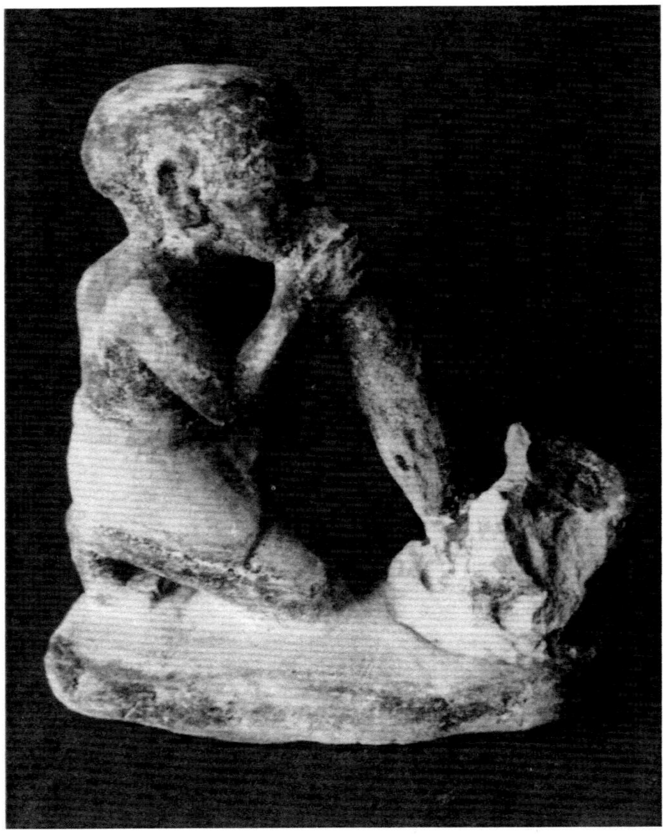

*Fig. 26. Terra-cotta figure of a man blowing into a bellows (mouth blower). From Giza, Egypt, early third millennium* B.C. *Chicago, University of Chicago, Oriental Institute Museum, no. OI 10631. Photo: Courtesy of the Oriental Institute of the University of Chicago.*

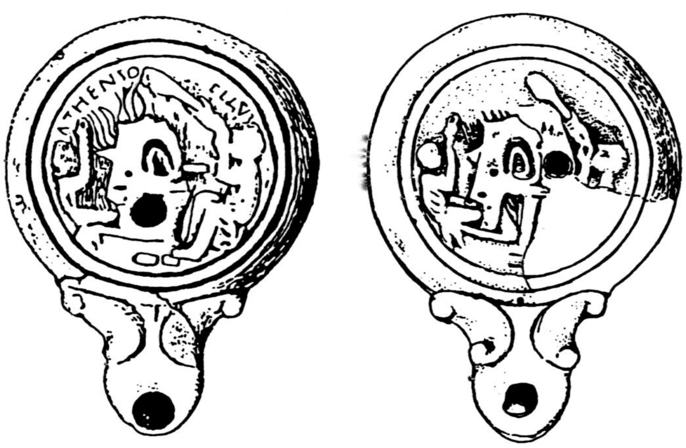

*Fig. 27. Two Roman lamps.* Left: *found at Asseria, Croatia, middle to second half of the first century* A.D. *Spalato, Croatia, Museum, no. 1094/30;* Right: *found near Ferrara, Italy, second half of the first century* A.D. *Ferrara, Museo Archeologico Nazionale, no. 52196, L 11.5 cm, D 8.3 cm. At the right a glass blower with a blowpipe is seated before a small glass furnace. The furnace is dome-shaped with two openings. The exact purpose of each opening is not clear. The lower opening might be the working port or the stoking hole; the upper opening might be the working port or the annealing area. At the end of the short blowpipe held aloft by the glass blower (compare the posture of a modern glass blower, Figs. 6 and 32) can be seen the bulb-shaped paraison. At the left can be seen an assistant holding up an unidentified object. Drawings: Lorene Sterner.*

Giza, Egypt, as early as the third millennium B.C. (Fig. 26). Bellows made of other materials such as leather, fitted with ceramic mouthpieces, tuyères, were commonly used throughout the Mediterranean and Europe in antiquity.[21] Bellows were an indispensable tool in glass working because they supplied the oxygen necessary to increase the temperature of the fire.[22] A short ceramic pipe, perhaps one originally used as a bellows (mouth blower), might therefore have served as the first blowpipe.

The hypothesis that the first blowpipes were made of clay seems to be supported by the scene of a glass blower depicted on the disks of two clay oil lamps of the second half of the first century.[23] A glass blower seated in front of his furnace blows a long-necked bulb-shaped bubble (Fig. 27, left and right). The blowpipe, however, has a strange, bulbous shape that is difficult to interpret as an iron pipe not only because of the difficulty of making a bulbous iron tube but also because the pipe is so short that if it had been of metal it would have been much too hot for the glass blower to put it to his mouth. The shape resembles that of the clay bellows in the hands of the Egyptian statuette (Fig. 26). If this interpretation of the scene on the lamps is correct, glass blowers were still using ceramic blowpipes in the middle or second half of the first century, long after the first experiments with inflation.

A recent experiment at The Toledo Museum of Art Glass Studio has proven that blowing glass with a fired ceramic pipe is feasible.[24] An inexperienced person can shape a ceramic blowpipe in about ten minutes. The pipe was probably fired at a low temperature in order to preserve some elasticity in the clay, which absorbs the heatshock of the fire. The end of the pipe is heated until it glows and then coated with small glass chips (Fig. 28). When the glass coating melts, it acts as an adhesive and makes it possible for a preheated chunk of cullet to stick to the end of the pipe (Fig. 29). The glass is reheated (see Fig. 19), shaped by manipulation in the furnace (Figs. 30 and 31, see also Fig. 20 left), and then blown (Fig. 32).

Although the Toledo experiment proved that ceramic blowpipes could have been used in antiquity, they did not, of course, prove that this was the case. The only incontrovertible proof would be the excavation of ceramic blowpipes.

[21] On the construction and use of bellows in antiquity: C. Roden, "Glasrohrdusen," *Der Anschnitt* 40 (1988) 62-82, esp. 68; G. Weisgerber and C. Roden, "Römische Schmiedeszenen und ihr Gebläse," *Der Anschnitt* 37 (1985) 2-21; G. Weisgerber and C. Roden, "Griechische Metallhandwerker und ihr Gebläse," *Der Anschnitt* 38 (1986) 2-26.

[22] Fragments of two tuyères with a diameter of 2.0 cm were excavated at a first-century glass-blowing factory in Avenches, Switzerland, Morel et al. 1992, 6.

[23] See above *Glossaries: furnace.*

[24] On April 18, 1992, at The Toledo Museum's Seminar in Experimental Archaeology, "From Bellows to Blowpipe," organized by the author with generous support from the Museum Education Department of The Toledo Museum of Art, twenty Toledo area glass blowers and ceramists gathered to blow glass with ceramic blowpipes of their own fashioning. The pipes varied in length between 35 and 60 cm, were made of several different clay bodies, and fired at different temperatures. See E. Marianne Stern, "Art and Archaeology at The Toledo Museum of Art Glass Studio," *The Toledo Conference 1993, Glass Art Society Journal* (1993) 70-77; Stern 1994, 81-85.

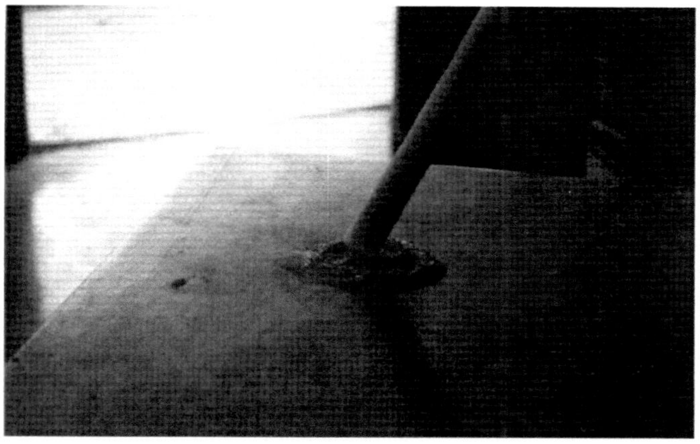

*Fig. 28. Before a new ceramic pipe will pick up a chunk of glass for glass blowing, the pipe must first be prepared. This is accomplished by heating the tip of the pipe until it glows, then coating the tip with crushed glass. The blowpipe is then heated further to melt the crushed glass. This step is only necessary the first time a ceramic pipe is used; once a vessel has been shaped and knocked off, a layer of glass remains on the pipe to which other chunks will adhere. Photo: Author.*

The use of ceramic pipes, however, could explain the rapid spread of the technique of glass blowing throughout the Roman empire, because clay was inexpensive and ceramic blowpipes were quickly and easily made. New insights are continually being made from excavations of ancient workshops. Ceramic cone-shaped tools, the function of which has not yet been identified, have recently come to light at two first-century factories.[25] Future excavations may uncover remains of first-century blowpipes that could shed light on the material and shape of the earliest blowpipes.

It is not clear when the first metal blowpipe was made. Knives, jacks, hammers, and various other tools used in the production of glass, ceramics, and metal were among finds dated first century B.C. to first century A.D. excavated at Malaya

[25] At Avenches (ancient Aventicum) in western Switzerland, Morel et al. 1992, 7, figs. 9, 10; and at Saintes, in southern France (A. Hochuli-Gysel, pers. com.). The cones have a bulging, flaring outline similar to that of the blowpipe depicted on the clay lamps (Fig. 27), but they could not have been used for blowing because they are solid; they have a deep cylindrical hollow on the wide, flat end.

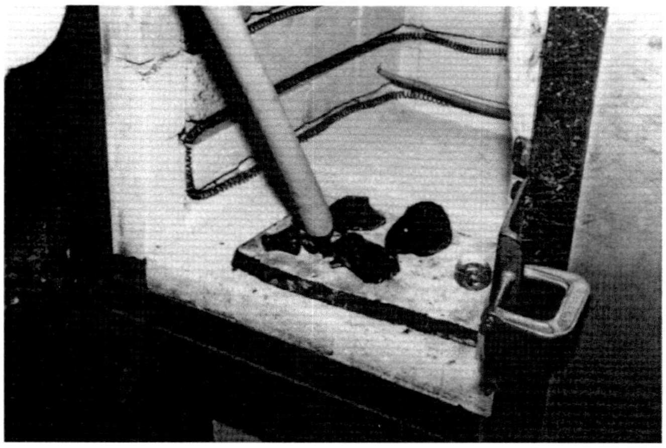

*Fig. 29. Picking up a chunk of cullet preheated in an annealing oven on the prepared tip of a ceramic pipe. Compare Fig. 18, which shows the same operation using a metal blowpipe. Photo: Author.*

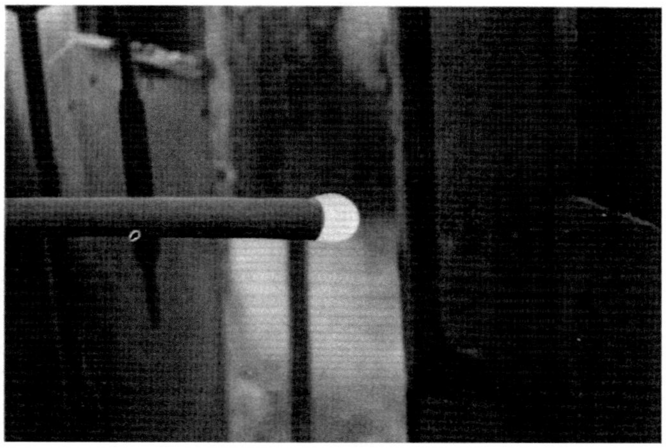

*Fig. 31. A chunk of glass shaped by rotating the ceramic blowpipe and ready to be blown. Photo: Lois Dorfman.*

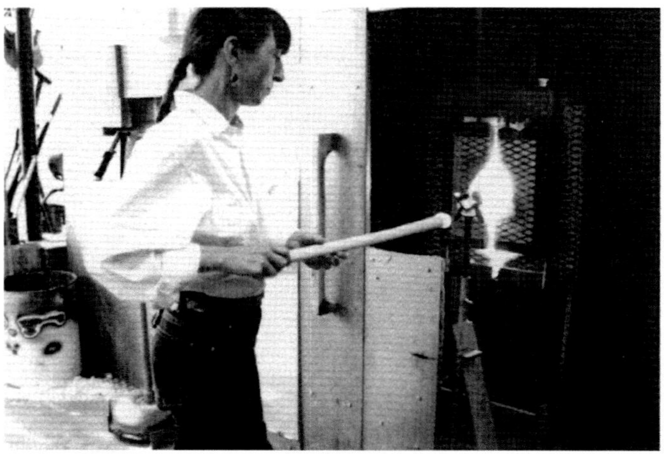

*Fig. 30. Reheating and shaping the chunk of glass on a ceramic blowpipe to a working temperature prior to blowing. The glow from the hot glass which helps the glass blower estimate the working condition of the chunk can be clearly seen in the photograph. Photo: Lois Dorfman.*

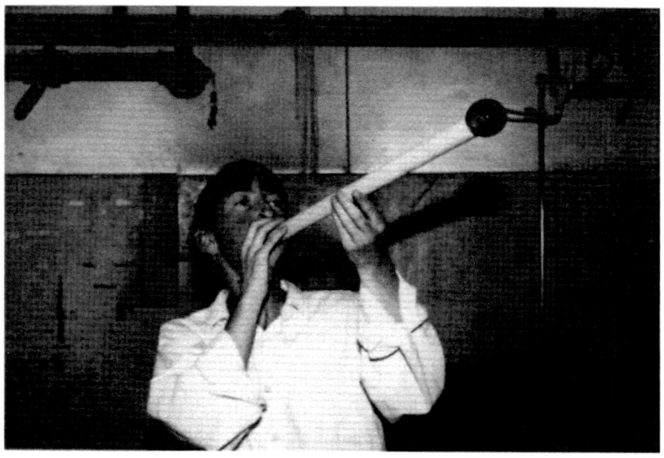

*Fig. 32. Blowing the initial bubble with a ceramic pipe. Compare the pose and the dimensions of the pipe seen here to those represented on Roman lamp disks in Fig. 27. Photo: Lois Dorfman.*

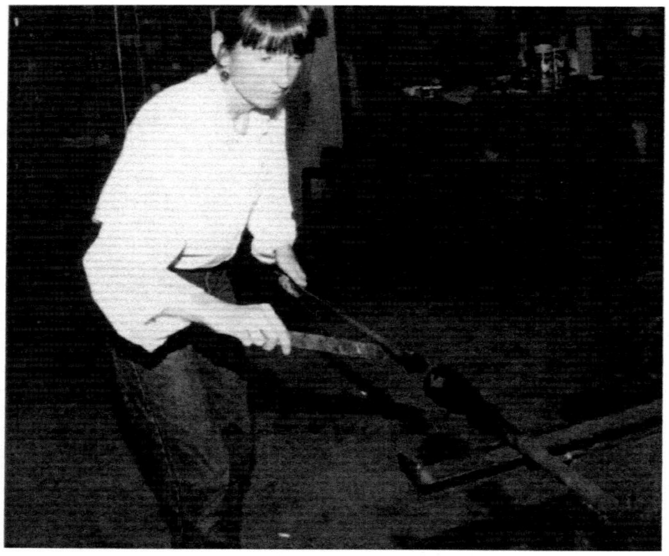

*Fig. 33. Attaching a pontil to a vessel. A file is used to guide the pontil to the center of the bottom of the vessel. Photo: Lois Dorfman.*

Kopanya in the Carpathian Mountains. They are apparently the remains of a blowpipe.[26] Remains of glass-working tools and iron rods that might have been solid (pontils) or hollow (blowpipes) were found in a third- to fourth-century context at Komarowa, in western Ukraine (see Fig. 8).[27] Hollow iron tubes that were probably blowpipes have been identified at Mérida, in Spain, and Salona, in Croatia.[28] Rust flakes in knock-offs excavated in a fourth-century glass factory at Jalame, Israel, show that iron pipes were used there too.[29] Iron blowpipes were a great improvement over ceramic ones. Not only were iron pipes much less prone to breakage, but they also allowed glass blowers more freedom in manipulating the glass and shaping it into vessels. They opened the way to the production of large and thick-walled vessels.

## THE INTRODUCTION OF HOT GLASS

The discovery that molten hot glass can be gathered on the end of a blowpipe must have come as a great surprise to a glass blower who was accustomed to gathering chunks. The exact circumstances of the discovery cannot be reconstructed, but it is possible to envisage probable scenarios. One scenario is a workshop accident in which chunks arranged for preheating were overheated and melted. In order to avoid the mess occurring after such accidents, glass workers probably arranged the chunks in small containers on a shelf inside the furnace.[30] Our imaginary glass blower saw that the chunk he was about to pick up had melted. He tried to save the glass by pushing the container away from the fire with the blowpipe he already held in his hand. When he noticed that molten glass adhered to his tool, the glass blower blew into his pipe for the fun of it and saw a bubble appear. At this point he might have tried it again and experimented with blowing bubbles from hot molten glass.

Molten hot glass used for gathering has a very low viscosity; it is drippy and difficult to manipulate — like thin honey running off a spoon (see Fig. 17). For a glass worker accustomed to chunk gathering, the challenge would have been to learn how to cool the outer skin of the glass before blowing or performing any other operation. This was exactly

26 Kotigoroshko 1989, 183-185, figs. 1 and 2 (blowpipe not ill.; dimensions not noted). For information on the blowpipe I am indebted to N. Sorokina.

27 Sorokina 1977b, 96, pl. I, no. 3. The iron rods are too corroded to determine whether they were hollow or solid (N. Sorokina, pers. com.).

28 Mérida: Price 1973, 80-84; Janet Lang and Jennifer Price, "Iron Tubes from a Late Roman Glassmaking Site at Mérida (Badajoz) Spain," *JAS* 2 (1975) 289-296. Salona: Susan Auth, "Roman Glass," in C. W. Clairmont, *Excavations at Salona, Yugoslavia (1969-1972)* (Park Ridge, N.J. 1976), 167.

29 Weinberg 1988, 35, 283.

30 Traces of crucibles that were probably used only for preheating chunks of glass have been found at a fourth-century glass factory at Jalame: Stern 1992a, 491-494.

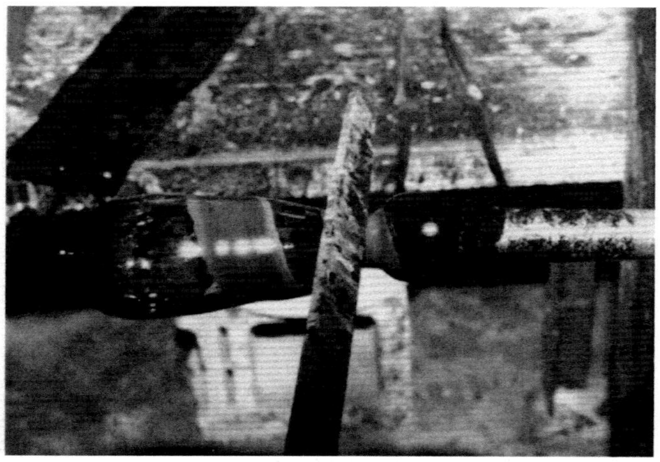

*Fig. 34. Attached to both the pontil (left) and the blowpipe (right), the neck of the vessel is scored with a file to prepare it for cracking off. Photo: Lois Dorfman.*

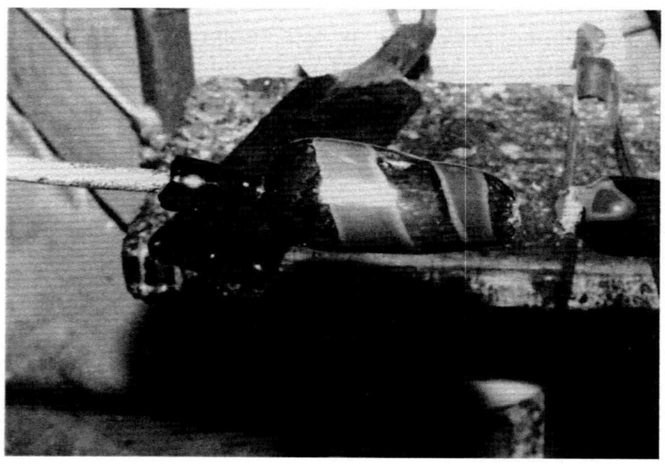

*Fig. 35. After cracking off the vessel from the blowpipe, the moil remains on the blowpipe (right) and the vessel remains attached to the pontil (left). Photo: Lois Dorfman.*

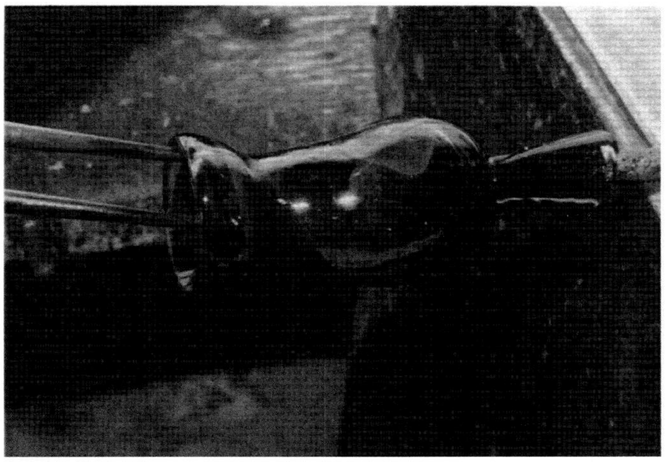

*Fig. 36. Manipulating a vessel while attached to a pontil. Pointed jacks are used to open the mouth of the vessel. Photo: Lois Dorfman.*

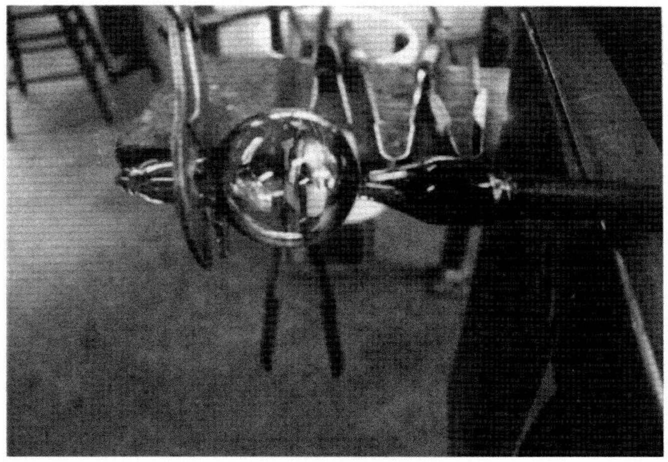

*Fig. 37. Using the blowpipe as a pontil. The glass blower grasps the vessel, cracks it off, and quickly turns it around. He holds the neck of the vessel with a clamping tool (here diamond shears) and reattaches the base to the ring of glass remaining on the blowpipe, so that the blowpipe can function as a pontil. Photo: Lois Dorfman.*

the opposite of what he was accustomed to doing, for after chunk gathering glass is still very viscous: it has to be heated further until the interior is soft enough to be worked.

Nevertheless, the potential of the bubble would probably have been apparent to a glass worker familiar with inflating chunks of glass to form vessels. Since blowing a bubble is done when glass is at its hottest, he may even have realized the advantage of gathering molten hot glass. Cooling the outer skin of a gather of hot glass takes less time than heating a chunk of glass and is both faster and more comfortable for the glass worker than standing in front of the furnace. When glass blowers began to blow larger vessels, other advantages would become apparent, such as the ease of dipping a paraison into hot glass in order to gather more glass on the pipe and the possibility of covering a bubble with a different color of glass by dipping it into a second pot of molten glass.

The discovery of hot glass was also an important factor in the development of making flat glass.[31] Flat glass made by pouring is the earliest archaeological evidence for the use of hot glass in antiquity. In Pompeii numerous fragments were found of rectangular glass window panes made by pouring molten hot glass on a tray with an upright rim.[32] Tool marks along the edges of the panes show that the glass was not fluid enough to fill the entire surface of the tray, but it was hot enough to be manipulated and pulled to the edge of the tray before it became too viscous. The finds from Pompeii are evidence that the use of molten hot glass was known in Italy before 79; it might have been introduced around the middle of the first century. The use of molten hot glass implies an improvement in furnace technology that made it possible to sustain high temperatures over a long period. Therefore it is tempting to assume that the improvements in glass making, mentioned by Pliny, *Natural History* 36, 194, related not only to the furnaces used for glass making, but also to those used for glass working (see *Glossaries: Glass Making*).

## THE PONTIL TECHNIQUE

Transferring a vessel from a blowpipe to a pontil in order to finish its mouth is always a precarious moment in the process of blowing. The pontil is attached to the underside of the vessel (Fig. 33), a score mark is made between the neck of the vessel and the end of the blowpipe (Fig. 34), and then the vessel is cracked off the pipe (Fig. 35). The bond between pontil and vessel has to be strong enough to hold the vessel on the pontil during manipulation (Fig. 36), while at the same time the bond needs to be loose enough to allow for knocking off the finished vessel from the pontil (see Figs. 10 and 11). To accomplish this, a preheated pontil is dipped in molten glass and pressed to the base of the vessel on a blowpipe. The pontil mark, the scar or rough surface left by such a pontil, is roughly circular (e.g. detail photographs at Cat. Nos. 120 and 121).

Although a pontil can be coated with glass by picking up a small chip of glass on the tip, the concept of transferring a vessel on a blowpipe to a second tool is more likely to have originated during work with hot glass. Dipping a preheated pontil in molten hot glass automatically coats the tip with a thin layer of glass that will adhere to both the pontil and a glass object that is hot. It is therefore proposed here that the development of the pontil technique resulted from the use of hot glass.

It is also possible to use the blowpipe itself instead of a pontil. After scoring the neck of a vessel, the glass blower grabs the vessel with a clamping tool, cracks it off the pipe, quickly turns the vessel around (Fig. 37), and presses the underside of the bottom against the blowpipe which still

[31] On earlier techniques of making circular blanks of flat glass for sagging: Stern 1991d; Stern 1994, 65-67.

[32] Haevernick 1955.

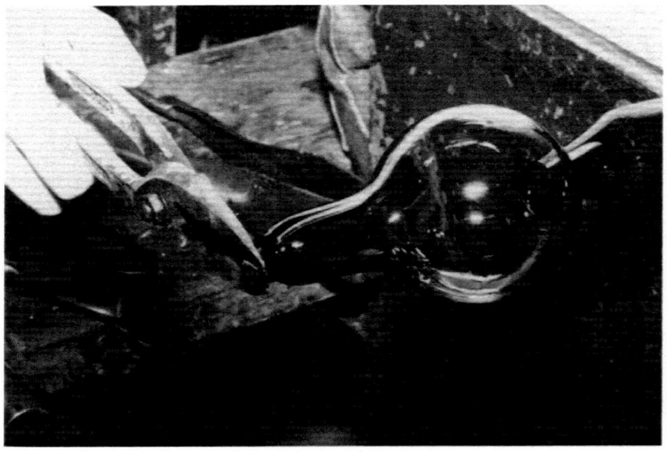

*Fig. 38. Trimming the mouth of the vessel with shears. The ring of glass which attaches the vessel to the blowpipe (functioning as a pontil) can be seen through the bottom of the vessel. Photo: Lois Dorfman.*

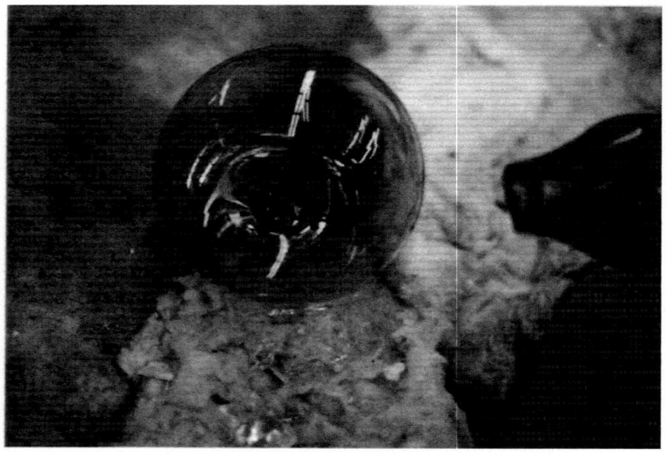

*Fig. 39. A ring-shaped mark, known as a ring pontil, can be seen on the underside of the vessel after it has been knocked off a blowpipe functioning as a pontil. (See also detail photographs of the undersides of the bases of Cat. Nos. 145 and 164). Photo: Lois Dorfman.*

retains a ring of glass from the other side of the score mark (Fig. 38).[33] The resulting scar on the underside of a vessel is ring-shaped (Fig. 39). Ring-shaped pontil scars which may or may not have been made by this technique[34] are found on many Roman vessels (e.g. Cat. Nos. 145, 146, and 164). At Cat. No. 145 the detail photograph of the underside of the base shows a circle of small gashes where the pontil was pulled away and the glass on the underside of the base was broken out. At Cat. No. 164 the detail photograph of the underside of the base shows a low, narrow, circular ridge of glass that was pulled off the pontil and remained stuck to the underside of the base.

Most first-century vessels do not have a pontil mark. The absence of a pontil mark cannot be explained by grinding and polishing. Although removing a pontil mark is common practice in studios today, this was rarely done in Roman times and has not been noted on first-century glass. No systematic investigation has yet been done to ascertain the first occurrence of pontil marks. In the Toledo Museum's collection pontil marks are not found on glass objects made before the second half of the first century.

Before the pontil technique was developed, a clamping tool must have been used to hold vessels while fire polishing the rim, see *Glossaries: clamp*.

## CONCLUSIONS

The development of glass blowing was not the result of a single event but rather the culmination of a series of discoveries that took place in the late Hellenistic and early Roman periods. These events can be reconstructed as follows. First, the discovery, in the early first century B.C., that glass can be inflated. This was followed immediately by tube blowing. Sometime between the second half of the first century B.C. and the early first century A.D., tube blowing led to the invention and development of the blowpipe. The use of a blowpipe was probably instrumental in the discovery that molten hot glass can be gathered and formed into objects. The earliest archaeological evidence for the use of hot glass comes from window panes excavated at Pompeii and therefore dates before 79. Working with hot glass prompted the development of the pontil technique, presumably in the second half of the first century. Some of these inventions and innovative uses of glass may have taken place simultaneously in more than one workshop. Present evidence suggests that inflation and the development of the blowpipe originated in the eastern Mediterranean, probably in glass centers along the Syro-Palestinian coast. The introduction of molten hot glass and the invention of the pontil technique might have taken place in Italy.

How and at what point mold blowing fits into this series of events is still a matter of conjecture. Mold blowing may have been practiced on a commercial basis as early as the first decade of the first century A.D., but the exact date of the first experiments with this technique and the circumstances which led to its development are not certain: see Chapter I: *Early Syro-Palestinian Mold-Blowing: Date.*

[33] The operation is performed so quickly that it is not necessary to reheat the glass at the end of the blowpipe.

[34] A ring pontil mark can also be made with a pontil. In order to do so, a gob of hot glass is gathered and marvered into a cone flaring out from the tip of the pontil. During marvering the pontil is tilted upward, causing the glass at the wide end of the cone to sag inward, thus leaving a circular ring of glass to attach to the bottom of the vessel. Today, such a ring pontil is often made if the vessel to be supported is very large or heavy. For an illustration of a ring pontil mark still attached to the vessel: Stern 1977a, pl. 8, no. 23, a pointed amphora made in Syria between the mid third and mid fourth centuries. Ring pontil marks made by the post technique are described by Mavis Bimson, "Ring 'Pontil Marks' and the Empontiling of a Group of Seventh-Century Anglo-Saxon Glass," *JGS* 22 (1980) 9-11.

# Mold Blowing

To blow glass vessels in a mold today, the glass blower first gathers and shapes a gob of glass, then inserts it into an open mold while one or more assistants closes the sections of the mold around the glass (Fig. 40). Insertion can be done either before or after starting an initial bubble in the paraison, but today it is usually done afterward because it is easier to control the bubble if it is started before the glass is in the mold. The blowpipe is kept in a vertical position in order to keep the glass at the end of the blowpipe centered during the operation. The glass blower slowly expands the glass until it fills the mold evenly. Because the glass is about the same thickness everywhere, the designs impressed on the exterior of the glass vessel appear in negative on the interior. Today the mold is usually pulled away from the glass as soon as the bubble has been expanded (Fig. 41). If the rim is to be fire polished, the vessel is reheated and transferred to a pontil.

In antiquity, the moment when the mold was pulled off might have varied according to workshop traditions and the shape of the vessel. Many Roman mold-blown drinking cups have rims that were lightly ground (e.g. Cat. Nos. 1 and 10). These cups were probably taken out of the molds as soon as the glass had been expanded; they were cracked off the blowpipe and set aside for cooling with an unworked rim. After annealing, the rim was ground.

The rims of most Roman mold-blown bottles were finished by reheating the neck of the vessel while manipulating and turning it in such a way that the glass at the rim collapsed inward to form a lip folded outward, upward, and inward. This could be done without touching the glass, just by using heat and gravity. When an ancient glass blower wished to make a fire-polished or folded rim, there were two options available for pulling off the mold. The first was to crack off the vessel from the blowpipe while it was still in the mold and leave it in the mold during fire-polishing. This would have had two advantages: first, the mold would have retarded the cooling of the glass so that there was more time to work on the rim; and second, the mold would have protected the bottle from damage to its shape and design during manipulation (e.g. Cat. Nos. 15-21, 23, and 24). The second option was to pull the mold off as soon as the glass had been expanded. If a vessel was to have handles, however, the mold had to be removed first because handles could only be applied after extraction from the mold, in which case there was always the danger of damage to design and shape while the glass blower held the vessel with a clamp in the fire. The clamping instrument that fitted around the body could also easily damage the glass if it became too hot (e.g. tool marks on Cat. Nos. 22, 26, 28-33, 40, 42, 49, 62, 78, 108, 161, and 162).

Glass vessels were probably not annealed in molds because differences in the shrinkage rate of mold materials and glass could cause the glass or the mold to crack.

## MOLD MATERIALS

A glass-blower's mold must be strong enough to withstand the heat and pressure of molten glass without the glass adhering to it. Ancient Roman glass-blower's molds were made of a variety of materials, but fired clay appears to have been used most widely. The large group of Hexagonal Bottles with High Relief was made in clay molds. Fragments of ancient clay glass-blowers' molds have been excavated in Corsica, Belgium, Germany, Hungary, and Ukraine.[1] Other

[1] Clay molds:

a) fired clay half-mold for making a grape-shaped bottle, from Macquenoise, Belgium, Chambon 1955, 3, frontispiece, 3; *Glass of the Caesars* 1987, 151-152;
b) fragments of two-part molds for making cylindrical cups, with impressions of charioteers and racing quadrigae, from Mariana, Corsica, Price 1973, 66, n. 4; Price 1976a, 119;
c) clay mold for a square bottle, Price 1976a, 119;
d) clay mold for the base of a square bottle inscribed in Greek with the proper name ΙΟΥϹΤΟϹ, *Ioustos*, and other letters, from Apulum, Roman Dacia (Hungary), Closca Baluta, "Fond de moule romain pour la fabrication de bouteilles carrées trouvé à Apulum (Dacie)," *AnnAIHV* 8 (1979), 112-114, figs. 1-2;
e) one-part clay mold for transferring designs for cutting facets in glass bowls, excavated among waste from glass working at Tanais, Russia, dated from the second half of the second century to about 240, Sorokina 1977c, 118-120, fig. 4;
f) one-part ribbed clay mold from Komarowo, western Ukraine, probably used for imparting only a ribbed pattern, M. Ju. Smisko, "Poselenie III-IV w.n.e. so sledami steklodelatelnogo proizvodsta us. komarov, cernovickoy oblasti," *Mat. Issled Arch. Prikarpattja i Volyni* 5 (1964) 67-80 (here Fig. 8, no. C).

*Fig. 40 (left). A glass blower blows into the blowpipe as two assistants press the halves of the mold together around the paraison. Photo: Author.*

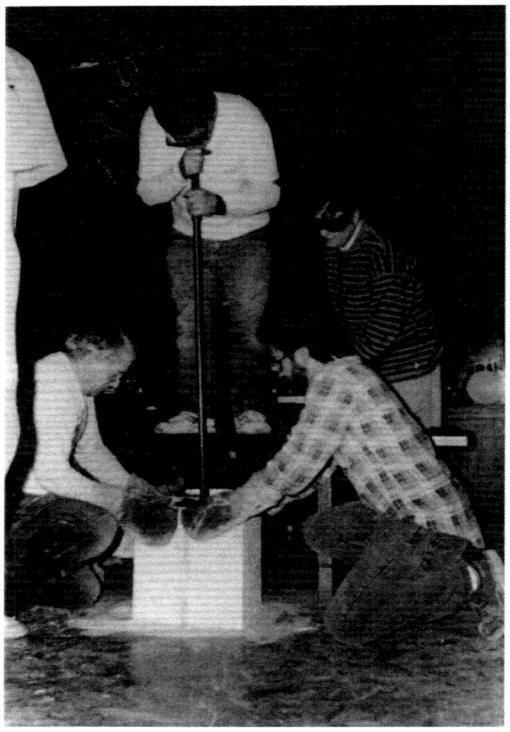

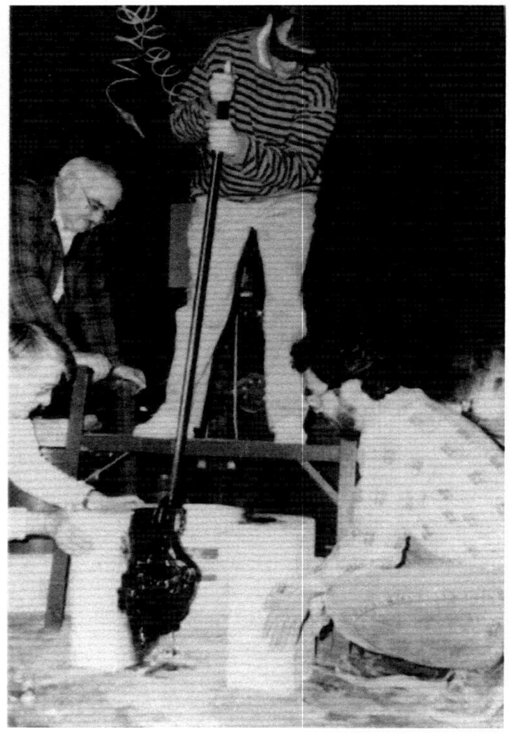

*Fig. 41 (right). After the bubble has been expanded and while the glass is still hot, the halves of the mold are pulled away from the vessel. Photo: Author.*

mold materials known from excavated contexts were stone,[2] plaster,[3] wood,[4] and sheet metal.[5] All these molds were light and prone to wear and breakage. Unlike modern molds, they did not represent a large financial investment on the part of the glass blower.

The molds used in nineteenth- and twentieth-century European and American glass factories were intricate custom-made objects cast from iron or bronze which did not wear out even with heavy use. Depending on the type of design carved on their interiors, these molds had two, three, or even four or more parts hinged together with sturdy, tapered iron pins.[6] Such durable molds had considerable value. The mold maker was a specialized craftsman, often an outside contractor but occasionally the most highly paid craftsman in a glass factory.[7]

## THE INVENTION OF THE MULTIPART GLASS-BLOWER'S MOLD

The makers of the earliest glass-blower's molds could have benefited from the experience of workshops producing molds for ceramic lamps, figurines, and vessels,[8] including how to avoid undercuts, and from metal workers how to build molds that could close so tightly that the hot glass could not escape during blowing.[9] But mold makers had no tradition on

The use of unfired open clay molds for casting glass chips has been documented as early as the fifth century B.C. in the workshop of Pheidias at Olympia: E. Kunze, "Olympia," in *Neue deutsche Ausgrabungen im Mittelmeergebiet und im vorderen Orient* (Berlin 1950) 284-289; A. Mallwitz and W. Schiering, *Die Werkstatt des Pheidias in Olympia*, I (Berlin 1964); A. Mallwitz and W. Schiering, "Die Funde der Phidiaswerkstatt," in *Deutsches Archäologisches Institut. Die Funde aus Olympia* (Athens 1980) 183ff; W. Schiering, "Gewandmatritzen aus der Werkstatt des Phidias in Olympia, bisherige Ergebnisse, Beobachtungen und Folgerungen," *Praktika tou XII diethnous synedriou klasikes archaiologias (Papers of the XII International Congress of Classical Archaeology, Athens, September 4-10, 1983)* Γ', 234-238; J. Letsch, Walter Noll, and W. Schiering, "Glasformgebung in Tonmatrizen: eine meisterliche Technologie der Werkstatt des Phidias in Olympia," *GlastechBer* 56.4 (1984) 96-105; Wolfgang Schiering, *Die Werkstatt des Pheidias in Olympia*, II: *Werkstattfunde* (Deutsches Archäologisches Institut, *Olympische Forschungen* XVIII, Berlin and New York 1991) 11-91, 123-129, pls. 1-50. Clay was used even earlier for bronze casting: Carol C. Mattusch *Greek Bronze Statuary: From the Beginnings through the Fifth Century B.C.* (Ithaca, N.Y. 1988) 54-58, sixth-century-B.C. clay investment molds from the Athenian Agora.

[2] Most stone molds appear to have been used for making the bases of square bottles, for example:

a) Augst, Switzerland, Römermuseum, no. 60.11062.-02, mold of "grünlicher Sandstein," the sides of which were originally thought to have been wooden panes, Max Martin, *Römermuseum und Römerhaus Augst (Augster Museumsheft* 4, Augst 1987) 62, no. 49, but have now been identified as stone panes, Rütti 1991, I, 163-164; fig. 103; pl. 218, 05-06;
b) Istanbul, Kocabaş Collection, *Kocabaş* 1984, 55, no. 96, fig. 45, with the name ΑΡΧΕϹΤΡΑΤΟΥ, *Archestratou*, retrograde, in Greek;
c) one-part limestone mold, probably for making fifth- to sixth-century bowls, found near Autun, France, Sapin 1985, 349; Foy et al. 1989, 66-67.

[3] A plaster mold for making grape-shaped bottles was excavated at Artashat, Armenia: G. Khachatarian, "Armenia and Syria — Centers of Glass-working," *Vjestnik Obschshestvennih Nauk* 7 (Yerevan 1983) 60-61, fig. 1 and 2.

[4] Cherry wood panels for the sides of a mold for square bottles were excavated at Saintes, France: Anne Hochuli-Gysel, in Foy et al. 1989, 58.

[5] A one-part hexagonal copper alloy mold for making sixth- to seventh-century glass vessels was excavated at Samaria: K. M. Kenyon, in G. M. Crowfoot and K. M. Kenyon, *Samaria-Sebaste III: The Objects* (London 1957) 450-451, no. 9. See Barag 1970c, 37-38, on molds of sheet metal used for Byzantine pilgrim vessels. See above, *Glossaries: Mold Construction Types*: MCT V A and Cat. No. 43.

[6] Luigi Zecchin, *Vetro e vetrai di Murano*, I (Venice 1987), fig. p. 37, illustrates the hinging of a three-part mold from Murano.

[7] Kenneth M. Wilson, *The Toledo Museum of Art, American Glass, 1760-1930* (New York 1994) chapters 3 and 4.

[8] Strong and Brown 1976, 78-80 (pottery); 94-98 (lamps); 105-108 (figurines).

[9] Strong and Brown 1976, 26-34.

which to rely for the construction of molds that could be quickly closed and opened. Glass blowing is the only craft that requires this feature. The construction of reusable multipart molds required ingenuity from both mold makers and glass blowers.

Two concerns are vital for the construction of glassblower's molds. The first is to close the mold tightly around a paraison before it is expanded and the second is to open the mold rapidly after blowing. These concerns affect the number and positioning of sections and require the construction of locking devices to prevent the glass from being forced through the mold edges under the pressure of the blowing process. Because most Roman molds seem to have been made of clay or plaster, they could not be hinged like modern molds. How the problem might have been solved in Roman times can be seen from surviving Islamic clay molds with small lug handles (Fig. 42).[10] These are two-part molds for blowing small bottles with decoration in relief very similar to Roman bottles like Cat. No. 55. The two vertical sections of the Islamic molds meet on the underside of the base (MCT VIII, see Fig. 15). At the top of each section is a semicircular channel carved through to the interior of the mold to form an opening through which the glass bubble was blown. The diameter of the channel, ca. 0.7 cm, is about the same size as the neck of Cat. No. 55. The thickness of the wall of the mold would have constricted the glass, leaving two short vertical mold seams at the base of the neck of the vessel,[11] a feature also seen on many small Roman mold-blown vessels (e.g. Cat. No. 55). On the outside, the mold halves have lugs that provide a grip for prying them apart to pull them off the glass after it was expanded. One of the halves has two vertical lugs, one at the top and one at the bottom, while the other has two horizontal lugs, one at either side of the mold, each placed at a different height. The lugs may have served as anchors for securing the mold halves together during blowing and finishing of the rim.

Eight basic types of mold construction were invented in the first century, MCTs I-VIII (see *Glossaries: Mold Construction Types* and Fig. 15). These first-century molds always included the base of the vessel. The earliest documented Mold Construction Type is MCT IV, used for Hexagonal Bottles with High Relief beginning perhaps as early as the year 13 (see Chapter I: *Hexagonal Bottles with High Relief: Date* and Remarks Cat. No. 15). MCTs I, III, V (used by Ennion), and VIII (used in the Workshop of the Floating Handles) were probably also created in the first quarter of the first century. MCT II may have been created around 25. MCTs VI and VII were in use by the third quarter of the first century, but they may have been invented somewhat earlier.

A third concern that affects the construction of glassblower's molds is that the paraison must fill a mold evenly during blowing. This influences the interior shape of the mold and is reflected in the circumference of the vessel at its widest point. Most first-century mold makers designed molds so that the greatest circumference of the interior would fit within a circle, to insure that the same amount of glass would be needed to fill the mold in all directions.

The cross section of the greatest circumference of the

[10] There appear to be at least eight halves of such molds in public or private collections. Five were published by Lamm as in the Sobernheim Collection, Berlin (Lamm 1929, 58, nos. 1-2, pl. 13, nos. 1-2), the Wasserman Collection, Berlin (Lamm 1929, no. 3), and the Berlin Antiquarium (Lamm 1929, 59, nos. 4-5). The Berlin Antiquarium pieces are now in the Museum für Islamische Kunst (nos. 30 660A and B), and one of the Sobernheim pieces (Lamm 1929, no. 2) is now in Tel Aviv (Museum Haaretz, no. 103958). In addition to these five, there are two additional mold halves in Tel Aviv (Museum Haaretz, nos. 104158 [here Fig. 42] and 104058). The Haaretz Museum mold halves are said to be from Hebron, but it is not clear whether this means they were found or acquired there. My thanks to Gusta Lehrer Jacobson for information and photographs of these pieces. Finally, a mold half in the Corning Museum of Glass (acc. no. 80.7.3) has recently been published (Kjeld von Folsach and David Whitehouse, "Three Islamic Molds," *JGS* 35 [1993] 149-153). The authors identify the Corning mold as from the Sobernheim Collection. Although the Corning mold half is similar in shape and design to one of the Sobernheim mold halves (Lamm 1929, no. 2), comparison of the Sobernheim piece and photographs of Museum Haaretz, no. 103958 clearly shows that the Tel Aviv piece is the one published by Lamm. Folsach and Whitehouse further conclude that the terra-cotta molds were not used for glass making but perhaps for casting objects in another medium such as plaster.

[11] Some of these Islamic mold sections appear to lack handles.

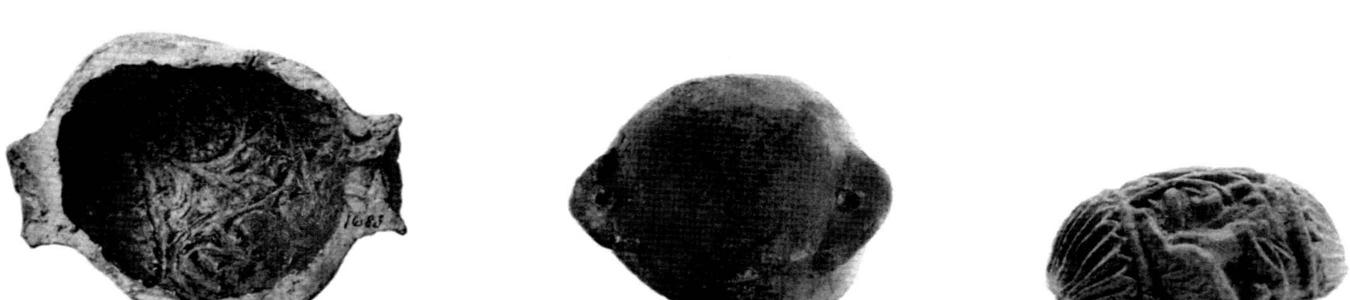

*Fig. 42.* Left: *Interior of one half of an Islamic terra-cotta mold. The semicircular channel at the top of the mold (upper right) would have served as one half of the opening through which the glass paraison would have been blown. Said to be from Hebron, Israel, exterior 10.3×7.5 cm; interior 7.0×6.0 cm; Th 0.8-0.5 cm. Tel Aviv, Eretz Israel Museum, no. MHG 104058.* Center: *Exterior of one half of an Islamic terra-cotta mold. The lug handles would have provided grips to pry the mold off the paraison once the glass had expanded. Said to be from Hebron, Israel, exterior 8.5×8.0 cm. Tel Aviv, Eretz Israel Museum, MHG 104158.* Right: *Modern plaster cast from the mold shown at center. Tel Aviv, Eretz Israel Museum, no. MHG 258. Photos: Courtesy Eretz Israel Museum, Glass Pavilion.*

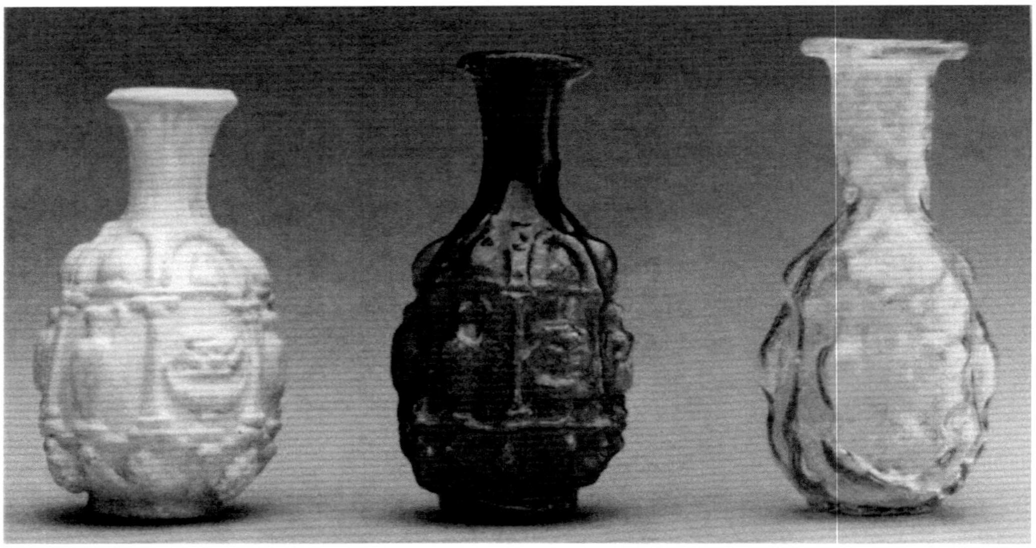

*Fig. 43. Diminution of crispness of detail and body size within one Series: Hexagonal Bottles with High Relief, Vessels Type.* From left: *Subseries A1, Generation A1a (Cat. No. 15), Subseries A3, Generations A3a (Cat. No. 26) and A3b (Cat. No. 33). Photo: Tim Thayer.*

earliest vessels is either circular (MCTs I-V, VII) or squarish (MCTs VIII B and VIII M). MCTs VI and VII A represent cautious experiments in filling an asymmetrical mold for a head-shaped vessel. MCT VI combines one half of a circular section with one half of a square one, but both halves fit within the same half circle; this mold was probably made in the second half of the first century (Cat. No. 140). MCT VII, introduced in the third quarter of the first century, was used for truncated conical beakers (Cat. Nos. 11 and 12) and for head-shaped vessels. MCT VII A was used for a cup shaped like a single head flask (Cat. No. 139); it has a slightly irregular circumference on the face side, but the greatest radius of each mold half was the same size, each fitting roughly within a half circle. Already in the early first century the Workshop of the Floating Handles experimented with mold constructions that were not symmetrical in all directions. The workshop used not only MCTs VIII A, VIII B, VIII E, and VIII H, which fitted into a circular circumference, but also MCT VIII F, which was lenticular.

## THE REPRODUCTION OF MOLDS

Many small glass bottles that appear to have been blown into the same mold vary not only in crispness and freshness of surface detail but also in size, which diminishes steadily over time (Fig. 43 and Remarks Cat. Nos. 15 [with Figs. 71 and 72], 26, and 33). This diminution in size can have only one explanation: the bottles were not blown in one and the same mold but rather into a succession of molds, each replacing its worn-out predecessor. Diminishing size is typical for the use of fired clay molds because clay shrinks when it is fired. The phenomenon of shrinking size is well known from terra-cotta figurines and lamps.[12]

[12] R. V. Nicholls, "Type, Group and Series: a reconsideration of some coroplastic fundamentals," *BSA* 47 (1952) 217-226; J. H. and S. H. Young, *Terracotta Figurines from Kourion in Cyprus* (Philadelphia 1955) 193.

The manufacturing process for glass-blower's molds can be reconstructed as follows. The mold maker, who may or may not have been the glass blower himself, made one or more clay molds from a model or archetype in wax or clay, possibly created by a specialized craftsman. While the molds were still new and unused, he also made one or more casts that could be used as secondary archetypes for this same Generation of molds (see Fig. 42, right, for a modern example). The glass blower used the first set of molds until they wore out or broke. When the molds became useless, he discarded them and made new molds either from the secondary archetypes or from one of the last bottles blown. At this point the mold maker might freshen the carving, which could modify the design to varying degrees. The whole process then started again with a second Generation of molds until the Type went out of production.

Because glass shrinks during annealing and clay shrinks during firing, the second-Generation molds were slightly smaller than their first-Generation ancestors. Consequently vessels blown into them were also slightly smaller. The size and shape of a vessel were also sometimes, but not always, affected when it was reheated and manipulated to form rims, handles, etc. And glass blowers sometimes gave the paraison an extra puff of air in order to prevent its collapsing or to correct the shape after extraction from the mold. Therefore, although bottles made in a mold of a younger generation can usually be distinguished, it is not possible to use shrinkage proportions alone to determine the number of Generations separating any given bottle from the archetype from which it was derived. An interesting example of the evolution of a fired clay mold is provided by the Generations of Hexagonal Bottles with High Relief (see Chapter I: *Syro-Palestinian Styles*, especially Figs. 46 and 54).

# COLOR PLATES

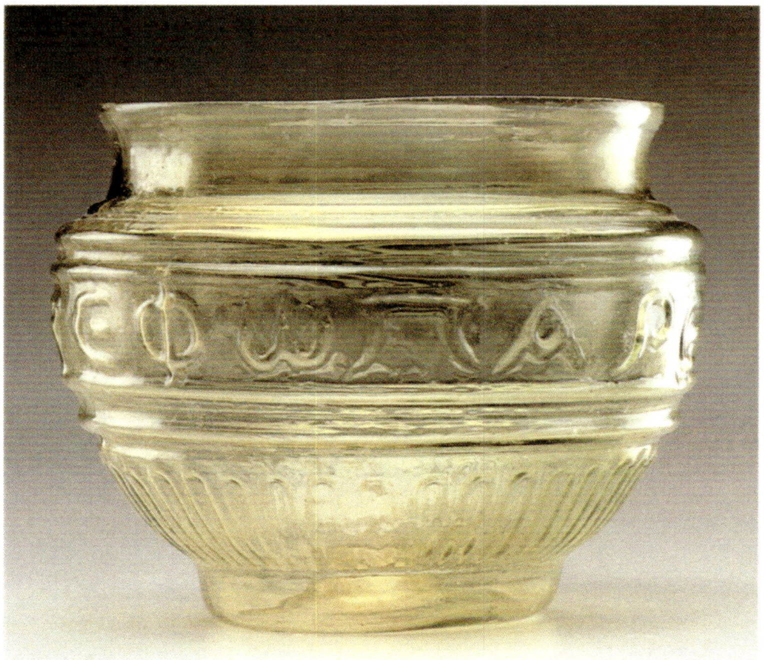

*1. Bulbous Cup with Mold-blown Inscription: Cat. No. 1.*

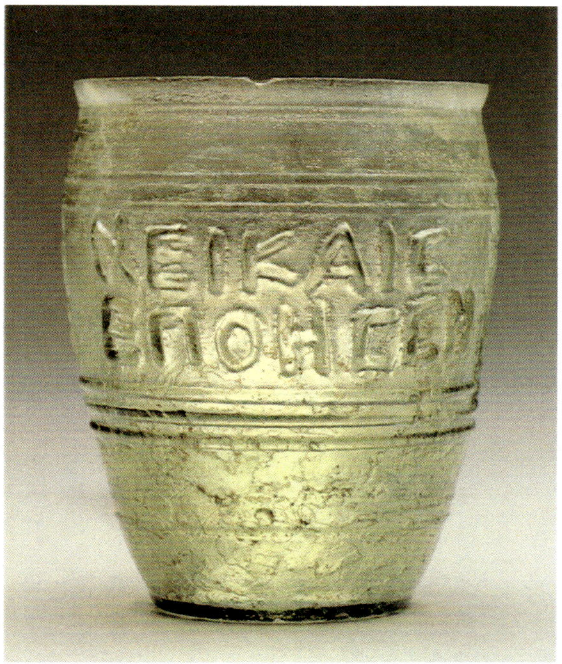

*2. Barrel-shaped Cup with Mold-blown Inscription (signed by Neikais): Cat. No. 5.*

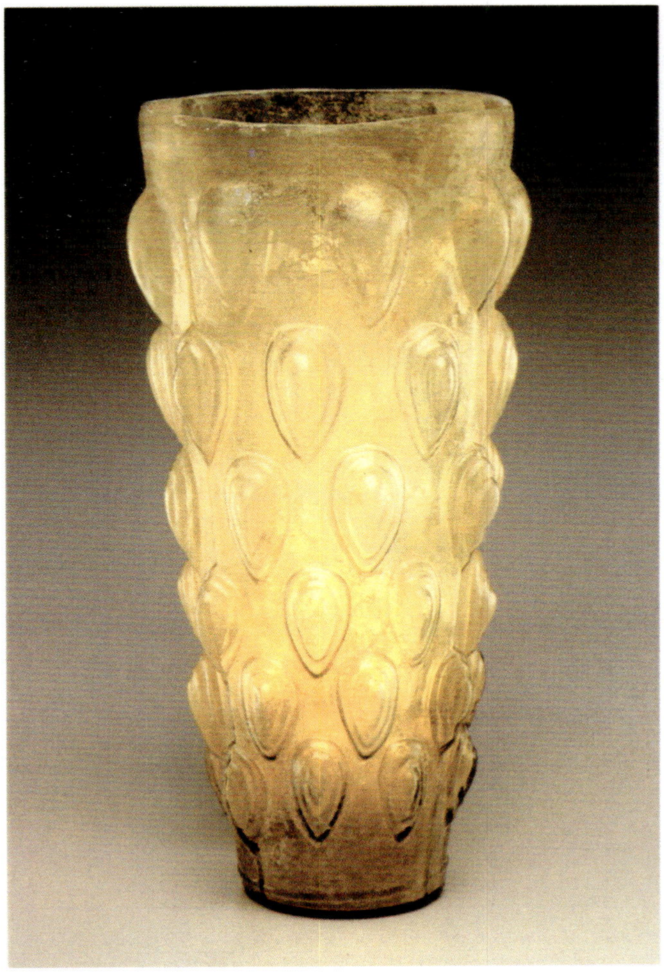

*3. Truncated Conical Beaker with Knot-shaped Knobs: Cat. No. 8.*

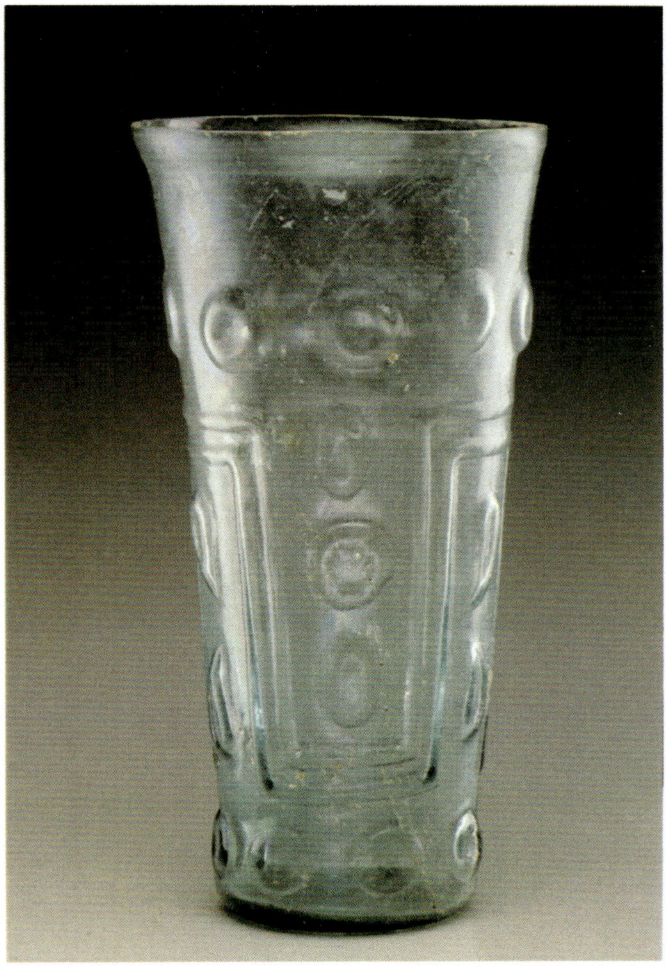

*4. Truncated Conical Beaker: Cat. No. 11.*

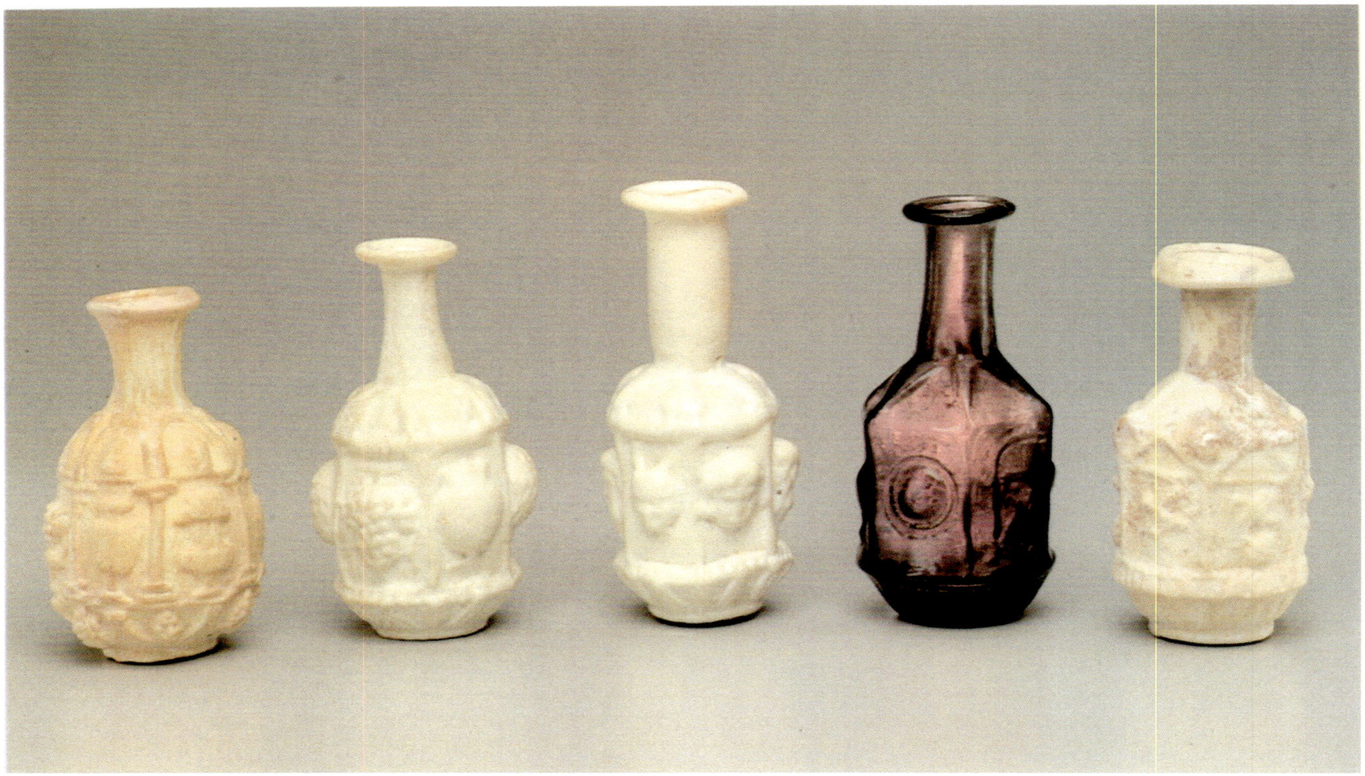

*5. Hexagonal Bottles. Left to right: Vessels Type, Series A1, Generation A1a: Cat. No. 15 (panels 3, 4); Fruit Type, Series A1, Generation A1a: Cat. No. 36 (panels 3, 4); Masks Type, Series A1, Generation A1a: Cat. No. 41 (panels 4, 5); Mixed Symbols Type, Series A1, Generation A1b: Cat. No. 43 (panels 3, 4); Bird Type, Series A1, Generation A1b: Cat. No. 45 (panels 5, 6).*

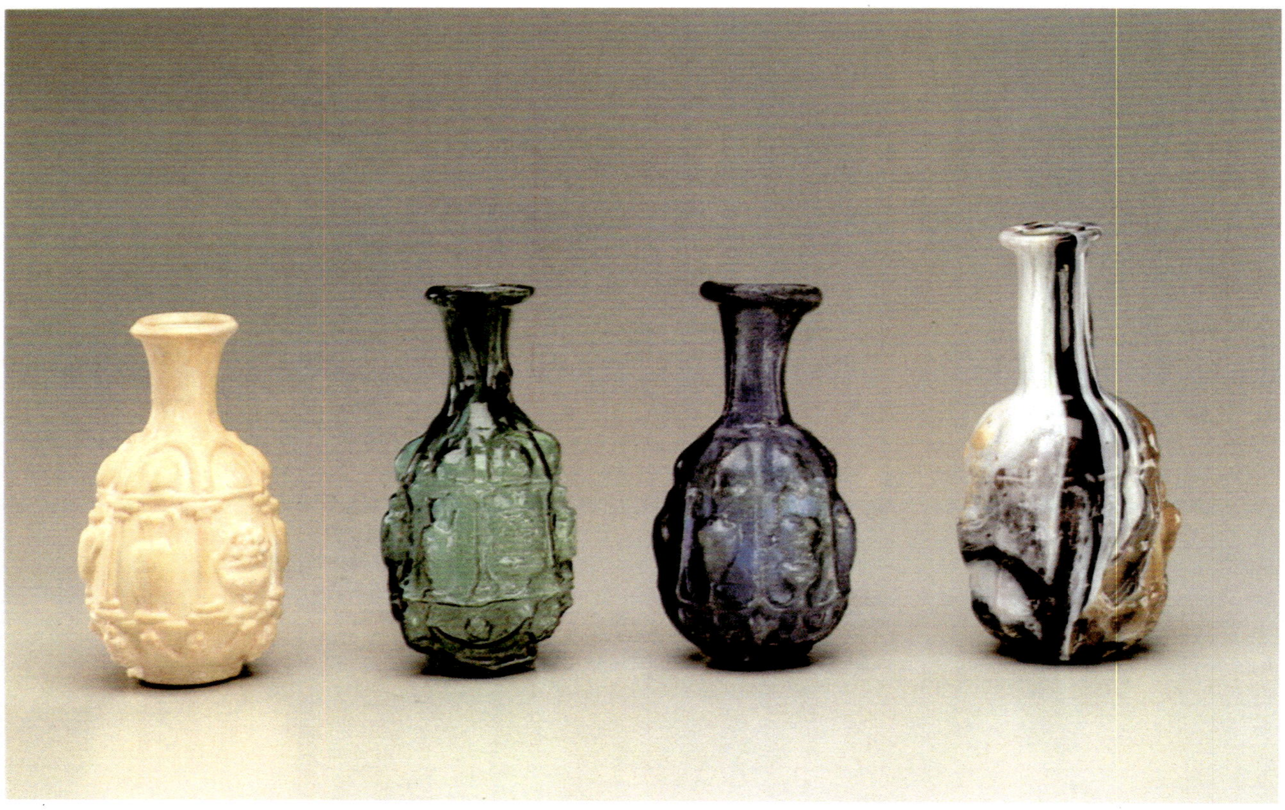

*6. Hexagonal Bottles, Vessels Type. Left to right (all with panels 1,2): Series A1, Generation A1a: Cat. No. 15; Series A3, Generation A3a: Cat. No. 26; Series A1, Generation A1c: Cat. No. 23; Series B1, Generation B1a: Cat. No. 34.*

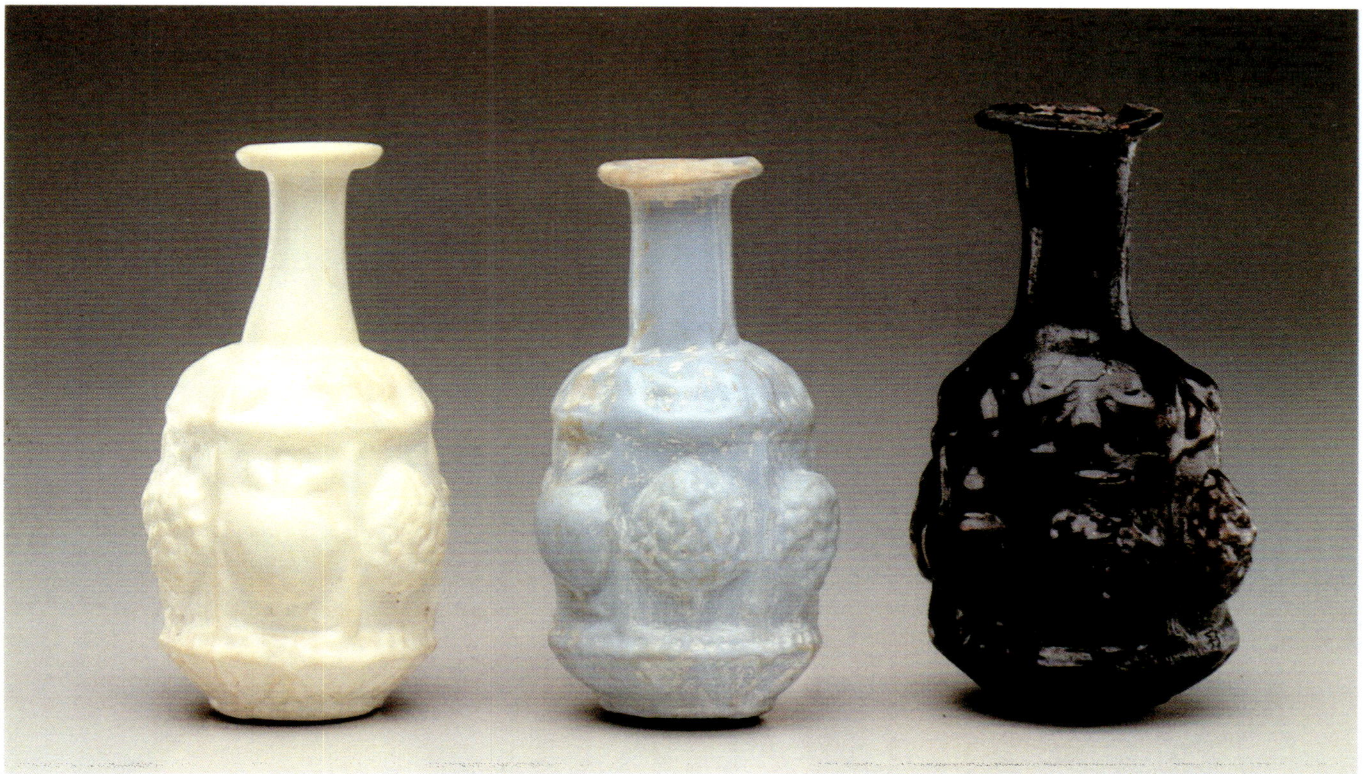

*7. Hexagonal Bottles, Fruit Type. Left to right: Series A1, Generation A1a: Cat. No. 36 (panels 6, 1, 2); Series A1, Generation A1a: Cat. No. 37 (panels 1, 2, 3); Series B: Cat. No. 40 (panels 4, 5, 6).*

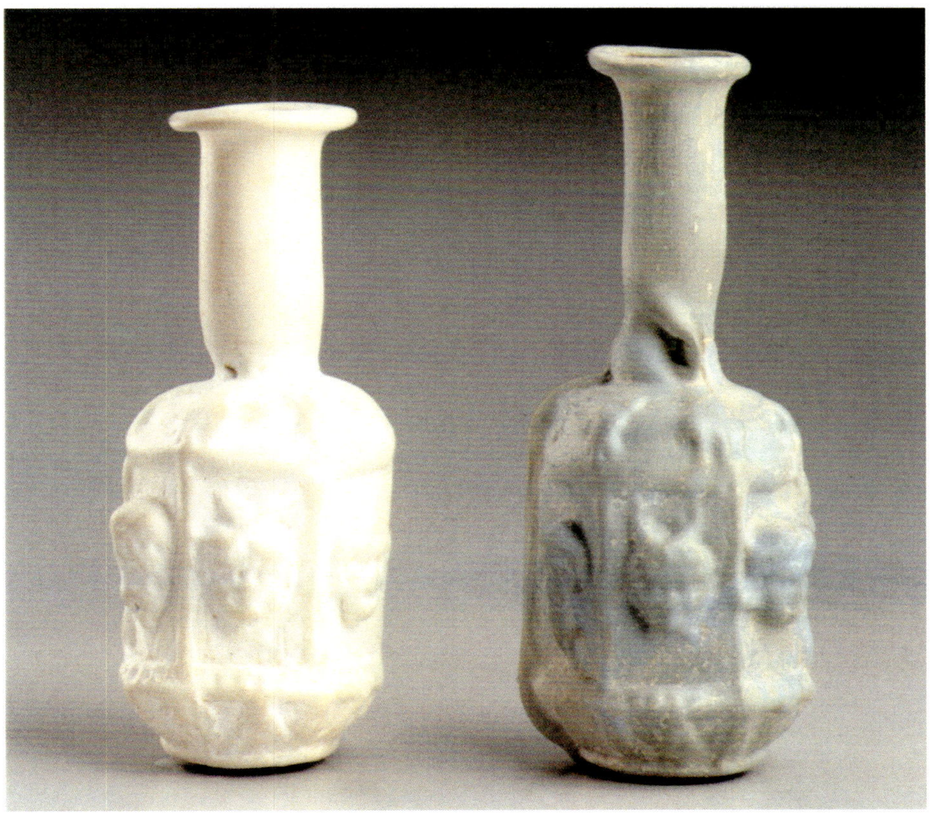

*8. Hexagonal Bottles, Masks Type. Left: Series A1, Generation A1a: Cat. No. 41 (panels 6, 1, 2); right: Series A, Generation A1a: Cat. No. 42 (panels 3, 4, 5).*

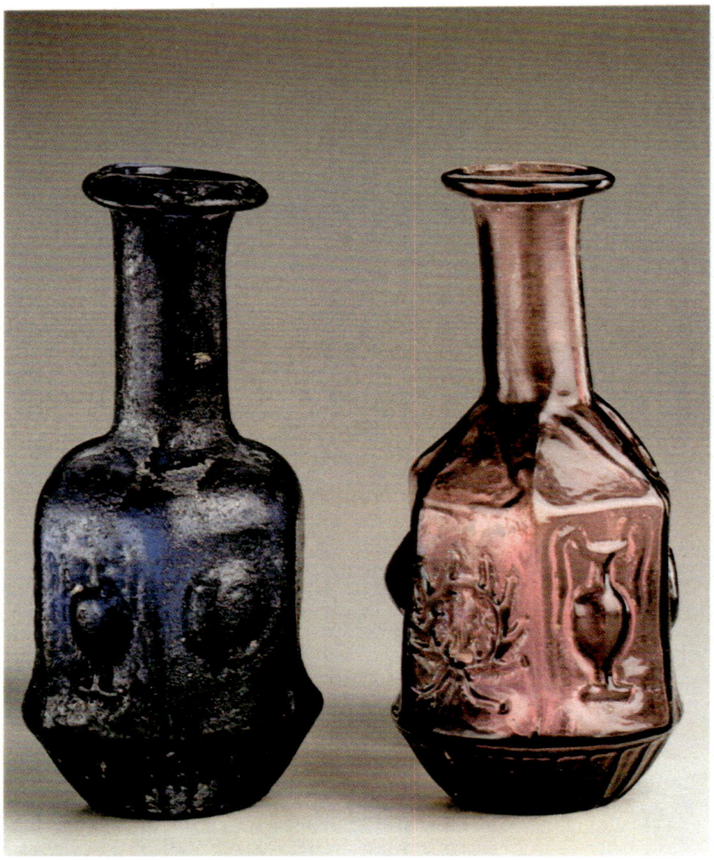

*9. Hexagonal Bottles, Mixed Symbols Type. Left: Series A1, Generation A1b-γγ: Cat. No. 44 (panels 5, 6); right: Series A1, Generation A1b: Cat. No. 43 (panels 1, 2).*

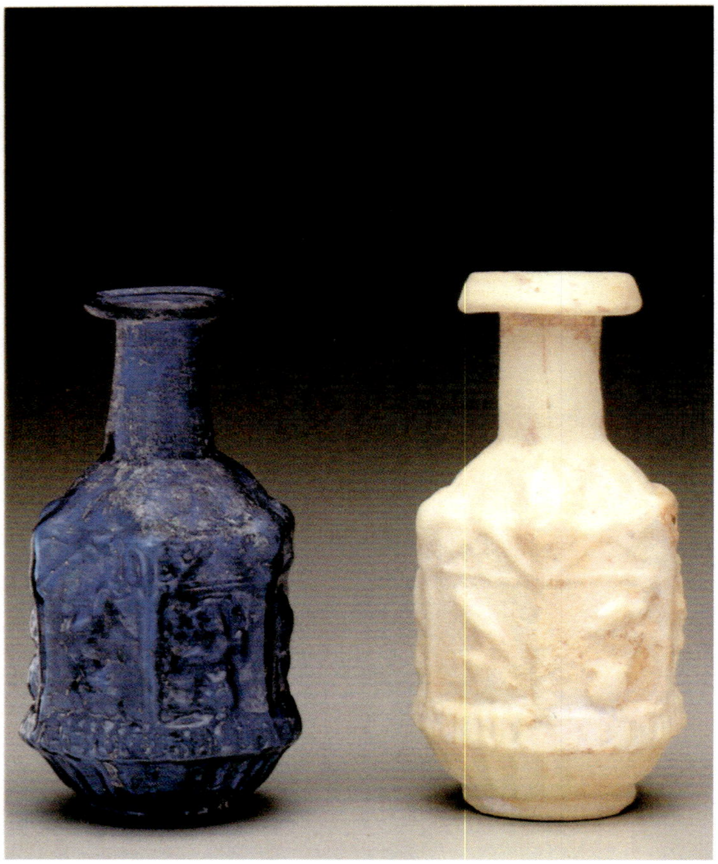

*10. Hexagonal Bottles, Birds Type. Left: Series A1, Generation A1c: Cat. No. 46 (panels 1, 2); right: Series A1, Generation A1b: Cat. No. 45 (panels 3, 4).*

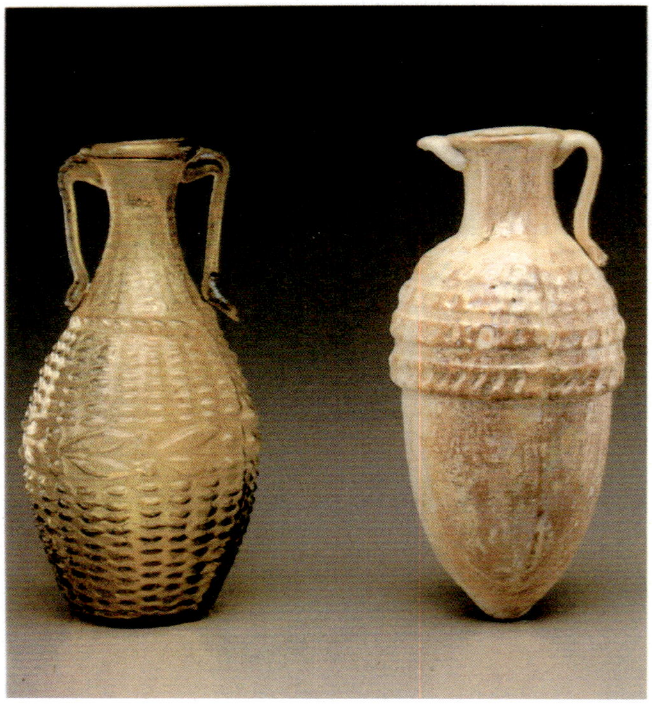

*11. Workshop of the Floating Handles. Left: Miniature Amphora in Wicker Basket: Cat. No. 59; right: Acorn-shaped Bottle: Cat. No. 113.*

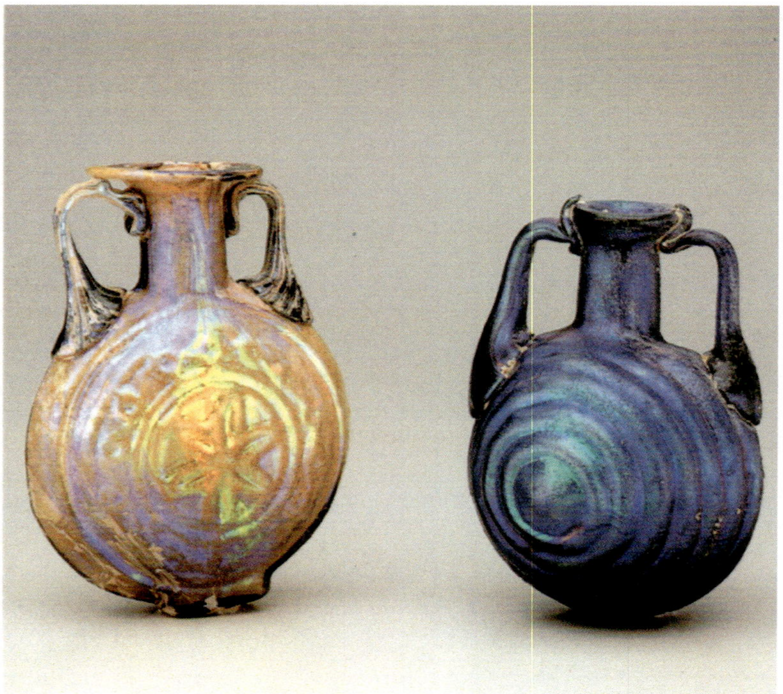

*12. Left: Lenticular Bottle with Two Handles: Cat. No. 60; right: Barrel-shaped Bottle with Two Handles: Cat. No. 70.*

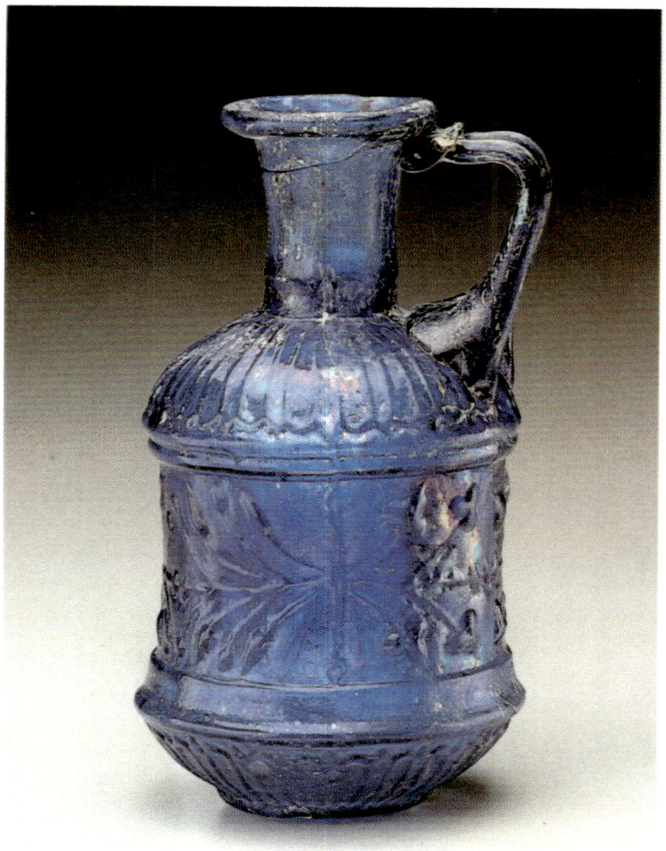

*13. Cylindrical Jug with Three Pairs of Floral Sprays: Cat. No. 76.*

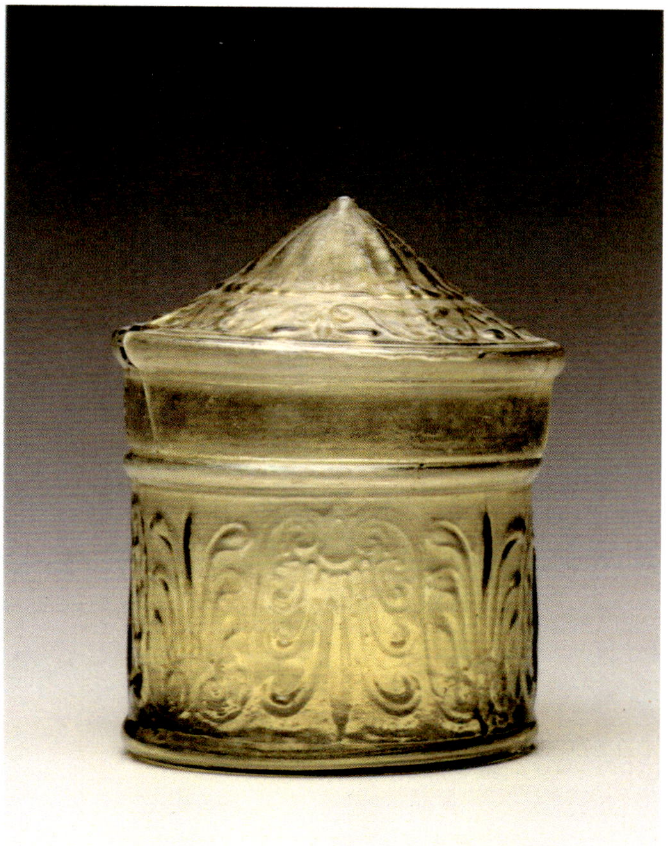

*14. Cylindrical Box and Conical Lid (Pyxis): Cat. No. 79.*

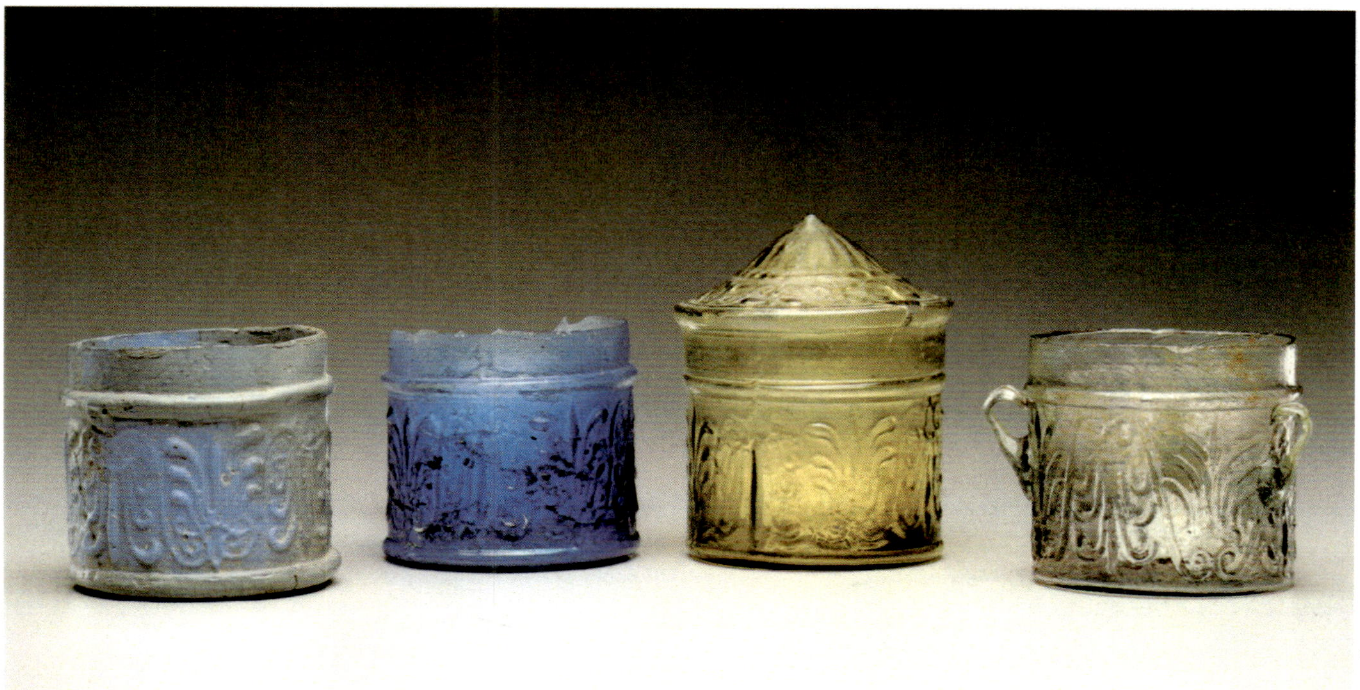

*15. Cylindrical Pyxides. Left to right: Cat. Nos. 82, 79, 80, and 81.*

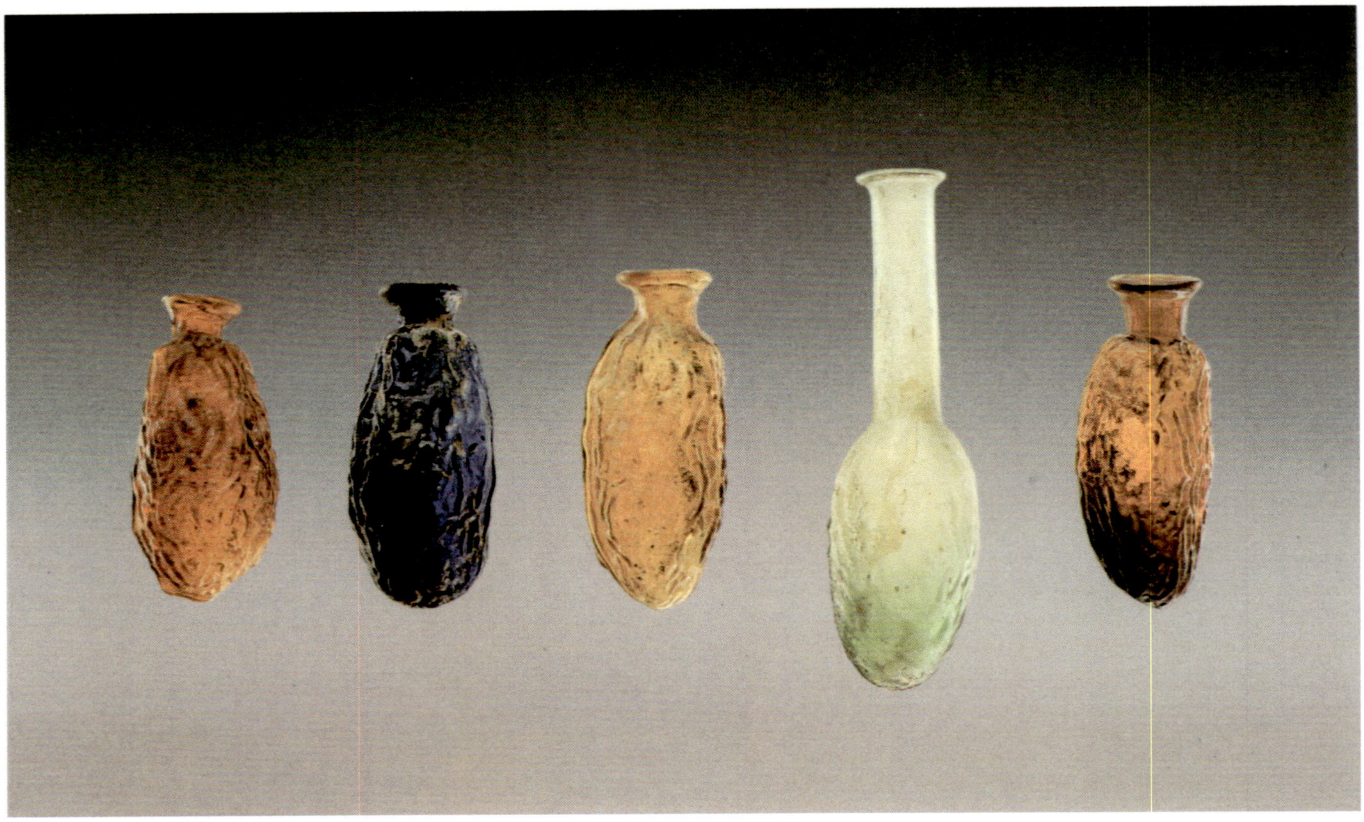

*16. Date-shaped Bottles. Left to right: Cat. Nos. 84, 89, 93, 103, and 104.*

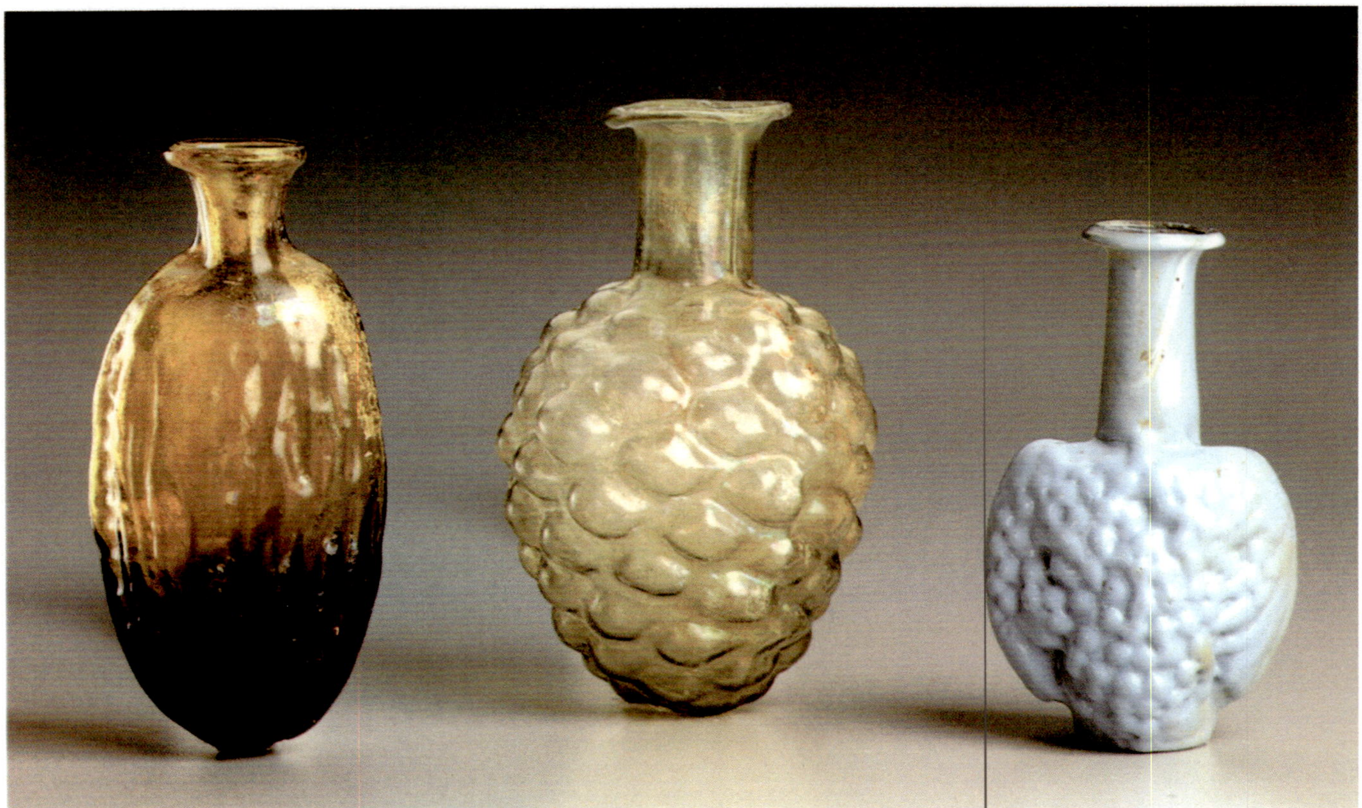

*17. Fruit-shaped Bottles. Left to right: Almond-shaped Bottle: Cat. No. 108; Pine Cone-shaped Bottle: Cat. No. 111; Grape-shaped Bottle: Cat. No. 109.*

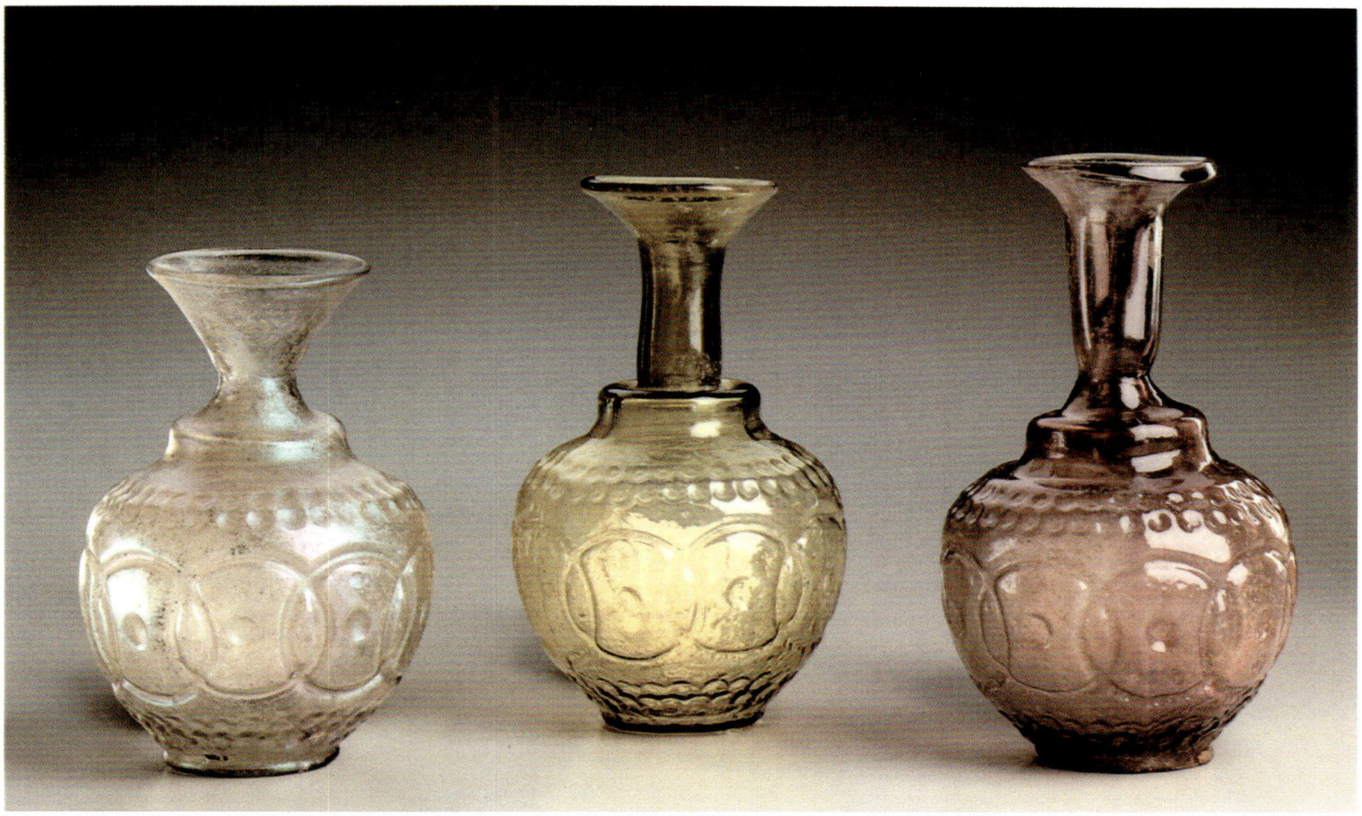

*18. Flasks with Mold-blown Designs in Sunken Relief. Left to right: Cat. Nos. 117, 115, and 116.*

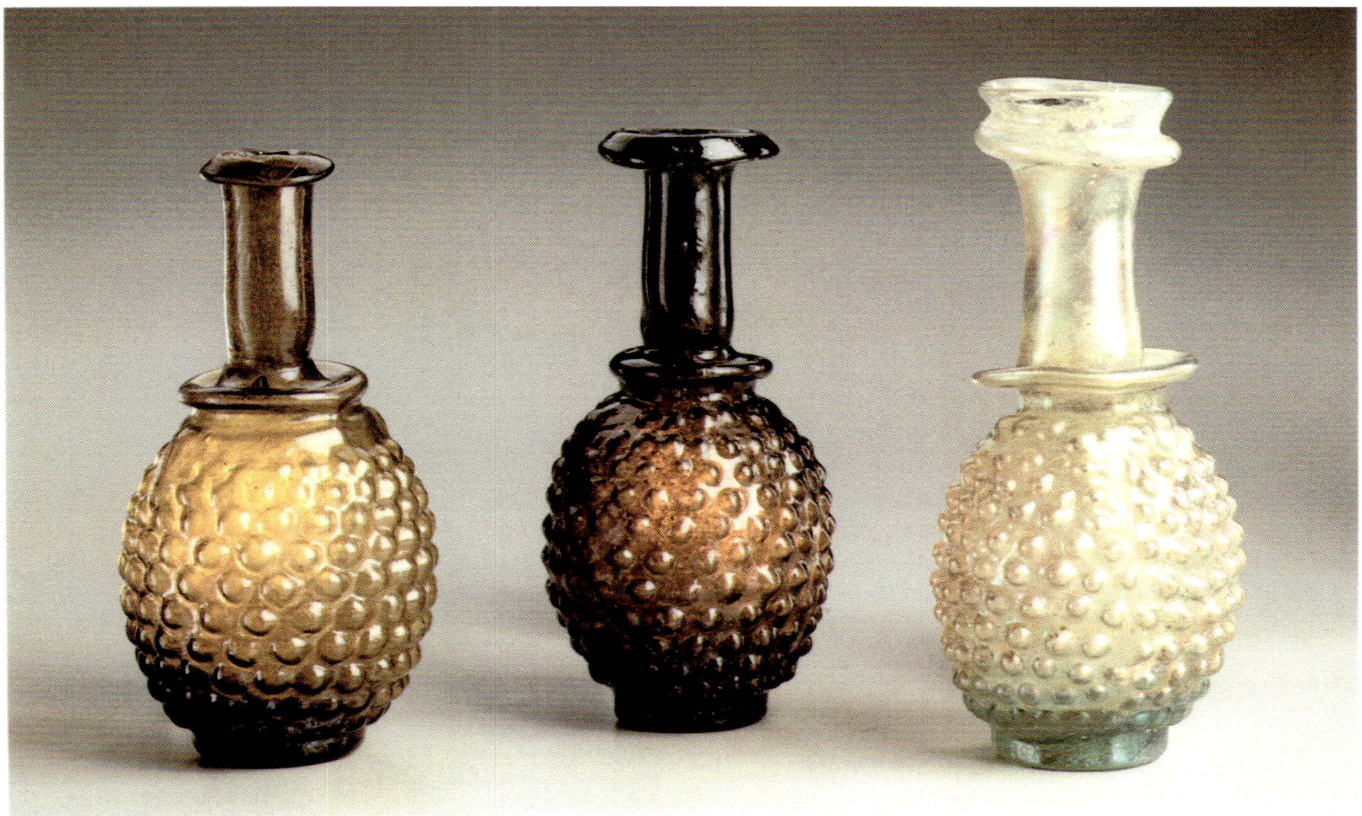

*19. Bulbous Bottles with Stylized Grape Pattern. Left to right: Cat. Nos. 120, 127, and 128.*

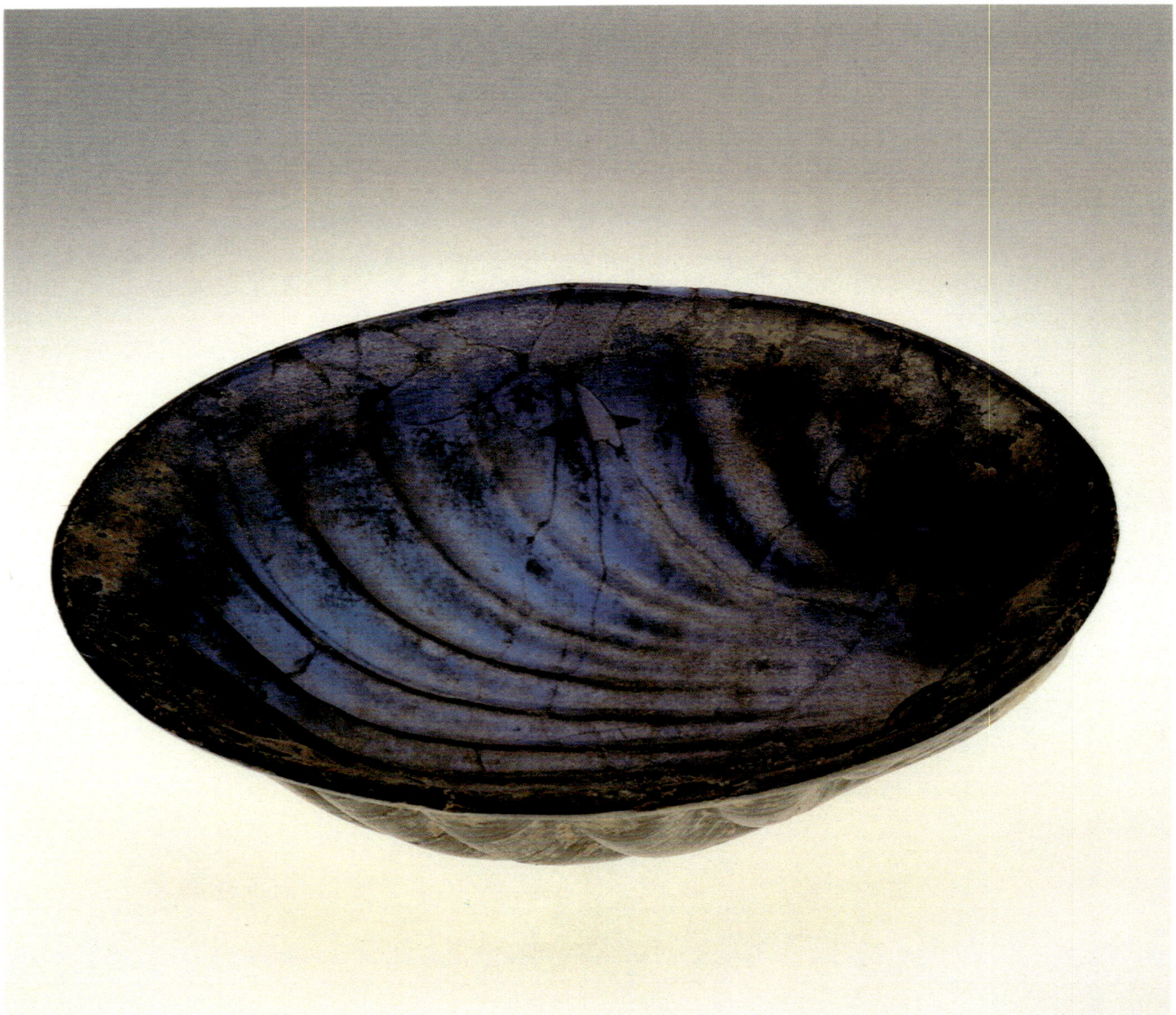

*20. Dish in the Shape of a Shell: Cat. No. 137.*

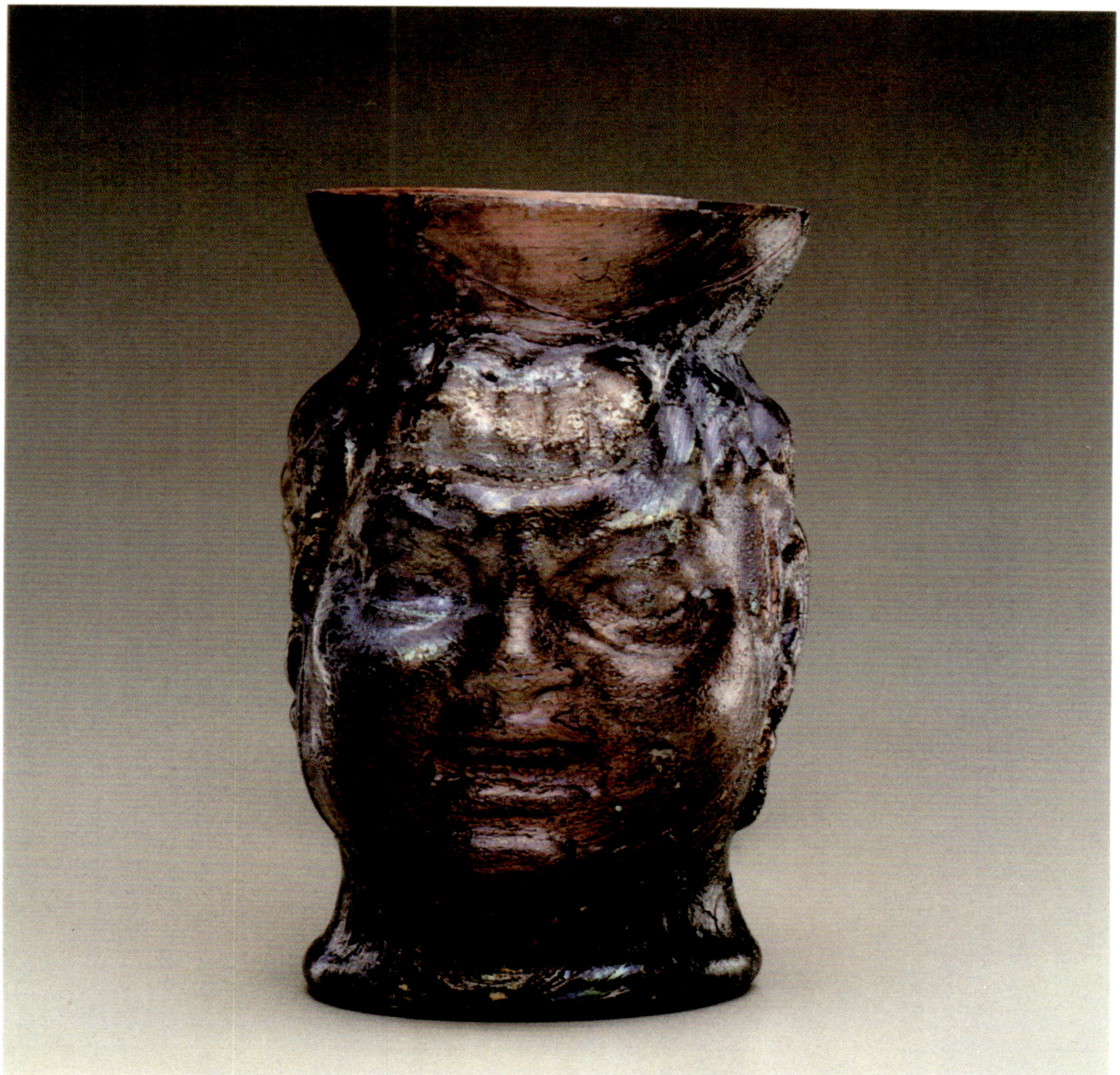

*21. Cup in the Shape of a Male Negroid Head: Cat. No. 139.*

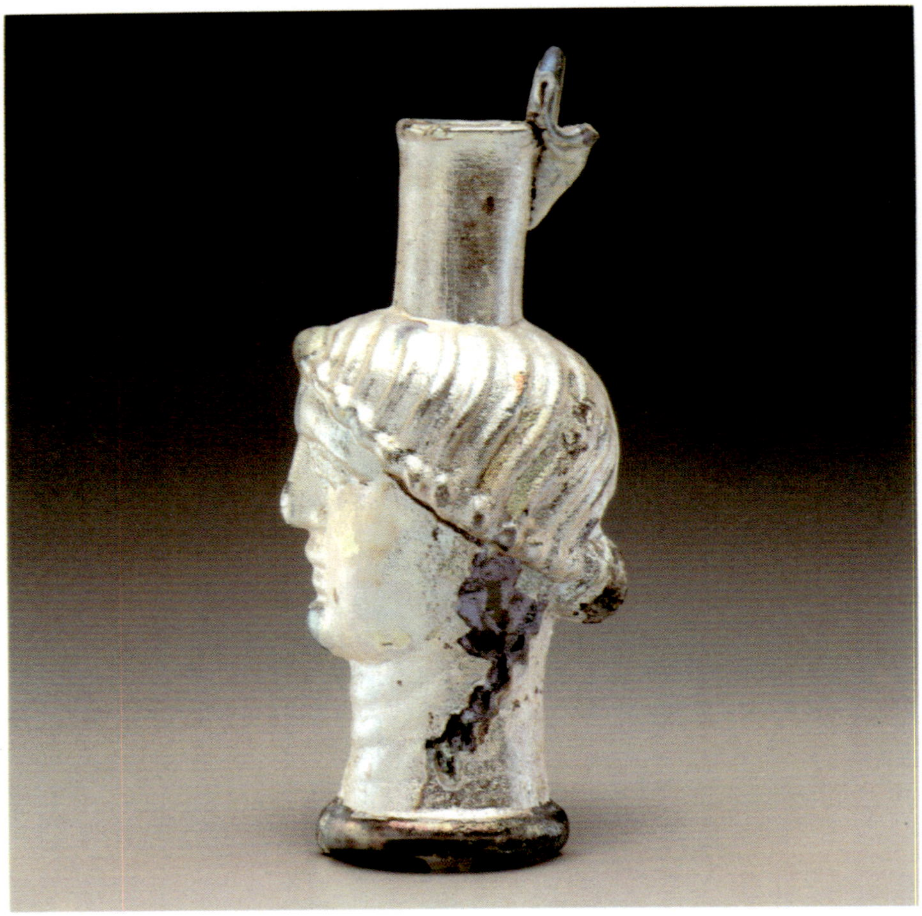

*22. Jug in the Shape of a Female Head, possibly Livia as Hera: Cat. No. 138.*

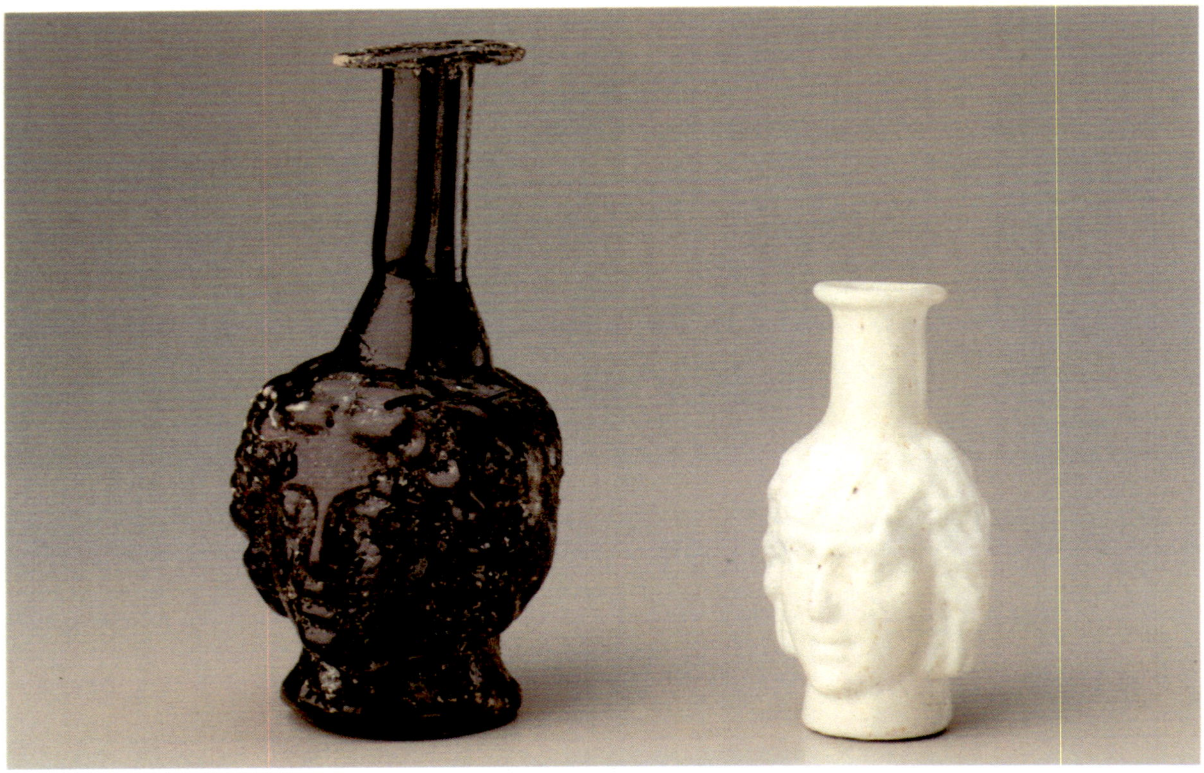

*23. Double Head-Shaped Vessels, Late First to End of Second Century. Left: Double Head-shaped Bottle, Two Similar Seductive Medusas: Cat. No. 142; right: Double Head-shaped Bottle, One Serious Face, One Smiling Face: Cat. No. 141.*

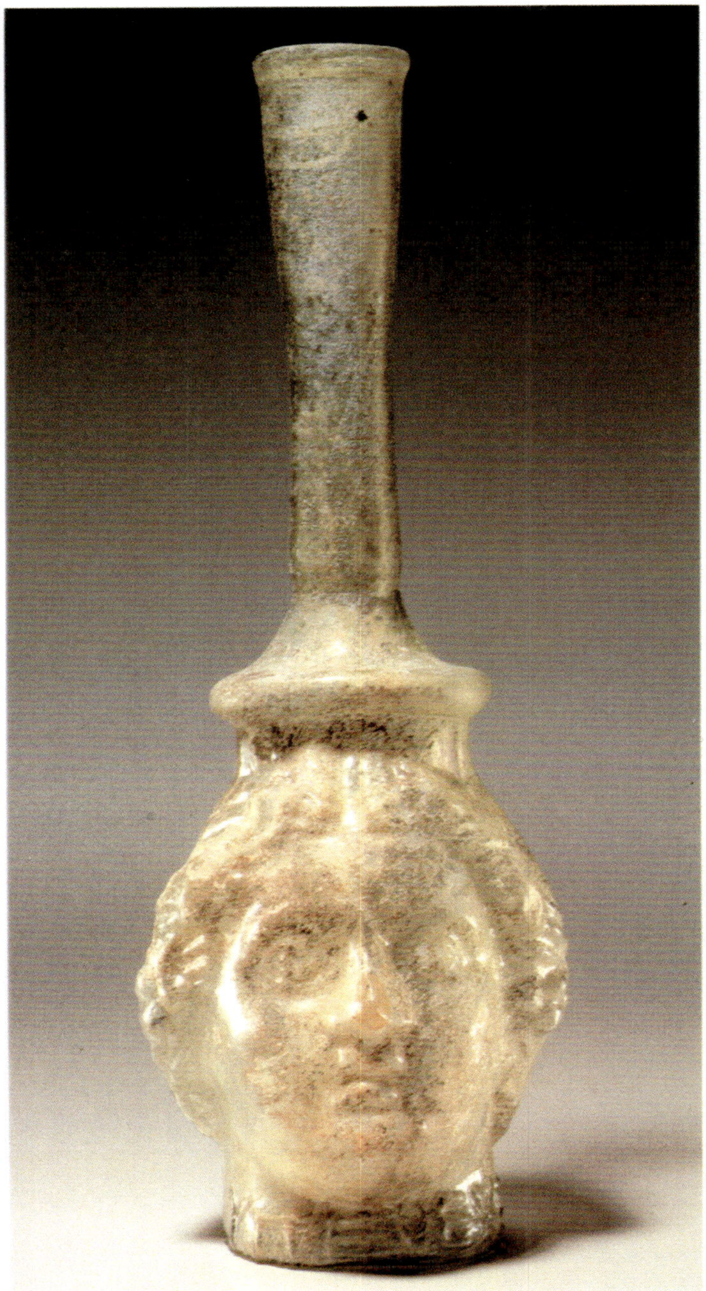

*24. Head Flask, Idealized Male Face with Mold-blown Inscription, possibly Dionysos: Cat. No. 147.*

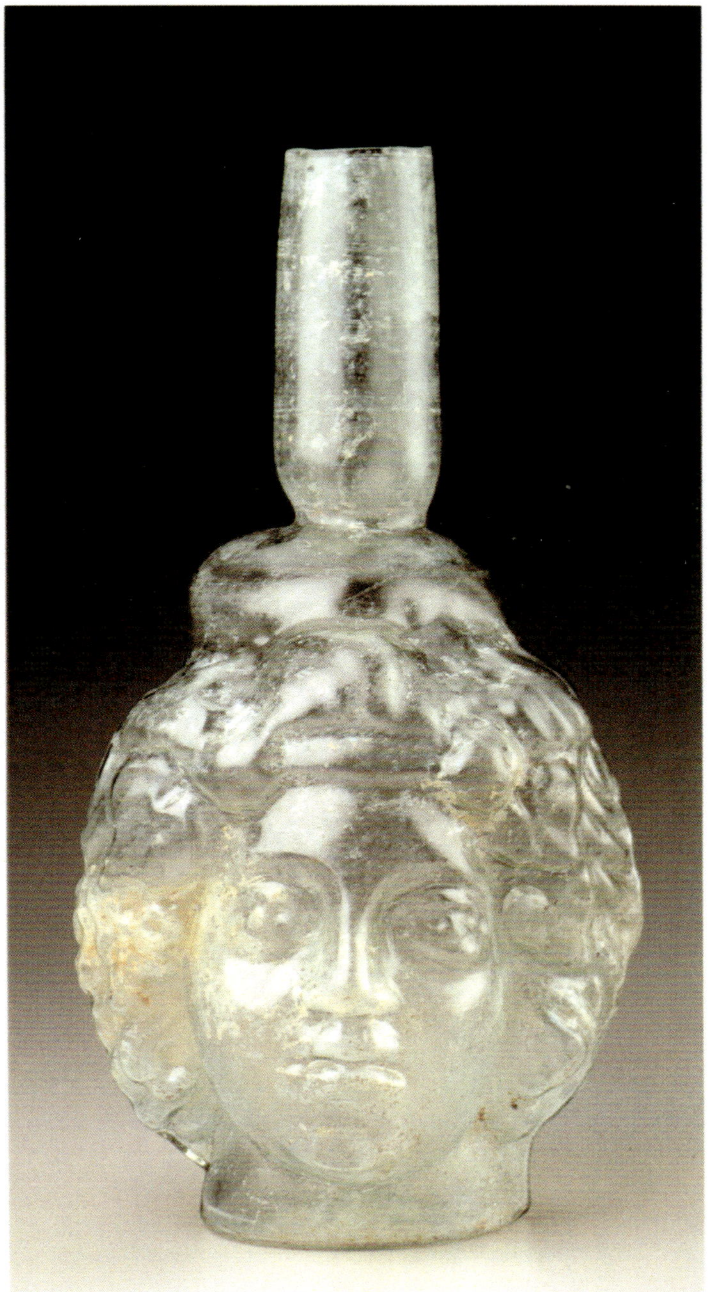

*25. Head Flask, possibly Antinous as Dionysos: Cat. No. 148.*

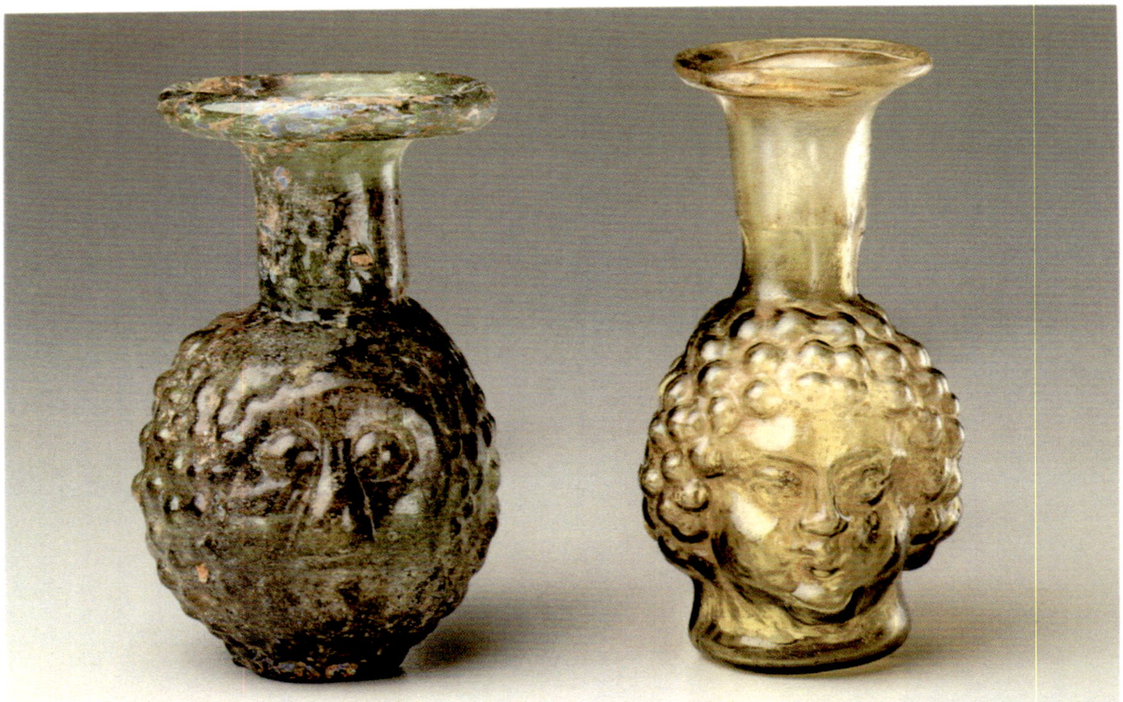

*26. Double Head-shaped Vessels with Knobby Hair, Third Century. Left: Double Head Flask with Internal Diaphragm (Sprinkler), Two Similar Stylized Droll Faces: Cat. No. 157 (Side A); Right: Double Head Flask, Two Similar Chubby Childlike Faces: Cat. No. 149 (Side B).*

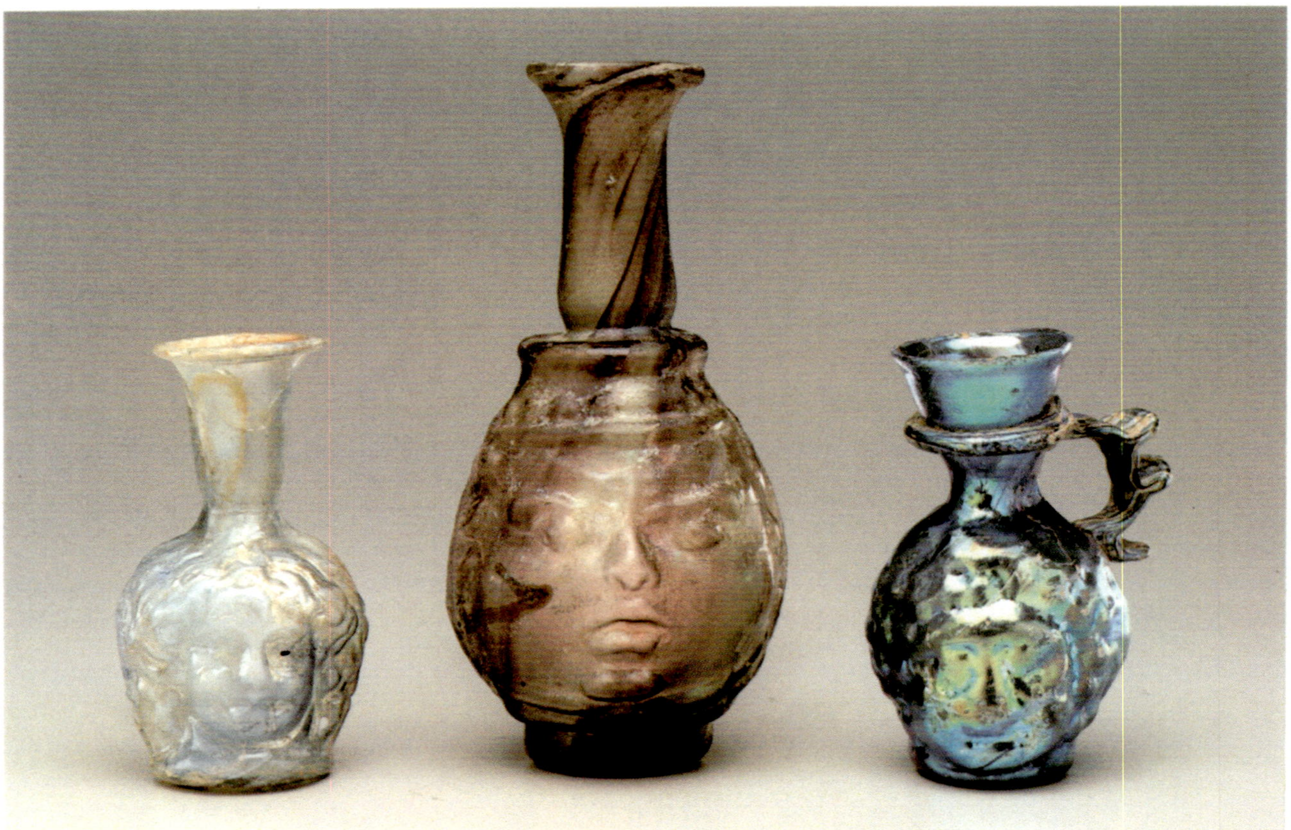

*27. Double Head-Shaped Vessels, Third or Fourth Century, and Double Head Flasks with Knobby Hair and One Handle, Fourth Century. Left to right: Double Head Flask, Two Similar Medusas with Wavy Hair: Cat. No. 158 (Side A); Double Head Flask, Two Similar Grotesque Faces: Cat. No. 160 (Side A); Double Head-shaped Vessel, Two Childlike Faces, One Serious Face, One Smiling Face, with Crimped Handle: Cat. No. 162 (Side A).*

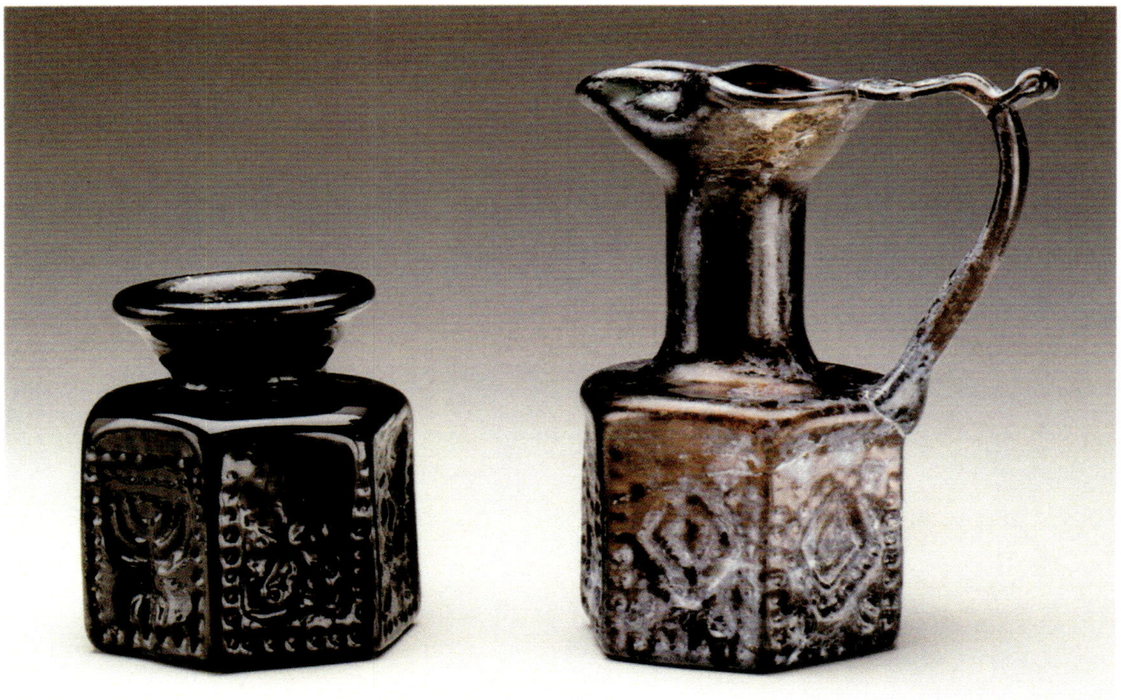

*28. Vessels with Jewish Symbols in Sunken Relief. Left: Hexagonal Jar: Cat. No. 173; Right: Hexagonal Jug: Cat. No. 174.*

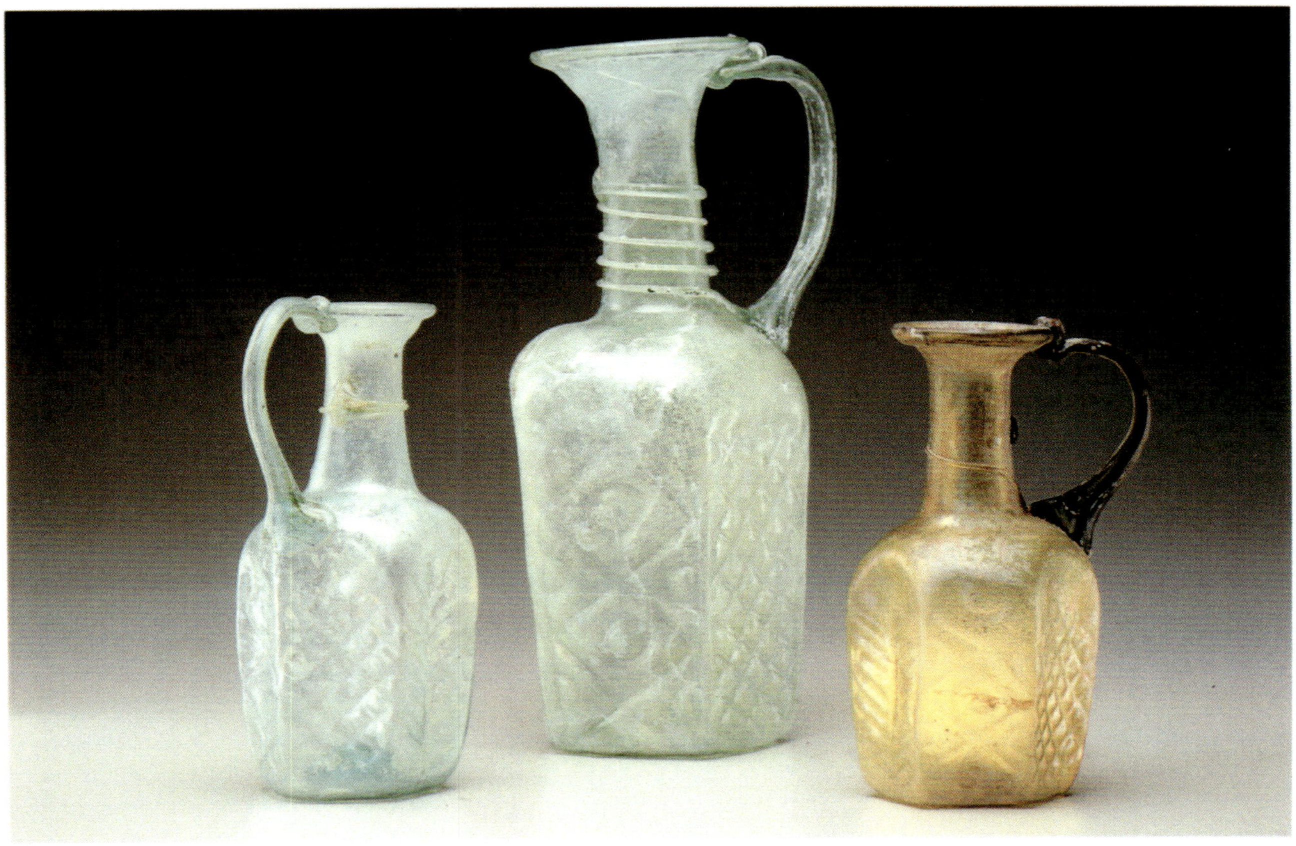

*29. Vessels with Symbols in Sunken Relief, perhaps Christian. Left to right: Hexagonal Jug: Cat. No. 180; Pentagonal Jug: Cat. No. 183; Hexagonal Jug: Cat. No. 181.*

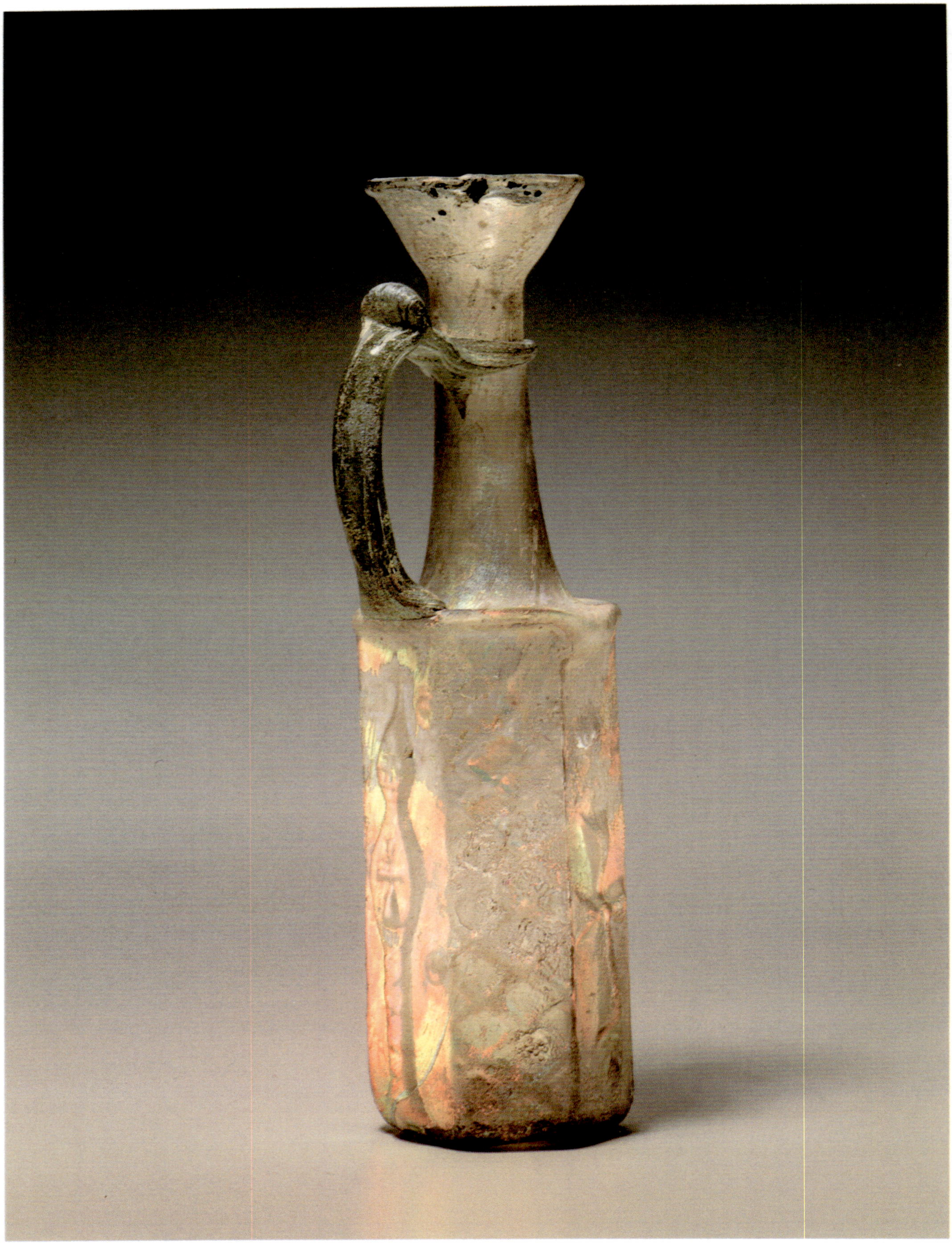

*30. Jug with Christian Symbols in Low Relief: Cat. No. 190.*

# I
# The First Century

## EARLY SYRO-PALESTINIAN MOLD BLOWING

Glass workers in the Syro-Palestinian area were not just the discoverers of but also the first to fully exploit all the possibilities of the new technique of glass blowing (see above: *Advances in Roman Glass Technology*). In Roman Egypt glass workers continued to produce fine mosaic glass inlays,[1] while the newly established glass industry in Rome and southern Italy was based on the older casting, sagging, and related heat-forming techniques inherited from the Hellenistic period.[2] The great tradition of blowing glass into molds that imparted shape, size, and decoration began around the coastal cities of the Roman provinces of Syria, Phoenicia, Palestine and flourished there and elsewhere[3] throughout the Roman period into Byzantine and Islamic times.

Long before the invention of glass blowing, the Syro-Palestinian coast was associated with glass making. The fame of the area is vividly documented by Pliny's well-known story about merchants on the beach near Akko, the colony of Ptolemais:

> There is a story that once a ship belonging to some traders in natural soda put in here and that they scattered along the shore to prepare a meal. Since, however, no stones suitable for supporting their cauldrons were forthcoming, they rested them on lumps of soda from their cargo. When these became heated and were completely mingled with the sand on the beach a strange translucid liquid flowed forth in streams; and this, it is said, was the origin of glass.[4]

Experiments with natron from Egypt and sand from the beach near Akko have proven that glass could have resulted from such a chance firing.[5] However, it is now generally accepted that the earliest glass dates from long before the setting of Pliny's story and was invented in Mesopotamia in the third millennium B.C. rather than on the Syro-Palestinian coast.[6]

Mold-blown vessels differ from glass vessels formed by other methods because multiple copies with integral decoration can be made that are closely similar to each other in shape, size, and decoration. The range of decorative designs and motifs in high or low relief or in raised outline is very wide. Vessels decorated with mold-blown designs are usually either tablewares, such as bowls, cups or beakers, or small bottles for unguents and scents. The quantities of decorated mold-blown glass produced during the first century A.D. outnumber by far mold-blown vessels produced in later Roman times.

Numerous vessels have survived that were made in one mold or in closely related molds. The information provided by studying these objects is invaluable for understanding and reconstructing early Roman glass-working techniques, production sequences, relations between workshops, and distribution patterns.

### *Date*

Very little is known about the exact date or circumstance of the beginning of mold blowing. Although it is now generally accepted that free blowing preceded mold blowing,[7]

[1] Stern 1993, 26.

[2] Grose 1989, 241-262 (with lit.).

[3] For place names used in this catalogue, see *Glossaries: Geographic Designations.*

[4] *Fama est adpulsa nave mercatorum nitri, cum sparsi per litus epulas pararent nec esset cortinis attollendis lapidum occasio glaebas nitri e nave subdidisse, quibus accensis, permixta harena litoris, tralucentes novi liquoris fluxisse rivos, et hanc fuisse originem vitri* (Pliny, *Natural History* 36.190); unless otherwise noted, all Latin quotations and English translations from Pliny are taken from the standard Loeb edition by D. E. Eichholz, *Loeb Classical Library*, vol. 419 (London and Cambridge, Mass. 1962, reprinted 1971, 1989).

[5] H. Löber, "Hatte Plinius der Ältere recht mit seinem Bericht über das Entstehen des Glases?," in T. E. Haevernick and A. von Saldern eds., *Festschrift für Waldemar Haberey* (Mainz 1976) 85-88.

[6] D. P. Barag, "The Origin of Glass," *IX Congrès internationale du verre, Communications artistiques et historiques* (Versailles, 27 September-2 October, 1971) Section BII, 183-189; D. P. Barag, in Oppenheim et al. 1970, 132-133, n. 15; Grose 1989, 45-48; Stern 1994, 86-88.

[7] For example, Grose 1989, 262; Israeli 1991, 47; Price 1991, 64.

the question is more complicated than has been previously recognized.[8] Evidently the inflation of tubes preceded any type of blowing into a mold. But glass tubes might have been expanded in molds before the development of the blowpipe. The chronological relationships among tube blowing, free blowing with the aid of a blowpipe, and using a blowpipe for blowing into a mold have not yet been established. The earliest identified mold-blown vessels are small bottles of a size that could have been made from tubes.[9] It has not been possible to determine whether they were made by tube blowing or by blowing with a pipe.

Three broad chronological groups of early mold-blown tablewares have recently been identified: the first dates about 35-45, the second about 45-61, and the third about 61-75. These groups are based on finds in Europe and the western Mediterranean.[10] Unfortunately there are no comparable dated finds from the eastern Mediterranean, let alone specifically from the Syro-Palestinian area. There are two reasons for this: first, the widespread custom in the eastern Mediterranean of using tombs for two or more generations[11] and, second, the shortage of closely datable military establishments. However, the scarcity of closely datable early finds does not mean that objects found in the eastern Mediterranean must be later than the dated finds in the West. In fact, the evidence discussed below suggests the opposite: that mold blowing began in the East by the first years of the first century.

The earliest finds in the West are sophisticated, decorated cups made by master blowers such as Ennion (see below: *Glass-Blowers' Signatures: Ennion*). Fragments of closely related mold-blown cups that are too small to be assigned definitely to Ennion have been found in early contexts, about A.D. 35-45, at Valkenburg, Netherlands,[12] and Kempten, Bavaria.[13] A fragment of a mold-blown jug with floral decoration was excavated at Magdalensberg, Austria.[14] Fragments of an oval pyxis with floral designs were excavated near Brescia.[15] It is impossible to determine how long before the period 35-45 these vessels were blown. They could have been relatively new or perhaps as much as a generation old, i.e. perhaps as much as thirty years old at the time of their disposal. The high quality of workmanship and the wide distribution clearly indicate that they were the products of a well-established industry. How long this industry took to establish is not known. Both the hypothesis that mold-blown wares were broken and discarded soon after they were made and the hypothesis that such wares were buried or discarded only after long years of use are speculative. When mold blowing first came into use it might well have been a technique closely guarded by the few workshops which practiced it. On the other hand, once the principles of the process were understood, mold blowing could have rapidly become common practice, since the technology is not overly complex.

Some mold-blown tablewares were probably made before the end of the first quarter of the first century. The existence of a commercially viable mold-blowing industry at this time is suggested by both the literary evidence (see below: *Multiregional Styles: The Spread of Styles and Designs*) and by the date of the earliest identifiable mold-blown vessels. These are small bottles decorated with designs in relief made in the Workshop of the Floating Handles (see below: *Workshop of the Floating Handles*), which may have begun production somewhat earlier in the first century than has heretofore been proposed.[16] One very popular type of mold-blown bottle with high relief may even have been created as early as the year 13 (see below: *Hexagonal Bottles with High Relief: Date* and Fig. 54). Thus, the earliest mold blowing on a commercial basis might date as early as the first decade of the first century. Experiments with blowing into molds may therefore have begun in the very first years of the century.

## *Origin*

Links with decorated mold-made pottery and the distribution of early dated finds of tablewares in the western part of the Roman empire have led to the suggestion that mold blowing may have been invented in Italy and from there spread eastward.[17] Until recently, however, it has generally been assumed that mold blowing was invented in the East, in part because it was thought to have preceded free blowing and it was logical to suppose that both techniques were invented in the same area. The argument that mold blowing preceded free blowing must now be reconsidered in the light of new discoveries about the history of inflation techniques (see above: *Advances in Roman Glass Technology: Inflation*), but research for this catalogue appears to confirm the hypothesis that mold blowing began in the East, somewhere along the Syro-Palestinian coast, probably in the vicinity of Sidon.

The earliest identifiable mold-blown vessels are small bottles made on the Syro-Palestinian coast. These bottles rarely reached the West, but they have been found throughout the eastern Mediterranean. The earliest mold-blown tablewares belong to types associated with eastern Mediterranean workshops: Ennion cups, jugs with floral decoration (Cat. No. 75), horizontally ribbed bowls and cups, and finely ribbed bowls (Cat. Nos. 13-14). The earliest datable mold-blown vessels found in the West appear to have been imports from the East.

## *The Role of Sidon*

Some modern discussions have associated Sidon primarily with the production of fine mold-blown glass. However, the exact

[8] Israeli 1991, 46-47, provides a concise survey of previously published opinions on this subject.

[9] See above: *Advances in Roman Glass Technology: Inflation*.

[10] Price 1991, 64-71, defines the groups as follows: first, Tiberian to early Claudian; second, Claudian to mid-Neronian; and third, mid-Neronian to early Flavian. The dates for these periods have been substituted here.

[11] Stern 1977a, 5-6.

[12] Van Lith 1978-79, 103, no. 327, pl. 22, from a context dated ca. 40-47.

[13] Fasold 1985, 213-214, no. 19, fig. 9.1, dated ca. 14-54.

[14] Czurda-Ruth 1979, 144-145, no. 1055, pl. 22, dated before 45.

[15] Lynn Passi Pitcher, *Sub ascia: Una necropoli romana a nave* (Modena 1987) 140-141, fig. 82, found in a late Tiberian to early Claudian context.

[16] Stern 1991a.

[17] Hayes 1975, 29-30; Price 1991, 71-72.

role of Sidon in glass making and glass working is elusive. The city's literary fame as a glass center has not been confirmed by archaeological excavations, and the toponymic Sidon does not occur on any inscription on a mold-blown vessel, although it does appear stamped on free-blown vessels (see below).[18] Nevertheless there are strong indications that some of the earliest mold-blown glass was produced in or near Sidon. The iconography of the hexagonal bottles of the Vessels Type suggests a link with Sidon (see Cat. No. 15), as does the iconography of the bottles belonging to the Birds Type (see Cat. No. 45).

A new interpretation of Pliny's phrase *argenti modo caelare* supports this view. Greek and Roman authors, writing in the first centuries B.C. and A.D., agree in assigning to Sidon the leading role in glass production (Strabo, *Geography* 16.2.25; Pliny, *Natural History* 5.76 and 36.193). The city profited from the availability of suitable glass-making sand[19] at the mouth of the river Belus which flows into the Mediterranean Sea about 70 km south of Sidon, just south of Akko (Acre, ancient Ptolemais) (Pliny, *Natural History* 5.75; 36.190; Josephus, *Jewish Wars* 2.10.2; Tacitus, *Histories* 5.7).[20] Strabo adds that, according to some, the Sidonians also had vitreous sands that were suitable for glass making (Strabo, *Geography* 17.758). Thus Sidon was primarily renowned for glass *making*. During the second century, the Greek philosopher Lucian (*Amores* 26) still quoted Sidonian glass as a standard of transparency.[21]

Sidon was also famous as a center of glass *working*, but its fame appears to have waned by the third quarter of the first century, for then Pliny wrote of it as something passé:

> From the lumps it (the glass) is melted again in the workshops (*in officinis*) and colored, and part is shaped by blowing, part *torno teritur*, and part *argenti modo caelatur*, Sidon being distinguished once for these workshops (*his officinis*), since it had even invented glass mirrors.[22]

Although the text poses a number of questions which will be addressed below, the wording of the Latin shows that, according to Pliny, Sidon had once been famous for workshops where glass was melted (*funditur*), colored (*tinguitur*), and worked. The phrase *his officinis* at the end of the passage obviously refers to the *officinis* just preceding: all these activities were done in Sidon and elsewhere, but especially in Sidon.

A curious object documenting Sidon's fame as a center for coloring glass is a small disk of opaque yellow glass stamped with a figure of the goddess of Victory, Nike, on one side and an inscription in Greek on the other.[23] The inscription reads in translation: "(Year) 188 / of Sidon / the holy and / (city) of refuge." The year 188 of the city of Sidon corresponds to A.D. 77/78. Inscription and image are proof that the object, and presumably the opaque yellow glass into which they were stamped, were made in Sidon, thus providing an illustration of Pliny's statement.[24]

The era of Sidon's fame as a glass-working center can be deduced from two items in Pliny's text: first, the mention of glass blowing as one of the techniques for which Sidon was famous provides a date about 90 B.C., when glass workers first discovered that glass can be inflated, and second, the use of *quondam*, once, indicates that Sidon's fame lasted no longer than until about 40 to 45, because Pliny's books on *Natural History*, completed in 77, were based on research undertaken since the early fifties. Thus Sidon's fame as a glass-working center included exactly that period in the history of glass working in which mold blowing was invented and perfected, the first century B.C. and early first century A.D.

According to Pliny, Sidon was famous for three glass-working techniques: *flatu figurare, torno terere*, and *argenti modo caelare*. *Flatu figurare*, "forming with breath," is obviously glass blowing. However, the other two techniques mentioned require some explanation.

*Torno terere* has often been interpreted as "turning on a lathe," *torno* being the turning wheel of a lathe to which the glass was attached and *terere* being the action of removing glass with a hand-held instrument.[25] Although some cast vessels of the first centuries B.C. and A.D. decorated with ridges and grooves were without question mounted on a wheel for cutting,[26] it is worth considering whether the term might refer to the actual production of vessels, like *flatu figurare*, rather than to a decorative technique. It has recently been suggested that the phrase might be translated as "smoothing on a wheel" or something similar and that it could refer to the production

[18] The only toponymic on a published mold-blown vessel is Cyprus, which occurs on a drinking cup with the inscription *Aristeas kuprios epoiei*, "Aristeas the Cypriot made (it)" (see below: *Glass-Blowers' Signatures: Aristeas*).

[19] The sand was suitable to glass making because it was not pure silica but actually mixed with about 9% lime which made the glass durable: Turner 1956, 276T-300T; Brill 1988, 266-269.

[20] Trowbridge 1930, 97-98.

[21] τὸ δ' ἄλλο σῶμα μηδ' ἀκαρῆ τριχὸς αὐταῖς ὑποφυομένης ἠλέκτρου, φασίν ἢ Σιδωνίας ὑέλου διαφεγγέστερον ἀπαστράπτει. ("But the rest of her person has not a hair growing on it and shines more pellucidly than amber, to quote the proverb, or Sidonian crystal.") *Lucian*, with a translation by M. D. MacLeod, 8 (Cambridge, Mass. 1979) 190-191.

[22] My thanks are due to David O. Ross, Department of Classical Studies, University of Michigan, Ann Arbor, for a new translation of and linguistic comments on this crucial passage which does indeed link Sidon to the glass-making and glass-working techniques described by Pliny: *ex massis rursus funditur in officinis tinguiturque, et aliud flatu figuratur, aliud torno teritur, aliud argenti modo caelatur, Sidone quondam his officinis nobili, siquidem etiam specula excogitaverat.* Pliny, *Natural History* 36.193.

[23] Israel, private collection, provenance unknown, D 2.1 cm, Dan Barag, "Recent Important Epigraphic Discoveries Related to the History of Glass-making in the Roman Period," *AnnAIHV* 11 (1985, published 1987) 111-115, fig. 4. The object might have functioned as a weight, a tessera (ticket), or some other unknown purpose.

[24] Opaque yellow glass is a prominent color in core-formed vessels and most types of mosaic glass including reticella or network glass, but it is not common in blown glass, whether free-blown or mold-blown. One cannot help but wonder for what Sidonian glass workers used opaque yellow around the year 77, unless it was for mosaic glass.

[25] Mutz 1972, 14-19.

[26] Mutz 1972, 173-176, figs. 506-519. For cutting, glass vessels were probably mounted vertically on a potter's wheel rather than horizontally on a lathe: see Stern 1994, 47-48.

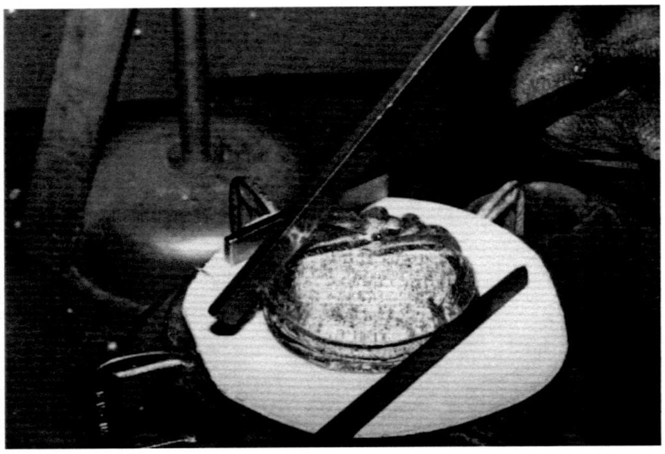

*Fig. 44. Reconstruction of making ribbed bowls on a potter's wheel. A glass blank heated to the proper temperature is struck with a strip of metal to form ribs. A horizontally held metal strip shapes the rim. A second horizontal strip was used here to prevent the mold from sliding. Only one horizontal strip would have been needed if the mold had been attached to the turning wheel. Photo: Author.*

of spiral reticella bowls which were made by spinning one or more glass threads around a convex half-mold, also called a former mold, set on a horizontally turning potter's wheel.[27] This interpretation is attractive because the Latin term *torno terere* could apply equally well to the production of glass ribbed bowls, which were also made on a potter's wheel according to recent research by Rosemarie Lierke.[28] She was able to reproduce ribbed bowls during an experiment at the Toledo Museum's Glass Studio.[29] A convex former mold was set on a potter's wheel and a hot disk of glass draped over it. The glass was struck repeatedly with a flat strip of metal, and indentations bordered by ribs were created. The rim of the bowl was made smooth by pressing the edge of the glass disk against the mold. By turning the mold on the potter's wheel, it was possible to quickly form all the ribs of the bowl and smooth the rim without having to reheat the glass (Fig. 44). The experiment confirmed Lierke's hypothesis that ribbed bowls could have been easily and quickly made.

During the first century B.C. many shapes of glass ribbed bowls were produced in Syro-Palestinian workshops, and many more in western shops.[30] If *torno terere* is interpreted as forming on a potter's wheel, this would be an adequate description of the production process of ribbed bowls which became one of the most widespread products of the Roman glass industry in Italy and other centers of the West. They were still being made throughout the Roman empire in large quantities into Pliny's time, long after the invention of blowing.[31]

*Argenti modo caelare* is usually translated as "engraving like silver." Engraved silver was imitated in glass by wheel-cut and abraded or scratched figural scenes. This decorative technique is not known on glass vessels dating to the first century B.C. and early first century A.D., although it is quite common from the middle of the second century on, when many glass vessels were so decorated.[32] *Caelare* as a technical term corresponds to the Greek *toreuein*, which not only means engraving but also embossing and chasing.[33] Embossing and chasing are, of course, impossible in glass, because glass cannot be hammered. It is worth considering whether Pliny might have had mold-blown glass in mind, the high and low relief designs of which resemble embossed metal.[34] No Latin term has been identified for mold blowing, although one would have expected Pliny to mention mold blowing in his list of techniques, since this was an important technique known to have been practiced on a large scale in the first century, not only in the Sidon area but also in Italy. If Pliny did not know how mold-blown designs were made, he may have thought (or even have been told by a glass blower who wanted to keep the technique secret) that glasses with designs in relief were made by chasing them like silver. Interpreting *argenti modo caelare* as mold blowing could explain Pliny's silence about this technique. However, Pliny's technical vocabulary throughout the *Natural History* is vague and imprecise.

The literary tradition regarding the prominence of Sidon as a glass-working center is supplemented by surviving signatures of glass blowers who added the toponymic "Sidonian" in Greek and in Latin to their names.[35] The signatures were stamped[36] in relief on the handles of free-blown drinking cups, *scyphi* (Greek, *skyphoi*), a shape popular in the

27 Lierke 1991, 315.

28 The basics of this production method are described by Lierke 1991, 316. R. Lierke, "›Aliud torno tiretur‹ Rippenschalen und die Spuren einer unbekannten Glastechnologie: Heisses Glas auf der Töpferscheibe," *AntW* (1993) 218-234; Stern 1994, 72-79.

29 Experiment undertaken May 30, 1992, with the support of the Department of Museum Education of The Toledo Museum of Art.

30 Gladys Davidson Weinberg, "Hellenistic Glass from Tel Anafa in Upper Galilee," *JGS* 12 (1970) 17-27; Gladys Davidson Weinberg, "Notes on Glass from Upper Galilee," *JGS* 15 (1973) 33-45. For a survey of eastern Mediterranean finds: Stern 1977a, 25-30, no. 4.

31 Roman ribbed bowls are found in almost every collection of ancient glass. The Toledo Museum of Art has numerous examples: Grose 1989, nos. 223, 228-242, 250-317, 384, and 394.

32 Harden 1969, 53-62. A few expertly *carved* glass vessels with figurative scenes exist: Athens, Nat. Mus., no. 23712, from a shipwreck off the island of Antikythera, ca. 80 B.C. (Gladys Davidson Weinberg, in Gladys Davidson Weinberg, Virginia R. Grace, G. R. Edwards, H. S. Robinson et al., "The Antikythera Shipwreck Reconsidered," *Transactions of the American Philosophical Society* 55, Pt. 3 (1965) 3-30, figs. 1-3); Athens, Nat. Mus., no. 16275, from a first-century A.D. grave on the Greek island of Siphnos (G. Mackworth Young, *BSA* 44 [1949] 90-91, pls. 33, 34).

33 *EAA* II, 252-253, s.v. *caelatura* (M. Cagiano de Azevedo); *EAA* VII, 919, s.v. *toreutica* (E. Simon). On these techniques: Strong and Brown 1976, 11-24.

34 Compare the story about Tiberius and the invention of malleable glass: Petronius, *Satyricon*, 51; Dio Cassius, 57.21.7; Pliny, *Natural History* 36.195. The story is discussed below: *Multiregional Styles: The Spread of Styles and Designs*.

35 Kisa 1908, III, 923-924; Fremersdorf 1938, 116-121; *Corinth XII*, 103, no. 650, a light blue handle preserving the name Aristoon (in Greek). Aristoon, Artas, and Philippos signed both in Latin and in Greek; the signatures of Neikoon and Eirenaios are known only in Greek.

36 The technique of stamping is also known from Sidon as can be seen by the stamped disk of opaque yellow glass discussed above; see above: note 23.

first centuries B.C. and A.D. Some of these glass blowers decorated the handles with impressions from coins; Artas seems to have used a coin of the emperor Augustus (27 B.C.-A.D. 14) and Eirenaios one of the emperor Caligula (37-41).[37] Because most handles with stamped signatures have been found in Rome, it has been suggested that these glass blowers had their workshops not in Sidon but in Rome or northern Italy.[38] It is tempting to speculate that these were the men who introduced glass blowing to Rome. The addition to their names of the toponymic Sidon was probably meant to convey an assurance of quality by invoking associations with the city's fame, not only as a glass-*making* center but also as a glass-*working* center.

## GLASS-BLOWERS' SIGNATURES

Greek and Roman authors do not mention the name of any glass worker. The names of glass blowers that have been preserved are known from epigraphic evidence only, either from signatures on glass objects or from inscriptions such as those found on tombstones bearing names and occupations. About 130 names of Roman glass blowers are known. Of these, five specialized in decorated mold-blown tablewares: Ennion, Aristeas, Neikais, Iason, and Meges. Their signatures, which appear in the common Greek formula *epoiei*, "made (it)," form part of the mold-blown decoration of their wares. All five were active in the Near East in the first century. The Toledo Museum collection includes one signed vessel, a beaker-shaped cup signed by Neikais (Cat. No. 5).

Although in the case of signatures on the undersides of mold-blown glass storage or transport bottles it is legitimate to question whether the signature refers to the maker of the container or to the maker of the contents,[39] in the case of mold-blown glass tablewares there is no such ambiguity. The signature can only refer to the vessel. Some doubts have been raised whether *fecit*, the Latin equivalent of *epoiei*, might in the case of a woman's name refer to the owner-manager of a workshop rather than to a glass blower, but this appears to be based on the bias that women could not blow glass.[40] Use of the formula by other craftsmen such as sculptors and potters is generally interpreted as referring to the craftsperson's pride in his or her achievement as maker or as owner of the workshop.[41] Glass workshop signatures are known, but these generally include the word *officina*, "workshop." Individual glass blowers occasionally added the name of the *officina* for which they worked.[42]

Of the names known for Roman glass blowers, two are definitely the names of women and one is probably a woman. The mold-blown signatures of Sentia Secunda of Aquileia and Ennia Fortuna appear on the undersides of square bottles.[43] The name Neikais, spelled with -ei-, appears to have been used by both men and women (see below: *Neikais* and Cat. No. 5). All three were active in the second half of the first century, Sentia Secunda and Ennia Fortuna in Italy and Neikais in the Near East. There is no obvious difference between the products of these three women and those of contemporary male glass blowers.

### *Ennion*

The most famous ancient glass blower is Ennion. He specialized in a wide variety of fine mold-blown tablewares. His signature has been preserved on more than thirty mold-blown vessels (Fig. 45). The signature itself is in Greek, ENNION ΕΠΟΙΗCΕ vel sim. (*Ennion epoiese*), "Ennion made it,"[44] but the name Ennion is not common in Greek and is probably a hellenized Semitic name.[45] Ennion may have been a Jew or a Phoenician, or he may have come from some other area with a Semitic population.[46] He did not add a toponymic to his name, which may indicate that he worked in his hometown throughout his career. In this volume Ennion's workshop is provisionally assumed to have been in Sidon and his style and technique are called "Sidonian."

Ennion's vessels stand out for the clarity and precision of their mold-blown designs, which are characteristic of late Hellenistic and early Roman toreutics.[47] But Ennion did not simply use existing metal vessels as the models for his tablewares. He designed his models especially for execution in glass, restricting his designs to shallow relief and outline, a procedure that guaranteed even filling of the mold when the paraison was expanded. These technical features also facilitated pulling the mold sections off the expanded but still hot glass quickly, without distorting shape or decorative design. Speed was necessary to prevent the glass from cooling too much before the handles were applied.

Pride in the technical achievements of his molds as well as the aesthetic qualities of his designs may explain why Ennion

[37] Isings 1957, Form 39; Kisa 1908, III, 704-706.

[38] Fremersdorf 1938, 116-121. See also below: *Multiregional Styles*.

[39] On signatures referring to contents: A. Frova, "Vetri con marchi," *JGS* 13 (1971) 38-44; J. Price, "Roman Unguent Bottles from Rio Tinto (Huelva) in Spain," *JGS* 19 (1977) 35; Taborelli 1983, 125; Biaggio Simona 1991, I, 162.

[40] F. K. Kiechle, "Die Struktur der gewerblichen Glaserzeugung in der frühen Kaiserzeit," *AnnAIHV* 6 (1973) 62-63: "doch kann man sich diese Geschäftsfrau schwerlich selber am heissen Glasofen werkend vorstellen" ("but one can hardly imagine this businesswoman/manager working in person at the hot furnace").

[41] For sculptors' signatures: K. T. Erim and J. Reynolds, "Sculptors of Aphrodisias in the Inscriptions of the City," in N. Basgelen and Mihin Lugal eds., *Festschrift für Jale Inan* (Istanbul 1989) 518.

[42] For examples: Trowbridge 1930, 116-117.

[43] Two square bottles in Linz, Austria, Oberösterreichisches Landesmuseum, Harden 1969, 49, 73, pl. IVB. The inscriptions differ slightly. One reads SENTIA SECUNDA FACIT AQUILEIAE, "Sentia Secunda made [it] in Aquileia." The other reads SENTIA SECUNDA FACIT AQ[*uileiae*] VITR[*earia*], "Sentia Secunda made [it] glassblower in Aquileia." Ennia Fortuna's name appears on the underside of the base of a glass bottle of a type known as a Mercury bottle: Trowbridge 1930, 122, no. B38; *CIL* 15, 2, 1, 6993; Kisa 1908, 929, no. 35.

[44] Harden 1935, 164-169, pl. 28.

[45] Lehrer 1979, 14. The name Ennion is also known on a third-century builder's inscription at Damascus, *SEG* II, 829.

[46] Lehrer 1979, 13-14. Other Semitic populations of the period included Syrians, Palmyrenes, and Nabataeans.

[47] On the connections between Ennion's designs and contemporary silver wares: Lehrer 1979, passim.

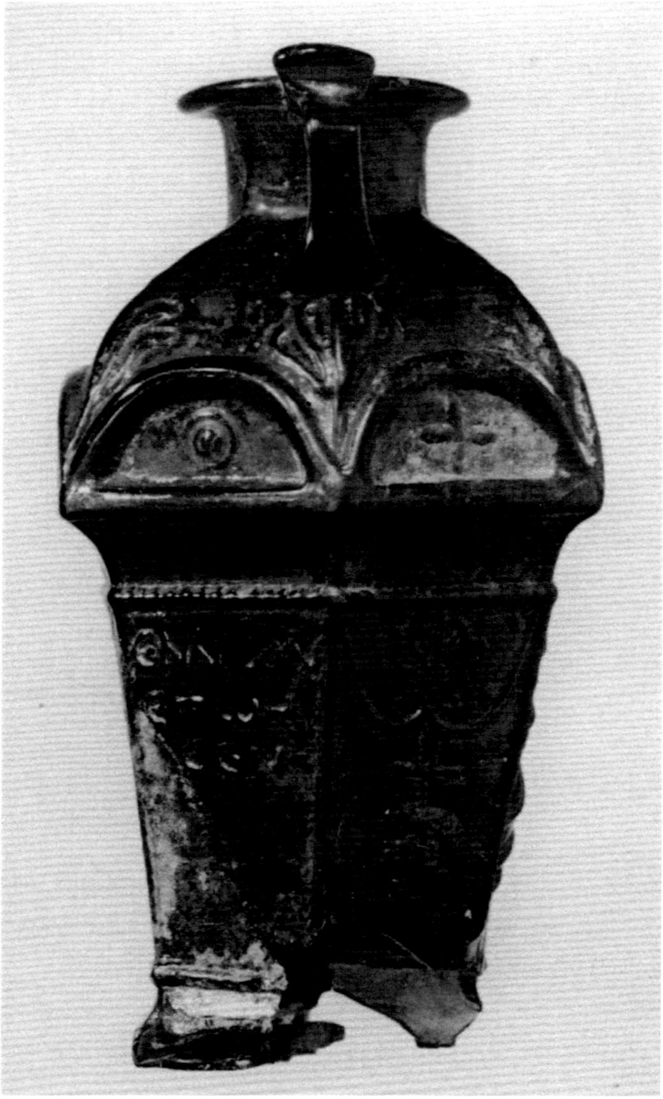

*Fig. 45. The decorative system used by Ennion on this six-sided bottle, consisting of panels decorated with thematically related objects and a shoulder decorated with palmettes and arches enclosing symbolic objects, is similar to the decorative system found on mold-blown Hexagonal Bottles with High Relief. The Hexagonal Bottles, however, differ from this bottle in having higher relief, smaller size, and a third zone of decoration around their bottoms. Bottle, signed by Ennion, H 14.3 cm, New York, The Metropolitan Museum of Art, Gift of Henry G. Marquand, 1881, no. 81.10.224. Photo: The Metropolitan Museum of Art.*

put his signature so prominently on his vessels. He designed models with decoration in the contemporary fashion, but these offered no risk when executed in the new technique of glass blowing. Although not large, his vessels are larger than most contemporary mold-blown glass vessels.

The earliest mold-blown tablewares preserved, or so we assume, are those signed by Ennion.[48] It is therefore tempting to infer that Ennion initiated the practice of mold blowing tablewares. Could he have begun his career as a silversmith? A silversmith who wished to make multiple copies of his designs for tablewares might have attempted to experiment with the new technique of mold blowing.

Vessels signed by Ennion were made in three different types of multipart molds, MCTs I, III, and V (see Fig. 15). MCTs I and V used a separate cup-shaped section for the lower part.[49] This construction must have included a sophisticated locking device in order to secure the vertical mold sections to the lower section. A cup-shaped lower section appears to have been characteristic of vessels made in Sidon or its immediate vicinity. It was also used by Neikais, Iason, and Meges (for MCT V E, see Cat. No. 5) and for two Series of unsigned cups with mold-blown inscriptions that recall Ennion (see Cat. No. 1). Ennion may have invented MCT III with its separate disk-shaped base portion.[50]

Ennion was a perfectionist who took great care to conceal the joins between mold sections, where small amounts of molten glass inevitably forced their way through, creating raised mold seams, by integrating them in the decorative design. Four unsigned jugs previously attributed to Ennion's workshop have recently been identified as ancient copies, because the mold seams between the vertical sections are not integrated into the design but run vertically between the palmettes.[51] Another idiosyncrasy was the way in which Ennion made handles on cups. Instead of applying a handle in the common manner at the lower end and pulling the bit up toward the rim, Ennion applied the handle near the rim and pulled it down, folding the lower end in and pressing it against the wall of the cup.[52]

The indistinctness of the mold-blown design of a cup in Pavia[53] led to the suggestion that the cup might be a local imitation made from a cast of a cup by Ennion himself.[54] The fact that this cup is slightly smaller than the rest of the most closely related cups suggests that it was blown into a clay mold made directly from a cup by Ennion. However, the characteristic way in which the handles are attached

[48] McClellan 1983, 73, fig. 2, a cup excavated at Corinth in a closed context with a coin of Caligula. See also further discussion below.

[49] Israeli 1983, 67-68, with reference to Ennion's jugs of Harden's Group A4i; see also Harden 1935, 168.

[50] For example, a signed amphora in St. Petersburg, Hermitage, no. P. 1852.54, Kunina 1973, 101-106, figs. 1-6.

[51] Israeli 1983, 68, n. 10, lists four examples, all of which probably had a trefoil mouth; a fifth example, not listed by Israeli, is in the Hermitage, see below: c. Just as on the jugs in Aquileia and Corning described by Israeli, the mold seams on all three jugs in the Hermitage run vertically between the palmettes: Kunina 1973, 107:

a) St. Petersburg, Hermitage, no. P 1896.32, from Kerch, transparent green, Kunina 1973, 106-111, figs. 7, 10-12; mentioned by Harden 1935, 186;
b) St. Petersburg, Hermitage, no. P 1896.33, from Kerch, transparent blue with colored flecks, Kunina 1973, 106-111, fig. 8; mentioned by Harden 1935, 186;
c) St. Petersburg, Hermitage, no. E. 560, find spot unknown, with opaque white flecks, Kunina 1973, 106-111, fig. 9;
d) Aquileia, Arch. Mus., no number, fragments of yellow glass, mouth not preserved, Calvi 1968, 98-99, fig. 2, 105, no. 245;
e) CMG, no. 65.1.2, formerly R. W. Smith Collection, find spot unknown, Smith 1957, no. 68.

[52] For example, CMG, no. 66.1.36, blue, *Glass of the Caesars* 1987, 164-165, no. 86; Lehrer 1979, 10-11, drinking cups types A1, B, and C.

[53] Pavia, Musei Civici, no. A243, transparent green, Maccabruni 1983, 33, no. 6, fig. 6, pl. 1.

indicates that it was blown in Ennion's workshop; the small size and indistinct design may be owing merely to Ennion following the standard practice of copying his own products.

Ennion may have begun production as early as the first quarter of the first century, for he had already established commercial contacts with the western half of the Roman empire by the second quarter of the century. Two fragments of a single mold-blown cup, each bearing part of the frame of a maker's signature in Greek, were excavated at Corinth in a closed context which included a coin of the Roman emperor Caligula (ruled 37-41).[55] The signatures are set in *tabulae ansatae*, rectangular frames with dove-tail side pieces, a convention known only for Ennion and Aristeas. The cup is assigned to Ennion because the two known examples of Aristeas's work bear only one *tabula ansata*, although the Corinth cup includes a palmette motif not otherwise known on Ennion cups with ansate frames. A cup preserving Ennion's signature in full was excavated in Italy with a coin of the emperor Claudius minted in 46.[56] Fragments of a cup with ivy and vine sprays were found in a closely dated Caligulan-early Claudian pit (37-ca. 45) at Fréjus, southern France.[57] Three fragments of a dark blue cup excavated at Kempten, Bavaria, in a Tiberian-Claudian context (14-54)[58] combine design elements known from two types of Ennion cups: a *tabula ansata* with part of the inscription ΜΝΗΘΗ Ο ΑΓΟΡΑΖΝΩ (*mnethe ho agorazno*),[59] "may the buyer be remembered," in three lines, and radiate tongues on the lower body. There is no way of knowing how old these vessels were at the moment of their disposal, but the fact that they were excavated at sites far from Sidon suggests Ennion's commercial success before the middle of the first century. The exported vessels are cups that have been proposed to belong to Ennion's later production,[60] but it is interesting to note that the only datable finds of shapes assumed to be early products, namely Ennion's jugs and hexagonal flasks, were excavated in and near a mansion destroyed by fire in 70 in Jerusalem.[61] Thus the vessels may have still been in use at this time, although it is possible they were made at a considerably earlier date.

Because all but two of thirteen drinking cups bearing Ennion's signature were found in Italy, it has long been believed that Ennion migrated from Sidon to northern Italy at some point in his career.[62] But recent finds of Ennion cups in Greece, Spain, Morocco, France, and perhaps southern Germany and Netherlands have led to various new hypotheses.[63] One suggestion is that it was not Ennion who moved, but rather his molds; in other words, that some kind of exchange of molds took place between workshops.[64] There is no way of proving or disproving an occasional or even a mass exchange of molds in antiquity. However, trade rather than migration or exchange appears to be a more convincing explanation because most Roman molds were inexpensive clay items that could easily be reproduced from an original model or from existing glass vessels (see *Mold Blowing: The Reproduction of Molds*). The distribution pattern of the several Ennion cups excavated in Italy appears to follow the Po valley inward from the Adriatic coast, a pattern that has also been observed for glazed ceramics with mold-formed decoration imported from Asia Minor in the first half of the first century.[65]

If Ennion's workshop was located in Sidon, a city known to have had extensive overseas trade connections and overland transport facilities, this could explain the wide distribution of Ennion's vessels throughout the Roman empire.[66] Two of Ennion's signed vessels come from Sidon.[67] Fragments of three were excavated in the Jewish Quarter of Jerusalem, and another is said to be from Jerusalem.[68] A fragment

54 G. Baravalli, "Un vetro di Ennione nel museo di Pavia," *Historia* 6 (1932) 362; Tammassia 1965, 3-4, fig. 1.

55 McClellan 1983, 73, fig. 2. Remains of a palmette and *tabula ansata* are preserved on one fragment; a portion of a second *tabula ansata* is preserved on the other.

56 Turin, Museo di Antichità, number unknown, from Vercelli, Harden 1935, 167, no. Aii4a; Calvi 1965a, 12, fig. 4; Lehrer 1979, drinking cup type C, 11-12, pl. 5. Other examples of cups made in this or closely related molds cannot be dated:

a) Newark, no. 50.1443, Auth 1976, 64-65, no. 58;
b) fragments excavated in the Athenian Agora, McClellan 1983, 72-73, fig. 1;
c) fragments from an imperial villa at Posilippo, Italy, R. T. Günther, *Pausilypon. The Imperial Villa near Naples* (Oxford 1913) 287, 289, fig. 191, cited in Price 1991, 65.

57 Price 1991, 65-66. The design is assigned to Harden 1935, type A2, i or ii; Lehrer 1979, drinking cup type B.

58 Fasold 1985, 213-214, no. 19, fig. 9.1.

59 *Mnethe*, with omission of the sigma, stands for *mnesthe*, a substandard spelling for the aorist optative with passive meaning, see at Cat. No. 5; *agorazno*, with metathesis of the n and o, stands for *agorazon*, the present participle of *agorazein*, "to buy."

60 Harden 1935, 165.

61 Avigad 1980, 107, 117, figs. 95-96, 112; Israeli 1983, passim.

62 That Ennion's workshop moved was first suggested by Harden 1935, 165. On migrating glass blowers, most recently: Price 1991, 72. A few chemical analyses have been performed on vessels signed by or attributed to Ennion, but these are of no help in distinguishing geographical origin since the composition of glass of the first centuries B.C. and A.D. excavated in Italy and the Near East does not show significant differences: McClellan 1983, 75-76, and 78, App. B.

63 To the cups listed by McClellan 1983, 76, and Price 1991, 65-66, add:

a) cup, Cadiz, Spain, Arch. Mus., no. 12.026, from excavations in the Area de la Avenida, Cadiz, Maria del Pilar Caldera de Castro, "Roman Glass in Southwest Spain: Some Notes on Trade Relations in the Early and Late Empire," *AnnAIHV* 11 (1988) 79-80, fig. 2;
b) fragmentary cup, Palermo, Regional Arch. Mus., no number, found at Solunto, Sicily, *BullAIHV* 9 (1981-83) 189, fig. 2;
c) fragment of a cup, Rome, Antiquarium Comunale, *BullAIHV* 9 (1981-83) 152;
d) cup, said to be from Jerusalem, currently on the art market, *Numismatics and Fine Arts*, auc. cat., 11 December 1991, lot 109.

64 McClellan 1983, 75-76.

65 Maccabruni 1983, 29 (with lit.).

66 Apart from the drinking cups already mentioned, most of which were found in Italy and the western Mediterranean, vessels signed by Ennion have also come to light in Cyprus, Panticapaeum (modern Kerch on the Crimea), Sidon (Harden 1935, 165-168), and Jerusalem (Lehrer 1979, 7 and 9).

67 Bowl, Yale, no. 1955.6.66, pale greenish blue, Matheson 1980, 43-45, no. 118; Harden 1935, 167, group A3; and bowl, Norfolk, Va., Chrysler Museum, no. 71.6779, greenish, H 6.7 cm, Harden, 1944-45, 89, pl. 8, no. 2; *JGS* 10 (1968) 180, fig. 6, 181.

68 On the excavated Ennion vessels: Israeli 1983, 65-69. A cylindrical cup, Type A, with angular handles as in Type C, currently on the art market, *Numismatics and Fine Arts*, auc. cat., 11 December 1991, lot 109, is also said to have been found in Jerusalem.

of an Ennion-type jug was excavated at Apamea, Syria.[69] This concentration of finds lends support to the hypothesis that his workshop was somewhere in the Syro-Palestinian area.

In addition to his maker's signature, Ennion often added a traditional Semitic blessing formula, "may the buyer be remembered" (see at Cat. No. 5).[70] The inscription itself, ΜΝΗ[C]ΘΗ Ο ΑΓΟΡΑΖΩΝ (*mnesthe ho agorazon*), is in Greek, the common language, koine, in the eastern half of the Roman empire, but the blessing would probably have been meaningful only to a buyer of Semitic origin. It does not occur on Ennion's jugs, amphoras, or six-sided bottles, but only and almost always on drinking cups, most of which were found in north Italy and the western Mediterranean. Perhaps this indicates that he produced these cups for export to Semitic communities.

### *Aristeas*

Aristeas appears to have specialized in fine mold-blown tablewares. He signed a mold-blown cup "Aristeas the Cypriote,"[71] which may indicate that his workshop was not in Cyprus.[72] Glass blowers using the toponymic Sidon appear to have done so only when they were active outside Sidon. Thus the most obvious explanation for the toponymic "the Cypriot" is that Aristeas did not work in Cyprus and added the toponymic perhaps to honor his homeland, to establish his credentials, or to differentiate himself from another artist with the same name.[73] This proposal is supported by the fact that there is no literary or epigraphic evidence for glass blowing in Cyprus during the first century. Nevertheless, an argument can be made for the hypothesis that in this case the toponymic refers to the location of Aristeas's workshop. First, the name Aristeas, though not uncommon in Palestine, Asia Minor, the Aegean islands, and Cyrenaica, has not been recorded previously from Cyprus.[74] Thus the fact that he was a Cypriot in spite of his uncommon name might have been something for Aristeas to point out. Second, technical and stylistic details of specific glass vessels have recently led to the suggestion that some luxury glass was being produced in Cyprus in the first century.[75]

[69] R. Donceel, "Un fragment de verre de l'atelier d'Ennion à Apamée," *RALouvain* 20 (1987) 36-49, fig. 1, from a pit filled around 70-80.

[70] The formula does not appear on either of the two cups signed by Aristeas.

[71] *Maxwell* 1979, 157-160, no. 280.

[72] Although it is accepted opinion among epigraphers that toponymics were omitted when a person was in his hometown, this was not invariably true. The Aphrodisian sculptor Polyneikes added the toponymic to his name in a signature found at Aphrodisias: Kenan T. Erim and Joyce Reynolds, "Sculptors of Aphrodisias in the Inscriptions of the City," in N. Basgelen and Mihin Lugal eds., *Festschrift für Jale Inan* (Istanbul 1989) 518, 524-525, pl. 202.

[73] A sculptor Aristeas from Aphrodisias in Asia Minor is known from two signatures on the bases of statues found at the villa of the Roman emperor Hadrian at Tivoli in Italy: *CIG*, III, 6140 a-b; E. Löwy, *Inschriften Griechischer Bildhauer* (Leipzig 1885) 259-260, no. 369; cf. M. Squarciapino, *La scuola di Afrodisia* (Rome 1943) 16, no. 27. Although the name Aristeas has not been found in Aphrodisias as a sculptor's signature, it is well attested in the city (Joyce Reynolds, pers. com.).

[74] Fraser and Matthews 1987, 62, s.v. *Aristeas.*

[75] N. P. Sorokina, "Wo wurde der Amphoriskos aus Pantikapaion geblasen und bemalt?," *AnnAIHV* 12 (1991) 59-66.

Only two vessels signed by Aristeas are known, one with and one without handles.[76] Both are drinking cups closely resembling those of Ennion in shape and style. Neither of Aristeas's cups carries the blessing formula "may the buyer be remembered." And, unlike Ennion, Aristeas applied handles in the usual way, first applied below then pulled up and attached to the rim.[77] Neither of Aristeas's cups was excavated in a dated context. An early first-century date is supported by his close association with Ennion and by the fact that a fragment of a mold-blown bottle bearing a stylized floral spray of the kind connected with his work was excavated at Magdalensburg, Austria, in a context that dates before the year 45.[78]

### *The Influence of Ennion and Aristeas*

The style of Ennion and Aristeas is called "Sidonian" in this catalogue because their work is characterized by a distinctly recognizable style and technique that appear to have influenced a number of other workshops in that area. Many more mold-blown vessels bearing Ennion's signature are preserved than examples by Aristeas, and the shapes made by Ennion consequently appear to be more varied than those of Aristeas.[79] But the products of the two glass blowers are similar in a number of stylistic and technical respects. Both used multipart molds with shallow intaglio on the interior to create decoration in low relief on the exterior of the finished vessels. Both placed their signatures in *tabulae ansatae*, rectangular frames with dovetail-shaped side pieces. The decorative designs of both are closely related to metalwork made and used throughout the Mediterranean in the late Hellenistic and early Roman periods. While Aristeas's designs are more stylized than those of Ennion,[80] the designs of both are predominantly linear and in raised outline with a marked preference for bands of geometric and floral designs. These designs inspired mold-blown vessels made in the Syro-Palestinian area with decoration in low relief, for example, Bottles with Handles (Cat. Nos. 53-63), Juglets (Cat. Nos. 75-78), and Pyxides (Cat. Nos. 79-83), as well as in high relief such as Juglets (Cat. Nos. 71-74).

Typical designs used by Ennion and Aristeas include various kinds of vertical fluting, but petal-shaped tongues with small spikes or leaves between the rounded ends have been noted only for Ennion (see at Cat. No. 49). The flutes are frequently placed in separate bands around the shoulder and the bottom of a cup, a disposition that became characteristic

[76] Cup, formerly Maxwell Collection, cited above, note 71; cup, in the Strada Collection, Pavia, Calvi 1965a, 9-12, figs. 1-3, 6-8.

[77] Compare Calvi 1965a, figs. 1-2, a cup by Aristeas, with fig. 4, a similarly shaped cup by Ennion.

[78] Czurda-Ruth 1979, 144-145, no. 1055, pls. 9, 22 (ill. upside down). One may also compare the mold-blown jug found in a first-century tomb at Yahmour, Harden 1944-45, 84-85, pl. 7. For other parallels: see Cat. No. 75.

[79] For an overview of the designs used by Ennion: Lehrer 1979, pls. I-VI; for Aristeas: Calvi 1965a, 9-16; *Maxwell* 1979, 157-160, no. 280.

[80] Calvi 1965a, 9-16; her observation was borne out by the Maxwell cup which was not known at the time.

of many types of small, unsigned mold-blown bottles with one or two handles (Cat. Nos. 53-59 and 71-74). In some cases the center frieze of these bottles suggests a more definite relationship with either Ennion or Aristeas. For example, inhabited scrolls and various types of stylized floral sprays (Cat. Nos. 75 and 76) have been connected with Aristeas's workshop.[81] Designs that may have been inspired by Ennion include bands of lozenges and hexagons arranged in a honeycomb pattern (Cat. Nos. 53 and 54),[82] bands of horizontal S-shaped scrolls (Cat. Nos. 57 and 58),[83] bands of alternating circumscribed palmettes with out-turned petals and free-standing closed palmettes (Cat. Nos. 79-82),[84] isolated floral motifs such as palmette-shaped plants with in- or out-turned petals (Cat. No. 48),[85] isolated symbolic motifs (Cat. Nos. 43-44),[86] suspended cult objects (Cat. Nos. 71-74),[87] and the use of arcades above panels (Cat. Nos. 15-35).[88] The exact relationship between small Hexagonal Bottles with High Relief (Cat. Nos. 15-47; see below: *Syro-Palestinian Styles*) and Ennion's Hexagonal Bottles[89] is not clear. Some of the smaller bottles may have been inspired by Ennion's bottle (see Fig. 45), which in that case would have been made before about 13 when the Hexagonal Bottles with High Relief started, or both Ennion and the Vessels Type were inspired by the same source.

Some designs and shapes seem to have appealed more to customers in the eastern Mediterranean, while others were preferred in Italy. Regional differences are reflected not only by the distribution patterns of signed vessels such as those made by Ennion and Aristeas, but also by the find spots of small unsigned bottles inspired by their designs. Popular in the eastern Mediterranean, for example, were small two-handled bottles with a central band of lozenges arranged in a honeycomb pattern (Cat. Nos. 53 and 54) and small one-handled bottles with panels showing cult objects (Cat. Nos. 71-74). Pyxides with bands of palmettes (Cat. Nos. 79-82) and small one-handled bottles with pairs of stylized floral sprays (Cat. Nos. 75-78) have been excavated at both eastern and western sites. In general, tablewares appear to have been exported more widely than small containers (see below: *Multiregional Styles*).

## *Neikais, Iason, and Meges*

Neikais, Iason, and Meges signed their names on barrel-shaped cups. The signatures of all three are accompanied by *epoiese*, the traditional Greek formula for indicating the maker. Iason and Meges are common men's names, but Neikais is not. The name has previously been interpreted as either a short spelling for N(e)ikaios or as N(e)ikias (with metathesis of the letters a and i), both of which are well-known Greek proper names.[90] However, the name Neikais, spelled with an -ei- as on these mold-blown cups, is now attested in Cyrenaica, where it appears to have been carried by both men and women.[91] Neikais may have been a woman glass blower, like Sentia Secunda and Ennia Fortuna who had workshops in Italy.

There is no proof that the workshop or workshops of Neikais, Meges, and Iason were in Sidon, but the distribution patterns of their vessels indicate an origin on the Syro-Palestinian coast. Of the six cups bearing Iason's name, one was found at Tel el Hoson, north of the Sea of Galilee, and another at Hasfin, in the Syrian Hauran, while two are said

[81] Calvi 1965a, 14-16. See also below: *Workshop of the Floating Handles.*

[82] Bands of lozenges occur on jugs and spherical bowls signed by Ennion:

Jugs:

a) Tel Aviv, no. MH 120058, Israeli 1964, 34-35, no. 1, figs. 1-3;
b) CMG, no. 59.1.76, formerly R. W. Smith Collection, *Glass of the Caesars* 1987, 166, no. 87;
c) MMA, no. 17.194.226, formerly Gréau Collection, Harden 1935, 168, A4ia; Lehrer 1979, 9, pl. 3, nos. 1-4;
d) fragments excavated in the Jewish Quarter of Jerusalem, Israeli 1983, 66-67, figs. 2-3.

Cf. Harden 1935, 168, group A4 i and ii; Harden 1944-45, 89-90; Lehrer 1979, 6-8, pl. I, nos. 1-2, 4-5.

Spherical bowls:

Harden 1935, 167, group A3; Harden 1944-45, 89; Lehrer 1979, 10, pl. IV, no. 1.

[83] A band of horizontal S-shaped scrolls occurs on Ennion's amphora in St. Petersburg, Hermitage, no. P. 1852.54, excavated at Panticapaeum (modern Kerch), Kunina 1973, 102-106, figs. 1-6; Harden 1935, 168, no. A5a; Lehrer 1979, 8-9, pl. II, nos. 1-2.

[84] A band of hanging palmettes, alternately circumscribed and free, occurs on Ennion's jugs:

a) Tel Aviv, no. MH 120058, Israeli 1964, 34-35, no. 1, figs. 1-3;
b) CMG, no. 59.1.76, *Glass of the Caesars* 1987, 166, no. 87;
c) MMA, no. 17.194.226, see above note 82c;
d) fragments excavated in the Jewish Quarter of Jerusalem, Israeli 1983, 66-67, figs. 2-3.

[85] Isolated floral motifs occur on three vessel shapes signed by Ennion:

a) amphora, St. Petersburg, Hermitage, no. P. 1852.54, excavated at Panticapaeum (modern Kerch), Kunina 1973, 102-106, figs. 1-3; Harden 1935, 168, no. A5a; Lehrer 1979, 8-9, pl. II, nos. 1-2;
b) hexagonal bottles, such as MMA, no. 81.10.224, Lehrer 1979, pl. III, 1-4; and fragments of two hexagonal bottles excavated in the Jewish Quarter of Jerusalem, Israeli 1983, 68-69, fig. 5;
c) one-handled drinking cup, Harden 1935, 165, group A1; Harden 1944-45, 88-89; Lehrer 1979, 11, drinking cup type A, pl. VI, nos. 1-2.

For fragments of similar cups from Spain and Morocco: Price 1973, 69, fig. 1, 1a-b.

[86] Isolated symbolic motifs such as crosses and circular objects occur on various vessels signed by Ennion:

a) hexagonal bottles, such as MMA, no. 81.10.224, Lehrer 1979, 9, pl. III, nos. 1-4; Harden 1935, 168-169, group A6 (here Fig. 45); and fragments of hexagonal bottles excavated in the Jewish Quarter of Jerusalem, Israeli 1983, 68-69, fig. 5;
b) one-handled drinking cups, such as Harden 1935, 165, group A1; Harden 1944-45, 88-89; Lehrer 1979, 11, drinking cup type A, pl. VI, nos. 1-2.

For fragments of similar cups from Spain and Morocco: Price 1973, 69, fig. 1, 1a-b.

[87] Suspended cult objects such as vessels, a syrinx, and a grape bunch occur on Ennion's hexagonal bottle in New York, MMA, no. 81.10.224, Lehrer 1979, 9, pl. III, nos. 1-4; Harden 1935, 168; group A6 (here Fig. 45); and on fragments of hexagonal bottles excavated in the Jewish Quarter of Jerusalem, Israeli 1983, 68-69, fig. 5.

[88] Compare the hexagonal bottles cited above in note 85.

[89] MMA, no. 81.10.224; Lehrer 1979; 9-10, pl. III (here Fig. 45); and fragments excavated in the Jewish Quarter of Jerusalem, Israeli 1983, 68-69, fig. 5.

[90] Harden 1935, 170, n. 15.

[91] Fraser and Matthews 1987, 328-329, s.v. *Nikaios, Nikais.*

to have come from "Syria."[92] Cups signed by Neikais have been excavated at Masada, Israel, and at Aleppo (see Cat. No. 5). Cups bearing Meges's name have been found in the vicinity of Nazareth and at Marion, Cyprus.[93] The hypothesis of a Syro-Palestinian origin is supported by the names of the glass blowers. All three names can be associated with the Syro-Palestinian coast. Meges was the name of a prominent Sidonian physician.[94] The tomb of a wealthy man named Iason was discovered in Jerusalem.[95] The name Neikais occurs in late first-century Jewish inscriptions at Taucheira-Arsinoe, a city in Cyrenaica where many Jews settled after the destruction of the Temple in Jerusalem in 70.[96]

The cups made by Neikais, Iason, and Meges are so similar to each other that they were probably made in one workshop, or in closely associated workshops, but the exact relationship among these glass blowers remains unclear. Did they work together in one workshop? Or did each have his own workshop? The differing forms of the letters in their inscriptions may speak against a common work place.

The facts that two of the names can be associated with Jews and that all three glass blowers put a traditional Semitic blessing formula on their cups may indicate that these glass blowers were Jews. Numerous references in Jewish sources such as the Talmud and Mishnah demonstrate that glass blowing was practiced by Jews.[97] However, there is no incontrovertible evidence that Neikais, Iason, and Meges were Jewish. Neikais and Iason, though common names among hellenized Jews already in the second century B.C.,[98] were also very common Greek names in the Aegean islands and Cyrenaica.[99] For a discussion of the cups by Iason, Meges, and Neikais and the blessing inscription, "may the buyer be remembered," see Neikais's cup, Cat. No. 5.

## SYRO-PALESTINIAN STYLES

The term "Syro-Palestinian styles" is used in this catalogue to describe vessels whose distribution is centered in those areas today divided among the countries of Israel, Jordan, Lebanon, and Syria, which during the Roman period were divided into the provinces of Judea, Palestina, Phoenicia, and Syria;[100] but they are found throughout the eastern Mediterranean. Most are small containers that probably held scented oils. Vessels in Syro-Palestinian styles were probably made in workshops along the Syro-Palestinian coast: in Sidon, possibly in other coastal cities such as Tyre and Akko (Acre), and perhaps in cities located farther inland.

### *Hexagonal Bottles with High Relief*
(Cat. Nos. 15-47)

Among the many types of early mold-blown wares, Hexagonal Bottles with High Relief form a distinctive group characterized by the shape of the body, the mold-blown decoration in high relief, and the use of three types of mold construction. As a group these bottles illustrate the extraordinary changes that affected the nascent technique of mold blowing as it was transformed from a local novelty into an empire-wide industry during the first half of the first century.

The decoration of Hexagonal Bottles in High Relief is arranged in three zones: the main frieze around the body and two zones of secondary decoration, one on the shoulder and one around the bottom. The main frieze usually has six panels, often making the body hexagonal in form. This decorative system is reminiscent of a large mold-blown bottle signed by Ennion (see Fig. 45).[101] This bottle's hexagonal body, division into panels decorated with thematically related objects, and decoration on the shoulder with palmettes and arches enclosing symbolic objects are all elements shared by the standard design of Hexagonal Bottles with High Relief. The main differences between these bottles and Ennion's bottle are their increased height of relief, their smaller size, and the addition of a third zone of decoration around their bottoms.

There are five Types of Hexagonal Bottles with High

[92] Harden 1935, 169, Group B. In all, six beakers with Iason's signature have been recorded, four from the Syro-Palestinian area and two of unknown provenance:

a) BM, no. 1913.5-22.21, from "Syria," bluish green, Harden 1935, 169, Ba;
b) formerly R. W. Smith Collection, from "Syria," greenish, Harden 1935, 169, Bb; Smith 1957, 59, no. 70;
c) MMA, no. 29.100.82, from Tell el Hoson, north of the Sea of Galilee, yellowish green, Harden 1935, 169, Bc;
d) Beirut, Nat. Mus., no. 1243, said to be from Hasfin, Hauran, olive green, Harden 1944-45, 90, Be.

From unknown provenance:

e) formerly Morgan Collection, Harden 1935, 169, Bd (perhaps now Yale no. 1955.6.58, Matheson 1980, 52-53, no. 133);
f) Switzerland, private collection, Luzern 1981, 80, no. 268.

A beaker in Tel Aviv, no. MHG 2168, *JGS* 11 (1969) 109, no. 3 is not ancient: Lehrer-Jacobson 1989, 42, no. 12.

[93] BM, no. 76.11-14.5, from Marion, Cyprus, yellowish green, Harden 1935, 170, Ca; MMA, no. A.G. 216, from near Nazareth, Israel, yellowish green, Harden 1935, 170, Cb.

[94] Harden 1935, 181, n. 38; *RE*, s.v. *Meges* 4. The name was not common in the Aegean islands, although it has been noted in Crete, Euboia, Rhodes, and Thera: Fraser and Matthews 1987, 301, s.v. *Meges*, and in Kerch: St. Petersburg, Hermitage, no. P. 1850.27, tombstone of Myrine, wife of Meges, G. Sokolov, *Antique Art on the Black Sea Coast* (Leningrad 1974) 122, no. 128, ill.

[95] Several Aramaic inscriptions were found in the tomb, one of which laments the deceased, Iason: Avi-Yonah 1984, 630-631, s.v. *Jerusalem*, The Tomb of Jason; L. Y. Rahmani, "Jason's Tomb," *IEJ* 17 (1967) 61-100.

[96] G. Lüderitz, *Corpus jüdischer Zeugnisse aus der Cyrenaika*, Beihefte zum Tübinger Atlas des Vorderen Orients, Reihe B (Geisteswissenschaft) no. 53 (Wiesbaden 1983) App. 16c.

[97] On Jewish glass blowers: D. Barag, *Encyclopedia Judaica* 7 (Jerusalem 1971) 603-611, s.v. *Glass*. See also Chapter IV: *The Fourth to Early Seventh Centuries*, esp. Cat. Nos. 169-177.

[98] For example, a hellenized Jew from Cyrene named Iason wrote a history of the exploits of Judas Maccabaeus in the second half of the second century B.C.

[99] Fraser and Matthews 1987, 229-230, s.v. *Iason*.

[100] The names and boundaries of the Roman provinces were as fluid as they are today. For example, under Hadrian, Judea was renamed Syria Palestina, from which the modern term Syro-Palestinian is derived, albeit with a wider geographic connotation. See also *Glossaries: Geographical Designations*.

[101] MMA, no. 17.194.226, formerly Gréau Collection, Harden 1935, 168, A4ia; Lehrer 1979, 9, pl. 3, nos. 1-4.

Relief. The main friezes feature one of five themes from which the name of each Type is derived: Vessels, Masks, Fruit, Mixed Symbols,[102] and Birds. Many of the finest bottles are blown in an opaque, glossy white glass that frequently has weathered to a dull, warm ivory white (e.g. Cat. No. 15). Some are opaque pale blue (e.g. Cat. No. 24) or green (e.g. Cat. No. 32) or even a mixture of colors imitating marble (Cat. No. 34). The designs are usually most legible on monochrome opaque vessels, but many were blown of translucent to transparent colored glass, the most frequently occurring color being dark blue (e.g. Cat. No. 27). Natural green glass examples are rare (Cat. Nos. 33 and 35). The elaborate mold-blown decoration and the use of colored glass may indicate that these bottles were luxury items.

The Toledo Museum's collection of Hexagonal Bottles with High Relief is important because there are so many examples with the same set of motifs. For the first time it has been possible to identify individual molds and generations of use and to record alterations and mutations of details of motifs and technique. These observations have shed new light on the development of the five Types and allowed reconstruction of a number of fascinating, complex interrelations among individual Types and Series (see Fig. 54). Identification of molds and generations is possible only by examination of the objects themselves; therefore when parallels in other collections are cited, it has usually not been possible to identify a specific mold or generation, although the Series can usually be determined.

### Mold Construction Types (MCTs II, IV A, V A)

Hexagonal Bottles in High Relief were made in three different MCTs which are easily recognized by the mold seams on the underside of the base (see Fig. 15 and *Glossaries: Mold Construction Types*). MCTs II and IV were made of clay, but MCT V A was probably metal. The MCT IV A mold appears to have been used earliest; therefore it is discussed here before MCT II.

MCT IV A was the most common type of mold used for blowing Hexagonal Bottles with High Relief. Although the mold seams are usually carefully concealed in the layout of the decorative design, they frequently form a Y-shaped pattern of ridges on the underside of the base, which is otherwise flat and undecorated. The earliest known bottles made in MCT IV A molds are Vessels Type, Generation A1a (see Cat. No. 15).

MCT II molds appear to have been used primarily but not exclusively within the Vessels Type (Series A3) (see Cat. No. 26).[103] The bottles made in MCT II molds are generally well made and the sections of the mold carefully aligned. The only bottle with high relief made in an MCT II mold for which a find spot is recorded is said to be from Sidon.[104]

MCT V A was made of metal. It consisted of two distinct sections, each made by a different technique (see Cat. No. 43). The mold sections were so well-aligned that it is difficult to distinguish the mold seams on the finished vessels. The earliest documented appearance of an MCT V A mold is in Series A3 of the Fruit Type. The MCT V A mold was probably not invented in the workshops producing Series A1 and A2 but in a third workshop that copied the design of a bottle of Fruit Series A1 into a metal mold, MCT V A. This third workshop then proceeded to design bottles of the Mixed Symbols Type (see Cat. No. 43), all of which were made in MCT V A molds. Because the decoration of the Mixed Symbols Type was probably created in Phoenicia, it is here assumed that the MCT V A mold also was invented in Phoenicia.

The use of a metal mold may have been prompted by the high breakage rate of clay molds. The degree of precision achieved with MCT V A molds was very high — no mold seams are visible — but the small number of vessels preserved indicates that this kind of mold did not gain wide popularity and was abandoned after a short period of use, perhaps because a decorated metal mold with a narrow opening was difficult to make and use.

Hexagonal Bottles with High Relief made in MCT VIII molds have been found only in the western part of the Roman empire. The use of an MCT VIII mold may therefore indicate that these bottles were blown in the West. The construction itself, however, was invented in the Near East, presumably in the Workshop of the Floating Handles, which appears to have experimented with a wide variety of MCT VIII molds (see below).

### Types

#### *Vessels Type*
(Cat. Nos. 15-35) (Fig. 46)

The fullest (but most idiosyncratic) discussion of the Vessels Type to date is that by Eisen, who named it the "Temple Series" because he thought the vessels appearing on the panels of the main frieze represented "sacred vessels in use in the temple," perhaps that of Sidonian Astarte.[105] Eisen focused on identifying the vessels appearing on the panels. His drawings indicate that he had probably seen examples of four different Series, two of which can be identified with Series represented by bottles in the Toledo collection. The bottle whose six vessels are illustrated in the top row of Fig. 47 belongs to Series A, the most common surviving Series.[106] Fig. 48 illustrates the vessels from the panels of a bottle of

[102] The so-called Athletic Symbols have here been renamed Mixed Symbols, as will be explained below.

[103] Compare a translucent blue bottle formerly in the Maxwell Collection (*Maxwell* 1979, 94-95, no. 151, here Fig. 49) which belongs to a Series of the Vessels Type that was created later than Generation A3b. For a possible example of an MCT II mold used in the Fruit Type: Isings 1975, 38, no. 663, fig. 99 (here Fig. 53 left).

[104] De Ridder 1909, 153, no. 273, pl. X.

[105] Eisen 1927, I, 233-248.

[106] Eisen 1927, I, fig. 114, top row (here Fig. 47).

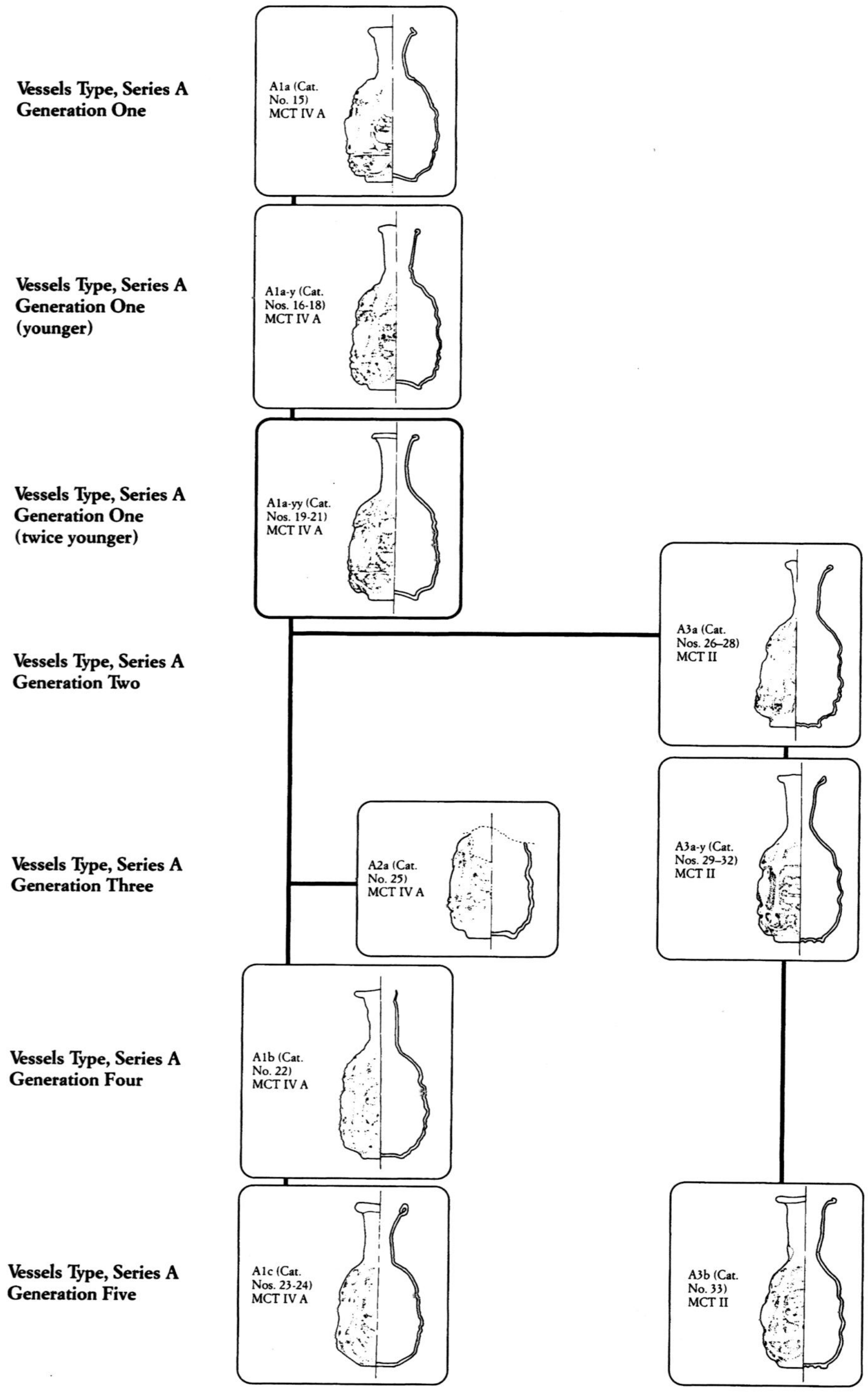


*Fig. 46. Hexagonal Bottles, Vessels Type, Series A1, 2, 3. Scale: 1:3.3. For an explanation of the changes by Generation, see Remarks Cat. No. 15. Chart designed by the author; drawings: Lorene Sterner and Timothy A. Motz.*

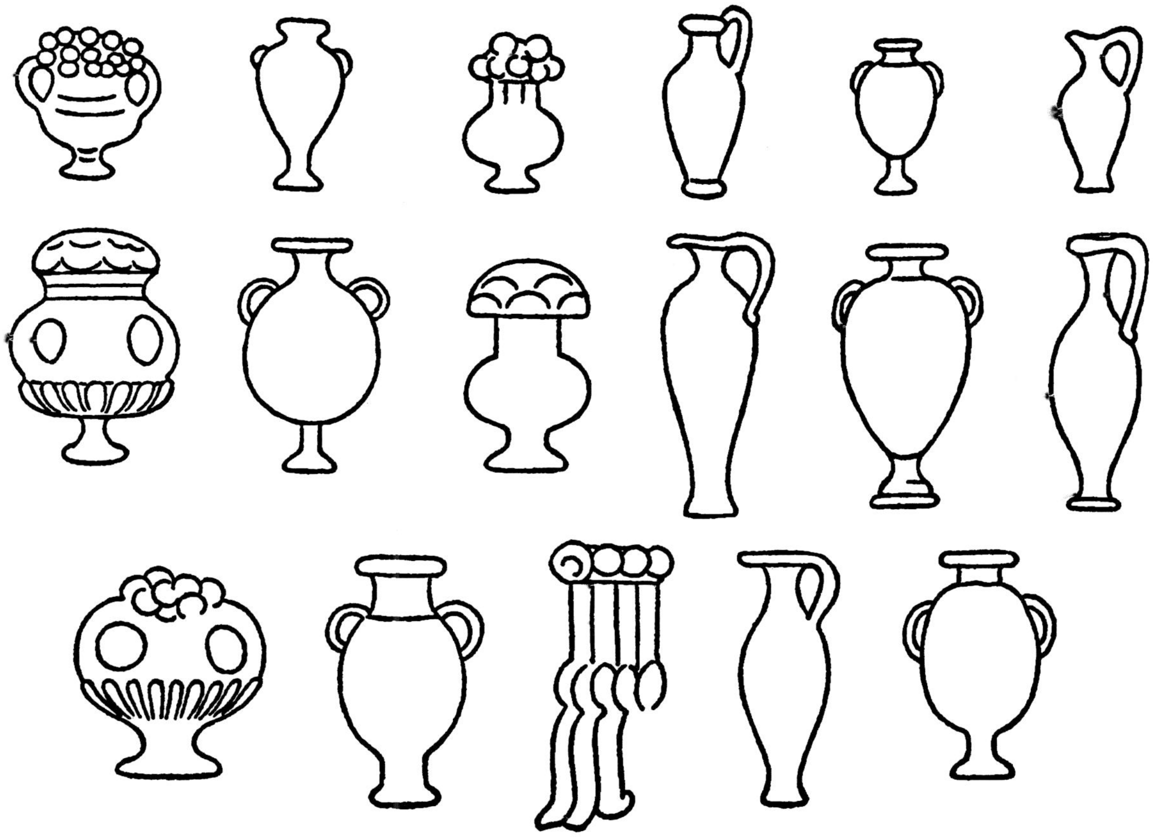

*Fig. 47. Vases represented in the panels of the Vessels Type. The top row shows the panels in Series A. The center row probably represents an unidentified Series D. The bottom row probably represents an unidentified Series E. Drawing: From Eisen 1927, I, p. 234, fig. 114.*

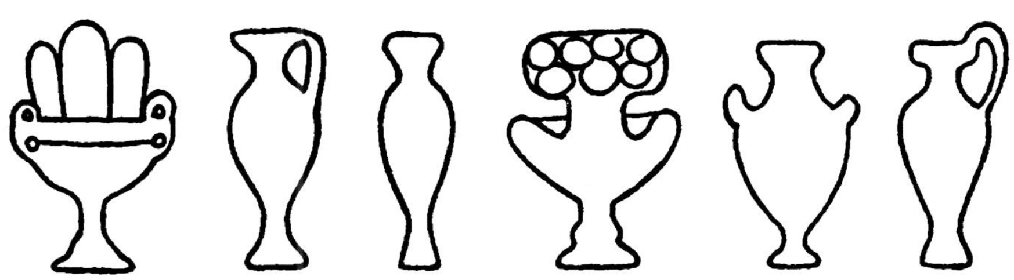

*Fig. 48. Vases represented in the panels of the Vessels Type. This row shows the panels in Series B. Drawing: From Eisen 1927, I, p. 247, fig. 115.*

Series B.[107] A third Series, C, is documented by a bottle formerly in the Maxwell Collection (Fig. 49).[108] The panel images in the middle row of Fig. 47[109] were reproduced by Eisen so much larger than those in the top row that they probably represent a fourth Series, D. The bottom row of Fig. 47[110] illustrates five panels only, perhaps from a five-sided bottle, which must represent a fifth Series, E.

Because Toledo has examples only of Series A and B, the discussion here is limited to those Series. Nineteen Toledo bottles belong to Vessels Type Series A; ten of these belong to Subseries A1, Generation A1a and its derivatives, Generations A1a-y through A1c (for a full discussion of the

[107] Eisen 1927, I, fig. 115 (here Fig. 48).

[108] *Maxwell* 1979, 94–95, no. 151 (here Fig. 49). De Ridder 1909, 154, no. 274, pl. X describes a bottle showing three footed jugs, one of which has the handle to the left, and three footed bowls; it is made of yellow glass and said to be from Tortosa on the coast north of Sidon. Compare Damascus, Nat. Mus., no. 13.121, an opaque white bottle from Tell Nabi Mind, southwest of Homs in Syria, Abdul Hak 1951, 113–114, no. 19.

[109] Eisen 1927, I, fig. 114, middle row (here Fig. 47).

[110] Eisen 1927, I, fig. 114, bottom row (here Fig. 47).

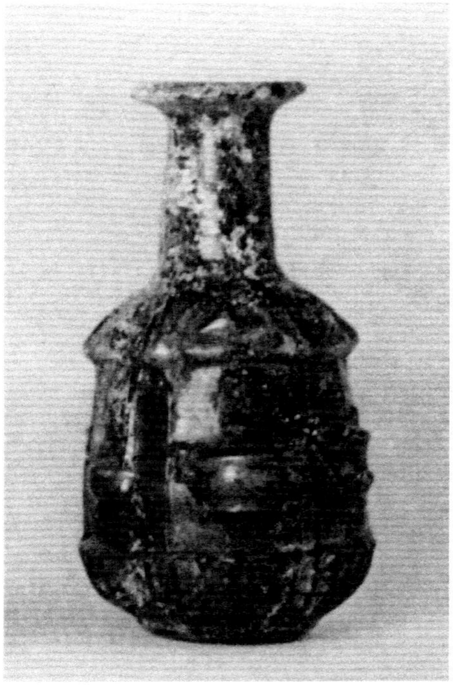

*Fig. 49. Hexagonal bottle, Vessels Type, Series C. H 8.6 cm, translucent blue, formerly Constable-Maxwell Collection. Photo and drawing: Reprinted with permission from* Maxwell *1979, 94, no. 151.*

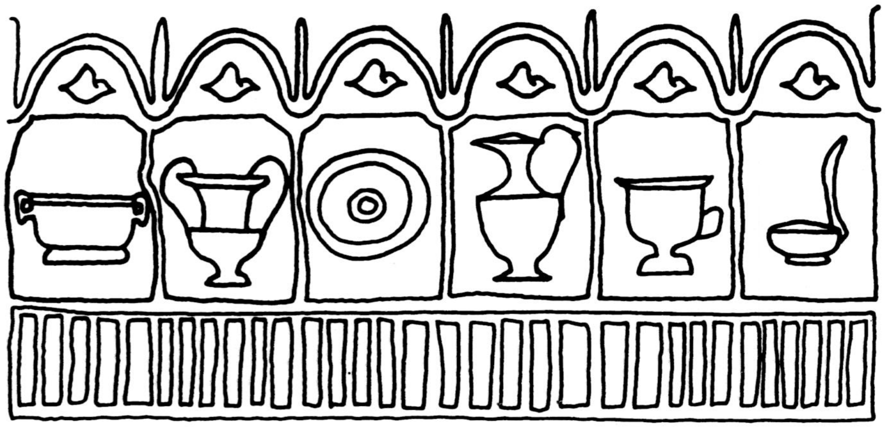

development of the Generations, see Cat. No. 15). All bottles belonging to Subseries A1 were made in three-part MCT IV A molds. One bottle belongs to Subseries A2 (Cat. No. 25) and was made in an MCT IV A mold with different sections. Eight bottles belong to Subseries A3 and its derivatives, all of which were made in four-part MCT II molds (see Cat. No. 26). It is possible that the workshop that made Subseries A3 also created Series C of the Vessels Type and Subseries A2 of the Fruit Type, because they were all made in MCT II molds. Two Toledo bottles belong to Vessels Type Series B; both were made in MCT IV A molds (Subseries B1) (see Cat. No. 34).

### *Fruit Type*

(Cat. Nos. 36-40) (Fig. 50)

It has been possible to identify two Series within the Fruit Type.[111] Four Toledo bottles belong to Series A, the most common Series surviving. One bottle belongs to Series B, which was probably made in a different workshop (see Cat. No. 40). Subseries A1 and B1 were both made in MCT IV molds.

The Toledo bottles of Fruit Type Series A belong to Subseries A1, Generation A1a, with a main frieze of two repeating sets of three fruits — pomegranate, cedar cone, and grape bunch (Cat. Nos. 36 and 37) — and Generation A1a-y, representing a younger generation without visible modifications (Cat. No. 38). Because Toledo does not have examples of later generations of Series A1, the entire Series is discussed here instead of at Cat. No. 36. In Generation A1b the egg-and-dart band below the panels of the main frieze was eliminated, but the panels still have two repeating sets of three fruits (Fig. 51).[112] Shoulder and bottom have a blurred version of the floral decoration characteristic of the preceding molds. The main frieze of Generation A1c has the same motifs as A1b bottles, two repeating sets of three fruits, but the mold maker replaced the triangular leaves around the bottom with stiff gadrooning (Fig. 52),[113] a motif that was probably copied from a Sidonian bottle of the Birds Type (see Cat. No. 45).

A bottle of Generation A1c seems to have provided the model for two new Subseries of the Fruit Type, Subseries A2 and A3; it is not clear whether they were made in different MCTs and thus presumably in different workshops. Subseries A2 was perhaps made in MCT V A molds, but the photograph shows a dot surrounded by a circle on the underside of the base (Fig. 53 left).[114] The main frieze has the traditional design of two repeating sets of three fruits, but an acorn appears in the place of a cedar cone, so the panels show pomegranates, grape bunches, and acorns. Subseries A3 was certainly made in MCT V A molds (Fig. 53 right).[115] The main frieze of Subseries A3 has two different sets of three fruits: one side is the same as Subseries A2, showing pomegranates, grape bunch, and acorn, but on the other side an acorn takes the place of the grape bunch, thus showing pomegranate, cedar cone, and acorn. The modeling of Subseries A2 and A3 is rather stiff, stylistically linking

[111] Lightfoot 1985, 123-125, gives an extensive list of parallels.

[112] Newark, no. 50.1457, opaque pale blue, Auth 1976, 198, no. 327, where it is erroneously described as having been made in a two-part mold.

[113] Tel Aviv, no. MHG 6558, Lehrer-Jacobson 1989, 41, ill.

[114] Schut Collection, Maastricht, Netherlands, Isings 1975, 38, no. 663, fig. 99. Isings's description suggests production in an MCT II mold, but the photographs sent by Mr. Schut suggest an MCT V mold. My thanks are owed to Mr. Schut for sending photographs.

[115] Newark, no. 50.1458, opaque white, Auth 1976, 69, no. 65.

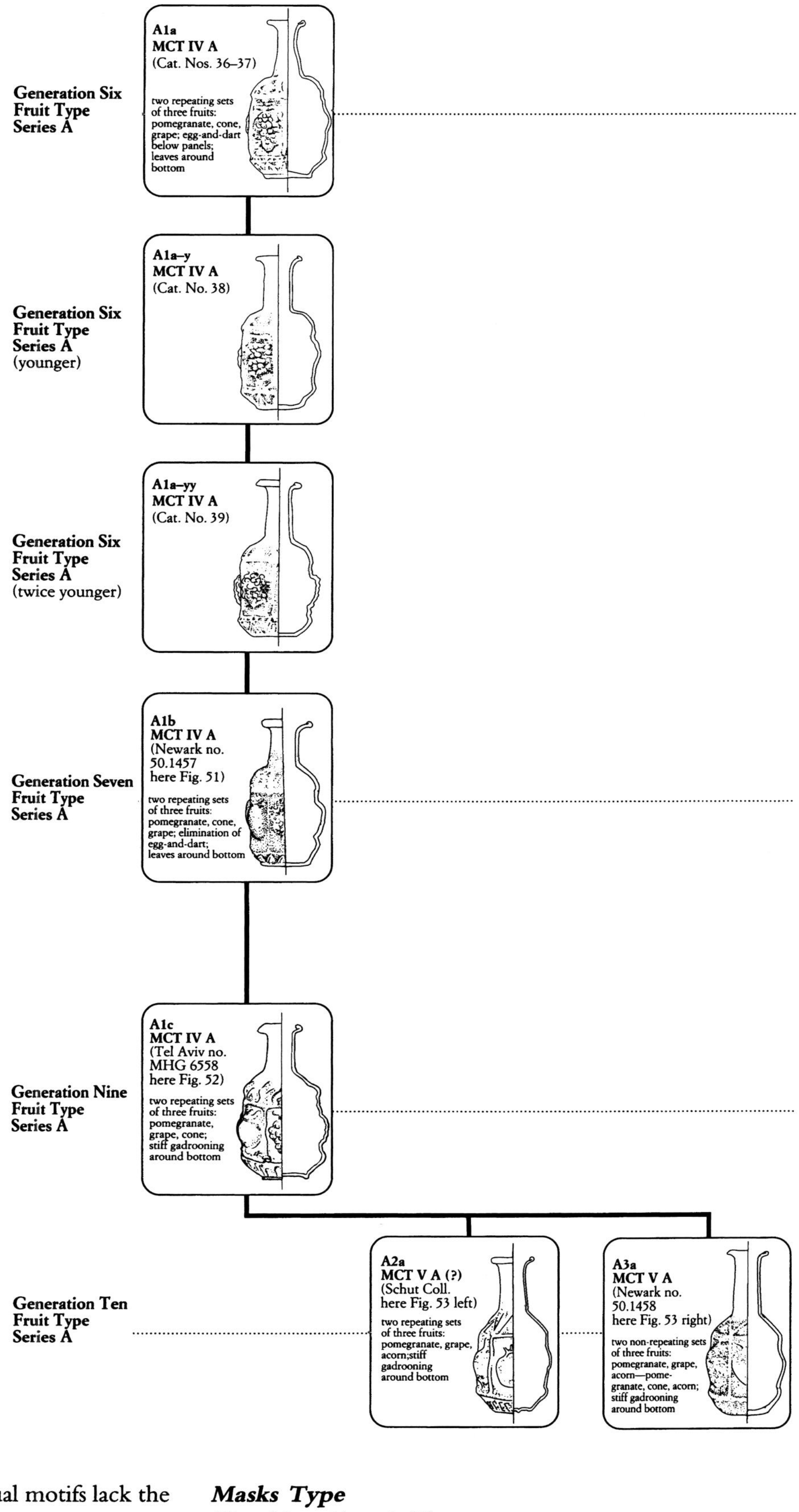


*Fig. 50. Hexagonal bottles, Fruit Type, Series A1, 2, 3. Scale 1:3.3. Chart designed by the author; drawings: Lorene Sterner and Timothy A. Motz.*

the fruits to Subseries A1, but the individual motifs lack the precision and delicacy of earlier generations.

Subseries A3 of the Fruit Type was probably made in the workshop that subsequently produced the Mixed Symbols Type, because MCT V A molds are otherwise known only in that Type.

### *Masks Type*

(Cat. Nos. 41 and 42)

The Masks Type includes both hexagonal and four-sided bottles.[116] The Toledo Museum has only two examples, and

[116] For a recent discussion of the Masks Type: Lightfoot 1987.

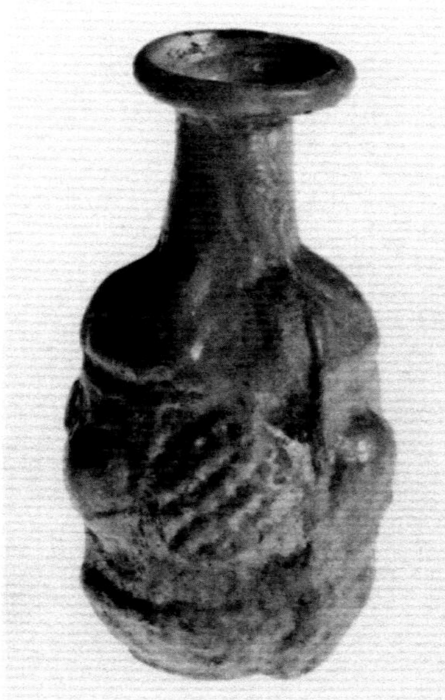

*Fig. 51. Hexagonal bottle, Fruit Type, Series A, Generation A1b. H 7.7 cm, opaque pale blue, Newark, N. J., Collection of the Newark Museum, The Eugene Schaefer Collection of Ancient Glass, no. 50.1457. Photo: Newark Museum.*

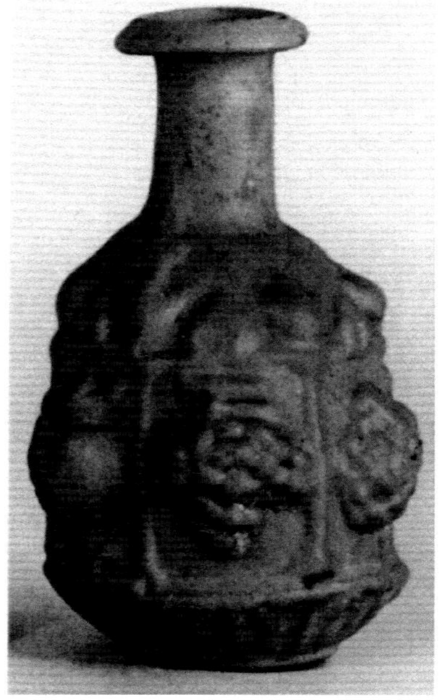

*Fig. 52. Hexagonal bottle, Fruit Type, Series A, Generation A1c. H 8 cm, opaque blue, Tel Aviv, Eretz Israel Museum, no. MHG 6558. Photo: Eretz Israel Museum, Glass Pavilion.*

it has not been possible to verify the exact appearance or even the number of different masks on individual bottles from published descriptions of other bottles belonging to the Masks Type. Consequently no attempt has been made to construct a chart for this Type. The Toledo bottles are assigned to Subseries A1. In order to avoid confusion with the terminology used in the rest of this chapter, the term "Group" is introduced here for bottles that appear to be related to each other but do not belong to the same Group as the Toledo bottles of Subseries A1, which is also Group 1.

The Toledo bottles show a different mask in each of the six panels (see Cat. No. 41). They were made in closely related molds, here assigned to Generation A1a because there appear to be no bottles extant that precede them. Subseries A1 was made in three-part MCT IV A molds.

A bottle of Subseries A1, or Group 1, was probably the model for the hexagonal bottles of Group 2, which shows two repeating sets of three masks.[117] A few hexagonal bottles

[117] Group 2: Found in a tomb at Knossos, Crete, transparent turquoise, Carington Smith 1982, 278-279, no. 62, fig. 6, pl. 39b; Wiesbaden, Landesmuseum, no. 2597, from a sarcophagus excavated at Heimersheim, Rheinhessen, translucent yellow, Lightfoot 1987, 20, no. 9, fig. 5, A-D. According to Spartz 1967, no. 26, the masks on the bottle from Heimersheim are the same as three of the masks on a four-sided bottle in Kassel, Staatliche Kunstsammlungen, no. G. 127 (see below, note 119f): Silenus, a female (perhaps a maenad), and a Gorgoneion.

Fig. 53. Left: *Hexagonal bottle, Fruit Type, Subseries A2. It is not clear whether this bottle was made in an MCT V A mold. The underside of the base appears to have a dot surrounded by a circle. Maastricht, The Netherlands, Schut Collection. Photo: Courtesy A. B. Schut.* Right: *Hexagonal bottle, Fruit Type, Subseries A3, made in an MCT V A mold. H 8.3 cm, opaque white, Newark, N. J., Collection of the Newark Museum, The Eugene Schaefer Collection of Ancient Glass, no. 50.1458. Photo: Newark Museum.*

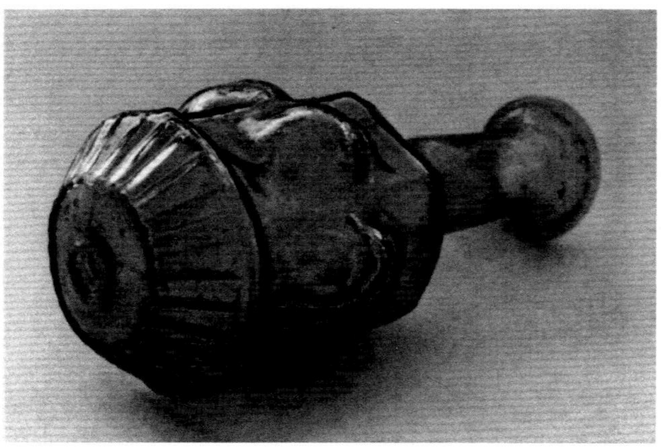

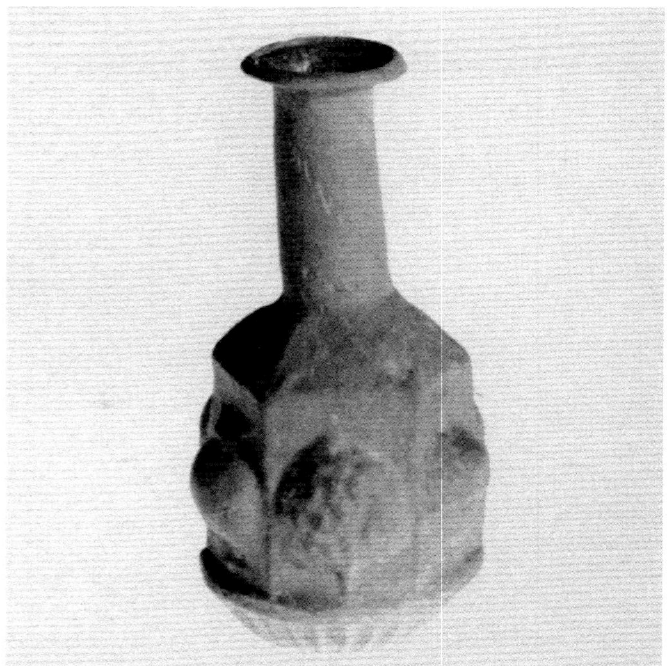

reported to have three repeating sets of two masks are assigned to Group 3.[118] Groups 2 and 3 lack secondary decoration on shoulder and bottom and do not have an egg-and-dart band below the panels with masks. Unfortunately, published descriptions are too vague to ascertain whether the masks themselves are identical within these Groups. Group 4 consists of four-sided bottles decorated with masks[119] and also lacking all secondary decoration. The style of Group 4 appears to be related to Group 3, a bottle of which may have inspired the decoration.

The farther west that bottles of the Masks Type are found, the simpler and more blurred is their decoration. This phenomenon suggests that a bottle of Group 1 was the model for making the archetype of the mold or molds used in the production of Group 2. It is not clear whether a bottle of Group 2 then became the model for Group 3 or whether Group 3 was derived independently from a bottle of Group 1.

In contrast to Group 1, bottles of Groups 2, 3, and 4 were made in two-part MCT VIII molds and not one example is reported to have been found in the Syro-Palestinian area. One bottle of Group 2 was found in Crete; another in Germany. Only one find spot has been recorded for Group 3: Herculaneum, in Italy. All bottles of Group 4 for which a provenance has been recorded were excavated in graves in southern France, western Switzerland, and Germany. An Italian origin has recently been suggested for the bottles blown in two-part molds, both four-sided and hexagonal.[120] This seems plausible since the use of an MCT VIII mold indicates a change of workshop and because none of the Bottles with High Relief excavated in the eastern Mediterranean appears to have been made in an MCT VIII mold.

A small group of four-sided bottles with masks dates to the nineteenth or early twentieth century.[121] They appear to be closely related to solid cylindrical blocks with two masks and cylindrical bottles with two or four masks are also of the same date and provenance.[122]

### *Mixed Symbols Type*

(Cat. Nos. 43 and 44) (see Fig. 54)

Bottles of the Mixed Symbols Type show a variety of objects in the panels of the main frieze. This is exceptional, because Hexagonal Bottles with High Relief usually show objects of a single kind. These objects have traditionally been interpreted as athletic symbols.[123] However, reexamination of the Type has revealed that these objects are definitely not athletic symbols and that a Dionysiac interpretation is more plausible. Therefore the name of this Type has here been changed to the "Mixed Symbols Type" (for a full discussion, see Cat. No. 43).

The Toledo Museum of Art has two bottles decorated with Mixed Symbols. They were blown into closely related molds belonging to Generations A1b and A1b-y. Although bottles of Generation A1b are the most common within this Type, there appears to have been an earlier Generation, represented by a bottle formerly in the Maxwell Collection, here named Generation A1a (see below Fig. 77).[124] Four panels of the bottles of Generation A1a have the same motifs as those of Generation A1b, but two panels are different. The earlier two panels show a radiating motif in which an X-shaped element with thickened ends is the basic feature. Although it is flanked by other objects, the X itself is identical to the double-ended thyrsoi on glass jugs showing cult objects (Cat. Nos. 71-74). In the more elaborate of the two panels, two small objects resembling the so-called strigil in panel 4 of Cat. No. 43 appear between the upper sections of the thyrsoi. The placing of these objects between thyrsoi, well-known Dionysiac symbols, indicates that they were not intended as athletic symbols, but that they too are Dionysiac symbols: the *pedum*, or shepherd's crook, which could also be used as a *lagobolon*, or hunting stick, for throwing at hares. The assemblage of objects depicted on the Maxwell bottle, Generation A1a, indicates that the Mixed Symbols in this Type originally had no connection with sports.

All bottles of the Mixed Symbols Type appear to have been blown into metal molds of MCT V A, which was probably invented in the workshop that originally created the design.

### *Birds Type*

(Cat. Nos. 45-47)

The bottles of the Birds Type stand somewhat apart from the majority of Hexagonal Bottles with High Relief because

118 Group 3: Ercolano, Antiquarium, no. E1238, from a wooden cabinet on the second floor of the Casa del Sacello di Legno, Herculaneum, opaque light green, Scatozza Höricht 1986, 52, no. 104, pls. 17 and 33; Sangiorgi 1914, no. 110, pl. 23, translucent dark blue.

119 Group 4 examples:

a) Saint-Germain-en-Laye, Musée des Antiquités Nationales, no. 24642, from Saintes, southern France, found in the sarcophagus of a young woman in the necropolis of Clousi, translucent brownish yellow, Lightfoot 1987, 11-16, 19, no. A4, fig. 2; H. Chew, "La tombe gallo-romaine de Saintes. Nouvel examen du matériel," *Antiquités Nationales* 20 (1988) 45 (cited by Morel et al. 1992, 17, n. 47);
b) Saint-Germain-en-Laye, Musée des Antiquités Nationales, no. 24641, provenance as for (a) above, translucent blue, Lightfoot 1987, 11-16, 19, no. A5, fig. 3;
c) from Avenches, ancient Aventicum, western Switzerland, excavated in the dump of a glass factory active from 40 to 70, perhaps for recycling, Morel et al. 1992, 14, fig. 19.11;
d) RGM, no. N6176, probably from Cologne, translucent blue-green, Lightfoot 1987, 11-16, 19-20, no. A6, fig. 4;
e) Bonn, Rheinisches Landesmuseum, no. 1448, found in a grave at Andernach, translucent dark green, Lightfoot 1987, 11-16, 19, no. A3;
f) Kassel, Staatliche Kunstsammlungen, no. G.127, perhaps from the Vogell Collection assembled in southern Russia and Ukraine, translucent green, Spartz 1967, no. 26, pl. 5, Lightfoot 1987, 11-16, 18, no. A2;
g) V&A, no. 1030-1868, translucent dark amber yellow, Lightfoot 1987, 11-16, 18, no. A1, fig. 1A-D;
h) Milan, Comune di Milano, Raccolte Archeologiche e Numismatiche, no. A1258, translucent pale blue, Lightfoot 1987, 11-16, 20, no. A7;
i) Bologna, Museo Civico, no. Rom. 32, translucent deep yellow, Lightfoot 1987, 11-16, 20, no. A8.

120 Lightfoot 1987, 21.

121 For example Smith 1957, no. 416; Axel von Saldern, "Originals, Reproductions, Fakes," *AnnAIHV* 5 (1970) 309-310.

122 Grose 1989, 383-384, fig. 184, p. 394, no. 713.

123 Eisen 1927, I, 252, fig. 120, calls this Type "Sidonian Six Sided Flasks with Palestra Objects;" Auth 1976, 69, no. 64, calls them "Athletic Symbols."

124 *Maxwell* 1979, 103, no. 176.

the main Series is decorated with low rather than high relief. The secondary decoration is also slightly different: the lower part of the body is always decorated with tongues or fluting instead of fillets or floral designs. The hexagonal shape of the bottles and the fact that they were blown in MCT IV A molds, the preferred construction type for Hexagonal Bottles with High Relief, are the reasons they are included in this section. Bottles of the Birds Type appear to have influenced the style and decoration of the latest generations of Bottles of the Vessels and Fruit Types (see Cat. No. 45 and Fig. 54).

Two Series have been identified, both of which are represented in the Toledo Collection: Series A, with decoration in low relief (Cat. Nos. 45 and 46), and Series B, with decoration in high relief (Cat. No. 47). The Birds Type has been discussed by Eisen, who already recognized the existence of two Series, but he was not able to identify the subjects of all individual panels.[125] Nor did he point out that both Series show the same subjects in the same order. In spite of extensive recutting of the mold of Series B, the two Series appear to be based on the same archetype. The correspondence between the two Series is discussed in Cat. No. 45.

### *Interrelations of Types*

(Fig. 54)

Fig. 54 shows the interrelations of Types and Series of Hexagonal Bottles with High Relief from Generation One to Twelve.[126] The earliest Series is Subseries A1 of the Vessels Type; Generation A1a of the Vessels Type is Generation One.

In Generation Five a bottle belonging to Generation A1c of the Vessels Type served as the starting point for the creation of two new Types: Fruit and Masks. The new archetype had a body that was more crisply hexagonal, and a new set of motifs for the panels of the main frieze was designed. The new archetype accentuated the vertical edges of the panels but omitted any reference to a column and modified the secondary decoration on shoulder and bottom. The arch design on the shoulder was simplified by eliminating the oval object under the arches and adding two curves on each side of the base of the legs of the arches so that they resemble inverted blossoms. Around the bottom of the bottles of Generation A1c of the Vessels Type is a rather indistinct pattern of triangular leaves encasing small oval fruit; for both the Fruit and Masks Types this was recarved to change the oval fruit to oval leaves, while all the leaves received a more linear definition, creating a new pattern of alternating triangular and oval leaves (Fig. 55).

The two Types that can be linked to the new archetype are Series A of the Fruit Type and Series A of the Masks Type. Similarities in the carving of certain motifs (see Cat. Nos. 36 and 41) in the panels of these two Series suggest that the design of the Fruit Type was created first (Generation Six) and the Masks Type then recut from the Fruit Type (Generation Seven). The bottles of Generation Six (A1a of the Fruit Type) are slightly larger than those of Generation Five (A1c of the Vessels Type), which can be explained in various ways (see Cat. No. 36). The form of the body of the Fruit Series is not as crisply hexagonal as that of the earliest bottles of the Masks Type, but the height of relief is greater. These observations suggest that Generation A1a of the Fruit Type was created shortly before Series A of the Masks Type, because there was room to cut away clay or wax from the fruit in the model to carve the masks.

After having been updated with gadrooning in Generation Nine (Fruit Type, A1c), Series A of the Fruit Type underwent at least two transformations in Generation Ten (Fruit Types A2a and A3a) before it inspired the creation of yet another Type, the Mixed Symbols Type in Generation Eleven.

Around the same time that the Mixed Symbols Type was created, or slightly earlier, Subseries A3 of the Vessels Type was updated with gadrooning around the bottom to create Series C, e.g. a bottle formerly in the Maxwell Collection (see Fig. 49).[127] The diminished size of the bottle and the evolution of the columns separating the panels of the main frieze into vertical ridges indicate that the creation of Series C took place at about the same time as Generation A1c or later of the Fruit Series was being produced, i.e. about Generation Nine of the Hexagonal Bottles with High Relief. Series C of the Vessels Type was decorated with a new set of vessels, including some that were not used previously on the Vessels Type. The Maxwell bottle with concentric circles on the underside of the base might have been blown into an MCT II mold. If that is correct, it might have been made in the same workshop as the bottles of Subseries A3 of the Vessels Series.

## Workshops and the Spread of Designs

The clay molds used to blow Hexagonal Bottles with High Relief had to be constantly freshened and replaced. This necessity, combined with the fact that a new archetype could easily be cast from a glass vessel made in a workshop far away, contributed to the spread of designs in a short period of time. As a result only the first Generation of bottles belonging to any given Series can be assigned to the workshop that created it. Each subsequent modification of the mold increases the possibility that the bottles blown in it were made elsewhere. A change in MCT within a Series clearly indicates production in a different workshop.

Bottles belonging to the Masks Type (Cat. Nos. 41 and 42) seem to indicate migration of designs from east to west. They were first created in Phoenicia, but they appear to have been copied in Italy using a differently constructed mold.

[125] Eisen 1927, I, 249.

[126] In order to avoid confusion between generations of Hexagonal Bottles in general and Generations of molds, Generations of molds are referred to with the numerical system explained above, while the generations of Hexagonal Bottles as a group are spelled out as on the chart (Fig. 54) as Generation One, Generation Two, etc.

[127] *Maxwell* 1979, 94-95, no. 151 (here Fig. 49).

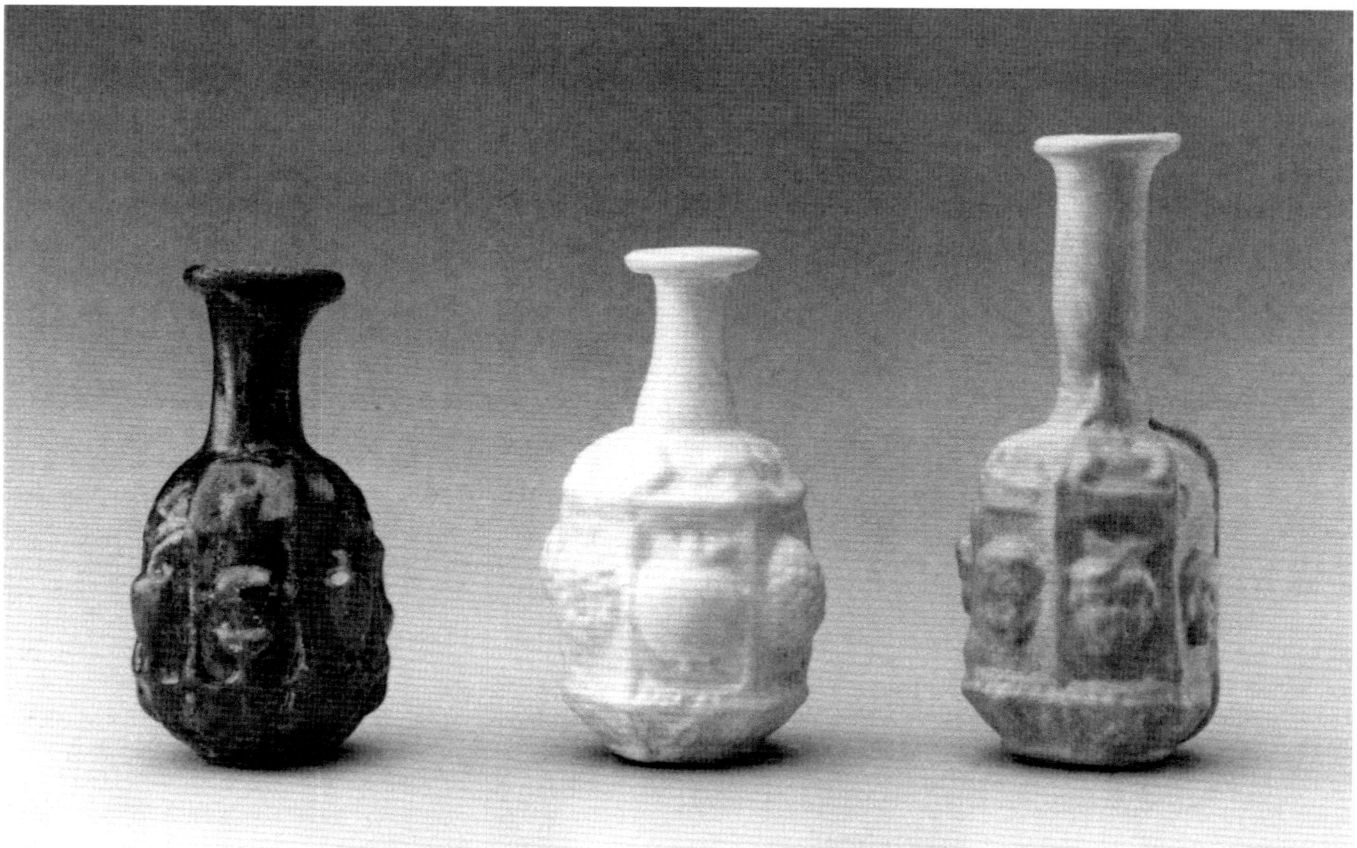

*Fig. 55. Hexagonal bottles of the Vessels, Fruit, and Masks Types, showing recarving of secondary decoration around bottom.* From left: *Vessels Type A1c (Cat. No. 23), Fruit Type A1a (Cat. No. 36), and Masks Type A1a (Cat. No. 42). Photo: Tim Thayer.*

The design used in the West was probably cast directly from a bottle blown in Phoenicia; but instead of making a three-part MCT IV A mold, such as was commonly used in the East, the Italian glass workshop made a two-part MCT VIII mold.

## Date

The dating of Hexagonal Bottles with High Relief is based on two criteria. The first consists of vessels found during controlled excavations to which an absolute date can be given independently for their context. The second criterion uses the interrelations of Types and Series to establish relative dating for the creation of various Series with respect to each other.

Very few bottles belonging to the Types discussed in this chapter have been found in datable contexts. In general, objects excavated in the western half of the Roman empire can be dated within a much more precise time span than those excavated in the eastern empire, because in the East tombs were used and reused, often for two or more generations. Thus there is a paradoxical situation in which bottles made and excavated in the West sometimes appear to be earlier in date than the similar eastern ones from which they may have been copied. Finds from the earliest datable published contexts are bottles of the Masks Type excavated in the West. One bottle belonging to Group 2 was discovered in a tomb dated to the second half of the first century,[128] while a bottle belonging to Group 3 must date from before 79 because it was excavated in Herculaneum[129] which was destroyed in that year. Of the four-sided bottles of Group 4 found in southwestern Gaul and the Rhineland, three were excavated in tombs dated to the middle of the first century.[130] The bottles were blown a few years earlier, at the latest around 40. This mid-century date is supported by the discovery of a bottle of the Mixed Symbols Type, Generation A1b (see Cat. No. 43), in a cremation deposit in Slovenia that was interred during the second quarter of the first century.[131]

[128] Carington Smith 1982, 278-279, no. 62, pl. 39b, fig. 6, from a tomb at Knossos, Crete. A bottle of Group 2 found in a third-century sarcophagus at Heimersheim, Germany, must have been an heirloom: Lightfoot 1987, 20, no. 9, fig. 5, A-D.

[129] Scatozza Höricht 1986, 52, no. 104, pls. 17, 33.

[130] For example:

a) Bonn, Rheinisches Landesmuseum, no. 1448, from a tomb at Andernach which also contained a coin of Claudius (41-54), C. Koenen, "Die vorrömischen, römischen und fränkischen Gräber in Andernach," *BJb* 86 (1888) 172, no. 12;

b, c) Saint-Germain-en-Laye, Musée des Antiquités Nationales, nos. 24642 and 24641, both from one burial in Saintes, dated mid first century, Lightfoot 1987, 19, nos. A4 and A5, figs. 2 and 3.

[131] From Ljubljana (ancient Emona), Slovenia, cremation grave 87, no. 3, one example of dark green glass, interred during the second quarter of the first century, Plesničar-Gec 1967, Y93; Petru 1974, 15, 28, figs. 1-2; Plesničar-Gec 1972, 217, grave 427, no. 3, pl. 205.3.

Evidence from mold-blown bottles and the relationships of their generations provides interesting data for speculation about the dates of these wares. For example, Fig. 54 shows how relative dating could be based on establishing the number of generations separating a given mold from Group 4 of the Masks Type and then estimating the number of years needed for the modifications observed in the development of decoration to take place (years per generation). A new generation can be recognized if a mold has been modified, but there is no way of determining the actual number of generations between modifications. Nor is it certain that all modifications have been recorded; there may be more than have been recognized thus far. In view of these uncertainties the duration of a generation has here been tentatively set at three years. This time span is used solely for the purposes of speculating about dates. It should not be taken to imply that this was the actual life span of an ancient glassblower's mold, which has not been established.

Generation A1a of the Vessels Type is the earliest known example of Hexagonal Bottles with High Relief. The smallest number of generations separating Generation A1a of the Vessels Type from Group 4 of the Masks Type is nine. Using three years as the duration of a mold generation and assuming that Group 4 of the Masks Type was blown no later than 40, this suggests a date some 27 years earlier for the creation of Subseries A1 of the Vessels Type, perhaps as early as 13. However, no known excavated bottles have provided a date earlier than 25/35.

## Distribution of Find Spots and Origin

Almost all the bottles for which find spots have been recorded were found in the eastern Mediterranean, but there is no concentration of finds in any particular area. Rather, the find spots appear to radiate from the Syro-Palestinian coast. A few bottles have been excavated in northern Israel, but most have been found at sites farther north, especially in western Syria along a line from Tell Nabi Mind (about 30 km southwest of Homs) to Gaziantep in eastern Turkey. These finds are balanced by bottles said to be from Tortosa on the Syrian coast, Cyprus, and Greece. A few have been excavated in Armenia and in the Crimea. Very few Hexagonal Bottles with High Relief have been recorded from Italy or the western Mediterranean, and those few are almost all copies probably made in the West.

This distribution pattern, combined with the fact that these bottles are not common in Israel, indicates that they were probably not made in a Palestinian glasshouse but somewhat

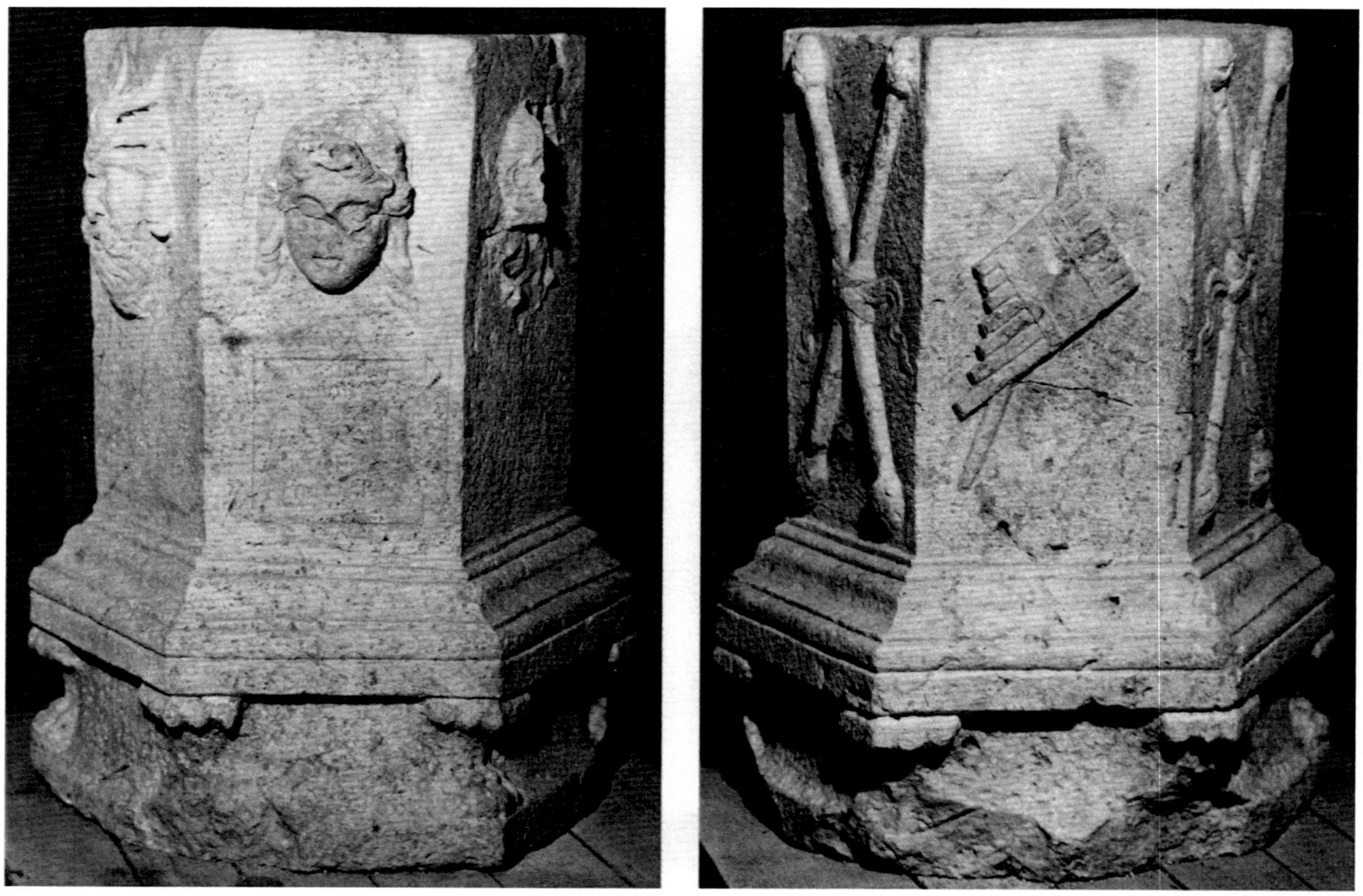

*Fig. 56. Hexagonal altar dedicated to the god Dionysos in the year 185 (A.D. 142), found at Beth Shean, Israel. The motifs appearing on this altar — masks, syrinx, pedum, and thyrsoi — are related to different aspects of Dionysos. The same motifs appear on Hexagonal Bottles with High Relief of the Masks Type and the Mixed Symbols Type as well as on Hexagonal Jugs with Dionysiac Symbols. Beth Shean Museum. Photo: Courtesy G. Foerster, Institute of Archaeology, Hebrew University, Jerusalem.*

farther north, presumably in one of the glasshouses of Phoenicia. The hexagonal shape and the depiction of a double-ended thyrsos, a type of thyrsos found predominantly in the Syro-Palestinian area, indicate strongly that these vessels were made in that area for a clientele that recognized these symbols. Therefore these bottles are assigned to unidentified workshops in Phoenicia, without ruling out the possibility that some, for example, the Vessels Type and Birds Type, were made in or near Sidon.

### Interpretation

It has repeatedly been claimed that the Hexagonal Bottles with High Relief are Jewish, made by Jewish glass blowers for a Jewish clientele, and connected with Jewish religious rites and festivals.[132] In this catalogue, the question of a Jewish interpretation is examined for each Type separately in the lead catalogue entries. The investigation is based on the assumption that market demand dictated in part the choice of motifs. To identify the clientele that would have purchased these bottles, the motifs decorating them are identified and interpreted against the background of cultural changes following the conquest of Alexander the Great in the years 333-323 B.C. In the Hellenistic and early Roman periods, approximately the third century B.C. to the third century A.D., numerous syncretisms of polytheistic cults flourished in the Near East. There also existed many Judaisms, the precise natures of which are not known, that had a majority or minority following. Although there were no "orthodox" Jews in antiquity, Jews continually struggled to preserve their heritage because daily life required interaction with their pagan neighbors. The attraction of Greek cultural life with its literary and artistic traditions was so strong that even Jewish priests and noble families began to adopt the Greek language and the Greek way of life.[133] Archaeological remains indicate that on the material level Jews and pagans used the same kinds of objects and furnishings.[134] It is against this varied background of religious beliefs and cultural changes that the Hexagonal Bottles with High Relief are considered.

The hexagonal shape appears to imitate religious architectural structures. The Vessels Type imitates an unidentified structure with six niches (see Cat. 15), while the Fruit, Masks, and Mixed Symbols Types probably imitate hexagonal altars such as one recently excavated at Beth Shean (ancient Nysa, Skythopolis) (Fig. 56).[135] An inscription on the altar

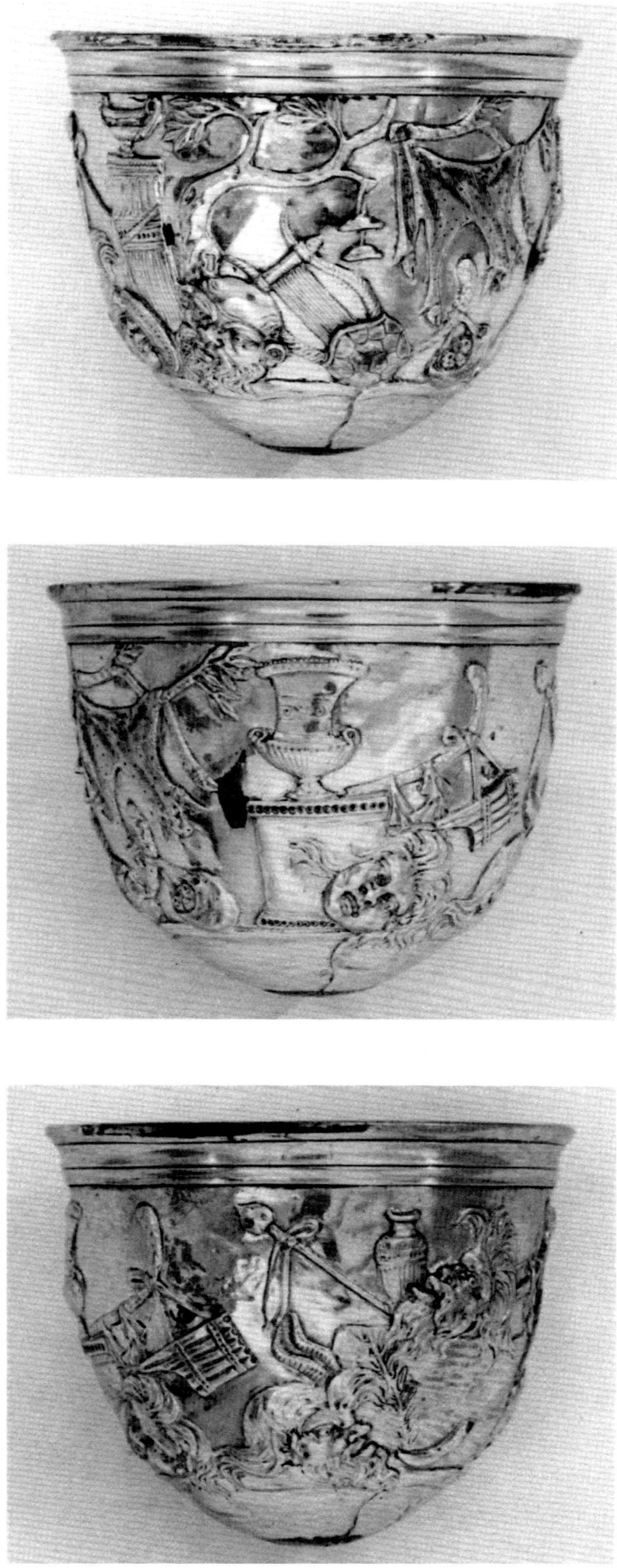

*Fig. 57. Silver cup, Roman, from Asia Minor, mid first century A.D. The symbols appearing on this cup — vessels, theatrical masks, musical instruments, and thyrsoi — evoke revelry associated with the god Dionysos and are paralleled on Hexagonal Bottles with High Relief. Toledo, Ohio, The Toledo Museum of Art, acc. no. 1961.9. Photo: The Toledo Museum of Art.*

132 A. Reifenberg, *Palästinensische Kleinkunst* (Berlin 1927) 69ff.; F. Neuburg, *Antikes Glas* (Darmstadt 1962) 51, figs. 41-42; A. Engle, "New Light on Sidonian Glassblowers," *Readings in Glass History* 21 (Jerusalem 1988) 73-91; Samuel Kurinsky, *The Glassmakers. An Odyssey of the Jews. The First Three Thousand Years* (New York 1991) 236. Goodenough 1953, I, 168, cites the Jewish interpretation with some hesitation and limits this to the Fruit Type.

133 Goodenough 1953, I, 13-14; Avi-Yonah 1984, 71-76. On the question when Hellenistic culture began to manifest itself: L. H. Feldman, "Hengel's Judaism and Hellenism in Retrospect," *JBL* 96 (1977) 371-382.

134 The question of interaction among Jews, pagans, and Christians in late antiquity has recently been put into perspective by Rutgers 1992, 101-118.

135 Foerster and Tsafrir 1988, 31, photo 18; Yael Israeli, "Marble Sculptures from Roman Israel," *Minerva* 3.5 (1992) 15-16, ill.

identifies it as an altar to Dionysos: "With good fortune, Seleukos, son of Ariston, [dedicated] as a thank offering to the god, the lord Dionysos, the founder, in the year 185" (year 185 of the Skythopolis era, or A.D. 142).[136] Dionysos is called the founder because of a local tradition that he founded the city and there buried his nurse Nysa. The six panels of the altar show Dionysiac motifs: 1, a mask of Dionysos above a *tabula ansata* with the dedicatory inscription; 2, broken; 3, crossed, double-ended thrysoi;[137] 4, syrinx and pedum; 5, crossed, double-ended thyrsoi; 6, a mask of Pan.

Hexagonal altars are not common in classical antiquity, but several are known from the Syro-Palestinian area and Asia Minor.[138] The motifs on the altar — masks, syrinx, pedum, and thyrsoi — are related to different aspects of Dionysos. The masks refer to Dionysos as god of the theater, while the thyrsoi, syrinx, and pedum are attributes of satyrs and maenads, his followers and worshippers. The same motifs occur on glass bottles of the Masks Type, the Mixed Symbols Type (Generation A1a), and on Hexagonal Jugs with Dionysiac Symbols (Cat. Nos. 71-74). However, Dionysiac and bucolic motifs need not always be interpreted as religious symbols in the late Hellenistic and early Roman periods. They are found on many objects where the original religious meaning had lost much of its cult association. These motifs often serve no more serious purpose than to evoke an atmosphere of joyous release and revelry associated with Dionysos, the god of wine. Dionysiac themes used in this way are well-known from luxury tablewares of the first centuries B.C. and A.D. A small silver cup with decoration in shallow relief, reported to be from Asia Minor (Fig. 57) and dated about the middle of the first century A.D.,[139] may serve as an example. The vessels and musical instruments evoke revelry associated with Dionysos. Similar iconography occurs on a pair of matching silver cups excavated in Hildesheim, Germany.[140]

The conclusion is that none of the Types of Hexagonal Bottles with High Relief is specifically Jewish. The decoration is either pagan or religiously neutral. The motifs on the Vessels and Mixed Symbols Types may have originally had polytheistic religious associations, but the themes of the Masks and Fruit Types were probably religiously neutral, appealing to local traditions in Phoenicia and Palestine that could be appreciated by gentile inhabitants of the cities on the coast, hellenized Jews, and other Semitic population groups living further inland. Each of these groups may have interpreted the motifs differently.

[136] The date on the altar was originally misread; it is now dated 142 (G. Foerster, pers. com.). My thanks are due to Professor Foerster for permission to illustrate the altar and for sending a description of the altar in English before its publication.

[137] On the double-ended thyrsos, known in antiquity as *thyrsolongchos*, see at Cat. No. 71.

[138] G. Foerster, pers. com.

[139] TMA, no. 1961.9, Oliver 1977, 116-118, no. 76 (with lit.); Lucia Pirzio Biroli Stefanelli, *L'Argento dei Romani* (Rome 1991) pls. 10-12.

[140] Berlin, Staatliche Museen, nos. 3779.13 and 3779.14, from Hildesheim, E. Pernice and F. Winter, *Der Hildesheimer Silberfund* (Berlin 1901) 37 ff., pls. 13 and 14; U. Gehrig, *Hildesheimer Silberfund* (Berlin 1967) 24-25, figs. 23-26; Oliver 1980, 130-131, no. 83.

The cultural and racial origins of the glass blowers producing Hexagonal Bottles with High Relief are a matter of conjecture.[141] The workshop that created the Masks Type, which was clearly intended for a hellenized clientele, also created the Fruit Type which was neutral in its associations. Examples of workshops catering to two or more separate population groups are not exceptional in antiquity. Workshops producing vessels with Christian and Jewish motifs have long been recognized in the fourth and early seventh centuries.[142] This phenomenon probably also existed in the first century.

### *The Workshop of the Floating Handles*
(Cat. Nos. 59, 62, 113, and 138)

Minor variations in the details of manufacture, such as the way rims, handles, and bases are formed or the number and position of mold seams, can be useful criteria for distinguishing the products of different glass workshops. When similar, unusual details are observed exclusively on vessel forms

[141] Lehrer 1979, 5, 14.

[142] On fourth-century gold glasses with Jewish and non-Jewish motifs made in one workshop: J. Engemann, "Bemerkungen zu römischen Gläsern mit Goldfoliendekor," *JAC* 11.2 (1967) 16-22. On early seventh-century pilgrim flasks with Jewish and Christian motifs: Barag 1970c; Barag 1971a; and below: Chapter IV and Cat. Nos. 169-192. On workshops producing sarcophagi for Jewish, Christian, and pagan customers in fourth-century Rome: Rutgers 1992, 101-108.

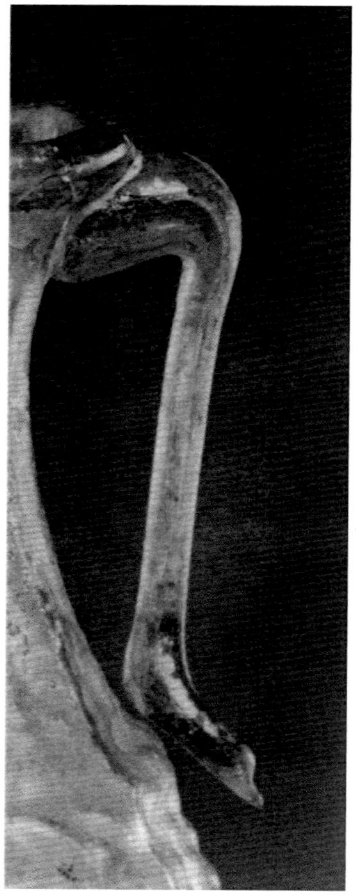

*Fig. 58. Workshop of the Floating Handles, enlarged detail of a miniature amphora encased in wickerwork (Cat. No. 59), showing the distinctive handles produced in the Workshop. Photo: Tim Thayer.*

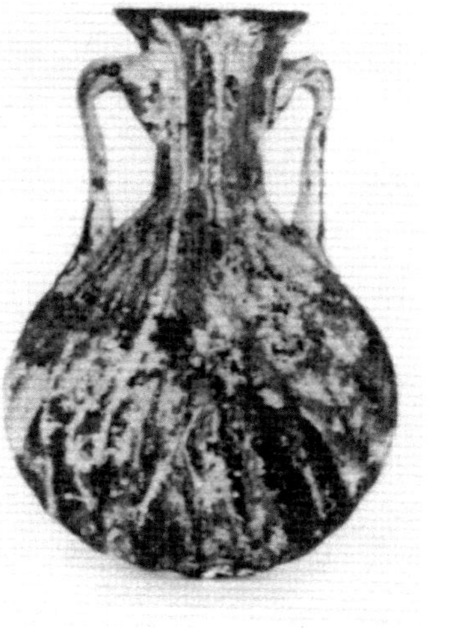

*Fig. 59. Workshop of the Floating Handles, Shell Type. H 6 cm, opaque purplish, Tel Aviv, Eretz Israel Museum, no. MHG 9358. Photo: Eretz Israel Museum, Glass Pavilion.*

found in the same area and dating from one period, it is likely that the objects were made in a single workshop. One Syro-Palestinian workshop that made small bottles has been identified by the peculiar way in which it made handles. It has been named "The Workshop of the Floating Handles" because the lower handle attachments do not adhere to the wall of its bottles.[143] Four of the mold-blown Types made in this workshop are represented by one vessel each in the collection of The Toledo Museum of Art (Cat. Nos. 59, 62, 113, and 138).

### Technique

Roman workshops usually made handles by applying a bit of glass to the shoulder or body of a vessel and then pulling it upward toward the neck or rim where it was attached. When a bit has been applied to the surface of the glass, it must be separated from the tool[144] on which it was brought to the vessel. Early Roman glass blowers usually did not cut the glass off the tool with shears; instead they pulled it out thin until it snapped. The excess glass was then folded back against and on top of the handle. In the Workshop of the Floating Handles, the glass blowers unusually and distinctively made handles the other way round. The hot bit was applied first to (or near) the rim and then pulled downward toward the shoulder or body. This technique was also unusual in that it used a tool, perhaps a type of shears or pincers, which separated the glass and then pinched it to form a small disk at the lower end of the handle. Occasionally the glass blower used the same tool to make a second pinch at the top of the handle for a vertical projection. Hot glass will adhere to a surface if that surface is also hot. However, this pinching

[143] Stern 1991a, 199-206.

[144] The tool was either a second blowpipe or a precursor of what later would be called a pontil.

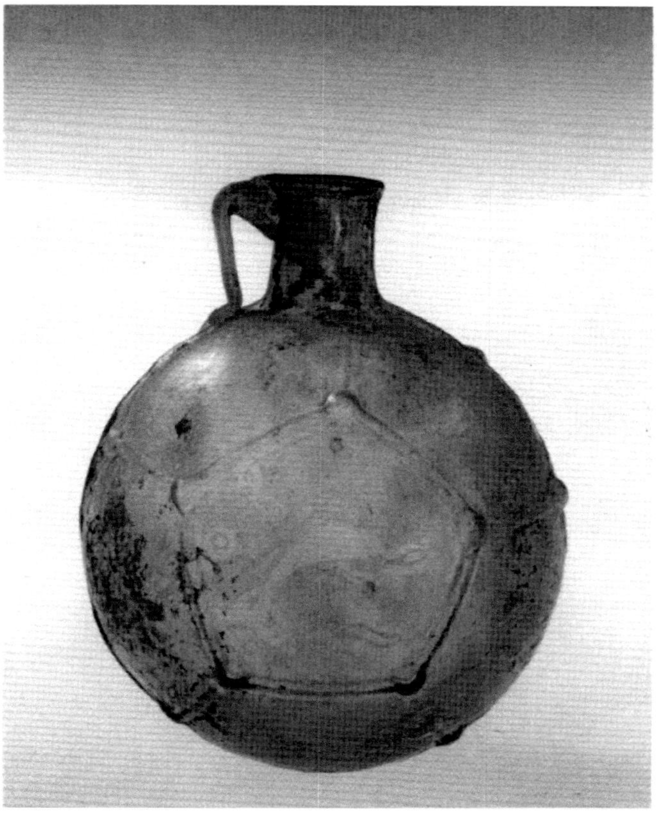

*Fig. 60. Workshop of the Floating Handles, Ball Type. H 8.2 cm, transparent greenish, Corning, New York, The Corning Museum of Glass, no. 68.1.45. Photo: The Corning Museum of Glass.*

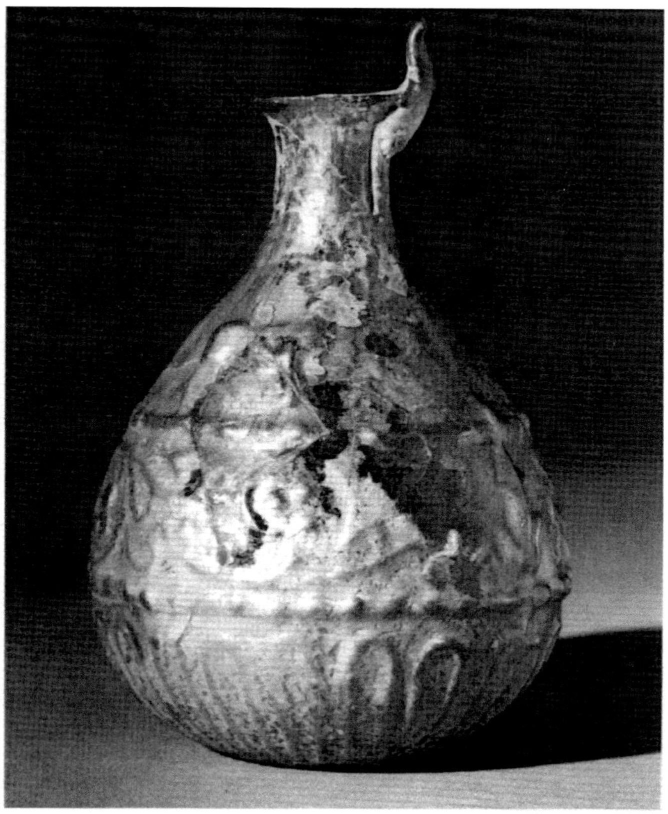

*Fig. 61. Workshop of the Floating Handles, Hunt-and-Scrolls Type. H 10.2 cm, greenish blue, Corning, New York, The Corning Museum of Glass, no. 74.1.22. Photo: The Corning Museum of Glass.*

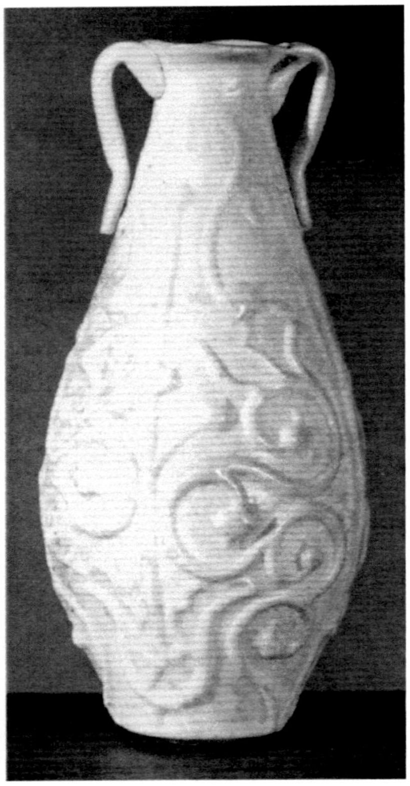

*Fig. 62. Workshop of the Floating Handles, Vertical Scrolls Type. H 9.8 cm, opaque white, Boston, Massachusetts, Museum of Fine Arts, no. 50.2280, Bequest of Charles B. Hoyt. Photo: Museum of Fine Arts.*

apparently cooled the glass at the lower tip of the handle too much for it ever to have been permanently attached to the shoulder or wall of the vessel. In effect, the lower end of the handle "floats" (Fig. 58). The distance between the lower end of the handle and the wall of the vessel is frequently wide enough to permit the handle to be slipped over a string or ribbon, but it seems highly unlikely that this was the intended purpose of the floating handles. They probably imitate objects in metal with similarly unattached handles (e.g., amphoriskoi and strigils).[145] The floating handles are so idiosyncratic and the tool used to make them so distinctive that these glass vessels must have been made in a single workshop.

All decorated bottles with floating handles were blown in two-part MCT VIII molds, a mold type that may perhaps have been invented in the Workshop of the Floating Handles. The glass blowers were also very adventurous in experimenting with the circumference of mold-blown vessels. The earliest vessels were blown in molds that fitted into a circular circumference (MCT VIII A; see Fig. 15 and Cat. No. 59), but later vessels were made both in molds with a lenticular circumference (MCT VIII F; see Cat. No. 62) and with asymmetrical halves (MCT VIII B; see Cat. No. 138).

## Description

Floating handles are known from ten Types of mold-blown vessels:[146]

1) Acorn Type, small bottles shaped like an acorn (Cat. No. 113);[147]

2) Shell Type, small bottles shaped like a shell (Fig. 59);[148]

3) Miniature Amphora Type, small bottles shaped like a miniature wickerwork-cased amphora with a laurel wreath around its middle (Cat. No. 59; see also Fig. 58)[149] (perhaps also a variant with the wreath in reverse[150] and another without a wreath);[151]

4) Ball Type, small globular bottles with a pattern of pentagons on the surface imitating a leather sports ball (Fig. 60);[152]

5) Hunt-and-Scrolls Type, small bulbous bottles decorated with a band of inhabited acanthus scrolls around the middle and a hunting scene on the shoulder (Fig. 61);[153]

145 For example, a silver amphoriskos, MMA, no. 03.24.5, from Bolsena, Italy, dated to the second half of the third century B.C., Oliver 1977, 56, no. 23, fig. 23.

146 The following does not attempt to be a complete catalogue of vessels with floating handles; its sole purpose is to draw attention to the existence of the group.

147 For example:

a) formerly R. W. Smith Collection, no. 860, acquired in Lebanon, opaque white, with remains of two floating handles, Smith 1957, 64-65, no. 86;
b) formerly Kofler-Truniger Collection, said to be from Syria, translucent green, with two floating handles (one broken off near the top), *Kofler-Truniger* 1985, 62-63, no. 95;
c) Tel Aviv, no. MH 9268, with one floating handle preserved;
d) Cat. No. 113, opaque white, with two floating handles (one broken off near the top).

148 For example:

a) Tel Aviv, no. MHG 9358, "probably bought in Italy," opaque purplish, H 6 cm (here Fig. 59);
b) Ravenna, Nat. Mus., no. 10026, from tomb at San Severo, transparent yellow, with two floating handles, Montanari 1967, 29, no. 38, ill. p. 24.

149 Numerous examples exist: see at Cat. No. 59.

150 Formerly Kofler-Truniger Collection, opaque white, with two floating handles, *Kofler-Truniger* 1985, 69, no. 112. For the same Type: Aquileia, Arch. Mus., no. 2893, perhaps found at Aquileia, handles and area of application not preserved, Calvi 1968, 105, no. 246, pl. 16, no. 4. It is not clear whether the wreath in reverse, appearing in the above-mentioned illustrations, is in the same direction on both sides of the bottles or in opposite directions on either side, as in Cat. No. 59.

151 Oppenländer Collection, no. 2471, pale bluish green, with two floating handles, von Saldern et al. 1974, 150, 155 ill., no. 432. Aquileia, Arch. Mus., no. R.C. 1032, pale bluish green, perhaps found at Aquileia, Calvi 1968, 105, no. 248, pl. 16, no. 3, belongs to the same Type but cannot be assigned to the Workshop of the Floating Handles because the remains of the handle appear to have the excess glass at the top folded back. Also the same Type: V&A, no. 1033-1868 (cited by Matheson 1980, 51, as a parallel for her no. 128).

152 For example:

a) Athens, Nat. Mus., no. 2701, from Thebes, Weinberg 1992, no. 97;
b) CMG, no. 68.1.45, transparent greenish colorless, with one remaining floating handle, H 8.2 cm, D (rim) 1.7 cm, (body) 6.4 cm (here Fig. 60).

The pattern of pentagons suggests an air-inflated leather sports ball. A miniature clay sports ball showing identical polygonal sections and a laced opening through which a bladder would have been inflated, was excavated in Samothrace, Dusenbery 1959, 168, fig. 8.

153 For example:

a) CMG, no. 74.1.22, formerly R. W. Smith Collection, no. 516, greenish blue, with remains of one floating handle, Smith 1957, 133, no. 252; *JGS* 18 (1976) 240, no. 3, ill. (here Fig. 61);
b) Switzerland, private collection, Luzern 1981, 79, no. 264, transparent blue green, with one floating handle;
c) Venice, Museo Vetrario, formerly Zara Museum, from the Dalmatian necropolis at Zadar, Croatia, translucent purplish blue, apparently without a handle, Calvi 1965a, 14, fig. 10;

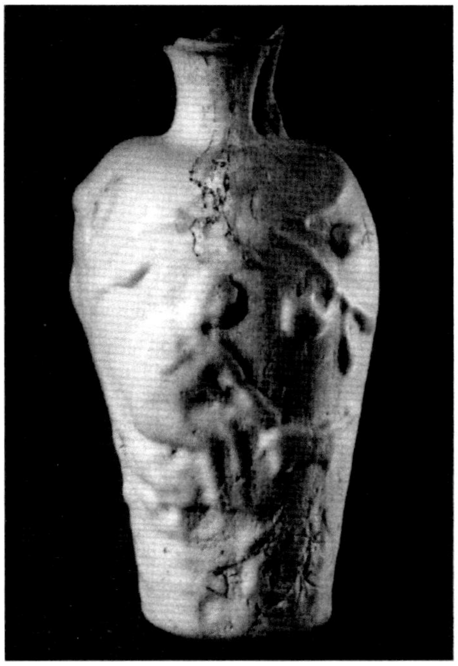

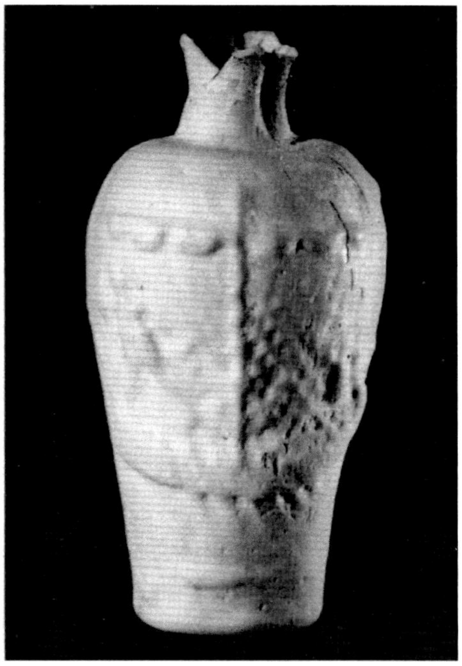

*Fig. 63. Workshop of the Floating Handles, Ajax Type. H 8.3 cm, opaque white, Corning, New York, Corning Museum of Glass, no. 62.1.16. Photo: The Corning Museum of Glass.*

6) Vertical Scrolls Type, tall bag-shaped bottles with a vertical design of scrolls and tendrils (Fig. 62);[154]

7) Ajax Type, tall bulbous bottles with scenes from the myth of Ajax (Fig. 63);[155]

8) Birds-and-Flowers Type, tall six-sided bottles with birds on the shoulder and floral patterns on the sides (Fig. 64);[156]

9) Medusa Type, small lenticular bottles with a Medusa mask on either side (Fig. 65; compare Cat. No. 62);[157]

10) Female Head-shaped Jugs Type (possibly Livia as Hera), head flasks in the shape of a female head with a *nodus* coiffure (Cat. No. 138 and Figs. 102 and 103).[158]

The shapes made in the Workshop of the Floating Handles thus constitute a diverse group, including molds with designs for shallow relief (Vertical Scrolls, Hunt-and-Scrolls, Ajax, Birds-and-Flowers), vessels shaped like natural objects (Acorn, Shell) or miniature everyday objects (Amphora, Ball), and head-shaped vessels, both the intermediate, lenticular type with a face in low relief on either side (Medusa) and the fully developed head flask reproducing a head in the round. The diversity is interesting in itself because it provides a glimpse into the variety of shapes made by a single workshop. In addition to mold-blown bottles, this workshop also made free-blown ones (Fig. 66).[159] All bottles were probably used to hold scented oils. This suggests that the workshop concentrated on the production of bottles with the same

d) MMA, no. 17.194.249, formerly Gréau Collection no. 1133, said to be from "Syria," H 9.0 cm, greenish, one handle preserved, Froehner 1903, I, 157, no. 1133; III, pl. 205, no. 4.

154 For example:

a) MFA, no. 50.2280, opaque white glass, with two floating handles, von Saldern 1968, no. 35 (here Fig. 62);

b) formerly Bachstitz Collection, no remains of handles, Eisen 1927, I, pl. 81.

155 This Type, previously thought to represent scenes from the Argonaut story (Matheson 1980, 49-50, no. 127), has recently been recognized as showing scenes of Ajax. An inscription identifies the man in the boat as Ajax: Özet 1993. Examples of this Type include:

a) CMG, no. 62.1.16, opaque white, *JGS* 5 (1963) 140-141, no. 3 (here Fig. 63);

b) Damascus, Nat. Mus., no. 12.813, from Tell Nabi Mind, opaque white, with two floating handles, Abdul Hak 1951, 113, no. 16, pl. 49, 1 (third from left), Al-'Ush et al. 1969, 104, fig. 39, cited as no. C 5573;

c) formerly Maxwell Collection, opaque white, with two floating handles, *Maxwell* 1979, 60-61, no. 90;

d) Bodrum, Turkey, Mus. of Underwater Archaeology, no. 2.17.91, from Stratonikeia, Turkey, transparent purple, Özet 1993, 142-145, fig. 13.

For the same Type, but without handles:

e) Yale, no. 1955.6.54, formerly Moore Collection, opaque white, Matheson 1980, 49-50, no. 127;

f) Houston, Museum of Fine Arts, no. 70-80, formerly R. W. Smith Collection, opaque white, upper part restored, *Bulletin of The Houston Museum of Fine Arts*, n.s. 2 (1972) 175, fig. 5.

A vessel of dark blue or purple glass with the remains of an opaque white handle, from the Dalmatian necropolis at Zadar, Croatia, now in Venice, Museo Vetrario, belongs to the same Type but cannot be assigned to the Workshop of the Floating Handles because the handle appears to have excess glass folded back on the top, Mariacher 1970, 16, fig. 4, 18.

156 BM, no. GR 1913.5-22.17, formerly Durighello Collection, transparent greenish, with two floating handles, reported to have been found at Beirut, *Glass of the Caesars* 1987, 161, no. 83.

157 See at Cat. No. 62.

158 Stern 1992b, 394-399, pl. 81, for examples, see at Cat. No. 138.

159 For example:

a) small amphora, Berlin, Staatliche Museum, no. M 107, blue, H 13.3 cm, from tomb D 47 at Miletos, Turkey, Rohde 1980, 158, no. 23, fig. 23 (here Fig. 66);

b) MMA, no. 17.194.157, formerly Gréau Collection no. 1182, find spot unknown, transparent blue, H 12.0 cm, Froehner 1903, I, 163, no. 1182; III, pl. 216, no. 5.

Remains of a first-century glass factory, active 40-70, producing free-blown and mold-blown wares have been excavated at Avenches (ancient Aventicum) in western Switzerland, Morel et al. 1992, 9-12. The production included Isings 1957, Forms 6, 8-11, 15, and 27, as well as mold-blown bottles shaped like a pine cone.

function rather than on production in one glass-blowing technique. This may suggest that the source producing the scented oils was nearby.

The vessels were made in a limited range of colors, with handles usually the same color as the body. Of the more than fifty vessels assigned to the workshop, twenty-one are natural bluish green, eleven opaque white, one opaque light blue, three translucent amber brown to dark yellow, one translucent purple, and one translucent dark green.[160] The colors divide unevenly among the Types. The majority of the vessels in the Miniature Amphora, Scrolls-and-Hunt, and Female Head-shaped Jug Types are in various hues of natural bluish green, but a few are made of translucent to transparent colored glass, the most notable being one in the Miniature Amphora Type and two each in the Medusa and Female Head-shaped Jug Types. Most, if not all, of the bottles of the Ajax type are opaque white.[161] Opaque white was also used for most of the vessels of the Acorn Type and a few of those of the Miniature Amphora Type; the sole representative of the Vertical Scrolls Type is also opaque white.

Applying the handles near the rim and pulling them down

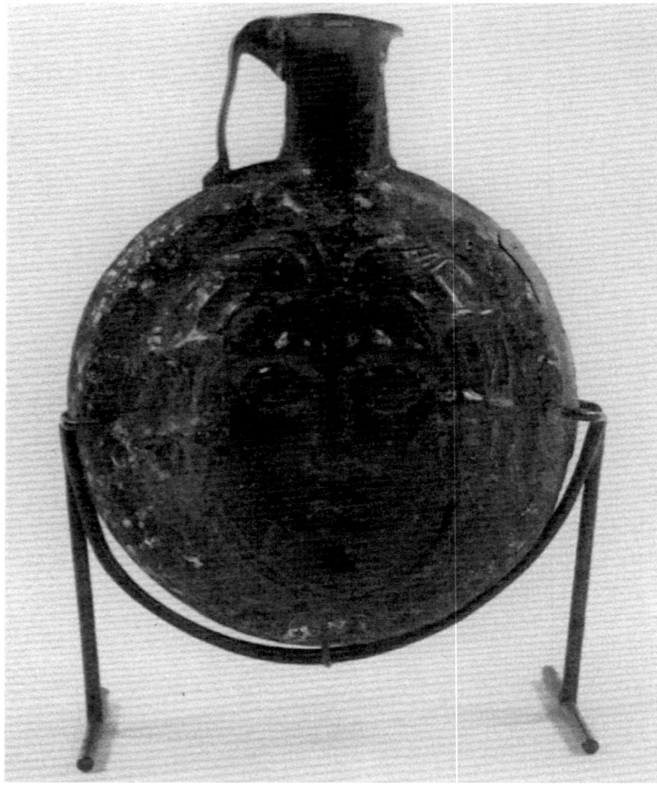

*Fig. 65. Workshop of the Floating Handles, Medusa Type. H 8.6 cm, translucent purple, Flushing, New York, The Godwin-Ternbach Museum of the Queens College of the City University of New York, no. 60-54. Photo: The Godwin-Ternbach Museum.*

is an idiosyncrasy also characteristic of Ennion's cups, but none of these has floating handles. Three bottles with floating handles are decorated with an acanthus scroll around the middle and a hunting scene on the shoulder (here called the Hunt-and-Scroll Type). The style of the acanthus scroll has been associated with the work of Aristeas.[162] The six-sided Birds-and-Flowers Type recalls Ennion's hexagonal bottle in the Metropolitan Museum of Art (see above Fig. 45).[163] These stylistic similarities and technical details suggest that all three workshops, those of Ennion, Aristeas, and the Workshop of the Floating Handles, were closely related; but the exact relationships among these three workshops are not clear. Ennion and Aristeas are known only for their tablewares, the Workshop of the Floating Handles only for small scent bottles.

### Date of the Workshop

One mold-blown bottle with floating handles has been found in a tomb in Italy datable to the second half of the first century,[164] and a free-blown example has been found in a first-

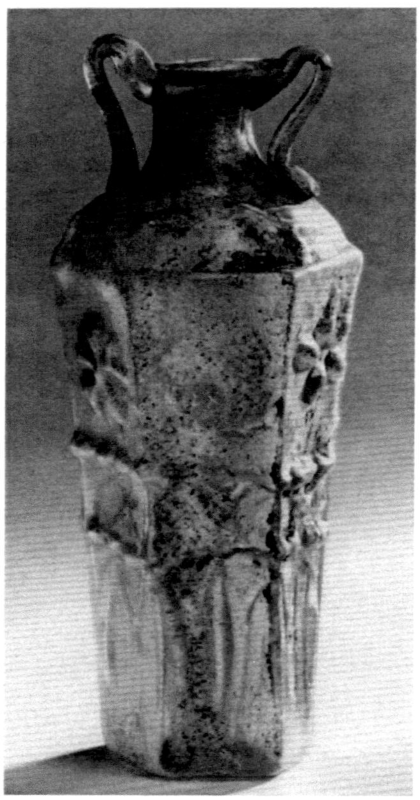

*Fig. 64. Workshop of the Floating Handles, Birds-and-Flowers Type. H 9.9 cm, transparent greenish, London, British Museum, no. GR 1913.5-22.17. Photo: Copyright British Museum.*

160 The only exceptions are two translucent, dark purplish blue vessels from the Dalmatian necropolis at Zadar, Croatia, now in Venice, Museo Vetrario: Calvi 1965a, 14, fig. 10, a bottle belonging to the Hunt-and-Scrolls Type, and Mariacher 1970, 16, fig. 4, 18, fig. 4, a bottle belonging to the Ajax Type. Neither bottle can be assigned with certainty to the Workshop of the Floating Handles, the first because it has no handles, and the second because it has normal handles.

161 The only exception appears to be the bottle of the Ajax Type mentioned above in note 160, which combines a dark purple or blue body with opaque white normal handles.

162 On the association of the Hunt-and-Scroll Type with Aristeas: Calvi 1965a, 14. On Aristeas: see above, *Glass-Blowers' Signatures: Aristeas.*

163 MMA, no. 81.10.224, Lehrer 1979, 9, pl. III, nos. 1-4.

164 Montanari 1967, 29, no. 38, Shell Type, from a tomb at San Severo that contained an amber ring dated to the Flavian period (69-98).

century tomb at Miletos, on the coast of Asia Minor.[165] From the themes chosen for decorating most mold-blown Types, it is possible to postulate a date in the first half of the first century. The head flasks of the Female Head-shaped Jugs Type, with their distinctive *nodus*-like hairstyle, were probably made before 29, possibly as early as 14 (see Cat. No. 138). Two Types are decorated with lush, inhabited acanthus scrolls, a motif that had become popular during the reign of Augustus.[166]

The colors used for the various Types may also provide a basis for relative dating. Opaque white glass appears to be characteristic for the earliest types of mold-blown bottles made in the Syro-Palestinian area (see Cat. No. 15). To judge from their opaque white color and shallow relief, the Ajax, Acorn, and Vertical Scrolls Types belong to the earliest production of the workshop, while the Medusa and Female Head-shaped Jug Types, which were made predominantly in natural bluish green glass — perhaps in imitation of silver — represent the latest years of activity. If the Female Head-shaped Jugs (Cat. No. 138) were made as early as the year 14 or soon after, this would posit a starting date for the activity of the Workshop of the Floating Handles somewhat earlier.

#### Location of the Workshop

Although only two bottles with floating handles have been found in controlled excavations, various alleged find spots point in a general way to an eastern Mediterranean origin. One of the free-blown bottles (see Fig. 66) was excavated at Miletos on the coast of Asia Minor.[167] Of three bottles belonging to the Acorn type, one is said to be from Syria, the second was acquired in Lebanon, and the find spot of the third, in Toledo, is unknown but probably Near Eastern (Cat. No. 113). One Miniature Amphora was excavated at Panticapaeum; one bottle of the Ball Type was found at Thebes, Greece; one of the Ajax bottles was excavated at Tell Nabi Mind, southwest of Homs in Syria, and fragments of a second one were found at Stratonikeia, Turkey. A six-sided bottle with birds and flowers is reported to have been found at Beirut. Five of the ten Female Head-shaped Jugs can be traced to the Near East. Western find spots are rare. A few bottles have been found in Italy,[168] but all are atypical in that they seem to have had a normal handle, applied from shoulder to rim, with the excess glass at the top folded back on top of the handle. At this time it is impossible to say whether the Workshop of the Floating Handles occasionally made bottles with normal handles. The possibility that the bottles from Italy were local imitations cannot be excluded.

Current opinion places the workshops of Ennion and Aristeas in the Sidon area.[169] This is also the most plausible location for the Workshop of the Floating Handles, principally because Sidon was the preeminent center of first-century glass blowing and would have been a magnet for enterprising craftsmen. That the glass blower or glass blowers of the Workshop of the Floating Handles were enterprising and innovative is shown by the diversity of their bottle shapes and subjects and the daring construction of their molds. Although the workshop began blowing bottles decorated with shallow relief in the style of Aristeas and Ennion, it went on to develop its own style. The Female Head-shaped Jugs also seem to be the earliest known glass head-shaped vessels. Glass head-shaped vessels are the most long-lived shape produced by mold blowing. The workshop may probably be credited with the invention of this important shape.

### *Date-shaped Bottles*
(Cat. Nos. 84-107)

Small bottles shaped like dates are among the most common first-century mold-blown glass vessels.[170] Their surfaces are

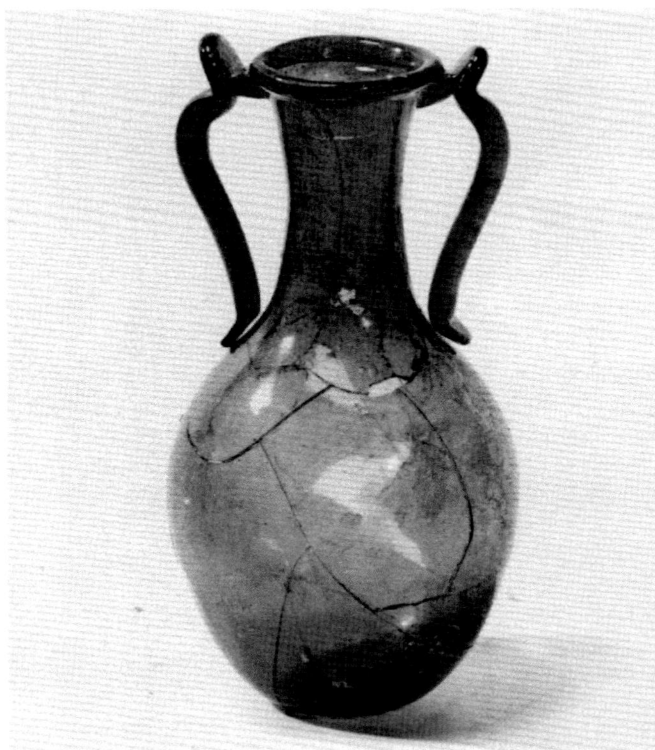

*Fig. 66. Workshop of the Floating Handles, free-blown glass amphora. From Miletos, grave D47, H 13.3 cm, translucent blue, Berlin, Staatliche Museen zu Berlin, Antikensammlung, no. M 107. Photo: Staatliche Museen, Antikensammlung.*

165 Berlin, Staatliche Museen, no. M 107, from Miletos tomb D 47, a small amphora of blue glass, H 13.3 cm, Rohde 1980, 158, no. 23, fig. 23.

166 The best-known example of the motif occurs on the Ara Pacis Augustae, erected at Rome between 13 and 9 B.C.: Simon 1967; P. Zanker, *Die Bilder des Augustus* (Munich 1987) 184-188 (English edition, *The Power of Images in the Age of Augustus* [*Jerome Lectures* 16, Ann Arbor, Mich. 1988] 179-183).

167 Rohde 1980, 158, no. 23, fig. 23.

168 The Shell Type, from Ravenna (Montanari 1967, 29, no. 38); an Ajax bottle, Venice, Museo Vetrario, from the Dalmatian necropolis at Zadar, Croatia (Mariacher 1970, 16, fig. 4, 18); and two bottles of the Miniature Amphora Type in the Museum at Aquileia (Calvi 1968, 105, nos. 246, 248, pl. 16) from unrecorded find spots, so it is not certain that they were discovered in the Aquileia area.

169 See above, *Glass-Blowers' Signatures: Ennion and Aristeas.*

170 Isings 1957, Form 78 d.

covered with a pattern of wavy ridges that imitates the wrinkled skin of a ripe date. On some vessels the wrinkles are rather stylized (Cat. Nos. 102 and 103), but most date-shaped bottles are naturalistic, not only in the wrinkling of the skin, but also in size and color, which is usually a shade of brown, although a few are known in blue (Cat. No. 89), green (Cat. No. 103), purple (Cat. No. 97), and "black" glass.[171] Different forms of date-shaped bottles suggest that they represent at least two species of dates, a plump type, comparable to modern Medjool dates (Fig. 67, below; see Cat. Nos. 84–92), the other lean and more elongated, like modern Empress dates (Fig. 67, above; see Cat. Nos. 104–107). However, observations on the technique of producing these glass vessels indicate that the vessels should not be too precisely related to actual species of dates.

### Technique

Most date-shaped bottles were blown in two-part molds with one continuous mold seam around the vessel, MCT VIII (Cat. Nos. 84–103 and 108), but a few were made in three-part molds, MCT IV B (Cat. Nos. 104–107). Some molds also formed the lower part of the neck (e.g. Cat. Nos. 84, 89, 93, and 97), but the necks and rims of most vessels were formed offhand. Occasionally the mold seam is ingeniously hidden in the wrinkles of the skin (Cat. Nos. 91 and 98), but this appears to be exceptional. When the parts of the mold were not well aligned, the glass filled the space between to form a distinct, protruding mold seam (Cat. Nos. 85, 86, and 93).

In the catalogue descriptions of the date-shaped bottles and the figure captions, the parts of the mold and the corresponding sides of the vessel are designated with the letters A and B if a two-part MCT VIII mold was used, and with A, B, and C for a three-part MCT IV mold. The three-part molds usually consist of one large section that formed half of the bottle (A), while the other half was divided into two approximately equal parts (B and C). The mold seam dividing the bottle into two equal halves is here called a *continuous* mold seam, the one between parts B and C a *central* mold seam. In measurements of diameter, the first measurement refers to the maximum distance measured from mold seam to mold seam (continuous mold seam), the second one to the maximum distance across the vessel measured between continuous mold seams.

The naturalistic appearance of many date-shaped bottles has led to the suggestion that the molds were cast directly from actual fruit.[172] This is plausible. The fruit could have been burned out of a plaster or clay mold. The mold must have been made from a durable material, because it had to be used more than once (see Cat. Nos. 84–88; 89 and 90; 93–95; 104–106).[173] Cat. Nos. 84–88 are particularly interesting because they demonstrate that bottles blown in the same mold can have different cross sections (compare Cat. Nos. 84 and 85) and even different profiles (Cat. No. 88, Remarks) although the pattern of wrinkles is identical. This was probably due to the process of manufacture. After the paraison was extracted from the mold, it was probably held with a clasping tool in order to finish the neck and rim. It was at this point in the production process that variations in cross section and profile occurred, but whether this happened by accident or intent is not clear.

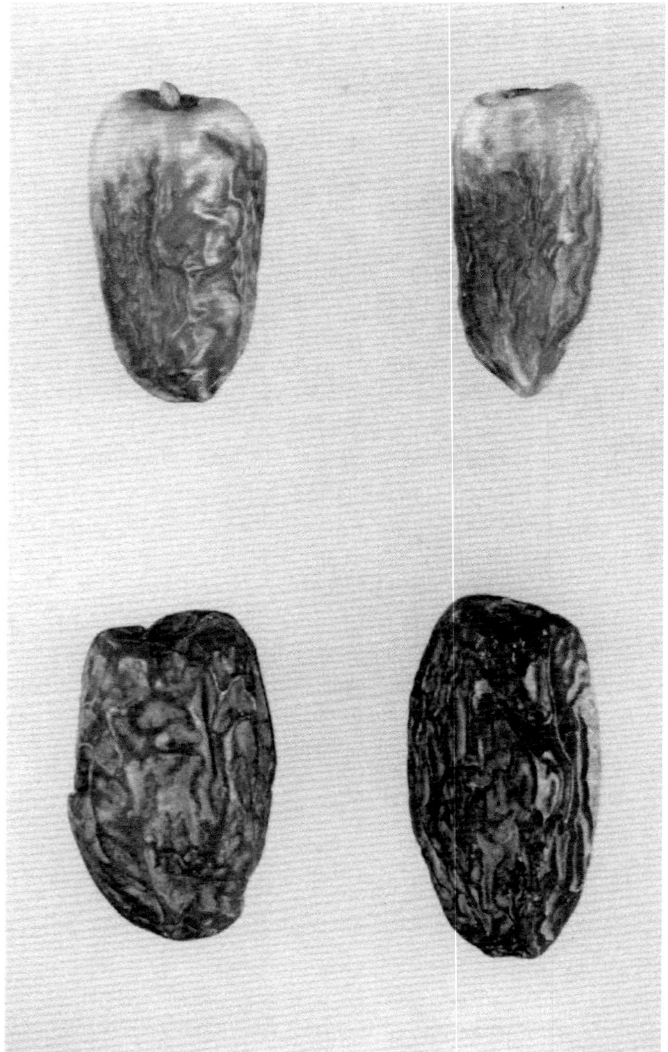

*Fig. 67. Varieties of modern dates.* Above: *Empress;* below: *Medjool. Photo: Photography Inc.*

### Date

These bottles were made from shortly before the middle of the first century until perhaps as late as the early second

[171] For example, Oppenländer Collection, nos. 23302, 2405, and 2337, von Saldern et al. 1974a, nos. 476–478; various examples in Tel Aviv, Haaretz Museum, case 18 (on exhibit in 1975). None of the date-shaped bottles in the Toledo Museum has the enigmatic layer of white glass noted on the interior of the neck of a few vessels in other collections: Carington Smith 1982, 280, nos. 65–67, fig. 6, pl. 39 f, g, and h; ROM, no. 918.5.6, Hayes 1975, 49, no. 87, pl. 7; Carnegie, no. 21342/8, Oliver 1980, 64, no. 60.

[172] *Holy Land* 1986, 256, at no. 137.

[173] Three date-shaped bottles from one mold were found in a tomb at Knossos: Carington Smith 1982, 280, nos. 65–67, fig. 6, pl. 39 f, g, h.

centuries. The earliest date-shaped bottle from a datable context was excavated at Trier, in western Germany, in a tomb dated to the Claudian period (41-54).[174] Examples from Pompeii[175] and Herculaneum[176] predate 79 when the eruption of Vesuvius destroyed these cities. Date-shaped bottles have been excavated in tombs used during the second half of the first century in Israel,[177] Crete,[178] and Croatia.[179] Examples from excavations in the Crimea[180] and in Syria[181] are dated to the first century and about 100 respectively. Date-shaped bottles are relatively sturdy vessels that seem to have been refilled and reused long after their original contents were consumed. Finds from tombs in Spain, sealed around the middle of the second century,[182] and at Aquileia on the northern coast of the Adriatic Sea, dated second half of the second century,[183] were probably heirlooms by the time they were interred. A few date-shaped bottles survived (or perhaps were rediscovered in antiquity) until they were interred several hundred years later as funerary gifts in fourth- and fifth-century graves excavated in Germany[184] and Israel.[185]

### Distribution of Find Spots and Origin

Two-thirds of the known date-shaped bottles have been found in the Near East. The others have surfaced all over the eastern Mediterranean and the West.[186] There is no real concentration of finds to indicate where the glasshouse(s) producing date-shaped bottles were located, but a relative dearth of finds from Israel may indicate that these bottles were not made in that region.[187] Their wide distribution suggests production in glass workshop(s) situated on the Syro-Palestinian coast, perhaps in one of the famous glass centers of Phoenicia. During the first century, Phoenicia was famous not only for its mold-blown glass but also for its date palms. In Greek, the geographical name Phoenike (Phoenicia) and the name for a date palm or date (*phoinix*) were homonymous. Ancient authors mention various species of dates, but the best were said to be those called *caryotae* from Phoenicia. According to Diodorus Siculus, writing in the second half of the first century B.C., *caryotae* excelled not only in sweetness and size but also in juiciness (Diodorus Siculus, *Bibliotheke Historike* 2.53.5). It is tempting to assume that the fame of the region's dates inspired glass workers in Phoenicia to blow vessels shaped like dates.

### Function and Interpretation

No remains of contents have been published from date-shaped bottles, but the relatively small size and narrow mouth suggest that they were designed to hold scented oils or medicines. Most date-shaped bottles have been found in tombs,[188] but their use as funerary gifts could have been secondary.

174 Trier, Rheinisches Landesmuseum, no. 05, 201e, from Trier, St. Matthias tomb 55, Goethert-Polaschek 1977, 96, no. 453, pl. 5, 55e.

175 *NSc* (1910), 382; compare also Naples, Mus. Nat. Arch., nos. 110223 and 13564, both from Pompeii, *Collezioni Napoli* 1989, 226, no. 47, 227, ill.

176 Ercolano, Antiquarium, no. E 2558, from Herculaneum, Decumano Massimo, Scatozza Höricht 1986, 52, Form 35, no. 105, pl. XXXIII.

177 Jerusalem, Israel Museum, nos. IDAM 78-131, 78-132, from Mishmar Haemek, *Holy Land* 1986, 255-256, no. 137.

178 Carington Smith 1982, 280, nos. 65-67, fig. 6, pl. 39 f, g, h.

179 Mikulčić 1974, 194, 201, 206, pl. 12, nos. 376 and 377, from Scupi, East Necropolis, grave 137 (dated "Domitian to Hadrian [81-138]" p. 200); Željko Rakníć, "Einige neue Funde der römerzeitlichen Brandgräber aus Zadar," *Diadora* 4 (1968) 214, pl. 2, from cremation burials/deposits at Zadar, Croatia, dated late first to early second century.

180 St. Petersburg, Hermitage, nos. P. 1866.32; P. 1899.71; P. 1909.6, three date-shaped bottles from Panticapaeum, Kerch, Kunina 1973, 128-130, fig. 21, nos. 1-3. One of these is illustrated in Kunina and Sorokina 1972, 167, fig. 10, 47, from a tomb that also contained an early first-century unguentarium of western type (perhaps the same one as illustrated in Sorokina 1967, 70, fig. 2, 14, from the Black Sea coast).

181 Clairmont 1963, 40, no. 150, pl. 21, from Dura Europos.

182 P. Paris et al., *Les Fouilles de Belo (Cadiz), 1917-21* II: *La Nécropole* (Bordeaux 1926) 80, no. 14, from Belo.

183 Aquileia, Mus. Arch., no. 50305, from Aquileia, tomb of Antestia Marciana, Brusin 1937, 190, and nos. 12661, 14951; Calvi 1968, 106, cat. nos. 251-253, pl. 17, no. 3.

184 Renata Pirling, "Spätrömische und Frühmittelalterliche Gläser aus Krefeld-Gellup," *AnnAIHV* 3 (1964) 97, fig. 29.2; Renata Pirling, *Das römisch-fränkische Gräberfeld von Krefeld-Gellup (Germanische Denkmäler der Völkerwanderungszeit*, ser. B, vol. 2, Berlin 1966) 106, grave 533, her Type 210, pl. 47, 14, from an early fourth-century grave at Krefeld-Gellep.

185 From a fifth-century tomb at Samaria, Crowfoot 1957, 413, no. 23, fig. 95, 23.

186 For a list of date-shaped bottles from recorded find spots: Stern 1977a, 44-46. To these add the datable finds discussed above from Mishmar Haemek (note 177), Crete (note 178), Herculaneum (note 176), Scupi (note 179), Zadar (note 179), and Trier (note 174), as well as finds that cannot be dated by their context from *Babylon*: BM, no. 91517, Barag 1985a, 93-94, no. 119; *Asia Minor*: Berlin, Antikensammlung, no. M 24, from Miletos, Rohde 1980, 161-162, no. 32, fig. 32; Adana, Bölge Museum, one example, unpublished, probably found in Cilicia, Stern 1984, 134; Stern 1989a, 123, n. 13; *Cyprus*: Cambridge, Fitzwilliam Mus., nos. GR.28.1876, GR.22.1877, both from Tremithus, *Fitzwilliam* 1978, 50, no. 97, ill.; the *Dalmatian coast*: from Nin Zadar, Damevski 1974, pl. IX 5; *Italy*:

a) Brescia, Museo dell'età romana, no. MR 1112, from the vicinity of Brescia, Calvi 1975, 214, fig. 16;
b) from Voghenza, northern Italy, tomb 4, F. Berti, "La necropoli romana di Voghenza," *Voghenza* (Ferrara 1984) 77-201, pl. XVII, fig. 124 (cited in Taborelli 1986-88, 142, n. 11);
c) from Aosta, via per le Gallie, unpublished (cited in Taborelli 1986-88, n. 12);;
d) from Melfi-Leonessa, tomb 21, C. Klein Andreau, "La romanizzazione," in *Civiltà Antiche del Medio Ofanto* (Napoli 1976) 30-36, fig. 14 (cited in Taborelli 1986-88, 142, n. 10);

and *Switzerland*: Zurich, Landesmuseum, no. 1024, from Baden (ancient Aquae Helveticae), Fünfschilling 1985, 118, no. 130, 155, no. 64.

187 Because all twenty-five date-shaped bottles in the Toledo Collection were acquired as part of the Curtis Collection and because Curtis is known to have bought most of his glass from Azeez Khayat, a dealer who excavated and collected in the vicinity of Nazareth and Haifa (on Khayat, see Grose 1989, 20-21), it might appear that this shape has a Palestinian origin. However, since the distribution pattern does not support Palestinian workshops, it seems prudent to abstain from citing the Toledo Museum examples as proof of Palestinian origin. Very few date-shaped bottles are listed in Khayat's sales catalogues (Khayat 1913, nos. 257 [ill.] and 312; Khayat 1919, nos. 45 [ill.] and 366). Unless one assumes that Khayat sold most of his date-shaped bottles privately to Curtis and other favored buyers (on the "Curtis Cabinet" in Khayat's inner office: Grose 1989, 20), it may be as valid to conclude that these bottles were not frequent finds in Galilee and southern Phoenicia. Date-shaped bottles are so attractive that Khayat and other dealers might have acquired them from unrecorded sources farther north or farther inland. Curtis is known to have also acquired glass through Lepke auctions in Berlin and elsewhere.

188 When found in tombs where the context has been recorded, date-shaped bottles were usually grave gifts for women and girls: Carington Smith 1982, 280, nos. 65-68, three of which were found with a female burial at Knossos, Crete; Brusin 1937, 190, from the tomb of a twelve-year-old girl at Aquileia. Date flasks also often occur in tombs in pairs.

In antiquity the fruit of the date-palm provided not only food and wine but also an important sweetening ingredient used in many recipes for aromatic wines, sauces for poultry and meat, and desserts (Apicius II, VI, and XIII; Pliny, *Natural History* 14.19.102, 107-108). On the use of dates for medicinal purposes, see Pliny, *Natural History* 23.51-52, and Dioscurides, *Materia Medica* I, 126. Various ancient authors mention unguents and scented oils made from dates.[189] The Roman poet Ovid, writing in the early first century, says that dates were common New Year's gifts in Rome "for the sake of the omen... that the event may answer to the flavor, and that the whole course of the year may be sweet, like its beginning" (Ovid, *Fasti*, I 185-188).[190] When an ambassador of Herod the Great, King of Judea, visited the emperor Augustus in Rome in 12 B.C., he brought fresh dates from Palestine as a gift.[191] The Roman custom of giving dates as a symbol of sweetness may explain the wide popularity of date-shaped bottles in the western as well as the eastern part of the Roman empire. A delicate glass bottle in the shape of a date filled with sweet date oil would have been a most appropriate gift on the occasion of a New Year, as well as on other occasions.

## MULTIREGIONAL STYLES

(Cat. Nos. 2-4, 8-10, 13 and 14, and 79-82)

Mold-blown vessels belonging to one and the same Type have frequently been excavated in widely distant places throughout the Mediterranean and northwestern Europe with no recognizable concentrations of finds in either the East or the West. Some styles are concentrated in the East, with the exception of Egypt,[192] or in the West. These Types constitute "multiregional" styles because they were apparently used and appreciated in many regions of the Roman empire and beyond.

A multiregional distribution pattern seems characteristic of the finest Sidonian-style mold-blown tablewares, such as those made by Ennion, which were probably made in the Syro-Palestinian area but are found throughout the Roman world. Unsigned cylindrical cups with a victory inscription, **ΛΑΒΕ ΤΗΝ ΝΙΚΗΝ**, *labe ten niken*, "seize the victory" (Cat. Nos. 2-4), have a wide distribution in the Mediterranean, but they too were probably made in the eastern part of the Roman empire. There are no indications that they were made in more than one workshop. Truncated conical beakers with knot-shaped bosses (Cat. Nos. 8-10) are found throughout the empire, but some variants in decoration and MCT appear to be concentrated in different regions, indicating that they were probably made at various locations in both the East and the West. Finely ribbed bowls with a pronounced shoulder also have a wide distribution (Cat. Nos. 13 and 14); future research may show that variants in shape and design predominate in different regions. Pyxides with floral decoration (Cat. Nos. 79-82) have no definite concentration in the East or the West.

It is largely a matter of conjecture why some styles and shapes became popular in many parts of the Roman world and others did not. However, one feature stands out: the glass vessels whose styles became multiregional are almost all ostentatious tablewares. This is true of not only mold-blown but also cast wares.[193] The only exception is the small group of pyxides with floral decoration, whose exact function has not been determined (Cat. Nos. 79-82). Fine glass tablewares might have been valued primarily as status symbols. Small mold-blown bottles for scented oils (like Cat. Nos. 15-70) were intended for personal use; thus they were less effective as objects for conspicuous consumption and remained tied to regional and local styles.[194] When luxury bottles for scented oils do show widespread distribution in East and West, it was probably owing to some symbolic meaning of the shape, such as date-shaped bottles (Cat. Nos. 84-107). The excavation of numerous fine glass vessels at Roman legionary camps[195] suggests that the chief attraction of mold-blown tablewares may have been to give the impression of culture and erudition at dinner parties. Spectacular mold-blown Syro-Palestinian tablewares would have been prime models for copying and imitating in Italy and the West, because during the early Roman empire, eastern Mediterranean styles were especially fashionable among certain classes of the nouveaux riches, as epitomized by the guests and host at Trimalchio's dinner.[196] Thus the creation of styles fashionable in East and West alike may have been primarily due to the desire to display established status symbols.

### *The Organization of the Glass Industry*

The ways in which mold-blown styles and designs were transferred from one region to another must have been closely linked to the organization of the industry which produced and sold the vessels. Historians have not yet succeeded in fitting the archaeological material into the known framework of Roman economic and social history.[197] A full discussion should include not only the facts about mold-blown vessels, but also what is known about other branches of the industry such as casting and free blowing, as well as legal, literary, and

[189] Theophrastos, *de Odoribus* 28; Pliny, *Natural History* 14.19.107-108.

[190] On dates as New Year's gifts: G. Heres, *FuB* 14 (1972) 185, 189. Compare the Jewish tradition of eating apples with honey at Rosh Hashana, Jewish New Year.

[191] Nikolaos of Damascus, fr. 136.F, in Jacoby, *Fragmente griechischer Historiker* (Berlin 1923) 90.

[192] On the slow adoption of glass-blowing techniques by Egyptian glass workers: Stern 1993, 26.

[193] On cast wares: Grose 1989, 242, who calls them "International" styles; Grose 1991, 12-18.

[194] Scents were usually imported in bulk, not in small luxury containers. Small containers for retail sale were probably locally made unguentaria.

[195] Van Lith and Randsborg 1985; Sophia M. E. van Lith, "First Century Cantharoi with a Stemmed Foot; Their Distribution and Social Context," in Newby and Painter 1991, 99-110.

[196] Petronius, *Cena Trimalchionis* (*Satyricon* 27-79), written before Petronius fell out of favor with Nero in 66 and committed suicide (as described by Tacitus, *Annales* 16.18).

[197] F. K. Kiechle, "Die Struktur der gewerblichen Glaserzeugung in der frühen Kaiserzeit," *AnnAIHV* 6 (1973) 53-64, does not take into account basic facts about surviving glass vessels and their find spots.

epigraphic sources pertaining to Roman commerce. Unfortunately very little information has been assembled. Any attempt at quantification or pattern construction remains tentative.

Research on other industries that used molds for reproduction, such as Roman terra-cotta lamps, has been confined largely to Italy and the western part of the Roman empire where conditions may have differed considerably from those in the Syro-Palestinian area during the first century. In the West, branch workshops and their managers, *institores*, were important factors in the distribution of signed terra-cotta lamps and the molds for making them. The find spots of surviving molds are evidence for the production of large numbers of terra-cotta lamps far away from their original center of production, but where and how the molds were made has not been established.[198]

In the case of Ennion (see above: *Glass Blowers' Signatures: Ennion*), it is not known what role, if any, subsidiary branches might have played in the Syro-Palestinian glass industry or in relations between workshops in the Syro-Palestinian area and Italy. The presence in Italy of Sidonian glass blowers such as Artas, Aristoon, and Eirenaios, who made the same type of drinking cups and stamped their names on the handles,[199] raises the question whether these craftsmen might have been organized in some type of cooperation or guild. The same question might be asked about Neikais, Meges, and Iason in the eastern Mediterranean, whose mold-blown glass cups imitate Italian ceramic drinking cups known as Aco beakers.[200]

It has been suggested that the transmission of designs on mold-blown wares might have taken place by the exchange of molds between workshops[201]. An occasional exchange of molds is conceivable, but in view of the relatively short life span of glass-blowers' molds and the ease with which new clay molds could be made, trade and copying from existing glass vessels were probably more important factors in the spread of mold-blown designs throughout the empire. Exchanging molds was common practice among late eighteenth- and early nineteenth-century American glasshouses.[202] But those molds were expensive metal molds that represented a substantial investment on the part of the glasshouse. The exchange of molds would have been much less attractive to Roman glass blowers whose molds were inexpensive clay items.

## *The Spread of Styles and Designs*

The new era of peace inaugurated by Augustus (27 B.C.-A.D. 14) insured safe travel for people, goods, and ideas over long distances. The unification and pacification of the Roman world made possible the miraculous speed with which not only styles and designs but also new inventions and techniques became known throughout the empire.

Three related factors contributed to the rise of multiregional styles of mold-blown glass. The first factor was expansion and increase in trade in glassware. The earliest wares with mold-blown designs were made on the Syro-Palestinian coast, from where they must have been exported to other regions. For example, the distribution pattern of Ennion cups in northern Italy follows the trade route along the Po River from the Adriatic Sea inland, along the same route followed by imported glazed wares.[203]

The second factor was the phenomenon of migrating glass workers.[204] Glass blowers from the Syro-Palestinian coast migrated not only to Rome but also to other parts of Italy and probably to the Roman provinces of northwest Europe. Glass workers could also travel with their trade. This was possible because an ancient glass worker needed only a few easily portable tools to set up shop. He could build a furnace anywhere wood was available for fuel. Raw glass would probably have had to be imported wherever he happened to work, because very few glass centers produced the colored glass from which most mold-blown vessels were made.

The third factor was wholly technical: the introduction of mold blowing in Rome, presumably by glass blowers from the Syro-Palestinian coast. Mold blowing appears to have been introduced in Rome during the reign of the Roman emperor Tiberius (14-37). Various Latin authors mention the invention in Tiberius's time of a glass that was malleable (*vitrum flexile*).[205] The most convincing explanation of what this flexible glass could have been is glass decorated with designs that resembled those commonly embossed in metal,[206] which is exactly what the mold-blown tablewares made by Ennion and Aristeas closely resemble. Pliny relates how Tiberius destroyed the entire workshop of the craftsman who "invented" this glass in order to avoid a devaluation of metals

198 W. V. Harris, "Roman Terracotta Lamps: The Organization of an Industry," *JRS* 70 (1980) 126-145.

199 Kisa 1908, III, 923-924; Fremersdorf 1938, 116-121; the names preserved are Artas, Aristoon, Eirenaios, Neikoon, Asinius Philippus, and Philippus. The name Annios Sidon occurs on a fragment found near Cadiz, Spain: Maria del Pilar Caldera de Castro, "Roman Glass in Southwest Spain. Some Notes on Trade Relations in the Early and Late Empire," *AnnAIHV* 11 (1988, published 1990) 79-80. On these glass blowers, see also *Early Syro-Palestinian Mold Blowing: The Role of Sidon*. On the literary evidence for the introduction of glass blowing in Rome and early deposits of blown glass in Rome and Sicily: David D. Grose, "Early Blown Glass: The Western Evidence," *JGS* 19 (1977) 9-29; Grose 1973, 31-52, discusses blown glass in Augustan deposits excavated at Cosa.

200 On Aco beakers: see at Cat. No. 5, note 18.

201 McClellan 1983, 75-76.

202 For instance, J. F. Magnus Amelung took molds to Pittsburgh from the Baltimore Glass Works that may have come in turn from his father's failed New Bremen Glass Manufactory (1785-1795): *Isaac Craig, his business activities, Pittsburgh, 1790-1804*, Carnegie Library, Pittsburgh; *General James O'Hara, account books and letters, 1790-1819*, Derring Papers, Historical Society of Western Pennsylvania, Pittsburgh. I am indebted to Kenneth M. Wilson for this information. See also Lowell Innes, *Pittsburgh Glass, 1797-1891* (Boston 1976) 19-29.

203 On glass and glazed wares excavated along this trade route: Maccabruni 1983, 29 (with lit.).

204 Price 1991, 72.

205 The evidence has been collected by Trowbridge 1930, 110-112: Petronius, *Satyricon* 51; Dio Cassius, *Roman History* 57.21.7; Pliny, *Natural History* 36.195; Isidore, *Origines* 16.16.6.

206 Pliny, *Natural History* 36.193, states that some glass vessels were made by *argenti modo caelare*, chasing (or engraving) like silver. On the interpretation of this passage, see above: *Early Syro-Palestinian Mold Blowing: The Role of Sidon*.

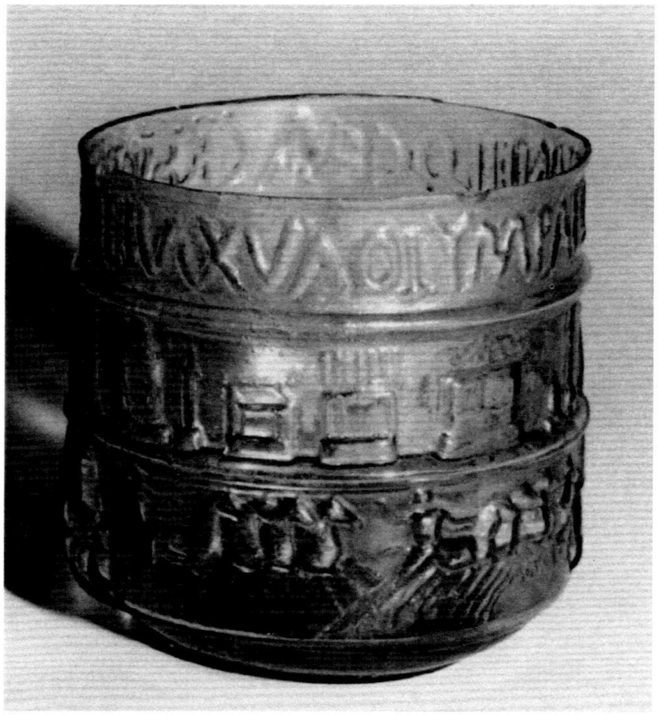

*Fig. 68. Beaker with circus scenes. From Colchester, England, H. 7.9 cm, light green, London, British Museum, no. BM PRB 1870.2-24.3. Photo: Copyright British Museum.*

such as bronze, silver, and gold.[207] The rejection of this sophisticated type of glass was perhaps inspired by the fear that metal vessels would become obsolete as tablewares, because metal has a disagreeable taste while glass has no taste.[208] By the time Pliny wrote his *Natural History*, completed in 77, "the use (of glass) (had) expelled silver and gold drinking vessels (from the market)."[209]

The hypothesis that mold blowing was introduced into Rome under Tiberius is supported by excavated finds. Among the earliest dated finds are two gladiator cups from Camulodunum, England, deposited between 49 and 60/61.[210] Cups decorated with circus scenes (Fig. 68) were a popular type that originated in the West; they have been excavated only in the western half of the empire. The technique of mold blowing was practiced at an early date not only in Rome but perhaps also in workshops located elsewhere in Italy and in the western half of the Roman empire.

Once the technique of mold blowing had become known in Rome and other areas of the Roman Empire, the transfer of styles and designs became a complex matter, because from then on trade was no longer the only possible way to disseminate mold-blown designs. Copying mold-blown vessels was relatively simple. A mold could be made directly from any imported vessel.[211] Local glass blowers were able to copy imported wares, and there are strong indications that they did indeed do so, for example, with Ennion's jugs, Hexagonal Bottles of the Masks Type (Groups 2 and 3), and some Types made in the Workshop of the Floating Handles. The copying of mold-blown designs began as an influence from the East to the West, but this probably soon became a two-way influence. A drinking cup showing a chariot race that may have been made in the East because it carries a fragmentary mold-blown signature in Greek, ... ΡΟϹ ΕΠΟΙΕΙ *[Alexand]ros epoiei*, "[Alexand]ros made it,"[212] appears to imitate Western circus cups.[213]

The role of the city of Rome in disseminating technology in this crucial period in the history of glass cannot be overestimated. The revival of the arts under Augustus and the lure of lucrative commissions in the city acted as a magnet to draw the most enterprising craftsmen from all over the empire to the capital.[214] At the same time, however, Rome acted as a centrifugal force, disseminating new ideas, ideologies, and techniques from the capital to the farthest frontiers.

An important agent in this dissemination was the Roman army. When legions were transferred, they were usually accompanied by all sorts of craftsmen, even glass workers for making window glass and supplying soldiers with tableware.[215] Frequent transfers engendered a cross fertilization of styles not only from Italy to outlying areas, but also from one province to another.[216] In short, three movements contributed to the spread of technology and designs in the early years of the Roman empire: a centripetal movement toward Rome, a centrifugal movement emanating from Rome, and a third movement among outlying provinces. These three movements combined to create truly multiregional styles.

[207] *Ferunt Tiberio principe excogitato vitri temperamento, ut flexile esset, totam officinam artificis eius abolitam ne aeris, argenti, auri metallis pretia detraherentur.* Pliny, *Natural History* 36.195.

[208] Price 1991, 64, suggests the story might be "a faint echo of the fears of manufacturers of more expensive forms of relief-decorated glass tablewares that the new discovery would supersede their own products."

[209] *Usus (vitri) vero ad potandum argenti metallo et auri pepulit.* Pliny, *Natural History* 36.199.

[210] Harden 1947, 300, nos. 50-51, pl. 86. Circus beakers have formed the subject of many studies, for example: Berger 1960, 56-67; Price 1973, 69-75; Ludwig R. Berger, "Neufund eines Gladiatorenbechers aus Vindonissa," *Vindonissa-Museum Jahresbericht, 1977* (1978) 62-68; Geneviève Sennequier, "Nouveau aperçu sur deux verres gallo-romains du musée des antiquités de Rouen: une coupe à course de chars et un gobelet à gladiateurs," *AnnAIHV* 7 (1977, publ. 1978) 67-94; Harden 1982; and Rütti et al. 1988. Price 1991, 67-68, gives a brief survey of datable finds. For another typically western type, see above *Hexagonal Bottles, Masks Type, Group 4.*

[211] Future mold research on mold-blown vessels from controlled excavations in the West may make it possible to distinguish between objects copied in the West and objects originating from Syro-Palestinian workshops.

[212] M. James Barrelet Collection, Paris, bought in Alexandria, Harden 1982, 31-34, figs. 1 and 2.

[213] Rütti 1988, 39-43, s.v. *Zirkusbecher* (with lit.). Jennifer Price, "Early Imperial Mold-Blown Glass Circus Cups: Some Problems of Production, Dating, and Distribution" (abstract), *AJA* 91 (1987) 315-316, shows that the distribution pattern of the various types of sports cups varies considerably within the western half of the Roman empire.

[214] On the role of Rome as a centripetal force: Stern 1977a, 29, 149-153, 155.

[215] Haevernick 1955; Clasina Isings, "Some of the 10th Legion's Glass at Nijmegen," in *Studies in Glass History and Design. Papers Read to Committee B Sessions of the VIIIth International Congress on Glass* (London 1968, published 1969) 9-11; Isings 1976; Sophia M. E. van Lith, "Römisches Glas aus Velsen," *OMROL* 58 (1977) 1-62; Isings 1979a; van Lith and Randsborg 1985.

[216] Stern 1977a, 152.

# CATALOGUE

## *Cups with Mold-blown Inscriptions* (Cat. Nos. 1-6)

### 1 **Bulbous Cup with Mold-blown Inscription**

Color plate 1, p. 51
Probably Syro-Palestinian, perhaps coastal region
Found at Pizzighettone, near Cremona, Italy
Probably mid first century
H 7.0 cm; max D 8.7 cm; D rim 7.5 cm; Th rim 0.1 cm; D base 4.7 cm
1967.6, Purchased with funds from the Libbey Endowment, Gift of Edward Drummond Libbey

Transparent dark yellowish green tint (10 Y 6/6). Medium thin glass. Pinprick and small spherical bubbles.

Blown into a three-part mold with two vertical sections joined to a cup-shaped base section (MCT V B); vertical mold seams from rim to second ridge below inscription on either side of **ΕΥΦΡΑΙΝΟΥ**. Relief crisp.

Flaring unworked rim. Concave lip. Squat bulbous body with complex profile. Base flat with three concentric raised circles around a central recessed knob.

At greatest diameter of body, frieze with an inscription in Greek bordered by three horizontal ridges above and two horizontal ridges below. The inscription reads: **ΕΥΦΡΑΙΝΟΥ // ΕΦΩ ΠΑΡΕΙ** (*euphrainou epho parei*), "Rejoice in that at which you are present." On lower part of body, a frieze of thirty-six upturned tongues in raised outline.

*Condition* Complete except for a few chips from rim. A few brown weathering spots on interior.

*Collections* Giorgio Sangiorgi, Rome (1913-65); Sergio Sangiorgi, Rome (1965-67).

*Published* Sangiorgi 1914, no. 103, pl. 19; Rendel Harris 1927, 291 (where the TMA vessel is incorrectly identified as one in the British Museum); A. Deissmann, *Licht vom Osten* (Tübingen 1923) 105, n. 1; Harden 1935, 174, Group G1iic; TMA, *Art in Glass* 1969, 24, ill.; Grose 1978, 79, fig. 15.

*Remarks* The inscription has been interpreted grammatically as a question: "what are you here for? be merry!"[1] However, the Greek phrase is probably not a question but a grammatically correct shortening of a longer expression: **ΕΥΦΡΑΙΝΟΥ ΕΠΙ ΤΟΥΤΩΙ ΕΦ ΩΙ ΠΑΡΕΙ** (*euphrainou epi toutoi eph hoi parei*), in which the demonstrative pronoun τουτωι (*toutoi*) is omitted and ἐπι ᾧ (*epi hoi*) is contracted to ἐφῳ (*epho[i]*); in Greek, the end iota is commonly omitted when capital letters are used.[2] The phrase happens to be found in the Greek koine text of the New Testament (Matthew XXII.50), a circumstance that prompted the hypothesis that glasses carrying this inscription were used at the Last Supper.[3] However, it was a well-known secular formula, meaning "delight in that at which you are present," enjoy the here and now — an appropriate exhortation at any meal or symposion.[4]

Two variants of this Type exist, their main distinguishing features being the contour of the walls and the relation of the height to the diameter. This vessel belongs to Harden's Group G1ii,[5] the tall variant with a complex profile and three horizontal ribs above the inscription instead of two ribs as in Harden's Group G1i,[6] which also has a more oval contour.

Cups of Harden's Groups G1i and G1ii have been found throughout the Mediterranean. Two appear to be from Lebanon, four were excavated in South Russia, one each from Asia Minor and Cyprus, three from Greece, two or three from Italy, and one from Tunisia.[7] Very small fragments probably belonging to one of Harden's Groups were excavated at Vindonissa, Switzerland,[8] and at Colchester, England.[9] This distribution is characteristic of first-century products associated with Sidon, such as glasses signed by Ennion. However, whether all cups belonging to Harden's Groups G1i and G1ii were actually made in one center from which they were exported, or whether some were copied by glass blowers working in glass centers elsewhere, as was evidently the case with Ennion's vessels, cannot be determined without further investigation of the molds in which they were blown.

The lower, fluted part of the body of most if not all of the cups of Harden's Groups G1i and G1ii was apparently blown in a separate cup-shaped mold section, the edge of which is marked by the second ridge below the band carrying the inscription. A similar use of a separate cup-shaped mold section for the fluted lower part of the body has been noted for a number of jugs signed by Ennion.[10] Cups blown in MCT V B molds were probably made in one of the workshops on the Syro-Palestinian coast, presumably in Sidon; the construction itself may have been invented in Ennion's workshop (see Fig. 15, MCTs I and V).

A glass cup of the same Type as Cat. No. 1 but belonging to Harden's Group G1v was discovered in a first-century cemetery on the Greek island of Siphnos.[11] Its date is probably an indication that all G1 Types were created before the third quarter of the first century.[12] A mid first-century date is supported by the fact that the shape occurs in Fine Ware Pottery used around the middle of the first century.[13] A large number of glass cups appears to have had a long life, not having been discarded before the second or third century.[14]

[1] Rendel Harris 1927, 289.

[2] My thanks are due to G. Neumann, Professor of Greek, University of Würzburg, and C. J. Ruijgh, Professor of Greek, University of Amsterdam,

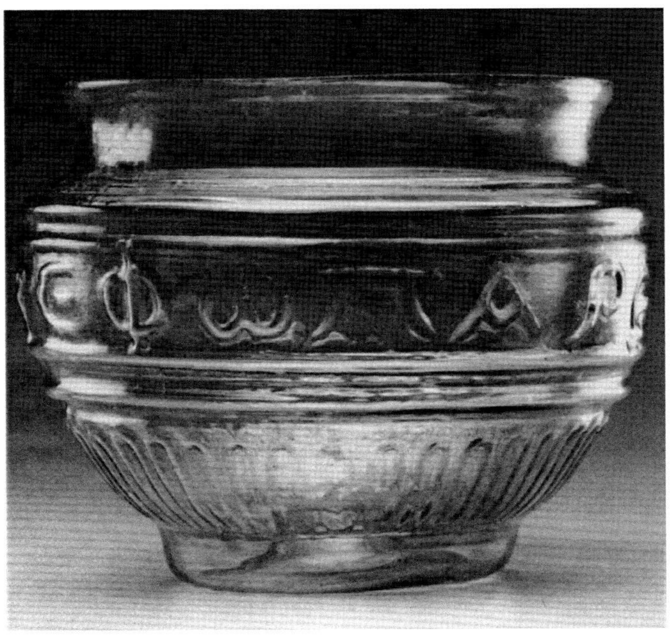

1 (1:1)

for discussing this phrase with me. Richter 1914, 85, translated: "Rejoice in that which Paris rejoiced" (i.e. women), but this is an impossible translation of the Greek.

3 Rendel Harris 1927, passim.

4 The inscription also occurs on mold-blown cups with an ovoid profile, for example, Birmingham, Museum and Art Gallery, on loan from Woodbrook College, from Kerch in the Crimea, Jennifer Price, "Roman Glass Other Than British," *BullAIHV* 8 (1977-1980) 47, fig. 14; Harden 1935, 175, Group G1iva.

5 Harden 1935, 174, Group G1ii:

a) BM, no. 1905.5-22.4, from Olbia, South Russia, yellowish green, Harden 1935, 174, Group G1iib;
b) Odessa, Museum, no. A51964, fragments from Berezan, Georgia, Sorokina 1962, 230, 231, fig. 15, bottom;
c) formerly Goluchow and Tyszkiewicz Collections, from Cyprus, colorless, Harden 1935, 174, Group G1iid;
d) Beirut, Nat. Mus., no. 1244, bought from D. Sarafian, colorless, Harden 1944-45, 92, Group G1iie;
e) Tunis, Bardo Mus., greenish, from the Roman cemetery at Sidi el Hani, Tunisia, Harden 1944-45, 292, Group G1iif;
f) Berlin, Antikenmus., no. 30219.212, formerly von Gans Collection, natural green, Harden 1935, 174, Group G1iia; Platz-Horster 1976, no. 62;
g) Oppenländer Collection, no. 2591, pale olive green, von Saldern et al. 1974, 159, no. 447;
h) CMG, no. 72.1.3, blue, *JGS* 15 (1973) 187, no. 5;
i) Athens, Agora Mus., nos. G138a and b, two fragments of a greenish cup belonging to Harden's Group G1i or ii, found at the bottom of a second-century well in the Athenian Agora, McClellan 1983, 74-75, fig. 5;
j) Delphi, Mus., Greece, light blue bowl found at Antikyra, Phocis, belonging to Harden's Group G1, not recorded which variant it is, McClellan 1983, 77.

6 Harden 1935, 174, Group G1i:

a) Berlin, formerly in Antiquarium of the National Museum, no. misc. 11 866, colorless, from Izmir (ancient Smyrna), Turkey, Harden 1935, 174, Group G1ia, pl. 25, no. e; lost during World War II according to Platz-Horster 1976, no. 62;
b) from Panticapaeum, fragments, Sorokina 1962, 230-231, fig. 15, top (not Sorokina 1967, 70-71, fig. 2, no. 16);
c) from Loo near Sochi in southern Russia, fragments, M. I. Ivanova and P. M. Goloubev, "Nachodki v Loo" ("Finds from Loo"), *SovArch* 1961, issue 3, 284f., fig. 4;
d) from Galaxidi, Greece, fragment [or a vessel], I. Threpsiades, *Ephem* 1972, 199, pl. 77, no. a;
e) in the Augusteum, a sanctuary for the imperial cult, at Cosa, Italy, was found a fragment preserving the letters ΦΡΑ, Grose 1973-76, 179;
f) Leiden, Netherlands, Rijksmus. van Oudheden, from Italy or North Africa, light grayish yellow, Harden 1935, 175, Group G1ib; L. J. F. Janssen, *BJb* 16 (1851) 72, no. 64, pl. 2; W. C. Bratt, *Glas uit de Oudheid: Verres de l'Antiquité* (Leiden 1962) no. 66, fig. 4; W. C. Braat, "The Glass Collection of the Rijksmuseum van Oudheden at Leiden," *OMROL* 44 (1963) 4-5, no. 10, pl. 35, no. 1; Salomonson 1975, 241, argues for a North African origin;
g) formerly Constable Maxwell Collection, *Maxwell* 1979, 174-175, no. 312;
h) Carnegie, no. 29490, pale blue green, Oliver 1980, 70, no. 64;
i) CMG, no. 59.1.78, formerly R. W. Smith Collection, no. 1329, acquired in Lebanon, greenish, Smith 1957, 60, no. 74;

See also previous note, nos. i and j.

7 In addition to Harden's list: Salomonson 1975, 241, with notes 27-33.

8 Vindonissa, Switzerland, Landesmuseum, no. 1928.684, Berger 1960, 49-50, no. 124, pl. 8.

9 Harden 1947, 299-300, no. 48, pls. 86-87.

10 Israeli 1983, 67-68, with reference to the jugs of Harden 1935, 168, Group A4i.

11 Athens, Benaki Mus., no. 43/70, from Siphnos, grave 6, no. 2, Clairmont 1977, 10, no. 32, pl. 3 (without mention of find spot); G. Mackworth Young, "Excavations in Siphnos," *BSA* 44 (1949) 86, pl. 28, nos. 3-5.

12 Harden 1944-45, 87-88.

13 For example, a cup from Hofheim, E. Ritterling, "Das frührömische Lager bei Hofheim i.T.," *AnnVereinNassauAltertumskunde* 34 (1904) pl. VI, no. 16.

14 For example, the fragmentary cups from Panticapaeum and Berezan, published by Sorokina 1962, 230-231, fig. 15, and the fragments from a second-century well in the Athenian Agora, McClellan 1983, 74-75.

## 2 Cylindrical Cup with Mold-blown Inscription

Probably Syro-Palestinian, probably made in Sidon
Reported to have been found near Sidon
First half of first century
H 6.5 cm; D rim (rec.) 6.6 cm; Th rim 1.2 cm; D base (rec.) 6.25 cm; max Th body 0.20 cm
1923.411, Gift of Edward Drummond Libbey

Transparent natural dusky yellow green tint (5 GY 5/2). Thin glass. A few small spherical and vertically elongated bubbles.
Blown into a three-part mold with two vertical sections joined to a disk-shaped base section (MCT VII); vertical mold seams from rim to base through palm leaves. Relief crisp.
Outsplayed unworked rim. Cylindrical body.[1]
On body, three friezes separated and bordered by single horizontal ridges. The central frieze contains an inscription in Greek. The upper and lower friezes each contain six stylized wreaths consisting of two concentric circles joined by radial lines; wavy lines below each wreath indicate the ends below the knot. Two diametrically opposed palm leaves divide the inscription into two almost equal parts: ΛΑΒΕ ΤΗ(Ν // Ν)ΕΙΚΗΝ (*labe ten neiken*), "seize the victory."
*Condition* Fragmentary; about two-thirds of the body, including most of the inscription and half of the rim, preserved. Iridescent film.
*Collection* Thomas E. H. Curtis.
*Published* Villefosse 1904, 278-280; Perdrizet 1904-1905, 291ff., Class 1B, nos. 5-8; Harden 1935, 178, Group K1iiie; erroneously cited as 1923.412 (Cat. No. 3) by Matheson 1980, 53, no. 134.
*Remarks* This cup belongs to a popular type of drinking cup bearing an inscription exhorting the user to "seize the victory." The motto probably refers to drinking contests,[2] as can be deduced from cups[3] in which this motto is accompanied by EICEΛΘΩΝ (*eiselthon*), an aorist participle of εἰcέρχεcθαι (*eiserchesthai*), "to enter the lists" in a contest.[4]
At least four distinct Series exist.[5] This Series (Harden's Group K1iii) is characterized by the disposition of the inscription in one line, with six stylized wreaths in the friezes above and below. Two variants are known, their major distinguishing feature being a reversed N in the word ΤΗΝ as in Cat. No. 3, which also clearly shows the stem of the Ε in ΝΕΙΚΗΝ extending up onto the horizontal ridge above the inscription. Both features are missing on this cup, which is probably from the variant with a normal N because the design of the wreath above the letters ΝΕ in the word ΝΕΙΚΗΝ differs slightly from that on Cat. Nos. 3 and 4.
Recorded find spots indicate that this Series originated in a Syro-Palestinian workshop: seven are from the Syro-Palestinian coast, seven from Cyprus, and one each from Greece, Panticapaeum, and Sardinia.[6] There appears to be no difference between the distribution of the two variants. The distribution pattern recalls that of bulbous cups inscribed ΕΥΦΡΑΙΝΟΥ ΕΦΩ ΠΑΡΕΙ (*euphrainou epho parei*) (see Cat. No. 1), ΚΑΤΑΧΑΙΡΕ ΚΑΙ ΕΥΦΡΑΙΝΟΥ (*katachaire kai euphrainou*), "rejoice and enjoy yourself,"[7] and other products of Sidonian workshops such as Ennion's and Aristeas's cups. This suggests that these cylindrical cups with victory inscriptions were made in or near Sidon.
No cups of this Type have been recorded from datable contexts, but a few examples of closely related Types have been found in contexts dated to the first half of the first century,[8] indicating that this Series was probably created before the middle of the first century.

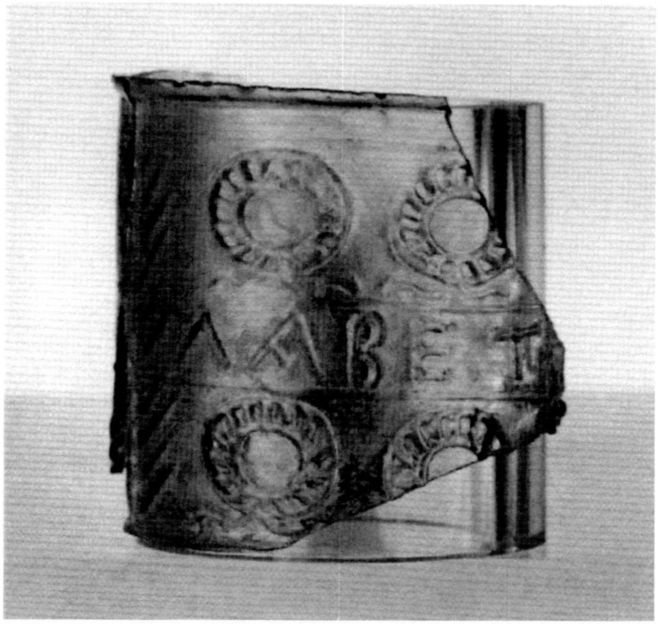

2 (1:1)

See also Cat. Nos. 3 and 4 from closely related molds. For a pyxis that may be related to this group of beakers, see Cat. No. 83.

[1] The base was probably flat, with a barely perceptible sunken dot in the center surrounded by a slightly raised ring, as can be seen on two better preserved examples in New York, MMA, nos. 81.10.214 and 32.128.10.

[2] Perdrizet 1904-5, 296-300, argues that cups with this inscription were containers for money given as prizes to victors in theatrical, gymnastic, or other contests.

[3] Paris, Louvre, no. MND 499; and Cagliari, Sardinia, Nat. Arch. Mus., *BullAIHV* 9 (1981-1983) 193-194, fig. 1.

[4] On εἰcέρχεcθαι (*eiserchesthai*), "entering a contest": Liddell-Scott, s.v. εἰcέρχομαι, II.

[5] Harden 1935, 176-180, Groups K1i, cylindrical beakers with the inscription ΛΑΒΕ ΤΗΝ ΝΙΚΗΝ, without an Ε in ΝΙΚΗΝ, "in four equal parts, each enclosed in a knotted oval wreath"; K1ii, cylindrical beakers with the inscription ΛΑΒΕ ΤΗΝ ΝΕΙΚΗΝ "in two parts, one of three lines, the other of two, each flanked on either side by knotted circular wreath, four in all"; K1iii, cylindrical beakers such as Cat. Nos. 2-4, with the inscription in one continuous frieze; and K2, splay-sided beakers, with the inscription ΛΑΒΕ ΤΗΝ ΝΙΚΗΝ, without an Ε in ΝΙΚΗΝ, "in two parts of two lines each, flanked on either side by knotted circular wreaths, four in all."

[6] From the Syro-Palestinian coast, *with reversed* N:

a) Cat. No. 4, from the vicinity of Sidon, dark manganese colored;
b) Cat. No. 3, from the vicinity of Sidon, probably with reversed N, natural yellowish glass;
c) MMA, no. 32.128.10, from Beisan, greenish, Harden 1944-45, 94, K1iiiq;
d) Yale, no. 1955.6.68, formerly Moore Collection, reported to be from Tyre, purple, Matheson 1980, 53, no. 134;
e) CMG, no. 55.1.4, formerly R. W. Smith Collection, no. 690, acquired in Lebanon, light amber, Smith 1957, 69, no. 96;

*probably with normal* N:

f) Cat. No. 2, from the vicinity of Sidon, natural yellowish glass;

*type of* N *unknown*:

g) Jerusalem, RM, no. 42.222, greenish, from a tomb at Joret-Ascalon, Harden 1944-45, 292, K1iiir.

From Cyprus, *with reversed* N:

a) BM, no. 1894.11-1.108, from Amathus, Cyprus, tomb 46, amber, Harden 1935, 178, K1iiia, pl. 26, no. c;
b) Rome, formerly Sangiorgi Collection, yellow, Harden 1935, 179, K1iii-i, Sangiorgi 1914, no. 104, pl. 20;
c) Nicosia, Cyprus Museum, no. unknown, yellowish, Harden 1935, 179, K1iiij; Vessberg 1952, 121, pl. III, nos. 7 and 8;
d) MMA, no. 81.10214 (formerly no. AG 214), greenish, probably from Idalion, Cyprus, Harden 1944-45, 93, K1iiin.

*with normal* N *or type of* N *unknown (not recorded or not preserved)*:

e) BM, no. 1884.12-28.20, amber, from Famagusta, Cyprus, Harden 1935, 178, K1iiib;
f) formerly Cesnola Collection, color not recorded, Harden 1935, 179, K1iiil;
g) formerly Djabra Pierides Collection, sea green, Harden 1935, 179, K1iiim.

From Greece, *with reversed* N: purple, Harden 1935, 178, K1iiid.
From Panticapaeum: St. Petersburg, Hermitage, no. P 1914.17, Kunina 1973, 133 and 134, fig. 23.
From Sardinia: Cagliari, Nat. Arch. Mus., no. 5136, from Cornus, colorless, type of N not recorded, Harden 1935, 179, K1iiik, pl. 276.
For others from unrecorded find spots: Harden 1935, 178-179; Harden 1944-45, 94; Auth 1976, no. 57; von Saldern et al. 1974, 162-166, no. 454; *Kofler-Truniger* 1985, 72-73, no. 118; Dobkin Collection, Jerusalem, *JGS* 13 (1971) 135, no. 8, ill.

[7] McClellan 1983, 76-77, n. 36, with addenda to Harden's lists, notes that six cups "come from Cyprus, three from Italy, two from Syria and Palestine, and one each from the Black Sea, Greece, and Switzerland."

[8] Examples known to the author:

a) Cosa, fragment of a cylindrical cup preserving the letters NH, excavated in a sealed deposit of about 40-45, Grose 1973, 33, 37-38, no. 8 (CE 2035), fig. 2. The letters are probably the last two letters of the Greek word ΕΥΦΡΟCΥΝΗ (*euphrosyne*), "happiness," as seen in two cylindrical cups b and c;
b) Pompeii, Antiquarium, no. 11307, from Pompeii, a cup with the entire word *euphrosyne* repeated on either side, Scatozza Höricht 1991, 79, fig. 13, no. a;
c) Düsseldorf, Kunstmuseum, no. P.1969-5, a cup from an unknown find spot, inscribed ΚΕΡΔΟC ΚΑΙ (*kerdos kai*), "profit and" on one side and ΕΥΦΡΟCΥΝΗ (*euphrosyne*) on the other, von Saldern 1974, 58, no. 38;
d, e) Beirut, Nat. Mus., nos. 108-109, found at Yahmour, Syria, two cylindrical beakers, Harden 1944-45, 82, 84, pl. 7, Group Fii;
f) Vindonissa, Mus., no. unknown and 1932.874, two fragments, excavated at Vindonissa, Switzerland, Berger 1960, 50, nos 117-118, pl. 8.

On this Series: Harden 1935, 171-173, Groups Fi, inscribed ΚΑΤΑΙΧΑΙΡΕ (*kataichaire*) [*sic*] ΚΑΙ ΕΥΦΡΑΙΝΟΥ (*kai euphrainou*), and Fii, correctly inscribed ΚΑΤΑΧΑΙΡΕ ΚΑΙ ΕΥΦΡΑΙΝΟΥ (*katachaire kai euphrainou*). A cylindrical beaker of Group Fi was excavated in a first-century cemetery on the Greek island of Siphnos: Harden 1944-45, 87-88, 90-91, pl. 18, no. 4. A related beaker was excavated in Sudan at Shirfadiq, necropolis no. 25, tomb no. 245, Säve-Söderbergh, "Preliminary Report of the Scandinavian Joint Expedition: Archaeological Investigations between Faras and Gemai, November 1961-March 1962," in *Kush* XI (1963) 66, pl. XII, no. B. Two cylindrical cups, one inscribed ΕΙCΕΛΘΩΝ ΛΑΒΕ ΤΗΝ ΝΙΚΗΝ (*eiselthon labe ten niken*), "enter [the contest] and seize the victory," the other inscribed ΚΑΤΑΧΑΙΡΕ ΚΑΙ ΕΥΦΡΑΙΝΟΥ (*katachaire kai euphrainou*), are in Cagliari, Sardinia, Nat. Arch. Mus., *BullAIHV* 9 (1981-1983) 193, 194, fig. 1.

## 3 Cylindrical Cup with Mold-blown Inscription

Probably Syro-Palestinian, probably made in Sidon
Reported to have been found near Sidon
First half of first century
H (pres.) 6.15 cm; D rim (rec.) 6.6 cm; D base (rec.) 6.0 cm; max Th body 0.22 cm
1923.412, Gift of Edward Drummond Libbey

Transparent dusky yellow green (5 GY 5/2). Thin glass. A few small spherical bubbles.

Blown into a mold with two vertical sections (MCT VII); vertical mold seams from rim to base through palm leaves. Relief crisp. Outsplayed unworked rim. Cylindrical body.

On body, three friezes separated and bordered by single horizontal ridges. The central frieze contains an inscription in Greek. The upper and lower friezes each contain six stylized wreaths consisting of two concentric circles joined by radial lines; wavy lines below each wreath indicate the knotted ends. Two diametrically

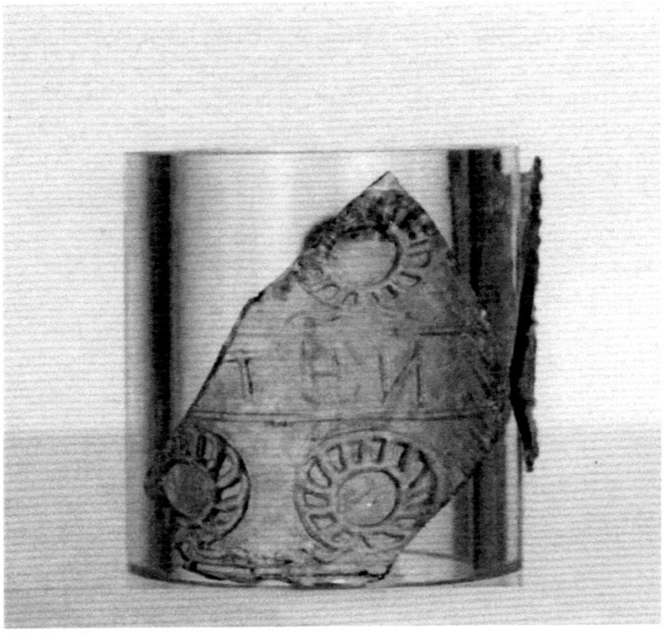
3 (1:1)

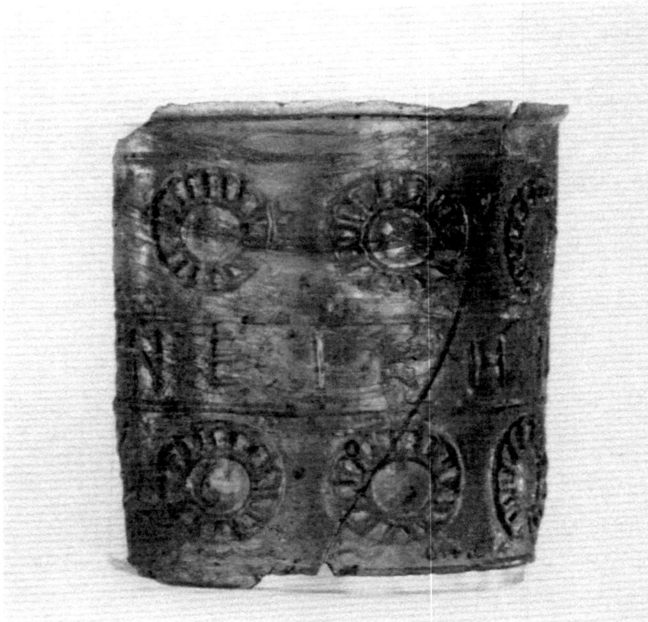
4 (1:1)

opposed palm leaves divide the inscription into two almost equal parts: (ΛΑΒΕ) THN (with reversed N) // NE(IKHN) (*labe ten neiken*), "seize the victory."
*Condition* Fragmentary; about one-third of the body preserved.
*Collection* Thomas E. H. Curtis.
*Published* Villefosse 1904, 278-280; Perdrizet 1904-5, 291ff., Class 1B, nos. 5-8; Harden 1935, 178, Group K1iiif and g; erroneously cited as acc. 1923.411 (Cat. No. 2) by Matheson 1980, 53, no. 134.
*Remarks* Same Type as Cat. No. 2, but the variant with a reversed N in the word THN (*ten*). An additional characteristic of this variant is that the stem of the E in NEIKHN (*neiken*) extends up onto the horizontal ridge above.[1] See also Cat. No. 4, probably the same variant.

[1] Cf. drawing, *Kofler-Truniger* 1985, 73, no. 118, which shows the reversed N in the word THN (ten) as well as the extension of the stem of the E in NEIKHN (neiken).

### 4 Cylindrical Cup with Mold-blown Inscription

Probably Syro-Palestinian, probably made in Sidon
Reported to have been found near Sidon
First half of first century
H (pres.) 6.7 cm; D rim (rec.) 6.6 cm; Th rim 0.15 cm; D base (rec.) 6.0 cm
1923.410, Gift of Edward Drummond Libbey

Translucent streaked dark manganese colored glass. Thin glass. A few small spherical bubbles.
Blown into a mold with two vertical sections (MCT VII); vertical mold seams from rim to base through palm leaves. Relief crisp.
Outsplayed unworked rim. Cylindrical body.
On body, three friezes separated and bordered by single horizontal ridges. The central frieze contains an inscription in Greek. The upper and lower friezes each contain six stylized wreaths consisting of two concentric circles joined by radial lines; wavy lines below each wreath indicate the knotted ends. Two diametrically opposed palm leaves divide the inscription into two almost equal parts: (ΛΑΒΕ THN) // NEIKHN (*labe ten neiken*), "seize the victory."
*Condition* Fragmentary; half of the body and rim preserved. Iridescent film.
*Collection* Thomas E. H. Curtis.
*Published* Villefosse 1904, 278-280; Perdrizet 1904-5, 291ff., Class 1B, nos. 5-8; Harden 1935, 178, Group K1iiih; included by Matheson 1980, 53, as a parallel to her no. 134 (reported to have been acquired from Azeez Khayat and excavated near Tyre).
*Remarks* Same Type as Cat. No. 2, but probably the variant with a reversed N in the word THN (*ten*), because the stem of the E in NEIKHN (*neiken*) extends up onto the horizontal ridge above it as on Cat. No. 3.

### 5 Barrel-shaped Cup with Mold-blown Inscription

Color plate 2, p. 51
Signed by Neikais
Probably Syro-Palestinian, coastal region
Reported to have been found in a tomb at Aleppo[1]
Probably mid first century
H 8.7 cm; max D 7.08 cm; D rim 6.6 cm; Th rim 0.07 cm; D base 4.0 cm
1930.5, Purchased with funds from the Libbey Endowment, Gift of Edward Drummond Libbey

Transparent natural grayish green tint (10 GY 5/2). Thin glass. Pinprick and small spherical bubbles, vertically elongated toward rim.
Blown into a three-part mold with two vertical sections joined to a cup-shaped base section (MCT V E). Vertical mold seams from rim through palm fronds to central ridge of the series of three ridges below the inscription.
Outsplayed unworked rim with constriction below. Barrel-shaped body. Flat base with circular depression around central boss.
At greatest diameter of body, frieze with an inscription in Greek and palm leaves, bordered by two horizontal ridges above and three horizontal ridges below. Near base, two fainter horizontal ridges. Diametrically opposed vertical palm leaves frame the inscription, which is arranged in two parts of two lines each: NEIKAIC / ΕΠΟΗCΕΝ (*Neikais epoesen*), "Neikais made (it)," on one side and on the other ΜΝΗCΘΗ / Ο ΑΓΟΡΑCΑC (*mnesthe ho agorasas*), "May the buyer be remembered."

*Condition* Complete except for a few chips from rim. Iridescent surface frosting.
*Collections* Fr. von Gans, Frankfurt (second collection) (before 1913); Gallerie Bachstitz, The Hague (1913 or later); F. Kouchakji, New York (1921 or later-1930).
*Published* W. Froehner, *Verre antique de la collection F. von Gans* (Paris 1913) 5-8, ill.; R. Zahn, *Galerie Bachstitz, 's Gravenhage* II (Berlin 1921), 56, no. 154, pl. 60; Eisen 1927, I, 272, pl. 59; Harden 1935, 170, Group D, no. a and pl. XXIV, nos. c and d; Riefstahl 1961, 37, ill.; Neuberg 1962, fig. 43; Harden 1964b, 22, fig. 4, nos. a and b; Labino 1968, 22, fig. 8; TMA *Guide*, 1966 (ill.); TMA, *Art in Glass*, 24, ill.; Burford 1972, ill. opp. p. 80.
*Remarks* Five vessels signed by Neikais are known. All five are barrel-shaped beakers like this one.[2] They are very similar, in shape, decoration, phrasing of the inscriptions, and construction of the molds in which they were blown, to beakers signed in Greek by two other glass blowers, Iason and Meges. The cups made by these three glass blowers obviously form a close group, but, as was the case with Ennion and Aristeas, the exact relationship between Neikais, Iason, and Meges remains unclear.[3]
Neikais is equivalent to Nikais, derived from νίκη (*nike*), "victory" or from the place name Nikaia. The alternative, substandard spelling -ei- to indicate a long -i- was common in many parts of the Greek-speaking world. The uncommon name has previously been interpreted as a misspelling of a Greek name Neikias or Neikaios,[4] but its discovery on inscriptions at Taucheira-Arsinoe in Cyrenaica, North Africa, establishes Neikais as a correct spelling.[5] The ending in -ais suggests a female name, but it appears to have been used by both men and women.[6] If Neikais was a female glass blower-entrepreneur, she may be regarded as the first woman known by name to have blown glass in the eastern Mediterranean. She joins Sentia Secunda, of Aquileia, and Ennia Fortuna in the ranks of identified ancient women glass blowers (see Chapter I: *Glass-Blowers' Signatures*).
The cups signed by Neikais, Iason, and Meges have opposite their signatures a second Greek inscription, ΜΝΗCΘΗ Ο ΑΓΟΡΑCΑC (*mnesthe ho agorasas*), "may the buyer be remembered." *Agorasas*, "buyer," is an aorist participle of *agorazein* "to buy" with the same practical meaning as the present participle *agorazon* appearing in other examples of this phrase.[7] *Mnesthe* is probably a substandard spelling for ΜΝΗCΘΕΙΗ (*mnestheie*), optative aorist passive of μιμνήσκεσθαι (*mimneskesthai*), "to be reminded, to remember," a verb with middle and passive voice in Greek. This has led to the grammatically possible translation of the phrase as "May the buyer remember (the glass blower)."[8] However, the fact that the inscription occurs also on glasses without a maker's signature was already reason for Harden[9] to interpret *mnesthe* as "to be remembered," passive of "to remember": The passive meaning "to be

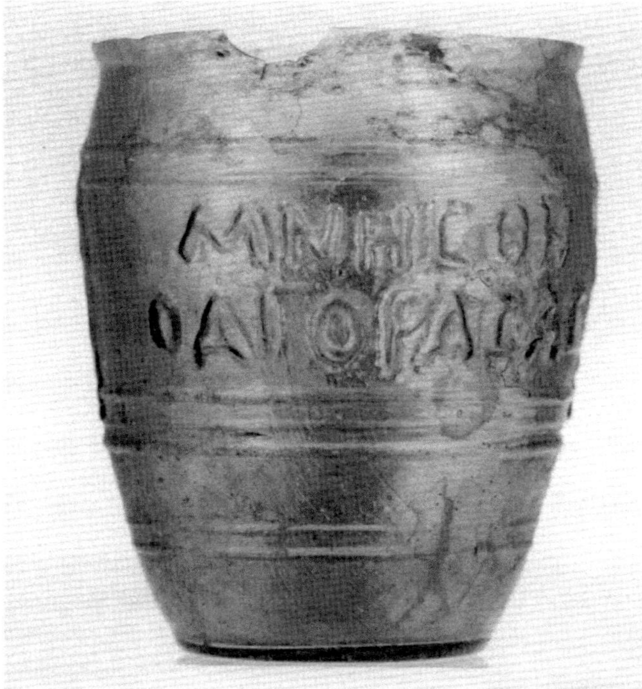

5 (1:1)

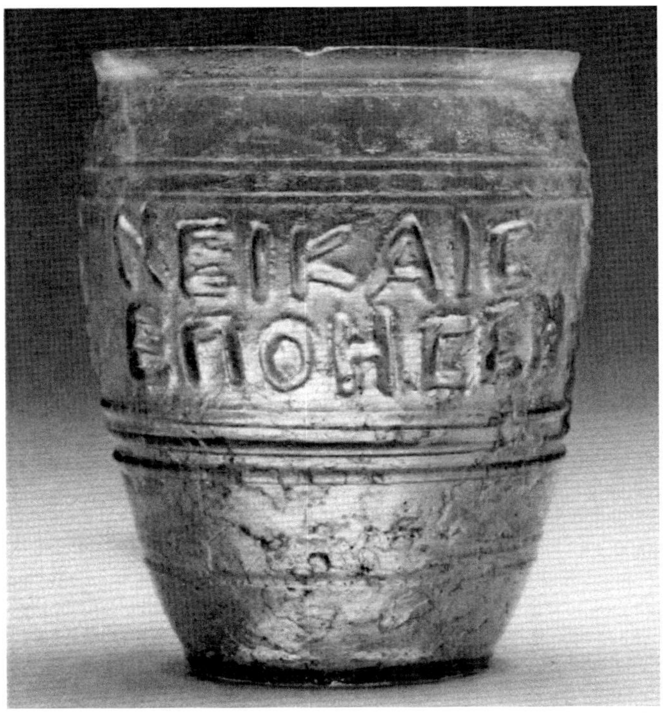


5 (1:1)

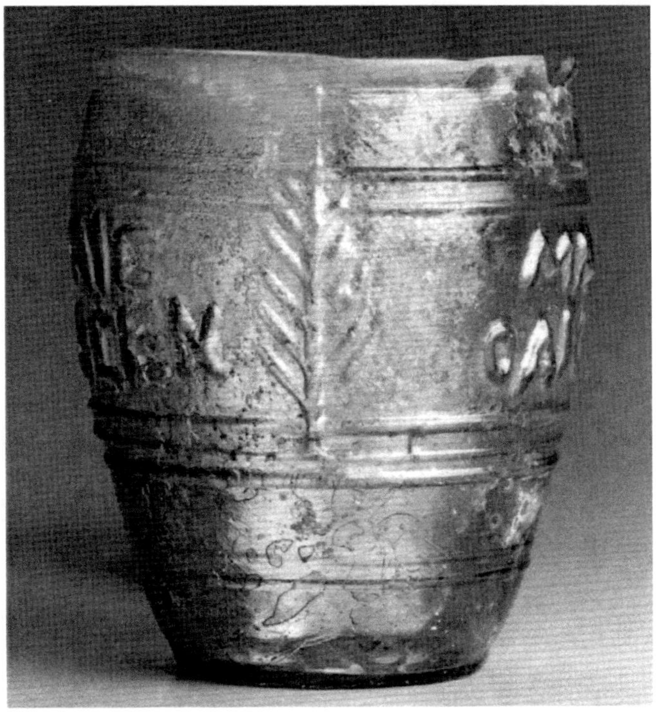

5 (1:1)

remembered" is most unusual for *mimneskesthai*. There are no parallels in classical or koine Greek. The only indication that the verb might have had this passive meaning appears in a fifth-century Greek text by the neo-Platonic philosopher Proklos (412-485). Proklos uses the perfect tense ὑπέμνηcται (*hypemnestai*) with the passive meaning "it has been observed." [10] This new passive perfect presupposes the previously undocumented existence of an aorist passive ὑπέμνηcθην (*hypemnesthen*) with the passive meaning "to be remembered" instead of the normal meaning "to remember." [11] with which one can connect the word ΜΝΗCΘΗ (*mnesthe*) in this inscription. The reason for the unusual use of *mnesthe* with this passive meaning is probably to be found in the fact that the phrase itself appears to be the translation of a shortened form of a typical Semitic blessing formula, the full form of which is "may [so-and-so] be remembered before the god [so-and-so]." Both the shorter and the longer versions are found on inscriptions by Semitic peoples such as Phoenicians, Hebrews, Nabataeans, and Syrians. [12] The grammatically surprising translation may be explained by the fact that Greek did not have a simple verb form to convey the meaning "to be remembered." The use of this passive form with passive meaning in devotional formulas might have been at the root of the rare use of *hypemnestai* with this passive meaning by Proklos.

The form and the language of the blessing inscriptions, the names of the glass blowers, [13] the find spots, and the use of MCT V all indicate an origin on the Syro-Palestinian coast for these cups. With seven cups from Syro-Palestinian find spots and one from Cyprus, the distribution appears to have been confined to a smaller area than many other types of Romano-Syrian glasses with mold-blown inscriptions.

Harden's suggestion [14] that the cups blown by Neikais, Iason, and Meges are degenerate types of later date than the other vessels with mold-blown inscriptions has not been confirmed by excavations. One of Neikais's cups was excavated at Masada, Israel, in a context not later than 66-73/74. [15] A mid first-century date for all three glass blowers is suggested independently by the wording of the maker's signature and blessing inscriptions, which are the same as those on Ennion's cups. The complicated construction of the three-part molds also indicates a relatively early date for this group. Similar barrel-shaped glass cups carry the inscription ΛΑΒΕ ΤΗΝ ΝΙΚΗΝ and circular wreaths [16] (see Cat. No. 2). The shape of the cups is related to first-century glass cups with wheel-cut incisions [17] (see also Cat. No. 6) and to ceramic cups of the type known as Aco beakers, which were made in the West from the late first century B.C. to about A.D. 25. [18]

1 Aleppo (ancient Beroia), northern Syria, on the caravan route between the Euphrates and the Mediterranean. After Alexander the Great's conquest of the Near East, Beroia was made a Greek city by Seleucus Nicator between 301 and 281 B.C.

2 Harden 1935, 170, Group D:

a) the Toledo beaker, cited as Group Da (acquired in 1930, not from the Bachstitz Gallery, The Hague, but from F. Kouchakji);
b) fragment excavated at Masada, Israel, Barag 1991, 139 (not ill.);
c) MMA, no. X21.186, formerly A.G. 463, fragmentary beaker, find spot not recorded, yellowish green glass, Harden 1935, 170, Group Db;
d) CMG, no. 70.1.8, *JGS* 13 (1971) 135, no. 6; the E at the end of ΜΝΗCΘΕ is an incorrect restoration for ΜΝΗCΘΗ;
e) Tel Aviv, no. MHG 138260, a pale yellow fragment preserving part of the signature ΝΕΙ(ΚΑΙC), Israeli 1964, 36, no. 3, fig. 5.

3 See Chapter I: *Glass-Blowers' Signatures: Neikais, Iason, and Meges.*

4 The name Neikais has been explained as either a shortened spelling for Neikai(o)s, or as Neikias, with metathesis of the letters A and I, both of which are well-known Greek proper names: Harden 1935, 170.

5 G. Lüderitz, *Corpus jüdischer Zeugnisse aus der Cyrenaika.* Beihefte zum Tübinger Atlas des Vorderen Orients, Reihe B (Geisteswissenschaft) Nr. 53 (Wiesbaden 1983) Appendix 16, no. c.

6 Fraser and Matthews 1987, 329, s.v. *Nikaios, Nikais.*

7 See Chapter I: *Glass-Blowers' Signatures: Ennion.*

8 Kisa 1908, III, 708, 716, since followed in many publications.

9 Harden 1935, 183, adopting an earlier proposal by G. Richter, "The Room of Ancient Glass," *BMMA* 6 (1911) 16.

10 Liddell-Scott, s.v. ὑπομιμνήcκω (*hypomimnesko*).

11 I am indebted to C. J. Ruijgh, Professor of Greek, University of Amsterdam, for discussing this expression with me.

12 Lehrer 1979, 13.

13 See Chapter I: *Glass-Blowers' Signatures.*

14 Harden 1935, 182; Harden, 1944-45, 88.

15 Barag 1991, 139.

16 Murano, Museo Vetrario, from the Dalmatian cemetery at S. Donato di Zara (modern Zadar, Croatia), construction of mold not described, Mariacher 1970a, 38-39, no. 15, side B; Mariacher 1970, 18, fig. 5, side A.

17 Isings 1957, Form 29.

18 On Aco beakers: J. Déchelette, *Les vases céramiques ornés de la Gaule romaine*, I (Paris 1904); A. Stenico, "Ceramica," in *Arte e civiltà romana nell'Italia settentrionale dalla Repubblica alla Tetrarchia*, II (Bologna 1965) 326-328; M. Vegas, "Aco-Becher," in *Rei Cretariae Romanae Fautorum Acta*, XI-XII (1969-1970) 107-124; M. P. Lavizzari Pedrazzini, "Bicchieri, ⟨tipo Aco⟩" in *Angera romana. Scavi nella necropolis 1970-1979* (Rome 1985) 373-387; L. Mazzeo Saracino, "Terra Sigillata nord-italica," *EAA Atlante delle forme ceramiche*, II (Rome 1985) 187-190; Maria Teresa Marabini Moevs, "Aco in Northern Etruria: The Workshop of Cusonius at Cosa," *MAAR* 34 (1980) 231-280.

## 6 Convex Cup with Mold-blown Inscription

Perhaps made in Italy
Probably second half of the first century
H 6.5 cm; D rim 6.84 cm; Th rim 0.11 cm; D base 4.0 cm
1923.413, Gift of Edward Drummond Libbey

Transparent natural grayish green tint (10 GY 5/2). Medium thin glass. Fabric cannot be determined because of weathering.

Blown into a three-part mold with two vertical sections and a disk-shaped base section (MCT VII C). [1] Vertical mold seams from rim to base on either side of the inscription ΕΥΦΡΑΙΝΟΥ. No mold seam visible on base. Relief crisp.

Rim lightly ground. Convex sides tapering toward base. Flat circular, undecorated base with a slightly raised central area of about 2.4×2.2 cm.

Near top of body, frieze with an inscription in Greek, bordered by one horizontal ridge above and two below. The inscription reads: ΕΥΦΡΑΙΝΟΥ // ΕΦΟΠΛΡΙ (*euphrainou epho parei*), "rejoice in that at which you are present." Near the base, two horizontal ridges.

*Condition* Intact; vertical strain crack in body. Iridescent film on exterior; areas of brown weathering and incipient pitting on interior.

*Collection* Thomas E. H. Curtis.

*Published* Richter 1914, 85, fig. 14; *BMMA* 9 (1914) 26; *ILN* December 25, 1926, 1268, fig. 4; Rendel Harris 1927, 291, no. 4, fig. 4; Harden 1935, 175, Group G 2a.

*Remarks* No parallels for this Type are known. The shape is a well-known first-century shape, [2] related to cups like Cat. No. 5, while the inscription imitates an inscription that occurs frequently on bulbous cups with a constricted rim (see Cat. No. 1). The maker of the mold was not literate in Greek, for there are three mistakes in the Greek: first, the spelling of ΕΦΟ with an *omikron* instead of an *omega*; second, an *alpha* without cross bar in ΠΑΡΕΙ; and third, the spelling of ΠΑΡΕΙ as ΠΑΡΙ with -Ι instead of -ΕΙ. The correct Greek would read: ΕΥΦΡΑΙΝΟΥ ΕΦΩ ΠΑΡΕΙ.

Most vessels blown in MCT VII molds have a decorated base. The facts that this cup has no decoration on its underside and that the maker of the mold was not literate in Greek suggest that it was

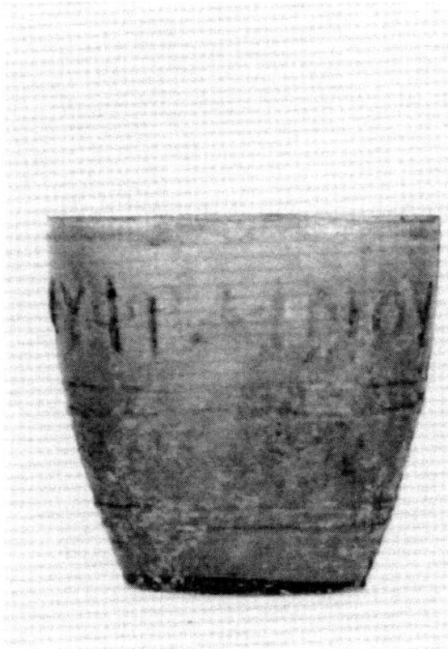

6 (1:2)

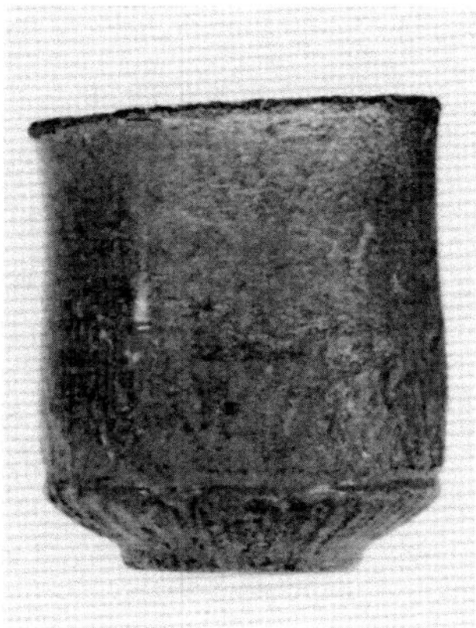

7 (1:2)

made in the second half of the first century, perhaps in the western part of the Roman empire where Greek was not the common language.

[1] Although the underside of the beaker is not decorated, a sharp edge at the junction of walls and base indicates that the mold was closed at the bottom, presumably with a separate disk-shaped section.

[2] Isings 1957, Form 29, free-blown, possibly imitating ceramic beakers known as Aco beakers, see Cat. No. 5, note 18.

## *Cup with Non-figural Designs in Low Relief* (Cat. No. 7)

### 7 **Cylindrical Beaker**

Eastern Mediterranean, probably Syro-Palestinian
First century
H 8.5-8.1 cm; D rim 7.6-7.2 cm; D base 4.2 cm
1923.443, Gift of Edward Drummond Libbey

Transparent decolorized or natural green. Exact color cannot be determined because of weathering. Medium thin glass.
Upper part of body free blown. Lower part of body blown into a three-part mold with two vertical sections joined to a cup-shaped base section (MCT V). The horizontal junction is between the ribs and the lozenge pattern. Relief indistinct.
Rim rounded and thickened in flame. Cylindrical body. Convex bottom. Base flattened.
On lower part of body, a band of contiguous lozenges or X's in raised outline. Around bottom, thirty-nine narrow ribs. On underside of base, three raised circles surrounding a central dot.
*Condition* Complete, but a large area of the exterior surface has been leached away. Remains of iridescence and mud brown encrustation on exterior. Golden iridescence and severe surface pitting on exterior and interior.
*Collection* Thomas E. H. Curtis.
*Published* Previously unpublished.
*Remarks* The design on the lower part of this beaker recalls designs on cups by Ennion and Aristeas,[1] but instead of fluting it has ribs like Cat. Nos. 13 and 14. Fragments of two cylindrical beakers with related designs were excavated in first-century contexts at Vitudurum, Switzerland.[2] The mold construction of the Toledo beaker, with two vertical sections joined to a cup-shaped base section, suggests it was made in the Syro-Palestinian area, as does the typical iridescent weathering.

[1] For example, two spherical bowls signed by Ennion:
a) Yale, no. 1955.6.66, formerly Moore Collection, Matheson 1980, 43-45, no. 118;
b) Norfolk, Va., Chrysler Museum, *JGS* 10 (1968) 181, no. 6.
Both have a pattern of lozenges above fluting. A cup signed by Aristeas formerly in the Maxwell Collection has a band of palmettes above fluting: *Maxwell* 1979, 157-160, no. 280, ill.

[2] Rütti 1988, 38-39, nos. 679-680, pls. 9, 30, and 33, the first dated to the second quarter of the first century, the second to the third quarter. Compare also two mold-blown beakers, one in the Altino (Venice), Archaeological Museum, from Altino, Bianca Maria Scarfì, *BullAIHV* 9 (1981-1983) 86, fig. 2, the other Yale, no. 1955.6.59, formerly Moore Collection, Matheson 1980, 52, no. 132, and a free-blown beaker in Bologna, Musei Civici, no. Rom. 15, Coll. Labella, Meconcelli Notarianni 1979, 50-51, no. 37.

## *Truncated Conical Beakers with Knot-shaped Knobs* (Cat. Nos. 8-10)

### 8 **Truncated Conical Beaker with Knot-shaped Knobs**

Color plate 3, p. 51
Probably Eastern Mediterranean
Mid to second half of first century
H 20.5 cm; max D 10.4 cm; D rim 9.4 cm; Th rim 0.22 cm; D base 5.3-5.13 cm
1923.490, Gift of Edward Drummond Libbey

Transparent to translucent dusky yellow (near 5 Y 6/4). Medium thin glass. A few small bubbles.
Blown into a four-part mold with three vertical sections joined to a disk-shaped base section (MCT III). Relief crisp.
Rim slightly ground with slight bulge below. Truncated conical body.
Flat base with two raised concentric rings and central boss.
A faint wheel-cut incision on exterior ca. 0.9 cm below rim. On body, six alternating rows of eight knot-shaped knobs in three-tiered relief diminishing toward base. Just above junction with base is a horizontal ridge; it is not a mold seam, since the vertical seams extend through this line to edge of base.
*Condition* Intact. Areas of buff weathering on exterior and interior; iridescent film and incipient pitting on exterior.
*Collection* Thomas E. H. Curtis.
*Published* Grose 1978, 78, fig. 15.
*Remarks* Knobbed beakers are one of the most common Types of mold-blown tableware.[1] The Type consists of many Series, all of which are characterized by straight walls that taper slightly toward the base and have as the principal element of their decoration alternating rows of usually oval knobs pointing downward like the knots of a tree trunk. The glass varies from transparent natural bluish green to decolorized clear, as well as transparent to translucent monochrome yellowish brown, purple, and blue. A few vessels have been finished as bottles[2] or jugs,[3] and one is finished as a drinking horn,[4] but most vessels are tall conical beakers. Two sizes are common: a large beaker about 20 cm high, like this one, and a smaller one about 12 cm high (see Cat. Nos. 9-10); but other sizes are known as well.[5] These beakers usually have a lightly ground rim, indicating that they were knocked off the blow pipe immediately after extraction from the mold and set aside for annealing.[6]

#### Decoration

The decoration occurs in many variations, the most common of which is alternating rows of pointed three-tiered knobs as on the three Toledo beakers,[7] or three-tiered knobs alternating with small circular bosses as on a beaker at Yale.[8] Other designs include alternating rows of plain pointed oval knobs,[9] sometimes alternating with small circular bosses,[10] and knobs or bosses framed by various linear patterns.[11] Rows whose knobs are arranged directly under each other are rare.[12] A few beakers have a unique decoration in which alternating knobs in the lowest row have been replaced by comedy masks;[13] others show a vine spray in addition to alternating rows of knobs and small circular bosses.[14]

### Technique

The Toledo beakers were blown in four-part molds consisting of three vertical sections and a fourth, disk-shaped base section. This type of mold construction appears to have been the one used most commonly for large beakers with three-tiered knobs.[15] But beakers blown in molds consisting of only two vertical sections for the walls are also known.[16] A few beakers have a pushed-in tubular base ring.[17] It has been suggested that these vessels were blown in open molds.[18] Theoretically, a mold without the disk-shaped base section could have been used (compare MCT Type IX, open at the base), but this is not the only possibility. Many beakers have a smooth area on the lower wall, just above the base.[19] After extraction from the mold, it would have been possible to reheat, expand, and tool the glass in this area without distorting the design above it.

### Function

Most of the beakers were probably drinking vessels, but it is possible that some were used as containers for medicine or medical instruments. One beaker was excavated with an array of medical instruments in the tomb of a physician.[20]

### Date

Finds at Pompeii and Herculaneum (see below) show that the Type existed when these cities were destroyed by the eruption of Mount Vesuvius in 79. This date is confirmed by dated finds from Switzerland,[21] the Netherlands,[22] and Turkey.[23] It is not clear when this Type went out of use.

### Distribution and Production Centers

Knobbed beakers were known in many parts of the Roman empire, but their distribution is uneven. They have been found in considerable numbers in some parts of Italy, in Campania[24] and at Cosa,[25] but they are rare in excavations in northern Italy.[26] Similarly, a few beakers have been found in tombs in the Ticino area, in southern Switzerland,[27] while they are relatively numerous at legionary sites at Vindonissa, Vitudurum, and Augst, in northern Switzerland.[28] At the sites of legionary camps along the Rhine in Germany, one fragment each has been excavated at Bonn and Asciburgium;[29] none has been reported from Cologne or Trier. On the other hand, fragments of knobbed beakers are relatively numerous in and around the legionary camp at Nijmegen in the Netherlands.[30] Finds have been reported also from England, France, Spain, and Portugal.[31] No beakers appear to have been found in northern Africa. Excavated finds from the Balkans[32] and the eastern Mediterranean[33] are not numerous, but their distribution does suggest at least one manufacturing center in the eastern Mediterranean.

The large variety in designs and technical details of the molds used indicates that this Type of vessel was made in more than one glass center. The most important criterion for distinguishing beakers made in different workshops should be the construction of the molds, but at this time not enough information is available. Another criterion may be the decoration itself. There appear to be many

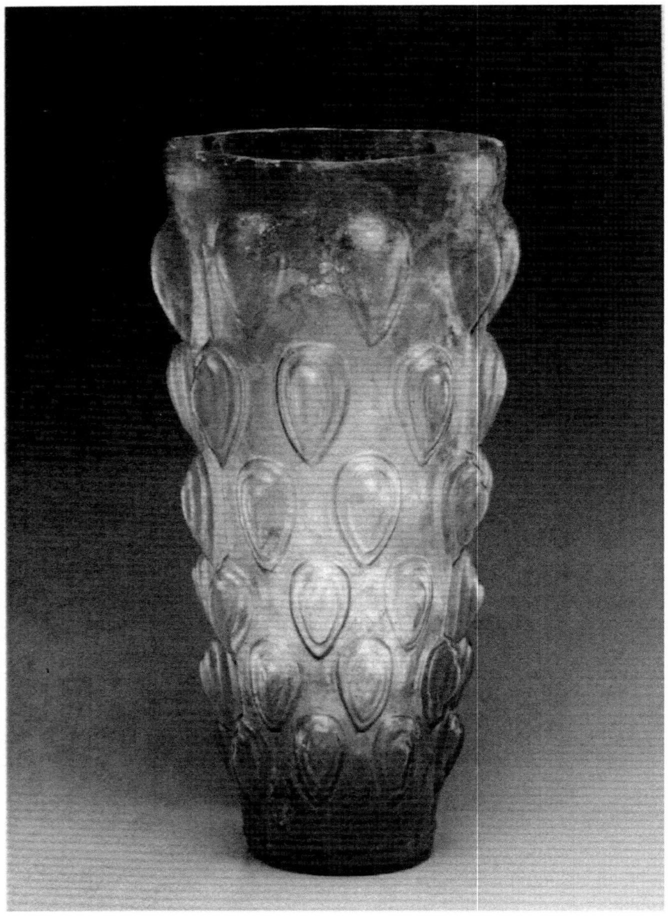

8 (1:2)

more variations in the decoration of beakers excavated in the West than in those from eastern sites. Beakers from the eastern Mediterranean appear to have been decorated almost exclusively with either three-tiered knobs or three-tiered knobs alternating with circular bosses, while plain knobs and knobs combined with linear patterns are common in the West.[34] It is therefore tempting to assign most of the Series with three-tiered knobs to eastern Mediterranean workshops, particularly those that were blown in four-part molds like the Toledo beakers. Designs with knobs and linear patterns were probably made only in the West, as were beakers with knobs in rows arranged directly under each other. The same western workshop (or workshops), which has not yet been located, may have made mold-blown beakers like Cat. No. 11, which are often found at the same western sites as the knobbed beakers and appear to have had largely the same distribution pattern. Vessels made in molds with only two vertical sections and vessels that were finished with a tooled base are also likely candidates for western workshops.

### Interpretation

Glass vessels of this Type have often been called "bossed," "lotus bud," or "almond" beakers.[35] Although it is possible to interpret smooth pointed knobs as rows of hanging lotus buds or almond nuts, three-tiered knobs with profiled edges are characteristic of knots in wood. Knots carved in alternating rows can be seen, for example, in the hanging clubs sculpted on a Roman altar dedicated to Hercules in 81 (Fig. 69).[36] Knots were an essential detail of Hercules's club, which was regarded as the trunk rather than the branch of an olive tree.[37] Knots are prominent in all Greek and Roman representations and imitations of Hercules's club.[38]

The importance of the knots is attested in a poem by the third-century-B.C. Greek poet Theokritos (17.31), where Hercules's club is said to be of iron cut to resemble knots of wood. The three-tiered knobs on the glass beakers indicate that the material imitated by these vessels must be wood,[39] presumably olive wood which is exceptionally knotty.

The glass beakers should therefore probably be associated with the Greek hero Herakles who was worshipped as Hercules in Rome.[40] In Rome, libations of wine were offered regularly to Hercules in a wooden skyphos, a two-handled drinking cup, that was said to have been left behind by Hercules himself.[41] Small bronze statuettes showing Hercules holding his skyphos in one hand and his club in the other were a regular feature in Roman household shrines.[42] A mold-blown glass bottle in a private collection is shaped like a club wound with ivy leaves (Fig. 70).

The shape of the knots on the glass vessels is very similar to decorative knots in a silver mount for a drinking horn shaped like Hercules's club found in Algeria in a Hellenistic tomb dated to the end of the second century B.C.[43] The conical shape of the glass beakers probably refers to Hercules's club.[44] In Greek and Roman tradition Hercules's vast appetite for food and wine was famous; he is shown in drinking contests with the wine god himself, Dionysos, on mosaics from Antioch in Syria.[45] Drinking from a glass beaker whose decoration evoked the presence of Hercules and his prowess in drinking would have been an appropriate way to ward off evil during a drinking bout when one was particularly vulnerable.

[1] Isings 1957, Form 31, now named Isings Form 31a by Berger 1981, 24, see at Cat. No. 11.

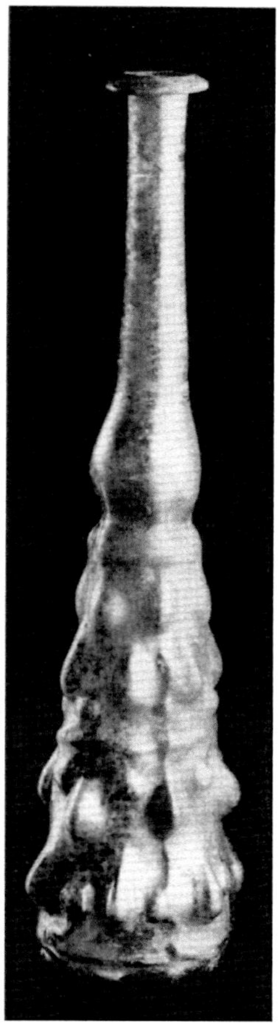

*Fig. 70. Bottle shaped like a Hercules club, Ernesto Wolf Collection. The shape of this bottle and the ivy leaves shown in relief wrapped around it suggest that this vessel is associated with Hercules. Photo: Courtesy Ernesto Wolf.*

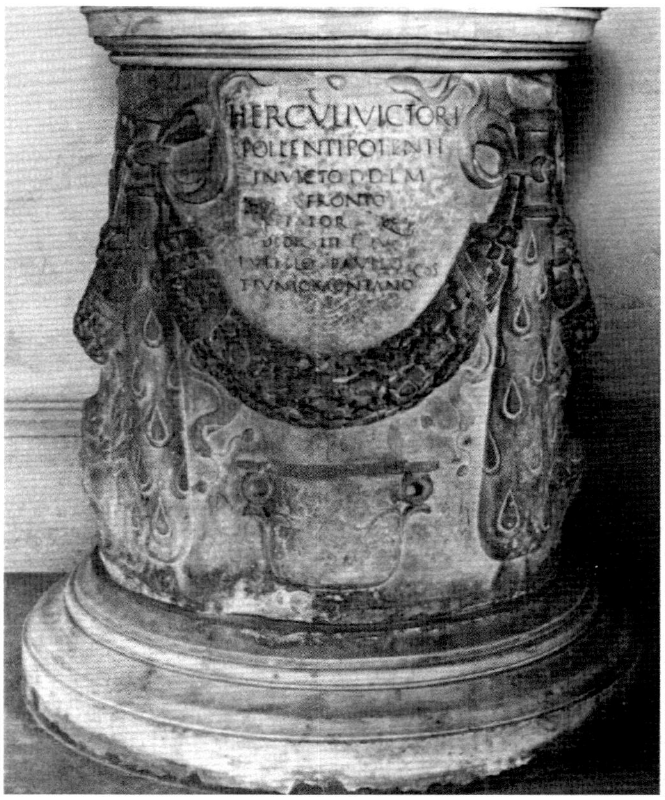


*Fig. 69. Circular marble altar dedicated to Hercules in A.D. 81. The knots seen on the Hercules clubs carved on this altar are similar in shape and arrangement to the bosses found on truncated conical beakers like Cat. Nos. 8-10, suggesting that these beakers should be associated with the god Hercules. Rome, Museo Capitolino, no. 1962. Photo: Oscar Savio Fotografo.*

[2] For example:
- a) Los Angeles County Museum of Art, M.88.129.51, formerly Hans Cohn Collection, tall-necked bottle, H 29.5 cm, pale green, with four alternating rows of eight knot-shaped bosses, von Saldern 1980b, 52, no. 45;
- b) ROM, no. 950.157.293, H 9.1 cm, amber, with two rows of eight alternating bosses, Hayes 1975, no. 86.

[3] For example, RGM, no. KL 38, formerly Loeffler Collection, no. 23, H 23.5 cm, yellowish green, *Glass of the Caesars* 1987, 162-163, no. 84; La Baume and Salomonson 1976, 37, no. 69 color pl. 4.

[4] BM, no. GR 1872.7-26.1, H 17.0 cm, amber, with a metal cap on the nozzle, *Masterpieces* 1968, 55, no. 66; Veronica Tatton-Brown, "Early Blown Glass: 1st century A.D.," in Tait 1991, 74-75, fig. 94, color.

[5] For example:
- a) Oxford, Ashmolean Mus., no. 1953.636, excavated at Kolophon in Asia Minor in the tomb of a physician, exceptionally large beaker of purple glass with six alternating rows of eight three-tiered knot-shaped bosses, Richard Caton, "Notes on a group of medical and surgical instruments found near Kolophon," *JHS* 34 (1914) 118, no. 16, pl. 12, no. 36;
- b) Loeffler Collection, from unknown origin, H 17.5 cm, natural green, also with six alternating rows of eight three-tiered knots, La Baume and Salomonson 1976, 36-37, no. 67.

[6] Exceptions are beakers with fire-polished rims:
- a) Hama, Syria, Mus., number unknown, transparent purple conical beaker with an everted rim, unpublished;
- b) Keszthely, Hungary, Balaton Mus., number unknown, from Keszthely, nearly cylindrical beaker decorated on the lower half of the body with only two rows of three-tiered knobs alternating with circular bosses, Barkóczi 1988, 94, no. 139, pl. 75.

[7] Large beakers with six rows of alternating three-tiered knobs:
- a) BM, no. 1913.5-22.18, formerly Durighello Collection (indicating a Syro-Palestinian provenance?), H 20.8 cm, said to have been made in a tripartite mold (referring to walls only?), *Masterpieces* 1968, no. 62;

b) formerly R. W. Smith Collection, no. 1368, probably from Asia Minor, H 20.0 cm, Smith 1957, 59, no. 71;
c) Bremen Collection, number unknown, natural olive green, H 21.0 cm, Bremen 1964, 181-184, no. 4a;
d) Oppenländer Collection, no. 2454, H 21.2 cm, light brown, von Saldern et al. 1974, no. 455.

For smaller vessels, see Cat. No. 9, with five rows of three-tiered knobs, and Cat. No. 10 with four rows.

[8] For example:
a) Yale, no. 1955.6.271, Matheson 1980, 49, no. 126;
b) from Panticapaeum, Sorokina 1962, 217, fig. 4;
c) Leiden, Rijksmuseum van Oudheden, no. B 1901/6.1, acquired from Jean Farah, said to have been found southeast of Tyre, Lebanon, H ca. 19 cm, transparent natural olive green;
d) from Podgrade, Croatia, light green, Damevski 1974, 84, pl. 12, no. 5;
e, f) Murano, Italy, Museo Vetrario, numbers unknown, said to be from the Nona necropolis at Zadar, Croatia, two beakers with four alternating rows of eight three-tiered knobs and small circular bosses.

[9] For example:
a) Vindonissa, Switzerland, Landesmuseum, eleven fragments, nos. 732, 9529, 9530, 1942: 429, and seven other pieces with unknown numbers, Berger 1960, 53, no. 129, pl. 8;
b) Zurich, Switzerland, Nat. Mus., no. 58920.08, from tomb S3 at San Materno near Ascona, dated about A.D. 100, H 14.8 cm, neutral bluish green, with five alternating rows of double, downward-pointing knobs, Donati et al. 1987, 86-87, no. 8, ill. 163; Carazzeti and Biaggio Simona 1988, 102, ill., 104, no. 124; Biaggio Simona 1989, 172, fig. 2; Biaggio Simona 1991, I, 97-98, no. 011.1.023, pl. 9, fig. 45;
c) Augst, Switzerland, Römermuseum, no. 78.13225, from Augst, Rütti 1991, II, 49, no. 991, pl. 45.

[10] Excavated at Nijmegen, Netherlands, natural bluish green, Isings 1980, 293, no. 14, fig. 7, no. 6.

[11] For example:
a) from Pompeii or Herculaneum, Spinazzola 1928, pl. 228, center row, sixth from right;
b) Verona, Italy, Mus. Arch., from Raldon near Verona, beaker from the same Series, Maria Carina Calvi, "I vetri romani dell'agro veronese," in *Il territorio Veronese in età romana. Convegno del 22-24 Ottobre 1971 (Atti Verona: Accademia di agricoltura, scienze e lettere* 9, Verona 1973) 214, fig. 5;
c) Oxford, Ashmolean Mus., number unknown, said to have been found at Stoke Bruerne, near Northhampton, England, Brown 1968, 405, fig. 74;
d) from Verulamium, England, fragment, Charlesworth 1972, 198, fig. 74, no. 3;
e, f, g) Vindonissa, Switzerland, Landesmuseum, nos. 1932: 4001, 12633, 3115, three beakers with various designs, from Vindonissa, Berger 1960, 53-54, nos. 131, 132, 134;
h) from Baden (ancient Aquae Helveticae), Switzerland, Fünfschilling 1985, 118, no. 128, pl. 12;
i, j, k, l) Augst, Switzerland, Römermuseum, nos. 62.7008, 37, 79.6802, 79.14393, from Augst, four beakers, Rütti 1991, II, 49, nos. 992-995, pl. 45;
m, n) from Nijmegen, Netherlands, fragments from two or three beakers, Isings 1980, 293, nos. 11, 12a and 12b, fig. 7, nos. 7, 9, and 10;
o) St.-Germain-en-Laye, Musée des Antiquités Nationales, from Néris (Allier), France, Lantier 1929, no. 23A;
p) BM, no. GR 1872.7-26.1 (see above note 4).

[12] For example:
a) Pompeii, Antiquarium, no. 12206, from Pompeii, Scatozza Höricht 1991, 82, fig. 16, no. a;
b) Bellinzona, Switzerland, Ufficio Cantonale dei Monumenti Storici, no. 134.58.143, from Locarno, Solduno, tomb 58.17, beaker with four rows of ten (?) knobs each, transparent natural green, Biaggio Simona 1991, I, 97-98, no. 134.2.060; Donati et al. 1987, 152-154, no. 143, color pl. 265;
c) Bologna, Museo Civico, no. Rom. 8 (Palagi 3357), possibly excavated in northern Italy but without documentation on find spot, transparent blue, with four rows of twelve knobs each, H 11.6 cm, Meconcelli Notarianni 1979, 46-47, no. 33, pl. 33;
d) Adria, Italy, Mus. Naz. Arch., no. Bocchi I 5, perhaps from the vicinity of Adria, transparent green, with four rows of eight knobs each, H 14.0 cm, G. M. Fogolari and B. M. Scarfí, *Adria Antica* (Venezia 1970) 86, no. 66, 5, fig. 66, no. 5.

[13] For example:
a) Split, Croatia, Arch. Mus., no. G1551, excavated at Podgrade near Benkovca (ancient Asseria), with four alternating rows of eight three-tiered knobs and small circular bosses above a fifth row consisting of four different comedy masks alternating with three-tiered knobs, H 13.5 cm, D base 4.2 cm, Rnjak 1979, 236, no. 476;
b) Loeffler Collection, no. 137, natural green, with the same design as the beaker from Podgrade, H 12.8 cm, La Baume and Salomonson 1976, 36, no. 66.

[14] For example:
a) Ercolano, Antiquarium, no. E2341, from Herculaneum, Scatozza Höricht 1986, 30, 19, pl. I, top center;
b) Pompeii, Antiquarium, no. 35115, from Pompeii, fragmentary beaker, Scatozza Höricht 1991, 82, fig. 16, no. b;
c) Vindonissa, Switzerland, Landesmuseum, no. 12633, from Vindonissa, beaker with remains of a wreath above oval bosses framed by diagonal lines, Berger 1960, 53, no. 132, pl. 8.

[15] From published descriptions it is not always clear whether a three-part mold refers to only the walls of a vessel or includes the disk for the base, in which case the walls would have been made with two equal mold halves.

[16] For example:
a, b) Murano, Italy, Museo Vetrario, numbers unknown, two beakers from the Nona necropolis at Zadar, Croatia, Giovanni Mariacher, *Vetri di Zara restaurati* (Venice 1966) 10-11, nos. 14-15, fig. 12;
c) Oxford, Ashmolean Museum, number unknown, the Newport Pagnell Beaker, from Stoke Bruerne, Brown 1968, 405, fig. 74;
d) see above note 9b;
e) see above note 13b.

[17] For example:
a) see above note 16c;
b) see above note 14a.

[18] See above note 16c.

[19] For example, beakers from Pompeii or Herculaneum, Spinazzola 1928, pl. 228.

[20] See above note 5a. On the function of beakers: Hilgers 1969, 255, s.v. *poculum*.

[21] Five of thirty-one fragments from Vindonissa were found in Flavian contexts, dated between 70 and 96: Berger 1960, 52-54. At Vitudurum, fragments of beakers with three-tiered knots were excavated in House 10, in a deposit belonging to the early periods of the site, which means that they were discarded before the middle of the century. However, this early date is not secure because the archaeological context shows that material from the second half of the first century was mixed with the finds in this area: see the discussion by Rütti 1988, 36. The earliest datable examples at Augst were found with pottery dated second half of first and early second century: Rütti 1991, II, 71.

[22] For example:
a) Nijmegen, Mus. Kam, no. XXA 98 E, from Cemetery E at Nijmegen, dated about 10-70;
b) from the legionary camp at Utrecht, fragment, in a context dated second half of first century. Both finds mentioned by Isings 1957, 46, Form 31.

[23] A fragment of a beaker decorated with small circular knobs between knot-shaped bosses was excavated at the fortress at Satala, built in 74 or 75: C. Lightfoot, pers. com.

[24] For example:
a) from Pompeii and Herculaneum, seven examples, each slightly different, Spinazzola 1928, pl. 228, center row;
b) from Herculaneum, Scatozza Höricht 1986, pl. 1, top row, center;
c, d) Pompeii, Antiquarium, nos. 12206, 35115, Scatozza Höricht 1991, 82, fig. 16, nos. a and b.

[25] Grose 1973-76, 179, mentions fragments of six bossed beakers from Cosa.

[26] For example:
a) see above note 11b, a single example with circular knobs framed by a wavy linear pattern imitating wood;
b, c) perhaps found in northern Italy, two beakers with rows of knobs above each other, see above notes 12c and d.

[27] For example:
a) see above note 9b;
b) see above note 12b.

[28] For example:
a) Vindonissa, Switzerland, Landesmuseum, thirty-one fragments from Vindonissa, Berger 1960, 52-54, nos. 129-136;

b) from Vitudurum (modern Oberwinterthur), Switzerland, fragments from five beakers, Rütti 1988, 165, nos. 661-665, pls. 8 and 33;
c) Augst, Römermuseum, nos. 78.13225, 62.7008, 37, 79.6802A, 79.14393, 61.1224B, 80.32029, 79.14382D, fragments of eight beakers from Augusta Rauricorum (modern Augst), Rütti 1991, II, 49, nos. 991-998, pl. 45, Forms AR 33.1 A, B, C, and D.

29 For example:
a) Bonn, Rheinisches Landesmuseum, no. 70.0534, from Bonn, Anna-Barbara Follmann-Schulz, *Die römischen Gläser aus Bonn* (*Bonner Jahrbucher* Beiheft 46, Cologne 1988) 90, no. 316;
b) Duisberg, Germany, Niederrheinisches Mus., no. A 74/152, from Asciburgium (modern Asberg), Germany, Sophia M. E. van Lith, *Glas aus Asciburgium* (*Funde aus Asciburgium* 10, Duisburg 1987) 47, no. 178, pl. 92.

30 For example:
a) see above note 22a;
b) eight fragments from the site of the military settlement at Nijmegen, Netherlands, Isings 1980, 292-293, nos. 11-14, fig. 7, nos. 6-10, p. 326, no. 1, fig. 36, no. 2, and p. 331, nos. 4-6, fig. 37, nos. 7, 2, and 5.

31 From *England*:
a) see above note 16c;
b) from Verulamium (modern St. Albans), seven fragments from two beakers, Charlesworth 1972, 198, 197, fig. 74, no. 3;

from *France*:
c) see note 11o;
d) from the Rhone Valley, Morin-Jean 1913, 139, fig. 186;
e) Metz, Mus., beaker from the cemetery of Marsbach, France, Isings 1957, 46, Form 31;
f) excavated at St. Cierges, fragments, Isings 1957, 46, Form 31;

*Spain*:
g) from Italica, M. Vigil Pascual, *El vidrio en el mundo antiguo* (*Bibliotheca Archaeologica* 7, Madrid 1969) 126, fig. 96;
h) eight more examples from Spain: Price 1973, 66-67;

*Portugal*:
i) Santiago de Cacém, Portugal, Mus. Municipal, green fragment of a cup from Mirobriga in southwest Portugal, Jorge de Alarcão, "Vidros romanos de museus do Alentejo e Algarve," *Conimbriga* 7 (1968) 37, no. 71, pl. 3;
j) two fragments from a beaker from Conimbriga, Jorge de Alarcão, *Fouilles de Conimbriga*, VI: *Céramiques et Verres* (Paris 1976) 168, 177, no. 45, pl. 35.

32 *Croatia* and *Slovenia*:
a, b) see above notes 8e, f;
c) Ljubljana, Slovenia, Nat. Mus., from Ljubljana (ancient Emona), Novotny-Schmid (Tito Street), grave 23, green, with four alternating rows of three-tiered knobs and circular bosses, Sonja Petru, *Emonske nekropole* (*odkrite med leti 1635-1960*) (Ljubljana 1972) 151, pl. 13, no. 21a;
d) see above note 8d.

*Hungary*:
e) see above note 6b.

*Romania*:
f) Constanta, Arch. Mus., no. 16003, from Neptun (Mangalia north), fragmentary beaker with remains of three rows of three-tiered knobs and circular bosses, pres. H 9.5 cm, Bucovala 1968, 52, no. 57.

33 *Syria* and/or *Palestine*:
a) Damascus, Nat. Mus., no. 15097, possibly from the Hauran, amber, with six alternating rows of three-tiered knobs, Zouhdi 1964, 53, no. 64, fig. 41;
b) see above, note 8c;
c) see above, note 6a;
d) see above, note 7a;
e) Beirut, formerly Kirchner-Schwarz Collection, probably acquired in Lebanon, *Kirchner-Schwarz* 1914, no. 107, pl. 2;
f) from Masada, Israel, Barag 1991, 139 (not ill.).

*Cyprus*:
g) Cambridge, Fitzwilliam Mus., no. GR.98.1876, formerly Cesnola Collection, from Golgoi, pale amber, with five alternating rows of two-tiered (?) knobs and circular bosses, H 11.0 cm, *Fitzwilliam* 1978, 33, no. 56.

*Egyptian Nubia*:
h) Oxford, Ashmolean Mus., no. 1912.1137, excavated at Faras, with five rows of three-tiered knobs, H 12.0 cm, P. Shinnie, *Meroe: A Civilization of the Sudan* (*Ancient Peoples and Places*, no. 55, New York and Washington 1967) 224, no. 83, fig. 83.

*Asia Minor*:
i) see above note 5a;
j) see above note 4;
k) see above note 7b;
l) fragment from the fortress at Satala (see above note 23).

*Kerch*:
m) from Panticapaeum, collection and number unknown, Sorokina 1962, 217, fig. 4; Sorokina 1967, 70, fig. 2, no. 7.

34 Conclusion confirmed by D. F. Grose [editor].

35 Bremen 1964, 181-184, no. 4a, with discussion.

36 Rome, Capitoline Mus. no. 1962: circular marble altar dedicated to Hercules by C. Ulpius Fronto, in 81, Helbig[4], II, no. 1425; *LIMC*, V, no. 2646.

37 *LIMC*, IV, 729.

38 For example, three glass bottles shaped like Hercules's club:
a) Besançon, France, Mus. des Beaux-Arts, mold-blown example of unknown provenance, Josiane Koltes, *Catalogue des collections archéologiques de Besançon*, VII: *La verrerie gallo-romaine (Annales littéraires de l'Université de Besançon 270, Paris 1982)* 62-63, no. 172, pl. 42;
b) Ernesto Wolf Collection (here Fig. 70);
c) Adana, Turkey, Bölge Mus., no. 56.43.1972/1070, from Kadirli, Cilicia, H 10.8 cm, bluish green, with applied opaque white knots, unpublished.

Compare:
d) MKG, no. 1968.3, bronze votive club, *AA* 89 (1974) 57, no. 9, fig. 9, nos. a-c;
e, f) Loeffler Collection, nos. N 118, N 118a, from North Africa, two ceramic bottles shaped like a club, with the knots pointed upward instead of downward, La Baume and Salomonson 1976, nos. 643 and 643a, pl. 70, no. 1.

39 As opposed to imitations of vessels of metal, ceramics, or vessels decorated with inlaid gems as for example, in Cat. No. 12. Wooden beakers, especially of beechwood, are frequently mentioned in Roman poetry: for example, Vergil, *Eclogae* 3, 36f; Ovid, *Metamorphoses* 8, 669; *Fasti* 5, 522; cf. Hilgers 1969, 260-261.

40 On the worship of Hercules in Rome: Simon 1990, 72-87.

41 Hercules's wooden scyphus is mentioned by Servius, *ad Aen.* 8, 278: Simon 1990, 76.

42 For example, a bronze statuette in Bologna, Museo Civico, no. 1012: Simon 1990, 76-77, fig. 97.

43 Constantine, Algeria, Mus., from Es Soumaa near El Khroub, L 9.5 cm, Heinz Günter Horn and Christoph B. Rüger eds., *Die Numider. Reiter und Könige nördlich der Sahara*, exh. cat., Rheinisches Landesmuseum Bonn (Cologne and Bonn 1979) 600-601, pl. 118. On Hercules's club as apotropaion for drinking: H. G. Horn, "Eine Weihung für Hercules Magusanus aus Bonn," *BJb* 170 (1970) 247-248.

44 The knobbed beakers were first associated with Hercules's club by Isings 1976, 353; Clasina Isings, "Roman Glass from Nijmegen," *AnnAIHV* 8 (1979, published 1981) 101-104, 105; Isings 1980, 282 and 292. So also Carnegie, no. 29490, Oliver 1980, no. 70, no. 65.

45 Two mosaics excavated at Antakya, Turkey (ancient Antioch-on-the-Orontes): Worcester, Mass., Worcester Art Museum, 1933.36, mosaic emblema, ca. 50-150, and Princeton, N.J., Princeton University Art Museum, Y1965.216, mosaic panel: ca. 250-300, D. J. Smith, "Mosaics," in M. Henig ed., *A Handbook of Roman Art* (Oxford 1983) color pls. 8 and 9. Both mosaics appear to copy one Hellenistic composition, Herakles drinks from a wooden skyphos; his club lies prominently in the foreground.

## 9 Truncated Conical Beaker with Knot-shaped Knobs

Probably Eastern Mediterranean
Mid to second half of first century
H 12.1 cm; max D 7.6 cm; D rim 6.25 cm; Th rim 0.15 cm; D base 3.96-3.95 cm
1923.492, Gift of Edward Drummond Libbey

Transparent to translucent dusky yellow. Thin glass. Fabric cannot be determined because of weathering.

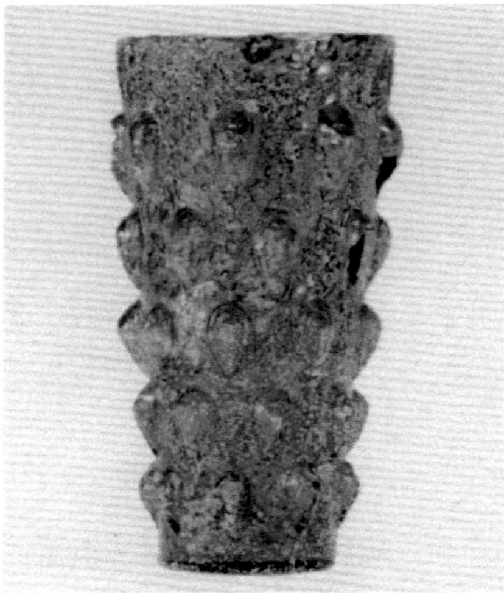

9 (1:2)

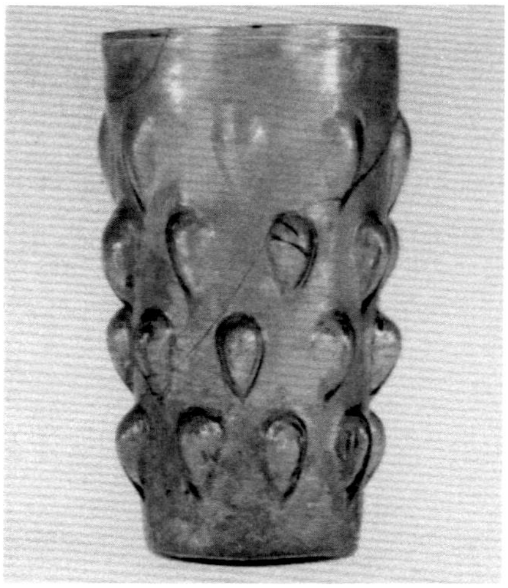

10 (1:2)

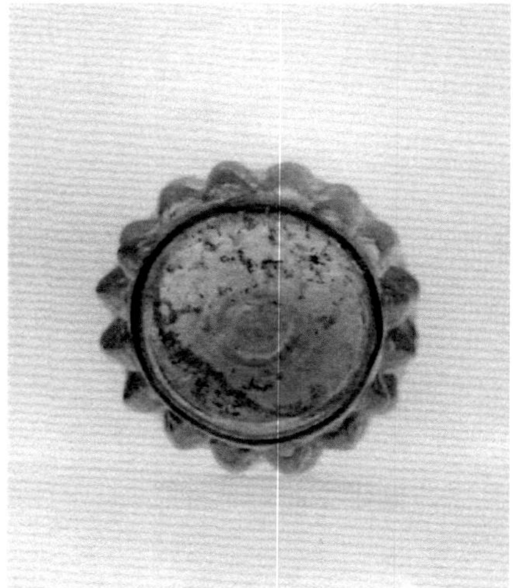

10, underside (1:2)

Blown into a four-part mold with three vertical sections joined to a disk-shaped base section (MCT III). Relief crisp.
Rim slightly ground. Truncated conical body. Flat base with two raised concentric rings around central boss.
On body, five alternating rows of eight knot-shaped knobs in three-tiered relief diminishing toward base.
*Condition* Complete; a few small chips missing from rim. Losses in four knots. Exterior covered with severely pitted golden iridescence; where this golden iridescence has flaked off, there is silver iridescence.
*Collection* Thomas E. H. Curtis.
*Published* Previously unpublished.
*Remarks* See Remarks Cat. No. 8.[1]

[1] Small beakers with five alternating rows of three-tiered knots:
- a) Newark, no. 50.1863, said to have been found at Caesarea, Israel, pale yellow green, Auth 1976, 76-77, no. 79;
- b) CMG, no. 59.1.79, formerly R. W. Smith Collection, no. 673, acquired in Lebanon, amber, Smith 1957, 59, no. 72, 60, ill.;
- c) formerly Beirut, Kirchner-Schwarz Collection, probably acquired in Lebanon, *Kirchner-Schwarz* 1914, no. 107, pl. 2;
- d) Oxford, Ashmolean Mus., no. 1912.1137, excavated in Faras, Egyptian Nubia, grave 2097; P. Shinnie, *Meroe. A Civilization of the Sudan* (*Ancient Peoples and Places*, no. 55, New York and Washington 1967) 224, no. 83, fig. 83;
- e) BM, no. GR 1892.3-13.50, said to come from the vicinity of Cyzicus, Turkey, on the south bank of the Bosporus, the top row of knots is single-tiered instead of three-tiered, Veronica Tatton-Brown, "Early Blown Glass: 1st Century A.D.," in Tait 1991, 74-75, fig. 96, right, color;
- f) formerly Paris, de Clerq Collection, find spot not mentioned, amber, de Ridder 1909, 240, no. 520, pl. 27;
- g) Loeffler Collection, no. 30, transparent light brown, La Baume and Salomonson 1976, no. 68, color pl. 4;
- h) Carnegie, no. 30082, Oliver 1980, 70, no. 65, 67, ill.;
- i) Baltimore, Walters Art Gallery, no. 47.574, Oliver 1980, 70;
- j) MMA, no. 29.100.80, Oliver 1980, 70;
- k) Saint Louis Art Mus., no. 422.19, Oliver 1980, 70.

### 10 Truncated Conical Beaker with Knot-shaped Knobs

Probably Eastern Mediterranean
Mid to second half of first century
H 12.4 cm; max D 7.2 cm; D rim 6.85 cm; Th rim 0.19 cm; D base 4.86-4.23 cm
1923.491, Gift of Edward Drummond Libbey

Transparent to translucent natural dusky yellow green tint. Medium thin glass. A few pinprick and small spherical and elongated bubbles.

Blown into a four-part mold with three vertical sections joined to a disk-shaped base section (MCT III). Relief indistinct.
Rim slightly ground. Truncated conical body. Flat base with two raised concentric rings and central boss.
A wheel-cut incision on exterior ca. 0.25 cm below rim. On body, four alternating rows of eight knot-shaped knobs, the top row consisting of single tiered knots, the lower three rows of knots in three-tiered relief.
*Condition* Complete; top broken and repaired. Exterior covered with iridescent film.
*Collection* Thomas E. H. Curtis.
*Published* Previously unpublished.
*Remarks* See Remarks Cat. No. 8.[1]

[1] Compare ROM, no. 950.157.40, from unknown find spot, H 12.6, yellowish green, Hayes 1975, 48, no. 83, pl. 7.

## *Truncated Conical Beaker with Isolated Figural Motifs* (Cat. No. 11)

### 11 Truncated Conical Beaker

Color plate 4, p. 51
Western Mediterranean, perhaps made in Italy
Second half of first century
H 14.5 cm; max D 7.5 cm; D rim 7.5 cm; Th rim 0.15 cm; D base 4.0 cm
1923.912, Gift of Edward Drummond Libbey

Transparent greenish gray tint (5 G 6/1). Medium thin glass. Pinprick and small vertically elongated bubbles.
Blown into a three-part mold with two vertical sections joined to a disk-shaped base section (MCT VII B). Vertical mold seams through battlement motif. Mold halves well aligned.
Horizontal, unworked rim with slight bulge below. Tall truncated conical body. Slightly concave base.
Below rim, two bands of fine wheel-cut incisions. On body, a mold-blown triple battlement motif of double lines bordered above by nine circular bosses with central depressions and below by eight similar bosses. In the space framed by the battlements, alternately a scallop between two oval bosses with central depressions and two asymmetrical leaf-shaped bosses with central depressions. On the underside of the base, a circular boss with a diameter of 1.5 cm surrounding a small central boss.
*Condition* Intact. Exterior covered with iridescent film.
*Collection* Thomas E. H. Curtis.

*Published* Grose 1978, 78, fig. 15; mentioned by Isings 1980, 331; Berger 1981, 24, n. 4, 26; von Saldern 1985, 31, fig. 12.
*Remarks* This beaker belongs to a small group of fine mold-blown vessels used as tablewares.[1] The delicate mold-blown patterns on the walls probably imitated *toreumata*, metal cups decorated with chased designs in relief.[2] The vessels are made of transparent decolorized or natural bluish green glass, presumably in imitation of silver, a few of a more strongly colored green. One vessel has cylindrical walls,[3] and another is a cup with a pushed-in base,[4] but most are conical beakers varying in height from about 12 to 14 cm. In recent years excavations have brought to light many new fragments, most of which are too small to permit reconstructing all the details of the decoration. The most remarkable feature of these beakers is the large diversity of related decorative patterns, the details of which appear to have differed slightly on each vessel, so that it is awkward to refer to each new design as a representative of a separate Series (see below). The beakers are closely related to truncated conical jugs decorated with similar designs.[5]

## Decoration

The decorative patterns consist of combinations of isolated floral and geometric motifs that are often drawn together by a battlement meander design as on this beaker and/or a floral wreath as on a beaker in Cologne.[6] In addition to a wreath or meander,[7] the most common motifs are dots, circles, ovals, asymmetrical leaf forms,[8] heart- and pelta-shaped[9] ornaments, shells, and rosettes. These motifs appear singly in rows, or alternating with another motif, or isolated within the frame of the battlement pattern. The shape of the vessels and the motifs decorating their walls connect this group on one hand with mold-blown beakers decorated with knot-shaped knobs, some of which carry a wreath in addition to knobs[10] (see Cat. No. 8), and on the other hand with beakers cut with high relief, bearing similar motifs[11] but usually without a wreath or meander pattern. The decoration of the Toledo beaker is very similar, if not identical, to that of a fragmentary beaker excavated at Pompeii.[12]

## Technique

In contrast to the knobbed beakers (Cat. Nos. 8-10) which have a lightly ground rim, the rims of beakers belonging to this group are often unworked. Beakers with a bulge below the rim, like this one, may have been knocked off the blowpipe with their moil, which was removed only after annealing. The large variety in decorative designs and the fact that most if not all examples of this Type appear to be unique suggest that the molds used for blowing were less durable than the clay molds used for other early mold-blown wares, including knobbed beakers. On the other hand, the motifs themselves are rarely as crisp as one would expect with a mold that was used only a few times. One explanation could be that small isolated motifs such as circles, leaf-shaped objects, shells, etc., were pressed in the mold with a hand-held clay stamp,[13] that wore off. The motifs on this beaker look as though they could have been made with four individual stamps, one for the circles, one for the ovals, one for the shells, and one for the asymmetrical leaf shapes.

## Date

This Type of beaker, and jugs blown in closely related molds, was still being made in 79, when Mount Vesuvius erupted. In a shop at Herculaneum a large number of glass vessels, including a related conical mold-blown beaker decorated with vertical scrolls, was

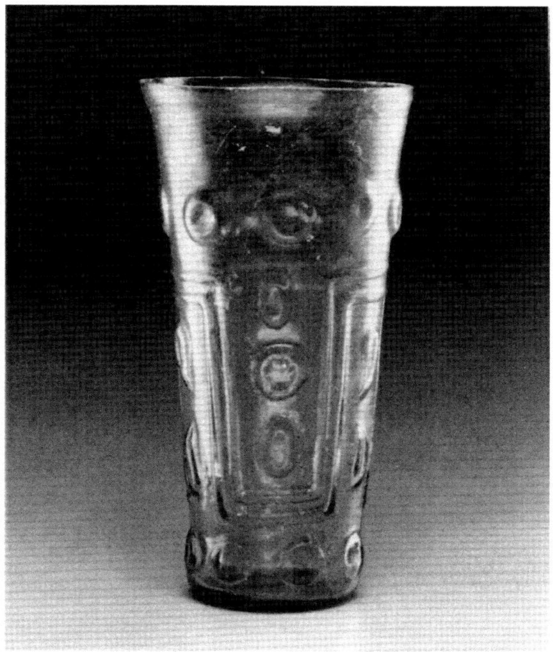

11 (1:2)

found still packed in straw and cloth.[14] However, finds in Switzerland and the Netherlands (see below) show that vessels of this Type were already being discarded around this time.

## Distribution and Origin

The distribution is similar to that of the western finds of beakers with knot-shaped knobs discussed at Cat. No. 8. Beakers and jugs are known from Pompeii and Herculaneum,[15] as are fragments from Solunto, Sicily,[16] Luni, northern Italy,[17] Switzerland,[18] and Nijmegen, Netherlands.[19] One fragment has been noted from Spain,[20] and another from Tunisia.[21] The distribution points to production in Italy, probably in Campania, because so many have been found at Pompeii and Herculaneum.[22] This is confirmed by the decoration, which has close links both to those mold-blown beakers with knot-shaped knobs that are found predominantly in the West (see Cat. No. 8) and to beakers with decoration cut in high relief which were probably made in the West. It is not clear whether a second production center existed in the eastern Mediterranean. One beaker is said to be from Syria and a related cup appears to have been found on Cyprus.[23] But most beakers with an Eastern Mediterranean find spot seem to have continuous floral designs[24] instead of isolated motifs.

1 Discussed by Isings 1957, 45-46, as Form 31; recognized as a separate form by Isings 1976; named Isings Form 31b by Berger 1981, 24.

2 Hilgers 1969, 261, s.v. *poculum*.

3 MMA, no. 17.194.232, formerly Gréau Collection, Froehner 1903, 156, no. 1125, pl. 203, no. 1; von Saldern 1985, 31, 38, fig. 15.

4 BM, no. GR 1892.5-26.1, with a row of circles and pelta-shaped objects below the rim and below that a row of masks, asymmetrical leaves, and floral motifs, von Saldern 1985, 31, 37, fig. 14; L. Palma di Cesnola, *Salaminia* (London 1882) 187, fig. 182.

5 Pompeii, Antiquarium, nos. 12104 and 12764, two jugs, from Pompeii, Scatozza Höricht 1991, 80, fig. 14, nos. a and b.

6 RGM, no. 1018, *Glass of the Caesars* 1987, 157, no. 79; La Baume 1973a, no. H4, pl. 45, no. 1.

7 The wreath and meander motifs are absent on a small number of beakers, including:

a) Vindonissa, Switzerland, Landesmuseum, no. 76.416.14, from Vindonissa, Berger 1981, 23-28, figs. 1, 3, and 5;

b) Naples, Mus. Naz. Arch., no. 111412, from Pompeii, yellow, with four horizontal rows of asymmetrical leaf motifs, hearts, dots, and pelta-shaped ornaments above a horizontal ridge just above the base, Berger 1981, 23, fig. 2.

8 The leaf-shaped designs have not yet been satisfactorily interpreted.

9 A *pelta* was originally a small, light shield in the shape of a half-moon that was used by Thracians and other peoples north of Greece. The shape became a popular motif in Hellenistic and Roman art; see C. Daremberg and E. Saglio, *Dictionnaire des antiquités grècques et romaines* 2, part C (Paris 1887) 1257-1258, s.v. *Clipeus*, V: La pelta.

10 For example, a beaker excavated at Herculaneum, Scatozza Höricht 1986, 19, pl. I, top center; Vindonissa, Switzerland, Landesmuseum, no. 12633, from Vindonissa, Berger 1960, 53, no. 132, pl. 8.

11 Axel von Saldern, "Römische Hochschliffgläser," *Jahrbuch des Museums für Kunst und Gewerbe Hamburg* 4 (1985) 27-30.

12 Pompeii, Antiquarium, no. 35083, from Pompeii, Scatozza Höricht 1991, 82, fig. 16, no. d.

13 Berger 1981, 26 with n. 7, remarks correctly that the stamps could have been impressed either directly in a negative mold or in the positive model from which the mold was cast.

14 For example:

a) Ercolano, Antiquarium, no. E 2356, from Herculaneum, H 14.5 cm, natural greenish, Alfonso de Franciscis, "Vetri antichi scoperti ad Ercolano," *JGS* 5 (1963) 137-139, figs. 1-4 (the mold-blown beaker is fig. 2, no. 3); Scatozza Höricht 1986, 39-40, no. 66, color pl. 7; perhaps identical with Eisen 1927, I, 293, fig. 130, pl. 63, no. c, noted as being from Pompeii.

In the same Series:

b) Oppenländer Collection, no. 22285, von Saldern et al. 1974, 166-167, no. 456, 169, ill.

15 For example:

a, b) see above note 5.
c) see above note 12.

Related mold-blown beakers:

d) Pompeii, Antiquarium, no. 12048, from Pompeii, Scatozza Höricht 1991, 81, fig. 15;
e) Pompeii, Antiquarium, no. 35096, from Pompeii, Scatozza Höricht 1991, 82, fig. 16, no. e;
f) see above note 7b;
g) see above note 14a.

16 Palermo, Italy, Mus. Arch. Regionale, two fragments of conical cups with ivy leaf and tendril decoration, Antoniana Giammellaro Spano and Francesca Spatafora, "Region: Sicilia," *BullAIHV* 9 (1981-1983) 189.

17 Antonio Frova, *Scavi di Luni* (Rome 1977) 288, no. 83, pl. 157, nos. 5, natural bluish, and 6, natural bluish green, fragments of two beakers, the first with a mold-blown base decorated with concentric circles and remains of a spiral, flower, and "drop" (asymmetrical leaf?) on the lower part of the wall, the second with a pushed-in base ring and remains of a spiral, and probably of "drops" on the lower part of the wall.

18 Fragments of seven beakers from excavations at four legionary settlements in northern Switzerland. From *Vindonissa*: three beakers:

a) Vindonissa, Landesmuseum, no. 76.416.14, almost complete beaker, with five horizontal rows of six asymmetrical leaf-shaped motifs only, Berger 1981, 23-28, figs. 1, 3, and 5;
b, c) Vindonissa, Landesmuseum, nos. 1954:40, 1923:40, fragments of two beakers unearthed during earlier excavations, the first, green, with remains of a horizontal ridge just above the base and a horizontal row of asymmetrical leaf-shaped ornaments alternating with dots bordered below and above by a row of dots only; the second a light green rim fragment preserving below a wheel-cut incision an asymmetrical leaf-shaped ornament and a dot, perhaps from a row of alternating leafs and dots, and below that a circular with central depression, Berger 1960, 54, nos. 137 and 138, pl. 8; Berger 1981, 25, figs. 6 and 7.

From *Vitudurum*:

d, e) wall fragments of two beakers, the first with floral spray, the second with remains of a battlement and ovals, Rütti 1988, 37, nos. G666-667, 165-166, pls. 8 and 33.

From *Aquae Helveticae* (modern Baden):

f) fragment, Fünfschilling 1985, 118, no. 129, pl. 12.

From *Augst*:

g) fragment preserving part of a battlement, a shell, and a wreath, Rütti 1991, II, 49, pl. 45, no. 999*.

19 Fragments of seven beakers from Nijmegen:

a) Nijmegen, Provincial Museum Kam, no. XXa 211a, lower half of a beaker of nearly colorless glass with a yellowish tinge, pres. H 8.6 cm, with a horizontal ridge just above the base, two straight laurel sprays joined in the middle by a round knob and at both ends by hanging fillets, *peltae* suspended at the ends of the sprays, and above this zone a meander encasing ovals and shells alternately, Isings 1976, 353-356, figs. 1 and 2;
b) excavated at Nijmegen in 1967, five fragments of clear glass with a bright green tinge, two of which join, in the upper row, raised rings with depressed centers, below that, part of a stalk, around the center a battlement pattern with remains of part of a shell which is not quite identical to that on beaker a) above and is less well blown, around the lower wall two branches of myrtle leaves with berries, from which a *pelta*-shaped object is suspended, below that a horizontal ridge, the rim itself polished and curved slightly inward, Isings 1976, 353-355; Isings 1980, 293, no. 7, figs. 6 and 7.1, 2, 3, and 5;
c) light green tinge, preserving an ivy leaf and part of a scroll, Isings 1980, 293, no. 5, fig. 5, no. 6;
d) colorless, preserving raised rings and a vine scroll, Isings 1980, no. 8, fig. 5, no. 7;
e) colorless wall fragment, preserving part of meander or rectangle pattern and traces of a scroll, Isings 1980, 331, no. 1, fig. 37, find 1/85, no. 1;
f) colorless wall fragment preserving traces of a vertical scroll, Isings 1980, no. 2, fig. 37, find 1/85, no. 4;
g) colorless fragment of lower side preserving a row of round bosses above a horizontal ridge, Isings 1980, no. 3, fig. 37, find 1/85, no. 4.

20 Gudiol Ricard, *Los vidrios Catalanes. Monumenta Cataloniae*, III (Barcelona 1936) 23, fig. 4, from Rubí, near Barcelona, "lower body and base of truncated conical beaker, moulded wreath encircling lower body with key pattern containing raised ovals above, rim and upper body missing" (cited in Price 1973, 68).

21 Tunis, Bardo Mus., no. 1790, perhaps from Sidi-El-Hani, lower part of beaker preserved, Mohammed Yacoub, "Les verres romains des musées de Sfax, de Sousse, et du Bardo," *BullAIHV* 6 (1971-72) 61, figs. 46 and 62, no. 21.

22 Isings 1976, 355; Scatozza Höricht 1991, 83.

23 A beaker in the RGM, no. 1018 (see above note 6) acquired from Kevorkian, London, is said to be from Syria. For the cup from Cyprus, BM, no. GR 1892.5-26.1, see above note 4.

24 For example, Cambridge, England, Fitzwilliam Mus., no. GR.99.1876, a beaker from Soloi, Cyprus, formerly Cesnola Collection, *Fitzwilliam* 1978, 32, no. 53.

## *Drinking Vessels with a Pronounced Shoulder* (Cat. Nos. 12-14)

### 12 Truncated Conical Beaker

Eastern Mediterranean
Mid to second half of first century
H 9.72 cm; max D 6.85 cm; D rim 5.85-5.75 cm; Th rim 0.1 cm; D base 4.0-3.9 cm
1923.913, Gift of Edward Drummond Libbey

Transparent to translucent dusky yellow green (near 5 GY 5/2, but more yellow). Medium thin glass. Numerous small spherical and vertically elongated bubbles.

Body blown into a three-part mold with two vertical sections joined to a disk-shaped base section (MCT VII).

Rim unworked with narrow bulge below. Pronounced shoulder. Truncated conical body. Flat base with four raised circles.

On body, a row of hanging triangles above seven alternating rows of twenty diamond-shaped bosses.

*Condition* Intact. Exterior covered with iridescent film.

*Collection* Thomas E. H. Curtis.

*Published* Previously unpublished.

*Remarks* Characteristic of this shape is the pronounced shoulder, also called "eingezogener Schulter," which also occurs on deep

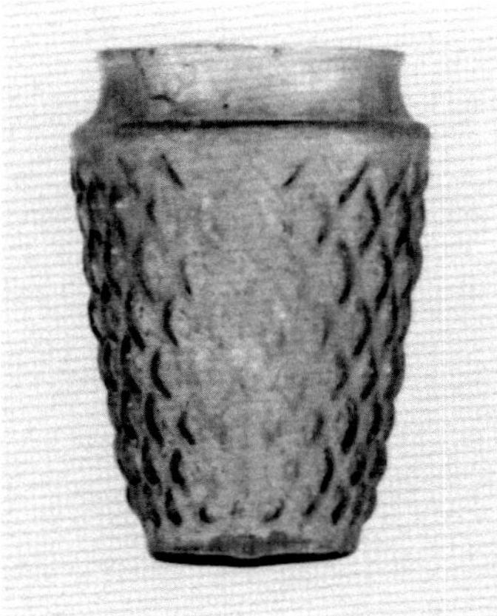
12 (1:2)

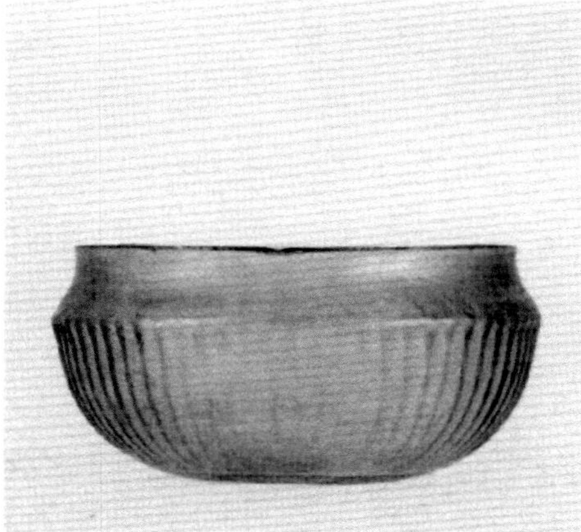
13 (1:2)

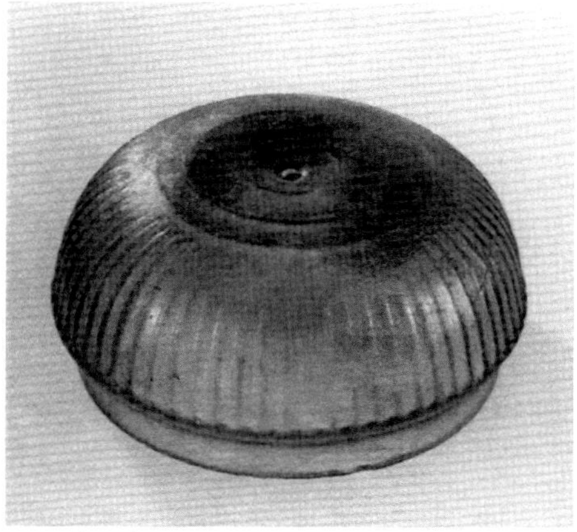
13, underside (1:2)

drinking bowls such as Cat. No. 13. Like those bowls, most beakers with a pronounced shoulder are decorated with mold-blown vertical flutes.[1] Decoration with diamond-shaped bosses is rare; excavated examples are known from Panticapaeum,[2] Ampurias,[3] and perhaps from Cologne.[4]

In northwest Europe, beakers with a pronounced shoulder decorated with fluting were already being discarded in the third quarter of the first century, so they could have been first made around the middle of the century. A similar production date may be assumed for the beaker with diamond-shaped bosses from Panticapaeum; it was found in a tomb that contained a locally made glass unguentarium of a type used in the second half of the first and early second century.

The bosses used as decoration for this beaker differ from the three-tiered knobs of beakers like Cat. No. 8 in that they appear to imitate not oval pointed knots in wood but facet-cut gems. They may be glass imitations of *pocula gemmata*,[5] gold and silver[6] cups decorated with gems, a royal custom especially in Syria according to the Roman author Cicero, writing in 70 B.C. (*In Verrem* 2.4, 62).[7] Triangular bosses imitating facet-cut gems appear on a bulbous beaker excavated at Fishbourne.[8] Oval gems cut cabochon style may be imitated by a few conical beakers with bosses and by a curious deep blue glass beaker blown into a silver mold with eight alternating rows of oval openings through which the glass bulges.[9]

[1] For example:
- a) Baden (ancient Aquae Helveticae), Switzerland, Fünfschilling 1985, 118, no. 127, pl. 12; Rütti 1988, 34, fig. 14;
- b) Aquileia, Italy, Mus. Arch., no. 12675, H 8.0 cm, colorless glass with reddish streaks, Calvi 1968, 101, 105, no. 250, pl. 16, no. 2;
- c-e) Augst, Römermuseum, nos. 72.4771, 37.3267, 58.2102, fragments of three examples from Augst, Switzerland, Rütti 1991, II, 47, pl. 44, nos. 983-985;
- f, g) unpublished examples from Benghazi, Libya, and Caerleon, Wales, cited in Price 1991, 72.

Ovoid beakers with pronounced shoulder and mold-blown fluting:
- h) RGM, no. 29.157, Fremersdorf VI, 1961, 42, pl. 66, from Cologne, Spiesergasse, natural bluish green, from a tomb that contained a coin of Nero (54-68);
- i) Valkenburg, no. 6003, fragment from Valkenburg, Z.H., Netherlands, dated ca. 47-69, van Lith 1978-79, 101, 103, no. 324, pl. 22.

[2] St. Petersburg, Hermitage, no. P.1912.15, from Panticapaeum, tomb 30, Kunina and Sorokina 1972, 163, fig. 8, no. 24; Kunina 1973, 138-139, fig. 26.

[3] Gerona, Spain, Mus. Arqueologico, nos. 767, 776 (?), from Ampurias, M. Oliva Prat, "Catalogo de los Vidrios Romanos de Ampurias del Museo Arqueologico di Gerona," *Anales del Instituto de Estudios Gerundenses* 6 (1951) 132, 134, pl. 28, fig. 2; Price 1973, 67.

[4] RGM, no. 758, said by Fremersdorf to be from Cologne, but according to La Baume the provenance is not known, Fremersdorf VI 1961, 50, pl. 98; La Baume 1973a, H 5, pl. 45, no. 2.

[5] On *pocula gemmata*, Hilgers 1969, 261, s.v. *poculum*.

[6] Silver cups decorated with gems are mentioned by the Roman poet Silius Italicus (lived ca. 26-101), *Punica*, 14.661.

[7] ... *pocula ex auro, quae, ut mos est regius et maxime in Syria, gemmis erant distincta clarissimis* (... golden cups, which, as is common with kings and especially those of Syria, were adorned with splendid jewels). Cicero, *The Verrine Orations* 2, translation by L. H. G. Greenwood (*Loeb Classical Library*, Cambridge, Mass., and London 1935) 2.4, 62.

[8] Donald B. Harden and Jennifer Price, "The Glass," in B. Cunliffe ed., *Excavations at Fishbourne 1961-1969: The Finds* (*Report of the Research Committee of the Society of Antiquaries of London* 27, London 1971), 338-339, no. 37, fig. 138, pl. 28, from a context dated to the years 43-75.

[9] BM, no. GR 1870.9-1.2, H 9.3 cm, said to have been bought in Florence but possibly originating from Brindisi in 1865, *Glass of the Caesars* 1987, 156, no. 78, color pl.; *Masterpieces* 1968, 59, no. 74. Its date is problematic: D. Harden, in *Glass of the Caesars*, 153-154, suggests a first-century date on account of the shape of the interior glass beaker. Veronica Tatton-Brown, "Early Blown Glass: 1st Century A.D.," in Tait 1991, 74-75, fig. 95, dates the piece "about A.D. 50-100;" while Fritz Fremersdorf, *Die römischen Gläser mit aufgelegten Nuppen in Köln* (*Die Denkmäler des römischen Köln* 7, Cologne 1962) 57, pl. 115, connects it with fourth-century vessels decorated with blue blobs. The nearest parallel for a surviving combination of metal and gems (perhaps of glass) is a stemmed cup made of electrum, a natural mixture of gold and silver, with large circular openings for gem inserts. The profile of the openings would not have allowed glass to be blown into them. This cup was found in a royal tomb of the fifth century in Hungary, near Szeged-Nagyszéksós. The beaker in the British Museum is unique in the history of ancient glass as the only extant example of blowing into an open-work mold.

## 13 Finely Ribbed Hemispherical Bowl with Pronounced Shoulder

Probably Eastern Mediterranean
Probably second quarter to mid first century
H 5.05 cm; max D 10.2 cm; D rim 9.2 cm; Th rim 0.18 cm
1923.425, Gift of Edward Drummond Libbey

Transparent pale green (?) glass. Medium thin glass. Pinprick and small spherical bubbles.

Rim free blown. Body blown into a vertically ribbed mold.[1] Rim lightly ground with concave lip below. Pronounced shoulder. Convex sides. Flat base with depression in center.

On sides, sixty-four vertical flutes bordered above by a faint ridge and extending to just above base. On underside of base, three raised concentric circles around a central boss.

*Condition* Intact. Exterior and interior covered with a transparent pale green, transparent pale blue, and transparent dark yellow iridescent film.

*Collection* Thomas E. H. Curtis.

*Published* Previously unpublished.

*Remarks* This bowl belongs to a group of vessels characterized by a pronounced shoulder ("Schalen mit eingezogener Schulter"). Most vessels in this group are bowls (e.g. Cat. No. 14), but some are tall and are classified as beakers (e.g. Cat. No. 12). The bowls occur in many variants, each of which appears to imitate metal bowls turned on a lathe.[2] Some glass bowls are steep, while others are shallow like Cat. No. 14. Three basic types are known. The first has a stepped omphalos at the bottom of the bowl which bears concentric raised rings on the underside, while the vertical ribs on the wall are bordered below by one or more horizontal ribs;[3] the profile is somewhat reminiscent of mold-blown cups with inscriptions such as Cat. No. 1. The second type is represented by this bowl: it does not have an omphalos, although the underside of the base is decorated with mold-blown concentric rings and is slightly depressed; the vertical ribs on the wall reach to the base without horizontal ribs below to offset them.[4] The third type is shallow and has a straight vertical rim and the shoulder is hardly pronounced (see Cat. No. 14).

### Date

Datable examples and fragments have been found in numerous first-century contexts (see below). Among the closest published parallels for the Toledo bowl are several from the north coast of the Black Sea at Panticapaeum and Kep, where they occur in tombs that have been dated independently to the second quarter to mid first century (see above note 4).

### Distribution and Origin

In most publications the three types of bowls and the beakers are discussed together, so it is usually not possible to distinguish between types. Vessels with a pronounced shoulder were used widely throughout the Roman empire, but find spots in the West predominate: Sidi Khrebish in Benghazi,[5] Tripolitania,[6] Spain,[7] Italy,[8] Switzerland,[9] Austria,[10] Germany,[11] Luxemburg,[12] Netherlands,[13] and England.[14] Find spots in Greece and the Eastern part of the Roman empire include Melos,[15] Knossos,[16] Cyprus,[17] western Anatolia,[18] Panticapaeum,[19] and Kep.[20] Finds from the Syro-Palestinian area are not well documented.[21] This distribution suggests that vessels with a pronounced shoulder were made in the West, and perhaps also in the East, but which were made where has not been established.

[1] MCT not possible to determine. No mold seams are visible. A faint horizontal mold-blown ridge (not indicated in profile drawing) marks the transition between ribs and lip undercutting the body of the bowl. This suggests either the use of a multipart mold that was opened in order to release the bubble after blowing, or that the glass blower tooled (pressed in) the shoulder after extracting the bubble from the mold.

[2] Mutz 1972, 66-69.

[3] For example:
- a) ROM, no. 924.14.4, Hayes 1975, 47-48, no. 82, with parallels from Cyprus, from western Anatolia, and from Vindonissa, Switzerland;
- b) fragment from Sidi Khrebish, Benghazi, Libya, Price 1985a, 294, nos. 43, 44, and 46, fig. 24, no. 3;
- c) from Vitudurum, Switzerland, base fragment, Rütti 1988, 34-35, no. 658, 165, no. G658, pl. 8;
- d) Augst, Römermuseum, no. 45.779, from Augst, Switzerland, Rütti 1991, II, 47, no. 974, pl. 44;
- e) Luxemburg, Musée d'Histoire et d'Art, from a tomb at Elvange-les-Hovelange, Luxemburg, Wilhelm 1969, 12, no. 8, pl. 57;
- f) Oberstimm, Germany, nos. 1975,165, 1975,270, 1975,305, fragments from Kastell Oberstimm, southern Germany, dated between 50 and 120, Hans Schönberger, *Kastell Oberstimm: Die Grabungen von 1968 bis 1971* (Berlin 1980) 280, nos. E21-22, fig. 88, pl. 108;
- g) Vindonissa, Landesmuseum, nos. 1919:231, 11287, from Vindonissa, Switzerland, Berger 1960, 55-56, nos. 141-142, notes the occurrence of a floral frieze below the vertical ribs of one fragment and a row of raised dots on another;
- h) Augst, Römermuseum, no. 69.6286, from Augst, Switzerland, with a row of raised dots below the ribs, Rütti 1991, II, 47, no. 980, pl. 44, 186;
- i) from Valkenburg, Z.H, the Netherlands, no. 6062, fragment with a row of raised dots, van Lith 1978-79, 101, no. 325, pl. 22.

[4] The closest parallels were found in the Crimea:
- a) St. Petersburg, Hermitage Museum, no. P.1910.43, from Panticapaeum, tomb 47, Kunina and Sorokina 1972, 158, fig. 6, no. 12, 159; Kunina 1973, 130-133, fig. 22, no. 1;
- b) from Kep, grave 88, Kunina and Sorokina 1972, 158, fig. 6, no. 42, 159;
- c) from Panticapaeum, tomb B (59-60), Kunina and Sorokina 1972, 166, fig. 10, no. 16.

All in contexts dated second quarter to mid-first century. See also:
- d) St. Petersburg, Hermitage, no. P.1902.140, from Panticapaeum, tomb 516 (328), Kunina 1973, 103-133, fig. 22, no. 2;
- e) St. Petersburg, Hermitage, no. P.1891.820, from Panticapaeum, excavated in 1891, Kunina 1973, 103-133, fig. 22, no. 3;
- f) St. Petersburg, Hermitage, no. E.681, said to be from near Nymphea in the Crimean Bosporus, Kunina 1973, 103-133, fig. 22, no. 3.

[5] See above note 3b; Price 1985a, 293, 294, nos. 43-46, fig. 24, no. 3.

[6] Jennifer Price, "Early Roman Vessel Glass from Burials in Tripolitania: A Study of Finds from Forte della Vite and other Sites now in the collections of the National Museum of Antiquities in Tripoli," in *Roman Tripolitania, Papers in Honour of Oliver Hackett* (*Society for Libyan Studies Occasional Papers* I, *BAR International Series* 274, Oxford 1985) 68, no. 3, fig. 6, no. 1, 69-70.

[7] Price 1973, 68, mentions fragments of at least twelve bowls found in Spain, mainly from sites on the Mediterranean coast.

[8] For example:
- a) from Valeggio, tomb 25, dated Claudian (41-54), Roffia 1979b, 113, fig. 2;
- b) Vigevano, Mus. Civico, no. ST. 21283, from Vigevano (Pavia), Roffia 1979b, 114, fig. 3;
- c) Torino, Mus., from Palazzolo Vercellese (Piemonte), Roffia 1979b, 114;
- d) Brescia, Mus. dell'età Romana, no. MR 170, from the region of Brescia, Calvi 1975, 214, fig. 20;
- e) from Cosa, twelve fragmentary ribbed bowls, Grose 1973-76, 179;
- f) Aquileia, Mus. Arch., no. 13027, perhaps from Aquileia, Calvi 1968, no. 249, pl. 16, no. 1.

[9] For example:
- a) Vindonissa, Landesmuseum, nos. 749, 750, 1919:231, 11287, 1934:830, from Vindonissa, Switzerland, from locations dated "Tiberian to early Claudian" (about 14-45?) to before 60/75, Berger 1960, 55-56, nos. 139-143;
- b) from Vitudurum (Oberwinterthur), Switzerland, fourteen fragments from periods I Early (7-35) through period III (70/80-110/120) but with most fragments dated between 35-70/80, Rütti 1988, 34-35, 165, nos. 647-660, pls. 8, 33;
- c) from Baden (ancient Aquae Helveticae), fragments from three beakers, Fünfschilling 1985, 118, nos. 124-126, pl. 11;
- d) Augst, Römermuseum, nos. 79.10233, 63.11871, 45.779, 80.34539, 67.16013, 80.29350, Aug-53-9, 45.1930, 69.6286, 62.11706A, 61.5749, from Augst, Switzerland, eleven fragments, found with ceramics dated 40-60 and 50-100, Rütti 1991, II, 47, nos. 972-982, pl. 44;
- e) Zürich, Nat. Mus., no. 58920.09, from Ascona, San Materno, grave 3, dated about 100, a unique example from the Ticino region, Biaggio-Simona 1991, I, 75-76, no. 011.1.017, pl. 4, fig. 37; Biaggio-Simona 1989, 172, fig. 3.

[10] From Magdalensberg, dated before 45, Czurda-Ruth 1979, 34, no. 318, pl. 2, 318.

[11] For example:
- a) Kempten (Bavaria), Germany, depot, no. LK 1941, from Kempten-Cambodunum, Fasold 1985, 212, no. 8, 211, fig. 7.6;
- b) see above note 3f.

[12] See above note 3e.

[13] From Valkenburg Z.H., the Netherlands, nos. 8706, 6003, 6062, 6869, in contexts dated period 2/3 (ca. 47-69), period 5 (ca. 120-to before 178), and period 6 (ca. 178-ca. 260), van Lith 1978-79, 101-103, nos. 323-326.

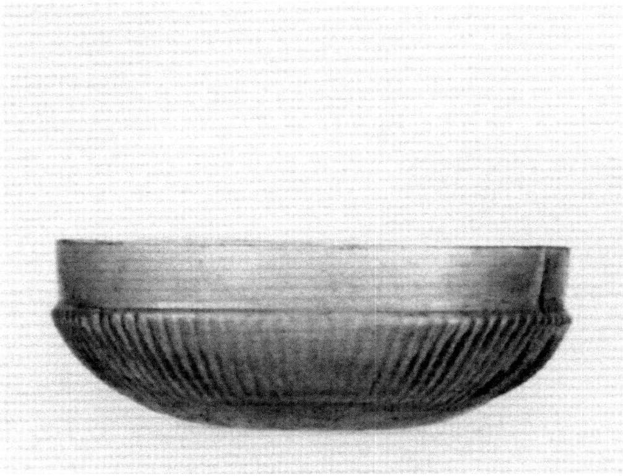

14 (1:2)

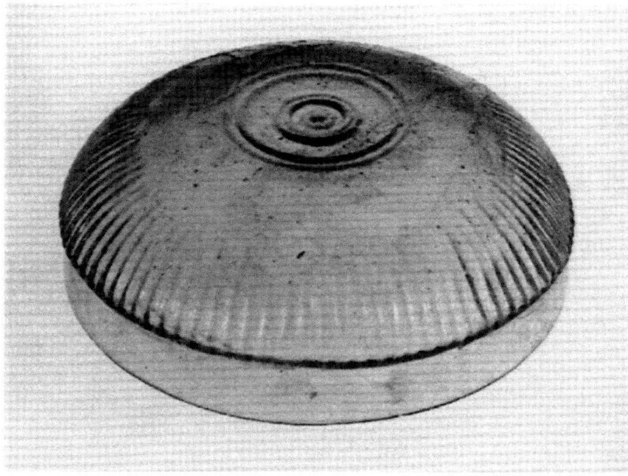

14, underside (1:2)

14 For example:

a) from Camulodunum, dated 10-65, Harden 1947, 300, no. 48;
b) from Verulamium, two examples, Charlesworth 1972, 196, no. 1, 197, fig. 74, no. 2, dated 60-75, and no. 3, dated 105-115;
c) London, Mus. of London, *London in Roman Times* (London Museum Catalogue, no. 3, London 1930) 122, fig. 42, no. 5 (cited by Berger 1960, 56).

15 BM, no. 1868.1-10.483, quoted by Calvi 1968, 101, n. 188.

16 Price 1988b, 32-33, fig. 3, nos. 25 and 26.

17 For example:

a) from "Kambi" Vasa, not dated independently, Harden 1958, 49, nos. 1 and 2, fig. 20, no. a, pl. 5, no. c;
b) Cambridge, Mus. of Archaeology and Ethnology, no. RC 23.725, (formerly Ransom Collection), from Cyprus?, made in a "3-piece mold (2 for sides, 1 for base)," *Fitzwilliam* 1978, 33, no. 55, fig. 55;
c) BM, no. 1912.10-25. 16, "from Cyprus," with four raised ridges on the underside of the base, covered with silvery weathering.

18 *JGS* 4 (1962) 140, no. 8, fig. 8.

19 See above note 4a and c.

20 See above note 4b.

21 For example:

a) V & A, no. 899-1902, from Jerusalem, quoted by Calvi 1968, 101, n. 188;
b) Princeton University Art Museum, no. 46-236, from Rakka, cited by Hayes 1975, 48, no. 82.

### 14 Finely Ribbed Shallow Bowl

Probably Eastern Mediterranean
Second quarter to mid first century
H 4.0 cm; D 11.02-10.95 cm; D rim 10.97 cm; Th rim 0.09 cm
1923.424, Gift of Edward Drummond Libbey

Transparent colorless (?) glass. Thin glass. Pinprick and small spherical bubbles.
Rim and body blown into a multipart vertically grooved mold with two vertical sections that probably included no more than the ribbed area.[1] Mold seams visible on rim only.
Straight rim, lightly ground. Convex sides. Base slightly concave. On sides, ninety-five short vertical ribs each ca. 2.5 cm long. On the underside of the base, around a central boss, three concentric circles, each bordered by a groove.
*Condition* Intact. Interior and exterior covered with transparent pale green and transparent dark yellow iridescent film.
*Collection* Thomas E. H. Curtis.
*Published* Previously unpublished.
*Remarks* See Remarks Cat. No. 13. Shallow bowls with fine ribs have been noted principally in the Mediterranean region, for example in Cyprus,[2] Cyrenaica,[3] northern Italy,[4] Ticino,[5] and southern France.[6] They are rarely found in the northwestern provinces.[7] Some bowls have an undecorated, plain base,[8] while others are decorated with concentric rings on the underside. Future research may show that bowls made in different molds were made in different regions.

1 MCT not possible to determine. No mold seam visible on the smooth area below the ribs or on the base.

2 Harden 1958, pl. 5, no. C.

3 From Sidi Khrebish, Benghazi, Libya, fragments of two vessels, Price 1985a, 293, nos. 43, 44, 294, fig. 24, no. 3.

4 From a tomb at Valeggio (Pavia), Gloria Vanacci Lunazzi, "Collections de verres en Italie. Region: Lombardia, Garlasco (Pavia) Museo Archeologico Lomellino," *BullAIHV* 9 (1981-83) 62-63; Aquileia, Italy, Museo, no. 13027, perhaps from Aquileia, Calvi 1968, 101, 105, no. 249, pl. 16, no. 1.

5 Zürich, Nat. Mus., no. 58920.09, from Ascona, San Materno, grave 3, dated about 100, Biaggio Simona 1991, I, 75-76, no. 011.1.017; Donati et al. 1987, 86, no. 9.

6 Jennifer Price, "The Aiguières and Argentière Sites at Fréjus (Forum Julii)," *AFAV* 2 (1987) 30-31, fig. 24.

7 Price 1991, 72.

8 For example, the bowl from Ascona, see above note 5.

## *Hexagonal Bottles with High Relief* (Cat. Nos. 15-47)

### ***Vessels Type***[1] (Cat. Nos. 15-35)

### 15 Hexagonal Bottle, Vessels Type, Series A

Color plates 5 and 6, p. 52; photographs, Figs. 43, 71, and 72
Phoenician
Probably first quarter of first century
H 7.25 cm; H body 5.0 cm; D 4.2-4.0 cm; D rim 2.15 cm; D base 2.3 cm
1923.472, Gift of Edward Drummond Libbey

Opaque white. Medium thin glass. Fabric cannot be determined because of weathering.
Neck and rim free blown. Body blown into a three-part mold of three vertical sections (MCT IV). Mold seams between panels 2 and 3, 4 and 5, and 6 and 1 meet off center on underside of base. Edges of mold between panels 1 and 6 and between panels 4 and 5 not well aligned. Relief crisp.
Flaring rim, folded outward and upward; tool mark along interior. Cylindrical neck with vertical crimp at its base. Bulbous body with greatest diameter midway between center and base. Low, offset base with concave underside.
On the shoulder, six pointed arches, each containing an unidentified, large dome-shaped object. On the body, six rectangular panels divided by columns with a broad abacus and torus capital, smooth shaft, and high double torus base. In the panels, six vessels, from

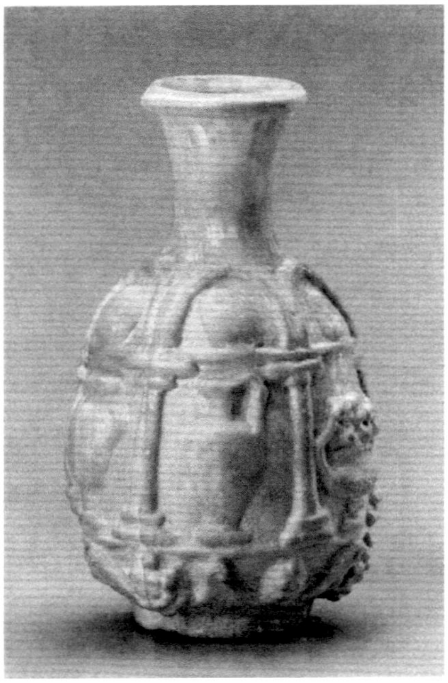

15, panel 1 (1:1)

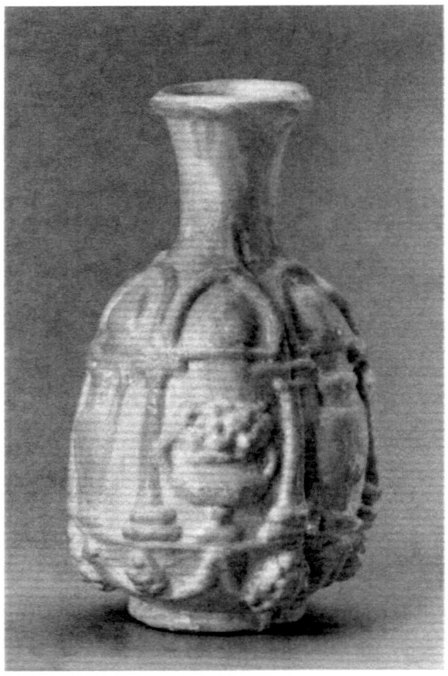

15, panel 2 (1:1)

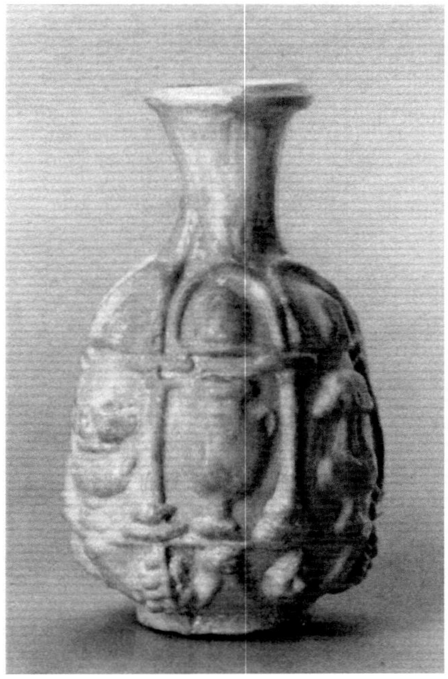

15, panel 3 (1:1)

left to right: 1, a spouted jug with high handle to the right and scalloped base; 2, a footed bowl with wide opening and two high, curved vertical handles from shoulder to rim, the opening containing three rows of rounded objects (fruit?); 3, a footed wide-mouthed vessel with two small vertical handles on the shoulder; 4, a footed bowl with tall cylindrical neck with vertical grooves, oblate body, and tall stem, the mouth filled with two rows of rounded dimpled objects (fruit?); 5, a footed jug with a round mouth and high handle to the right; and 6, a second footed, wide-mouthed vessel as in panel 3. Around the bottom, fillets suspended from the center of one panel to the center of the adjacent panel, with alternating large and small fruits with knobbed surfaces below each column, the larger ones on the mold seams and covering the fillet.
*Classification* Vessels Type, Series A1, Generation A1a.
*Condition* Intact. Exterior covered with weathering which ranges from grayish yellow (5 Y 8/4) to grayish orange (10 YR 7/4).
*Collection* Thomas E. H. Curtis.
*Published* Previously unpublished.
*Remarks* One of seven bottles in the Toledo Museum collection from closely related molds (Cat. Nos. 15-21).[2] These bottles belong to the Vessels Type, the most common Type of early mold-blown glass vessels preserved from antiquity. The decoration has three zones. The main zone, on the body, is divided into a frieze of six panels, each depicting a vessel in high relief. Series A of this Type is characterized by showing in a specific order: two footed jugs, two footed bowls (probably filled with fruit), and two wide-mouthed vessels. Although each shape is represented twice, the pairs are not identical and the order in which they occur does not form two repeating sets. The elaborate secondary decoration consists of arches enclosing an unidentified object above each panel and fillets with fruit below the panels.
Series A is by far the most common of the Vessels Type. Three Subseries can be recognized in Series A. Of these, Subseries A1 (Cat. Nos. 15-25) the most common, was made in three-part molds consisting of three vertical sections, from shoulder to underside of base (MCT IV A). The mold seams form a Y-shaped pattern on the underside of the base, with the stem of the Y belonging to the seam between panels 4 and 5 (see Fig. 15). Most rims were probably finished while the bottles were still in the mold, because crimp marks from a clamp are rare. Subseries A2 was made in MCT IV A molds like Subseries A1, but the section divisions were positioned differently (see Cat. No. 25). Subseries A3 was made in four-part molds of MCT II (Cat. Nos. 26-33), these bottles were finished after removal of the mold; all have crimp marks at the base of the neck from being held with a clamp.
Because the molds were probably made of clay, they wore out and had to be replaced frequently. During the process of manufacture, the molds into which the bottles were blown became progressively smaller, while the design became less detailed and was periodically modified (see *Mold Blowing: The Reproduction of Molds*). Bottles belonging to Subseries A1 provide a revealing example of the process (see Chapter I: *Syro-Palestinian Styles: Hexagonal Bottles with High Relief*). The earliest are those of Generation A1a (Fig. 71). By the time molds of Generation A1c were made, the columns dividing the panels of the main frieze had degenerated to simple vertical ribs, while the fillets with fruit around the bottom of the bottles had gone through many small modifications, gradually becoming less detailed until the design was finally redefined as triangular leaves alternating with small oval fruit. The modifications leading to these transformations can be followed step by step in the bottles of the Toledo Museum collection (Fig. 72).

*Generation A1a*: A characteristic of the design of the first generation molds is that the base of the arches above panels 1 and 2 forms one continuous architrave above the columns while all other arches have separate base lines. All columns have a double torus base and a capital with a torus and abacus. These characteristics are clearly seen in this bottle which is also one of the most bulbous and crisp bottles known from Generation A1a.[3]

*Generations A1a-y and A1 a-yy*: Cat. Nos. 16-21 are slightly more slender (greatest diameter from 4.1×3.9 to 4.1×3.7 cm), while the vessels shown in the panels are also more slender. They belong to younger generations of A1a in which no modification of the molds took place. The Toledo bottles are opaque white, the most common color for bottles of Generation A1a. Certain details disappeared from the surface of these bottles, but probably only because of wear and tear of the molds during blowing.

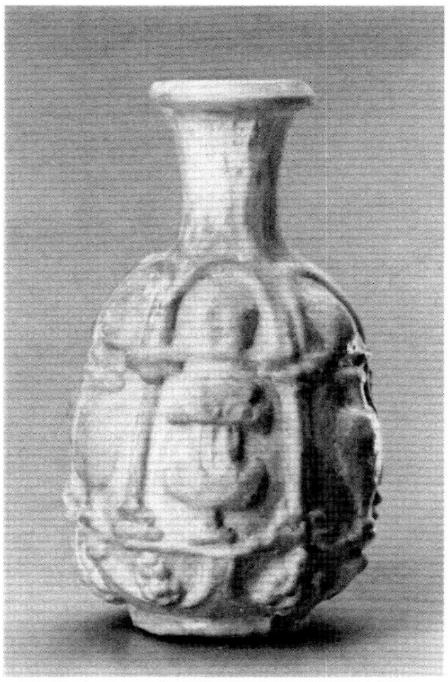

15, panel 4 (1:1)

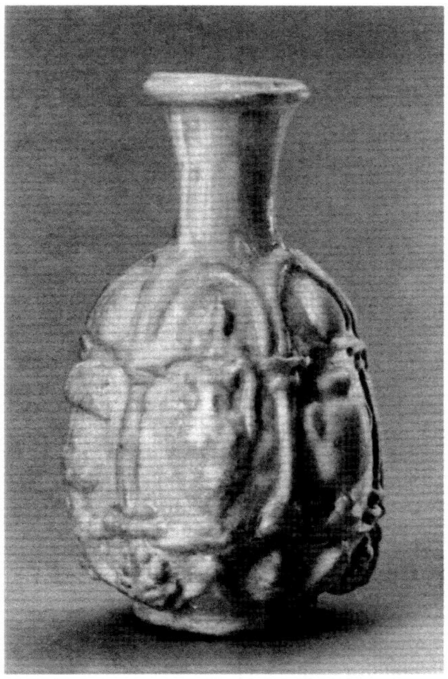

15, panel 5 (1:1)

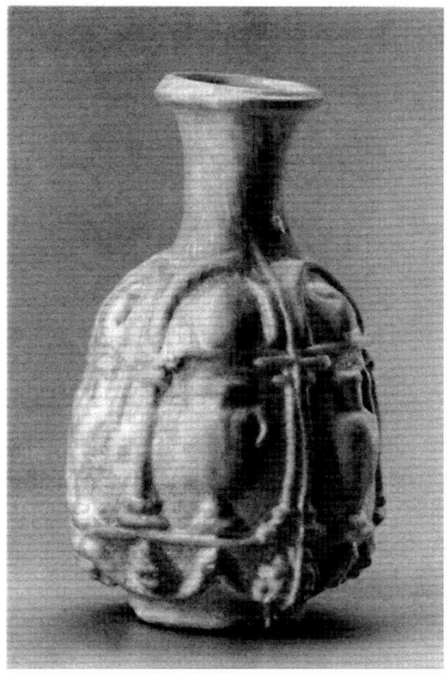

15, panel 6 (1:1)

*Generation A1a-y*: The capital of the column between panels 1 and 2 has lost the torus element (see Cat. Nos. 16-18).

*Generation A1a-yy*: In addition to the above, all the arches on the shoulder now have separate baselines (see Cat. Nos. 19-21).

*Generations A1b and A1c*: Cat. Nos. 22-24 have not only a modified design but are also made of translucent to transparent colored glass, usually blue, instead of opaque white glass. These two significant deviations may indicate that these bottles were made in a different workshop that used a bottle of Generation A1a-yy as the model for its archetype. The bottles are a fraction smaller than those of Generation A1a.

*Generation A1b*: The legs of all arches flow together; the architrave is missing. All columns have lost the torus from their capitals; the base of the column between panels 1 and 2 has become an indistinct triangle. The difference in size between the large and small fruit in the fillet has become much less obvious, the surface of the larger fruit has only a few bumps, and the smaller fruit has become smooth (see Cat. No. 22).

*Generation A1c*: In addition to the features of Generation A1b, the arches have become indistinct and the dome-shaped objects framed by them now look like oval leaves. The columns between the panels have become vertical ribs without any indication of capitals and with slightly thickened ends where the bases would have been. The fillet around the bottom has been redesigned to form pointed leaves alternating with uniform small oval fruits with smooth surfaces (see Cat. Nos. 23 and 24).

*Subseries A2* (see Cat. No. 25) and *Subseries A3* (see Cat. No. 26) began between Generations A1a-yy and A1b. Like bottles of Generations A1b and A1c, the earliest bottles of Subseries A2 and A3, Generations A2a and A3a, are usually made of translucent to transparent dark blue glass. The change in MCT indicates that they were probably produced in different workshops.

Most of the bottles belonging to Series A appear to belong to Subseries A1, but it is often not possible to distinguish this, since most publications do not describe the base.

### Distribution

Bottles belonging to Series A have a wide distribution throughout the eastern Mediterranean, ranging from the Black Sea in the north to Galilee and southern Phoenicia in the south, and from the Greek islands and Cyprus in the west to Dura Europos in the east.[4] Few if any bottles of Series A have been recorded from Italy or other find spots in the western part of the Roman empire.[5]

### Date

Very few bottles of the Vessels Type have appeared in controlled excavations that provide datable contexts. One bottle belonging to Series A (probably Generation A1a-y) was excavated at Mishmar Haemek in northern Israel, in a family vault that was used throughout most of the first century.[6] Another bottle of Series A, Generation A1b, was excavated in Armenia, where it was found in a tomb with other glass vessels typical of the first half of the first century.[7] This Armenian find indicates that Series A of the Vessels Type was designed sometime before the middle of the first century. A fragment belonging to one of the latest generations was found at Knossos in a Neronian (54-68) or Flavian (69-96) context.[8] Relative dating, based on the number of generations of molds separating the earliest bottles of Series A from a group of bottles of the Masks Type excavated in graves that can be dated around A.D. 40,[9] seems to indicate a date for the inception of Series A around the year 13 (see Chapter I: *Syro-Palestinian Styles: Hexagonal Bottles in High Relief: Date* and Fig. 54). The large number of vessels surviving from this Series, including twenty-one in Toledo's collection (Cat. Nos. 15-35), indicates that the Type must have been popular. The various Series and Generations further suggest that the Type was made over an extended period of time, possibly through the third quarter of the first century.

### Interpretation

The shapes of the vessels appearing in the panels of the bottles are characteristic of late Hellenistic metalwares popular throughout the Roman empire. The footed bowl on panels 2 and 4, the jug on panels 1 and 5, and the tall wide-mouthed vessel on panels 3 and

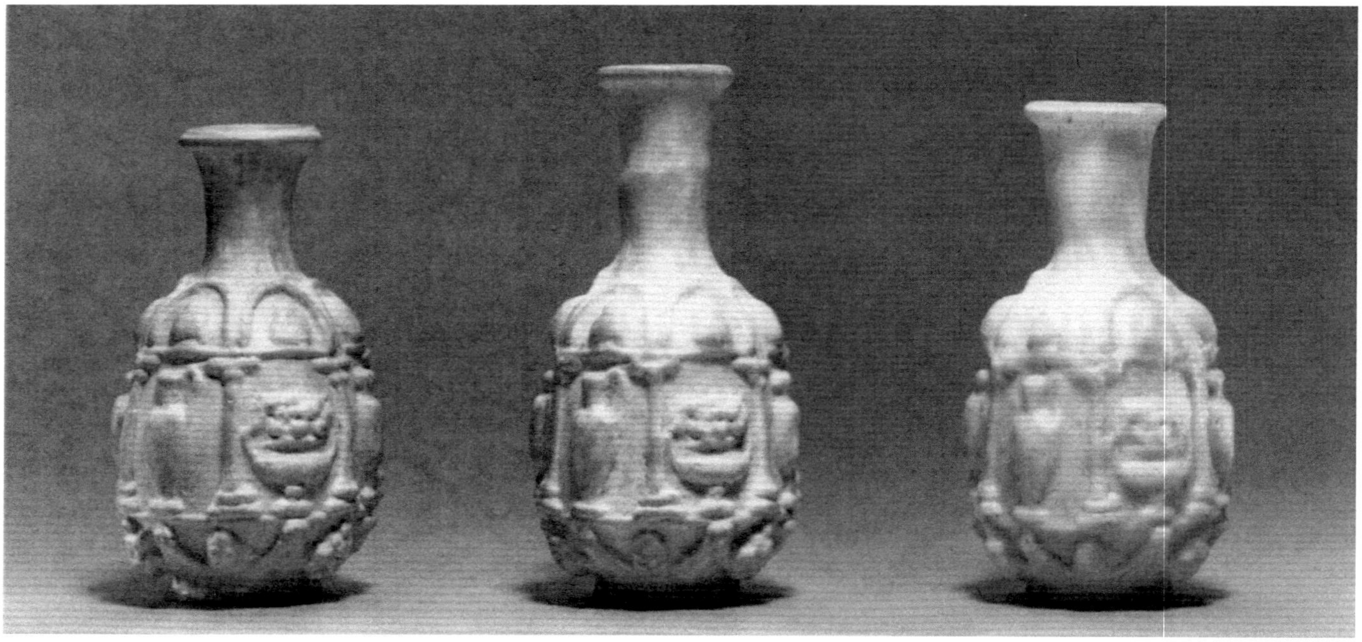

*Fig. 71. Hexagonal bottles, Vessels Type, Subseries A1, showing wearing out of mold within one Generation.* From left: *Generations A1a (Cat. No. 15), A1a-y (Cat. No. 17), and A1a-yy (Cat. No. 21). Photo: Tim Thayer.*

6 (which may be a hydria for water with the vertical handle turned to the back or an amphora for wine) are shapes that occur in a variety of settings similar to those on a Roman silver cup in Toledo with scenes that are unmistakably Dionysiac (see Fig. 57).

The bowls in panels 2 and 4 are filled with small round objects that could be fruit or nuts. This sets them apart from vessels in Dionysiac scenes where the contents are usually not shown, although a bowl of fruit or nuts is not an uncommon motif in first-century secular [10] and funerary [11] contexts. The knobbed fruit in the loops of the fillet shown on the bottom of Series A has been interpreted as a bunch of grapes, but this seems unlikely because the fruits in the fillet of Series A are shown as an oval shape with its point turned upward, while grapes are most commonly represented with the narrow end of the bunch pointed downward (see Cat. Nos. 36-40, Fruit Series, and Cat. Nos. 109, 110, 118, Bottles Shaped like a Grape Bunch). The outline of the shape resembles more closely that of a cedar cone (see Cat. No. 36).

The detailed rendering of the vessels depicted in the panels of the glass bottles, the uniformity of the overall design of the bottles throughout many generations of molds, the large number of bottles preserved, and the fact that three or more workshops were involved in their production are curious phenomena suggesting that bottles of the Vessels Type might have had some special meaning for their buyers. Various elements appear to suggest a religious function or a recurring special event for the production or use of the Vessels Type. Shape and decoration of the earliest bottles evoke a hexagonal architectural structure with vessels set in niches or aediculae. Six carefully rendered columns support an architrave surmounted by an arch above each aedicula. A fillet with fruit is suspended below each aedicula. Arches, columns, and fillets with or without fruit are traditional elements of the classical visual vocabulary for enhancing or sanctifying an object or for indicating a niche or a sacred locale such as a precinct or temple. [12]

The vessels appearing in the aediculae have been interpreted as sacred vessels, but opinions are divided about whether they represent Jewish or pagan sacred vessels. The Vessels Type was originally named the Temple Series because the vessels shown in the panels were thought to represent ritual objects from the temple of Sidonian Astarte; the dome-shaped object in the arches above the panels was interpreted as the cone or stone under which the deity was worshipped. [13] It is not known what shapes of cult vessels were used in Graeco-Roman temples, but there is no evidence to suggest that they differed from the slender shapes of metal vessels which were popular throughout the Roman empire in the late Hellenistic and early imperial period and are represented in the main panels of the Vessels Type.

A Jewish interpretation ascribes the vessels appearing in the aediculae to the Second Temple in Jerusalem, destroyed by the Romans in the year 70. [14] The problem with this hypothesis is that known depictions of Jewish ritual vessels do not share the characteristic shapes represented in the panels of the Vessels Type. Jewish temple vessels are known from coins struck in the years 66-70 during the revolt against Rome. A tall-stemmed cup with a globular body and wide horizontal rim appears on Jewish shekels. It is usually identified as the Omer cup, a temple vessel used on the second day of Passover to offer a measure of barley. [15] Another temple vessel appearing on coins from this period has a hemispherical, fluted body, two upward curving handles, and a tall neck becoming wider at the rim which is decorated along the edge with large beads; the mouth is covered with a tall, fluted, conical lid with a small knob at the top. [16] Two other types of vessels occur on Jewish gold glasses of the fourth century. The one depicted most frequently has a low disk-shaped foot, spherical body, narrow neck, and two angular handles from shoulder to rim; [17] the other is a wide-mouthed vessel with a tall flaring neck, globular body, conical foot, and two tall handles curving from shoulder to rim. [18] The contents is not shown in any of these Jewish vessels, and all are characterized by globular body shapes, suggesting that Jewish ritual vessels did not look like contemporary Hellenistic wares. Since not one of the Jewish shapes known from depictions of temple vessels is represented in the Vessels Type, it is legitimate to conclude that the glass bottles of the Vessels Type do not depict Jewish ritual vessels.

Although an interpretation as Jewish vessels can be discarded, it is difficult to prove or disprove a connection with a specific pagan

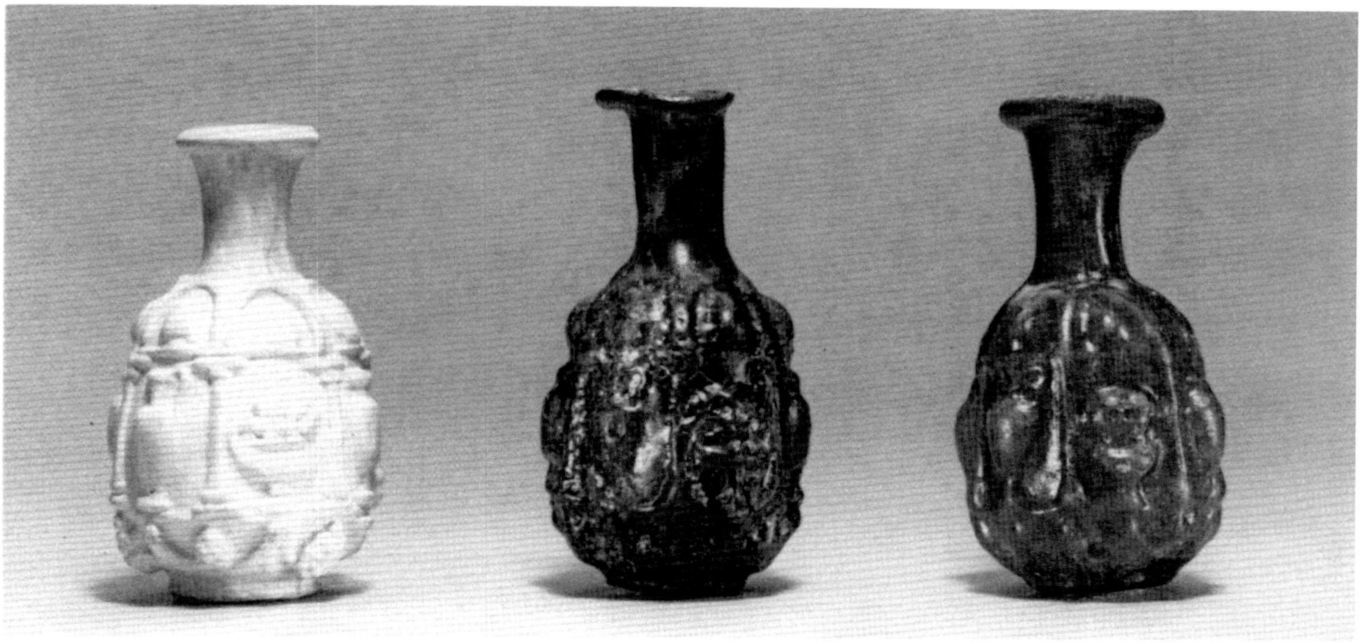

*Fig. 72. Hexagonal bottles, Vessels Type, Subseries A1, showing the development of the border of leaves around the bottoms of the vessels. Note also the diminution of size.* From left: *Generations A1a (Cat. No. 15), A1b (Cat. No. 22), and A1c (Cat. No. 23). Photo: Tim Thayer.*

temple or cult association. The same vessels as those depicted on the bottles occur on an architectural element from Sidon, an indication that the motifs were in some way connected with this area.[19] There were many local cults and trade societies in Phoenicia, each of which had their own rituals and sacred objects. The bottles may well have been associated with one of these.

[1] See Chapter I: *Syro-Palestinian Styles: Hexagonal Bottles with High Relief.*

[2] An eighth bottle originally received by The Toledo Museum of Art as part of the Curtis Collection (TMA no. 1923.469) was acquired by exchange by the Kelsey Museum of Archaeology, The University of Michigan, Ann Arbor, no. 77.2.2, Jay J. Zumeta, "A Recent Acquisition of Ancient Glass at the Kelsey Museum," *Bulletin of the Museums of Art and Archaeology, The University of Michigan* 2 (1979) 1-10.

[3] The same characteristics can be seen on the bottle *Kofler-Truniger* 1985, 69, no. 111.

[4] For examples see:

a) St. Petersburg, Hermitage, no. PAN 599, opaque white, from Panticapaeum, Kunina 1973, 113-114, fig. 14;
b) Yerevan, State History Mus. of Armenia, no. 2112/1, from Verin Artashat, Armenia, transparent dark blue, probably Generation A1b, Arakelian et al. 1969, 54, no. 105, fig. 105, pl. 5;
c) Damascus, Nat. Mus., no. 18.814, from Aleppo, opaque white, probably Generation A1a-y or younger, Abdul Hak 1951, 114, no. 20, pl. XLIX, no. 1; Zouhdi 1969, fig. 39, right;
d) from Dura Europos, tomb 24, loc. XVI, opaque pale blue, Generation A1a-y or younger, Clairmont 1963, 39, no. 146, pl. XXI;
e) Cambridge, Fitzwilliam Museum, no. GR 23.1876, from Cyprus, opaque white, probably Generation A1a-y, *Fitzwilliam* 1978, 32, no. 54a, ill. p. 44;
f) BM, no. S 179, reported to be from the Greek archipelago, opaque pale blue, probably Generation A1a-y or younger, Nesbitt 1871, 30, no. 179, pl. 5, fig. 2;
g) formerly de Clercq Collection, reported to be from Sidon, blue, de Ridder 1909, 153, no. 273. pl. X;
h) Yale, no. 1955.6.53, reported to be from Mount Carmel, Israel, opaque grayish green, probably Generation A1a-y or younger, Matheson 1980, 46, no. 122, fig. 122;
i) Israel Department of Antiquities and Museums, no. 78-110, from Mishmar Haemek, glossy translucent white, probably Generation A1a-y or younger, *Holy Land* 1986, 257-258, no. 138, ill.;
j) judging from illustrations, a fragment from Knossos, Crete, belongs to a late mold, perhaps Generation A1c or younger, Price 1988b, 33-44, fig. 3, no. 27.

[5] A bottle in Murano, Museo Vetrario, reported to be from the necropolis of Nona at Zadar, Croatia, natural bluish green, is a very late example of Series A, like Cat. Nos. 23, 24, and 33, with leaves instead of fillets around the bottom. A bottle excavated in northern Italy appears to belong to Series B or a related Series: Giuliana M. Facchini, "Le verre en Italie," *Bull-AIHV* 9 (1981-1983) 207, fig. 1; Giuliana M. Facchini, "Roman Glass in an Excavational Context: Angera (VA)," *AnnAIHV* 11 (1988, published 1990) 108-109, fig. 2, no. 2.

[6] D. Barag, pers. com.; cf. *Holy Land* 1986, 255-256.

[7] A bottle from Verin Artashat (see above note 4b) found with other first-century glass, Arakelian et al. 1969, 33, 43, nos. 10, 11, 62. See also Cat. No. 22.

[8] See above note 4j.

[9] Lightfoot 1987, 11-12, 19, figs. 2-3. The bottles found in datable contexts include:

a) Bonn, Rheinisches Landesmuseum, no. 1448, found at Andernach in grave 31, translucent dark green, H 7.7, Lightfoot 1987, 19, no. 3;
b, c) Saint-Germain-en-Laye, Musée des Antiquités Nationales, nos. 24642 and 24641, from Saintes, found in the sarcophagus of a young woman in the necropolis of Clousi, translucent brownish yellow and translucent blue, H 6 cm and 7.4 cm, Lightfoot 1987, 19, nos. 4 and 5.

[10] For example in Pompeian Second Style wall paintings at Oplontis, about 30 B.C.

[11] For example Wolfgang Binsfeld, Karin Goethert-Polaschek, Lothar Schwinden, *Katalog der römischen Steindenkmäler Trier, Götter und Weihedenkmäler* (*Trierer Grabungen und Forschungen*, vol. XII, no. 1; *Corpus Signorum Imperii Romani. Corpus der Skulpturen der römischen Welt, Deutschland*, vol. IV, no. 3, Mainz am Rhein 1988).

[12] An arch need not always be understood as an abbreviated form of a temple or shrine; nor must objects appearing under an arch represent cult items.

[13] Eisen 1927, I, 233: "The circumstance that the forms of the illustrated vases are but six, rarely repeated, and always succeeding each other in the same order, suggests that they represented sacred vessels in use in the temple, each standing under a tympanum in a niche or between columns precisely as depicted. What temple was actually represented I do not know; but it is not unlikely that it was the temple of the Sidonian Venus, Ashtoreth. This theory is suggested because the cone standing in each tympanum might represent the cone under which the deity was worshiped. It is known that

several mythological deities were worshiped as a stone or cone fallen from heaven, and this might have been one of them." Eisen identifies the individual vessels as follows, pp. 234 and 247: "First type [compare panel 2 in this catalogue]. A cantharus or crater to hold fruit, with low stem and narrow foot, as in use in the first century B.C. Two large conspicuous handles. Base sometimes decorated with a lotus corolla. The cavity was undoubtedly used to hold fruits or sacred loaves. Possibly the fruit rested on a platter placed over the mouth. Second type [compare panel 3 in this catalogue]. The form is that of an amphora with foot-stand, or a jar. The base varies in the three specimens [here Fig. 47], but generally it is like that in the upper row [compare this bottle, Cat. No. 15]. The form is more or less repeated in the fifth row [panel 6 in this catalogue]. Holmos. Third Type [compare panel 4 in this catalogue]. A wide jar with cylindrical neck and narrow foot. In the third row [here Fig. 47, bottom row] the body is fluted, but this is an isolated and unique occurrence. The vase holds fruit or loaves. Fourth type [compare panel 5 in this catalogue]. A tall, slender wine or oil flask, but slightly varied in the different specimens but most generally, like the one in the two lower rows [here Fig. 47], each pitcher has a spout. Fifth type [compare panel 6 in this catalogue]. Jar with narrow neck, and low narrow foot-stand. Holmos. Sixth type [compare panel 1 in this catalogue]. Pitchers with more or less distinct spout, more or less repeating the fourth type." Eisen attempts to identify the vessels as follows, p. 247: "In a general way these vases must be such as were used in Bacchic rites or in those of the Sidonian Ashtoreth. At least one of the vases can be shown to have been in use several centuries before the time when the temple series was made. This is the 'holmos' used in the Eleusinian rites, as seen depicted on the scene with the idol standing on the table or altar. On each side of Bacchus stands a holmos vase of the same form as represented by the heavy, two-handled vase on the flask. Female attendants are in the act of mixing the wine, one pouring it into one of the two flasks or jars, the other one dipping it out by means of a trulla. Compare: Daremberg et Saglio, fig. 2424, under *Dionysos*. Also Museo Borbonico, XII, pl. XXI."

14 A. Engle, "The Sidonian Glassmakers' Market," *Readings in Glass History*, no. 11 (1980) 30.

15 For example, *Holy Land* 1986, 227-229, no. 116, figs. 103-106; Goodenough III, figs. 685, 689. Goodenough I, 276, interprets the cup as a wine cup. On Jewish coins: Y. Meshorer, *Ancient Jewish Coinage*, 2 vols. (New York 1982).

16 *Holy Land* 1986, 226, fig. 99, a bronze prutah, struck in 68.

17 For example, Jerusalem, Israel Museum, no. IMJ 66.36.14, previously in the collection of Count Goluchow, *Holy Land* 1986, no. 146, fig. 146 (color); Goodenough III, figs. 964-967, 969-970, 974.

18 For example, Vatican, Museo Sacro, no. 479, from the cemetery of SS. Pietro e Marcellino, Rome (1882), C. R. Morey, *The Gold-Glass Collection of the Vatican Library* (Vatican City 1959) 27, no. 116, pl. XX. Goodenough III, fig. 978, reproduces a drawing on which two vessels appear, one with a smooth body and neck, the other with a fluted body and neck. Neither vessel has a lid. A large seven-branched candlestick identifies the subject beyond doubt as Jewish.

19 D. Barag, pers. com.

## 16 Hexagonal Bottle, Vessels Type, Series A

Phoenician
Probably first quarter of first century
H 7.67 cm; H body 5.14 cm; D 4.1-3.76 cm; D rim 1.9 cm; D base 2.35 cm
1923.467, Gift of Edward Drummond Libbey

Opaque white. Medium thin glass. Fabric cannot be determined because of weathering.
Neck free blown. Body blown into a three-part mold of three vertical sections (MCT IV). Mold seams between panels 2 and 3, 4 and 5, and 6 and 1 meet off center on underside of base. Relief moderately crisp. Edges of mold not well aligned.
Flaring rim, folded outward and upward; tool mark along interior. Cylindrical neck. Bulbous body with greatest diameter halfway between center and base. Low, offset base with concave underside.
Decoration like Cat. No. 15.
*Classification* Vessels Type, Series A1, Generation A1a-y.
*Condition* Intact. Buff-colored weathering and severe surface pitting over large areas of exterior.
*Collection* Thomas E. H. Curtis.
*Published* Previously unpublished.
*Remarks* See Remarks Cat. No. 15. One of seven bottles in the Toledo Museum collection from closely related molds (Cat. Nos. 15-21), but a younger Generation in which the capital of the column between panels 1 and 2 has lost its torus. Same Generation as Cat. Nos. 17 and 18.

## 17 Hexagonal Bottle, Vessels Type, Series A

Photograph, Fig. 71
Phoenician
Probably first quarter of first century
H 8.15 cm; H body 5.13 cm; D 4.1-3.9 cm; D rim 2.1 cm; D base 2.25 cm
1923.470, Gift of Edward Drummond Libbey

Opaque white. Medium thin glass.
Neck free blown. Body blown into a three-part mold of three vertical sections (MCT IV). Mold seams between panels 2 and 3, 4 and 5, and 6 and 1 meet off center on underside of base. Relief moderately crisp. Edges of mold carefully aligned.
Flaring rim, folded outward and upward; tool mark along interior. Tall cylindrical neck with unintentional bulge at midpoint. Bulbous body with greatest diameter halfway between center and base. Low, offset base with concave underside.
Decoration like Cat. No. 15.
*Classification* Vessels Type, Series A1, Generation A1a-y.
*Condition* Two long narrow holes along edge of base. Thin layer of ivory-colored weathering on exterior; incipient surface pitting.
*Collection* Thomas E. H. Curtis.
*Published* Previously unpublished.
*Remarks* See Remarks Cat. No. 15. One of seven bottles in the Toledo Museum collection from closely related molds (Cat. Nos. 15-21). Same Generation as Cat. Nos. 16 and 18.

## 18 Hexagonal Bottle, Vessels Type, Series A

Phoenician
Probably first quarter of first century
H 8.2 cm; H body 5.1 cm; D 4.1-3.8 cm; D rim 2.1 cm; D base 2.3 cm
1923.468, Gift of Edward Drummond Libbey

Opaque white. Medium thin glass.
Neck free blown. Body blown into a three-part mold of three vertical sections (MCT IV). Mold seams between panels 2 and 3, 4 and 5, and 6 and 1 meet off center on underside of base. Relief moderately crisp.
Everted rim, folded outward, upward, and inward. Tall cylindrical neck. Bulbous body with greatest diameter halfway between center and base. Low, offset base with flat underside.
Decoration like Cat. No. 15.
*Classification* Vessels Type, Series A1, Generation A1a-y.
*Condition* Intact. Interior of neck covered with ivory-colored weathering, dark greenish brown layer, and severe surface pits. Weathering on exterior has been removed, leaving a mealy off-white surface.
*Collection* Thomas E. H. Curtis.
*Published* Previously unpublished.
*Remarks* See Remarks Cat. No. 15. One of seven bottles in the Toledo Museum collection from closely related molds (Cat. Nos. 15-21). Same Generation as Cat. Nos. 16 and 17. On the base, a rectangular sticker with first and last of four digits legible: 2.7 [7 or 1].

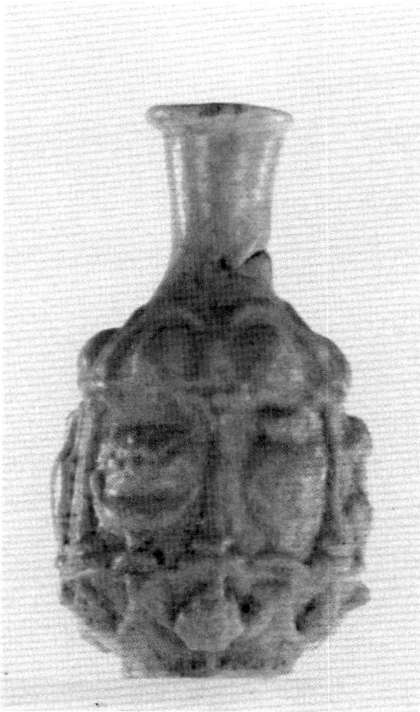

16, panels 2 and 3 (1:1)

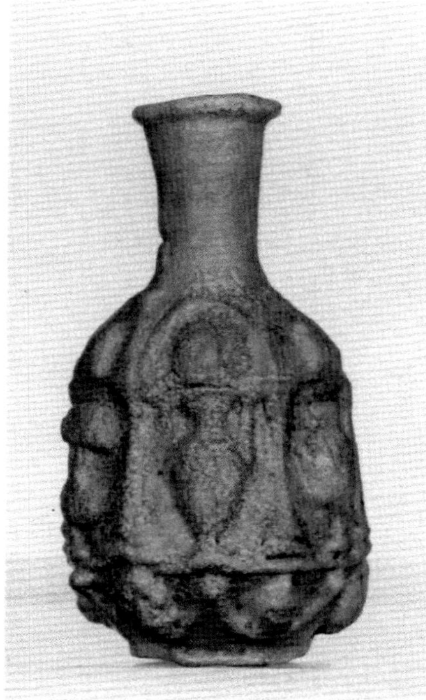

16, panel 5 (1:1)

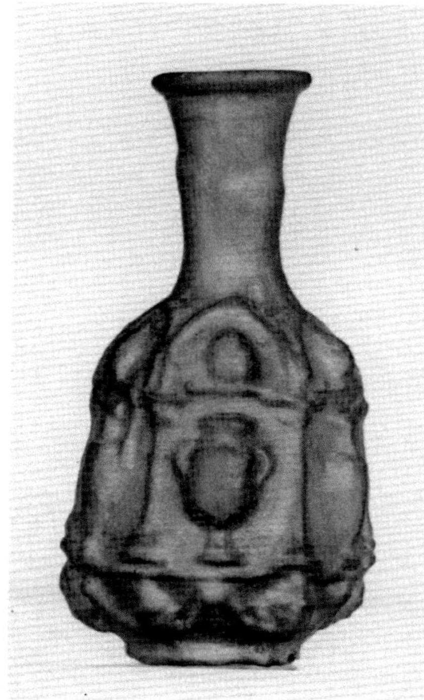

17, panel 6 (1:1)

19 **Hexagonal Bottle, Vessels Type, Series A**

Phoenician
Probably first half of first century
H 7.7 cm; H body 5.0 cm; D 4.1-3.9 cm; D rim 2.1 cm; D base 2.2 cm
1923.471, Gift of Edward Drummond Libbey

Opaque white. Medium thin glass. Fabric cannot be determined because of weathering.
Neck free blown. Body blown into a three-part mold of three vertical sections (MCT IV). Mold seams between panels 2 and 3, 4 and 5, and 6 and 1 meet off center on underside of base. Relief crisp.
Flaring rim, folded outward and upward; tool mark along interior. Cylindrical neck. Bulbous body with greatest diameter halfway between center and base. Low, offset base with concave underside.
Decoration like Cat. No. 15.
*Classification* Vessels Type, Series A1, Generation A1a-yy.
*Condition* Intact. Exterior covered with opalescent weathering; where this has flaked off, the surface is yellowish gray.

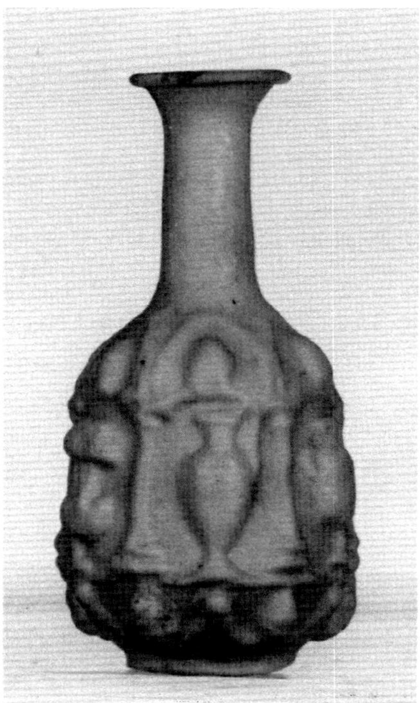

18, panel 5 (1:1)

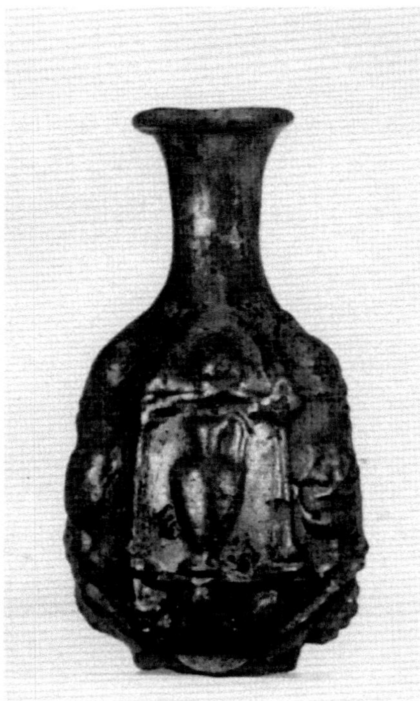

19, panel 1 (1:1)

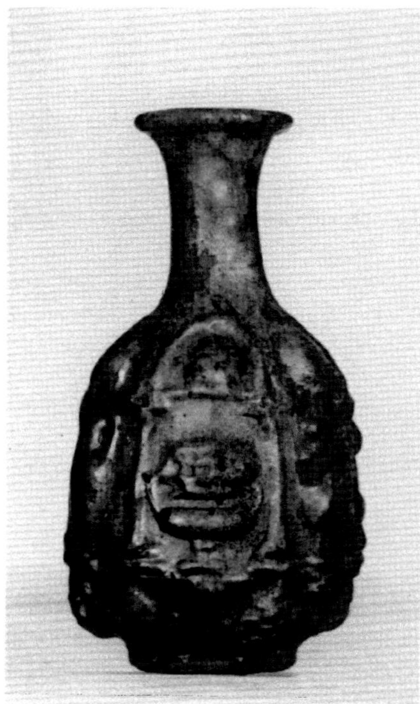

19, panel 2 (1:1)

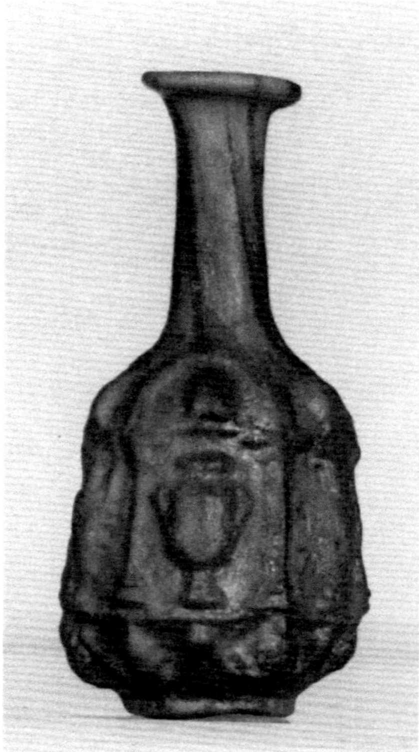

20, panel 6 (1:1)

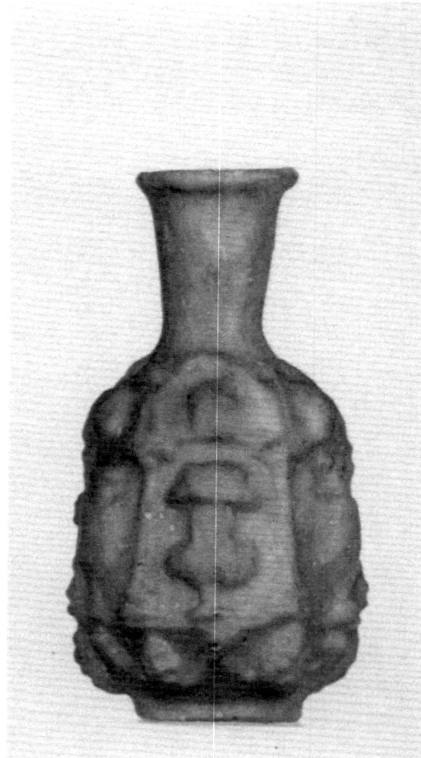

21, panel 4 (1:1)

*Collection* Thomas E. H. Curtis.
*Published* Previously unpublished.
*Remarks* See Remarks Cat. No. 15. One of seven bottles in the Toledo Museum collection from closely related molds (Cat. Nos. 15-21). Same mold as A1a-y, but a younger generation in which, in addition to the losses recorded at Cat. No. 16, the architrave above the columns has been freshened in such a way that each arch has its own base line. Same Generation as Cat. Nos. 20 and 21.

### 20 Hexagonal Bottle, Vessels Type, Series A

Phoenician
Probably first half of first century
H 8.8 cm; H body 5.15 cm; D 4.1-3.8 cm; D rim 2.1 cm; D base 2.25 cm
1923.465, Gift of Edward Drummond Libbey

Opaque white. Medium thin glass. Fabric cannot be determined because of weathering.
Neck free blown. Body blown into a three-part mold of three vertical sections (MCT IV A). Mold seams between panels 2 and 3, 4 and 5, and 6 and 1 meet off center on underside of base. Relief indistinct.
Everted rim, folded outward, upward, and inward. Tall cylindrical neck. Bulbous body with greatest diameter halfway between center and base. Low, offset base with concave underside.
Decoration like Cat. No. 15.
*Classification* Vessels Type, Series A1, Generation A1a-yy.
*Condition* Intact. Streaked ivory-colored weathering and iridescence on exterior; surface pitting.
*Collection* Thomas E. H. Curtis.
PUBLISHED Previously unpublished.
*Remarks* See Remarks Cat. Nos. 15 and 19. One of seven bottles in the Toledo Museum collection from closely related molds (Cat. Nos. 15-21). Same Generation as Cat. Nos. 19 and 21.

### 21 Hexagonal Bottle, Vessels Type, Series A

Photograph, Fig. 71
Phoenician
Probably first half of first century
H 7.6 cm; H body 5.1 cm; D 4.1-3.7 cm; D rim 2.1 cm; D base 2.2 cm
1923.466, Gift of Edward Drummond Libbey

Opaque white. Medium thin glass. Fabric cannot be determined because of weathering. Neck free blown. Body blown into a three-part mold of three vertical sections (MCT IV). Mold seams between panels 2 and 3, 4 and 5, and 6 and 1 meet off center on underside of base. Relief indistinct. Edges of mold carefully aligned. Flaring rim, folded outward and upward; tool mark along interior. Cylindrical neck. Bulbous body with greatest diameter halfway between center and base. Low, offset base with concave underside.
Decoration like Cat. No. 15
*Classification* Vessels Type, Series A1, Generation A1a-yy.
*Condition* Intact. Surface pits and traces of iridescent film.
*Collection* Thomas E. H. Curtis.
*Published* Previously unpublished.
*Remarks* See Remarks Cat. Nos. 15 and 19. One of seven bottles in the Toledo Museum collection from closely related molds (Cat. Nos. 15-21). Same Generation as Cat. Nos. 19 and 20.

### 22 Hexagonal Bottle, Vessels Type, Series A

Photograph, Fig. 72
Phoenician
Mid first century
H 7.3 cm; H body 5.0 cm; D 4.1-3.7 cm; D rim 2.1-1.85 cm; D base 2.2 cm
1923.475, Gift of Edward Drummond Libbey

Translucent grayish blue. Medium thin glass. Fabric cannot be determined because of weathering.

Neck free blown. Body blown into a three-part mold of three vertical sections (MCT IV). Mold seams between panels 2 and 3, 4 and 5, and 6 and 1 meet off center on underside of base. Relief crisp, but difficult to distinguish because of weathering. Vertical tool mark at base of neck.
Flaring rim, partially folded outward, upward, and inward. Cylindrical neck. Bulbous body with greatest diameter halfway between center and base. Low, offset base with irregular flat underside.
On the shoulder, six arches, each containing an unidentified, large egg-shaped object. On the body, six rectangular panels divided by columns, each with an abacus, smooth shaft, and high double torus base. In the panels, six vessels, from left to right: 1, a spouted jug with high handle to the right; 2, a footed bowl with wide opening and two vertical handles from shoulder to rim, the opening containing three rows of rounded objects (fruit?); 3, a footed wide-mouthed vessel with two small vertical handles on the shoulder; 4, a footed bowl with tall cylindrical neck with vertical grooves, oblate body, and tall stem, the mouth filled with two rows of rounded objects (fruit?); 5, a footed jug with a spout and a high handle to the right; and 6, a second footed, wide-mouthed vessel as in panel 3. Around the bottom, fillets suspended from the center of one panel to the center of the adjacent panel, with alternating large and small fruits below each column, the larger ones with a knobbed surface, the smaller ones perhaps smooth.
*Classification* Vessels Type, Series A1, Generation A1b.
*Condition* Intact. Exterior covered with blue-green and golden iridescence and areas of thick pinkish crust; severe surface pitting. Iridescence appears to have been treated to prevent flaking.
*Collection* Thomas E. H. Curtis.
*Published* Previously unpublished.
*Remarks* See Remarks Cat. No. 15. The archetype for this mold may have been made after a bottle of Generation A1a-yy or, more probably, an even younger generation, for Generation A1b is younger than Subseries A2, Generation A2a (see Cat. No. 25 and Fig. 46). The legs of the arches on the shoulder flow together, the architrave appears to be missing, and all columns have lost the torus. The base of the column between panels 1 and 2 has become one large indistinct triangle. The size of the large fruit between the fillets has decreased; the small fruit has lost the knobs on its skin. The recutting of the mold and the fact that the bottle is made of translucent to transparent blue glass may indicate that it was made in a different workshop than the bottles of Generation A1a. A transparent blue bottle excavated in a first-century grave in Armenia may belong to the Generation A1b.[1]

[1] Yerevan, State History Mus. of Armenia, no. 2112/1, from Verin Artashat, bluish, Arakelian et al. 1969, 54, no. 105, fig. 105, pl. 5.

### 23 Hexagonal Bottle, Vessels Type, Series A

Color plate 6, p. 52; photographs, Figs. 55 and 72
Phoenician
Probably mid first century
H 7.7 cm; H body 4.85 cm; D 4.1-3.7 cm; D rim 2.5 cm; D base 2.2 cm
1923.476, Gift of Edward Drummond Libbey

Translucent royal blue. Medium thin glass. Pinprick and small spherical bubbles in body and medium to large linear bubbles in neck. Neck free blown. Body blown into a three-part mold of three vertical sections (MCT IV). Mold seams between panels 2 and 3, 4 and 5, and 6 and 1 meet at center of underside of base. Relief moderately crisp.
Flaring rim, folded outward, upward, and inward. Cylindrical neck. Ovoid body. Low, offset base with flat underside.
On the shoulder, six rounded arches, each containing an unidentified, smooth, ovoid object. On the body, six rectangular panels divided by degenerated columns with rudimentary bases. In the panels, six vessels, from left to right: 1, a spouted jug with high handle to the right; 2, a slender footed bowl with wide opening and two large curving handles from hip to rim, the opening containing a semicircular mound of fruit (or covered with a semicircular bowl?); 3, a footed jar with slender body and two very small handles on the shoulder; 4, a footed bowl with tall cylindrical neck and oblate body with two small vertical handles, the mouth filled with unidentified contents; 5, a footed jug with a spout and handle to the right; and 6, a footed jar with two small vertical handles

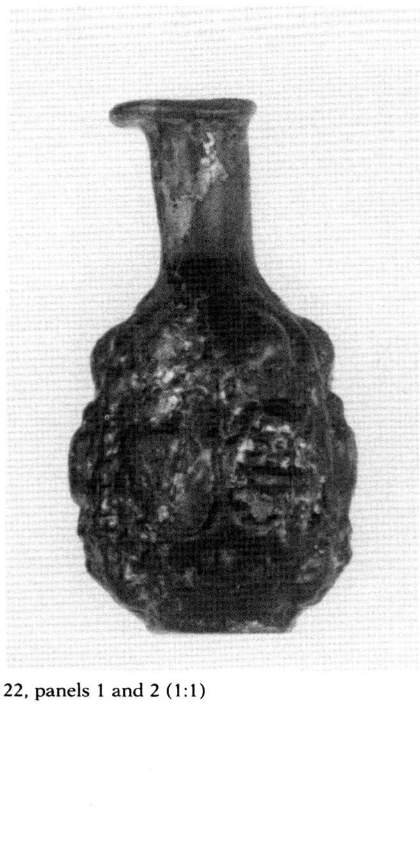
22, panels 1 and 2 (1:1)

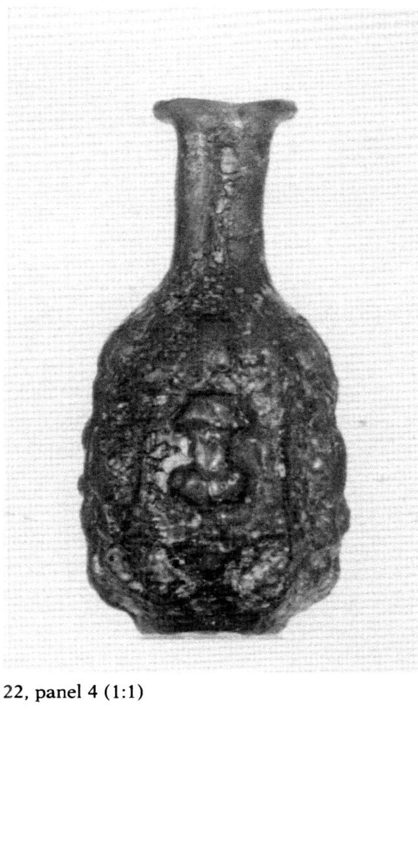
22, panel 4 (1:1)

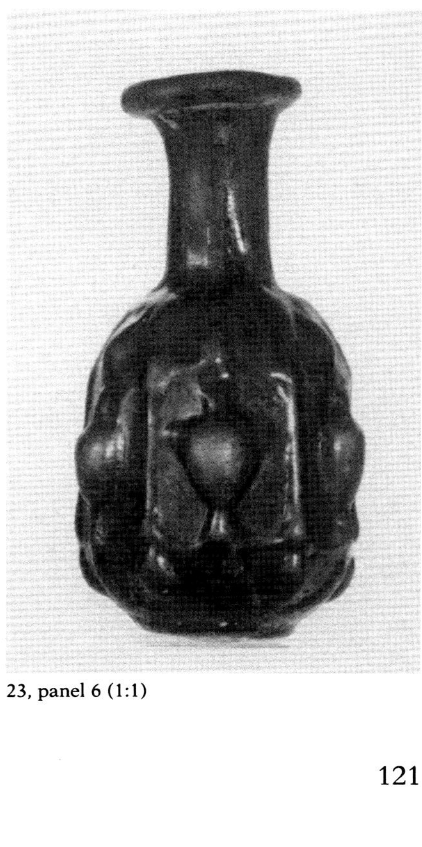
23, panel 6 (1:1)

on the shoulder. Around the bottom, six triangular leaves with tips centered beneath panels alternating with an irregularly shaped ovoid fruit below each column.
*Classification* Vessels Type, Series A1, Generation A1c.
*Condition* Intact. Areas of iridescent film; surface pits.
*Collection* Thomas E. H. Curtis.
*Published* Previously unpublished.
*Remarks* See Remarks Cat. No. 15. One of two bottles in the Toledo Museum collection from closely related molds (Cat. Nos. 23 and 24). Generation A1c is a simplified version of the more detailed molds of the previous bottles. It is younger than Subseries A3, Generation A3a-y (see Cat. No. 29 and Fig. 46); it seems to be of the same generation as Generation A3b (see Cat. No. 33 and Fig. 46). The arches on the shoulder are indistinct; the oval object below looks like an oval leaf. The columns dividing the panels of the main frieze have become distorted vertical ribs with slightly thickened lower ends where the base would have been; there is no remnant of a capital. The architrave is missing. The fillet around the bottom has been recut in such a way that solid triangular petals or leaves replace the earlier suspended fillet motif; all the fruits are smooth and small. Peculiarities of this mold include the ovoid object between and below panels 3 and 4 and the petals under panels 4 and 6, all of which are off center to the right.

### 24 Hexagonal Bottle, Vessels Type, Series A

Phoenician
Probably mid first century
H 8.6 cm; H body 4.85 cm; D 4.05-3.7 cm; D rim 2.5 cm; D base 2.2 cm
1923.478, Gift of Edward Drummond Libbey

Translucent to opaque patchy pale bluish gray. Medium thin glass. Linear bubbles in neck.
Neck free blown. Body blown into a three-part mold of three vertical sections (MCT IV). Mold seams between panels 2 and 3, 4 and 5, and 6 and 1 meet at center of underside of base. Relief indistinct. Vertical crimps at base of neck.
Flaring rim, folded outward, upward, and inward. Cylindrical neck. Ovoid body. Low, offset base with flat underside.
Decoration like Cat. No. 23.
*Classification* Vessels Type, Series A, Generation A1c.
*Condition* Body cracked. Hole in body and another small hole in base. Portion of panel 3 restored with blue iridescence added to surface.
*Collection* Thomas E. H. Curtis.
*Published* Previously unpublished.
*Remarks* See Remarks Cat. Nos. 15 and 23. One of two bottles in the Toledo Museum collection from closely related molds (Cat. Nos. 23 and 24).

### 25 Hexagonal Bottle, Vessels Type, Series A

Phoenician
Probably mid first century or earlier
H (pres.) 5.3 cm; H body (pres.) 5.3 cm; D 4.0-3.6 cm; D base 2.2 cm
1977.26 (acquired with the Curtis Collection in 1919), Gift of Edward Drummond Libbey

Translucent grayish blue. Medium thin glass. Pinprick and small spherical bubbles.
Neck free blown. Body blown into a three-part mold of three vertical sections (MCT IV). Mold seams between panels 1 and 2, 3 and 4, and 5 and 6 meet off center on underside of base. Relief crisp.
Ovoid body. Low, offset base with flat underside.
On the shoulder, six pointed arches, each containing an unidentified, large, smooth, egg-shaped object. On the body, six rectangular panels divided by columns, each with an abacus and torus capital, smooth shaft, and high double torus base. In the panels, six vessels, from left to right: 1, a spouted jug with angular handle to the right; 2, a footed bowl with wide opening and two curving handles from hip to rim, the opening containing three rows of rounded objects (fruit?); 3, a footed wide-mouthed vessel with two small handles on the shoulder; 4, a slender footed bowl with tall cylindrical neck with vertical grooves, oblate body, and tall stem, the mouth filled with two rows of rounded dimpled objects (fruit?); 5, a footed jug with a spout and high handle to the right; and 6, a second footed wide-mouthed vessel as in panel 3. Around the bottom, fillets suspended from the center of one panel to the center of the adjacent panel, with a large fruit with a knobbed surface alternating with a small drop-shaped smooth fruit, the smaller fruit on the seams.
*Classification* Vessels Type, Series A2, Generation A2a.
*Condition* Fragmentary. Neck and one-quarter of body missing. Incipient surface pitting.
*Collection* Thomas E. H. Curtis.
*Published* Previously unpublished.
*Remarks* See Remarks Cat. No. 15. The mold seams of Subseries A2 are one panel off from those in Series A1. Features such as the use of a mold with different section divisions and translucent blue glass instead of opaque white may indicate that this bottle was blown in a different workshop than Subseries A1. This Subseries A2, Generation A2a, is not only younger than Generation A1a-yy (see Cat. No. 19 and Fig. 46) but also younger than Subseries A3, Generation A3a (see Cat. No. 26 and Fig. 46). However, it is older than Subseries A1, Generation A1b (see Cat. No. 22 and Fig. 46). All arches above the columns have separate base lines. The capitals of the columns between panels 6 and 1 and between 4 and 5 have lost the lower torus; the small fruit in the fillet around the base has lost the knobs in the surface, and there are fewer knobs in the large fruit. The mold seams pass through the smaller fruit, which is nearly drop-shaped. The peculiarities of this mold include the fruit below the column between panels 2 and 3 being slightly off center to the left, while the festoons below the panels protrude less than in Subseries A1, Generation A1a, though the fruit protrudes more. Panel 2 is broader than in Subseries A1, Generation A1a, and the vessel shown in it is better centered, while the right handle is more clearly defined and its stem taller.

### 26 Hexagonal Bottle, Vessels Type, Series A

Color plate 6, p. 52; photograph, Fig. 43
Phoenician
Probably first half of first century
H 7.46 cm; H body 5.24 cm; D 4.0-3.5 cm; D rim 2.2 cm; D base 2.24×2.45 cm
1923.531, Gift of Edward Drummond Libbey

Translucent dark yellowish green (10 GY 4/4) and moderate blue green (5 BG 4/6). Opaque reddish brown streaks in neck and rim. Batch not well melted; possibly same batch as Cat. No. 31. Medium thin glass. Numerous small spherical bubbles in body and small to large linear and pinprick bubbles in neck.
Neck free blown. Body blown into a four-part mold of three vertical sections joined to a disk-shaped base section (MCT II). Mold seams between panels 2 and 3, 4 and 5, and 6 and 1. Relief crisp on panels 1, 2, and 3, but difficult to distinguish because of the transparency of the glass. The bottle sags toward panel 5 where the relief is blurred. Horizontal bulge and vertical crimp marks

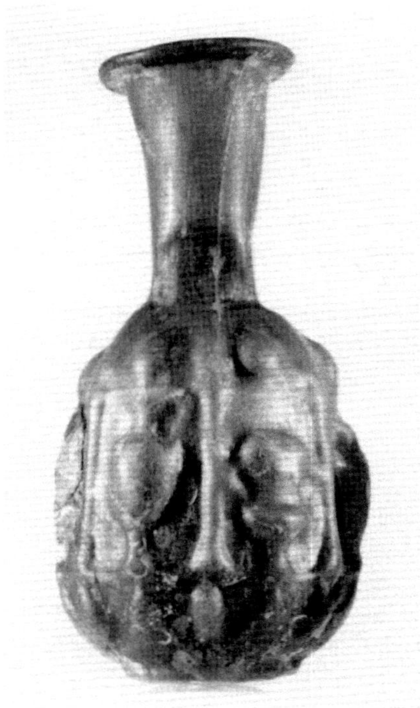

24, panels 1 and 2 (1:1)

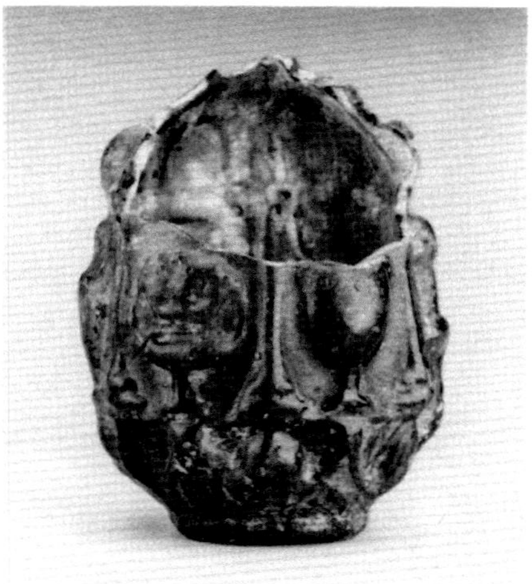

25, panels 2 and 3 (1:1)

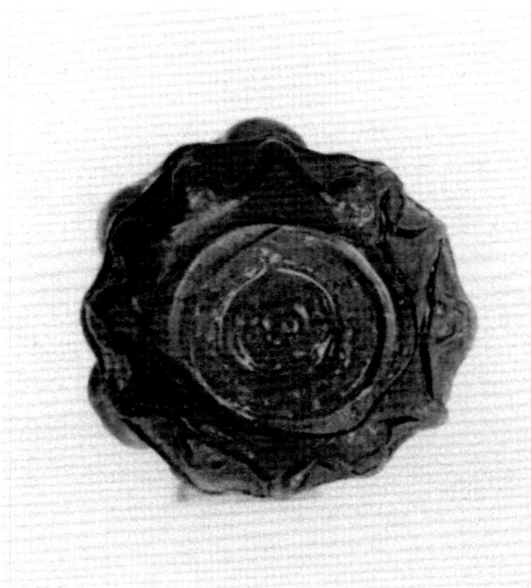

26, underside (1:1)

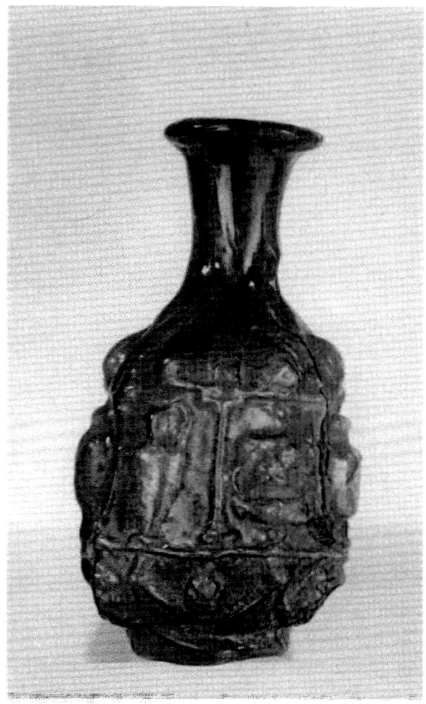

26, panels 1 and 2 (1:1)

at base of the neck from tool used to hold the bottle while finishing the rim.

Flaring rim, folded outward, upward, and inward. Cylindrical neck with slight downward taper. Sloping shoulder. Bulbous body with greatest diameter halfway between center and base. Low, offset base with flat underside with two raised concentric circles.

On the shoulder, six pointed arches, each containing an unidentified, large, egg-shaped object. On the body, six rectangular panels divided by columns, each with an abacus and torus capital, smooth shaft, and high double torus base. In the panels, six vessels, from left to right: 1, a spouted jug with handle to the right; 2, a footed bowl with wide opening and two curving handles from shoulder to rim, the opening containing three rows of rounded objects (fruit?); 3, a footed wide-mouthed vessel with two vertical handles on the shoulder; 4, a footed bowl with tall cylindrical neck with vertical grooves, oblate body, and tall crooked stem, the mouth filled with two rows of rounded objects (fruit?); 5, a footed jug with a round mouth and high handle to the right; and 6, a second footed wide-mouthed vessel as in panel 3. Around the bottom, fillets suspended from the center of one panel to the center of the adjacent panel, with alternating large and small fruits with knobbed surfaces below each column, the larger ones on the seams and covering the fillet.

*Classification* Vessels Type, Series A3, Generation A3a.

*Condition* Body severely cracked with a few small losses. Interior incrusted.

*Collection* Thomas E. H. Curtis.

*Published* Previously unpublished.

*Remarks* See Remarks Cat. No. 15. One of seven bottles in the Toledo Museum collection from closely related molds (Cat. Nos. 26-32). The design of concentric circles on the underside of the base imitates contemporary metalwares turned on a lathe,[1] but it is purely decorative, as can be seen by comparison with Cat. Nos. 15-25, which do not have these circles. The diminishing size of the bottles belonging to this Type is evidence that the molds were made of clay. Series A3 was made in four-part MCT II molds with two characteristic short diagonal mold seams on the side of the base. The position of these mold seams suggests that the disk-shaped base section of the mold was permanently attached to the vertical section containing panels 5 and 6 and that the other two vertical sections were moved separately to enclose the bubble. The glass expanded at the bottom of the bubble and forced the movable sections of the mold apart slightly so that a ridge of glass marks these two short diagonal mold seams.

Subseries A3, Generation A3a, is younger than Subseries A1, Generation A1a-yy, but older than Subseries A2, Generation A2a (see Fig. 46). All the arches in Subseries A3, Generation A3a have separate bases. The capitals between panels 1 and 2, 2 and 3, and 5 and 6 have lost their toruses, the capital between panels 2 and 3 has lost its abacus, and the base of the column between panels 2 and 3 is an indistinct triangle. There is only a slight difference in size between small and large fruit in the fillet around the bottom; both have knobbed surfaces. Cat. No. 26 is the most crisp example in Toledo of Generation A3a, but Cat. Nos. 27 and 28 are probably of the same Generation. Cat. Nos. 29-32 represent a younger generation.

The peculiarities of this Generation A3a mold include the fruit between panels 4 and 5 being off center to the right and the egg-shaped objects above panels 2 and 3 appearing to have a dimpled surface.

Most of the bottles belonging to Subseries A3 are translucent to transparent dark blue, and all but one (Cat. No. 27) have vertical crimps at the base of the neck from the clamp used to hold the bottle while finishing the rim.[2] These details, and the fact that the molds were four-part MCT II molds, indicate that the bottles of Subseries A3 were made in a different workshop than those of Subseries A1. The workshop using MCT II molds may have operated in Sidon. If this piece represents the first generation of bottles of Subseries A3, the archetype used to make the mold may have been modeled from a bottle of Subseries A1, Generation A1a-yy, or an undocumented younger generation.

[1] Mutz 1972.

[2] Bottles of Vessels Type, Subseries A1 rarely have tool marks at the base of the neck: see Cat. Nos. 15, 22, and 24.

## 27 Hexagonal Bottle, Vessels Type, Series A

Phoenician
Probably first half of first century
H 8.5 cm; H body 5.15 cm; D 4.0-3.5 cm; D rim 2.5-2.3 cm; D base 2.2 cm
1923.480, Gift of Edward Drummond Libbey

Translucent dark purplish blue. Large pale blue and colorless areas resulting from an imperfectly melted batch. Medium thin glass. Clusters of small spherical and large oval bubbles in body and small to very large linear bubbles in neck.

Neck free blown. Body blown into a four-part mold of three vertical sections joined to a disk-shaped base section (MCT II). Mold seams between panels 2 and 3, 4 and 5, and 6 and 1. Relief crisp but difficult to distinguish because of the transparency of the glass. Flaring rim, folded outward, upward, and inward. Cylindrical neck. Sloping shoulder. Bulbous body with greatest diameter halfway between center and base. Low, offset base with flat underside with two raised concentric circles.

Decoration like Cat. No. 26.

*Classification* Vessels Type, Series A3, Generation A3a.

*Condition* Intact. Annealing crack at top of neck.

*Collection* Thomas E. H. Curtis.

*Published* Previously unpublished.

*Remarks* See Remarks Cat. Nos. 15 and 26. One of seven bottles in the Toledo Museum collection from closely related molds (Cat. Nos. 26-32). Same Subseries and same Generation as Cat. Nos. 26 and 28. A rare bottle in Subseries A3 without vertical crimps at the base of the neck.

## 28 Hexagonal Bottle, Vessels Type, Series A

Phoenician
Probably first half of first century
H 8.2 cm; H body 5.2 cm; D 4.05-3.6 cm; D rim 2.45 cm; D base 2.35 cm
1923.477, Gift of Edward Drummond Libbey

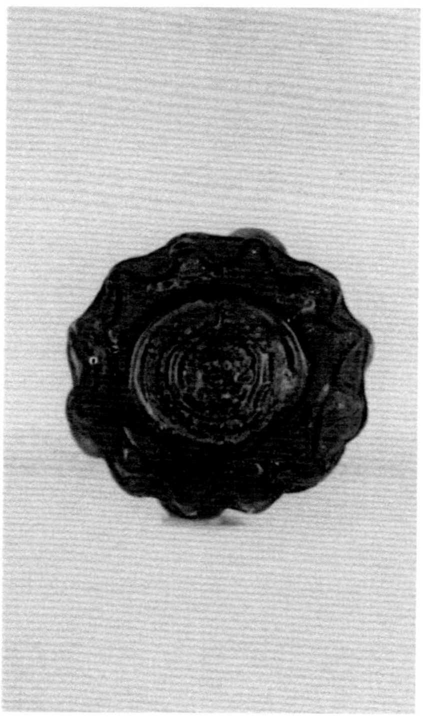

27, underside (1:1)

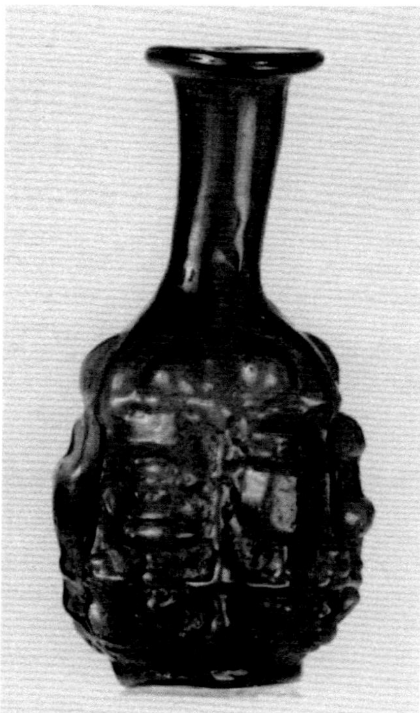

27, panels 2 and 3 (1:1)

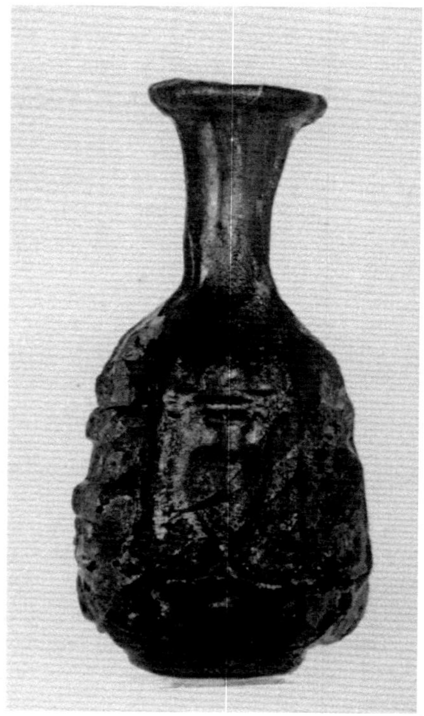

28, panel 5 (1:1)

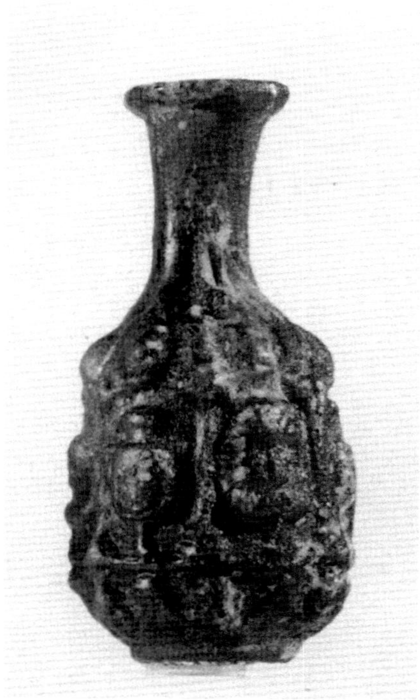

29, panels 3 and 4 (1:1)

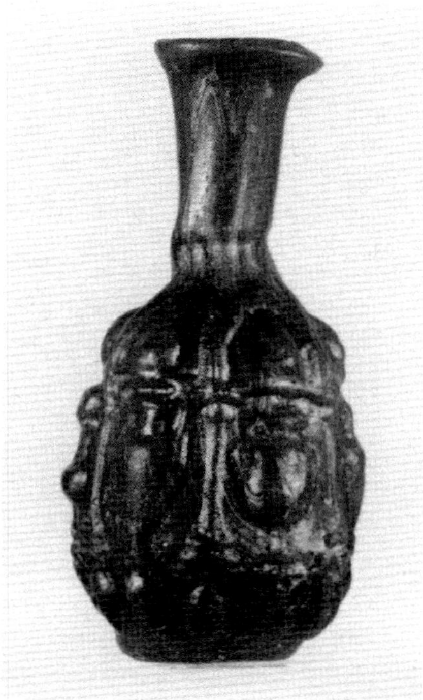

30, panels 1 and 2 (1:1)

Translucent streaked grayish blue. Medium thin glass. Numerous small oval and linear bubbles in neck.
Neck free blown. Body blown into a four-part mold of three vertical sections joined to a disk-shaped base section (MCT II). Mold seams between panels 2 and 3, 4 and 5, and 6 and 1. Relief crisp but difficult to distinguish because of weathering. Crimps at base of neck from tool used to hold vessel while finishing rim.
Flaring rim, folded outward, upward, and inward. Cylindrical neck with slight downward taper. Sloping shoulder. Bulbous body with greatest diameter halfway between center and base. Low, offset base with flat underside with two raised concentric circles. Decoration like Cat. No. 26.
*Classification* Vessels Type, Series A3, Generation A3a.
*Condition* Intact. Exterior covered with reddish brown crust and purple iridescence; severe surface pitting.
*Collection* Thomas E. H. Curtis.
*Published* Previously unpublished.
*Remarks* See Remarks Cat. Nos. 15 and 26. One of seven bottles in the Toledo Museum collection from closely related molds (Cat. Nos. 26-32). Same Subseries and same Generation as Cat. Nos. 26 and 27.

### 29 Hexagonal Bottle, Vessels Type, Series A

Phoenician
Probably mid first century or earlier
H 7.9 cm; H body 5.1 cm; D 4.05-3.6 cm; D rim 2.4 cm; D base 2.4 cm
1923.479, Gift of Edward Drummond Libbey

Translucent streaked grayish blue. Medium thin glass. Fabric cannot be determined because of weathering.
Neck free blown. Body blown into a four-part mold of three vertical sections joined to a disk-shaped base section (MCT II). Mold seams between panels 2 and 3, 4 and 5, and 6 and 1. Relief indistinct. Vertical crimps at base of neck from tool used to hold vessel while finishing rim.
Flaring rim, folded outward, upward, and inward. Cylindrical neck with slight downward taper. Sloping shoulder. Ovoid body. Low, offset base with flat underside with two raised concentric circles. Decoration like Cat. No. 26.
*Classification* Vessels Type, Series A3, Generation A3a-y.
*Condition* Hole in shoulder. Golden green iridescence; severe surface pitting.
*Collection* Thomas E. H. Curtis.
*Published* Previously unpublished.
*Remarks* See Remarks Cat. Nos. 15 and 26. One of seven bottles in the Toledo Museum collection from closely related molds (Cat. Nos. 26-32). Same Subseries as Cat. Nos. 26-28, but a younger Generation in which the body is ovoid and the details of the design are less distinct. Same Subseries and same Generation as Cat. Nos. 30-32.

### 30 Hexagonal Bottle, Vessels Type, Series A

Phoenician
Probably mid first century or earlier
H 8.38 cm; H body 5.1 cm; D 4.0-3.5 cm; D rim 2.25 cm; D base 2.25 cm
1923.474, Gift of Edward Drummond Libbey

Translucent to opaque streaked dark greenish gray (near 5 GY 4/1) and grayish olive green (5 GY 3/2). Batch not well melted. Medium thin glass. Fabric cannot be determined because of weathering.
Neck free blown. Body blown into a four-part mold of three vertical sections joined to a disk-shaped base section (MCT II). Mold seams between panels 2 and 3, 4 and 5, and 6 and 1. Relief indistinct. Vertical crimps at base of neck from tool used to hold vessel while finishing rim.
Flaring rim, folded outward, upward, and inward. Cylindrical neck with slight downward taper. Sloping shoulder. Ovoid body. Low, offset base with flat underside with two raised concentric circles. Decoration like Cat. No. 26.
*Classification* Vessels Type, Series A3, Generation A3a-y.
*Condition* Intact. Remains of reddish brown weathering; purple and golden iridescence where the weathering has flaked off; incipient pitting.

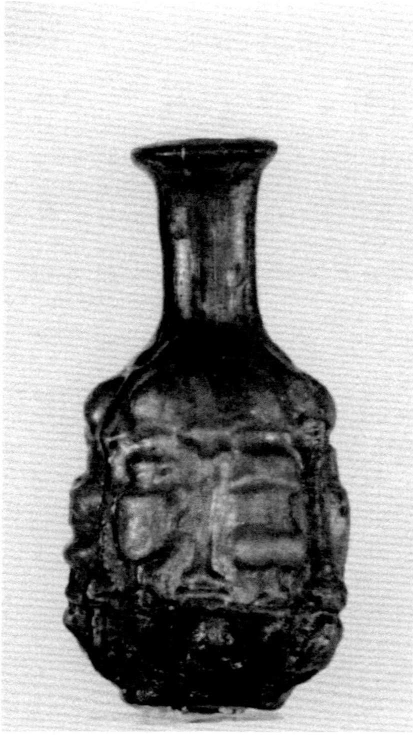

31, panels 3 and 4 (1:1)

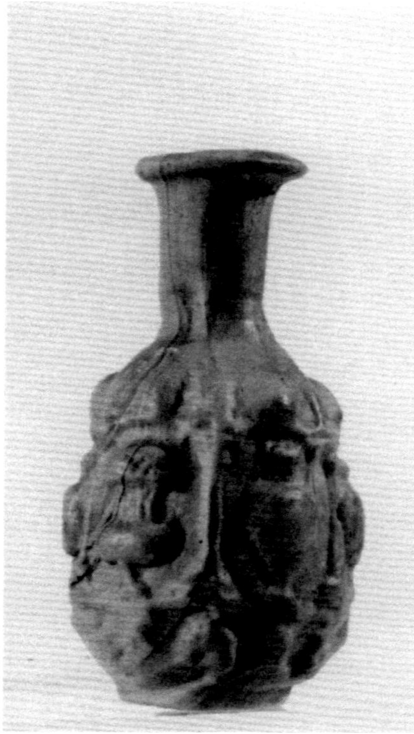

32, panels 4 and 5 (1:1)

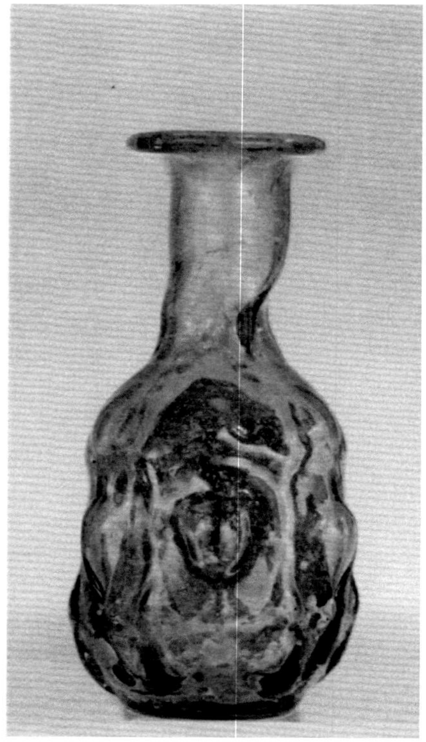

33, panel 6 (1:1)

*Collection* Thomas E. H. Curtis.
*Published* Previously unpublished.
*Remarks* See Remarks Cat. Nos. 15 and 26. One of seven bottles in the Toledo Museum collection from closely related molds (Cat. Nos. 26-32). Same Subseries and same Generation as Cat. Nos. 29, 31, and 32.

31 **Hexagonal Bottle, Vessels Type, Series A**

Phoenician
Mid first century or earlier
H 7.8 cm; H body 5.1 cm; D 4.0-3.7 cm; D rim 2.0 cm; D base 2.2 cm
1923.473, Gift of Edward Drummond Libbey

Translucent dark yellow green (10 GY 4/4) and moderate blue green (5 BG 4/6). Opaque reddish brown streaks in neck and rim. Batch not well melted; possibly same batch as Cat. No. 26. Medium thin glass; numerous small spherical bubbles in body and small to large linear and pinprick bubbles in neck.
Neck free blown. Body blown into a four-part mold of three vertical sections joined to a disk-shaped base section (MCT II). Mold seams between panels 2 and 3, 4 and 5, and 6 and 1. Relief indistinct. Vertical crimps at base of neck from tool used to hold vessel while finishing rim.
Flaring rim, folded outward, upward, and inward. Cylindrical neck with slight downward taper. Sloping shoulder. Ovoid body. Low, offset base with flat underside with two raised concentric circles. Decoration like Cat. No. 26.
*Classification* Vessels Type, Series A3, Generation A3a-y.
*Condition* Intact. Iridescent areas; incipient surface pitting.
*Collection* Thomas E. H. Curtis.
*Published* Previously unpublished.
*Remarks* See Remarks Cat. Nos. 15 and 26. One of seven bottles in the Toledo Museum collection from closely related molds (Cat. Nos. 26-32). Same Subseries and same Generation as Cat. Nos. 29, 30, and 32.

32 **Hexagonal Bottle, Vessels Type, Series A**

Phoenician
Probably mid first century or earlier
H 7.67 cm; H body 5.14 cm; D 4.1-3.7 cm; D rim 2.35 cm; D base 2.2 cm
1923.481, Gift of Edward Drummond Libbey

Opaque streaked light greenish gray. Medium thin glass. Small spherical bubbles in body (burst on the surface) and linear bubbles in neck.
Neck free blown. Body blown into a four-part mold of three vertical sections joined to a disk-shaped base section (MCT II). Mold seams between panels 2 and 3, 4 and 5, and 6 and 1. Relief indistinct. Vertical crimps at base of neck from tool used to hold vessel while finishing rim.
Flaring rim, folded outward, upward, and inward. Cylindrical neck with slight downward taper. Sloping shoulder. Ovoid body. Low, offset base with flat underside with two raised concentric circles. Decoration like Cat. No. 26.
*Classification* Vessels Type, Series A3, Generation A3a-y.
*Condition* Intact. Remains of buff-colored weathering; surface pits.
*Collection* Thomas E. H. Curtis.
*Published* Previously unpublished.
*Remarks* See Remarks Cat. Nos. 15 and 26. One of seven bottles in the Toledo Museum collection from closely related molds (Cat. Nos. 26-32). Same Subseries and same Generation as Cat. Nos. 29-31.

33 **Hexagonal Bottle, Vessels Type, Series A**

Photograph, Fig. 43
Phoenician
Probably mid first century
H 8.0 cm; H body 5.1 cm; D 4.0-3.5 cm; D rim 2.8 cm; D base 2.2 cm
1923.530, Gift of Edward Drummond Libbey

Transparent pale green tint (10 GY 6/2). Medium thin glass.

Numerous pinprick and small vertically elongated bubbles in neck; some large spherical bubbles in body.
Neck free blown. Body blown into a four-part mold of three vertical sections joined to a disk-shaped base section (MCT II). Mold seams between panels 2 and 3, 4 and 5, and 6 and 1. Relief indistinct and difficult to distinguish because of the transparency of the glass. Vertical crimps at base of neck from tool used to hold vessel while finishing rim.
Flaring rim, folded outward, upward, and inward. Cylindrical neck with slight downward taper. Sloping shoulder. Slender ovoid body. Low, offset base with flat underside with two raised concentric circles.
On the shoulder, six arches, each containing an unidentified, small, smooth, egg-shaped object. On the body, six rectangular panels divided by degenerated columns, each with a rudimentary base. In the panels, six vessels, from left to right: 1, a spouted jug with handle to the right; 2, a footed bowl with wide opening and two handles from shoulder to rim, the opening containing one (?) row of rounded objects (fruit?); 3, a footed vessel with two very small handles on the shoulder; 4, a slender footed bowl with tall cylindrical neck with vertical grooves, oblate body, and tall crooked stem, the mouth filled with two (?) rows of rounded objects (fruit?); 5, a footed jug with handle to the right; and 6, a second footed vessel as in panel 3. Around the bottom, six triangular leaves with tips centered beneath panels alternating with a small ovoid fruit below each column.
*Classification* Vessels Type, Series A3, Generation A3b.
*Condition* Intact. Small dimples on surface, probably from grit on interior of mold.
*Collection* Thomas E. H. Curtis.
*Published* Previously unpublished.
*Remarks* See Remarks Cat. Nos. 15 and 26. This vessel was made in the same type of four-part mold (MCT II) as Cat. Nos. 26-32, but the decorative design was modified in such a way that the columns between the panels have degenerated to vertical ribs with thickened ends where the base would have been, while there is no remnant of a capital. The fillet around the bottom has been recut to form six solid triangular petals or leaves alternating with a small smooth fruit.
Subseries A3, Generation A3b, may be the same Generation as Subseries A1, Generation A1c (see Fig. 46), but it was probably blown in the same workshop as the other bottles belonging to Subseries A3. Like most of the bottles of Generation A3a, this one has tool marks at the base of the neck. The use of natural green glass instead of colored glass is rare and confirms a relatively late date for the creation of Generation A3b.

## 34 **Hexagonal Bottle, Vessels Type, Series B**

Color plate 6, p. 52
Probably Eastern Mediterranean
Probably mid to second half of first century
H 8.8 cm; H body 5.4 cm; D 4.7-4.0 cm; D rim 2.1 cm; D base 2.5 cm
1923.430, Gift of Edward Drummond Libbey

Polychrome glass composed of a mixture of colors ranging from opaque to translucent brownish black (5 YR 2/1), light brownish gray (near 5 YR 6/1), white (or light gray), and moderate yellowish brown glass (near 10 YR 5/4). Interior predominantly opaque white. Medium thin glass.
Neck free blown. Body blown into a three-part mold of three vertical sections (MCT IV). Mold seams between panels 2 and 3, 4 and 5, and 6 and 1 meet at center of underside of base. Relief moderately crisp.
Rim folded outward, upward, inward and flattened. Tall cylindrical neck. Convex shoulder and convex bottom joined by a nearly cylindrical body. Base flat.
On the shoulder, six rounded arches, each containing an unidentified dome-shaped object with a depression in the center of its bottom. On the body, six rectangular panels divided by degenerated columns, each with a rudimentary capital and base. In the panels, six vessels, from left to right: 1, a footed jug with spout to the left and high handle to the right; 2, a wide, stemmed bowl with two double looped handles, containing three tall upright dimpled objects with round tops; 3, a footed jug with spout to the left and angular handle to the right; 4, a footed jug with spout to the left and handle to the right; 5, a footed bowl with tall cylindrical neck, oblate body with two small vertical handles, the mouth filled with three rows of small round objects (fruit?); and 6, a footed vessel with two small handles on the shoulder. Around the bottom, six triangular petals or leaves with tips centered beneath the panels, alternating with a large ovoid fruit or leaf beneath each column.
*Classification* Vessels Type, Series B1, Generation B1a.
*Condition* Intact. Some light brown incrustation on exterior; surface pits.
*Collection* Thomas E. H. Curtis.
*Published* Previously unpublished.
*Remarks* One of two bottles in the Toledo Museum collection from related molds (Cat. Nos. 34 and 35). These bottles are assigned to Series B because the vessels shown in the panels differ from those of Series A (Cat. Nos. 15-33). Series B is characterized by showing in a specific order three footed jugs with handles to the right; two footed bowls, the one in panel 5 filled with small round fruit as in panels 2 and 4 of Series A, the other in panel 2 filled with three tall upright dimpled objects that have been interpreted as loaves;[1] and one footed jar with two small handles. The main differences from Series A are the order and number of times each type of vessel is represented in Series B and the change of contents of the footed bowl in panel 2. However, Series B is so closely related to Series A that it might have been derived from the same archetype (see Remarks Cat. No. 15).
Certain details in the representation of the vessels and in the secondary decoration around the base suggest a close relationship between Series B and Generation A1c of Subseries A1 (see Cat. Nos. 23 and 24). For instance, the addition of small handles to the footed bowl with oblate body and tall cylindrical neck is common to both panel 5 of Series B and panel 4 of Subseries A1, Generation A1c, while the triangular leaves alternating with oval fruit or leaves around the bottom of the bottles are also very similar. The columns dividing the panels of Series B still having rudimentary capitals and bases and the larger size of the bottles of Series B suggest that Series B was created before Generation A1c of Subseries A1.
Series B is a small Series which may not have been made in Phoenicia. No examples have been published from the Syro-Palestinian area, but two examples are reported to be from Cyprus.[2]
A few small fragments of an opaque blue bottle have been excavated in a first-century cemetery at Angera in northern Italy;[3] too little remains to be absolutely sure of the Series to which it belongs, but what survives appears to agree with Series B of the Vessels Type.
The multicolored glass of this bottle, with its mostly white interior, was probably not made by a complicated casing process. Experiments by the author have shown that a similar result can be achieved by picking up small chunks of three or four different colors on the end of the blow pipe; heating, pinching, and shaping them; then blowing out the bubble. The color which happens to

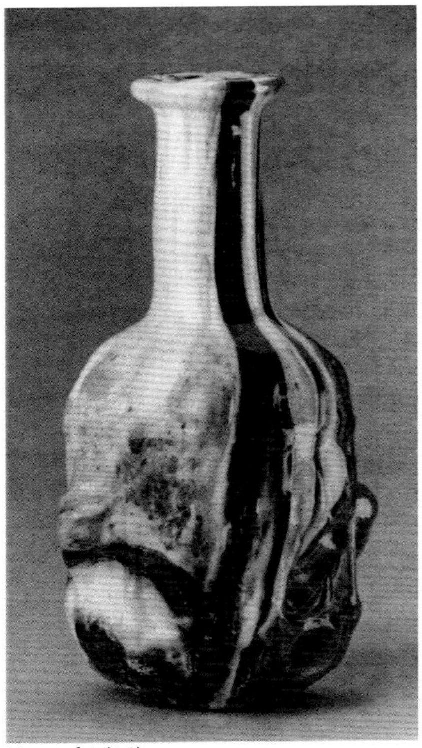

34, panel 1 (1:1)

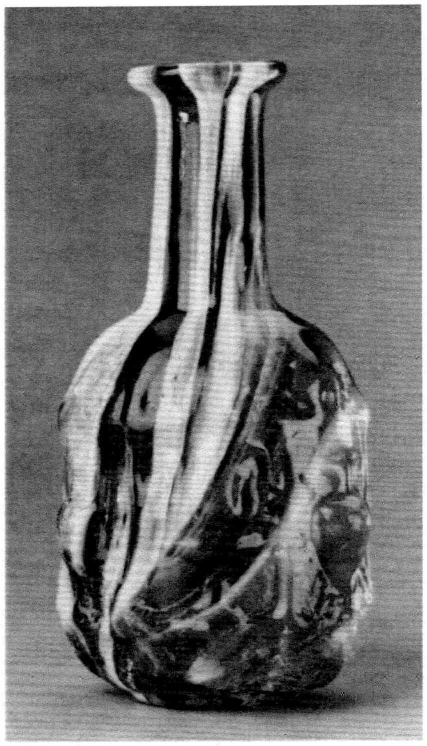

34, panel 2 (1:1)

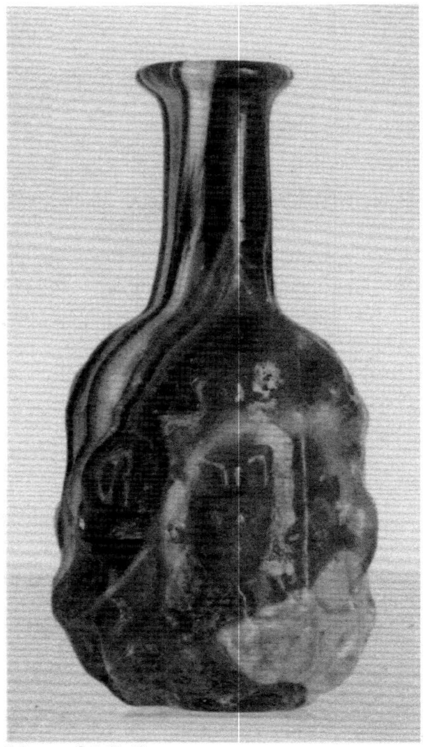

34, panel 3 (1:1)

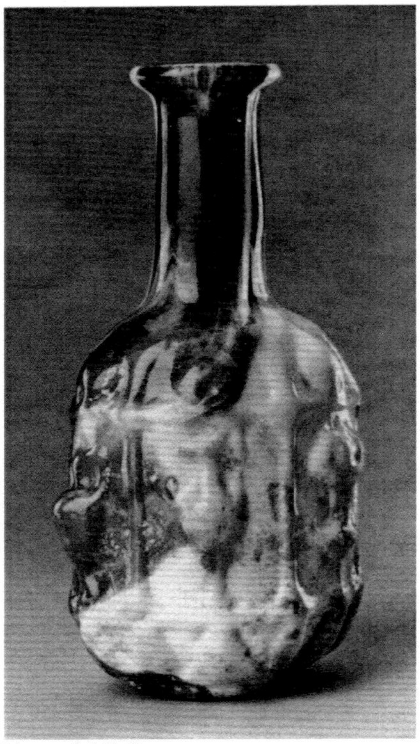

34, panel 4 (1:1)

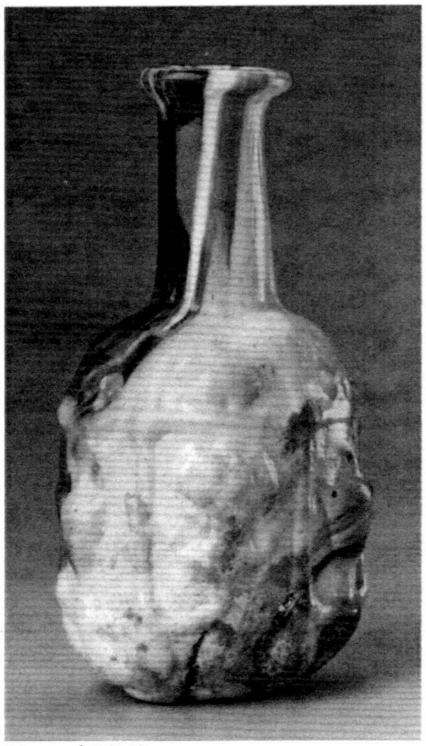

34, panel 5 (1:1)

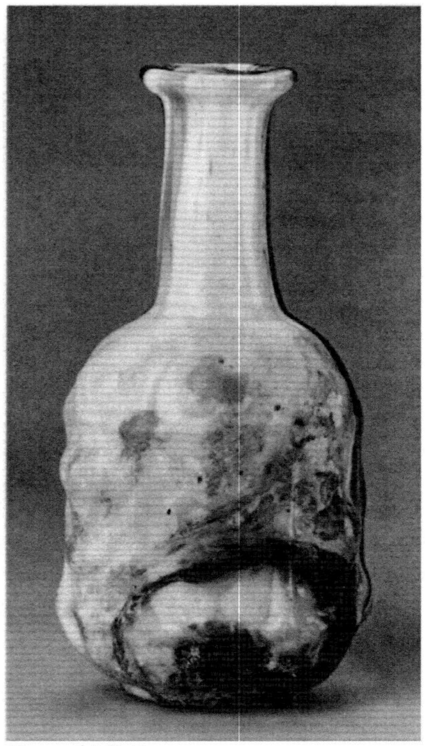

34, panel 6 (1:1)

cover the hole of the blowpipe spreads unevenly over the entire interior of the vessel.

[1] Eisen 1927, I, 248, fig. 115 (here Fig. 48).

[2] For example:

a) Cambridge, Fitzwilliam Museum, no. GR.27.1876, from Larnaca, Cyprus, transparent light blue, *Fitzwilliam* 1978, 32, no. 54b, ill. 44;

b) Nicosia, Cyprus, Mus., no. B1709, find spot unknown but presumably from Cyprus, Vessberg 1952, 130-131, pl. VI, 21.

[3] Giuliana M. Facchini, "Le verre en Italie," *BullAIHV* 9 (1981-1983) 207, fig. 1.

## 35 **Hexagonal Bottle with Handle, Vessels Type, Series B**

Perhaps Eastern Mediterranean
Probably mid to second half of first century
H 8.7 cm; H W handle 9.37 cm; H body 5.4 cm; D 4.5-3.9 cm; D rim 1.7 cm; D base 2.4 cm
1923.529, Gift of Edward Drummond Libbey

Transparent pale green tint (10 G 6/2). Translucent similarly colored handle. Medium thin glass. Clusters of pinprick bubbles in body and numerous small to medium-sized linear bubbles in neck. One large linear bubble from near rim to base of neck.
Neck free blown. Body blown into a three-part mold of three vertical sections (MCT IV). Mold seams between panels 2 and 3, 4 and 5, and 6 and 1 meet at center of underside of base. Relief moderately crisp but difficult to distinguish because of the transparency and glossiness of the glass. Handle irregularly attached at rim with a large fold against neck.
Rim rounded and thickened in flame; tool mark on interior. Tall cylindrical neck. Convex shoulder and convex bottom joined by a nearly cylindrical body. Base flat. Coil handle from shoulder to rim with projecting loop above. Handle positioned above panel 6. Decoration like Cat. No. 34.
*Classification* Vessels Type, Series B1, Generation B1a-y (younger Generation).
*Condition* Neck broken and repaired. Numerous annealing (?) cracks in neck. Interior incrusted.
*Collection* Thomas E. H. Curtis.
*Published* Previously unpublished.
*Remarks* See Remarks Cat. No. 34. The addition of a handle is not common in the Vessels Type. This feature, together with the unusual finishing of the rim, the rare use of natural green glass, and the diminished size, indicating a younger Generation than Cat. No. 34, suggest that this bottle was not made in the same workshop as Cat. No. 34. The glass blower may have used a bottle of Generation B1a as a model.

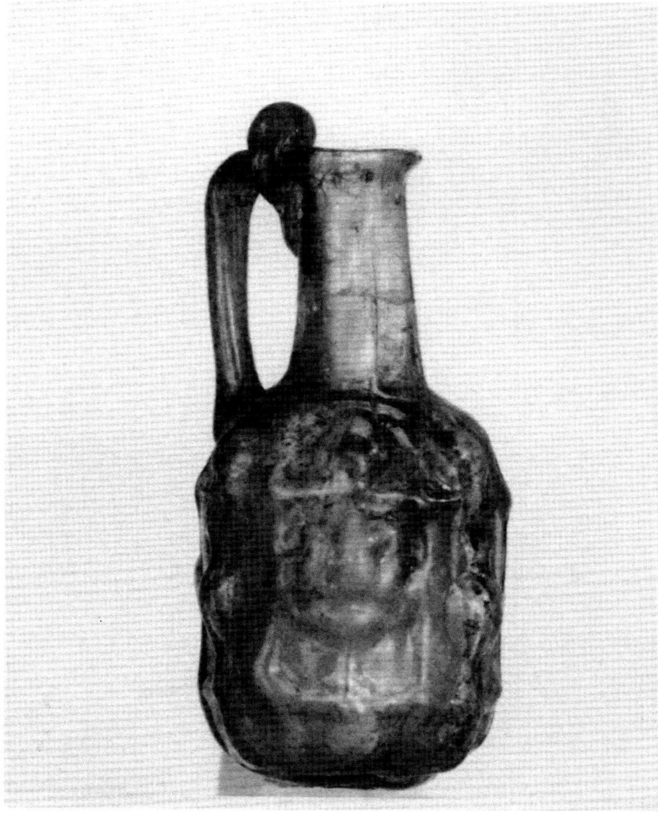
35, panel 2 (1:1)

### *Fruit Type*[1]
(Cat. Nos. 36–40)

#### 36 Hexagonal Bottle, Fruit Type, Series A

Color plates 5 and 7, pp. 52 and 53; photograph, Fig. 55
Phoenician, possibly made in Sidon
Perhaps second quarter of first century
H 7.93 cm; H body 5.1 cm; D 4.55–3.8 cm; D rim 2.0 cm; D base 2.35 cm
1923.433, Gift of Edward Drummond Libbey

Opaque white. Medium thin glass.
Neck free blown. Body blown into a three-part mold of three vertical sections (MCT IV). Mold seams between panels 2 and 3, 4 and 5, and 6 and 1 meet at center of underside of base. Relief crisp.
Rim folded outward, upward, and inward. Tapering neck. Convex circular shoulder and convex circular bottom joined by a hexagonal body. Low, offset base with flat underside.
On the shoulder, in blurred relief, an inverted (?) floral motif (pomegranates, lotus, or rounded leaves?). On the body, six rectangular panels divided by thin vertical ribs; bordered below by a narrow band with a simplified egg-and-dart pattern; each panel contains a fruit. In the panels, from left to right: 1 and 4, a pomegranate; 2 and 5, a cedar cone; 3 and 6, a bunch of grapes. Around the bottom, six triangular leaves with double border centered beneath the panels, alternating with slender oval petals beneath each rib.
*Classification* Fruit Type, Series A1, Generation A1a.
*Condition* Intact. Traces of iridescent film.
*Collection* Thomas E. H. Curtis.
*Published* Lightfoot 1985, 124, no. 14.
*Remarks* One of four bottles in the Toledo Museum collection from closely related molds (Cat. Nos. 36–39). These bottles belong to the so-called Fruit Type,[2] a relatively small group of hexagonal bottles with a fruit in high relief in the center of each panel. The archetype of Series A had two sets of three repeated but not identical fruits and elaborate secondary decoration of floral motifs around shoulder and bottom. The shoulder decoration consists of indistinct, irregularly shaped bulges above the vertical ribs separating the panels. The panels are bordered below by a narrow band with a simplified egg-and-dart pattern.
With at least fourteen bottles surviving, Subseries A1 is the most common Subseries of Series A.[3] About half of the surviving bottles of Subseries A1 are translucent blue, four are opaque white, and two opaque blue. It is impossible to determine to which generations they belong, but they were probably all blown in three-part MCT IV A molds. The development of the molds in Generations A1b and A1c is discussed in Chapter I: *Syro-Palestinian Styles: Hexagonal Bottles with High Relief: Fruit Type* because there are no bottles of these Generations in the Toledo Museum collection.
The four Toledo bottles show the same minor differences in the rendering of the fruit within each of the two sets of decoration. For example, the base line of the pomegranate in panel 1 extends farther to the right than to the left, while the base line of the pomegranate in panel 4 extends farther to the left than to the right and the fruit itself is placed higher above the base line. The cedar cones in panels 2 and 5 are not differentiated from each other. The grape bunch in panel 3 touches the base of the panel and has long, heavy side bunches with a fat tendril on either side trailing down to the base of the panel. In contrast, the side bunches in the grape bunch of panel 6 are scarcely differentiated from the main bunch, while the tendrils extending down on either side are extremely thin and curl outward before reaching the base of the panel. The mold-maker's style is stiff and formal; the fruits are stylized and shown in strict symmetry, each centered in its panel.

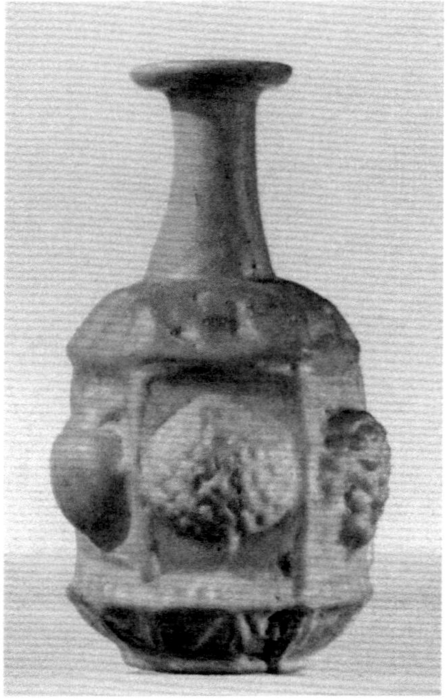

36, panel 2 (1:1)

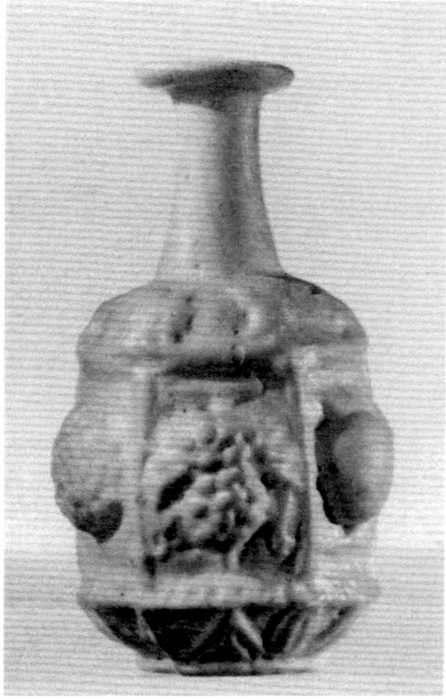

36, panel 3 (1:1)

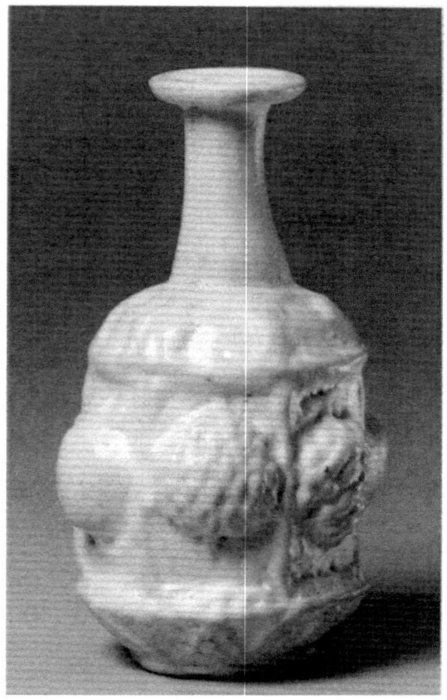

36, panel 5 (1:1)

### Origin and Distribution

The secondary decoration of Subseries A1, Generations A1a to A1a-yy, of the Fruit Type is identical to that of Generation A1a of the Masks Type (see Cat. No. 41). The simplified egg-and-dart band used as a border below the panels is also common to Series A of both Types. The composition of the panels in both Types is also similar, with one large round object in the center of each panel, recalling the composition of metopes on friezes above contemporary Palestinian tombs. [4]

The grape bunches of Series A of the Fruit Type are reminiscent of two of the masks in Series A of the Masks Type (see Cat. No. 41): the outline and overall impression of the grape bunch in panel 3 are similar to those of the youthful mask in panel 4, while the grape bunch in panel 6 is reminiscent of the Medusa mask in panel 1. These similarities suggest that the motifs of the Masks Type may have been recut from the motifs of the Fruit Type. These similarities further suggest that the archetypes for Series A of both Masks and Fruit Type were made in one workshop at around the same time. [5]

The find spots noted for bottles belonging to Series A of the Fruit Type are inland Syria, [6] eastern Turkey, [7] and Lebanon. [8] Most of these bottles appear to belong to Subseries A1, Generations A1a and A1a-y. The archaeological evidence is insufficient to locate the workshop that created Series A of the Fruit Type, but it was most probably somewhere in northern or central Phoenicia, possibly in Sidon. This proposal is supported by an otherwise tantalizingly uninformative remark that four bottles decorated with fruit in the panels, most of which are said to show two repeated sets of pomegranate, cedar cone, and grape bunch, were excavated at Tortosa (on the Syrian coast, in northern ancient Phoenicia) and one in Cyprus. [9]

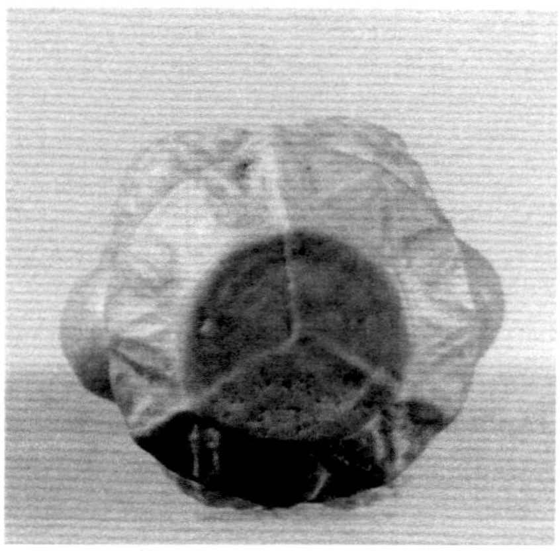

36, underside (1:1)

### Date

None of the bottles belonging to the Fruit Type can be dated by its context. A date in the second quarter of the first century for the earliest bottles of Subseries A1, Generation A1a, appears probable because that is the time when the archetype of Series A of the Masks Type was created (see Cat. No. 41 and Chapter I: *Syro-Palestinian Styles: Hexagonal Bottles with High Relief: Relations of Types*). The increased use of transparent glass instead of opaque glass for bottles of the Fruit Type also seems to reflect the general trend in first-century glass in which opaque colors were gradually supplanted by transparent ones, to be finally superseded by decolorized glass in the last decades of the century. In view of the small numbers of the Fruit Type that survive, production probably did not extend over many years.

### Interpretation

Unlike bottles of the Vessels Type (Cat. Nos. 15-35), which appear to imitate a six-sided architectural structure with columns, niches, and fillets, the six panels of the Fruit Type are separated by thin vertical ridges that do not evoke an association with columns or pillars. The shape of the body is much more plainly hexagonal. The design recalls six-sided pagan altars such as an altar to Dionysos, recently excavated at Beth Shean, Israel (see Fig. 56 and Chapter I: *Syro-Palestinian Styles: Hexagonal Bottles with High Relief: Interpretation*). However, there is no evidence of a ritual or religious use of the glass bottles. The bottles of the Fruit Type share this simple hexagonal body shape with those of the Masks Type (Cat. Nos.

41 and 42) and the Mixed Symbols Type (Cat. Nos. 43 and 44). The attraction of the Fruit Type appears to have been the depiction of local fruits for which Phoenicia and Palestine were famed. Of the fruit shown in the panels of Generation A1a, the grape bunch and pomegranate appear to be constants. The third fruit has been called a pine cone,[10] a custard-apple,[11] and an etrog, a citrus fruit indigenous to the Syro-Palestinian Coast.[12] On a few later bottles the position of this fruit is occupied by an acorn.[13] Although the etrog played an important role in the Jewish Festival of Tabernacles, Sukkoth, and had been cultivated along the Syro-Palestinian coast at least since the Hasmonean period (152-137 B.C.),[14] the swelling shape of the fruit in panels 2 and 5 does not strongly resemble an etrog (Fig. 73) or known ancient represention of etrogs which are depicted either as a beet root or a heart-shaped fruit with the end pointing downward[15] or as a citron with a point at either end.[16] A cone fruit therefore remains the most probable identification. Many kinds of cone-bearing trees were native to the Syro-Palestinian coast, the most famous being the cedars of Lebanon which have squat, swelling cones (Fig. 74) that look exactly like the fruit on these bottles.[17] Most cone fruits grow hanging downward, but cedar cones grow erect, as indicated by the short upright stem shown under the cones depicted in the Fruit Type. The carefully rendered, small, regularly placed scale-shaped elements on the surface of the cone are identical to representations of the finials of thyrsoi on a Roman silver cup in Toledo (see Fig. 57). The finial of a thyrsos originally consisted of ivy leaves but these were sometimes replaced by a cone fruit in Hellenistic and Roman times.[18]

The grape bunch, cone fruit, and pomegranate could have both polytheistic and Jewish associations. In Hellenized pagan circles a grape bunch symbolized the source of wine and a cone could stand for the finial of a thyrsos, both of which were sacred to the Greek god Dionysos who was worshipped throughout the Mediterranean world. However, in Phoenicia a cedar cone also had local importance as a symbol of the cedar forests which then covered extensive areas of the Lebanon mountains. The trees provided fragrant oil and resin, while their aromatic wood was the most famous and prized export of ancient Phoenicia. According to the Bible, the ancient Sidonians were experts in felling the giant trees (*I Kings* 5: 20, 28). As fertility symbols pomegranates not only played an important role in Greek mystery rites but were also sacred to the Phoenician goddess Astarte.

In Jewish tradition the grape vine was one of the seven choice fruits of Israel (*Deuteronomy* 8:8; *Numbers* 13:23). The three-lobed bunch of grapes is a common motif on Jewish tombs in Palestine and also occurs often on Jewish coins of the first century.[19] The cedars of Lebanon are frequently praised in the Bible as "the trees of the Lord" (*Psalms* 104:16) because of their beauty, hardiness, and longevity. The timber of Lebanon cedars was used in the construction of both the First and Second Temples in Jerusalem (*Ezra* 3:7). Pomegranates grew in nearly every part of Palestine. Like the vine, the pomegranate was one of the seven choice fruits of the land, symbolizing its fertility (*Deuteronomy* 8:8; *Numbers* 13:23); its beauty is remarked in the Song of Songs (*Songs* 4:3,13; 6:7; 6:11) as well as the delicious taste of its juice (*Songs* 8:2). The capitals of the pillars of the First Temple are said to have been decorated with pomegranates (*I Kings* 7:18, 42), and pomegranates also appear in fourth- to sixth-century synagogues in Palestine.[20] Three joined pomegranates appear on Jewish silver coins struck in the years 66-67 during the Jewish Revolt against Rome.[21]

Thus the fruits depicted on Series A of the Fruit Type could have appealed to both Jews and non-Jews. The strong associations with local fruits would have made Series A equally attractive to all inhabitants of Phoenicia and Palestine because customers could interpret the fruits as they wished.

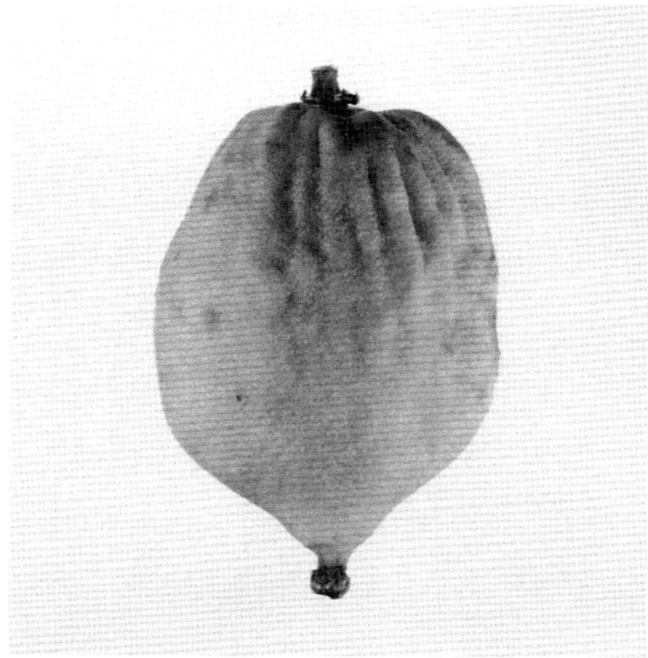

*Fig. 73. Etrog. Photo: Photography Inc.*

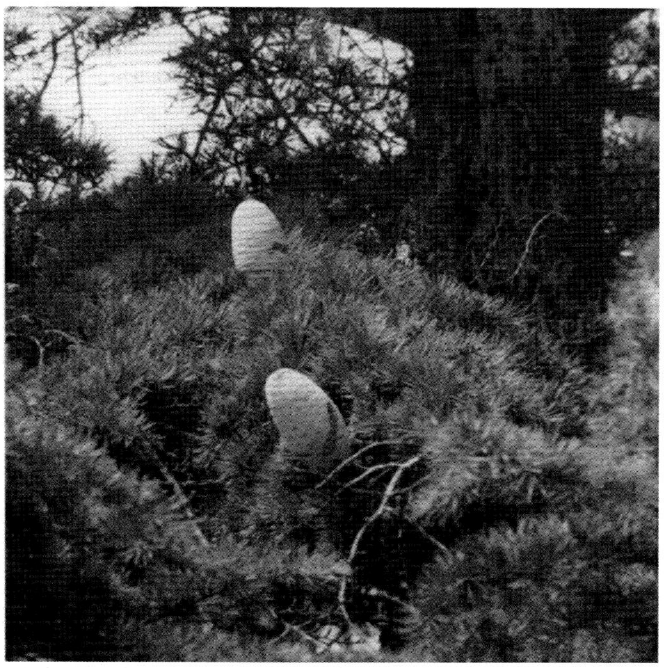

*Fig. 74. Immature cones on a cedar of Lebanon tree in the Stranahan Arboretum of The University of Toledo, Toledo, Ohio. Photo: Timothy A. Motz.*

[1] See Chapter I: *Syro-Palestinian Styles: Hexagonal Bottles with High Relief: Fruit Type.*

[2] On this Type: Lightfoot 1985, 123-125, nos. 1-22, who calls it the "Fruit Series." Eisen 1927, I, 252, named this Type "Flasks with Fruit Trees and Wreaths."

[3] Lightfoot 1985, 123-124, lists the following parallels:

a) Gaziantep, Turkey, Gaziantep Mus., number unknown, translucent cobalt blue, Lightfoot 1985, 123-124;

b) Damascus, Nat. Mus., no. 8521, opaque white, Lightfoot 1985, 124, no. 1;
c) formerly Constable-Maxwell Collection, no number, opaque blue, Lightfoot 1985, 124, no. 5; *Maxwell* 1979, 63, no. 95;
d) formerly Constable-Maxwell Collection, no number, translucent blue, Lightfoot 1985, 124, no. 7; *Maxwell* 1979, 63, no. 97;
e) formerly Kofler-Truniger Collection, no number, opaque grayish blue, Lightfoot 1985, 124, no. 8; *Kofler-Truniger* 1985, 68, no. 108;
f) Oppenländer Collection, no. 2492, translucent blue, Lightfoot 1985, 124, no. 9; von Saldern et al. 1974, 143-144, no. 406;
g) Oppenländer Collection, no. 2493, weathered blue, Lightfoot 1985, 124, no. 10; von Saldern et al. 1974, 144, no. 407;
h) New York, Harvey Rothenberg Collection, no number, opaque white, Lightfoot 1985, 124, no. 11; *JGS* 20 (1978) 119, no. 5, ill.;
i) Switzerland, private collection, no number, dark blue, Lightfoot 1985, 124, no. 12; Luzern 1981, 76, no. 241;
j-m) here Cat. Nos. 36-39.

Lightfoot 1985, nos. 4 and 6, do not belong to Series A; there has been no opportunity to check nos. 2 and 3. The description of no. 11 is not sufficient to say whether it belongs to Generation A1a or A1b. To the bottles listed by Lightfoot, add:

n) Damascus, Nat. Mus., no. C5739, from Tell Nabi Mind (about 25 km southwest of Homs, Syria), opaque white, Zouhdi 1969, 105, no. 4;
o) Damascus, Nat. Mus., no. 8521, from Dura Europos, tomb 24, loc. XIV, Clairmont 1963, 39, no. 147, pl. XXI;
p) Paris, Louvre, number unknown, purchased in Beirut, white, Goodenough I, 168; III, fig. 386.

[4] Goodenough I, 79-84; III, figs. 25-32.

[5] The conclusion that vessels belonging to the Fruit Type and Masks Type originated from the same workshop was anticipated by Lightfoot 1985, 127.

[6] See above note 2n, o.

[7] See above note 2a.

[8] See above note 2p.

[9] Froehner 1879, 63, n. 3. Unfortunately, the description is so general that the vessels referred to could also belong to Series B of the Fruit Type; see Cat. No. 40.

[10] Most recently in Lightfoot 1985, 123.

[11] See above note 2e. In spite of the clear resemblance to a custard apple or paw paw, this interpretation is impossible because that fruit is indigenous to Central America and the eastern United States and was not known in classical antiquity.

[12] A. Engle, "New Light on the Sidonian Glassmakers," *Readings in Glass History* 21 (Jerusalem 1988) 77, fig. 45C. Goodenough I, 168, n. 49, quotes the opinion of "others" with some reservation about this identification.

[13] Auth 1976, 69, no. 65 (a bottle blown into a three-part mold of MCT V A [Subseries A2, Generation A2a]); Isings 1975a, 38-39, no. 663, fig. 99 (here Fig. 53).

[14] The earliest mention of the etrog as a citron in combination with a Jewish ceremony occurs in Flavius Josephus, *Jewish Antiquities*, 13.372, who relates that the Jewish king Alexander Jannaeus (103-76 B.C.) was pelted with etrogs during the festival of Sukkoth. The etrog appears also on Jewish coins dated 69/70, Goodenough I, 276; III, fig. 689. On the etrog as a Jewish cult symbol: Goodenough IV, 145-166; Charles B. Chavel, *Ramban (Nachmanides). Commentary on the Torah. Leviticus* (translated and annotated) (New York 1974) 396ff.

[15] Goodenough IV, 146; III, figs. 651, 689, 704, 708, 711, 720; Hachlili 1988, figs. 56a, 60.

[16] Goodenough IV, 146; III, fig. 715; Rachel Hachlili, *Ancient Jewish Art and Archaeology in the Land of Israel* (*Handbuch der Orientalistik*, vol. 1, pt. 2, *Die Denkmäler*, pt. B, *Vorderasien*, sect. 4, Leiden 1988) fig. 61.

[17] The cedar of Lebanon tree has both male and female cones; the male cones are small and slender and are often found on the lower branches of the tree, while the female cones are large (about 10 cm long) and squat and are often found on the upper branches (my thanks to David C. Michener of the Matthei Botanical Gardens at the University of Michigan for explaining this). It is the female cedar of Lebanon cone which appears on hexagonal bottles of the Fruit Type.

[18] F. G. Von Papen, *Der Thyrsos in der griechischen und römischen Literatur* (Diss. Bonn, 1905). The silver cup in Toledo was reportedly found in Turkey, where in antiquity there were large forests of Lebanon cedars. The use of a cedar cone instead of a pine cone would therefore have been understandable in this area. See also the *thyrsolongchoi* depicted on Cat. Nos. 71-74.

[19] Goodenough III, figs. 21-24, 677, 694-697. Compare also the three-lobed grape bunches depicted on Jewish sarcophagi of the first half of the first century, e.g. *Holy Land* 1986, 222-223, no. 114 (with references to additional, similarly decorated sarcophagi). Small bottles in the shape of a three-lobed grape bunch were also made in mold-blown glass: see Cat. Nos. 109 and 110.

[20] Goodenough III, figs. 563 (carved screen), 600 (fresco), 635 (mosaic floor). On the meaning of the pomegranate in the Near East: F. Muthmann, *Der Granatapfel, Symbol des Lebens in der alten Welt* (Bern 1982).

[21] *Holy Land* 1986, 227-229, no. 116, figs. 103-106.

## 37 Hexagonal Bottle, Fruit Type, Series A

Color plate 7, p. 53
Phoenician, possibly made in Sidon
Probably second quarter of first century
H 7.7 cm; H body 5.0 cm; D 4.5-3.8 cm; D rim 2.4 cm; D base 2.3 cm
1923.434, Gift of Edward Drummond Libbey

Opaque pale blue (near 5 PB 7/2). Medium thin glass.
Neck free blown. Body blown into a three-part mold of three vertical sections (MCT IV). Mold seams between panels 2 and 3, 4 and 5, and 6 and 1 meet at center of underside of base. Relief moderately crisp.
Rim folded outward, upward, and inward. Tapering neck. Convex circular shoulder and convex circular bottom joined by a hexagonal body. Base flat.
Decoration like Cat. No. 36.
*Classification* Fruit Type, Series A1, Generation A1a.
*Condition* Intact. Interior and exterior of rim covered with buff-colored weathering which continues over about one-fourth of exterior surface.
*Collection* Thomas E. H. Curtis.
*Published* Lightfoot 1985, 124, no. 15.
*Remarks* See Remarks Cat. No. 36. One of four bottles in the Toledo Museum collection from closely related molds (Cat. Nos. 36-39).

## 38 Hexagonal Bottle, Fruit Type, Series A

Phoenician, possibly made in Sidon
Probably mid first century
H 7.85 cm; H body 5.0 cm; D 4.45-3.8 cm; D rim 2.2 cm; D base 2.3 cm
1923.429, Gift of Edward Drummond Libbey

Opaque white. Medium thin glass.
Neck free blown. Body blown into a three-part mold of three vertical sections (MCT IV). Mold seams between panels 2 and 3, 4 and 5, and 6 and 1 meet at center of underside of base. Relief moderately crisp.
Rim folded outward, upward, and inward. Tapering neck. Convex circular shoulder and convex circular bottom joined by a hexagonal body. Base flat.
Decoration like Cat. No. 36.
*Classification* Fruit Type, Series A1, Generation A1a-y.
*Condition* Small hole in shoulder. Traces of iridescent buff-colored weathering; surface pitting. Most of weathering has been removed, leaving areas of a white surface.
*Collection* Thomas E. H. Curtis.
*Published* Lightfoot 1985, 124, no. 13.
*Remarks* See Remarks Cat. No. 36. One of four bottles in the Toledo Museum collection from closely related molds (Cat. Nos. 36-39).

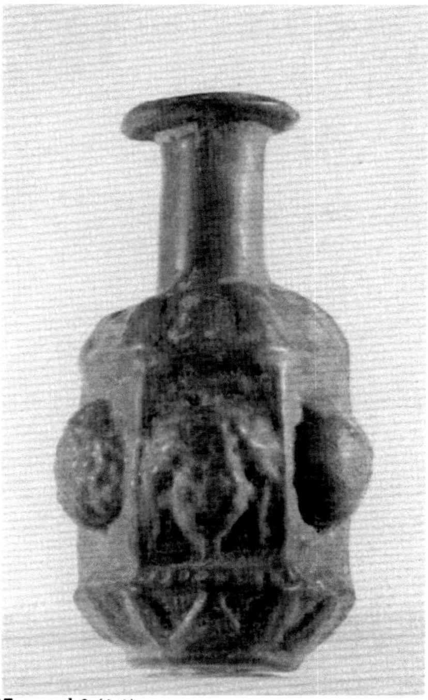

37, panel 3 (1:1)

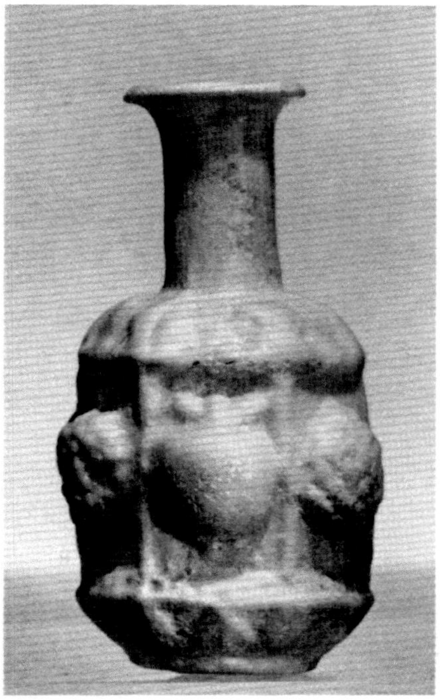

38, panel 1 (1:1)

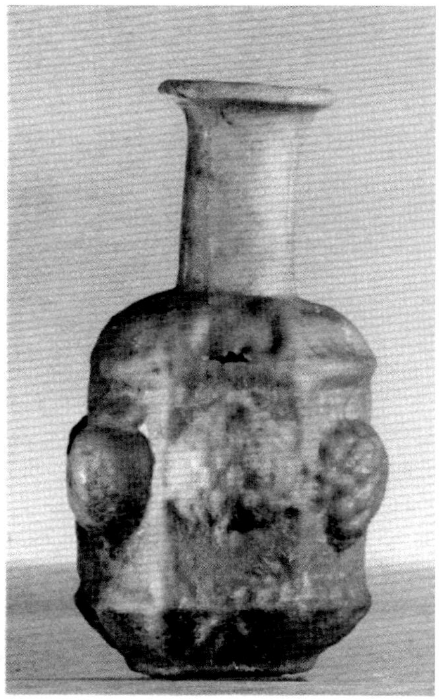

38, panel 5 (1:1)

39 **Hexagonal Bottle, Fruit Type, Series A**

Phoenician, possibly made in Sidon
Probably second half of first century
H 8.0 cm; H body 5.4 cm; D 4.45-3.2 cm; D rim 2.3 cm; D base 2.3 cm
1923.435, Gift of Edward Drummond Libbey

Possibly transparent pale blue, but color difficult to determine because of weathering. Medium thin glass. Fabric cannot be determined because of weathering.
Neck free blown. Body blown into a three-part mold of three vertical sections (MCT IV). Mold seams between panels 2 and 3, 4 and 5, and 6 and 1 meet at center of underside of base. Relief moderately crisp.
Rim folded outward, upward, and inward. Cylindrical neck. Convex circular shoulder and convex circular bottom joined by a hexagonal body. Base flat.
Decoration like Cat. No. 36.
*Classification* Fruit Type, Series A1, Generation A1a-yy.
*Condition* Intact. Exterior and interior of mouth covered with purple, golden, and green iridescence. Where this iridescence has flaked off, e.g. on the neck, severe surface pitting.
*Collection* Thomas E. H. Curtis.
*Published* Lightfoot 1985, 124-125, no. 16.
*Remarks* See Remarks Cat. No. 36. One of four bottles in the Toledo Museum collection from closely related molds (Cat. Nos. 36-39).

40 **Hexagonal Bottle, Fruit Type, Series B**

Color plate 7, p. 53
Phoenician, possibly made in Sidon
Probably second half of first century
H 8.5 cm; H body 5.26; D 4.9-4.1 cm; D rim 2.6 cm; D base 2.3 cm
1923.436, Gift of Edward Drummond Libbey

Translucent dark brownish manganese colored glass. Medium thin glass. Fabric cannot be determined because of weathering.
Neck free blown. Body blown into a three-part mold of three vertical sections (MCT IV). Mold seams between panels 2 and 3, 4 and 5, and 6 and 1 meet at center of underside of base. Relief high but indistinct. Two vertical crimps at base of neck from tool used to hold vessel while finishing rim.
Rim folded outward, upward, and inward. Cylindrical neck. Convex circular shoulder and convex circular bottom joined by a hexagonal body. Base flat.
On the shoulder, in blurred relief, a floral motif (heart-shaped lotus buds above center of panels, alternating with pomegranates or palmettes between panels). On the body, six rectangular panels divided by thin vertical ribs with slightly broadened ends that make them resemble degenerated columns with capitals and bases; each panel contains a fruit. In the panels, from left to right: 1 and 4,

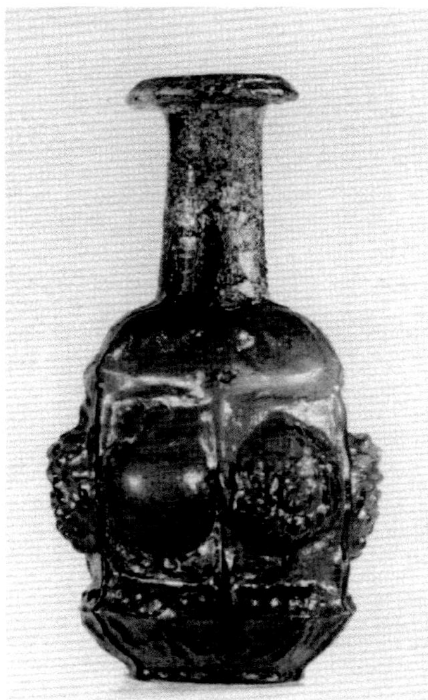

39, panels 1 and 2 (1:1)

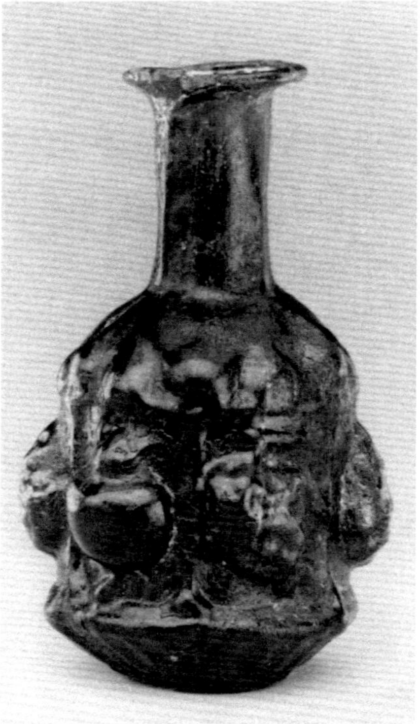

40, panels 1 and 2 (1:1)

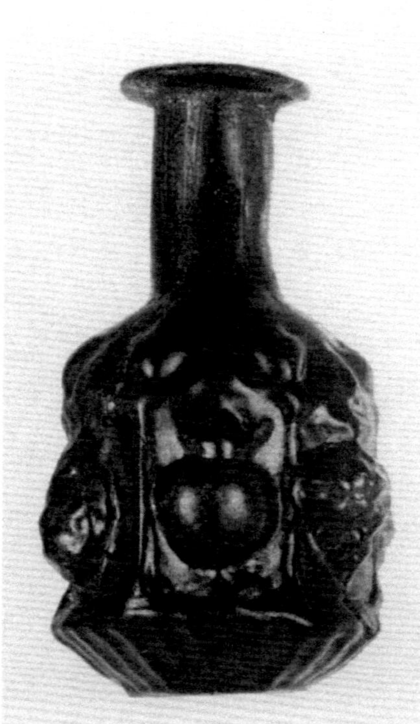

40, panel 4 (1:1)

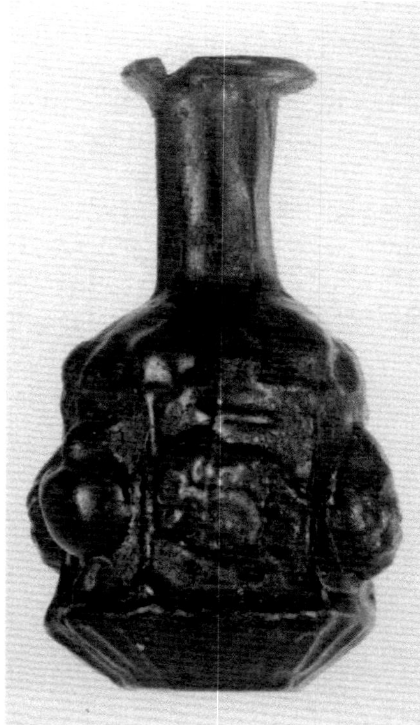

40, panel 5 (1:1)

a pomegranate; 2 and 5, a bunch of grapes; 3 and 6, a cedar cone. Around the bottom, twenty-seven upturned flutes.
*Classification* Fruit Type, Series B1.
*Condition* Small chip from rim restored. Severe surface pitting on exterior. Exterior has been coated with a solution of acryloid B72 and toluene to consolidate outer crust of glass.

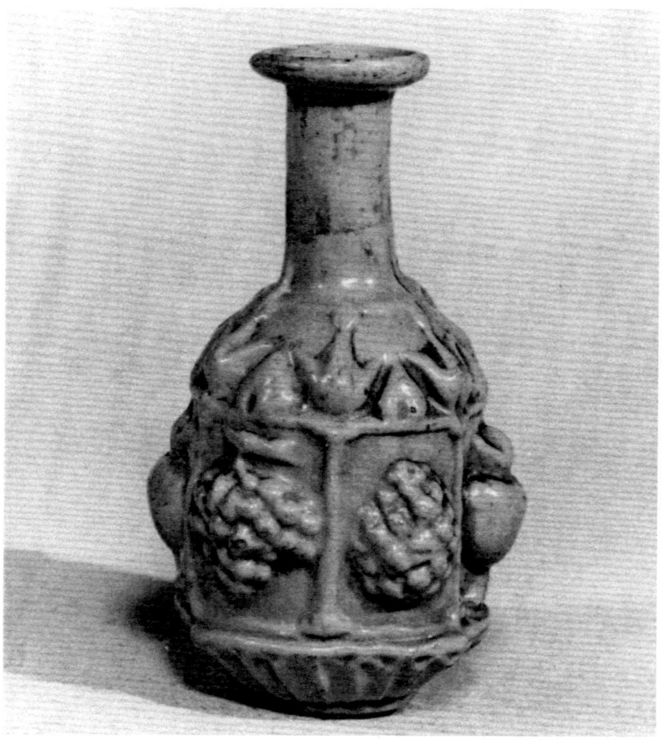

*Fig. 75. Hexagonal bottle, Fruit Type, Series B. This bottle may have been blown in the same mold as Cat. No. 40, but because the design is crisper and a mold crack seen in Cat. No. 40 is not present, it is likely that this bottle was produced earlier. H 9.2 cm, opaque pale blue, Pavia, Musei Civici, no. A219. Photo: Musei Civici, Pavia.*

*Collection* Thomas E. H. Curtis.
*Published* Eisen 1927, I, 253, fig. 121, pl. 52; Lightfoot 1985, 125, no. 22.
*Remarks* Series B of the Fruit Type differs from Series A (see Remarks Cat. No. 36) not only in the order of the fruits and style of their rendition but also in significant details of the secondary decoration. The floral motif on the shoulder is more elaborate than that of Series A, but the most striking difference lies in the use of gadrooning instead of triangular leaves around the bottom of the vessels of Series B. Below the panels, the egg-and-dart border characterizing Series A is omitted, but the ribs separating the panels clearly reveal their original conception as columns (see Remarks Cat. No. 15). The mold was also slightly larger at its widest point than the molds used for Series A.
Like Series A (Cat Nos. 36-39), the vessels of Series B show two sets of three repeated, but not identical, fruits. The pomegranate in panel 1 has a short thick stem, curving to the left, which is so faint in panel 4 that it appears to be missing. The fat grape bunch in panel 2 makes a sweeping curve from left to right, pointing its tip toward the lower right corner of the panel; the side bunches are only faintly indicated. In contrast, the large grape bunch in panel 5 has clearly rendered side bunches and is arranged vertically along the axis of the panel. The cedar cone in panel 3 is slightly thinner and more pointed than that in panel 6; both are characterized by a pattern of irregular surface bumps or scales.
This mold maker's style is much less stiff than that of the maker of the molds for Series A. In Series B the fruits are rendered in a lively manner, appearing to burst out of the space allotted to them even though the mold maker varied the widths of the panels in order to accommodate his design (2.0 cm for panels 1 and 2; 1.75 cm for panel 3).
Very few vessels can be assigned to Series B, therefore no attempt has been made to classify them. An opaque pale blue bottle in Pavia[1] (Fig. 75) may have been blown in the same mold as this

Toledo bottle. The photograph of the bottle in Pavia shows clearly that the mold was in much better condition when that bottle was blown: not only is the design crisper, but a small crack in the mold, visible on the Toledo bottle along the base of panel 4, had not yet formed. When this bottle was blown, hot glass seeped through the crack, forming a short, upward curving protrusion. [2]

ORIGIN AND DATE

Nothing definite is known about the find spots of the bottles assigned to Series B. A Near Eastern find spot is likely for the Toledo bottle because it entered the collection as part of the Curtis Collection, but the possibility that the bottle in Pavia was found in one of the cemeteries around Pavia "cannot be excluded." [3] In view of the fact that both bottles were probably blown in the same mold, production in an eastern Mediterranean glass house seems likely for both vessels.

The motifs of Series A and Series B of the Fruit Type are so similar that they reflect a very special relationship. This is not simply a relationship of copying and degenerating molds, as is the case for the molds of the Masks Type (see Cat No. 41). The independent, vigorous style of Series B and the different secondary decoration are too far from that of Series A to assume that they were made in the same glass house. However, the common use of MCT IV A molds and the subjects of the panels in both Series suggest the workshops were located in the same general area. Series A was made in a Phoenician glass house, possibly in Sidon, around the middle of the first century (see Cat. No. 36). A similar origin and date seems probable for Series B.

[1] Pavia, Musei Civici, no. A219, opaque pale blue, Maccabruni 1983, 34-35, no. 8, fig. 8, color pl. 2, top right.

[2] A so-called flash, in the language of American contemporary glass makers.

[3] Maccabruni 1983, 32.

## *Masks Type* [1]

(Cat. Nos. 41 and 42)

### 41 Hexagonal Bottle, Masks Type, Series A

Color plates 5 and 8, pp. 52 and 53
Phoenician
Probably second quarter of the first century
H 8.8 cm; H body 5.0 cm; D 3.95-3.6 cm; D rim 2.6 cm; D base 2.3 cm
1923.431, Gift of Edward Drummond Libbey

Opaque white. Thick glass.

Neck free blown. Body blown into a three-part mold of three vertical sections (MCT IV). Mold seams between panels 2 and 3, 4 and 5, and 6 and 1 meet at center of underside of base. Relief moderately crisp. Edges of mold carefully aligned.

Rim folded outward, upward, and inward. Tall cylindrical neck with constriction at its base. Convex circular shoulder and convex circular bottom joined by a hexagonal body. Base flat.

On the shoulder, in blurred relief, an inverted floral motif (a pomegranate or a lotus?). On the body, six rectangular panels, bordered below by a narrow band with a simplified egg-and-dart pattern; each panel contains a frontal theatrical mask. In the panels, from left to right: 1, Medusa, Rondanini type [2] but with more upright wings; 2, a youthful face with half-open mouth and hair worn close over the ears (a slave?); 3, a man with long pointed beard (an old man); 4, Pan or paniskos (a youthful human face with goat horns and split beard); 5, a man or woman with short hair curling up at ears (a mourning mask?); and 6, a man with medium-length hair and medium-length curly beard (a king?). Around the bottom, alternating rounded and pointed leaves.

*Classification* Masks Type, Series A1, Generation A1a.

*Condition* Intact. Remains of dull white weathering on exterior and yellowish brown (weathering?) on interior.

*Collection* Thomas E. H. Curtis.

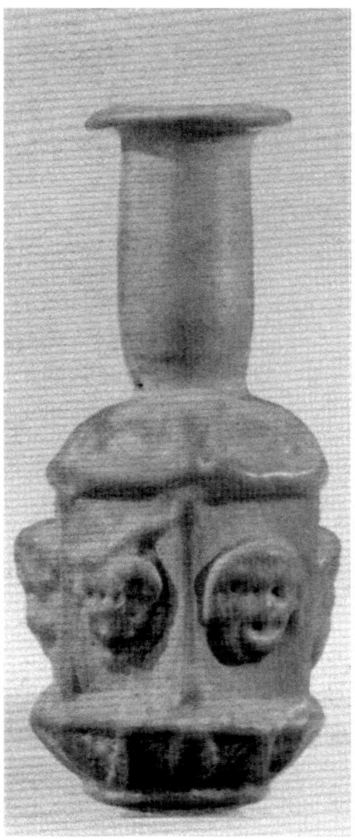

41, panels 1 and 2 (1:1)

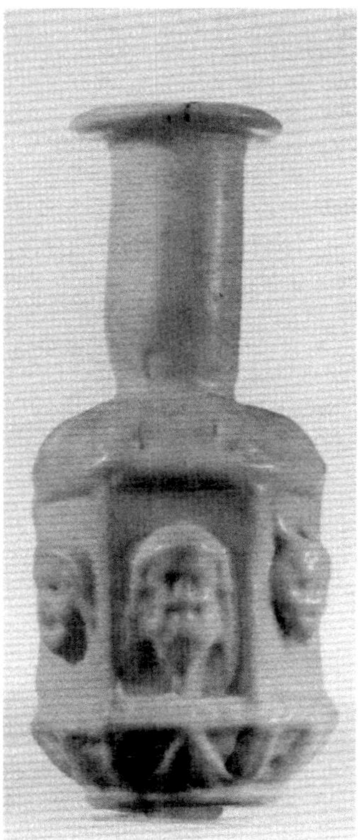

41, panel 3 (1:1)

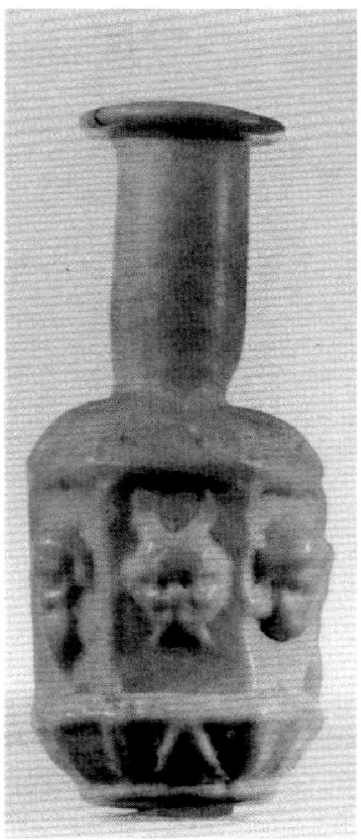

41, panel 4 (1:1)

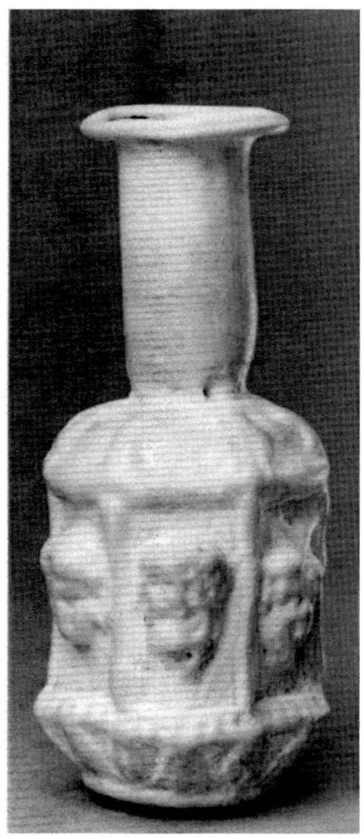

41, panel 6 (1:1)

*Published* Richter 1914, 85, fig. 10; Lightfoot 1987, 21, no. 11, fig. 6 (incorrectly describing the panels as showing two repeated sets of three masks).

*Remarks* One of two bottles in the Toledo collection from closely related molds (see also Cat. No. 42). These bottles belong to the so-called Masks Type,[3] a small group of hexagonal bottles with a mask in high relief in the center of each panel. Series A is characterized by six different masks and elaborate secondary decoration on shoulder and bottom. The panels have a lower border of a narrow band with a simplified egg-and-dart pattern. The shoulder decoration has not yet been satisfactorily identified; it has indistinct bulges where the glass did not completely fill the mold during blowing. Series A seems to be the most numerous of this Type; in addition to the two bottles in Toledo, at least five more bottles are known.[4] They appear to belong to Generation A1a and A1a-y and were probably all blown into MCT IV A molds. The only colors noted thus far are opaque white and opaque pale blue.

Similarities between the Medusa mask in panel 1 and the grape bunch in panel 6 of Series A of the Fruit Type and between the youthful mask in panel 4 and the grape bunch in panel 3 of the Fruit Type suggest that these masks may have been recut from the grape bunches (see Remarks Cat. No. 36 and Chapter I: *Syro-Palestinian Styles: Hexagonal Bottles with High Relief: Interrelations*). The archetypes for both Series were probably created by one craftsman at about the same time.

The indistinct descriptions of the masks in most publications are probably the result of difficulties in identifying them when the relief is blurred.[5] The Toledo bottles reveal for the first time that although the masks are arranged in two symmetrical sets of three, each of the six masks on the panels of Series A is different. The sets consist of one winged and one horned mask (panels 1 and 4), one youthful and one female mask with short hair (panels 2 and 5), and two bearded masks, one with a long beard, the other with a curly medium length beard (panels 3 and 6). Either set could have been the model for Group[6] 2 with its two repeated sets of three masks.[7] A few bottles said to have three repeated sets of two masks are assigned to Group 3.[8] Groups 2 and 3 lack secondary decoration on shoulder and bottom and also lack an egg-and-dart band as a lower border on the panels with masks. Unfortunately, descriptions are too vague to determine whether the masks themselves are identical within each of these tentative Groups. A small group of four-sided bottles decorated with masks, Group 4 was probably inspired by bottles of Group 3.[9]

### Origin and Distribution

Series A of the Masks Type was produced in the same workshop as Series A of the Fruit Type (see Remarks Cat. No. 36). Most bottles of Series A of the Masks Type have been unearthed in the eastern Mediterranean. One was found about 30 km southwest of Homs, Syria; one was excavated in Eretria, Greece; and three are from unrecorded find spots but probably of eastern Mediterranean provenance because of their history and present location. A Near Eastern find spot is also likely for the two bottles in Toledo because they were originally in the Curtis Collection.

### Date

None of the bottles belonging to Series A can be dated by archaeological context. However, a date in the second quarter of the first century is suggested by the relationship with the four-sided bottles of Group 4, whose design was probably inspired by bottles of Group 3. The Masks Type went through at least three major changes between the creation of the archetype for Series A and the beginning of the production of the four-sided bottles. The latter were probably blown around the year 40 (see Chapter I: *Syro-Palestinian Styles: Hexagonal Bottles with High Relief: Date*); consequently Series A might have been created around the year 30, or slightly earlier. The relatively small number of examples that survives from Series A suggests that it quickly went out of production, presumably around the middle of the first century.

### Interpretation

Like bottles of the Fruit Type (Cat. Nos. 36-40) the simple hexagonal shape of bottles of the Masks Type appears to imitate altars such as one recently excavated at Beth Shean, Israel (see *Syro-Palestinian Styles: Hexagonal Bottles with High Relief: Interpretation* and Fig. 56). Like the panels of this stone altar dedicated to Dionysos, which are decorated with masks and Dionysiac symbols, the six panels of the glass bottles abut each other without columns or any other type of dividing element. It is possible that the bottles of the Masks Type were in some way associated with a cult of Dionysos. But unlike the masks on the altar, which represent the god and his follower Silenos (or Pan?), the masks on the glass bottles can be identified as theatrical masks, which are associated with Dionysos only indirectly by virtue of his being the god of theatrical performances.

The six masks of Series A do not have the high headdress, or *onkos*, typical of late Hellenistic and Roman tragic masks used in performances of tragedies and satyr dramas, nor do they show the exaggerated features of Hellenistic and Roman comic masks. Instead, these masks have closed mouths[10] and beautiful, symmetrical facial expressions, features that characterize them as the masks worn by pantomime actors who did not speak but danced while an unseen chorus accompanied by an orchestra sang a text.[11] The themes of pantomimes were usually taken from Greek mythology. Some dancers specialized in light themes akin to satyr drama in which Pan led a troup of satyrs in teasing and making fun of the heroes of the myths. The masks shown in Series A of the Masks Type include a Medusa mask in panel 1 and a youthful Pan in panel 4. The bearded mask in panel 6 would have been appropriate for a king. This suggests that the masks depicted in Series A might have been inspired by a pantomime performance on the theme of Perseus killing Medusa.

The decoration of the bottles of the Masks Type was designed to please a theater-loving clientele. In the first century A.D. this would have included people from most population groups living along the Syro-Palestinian coast. Many Jews who were attracted by the Greek way of life had become so Hellenized that rabbis found it necessary to repeat warnings against attending Greek theaters and games.[12] Although pantomime had a long tradition in Syria,[13] the find spots of bottles belonging to the Masks Type suggest that this Type of hexagonal bottle with high relief had more success in the West than in its area of origin. The success of the Type in the West may have been related to the formal introduction of pantomime in 22 B.C. to Rome where this form of theater immediately became very popular and was actively supported by Roman emperors such as Augustus (14 B.C.-A.D. 27), Caligula (27-41), Claudius (41-54), and Nero (54-68), to name only the most eminent patrons of the early first century.

[1] For the Masks Type, see also Chapter I: *Syro-Palestinian Styles: Hexagonal Bottles with High Relief: Masks Type.*

[2] E. Buschor, *Medusa Rondanini* (Stuttgart 1958) 11-12, pl. 1. See also Cat. No. 142, which shows two faces of Medusa Rondanini type.

[3] Named the "Mask Series" by Lightfoot 1987, 13.

[4] Published descriptions are too vague to identify the number and type of masks, but from what is visible in the accompanying photographs, masks

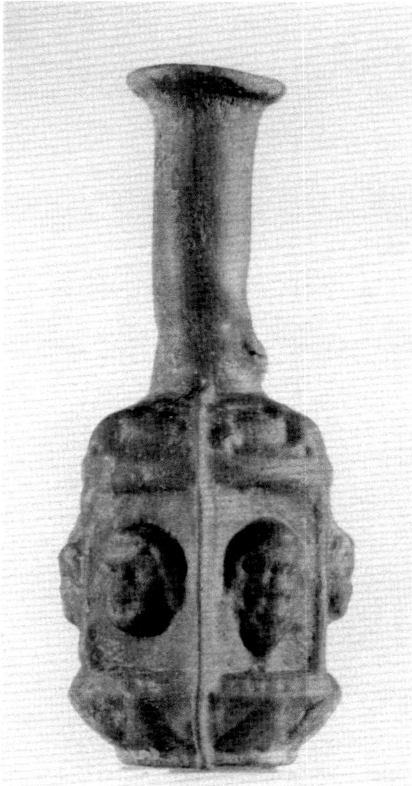

42, panels 2 and 3 (1:1)

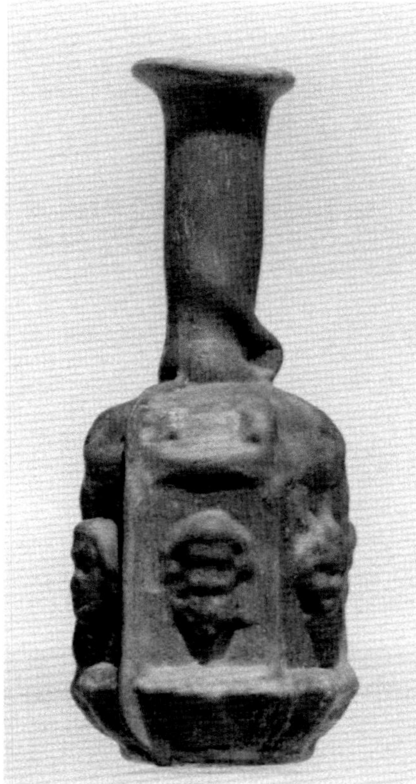

42, panel 3 (1:1)

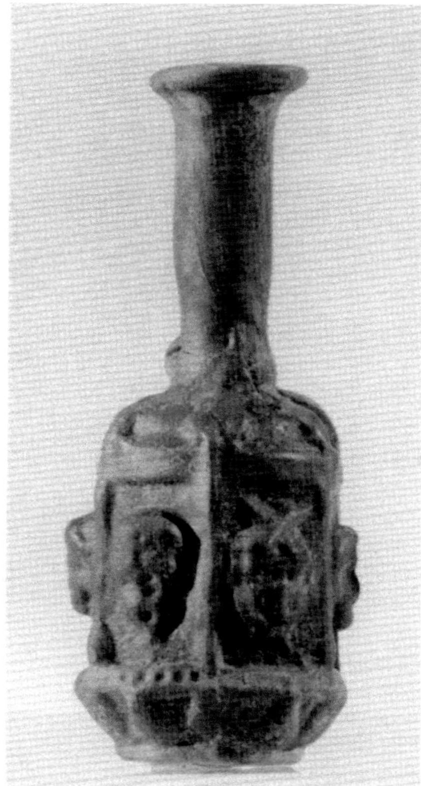

42, panels 6 and 1 (1:1)

and secondary decoration of the following bottles appear to be identical to the Toledo bottles:

a) Damascus, Nat. Mus., no. C5740, from Tell Nabi-Mind (ancient Qadish, appears on the map in Al-'Ush et al. 1969 as Tall An-Naby-Mand), opaque white, Abdul Hak 1951, 113, no. 17 (where the inv. no. is given as 13.122), pl. XLIX left; Zouhdi 1969, 104–105, no. 3, fig. 39 left;

b, c) Tel Aviv, nos. MH 6758, opaque white, and MH 6858, opaque blue, Israeli 1964, 39, nos. 7a and 7b, figs. 10 and 11, both from the same mold;

d) Oppenländer Collection, no. 2404, formerly Mavrogordato Collection, opaque white, von Saldern et al. 1974, 143, no. 408;

e) Athens, Nat. Mus., no. 2906, from Eretria, opaque white, probably belongs to Series A (it is too fragmentary and the relief too blurred to say anything definite, but it has the simplified egg-and-dart band and floral decoration on the bottom that are characteristic of Series A, and it was blown into a three-part mold), Weinberg 1992, 125, no. 96.

5 Eisen 1927, I, 253, fig. 123, identified the masks as etrogs.

6 The term "Group" is introduced here because there has been no opportunity to verify the appearance of the masks on these bottles. Consequently, although some of them might belong to Series A, it has been impossible to determine the exact relationship between these Groups and Series A or of one Group to another.

7 See Chapter I: *Hexagonal Bottles in High Relief: Masks Type, Group 2.*

8 See Chapter I: *Hexagonal Bottles in High Relief: Masks Type, Group 3.*

9 See Chapter I: *Hexagonal Bottles in High Relief: Masks Type, Group 4.*

10 The mouth of the youthful mask in panel 2 is half open.

11 On pantomime masks: M. Kokolakis, *Pantomimus and the Treatise Peri Orcheseos* (Athens 1959) 36f. On stage arrangements for pantomime performances: E. Simon, *Augustus. Kunst und Leben in Rom um die Zeitenwende* (Munich 1986) 217-219.

12 Avi-Yonah 1984, passim. However, masks do not appear to have had any specific Jewish relevance among Jewish communities living along the Syro-Palestinian coast. Masks are carved on a few late Roman Jewish sarcophagi discovered in catacombs at Rome (Goodenough III, figs. 787 and 789, from the catacomb Randanini), but were not common motifs on funerary monuments in Palestine (not one example from Palestine is cited in Goodenough III). On the catacombs of the Vigna Randanini: Leonard Rutgers, "Ein in situ erhaltenes Sarkophag-fragment in der jüdischen Katakombe an der Via Appia," *Jewish Art. Journal of the Center for Jewish Art,* 14 (The Hebrew University Jerusalem, 1988) 16-27; Leonard Rutgers, "Überlegungen zu den jüdischen Katakomben Roms," *JbAntC* 33 (1990) 140-157; Leonard Rutgers, "The Jewish Catacombs of Rome Reconsidered," *Tenth World Congress of Jewish Studies* Division B, vol. II (Jerusalem 1990) 29-36.

13 M. Bieber, *The History of the Greek and Roman Theater* (Princeton 1961) 236; L. Robert, "Pantomimen im griechischen Orient," *Hermes* 65 (1930) 106ff.

### 42 Hexagonal Bottle, Masks Type, Series A

Color plate 8, p. 53; photograph, Fig. 55
Phoenician
Probably second quarter of the first century
H 9.63 cm; H body 5.1 cm; D 3.9-3.4 cm; D rim 2.2 cm; D base 2.5-2.3 cm
1923.432, Gift of Edward Drummond Libbey

Opaque pale blue (5 B 6/2). Medium thin glass.
Neck free blown. Body blown into a three-part mold of three vertical sections (MCT IV). Mold seams between panels 2 and 3, 4 and 5, and 6 and 1 meet at center of underside of base. Relief moderately crisp. A sharp, protruding mold seam between panels 2 and 3 indicates that the edges of the mold sections were not tightly closed at this point. Deep irregular crimp at base of neck.
Rim folded outward, upward, and inward. Tall cylindrical neck with constriction at its base. Convex circular shoulder and convex circular bottom joined by a hexagonal body. Base flat.
Decoration like Cat. No. 41.
*Classification* Masks Type, Series A, Generation A1a.
*Condition* Intact; deep scratches at base of neck. Thin layer of buff-colored weathering on exterior.
*Collection* Thomas E. H. Curtis.
*Published* Lightfoot 1987, 21, no. 12, fig. 7 (incorrectly describing

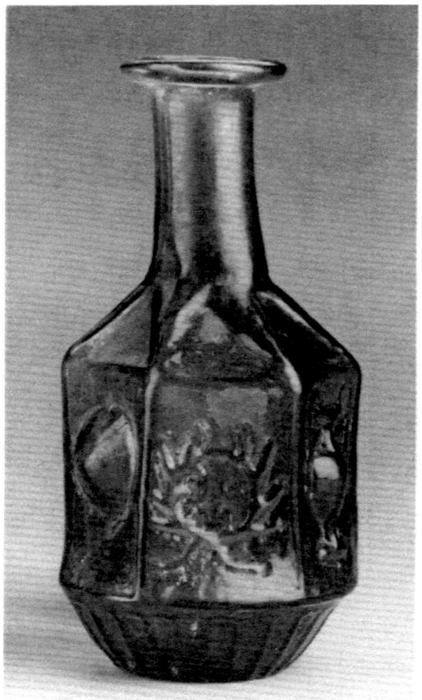

43, panel 1 (1:1)

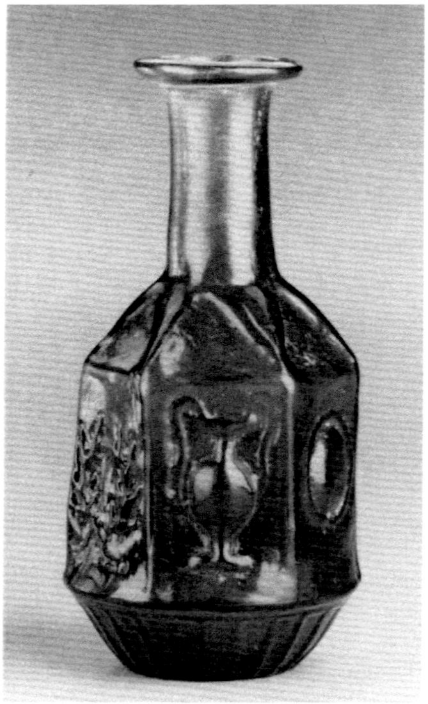

43, panel 2 (1:1)

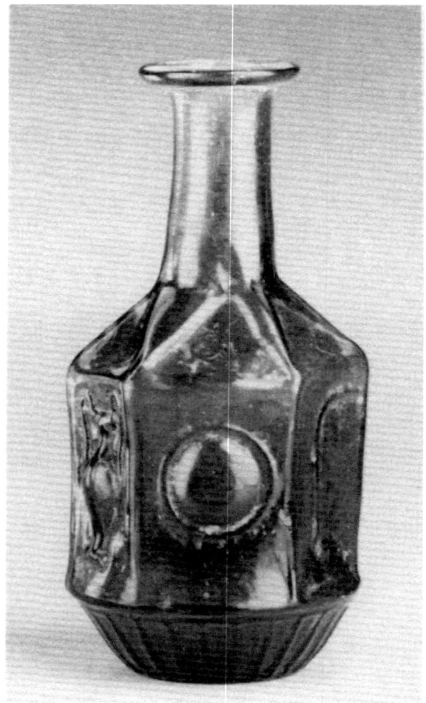

43, panel 3 (1:1)

the panels as showing two repeated sets of three masks and the shoulder as having six radiating ribs).
*Remarks* See Remarks Cat. No. 41. One of two bottles in the Toledo Museum collection from closely related molds.

### *Mixed Symbols Type*[1]
(Cat. Nos. 43 and 44)

#### 43 **Hexagonal Bottle, Mixed Symbols Type, Series A**

Color plates 5 and 9, pp. 52 and 54; photograph, Fig. 3
Probably Phoenician, possibly made in Sidon; or perhaps Galilean, possibly made in Nysa-Skythopolis
Second quarter to mid first century
H 8.5 cm; H body 5.4 cm; D 4.0 cm; D rim 2.3 cm; D base 2.3 cm
1923.464, Gift of Edward Drummond Libbey

Translucent manganese colored glass. Medium thin. Vertical linear bubbles in neck.
Neck free blown. Body blown into a three-part mold of two vertical sections joined to a cup-shaped base section that formed the base and the tongues around the bottom (MCT V A). Vertical mold seams perhaps between panels 6 and 1 and panels 3 and 4 (see Remarks). Relief moderately crisp. Mold sections exceptionally well aligned. Dot-shaped punch marks around outlines of relief. Three rings of circular concentric chill marks on underside of base (see Fig. 3). Rim folded outward, upward, and inward. Cylindrical neck. Steep downward sloping hexagonal shoulder and short upward sloping circular bottom joined by a hexagonal body. Flat base with central small recessed cylindrical cavity with a tiny round bump in center (see Fig. 3).
On the shoulder, six downturned rays pointing toward corners of panels. On the body, six rectangular panels; each panel contains a different object. In the panels, from left to right: 1, a wreath, or crown, of olive leaves; 2, a footed amphora with globular body and two high handles; 3, a convex, circular object with flat edge (a patera or tympanon); 4, a shepherd's crook; 5, a spouted, footed jug with a high handle to the left; and 6, a conical boss with ten radiate markings. Around the bottom, twenty-five upturned tongues.
*Classification* Mixed Symbols Type, Series A1, Generation A1b.
*Condition* Intact. Areas of milky iridescent film on exterior.
*Collection* Thomas E. H. Curtis.
*Published* Previously unpublished.
*Remarks* One of two bottles in the Toledo Museum collection from closely related molds (see also Cat. No. 44).[2] Series A has six different objects and geometric secondary decoration of stiff gadrooning on the bottom and long, pointed rays on the shoulder. The Toledo bottles are assigned to Generation A1b because they appear to have had at least one predecessor (see Fig. 77) which was probably the archetype and is therefore assigned to Generation A1a[3]. Generation A1b differs from Generation A1a in showing a wreath or crown in panel 1 and a shepherd's crook in panel 4 instead of the *thyrsolongchoi*, double-ended thyrsoi, decorating these panels in Generation A1a (see Fig. 54). In addition, the circular objects in panels 3 and 6 were interchanged. Although bottles of Generation A1b are the most common within this Series, only a few have been published.[4] Most, if not all, examples of the Mixed Symbols Type were made of transparent or translucent colored glass.

#### Technique

The most notable feature of the Mixed Symbols Type is that it appears to have been made in metal molds instead of a clay ones. This is indicated by a number of features: chill marks and a cylindrical cavity with a round bump in the center of the underside of the base of the glass bottles, punch marks around the symbols in the panels, and overblows between the triangular forms on the shoulder.
MCT V A molds were very well constructed. The sections fitted together so well during blowing that mold seams are extremely difficult to identify. The fact that each set of three panels has a different height, panels 1, 2, and 3 being 3.1 cm, while panels 4, 5, and 6 are 3.3 cm, suggests that the mold might have had two

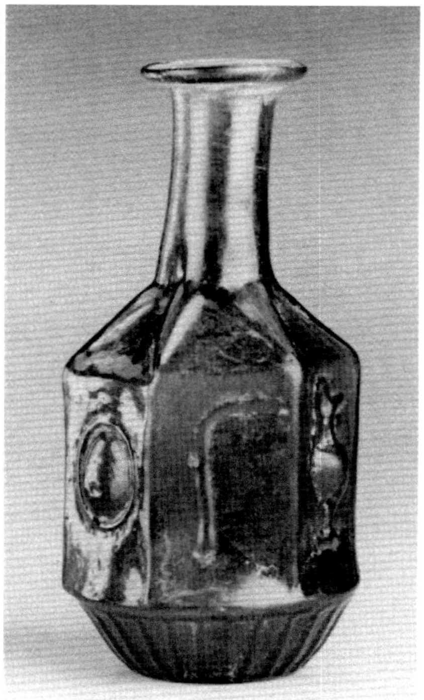
43, panel 4 (1:1)

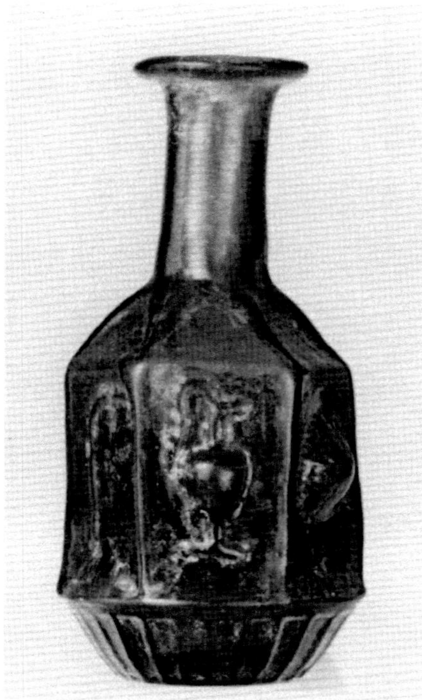
43, panel 5 (1:1)

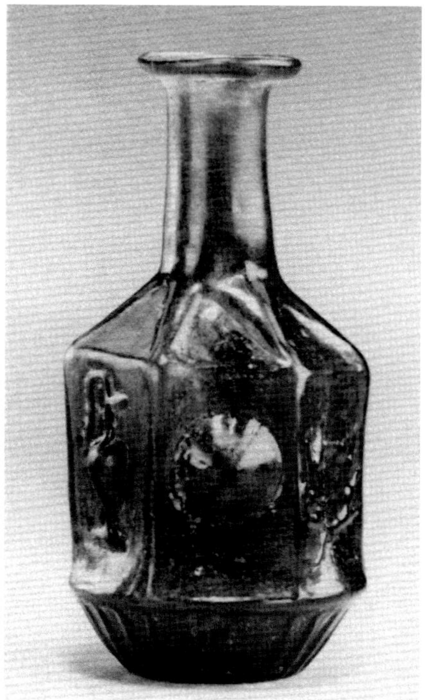
43, panel 6 (1:1)

vertical sections (but see below). The high quality of the construction of the molds contrasts sharply with the crudeness of the execution of the designs on the panels. The motifs are simply outlined and hollowed out without modeling or detailing of the surface. MCT V A molds must have been invented in an enterprising workshop with great technical skills but little artistic sophistication.
The molds consisted of two distinct parts, each made by a different technique. The cup-shaped bottom section was probably turned on a lathe at some point during its manufacture. In the center of the interior of the metal cup was a short cylindrical projection with a concave upper surface where the point of a lathe's mandrel pressed the metal against the lathe spindle on the other side.[5] The central projection in the mold cup produced a cavity with a small rounded bump at the bottom on the underside of the glass bottle. After the metal cup was removed from the lathe, decorative ribs were hammered into its wall.
The sides and shoulder of the mold were probably made from a rectangular sheet of metal folded to form six rectangular panels, each topped by a triangle (Fig. 76). The craftsman laid the sheet on pitch or leather with side A down and executed the symbols in repoussé from side B; he also added details such as radiate markings to the symbol in panel 6. Then he turned the sheet of metal over, with side A up, and gave the design its final definition by punching around the edges of the symbols. The punch marks on side A protruded as small bumps on side B, which was to be the interior of the mold. Apparently, the craftsman became confused about which side was up when he executed the wreath in panel 1 of Generation A1b. After having completed all the symbols, he folded the sheet along the vertical edges of the panels with side B on the interior, and he bent the triangles at the top inward. He must have created a locking device between the metal sheet and the cup-shaped bottom section, because all bottles blown in this mold appear to have their mold seams in the same position above the cup. When the metal mold was filled with glass, the symbols were reproduced in positive and the bumps from punching appeared as little cavities or dots; on the shoulder, glass escaped upward, forming overblows between the metal triangles.

## Date

A bottle excavated in Slovenia, in a cremation deposit that was interred during the second quarter of the first century,[6] provides a date before which Generation A1b of the Mixed Symbols Type must have been created. Judging from the small number of surviving vessels, the Type was made during only a brief period, shortly before and around the middle of the first century.

## Origin

Thus far, the only bottle belonging to the Mixed Symbols Type for which a find spot is known is the bottle excavated in Slovenia. Archaeological evidence is therefore not sufficient to locate the origin of this Type. However, iconographical details point to an origin in Phoenicia or Galilee. Two panels of the Maxwell bottle (Figs. 77 and 78),[7] assigned to Generation A1a, show double-ended thyrsoi, or *thyrsolongchoi*, a type of thyrsos that was not common in Greece and the Western Mediterranean but very common in Phoenicia and northern Palestine from the Hellenistic period (see Cat. No. 71). The thyrsoi on the Maxwell bottle and the Jugs with Cult Objects (Cat. Nos. 71-74) cross to form an X-shaped motif. This rare motif is found on two panels of a second-century altar to Dionysos excavated at Beth Shean (ancient Nysa-Skythopolis), Israel (see Fig. 56). It is tempting to speculate that crossed *thyrsolongchoi* and glass vessels decorated with this motif might have been associated with Nysa-Skythopolis, which remained a polytheistic city in Galilee throughout antiquity. The bottles were probably made in that area, possibly in Sidon or, possibly, in Nysa-Skythopolis.
The construction of MCT V A molds is so unique that we may assume that all bottles made in this type of mold were made in the same workshop. The earliest evidence for the existence of this workshop is a bottle of the Fruit Type made in an MCT V A mold (see Fig. 53 right).[8] The stiff upturned tongues around the bottom of the Mixed Symbols Type were copied from a bottle of the Fruit Type (Subseries A3); the steep slope and hexagonal form of the shoulder are also similar.

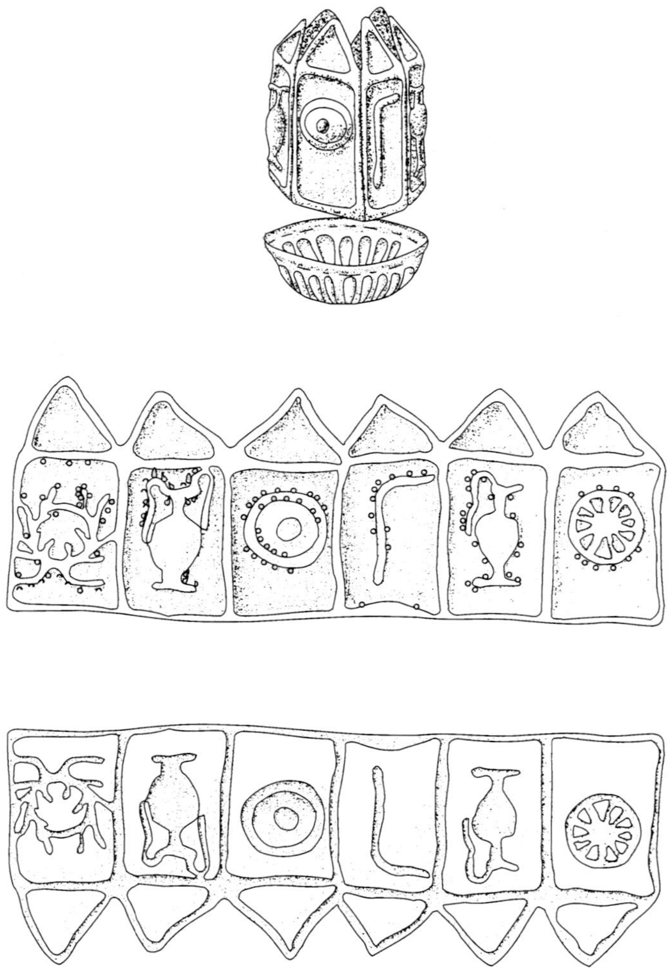

*Fig. 76. Tentative reconstruction of a two-part mold for a hexagonal bottle, Mixed Symbols Type, Series A1, Generation A1b (MCT V A).* Bottom: *The one-piece wall of the mold was a sheet of metal; this sheet was worked in repoussé from Side B (the interior).* Middle: *the sheet was then turned over and punched on Side A (the exterior side of the mold) to define the shapes.* Top: *The sheet of metal was folded into a hollow hexagon, and the triangular portions at the top were folded inward to produce the shoulder of the vessel. The base section of the mold was probably turned on a lathe, after which the decorative ribs were hammered into the surface. Drawings: Author and Timothy A. Motz.*

## Interpretation

Analysis of the motifs on the panels of the Maxwell bottle (see Fig. 78) shows that the symbols on the first Generation of the Mixed Symbols Type were associated with Dionysos. Two of its panels show crossed *thyrsolongchoi*; one of these panels also incorporates two small shepherd's crooks in the design. The amphora, jug, and patera or tympanon on the Maxwell bottle and on the Generation A1b bottles (Cat. Nos. 43 and 44) also belong to the realm of Dionysos.

The only object that is difficult to identify is the conical boss with radiate markings shown in panel 3 of Generation A1a and in panel 6 of Generation A1b. The Dionysiac connotations of the symbols in the other five panels of Generation A1a suggest that the conical object was associated with Dionysos or a syncretistic deity of the Near East. Although the radiate markings suggest the spokes of a wheel, this object does not resemble representations of the wheel of Nemesis [9] or the chariot of Astarte [10] or other celestial chariots [11] which occur in combination with Dionysiac symbols in the Near East, because these wheels are not shown as a conical elevation. Nor does it resemble representations of cymbals, a musical instrument with a roughly conical shape commonly associated with Dionysos throughout the Mediterranean, because cymbals are usually depicted in side view (see Fig. 57). In fact, the conical object on the glass bottles does not resemble any of the numerous motifs and symbols depicted on monuments and objects found in the Near East. The conical boss is here tentatively interpreted as a shield boss or shield. Shields play a role in myths about Dionysos. First, when Dionysos was a baby, the Korybantes who took care of him drummed on their shields in order to conceal the voice of the crying infant. Second, for his legendary campaign against the Indians, Dionysos was given a shield made by the blacksmith of the gods, Hephaistos. [12]

Another interpretation, suggested by Erika Simon, [13] is that these conical objects represent sacrificial cakes, such as for example a *pyramis*, shaped like a cone, or a *plakous*, shaped like a conical mallow seed. Eight-part sacrificial cakes are found in connection with Dionysos and Sabazios, [14] originally a Thrako-Phrygian god, regarded sometimes as identical with Dionysos, sometimes as identical with Zeus. The cakes are rendered as flat circular disks with grooves and dots indicating the eight parts, but in reality they were somewhat conical. [15] If the object depicted in panel 6 of the Mixed Symbols Type is such a cake, this would also agree with the Dionysiac symbolism of the Mixed Symbols Type.

The bottles of Generation A1b (Cat. Nos. 43 and 44) differ from the Maxwell bottle assigned to Generation A1a by the substitution of two new panels in place of the panels with the thyrsoi. The thyrsoi in panel 1 were replaced by an olive wreath and in panel 4 by a shepherd's crook or pedum, previously mistaken for a giant strigil. [16] Comparison with the Maxwell bottle shows that the so-called strigil is a larger version of the two shepherd's crooks appearing between the crossed thyrsoi on the Maxwell bottle. A large shepherd's crook also appears on one side of the hexagonal altar to Dionysos from Beth Shean (see Fig. 56).

The connection with Dionysos is less obvious in Generation A1b bottles. Perhaps this was a deliberate decision by the workshop to widen the circle of prospective customers.

[1] See also Chapter I: *Hexagonal Bottles with High Relief: Mixed Symbols Type.*

[2] Eisen 1927, I, 252, and fig. 120, calls this Type "Sidonian Six Sided Flasks with Palestra Objects"; Auth 1976, 69, no. 64, calls them "Athletic Symbols." The Type has been renamed Mixed Symbols Type in this catalogue.

[3] See *Maxwell* 1979, 103, no. 176, ill. (here Figs. 77 and 78).

[4] See the following examples:
- a) from a cremation in Ljubljana (ancient Emona), Slovenia, dark green, interred during the second quarter of the first century, Plesničar-Gec 1967, grave 87, no. 3; Plesničar-Gec 1972, 217, grave 427, pl. 205.3; Petru 1974, 28, figs. 1-2;
- b) Newark, no. 50.1456, transparent light green, Auth 1976, 69, no. 64;
- c) V&A, no. unknown, no color indicated, Honey 1946, pl. 5A;
- d) Switzerland, private collection, transparent manganese colored, Luzern 1981, 77, no. 245.

Unfortunately, the publications do not describe the undersides of the bases of these bottles, so it is impossible to determine whether they were blown into an MCT V A mold like the Toledo bottles. However, the fact that bottles of Generations A1a and A1b were blown into the same types of mold suggests that all bottles belonging to the Mixed Symbols Type were made in MCT V A molds.

[5] On raising around a planishing stake and lathe techniques: Mutz 1972.

[6] See above note 4a.

[7] See above note 3.

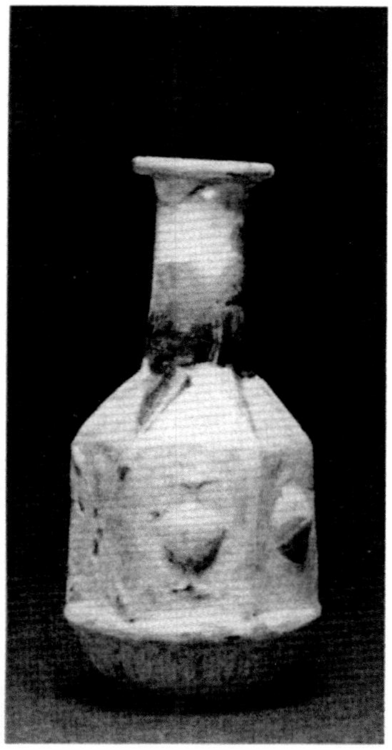

*Fig. 77. Hexagonal bottle, Mixed Symbols Type, Generation A1a. H 9 cm, almost colorless, formerly Constable-Maxwell Collection. Photo: Reprinted with permission from* Maxwell *1979, 103, no. 176.*

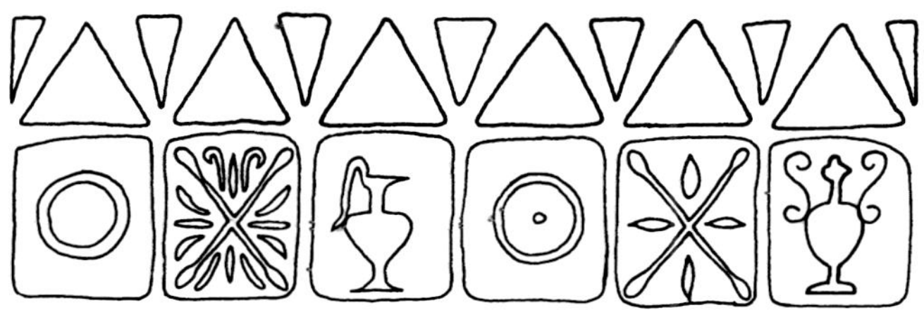

*Fig. 78. Drawing of hexagonal bottle in Fig. 77. Drawing: Reprinted with permission from* Maxwell *1979, 103, no. 176.*

[8] Newark, no. 50.1458, opaque white, Auth 1976, 69, no. 65; see also Chapter I: *Hexagonal Bottles with High Relief: Fruit Series*, Subseries A2, Generation A2a.

[9] On the wheel of Nemesis: J. Leibovitch, "Le Griffon d'Erez et le sens mythologique de Némésis," *IEJ* 8.3 (1958) 141-148, pls. 25-29; the wheel occurs in combination with a double-ended thyrsos on a funerary relief reported to be from Alexandria, Bologna, Museo Civico, Leibovitch 1958 145, fig. 2, K. Schauenburg, "Pluton und Dionysos," *JdI* 68 (1953) 70, fig. 23.

[10] The chariot of Astarte appears in the sky between Apollo and Dionysos on numerous coins: *LIMC* III.1, *Dionysos in periferia orientali*, 516, no. 9c; 521, no. 716, from Sidon.

[11] Bernard Goldman, "The Celestial Chariot East and West," *Bulletin of the Asia Institute*, n.s. 2 (1988) 87-105.

[12] Nonnos, *Dionysiaka* 25.337-565.

[13] E. Simon, pers. com.

[14] E. N. Lane, *The Other Monuments and Literary Evidence, Corpus Cultus Iovis Sabazii*, II (Leiden 1985) 40, no. 84 (brought to my attention by J. Derksen, Amsterdam). On similar cakes in Dionysiac context, see below note 15. Perhaps one may also compare a small conical protuberance with radiate markings on a bronze bust of Sabazios in the Vatican Museum: M. P. Nilsson, *Geschichte der griechischen Religion*, II: *Die Hellenistische und Römische Zeit* (Munich 1961) pl. 14, 1; F. Cumont, *RA* 1 (1892) 186ff., pl. 10.

[15] A terra-cotta vessel with Dionysiac objects in relief, including eight-part circular loaves or sacrificial cakes, is illustrated by M. J. Vermaseren, *The Hands, Corpus Cultus Iovis Sabazii*, I (Leiden 1983) 8-9, no. 15 bis a, pls. XIII-XIV (brought to my attention by J. Derksen, Amsterdam).

[16] Eisen 1927, I, 252, fig. 120; Newark, no. 50.1456, Auth 1976, 69, no. 64.

### 44 **Hexagonal Bottle, Mixed Symbols Type, Series A**

Color plate 9, p. 54
Probably Phoenician, possibly made in Sidon; perhaps Galilean, possibly made in Nysa-Skythopolis
Probably second quarter to mid first century
H 8.68 cm; H body 4.9 cm; D 3.95 cm; D rim 2.45 cm; D base 2.4 cm
1923.463, Gift of Edward Drummond Libbey

Translucent dark blue. Medium thick glass. Thick base.
Neck and shoulder free blown. Body blown into a three-part mold of two vertical sections joined to a cup-shaped base section that formed the base and the tongues around the bottom (MCT V A). Vertical mold seams perhaps between panels 6 and 1 and panels 3 and 4 (see Remarks Cat. No. 43). Relief indistinct. Mold sections exceptionally well aligned. Dot-shaped punch marks around outlines of relief. Three concentric rings of circular chill marks on underside of base (see detail photograph).
Rim folded outward, upward, and inward. Cylindrical neck with curved transition to shoulder. Round circular shoulder and short upward-sloping circular bottom joined by a hexagonal body. Flat base with central small recessed cylindrical cavity with a tiny round bump in center (see detail photograph).
Shoulder undecorated. On the body, six rectangular panels; each panel contains a different object. In the panels, from left to right: 1, a wreath, or crown, of olive leaves; 2, a footed amphora with globular body and two high handles; 3, a convex, circular object with flat edge (a patera or tympanon); 4, a shepherd's crook; 5, a spouted, footed jug with a high handle to the left; and 6, a conical boss with ten radiate markings. Around the bottom, twenty-five upturned tongues.
*Classification* Mixed Type, Series A1, Generation A1b-yy.
*Condition* Intact. Iridescent film and incipient surface pitting on exterior.
*Collection* Thomas E. H. Curtis.
*Published* Previously unpublished.
*Remarks* See Remarks Cat. No. 43. One of two bottles in the

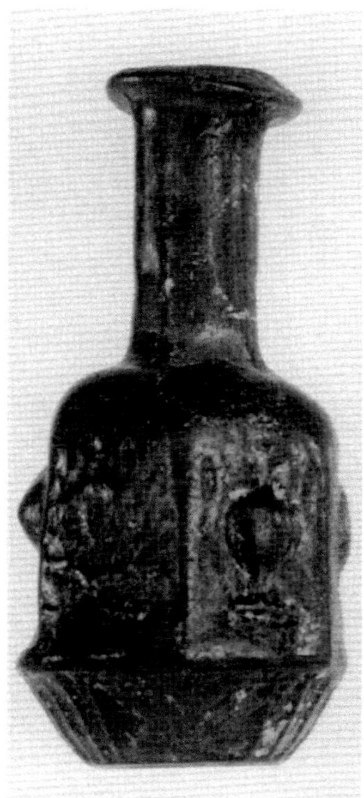

44, panels 1 and 2 (1:1)

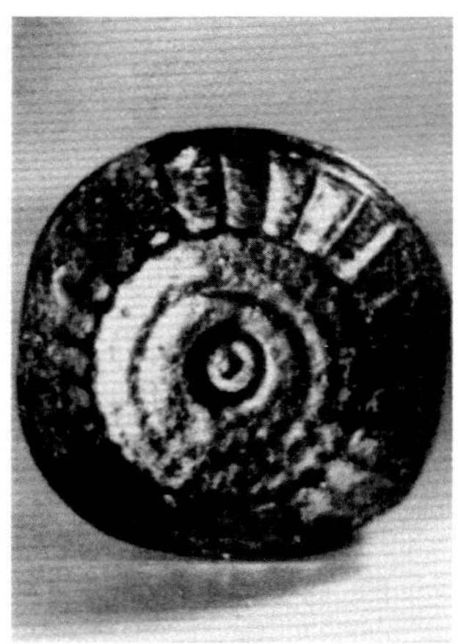

44, underside (1:1)

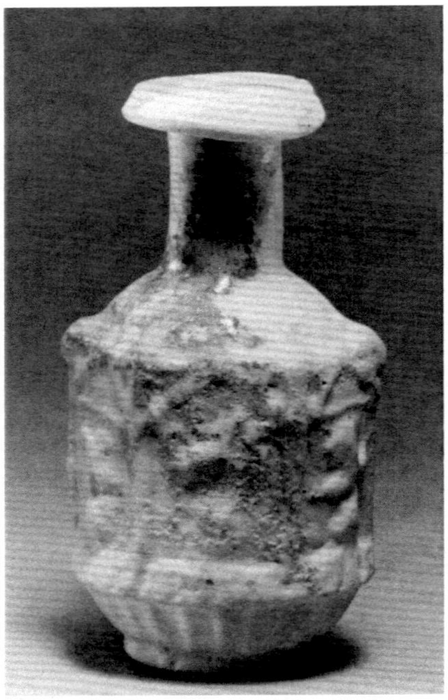
45, panel 1 (1:1)

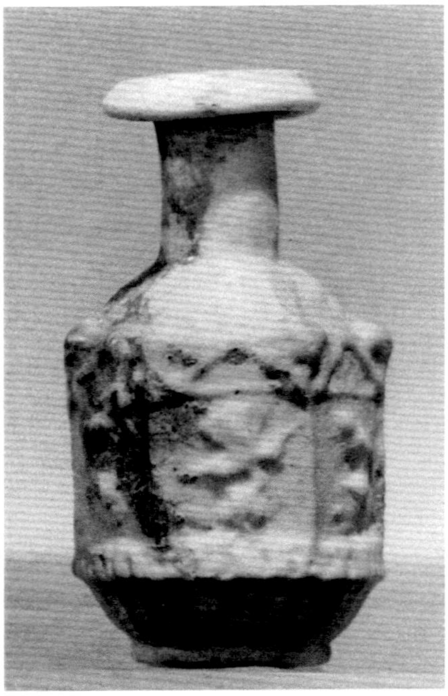
45, panel 2 (1:1)

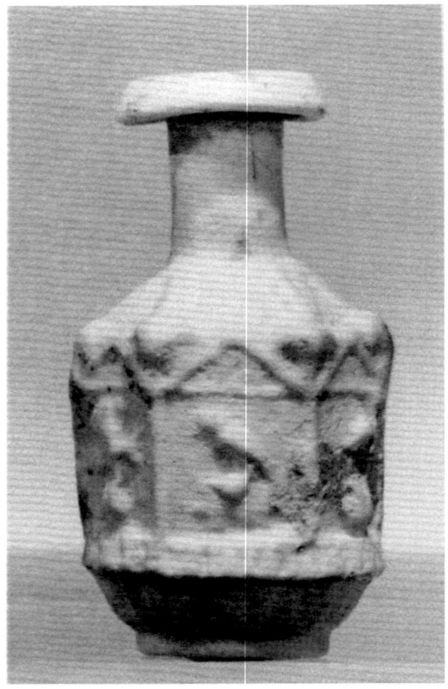
45, panel 3 (1:1)

Toledo Museum collection from closely related molds. The crude execution of this vessel is probably due to a combination of deterioration (or a bad copy?) of the mold and a problem with filling the mold while blowing. The glass blower may have had too much glass on his blowpipe to blow out the bubble. The entire shoulder design is missing, as well as the handles and rim of the amphora in panel 2, the horizontal part of the shepherd's crook in panel 4, and the spout and handle of the jug in panel 5. However, the portions of the design that are preserved are identical to Cat. No. 43.

### *Birds Type*[1]

(Cat. Nos. 45-47)

#### 45 Hexagonal Bottle, Birds Type, Series A

Color plates 5 and 10, pp. 52 and 54
Probably Phoenician, perhaps made in Sidon or vicinity
Probably second quarter of first century
H 7.7 cm; H body 5.25 cm; max D 4.15 cm; D rim 2.2 cm; D base 2.45 cm
1923.437, Gift of Edward Drummond Libbey

Opaque white. Medium thick glass.
Neck free blown. Body blown into a three-part mold of three vertical sections (MCT IV A). Mold seams between panels 6 and 1, 2 and 3, and 4 and 5 meet at center of underside of base. Relief moderately crisp. Edges of mold carefully aligned.
Rim folded outward, upward, and inward. Cylindrical neck. Downward sloping shoulder and upward sloping bottom joined by a hexagonal body. Low, offset base with flat underside.
On the shoulder, six downturned rays or pointed petals. On the body, six almost square panels, each topped with a triangular pediment, divided by ribs or posts; bordered below by a band with egg-and-dart pattern. Between the pediments, unidentified spherical objects. Each panel contains a bird or winged insect in a different pose perched on or flying over a pot-shaped nest or rock. In the panels, from left to right: 1, a bird with a long spoon-shaped bill (perhaps a swan)[2] to the right on a nest; 2, a simplified butterfly to the right grasping a rock(?) between its feet;[3] 3, a bird (perhaps a falcon) to the left perched on an irregularly shaped rock; 4, a bird (perhaps an ibis) to the right perched on a pot;[4] 5, a bird (a small raptor) with diagonally outspread wings flying to left over a rock or nest;[5] and 6, a small songbird swooping down to left to feed invisible young in its nest. On the bottom, twenty-six upturned tongues.
*Classification* Bird Type, Series A1, Generation[6] A1b.
*Condition* Complete; small hole on shoulder. Areas of thick purplish brown weathering on exterior.
*Collection* Thomas E. H. Curtis.
*Published* Richter 1914, 83, fig. 8.
*Remarks* One of three bottles in the Toledo Museum collection from related molds (Cat. Nos. 45-47). This bottle belongs to the same Subseries and possibly the same Generation as Cat. No. 46, but its mold was carved by a craftsman with better understanding of the anatomy of birds. Cat. No. 47 belongs to Series B. The bottles belong to the Birds Type, a relatively small group of hexagonal bottles. The decoration is divided into the three zones customary with hexagonal mold-blown bottles, but the relief of Series A is low rather than high. The six panels of the main decoration show six different scenes in a set order, each with a bird and an object such as a rock, nest, or pot. These scenes are depicted as if seen from within a temporary wooden structure with posts and beams, or from within a tent (see Remarks Cat. No. 47).
The secondary decoration of the archetype of Series A consists of floral motifs on the shoulder and fluting around the bottom, while the panels are bordered below by a band of egg-and-dart pattern. The earliest Generation, A1a, is represented by a bottle in Tel Aviv (Fig. 79); it has pointed leaves alternating with palmettes with inturned petals on the shoulder.[7] The Toledo bottles have no palmettes on the shoulder, and palmettes are not mentioned for any of the other bottles of this Series. The exact relationship between Generations A1b (this bottle) and A1c (Cat. No. 46) is not clear. The molds appear to have been carved by different hands, but they may belong to the same generation, so they are here referred to as Molds instead of as generations. Bottles blown in Molds A1b and A1c are too similar to be distinguished from descriptions or photographs. Generation A1d is not represented in the Toledo Museum collection (Fig. 80),[8] but the decoration of the

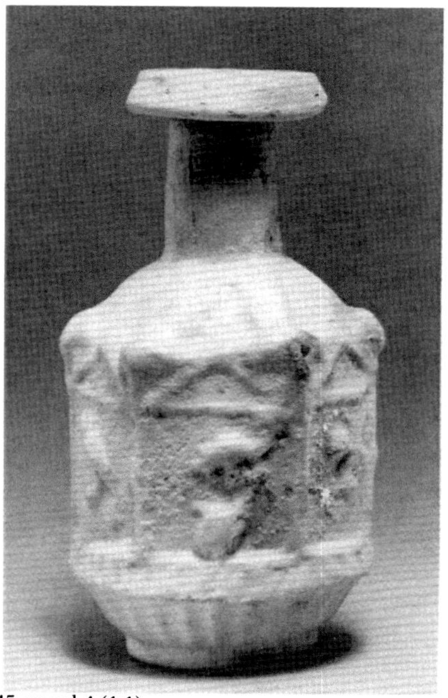

45, panel 4 (1:1)

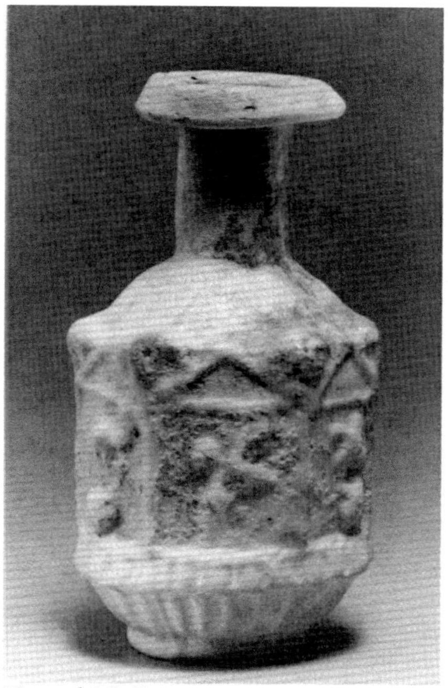

45, panel 5 (1:1)

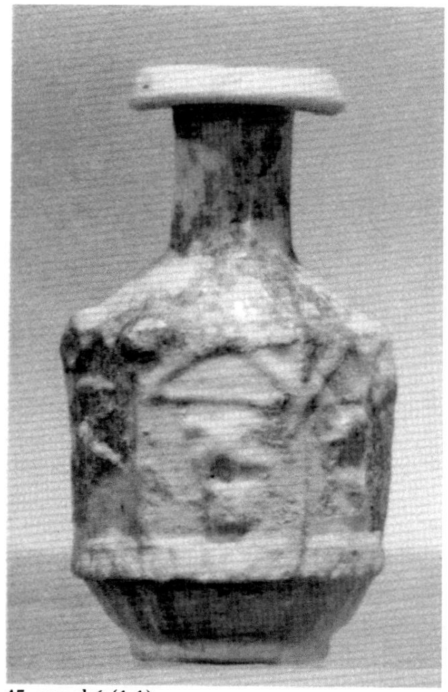

45, panel 6 (1:1)

shoulder has been recarved to form ribs or petals radiating from the base of the neck.

With at least twelve bottles surviving, Series A was the most common of the Birds Type. Most bottles were blown into Molds A1b or A1c, which differ primarily in the style of carving and the number of tongues on the lower body, Mold A1b having twenty-six tongues and Mold A1c, twenty-seven.[9] Where it has been possible to check, the molds used for Series A were all standard three-part molds of MCT IV A. Most of the bottles are opaque white, one opaque blue, and a few transparent to translucent blue.

### Distribution and Origin

None of the bottles of the Birds Type was found during a controlled excavation. One bottle of Series A, Mold A1b or A1c, is said to have been found at Amrit (ancient Marathos)[10] on the Syrian coast,[11] and two were possibly unearthed in Turkey.[12] One bottle of Series B was acquired at Aleppo.[13] Although these find spots suggest an eastern Mediterranean origin for Series A, they are not sufficient to indicate where the workshop or workshops might have been. Stylistic, technical, and iconographical features, however, point toward an origin on the Syro-Palestinian coast, perhaps in or near Sidon. The attempt at variation in the rendering of the birds and motifs such as the triangular pediments in the upper part of the panels, tongues around the bottom of the bottles, and palmettes on the shoulder of Generation A1a, as well as the shallow relief, are all features that reflect Ennion's influence. On the other hand, the poor quality of design and execution indicates that Series A was not made by Ennion but by another workshop in the area. All bottles belonging to the Birds Type appear to have been blown in MCT IV A molds, a mold type which does not appear to have been used frequently outside the Syro-Palestinian area where they were the most common type of mold construction for bottles with high relief. The iconography also suggests a connection with Sidon (see below).

### Date

The influence of Ennion's workshop on other glass blowers working in Phoenicia was probably felt from around the year 25 at the latest (see Chapter I: *Glass-Blowers' Signatures: Ennion*). Therefore, the creation of the Birds Type is tentatively dated to the second quarter of the first century. This date agrees with the hypothesis (presented in Chapter I: *Syro-Palestinian Styles: Hexagonal Bottles with High*

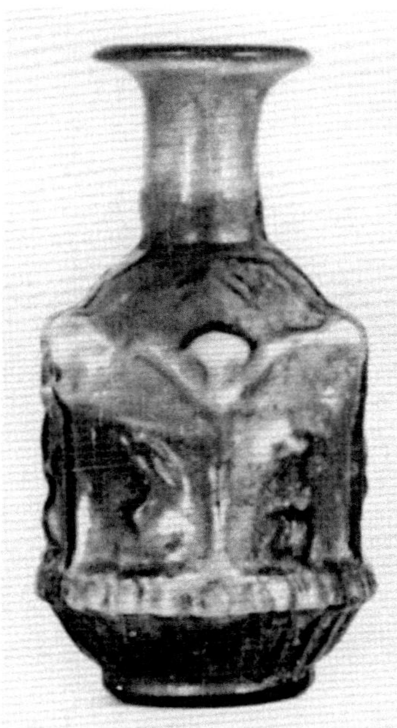

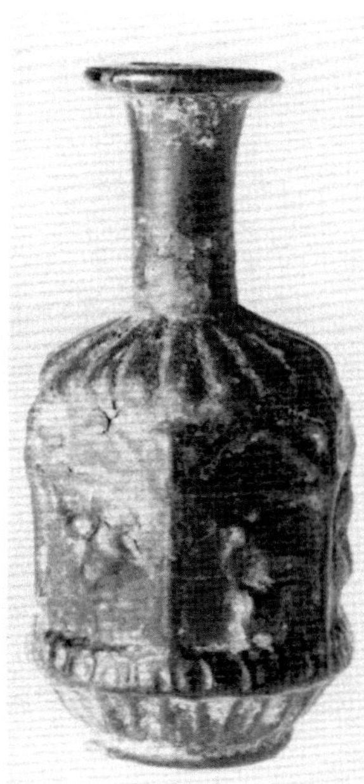

*Fig. 79. Hexagonal bottle, Birds Type, Series A, Generation A1a. This archetype has pointed leaves alternating with palmettes with inturned petals on the shoulder. H 7.7 cm, deep blue, Tel Aviv, Eretz Israel Museum, no. MHG 6658. Photo: Eretz Israel Museum, Glass Pavilion.*

*Fig. 80. Hexagonal bottle, Birds Type, Series A, Generation A1d. The decoration of the shoulder has been recarved from downturned rays (Generations A1b and A1c) to become round-tipped ribs or petals radiating from the base of the neck. H 7.7 cm, translucent blue, Newark, N.J., Collection of the Newark Museum, The Eugene Schaefer Collection of Ancient Glass, no. 50.1454. Photo: Newark Museum.*

*Relief: Date*) that before the middle of the first century, perhaps around the year 40, bottles of the Birds Type were the model for the fluting around the bottom of two updates in Generation Nine of Hexagonal Bottles: Subseries A1, Generation A1c, of the Fruit Type and Series C of the Vessels Type (see Fig. 54).

### Interpretation

The relief of the Toledo bottles is so distinct that for the first time it has been possible to identify the subject of each panel, and it can now be seen that Series B (Cat. No. 47) [14] shows the same subjects as Series A (Cat. Nos. 45 and 46) and in the same order, but in high relief. The person who made the model for Series B was a more accomplished craftsman. For identification of the birds in Series A, see also Cat. No. 46, where the relief is more crisp than on this bottle. The birds are the same as those described and depicted by Eisen. [15]

The correspondence between the birds of Series A and Series B is most obvious in panels 5 and 6, which will therefore be discussed first. Panel 5 shows a little raptor flying to the left above a rock or nest. The main difference between this panel in Series A and B is the height of the relief. The relief of the center portion of the body of the bird in Series B stands up very high, while the tip of its lower wing flows into the nest or rock below; the bird itself is larger than in Series A. In panel 6 bird and nest are located in the right half of the panel in Series A and are pressed against the left edge of the panel in Series B. In Series A the songbird appears to be swooping down toward the nest with wings spread horizontally, while in Series B the bird perches on the rim of the nest, with its head to the left; the nest is larger and the eager head of a hungry young bird is clearly indicated.

The correspondence of panels 1 through 4 is less obvious. Panel 1 is the only panel of Series A that can compare to Series B in quality of design. In Series A the swan is perched on a pot-shaped object or nest which has been replaced in Series B by a decorative three-lobed bush; the bird in Series B is smaller and lacks the distinctive spoon-shaped bill of the bird in Series A. Panel 2 of Series A shows a simplified butterfly with antennae on its head to the right, grasping a rock (?) between its feet; [16] in Series B the butterfly has been transformed into a bird in a very unnaturalistic pose: the butterfly's three wings have become one large front (left) wing and one small (right) wing. In Series B the indistinct object (rock?) of Series A has been changed into a pot-shaped nest with its opening to the left. The main difference between the birds in panel 3 is that in Series A the bird sits isolated in the center of the panel, while the bird in Series B is smaller and has been moved to the left against the post, as though it were pecking the wood and pulling out a bug, but the pose of both birds is the same. Panel 4 of Series A shows a bird (perhaps an ibis) to the right, perched on a pot; the curve of the underside of its body and neck can still be seen in the robust bird (swan) of Series B perched on a semicircular mound-shaped nest rendered in extremely high relief.

Birds are a popular motif on Augustan silver, [17] but in the case of these glass bottles, which were probably blown in or near Sidon, the subject may have been chosen in order to evoke the name Ornithopolis, "Bird City." [18] Located between Sidon and Tyre on the Phoenician coast, this city is mentioned by various ancient authors (Strabo, *Geography* 16.2.24; Pliny, *Natural History* 5.17.76) and is said to have belonged to Sidon (Skylax, 104; *Geographi Graeci Minores* I.78).

[1] See Chapter I: *Syro-Palestinian Styles: Hexagonal Bottles with High Relief: Birds Type.*

[2] This bird is described by Eisen 1927, I, 248, fig. 116, as a stork. However, a stork does not bend its neck the way the bird in panel 1 does. The short legs of this bird and the curve between forehead and bill are features that are more typical of a swan. On the other hand, the object below this bird does not resemble a swan's nest, which is usually shaped like a semicircular mound. For assistance in identifying the birds in Cat. Nos. 45-47 my thanks go to Dr. H.-J. Boehr, Wiesbaden.

[3] Two antennae, three of four upright curving wings, and two of six legs are shown. A similar insect appears on a Hellenistic silver cup in Toledo, TMA, no. 1961.9 (here Fig. 57).

[4] The pose and the line of the back of the bird's head suggest this is an ibis.

[5] In the drawing, the outlines of the wings are rendered too rounded.

[6] The relationship between Generations A1b and A1c is not clear: see Remarks.

[7] Tel Aviv, no. MHG 6658, deep blue, Israeli 1964, 38, no. 6, fig. 9 (here Fig. 79).

[8] Newark, no. 50.1454, deep blue, Auth 1976, 198, no. 325 (here Fig. 80). See also de Ridder 1909, 153, no. 272 (not ill.). Both from unknown find spots.

[9] It has not been established whether the following bottles were made in Mold A1b or A1c:

- a, b) Istanbul, H. Kocabaş Collection, possibly unearthed in Turkey, both opaque white glass, *Kocabaş* 1984, 24, 55, nos. 97 and 98, figs. 46a and b;
- c, d) Switzerland, private collections, both opaque white glass, Luzern 1981, 77, nos. 243-244, ill.;
- e) formerly Constable-Maxwell Collection, opaque white, *Maxwell* 1979, 64-65, no. 99, ill.;
- f, g) Oppenländer Collection, no. 2533, opaque white and no. 22216, opaque pale blue, von Saldern et al. 1974, 143-144, nos. 410, ill., and 411 (not ill.);
- h) formerly de Clercq Collection, reported to be from Amrit (ancient Marathos), on the Syrian coast, about 5 km south of Tortosa, opaque white, de Ridder 1909, 152, no. 271, pl. X.

[10] See above note 9h. The find spot of a bottle in Damascus, Nat. Mus., no. 12.022, Abdul Hak 1951, 114, no. 22, is unknown, but it was probably somewhere within the borders of the Arab Republic of Syria because most of the material in the Damascus Museum is from Syria; from the description it is not clear whether this bottle belongs to Series A or Series B.

[11] Marathos was the principal mainland city of the federation of Arados, a small island off the Phoenician coast, during the Hellenistic period.

[12] See above note 9a and b.

[13] Philadelphia, The University Museum, no. MS 5009 (unpublished).

[14] Series B corresponds to Eisen 1927, I, 249, and fig. 117, "nesting birds without a guardian." The numbering of the panels of the Toledo bottle is the same as that of the panels in Eisen's fig. 117.

[15] Series A corresponds to Eisen 1927, I, 249, fig. 116, "nesting birds and a seated guardian." Panel 1 of the Toledo bottles corresponds to panel 3 in Eisen's fig. 116; panel 2, showing a butterfly with antennae on its head and a rock(?) between its feet is Eisen's panel 4, the so-called guardian. Eisen's drawings are badly distorted.

[16] This is the panel identified as a guardian or bird-watcher by Eisen 1927, I, 249, fig. 116, fourth panel from left.

[17] For example, Oliver 1977, nos. 96 and 97.

[18] A connection with Ornithopolis was suggested by Eisen 1927, I, 249, who named the Birds Type the "Ornithopolis Series."

## 46 Hexagonal Bottle, Birds Type, Series A

Color plate 10, p. 54
Probably Phoenician, perhaps made in Sidon or vicinity
Probably second quarter of first century
H 7.47 cm; H body 5.1 cm; max D 4.05 cm; D rim 2.2 cm; D base 2.4 cm
1923.462, Gift of Edward Drummond Libbey

Translucent royal blue. Medium thin glass.

Neck free blown. Body blown into a three-part mold of three vertical sections (MCT IV). Mold seams between panels 6 and 1, 2 and 3, and 4 and 5 meet at center of underside of base. Relief moderately crisp. Edges of mold carefully aligned.

Rim folded outward, upward, and inward. Cylindrical neck.

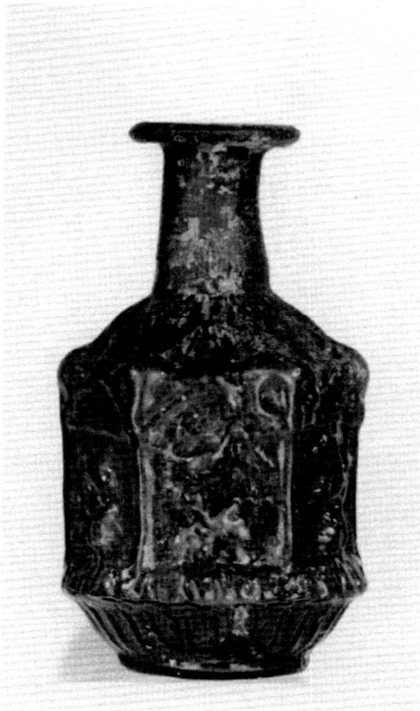

46, panel 1 (1:1)

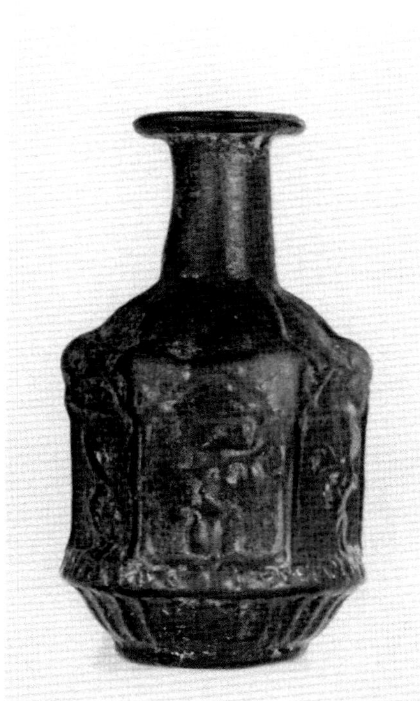

46, panel 4 (1:1)

Downward sloping shoulder and upward sloping bottom joined by a hexagonal body. Low, offset base with flat underside.
On the shoulder, six downturned rays or pointed petals. On the body, six almost square panels, each topped with a triangular pediment, divided by ribs or posts; bordered below by a band of egg-and-dart pattern. Between the pediments, unidentified spherical objects. Each panel contains a bird or winged insect in a different pose perched on or flying over a pot-shaped nest or rock. In the panels, from left to right: 1, a bird with a long spoon-shaped bill (perhaps a heron rather than a swan)[1] to the right on a nest; 2, a fat butterfly to the right with three upright curving wings and antennae on its head, grasping a rock (?) between its feet; 3, a small bird to the left perched on an irregularly shaped rock;[2] 4, a bird to the right perched on a pot;[3] 5, a bird (probably a small raptor) with diagonally outspread wings flying to left over a rock or nest; and 6, a small songbird swooping down to the left[4] to feed invisible young in its nest. On the bottom, twenty-seven upturned tongues.
*Classification* Birds Type, Series A1, Generation A1c.
*Condition* Intact. Remains of iridescent film; incipient frosting.
*Collection* Thomas E. H. Curtis.
*Published* Previously unpublished.
*Remarks* See Remarks Cat. No. 45. One of three bottles in the Toledo Museum collection from related molds (Cat. Nos. 45-47). This bottle belongs to the same Subseries and possibly the same Generation as Cat. No. 45, but the mold was carved by a craftsman with little understanding of the anatomy of birds. There are twenty-seven tongues around the bottom instead of twenty-six.

[1] Although the subject is clearly the same as in panel 1 of Cat. No. 45, the long legs of this bird are more characteristic of a heron than of a swan.

[2] The bird is very similar to the one in Cat. No. 45, panel 3, but its head is rather small and its neck and legs too long for a falcon.

[3] The bird in this panel appears to be the same as the one in Cat. No. 45 (perhaps an ibis), but the proportions here are no longer natural.

[4] The pose of this bird can also be interpreted as perching on the edge of a nest like the bird in Cat. No. 47, panel 6.

### 47 **Hexagonal Bottle, Birds Type, Series B**

Probably Phoenician, perhaps made in Sidon or vicinity
Probably mid first century
H 7.25 cm; H body 5.1 cm; max D 3.8 cm; D rim 2.0 cm; D base 2.3 cm
1923.461, Gift of Edward Drummond Libbey

Translucent blue. Medium thin glass. Fabric cannot be determined because of weathering.
Neck free blown. Body blown into a three-part mold of three vertical sections (MCT IV). Mold seams between panels 1 and 2, 3 and 4, and 5 and 6 meet at center of underside of base. Relief moderately crisp. Edges of mold carefully aligned.
Rim irregularly folded outward, upward, and inward. Cylindrical neck. Downward sloping shoulder and upward sloping bottom joined by a hexagonal body. Flat base.
On the shoulder, six rounded arches. On the body, six rectangular panels, divided by timber framework consisting of posts and beams with diagonal supports branching out near the top (forming the Greek letter Ψ [*psi*]) and meeting above the middle of the panels. Indication of the crossbeams connecting the bases of the diagonal supports is indistinct, making the main fields of the panels nearly square. The panels are bordered below by a smooth band slanting outward. Each panel contains a bird in a different pose perched on or flying over a pot-shaped nest or rock. In the panels, from left to right: 1, a bird with its head to the right (perhaps a swan)[1] perched on the center of a trefoil bush; 2, a long-necked bird swooping down toward left and sticking its long beak into a pot lying on its side;[2] 3, a small bird to the left perched on an unidentified object and touching the frame of the panel with its beak; 4, a large bird with a long neck (a swan) perched on a semicircular object, perhaps a mound-shaped nest; 5, a bird (a small raptor) flying to left with outspread wings over a nest or rock; and 6, a songbird with wings extended upward perched on the edge of a pot-shaped nest and bending its head down to the left to feed a young bird lifting its head up out of the nest. On the bottom, twenty-six upturned tongues.

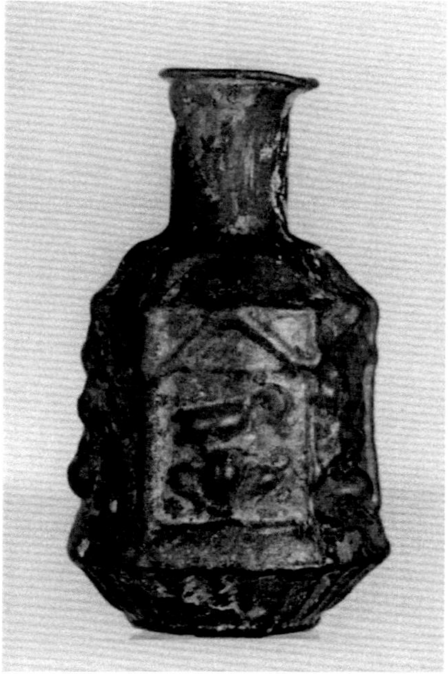
47, panel 1 (1:1)

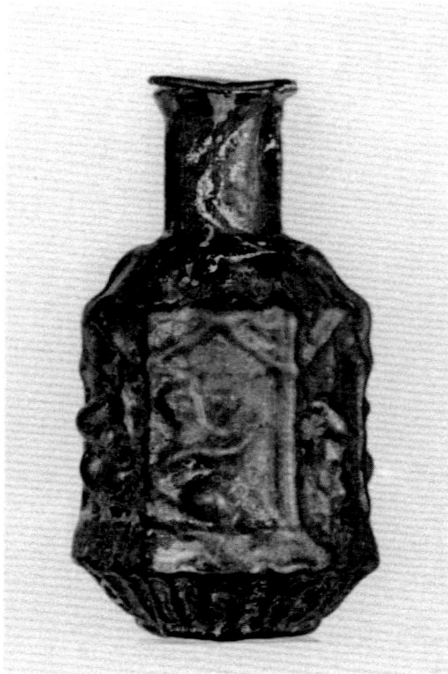
47, panel 2 (1:1)

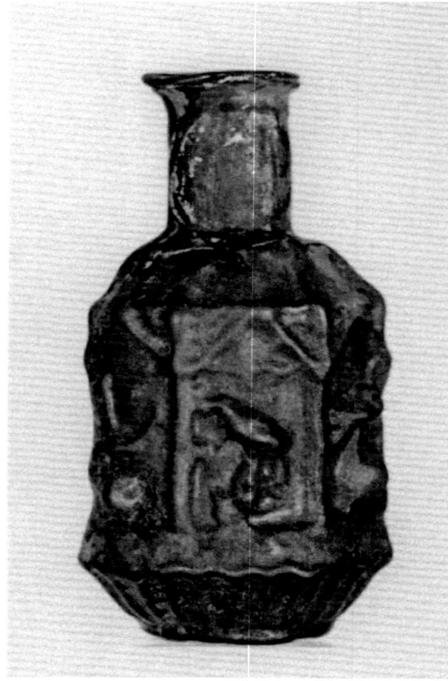
47, panel 3 (1:1)

*Classification* Birds Type, Series B1, Generation B1a.
*Condition* Intact. Surface areas of buff-colored weathering and surface pitting.
*Collection* Thomas E. H. Curtis.
*Published* Previously unpublished.
*Remarks* See Remarks Cat. No. 45. One of three bottles in the Toledo Museum collection from related molds (Cat. Nos. 45-47). This bottle belongs to Series B,[3] which shows the same subjects in the same order as Series A, however, there are considerable differences between the two Series, both in the execution of the decoration and in mold construction. The mold seams of Series B are one panel off from the mold seams of Series A. In contrast to the low relief of Series A, the relief of Series B is high. The craftsman who created the molds for Series B was an accomplished draftsman and artist who made very effective designs but had little knowledge of the behavior of birds (compare, for example, panel 2). The smooth band bordering the panels below still bears traces of the egg-and-dart pattern characteristic of Series A, indicating clearly that the model for the archetype of Series B was taken from a bottle of Series A. The mold maker adapted the subsidiary decoration to the subject of the main frieze. He recut the pointed leaves on the shoulder of Series A to form a pattern of arches such as was common in the Vessels Type (see Cat. No. 15). He also transformed the triangular pediments divided by posts into a design that was not only more graphic but also more in agreement with the outdoor subject of the birds in the panels. By thickening the angles at the bases of the triangles, he transformed them into a construction of wooden beams and posts that appears to be bound together with rope, so that the bases of the triangular pediments have become crossbeams. The result evokes a light architectural structure, perhaps window frames or wooden posts to support a tent, opening to a view of bird life. He smoothed the egg-and-dart below the panels to form a base for this structure.
For the correspondence between the subjects of the panels of Series A and Series B, see Remarks Cat. No. 45.

#### Origin and Workshop

Although no bottles of Series B have been found during controlled excavations, what is known about their origins points to the eastern Mediterranean.[4]
The style of Series B is so different from that of Series A that one may assume the two Series were produced by different workshops. Workshop affinities appear to be closest with Series C of the Vessels Type (see Fig. 48), for example in the depth of the mold, the variation of the scenes, and the arches on the shoulder.

[1] For a clearer rendering of a swan, see panel 4 of Cat. No. 45.
[2] The pose of this bird cannot be paralleled in nature. No known bird acts this way. It could only have been conceived by someone who did not know bird behavior.
[3] Series B corresponds to Eisen 1927, I, 249, fig. 117. The panels in the Toledo bottles have the same numbering as Eisen's fig. 117; in both cases panel 1 shows a bird on a three-lobed bush.
[4] For example:
a) BM, no. 135009, translucent dark blue, formerly Durighello Collection, which contained for the most part finds from southern Phoenicia and Galilee (cf. Durighello 1911, foreword), Barag 1985a, 102, no. 148;
b) Philadelphia, The University Museum, no. MS 5015 (unpublished), acquired at Aleppo;
c) CMG, no. 61.1.16, formerly R. W. Smith Collection, no. 812, Smith 1957, 61, no. 78, ill.;
d) formerly Kofler-Truniger Collection, opaque white, Luzern 1981, 77, no. 242, ill., *Kofler-Truniger* 1985, 71, no. 115, ill.;
e) Oppenländer Collection, no. 2415, translucent blue, von Saldern et al. 1974, 143, no. 409, ill.

## *Cylindrical Bottle with Low Relief* (Cat. No. 48)

### ***Plants Type*** (Cat. No. 48)

#### 48 Cylindrical Bottle, Plants Type

Syro-Palestinian, perhaps made in Sidon
Probably second to third quarter of first century
H 8.2 cm; H body 5.5 cm; max D 4.34 cm; D rim 2.6 cm; D base 2.42 cm
1923.439, Gift of Edward Drummond Libbey

Opaque white. Medium thin glass.

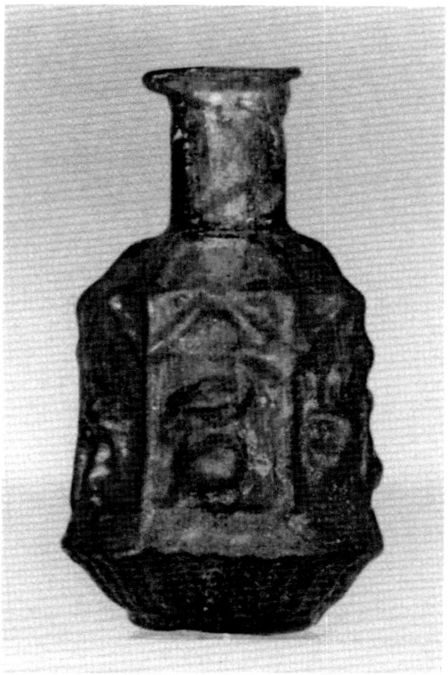
47, panel 4 (1:1)

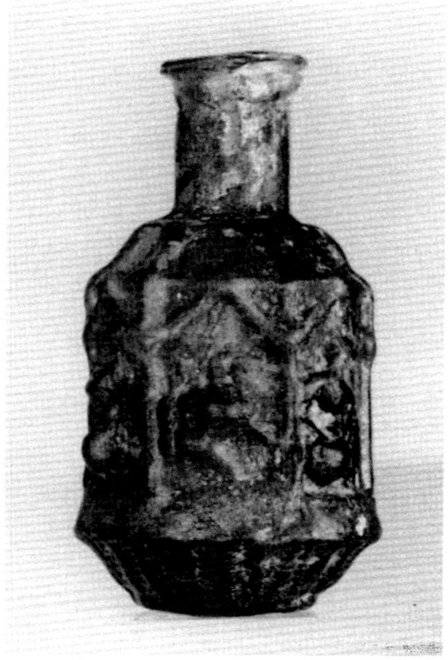
47, panel 5 (1:1)

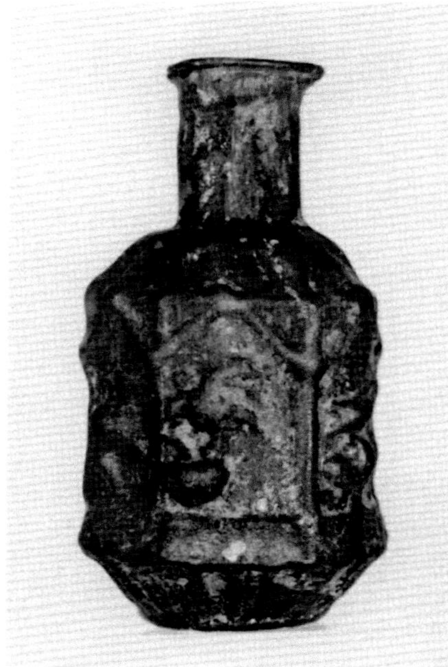
47, panel 6 (1:1)

Neck free blown. Body blown into a three-part mold of two vertical sections joined to a cup-shaped base section (MCT V C). Mold seams through centers of shoulder palmettes with inturned petals and between panels 6 and 1 and panels 3 and 4.

Rim folded outward, upward, inward, then flattened. Cylindrical neck with slight downward taper. Convex shoulder and convex bottom joined by a cylindrical body. Flat base with three raised concentric circles around a small central boss on its underside.

On the shoulder, two inverted seven-leaf palmettes with inturned petals (asphodelus) alternating with inverted five-leaf palmettes with outturned petals rising from inturned petals. On the body, six square panels framed by beaded ribs topped alternately with triangular pediments and arches in beaded outline; each panel contains a foliate motif. In the panels, from left to right: 1 and 4, a seven-leaf palmette with five outturned petals rising from two inturned petals; 2 and 5, a seven-leaf palmette with inturned petals (asphodelus); and 3 and 6, a plant with seven upsloping side branches (acanthus).[1] On the bottom, thirty upturned contiguous tongues.

*Condition* Complete; small hole in shoulder. The glass has degenerated throughout to a chalky pinkish white material with pink iridescent areas on the exterior.

*Collection* Thomas E. H. Curtis.

*Published* Previously unpublished.

*Remarks* This bottle belongs to a family of vessels that includes small bottles and pyxides. The bottles are cylindrical, the pyxides usually octagonal.[2] Characteristic is the use of beaded lines to form the decoration of rectangular panels framing isolated motifs and capped by alternating triangular and lunette-shaped pediments.

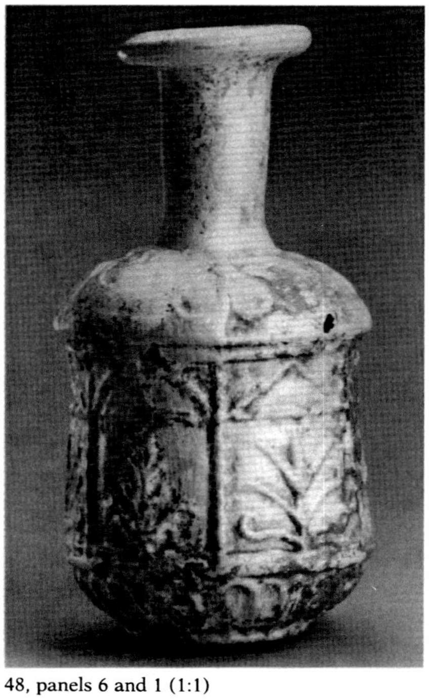
48, panels 6 and 1 (1:1)

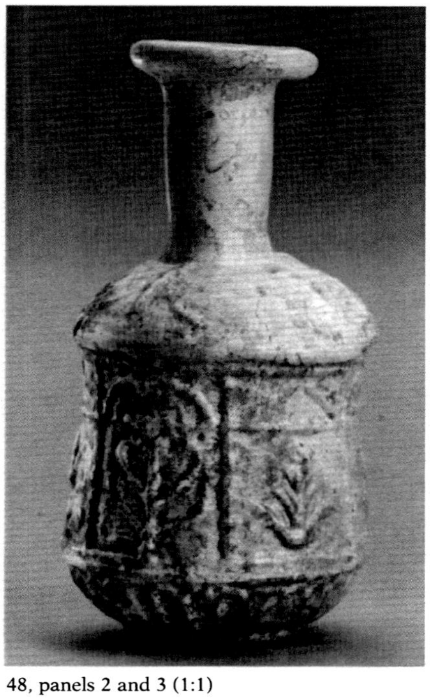
48, panels 2 and 3 (1:1)

Decoration

The most common motifs are stylized palmettes, but more elaborate compositions also occur. The earliest (?) vessels of this family not only show other motifs in addition to palmette-shaped plants,[3] but also have a distinctive complex profile[4] reminiscent of jugs with paired floral sprays (compare Cat. No. 75). Later examples such as this bottle[5] have not only a much less distinct profile but also a reduced repertoire of motifs. An even later variant has only two alternating plant motifs and gadroons on the shoulder.[6]
The stylized palmettes are similar to those on at least three shapes of vessels signed by Ennion.[7] However, Ennion's vessels do not have beaded lines. Nor are animals, such as appear as shoulder decoration on some bottles belonging to this family,[8] known in Ennion's repertoire. It is therefore unlikely that Ennion's workshop created any of the molds used in the production of this family,[9] especially because he appears to have produced only larger tablewares. Although the choice of motifs may have been inspired by Ennion, the motifs themselves are part of the Hellenistic heritage upon which all craftsmen in the early Roman empire drew.[10]

Technique

The arrangement of the decoration in panels connects this family of Plants Type vessels with Hexagonal Bottles with High Relief (Cat. Nos. 15-47), but the MCT V C mold sets them apart and suggests a closer connection to Hexagonal Jugs with Dionysiac Symbols (Cat. Nos. 71-74). The beaded lines are reminiscent of metalwork and could indicate that the designs were originally chased into metal sheets with a pointed tool, as was done to create the molds for the Mixed Symbols Type (Cat. Nos. 43 and 44).

Date

No examples have been published from datable contexts, but the MCT V mold was used throughout the first century, and vessels decorated with concentric rings on the underside appear to have been most common in the second and third quarters of the century.

Distribution and Origin

Most of the bottles that are direct parallels come from unknown find spots. However, find spots for other members of the family point to an eastern Mediterranean origin. Two pyxides were found in Athenian tombs,[11] and one is reported to be from Sidon.[12] This suggests that the workshop producing this family was located in an internationally oriented trade center such as Sidon.

[1] My thanks are owed to Sam Segal, Amsterdam, for identifying the plant motifs.

[2] On the bottles: Matheson 1980, 45-46, nos. 119-120, with lit. For the pyxides: Matheson 1980, 46, no. 121, with lit. The complex profile of the pyxides recalls cylindrical glass pyxides made in Greece: Weinberg 1959, 11-22; Weinberg 1992, nos. 39-43; Stern and Schlick-Nolte 1994, 79-81. Small pyxides of this shape were made from the third century B.C. onward not only in glass but also in silver, wood, and clay: for example, Oliver 1977, 53, no. 21.

[3] For example:
a) formerly Constable-Maxwell Collection, *Maxwell* 1979, 102, no. 175;
b) Yale, no. 1930.423, Matheson 1980, 46, no. 120, with more parallels.

The six panels of these bottles show four different plant motifs, a vessel, and a pair of strigils in the main frieze; the shoulder shows animals divided into pairs.

[4] For example:
a) see note 3b (bottle);
b) Yale, no. 1955.6.56, Matheson 1980, 46, no. 121 (pyxis).

[5] For example:
a) Yale, no. 1955.6.52, Matheson 1980, 45, no. 119 (Eisen 1927, I, 253, fig. 122, pl. 52d), with parallels;
b) Philadelphia, University Mus., no. MS 5014, reported to be from Aleppo (unpublished).

[6] For example:
a) formerly Constable-Maxwell Collection, *Maxwell* 1979, 52, 71;
b) Tel Aviv, no. MHG 6958, Israeli 1964, 37, no. 5, fig. 8.

The plants rise from petals with outturned tips instead of inturned ones as on the Toledo bottle.

[7] For discussion and connections with Ennion: Israeli 1964, 37, no. 5, fig. 8. Isolated floral motifs occur on amphorae:
a) St. Petersburg, Hermitage, no. P 1852.54, an amphora bearing Ennion's signature excavated at Panticapaeum, Kerch, Kunina 1973, 102-106, figs. 1-3; cf. Harden 1935, 168, group A5; Lehrer 1979, 8-9, pl. II, nos. 1-2;

on hexagonal bottles such as:
b) MMA, no. 81.10.224, Lehrer 1979, 9-10, pl. III, 1-4;
c) fragments of hexagonal bottles excavated in the Jewish Quarter in Jerusalem, Israeli 1983, 68-69, fig. 5; cf. Harden 1935, 168, group A6;

and on one-handled drinking cups, cf. Harden 1935, 165, group A1; Lehrer 1979, 11, drinking cup Type A, pl. VI, nos. 1-2. For fragments of similar cups from Spain and Morocco: Price 1973, 69, fig. 1, nos. 1a-b.

[8] See above note 3b.

[9] *Pace* Harden, in *Maxwell* 1979, 10, referring to lot no. 175.

[10] Compare the arrangement of panels on Oxford, Ashmolean Mus., no. 1889.16g, bought in Arezzo, a fragment of Arretine ware, A. C. Brown, *Catalogue of the Italian Terra-Sigillata in the Ashmolean Museum* (Oxford 1968) 31, no. 89, pl. 21, fig. 1.

[11] Athens, Nat. Arch. Mus. no. 2779, from Athens, the Kerameikos, and no. 2764, from a tomb in Athens, Stadiou Street, Weinberg 1992, 123-124, nos. 94 and 95.

[12] For example:
a) BM, no. GR 1893.10-16.1, reported to be from Sidon, opaque white, *Masterpieces* 1968, 52, no. 59; Tait 1991, 70-71, fig. 89, color;
b) see note 4b, with additional parallels from unknown find spots.

## *Carinated Bottle*
(Cat. No. 49)

### ***Carinated Bottle***
(Cat. No. 49)

### 49 **Carinated Bottle**

Eastern Mediterranean, probably Syro-Palestinian
Mid first century
H 8.9 cm; H body 5.6 cm; max D 5.7 cm; D rim 2.28-2.15 cm; D base 3.36 cm
1923.444, Gift of Edward Drummond Libbey

Transparent to translucent dark blue. Thin glass. A few small spherical bubbles in body and elongated vertical bubbles in neck. Neck free blown. Body blown into a three-part mold of two vertical sections joined to a cup-shaped base section (MCT V). Rim folded outward, upward, inward, and downward. Tall cylindrical neck with two tool marks at its base. Upper part of body conical; bottom part of body rounded. Flat base with two raised concentric circles around a small central depression.
On the body, two faint grooves, each bordered by two faint ridges, at 1.5 cm and 1.1 cm above carination. On the bottom, twenty-nine petal-like tongues alternating with darts in raised outline.
*Condition* Intact. Exterior covered with iridescent film.
*Collection* Thomas E. H. Curtis.
*Published* Previously unpublished.
*Remarks* Very few parallels for this bottle are known;[1] not one appears to have been found in a controlled excavation. The delicate petal-like tongues, with small darts between the tongues, recall the

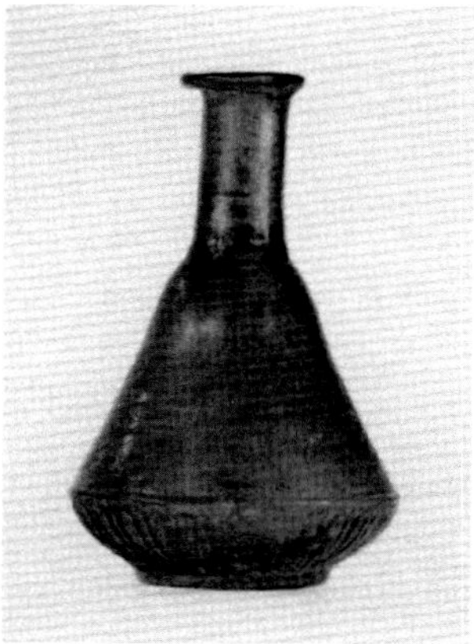

49 (1:2)

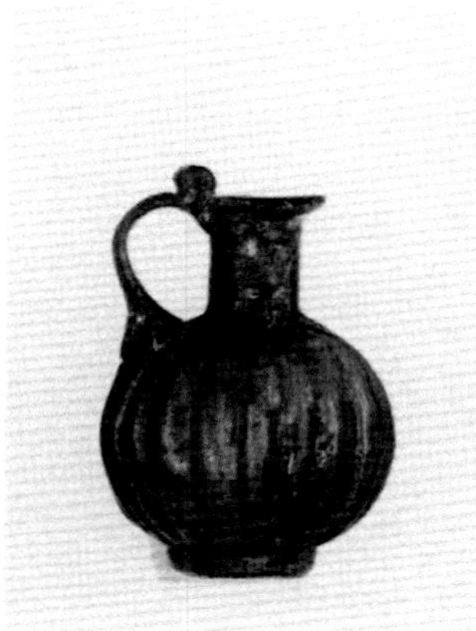

50 (1:2)

work of Ennion[2] who probably derived the design from contemporary Hellenistic metalware.[3] This petal-like tongue pattern also occurs on two cylindrical glass jugs with floral motifs which are here dated to the first half of the first century (Cat. Nos. 75 and 76). The use of an MCT V mold for this bottle recalls bottles such as Cat. No. 48, but there is no indication that this bottle was blown in a metal mold. The concentric rings on the underside of the base, petal-like fluting, and the use of an MCT V mold are all features that suggest production in the second or third quarter of the first century, most probably on the Syro-Palestinian coast.

[1] For example:
a) Nicosia, Cyprus Mus., no. D 1629, H 9.0 cm, light green, with one handle, Vessberg 1952, 130, pl. VI.18;
b) formerly Constable Maxwell collection, *Maxwell* 1979, no. 91, H 8.8 cm, bluish green, without a handle.

[2] For example, Ennion's drinking cups, Harden 1935, 167, Group A2ii4:
a) Turin, Arch. Mus., Lehrer 1979, 11, nos. V, 1-2, pl. V, figs. 1-2, Calvi 1965a, 11, figs. 4, 12;
b) Newark no. 50.1443, Auth 1976, 64-65, no. 58.

Ennion's jugs, Harden 1935, 168, Group A4i:
c) MMA, no. 17.194.226, Lehrer 1979, pl. 1, 4;
d) Tel Aviv, no. MH 120058, Lehrer 1979, 6-8, pl. 1, 1; 2 (Israeli 1964, 34-35, no. 1, figs. 1-3);
e) fragments of two jugs excavated in the Jewish Quarter of Jerusalem, Israeli 1983, 65-69, figs. 2-4.

[3] Compare:
a) MFA, no. 58.319, from Asia Minor, silver bowl, Oliver 1977, 40, no. 10 (with lit.)

For the floral origin of the fluted design, compare:
b) Paris, Louvre, no. unknown, provenance unknown, silver cup, J. Charbonneaux, *Revue du Louvre* 10 (1960) 133-135, color pl.
c) Berlin, Staatliche Museen, Antikenabteilung, no. 31 425, reported to be from Nihawand, Iran, silver bowl, Oliver 1977, 76, no. 41;
d) Stockholm, Medelhausmuseet, no. MM 1974:37, from an unknown find spot, Oliver 1977, 77, no. 42.

## *Bottles with One or Two Handles*
(Cat. Nos. 50-70)

### ***Spherical Bottles with Vertical Ribs***
(Cat. Nos. 50 and 51)

#### 50 **Melon-shaped Bottle with One Handle**

Eastern Mediterranean, probably Syro-Palestinian
Probably second half of the first century
H 6.7 cm; H body 4.65 cm; max D 5.2-4.8 cm; D rim 2.34 cm; D base 2.3 cm
1923.684, Gift of Edward Drummond Libbey

Translucent manganese colored glass. Similarly colored handle. Medium thin glass. A few small spherical bubbles in neck and one large vertically elongated bubble near handle.
Rim tooled. Neck and body blown into a two-part mold of two vertical sections (MCT VIII). One continuous mold seam around body and base extends onto either side of neck and follows line of vertical ribs on body. Edges of mold not well aligned.
Rim partly folded outward, upward, and inward with area of extreme overlap. Cylindrical neck. Melon-shaped body. Circular flat base. Bifurcated handle applied to shoulder and attached to rim with projecting thumbrest above. Handle positioned adjacent to seam.
Around the body, twenty-two vertical ribs, wider at middle and tapering toward ends.
*Condition* Intact. Exterior covered with purplish and rainbow iridescence; surface pitting.
*Collection* Thomas E. H. Curtis.
*Published* Previously unpublished.
*Remarks* This bottle belongs to a Type that was usually finished with one handle[1] and was often made of manganese colored purplish glass.[2] Various Series exist, two of which are represented in the Toledo Museum collection (see Cat. No. 51, which is of the same Type but a different Series). The shape of the ribs may imitate a melon or squash.
Although excavated finds are numerous in first-century tombs in Armenia,[3] Georgia,[4] and Panticapaeum, Kerch,[5] the bottles are regarded as "Syrian" (i.e. Syro-Palestinian) imports. The type has not been recorded from sites in the West.[6]

[1] For an example like this bottle but with two handles, see below note 4c. For an example without handles, Oppenländer Collection, no. 2211, von Saldern et al. 1974, 146, ill., 155, no. 428.

[2] For example, Oppenländer Collection, nos. 22241 and 22200, von Saldern et al. 1974, 147, ill., 155, nos. 426 and 427.

[3] From Armenia:
a) Yerevan, State Hist. Mus., no. 1964/85, from Garny, tomb 6, light violet, H 7.5 cm, found with a small bottle like Cat. No. 54, Arakelian et al. 1969, 57, no. 116;
b) Yerevan, State Hist. Mus., no. 2100/79, from Garny, tomb 79, opaque blue, H 7.0 cm, Arakelian et al. 1969, 57, no. 117, color pl. 7;
c) Moscow, Hist. Mus., no. 47733 OP, $\frac{567}{VXII}$, from the district of Artashat, Arakelian et al. 1969, 58, no. 118.

[4] From Georgia:
a) Tblisi, Mus., no. 61:3464, from Urbnis, tomb 187, light purple, H 6.8 cm, Saginashvili 1970, 50, 95, no. 5;
b) Tblisi, Mus., no. 62:3737, from Urbnis, tomb 207, light purple, H 7.4 cm, Saginashvili 1970, 50, 95, no. 6;
c) Tblisi, Mus., no. 62:3724, from Urbnis, tomb 205, H 7.9 cm, two handles, Saginashvili 1970, 50, 94, no. 4.

[5] Two examples are cited by Arakelian:
a) Moscow, Hist. Mus., no. 34405 A37/11, from Panticapaeum, Arakelian et al. 1969, 58;
b) St. Petersburg, Hermitage, no. E. 1249, from Panticapaeum, Arakelian et al. 1969, 58 (Hermitage number is perhaps PAN 590, cf. Kunina 1973, 123-124, fig. 18).

[6] A related type with two handles in Venice, Museo Vetrario, no. I.G. VE 382, from the necropolis of Nona, Zadar, Croatia, has a small rectangular base like Cat. Nos. 60 and 61.

#### 51 **Melon-shaped Bottle with One Handle**

Eastern Mediterranean, probably Syro-Palestinian
Probably second half of first century
H 8.05 cm; H body 4.9 cm; max D 5.25-4.9 cm; D rim 2.9 cm; D base 3.6-3.5 cm
1923.685, Gift of Edward Drummond Libbey

Translucent patchy manganese colored glass. Translucent streaked

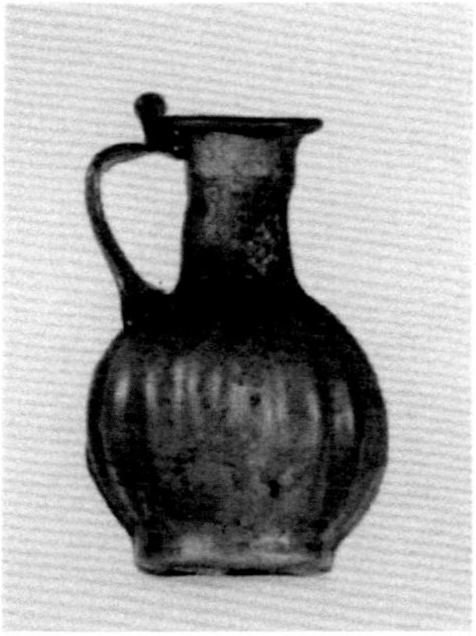

51 (1:2)

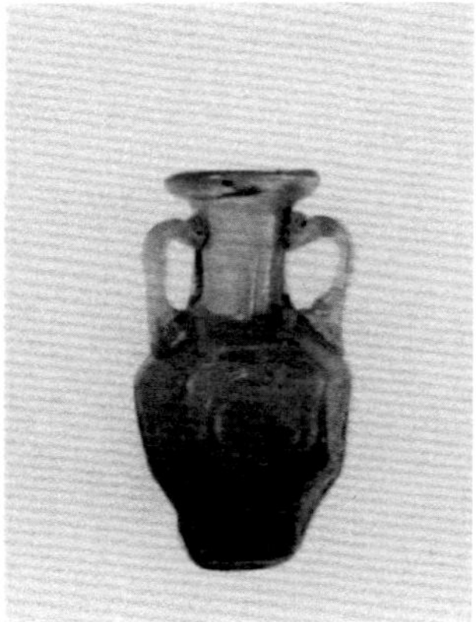

52 (1:2)

grayish green handle. Medium thin glass. Fabric cannot be determined because of weathering.
Rim tooled. Neck and body blown into a two-part mold of two vertical sections (MCT VIII C). One continuous mold seam around body and base extends onto either side of neck and follows line of vertical ribs on body. Edges of mold not well aligned. Tip of handle drawn out thin.
Rim folded outward, upward, and inward. Concave neck with bulge above edge of mold. Melon-shaped body. Circular flat base. Bifurcated handle applied to shoulder and attached to rim with projecting thumbrest above. Handle centered over mold seam. Around body, seventeen vertical ribs, wider at middle and tapering toward ends.
*Condition* Intact. Exterior covered with areas of purplish and rainbow iridescence; areas of severe surface pitting. Interior incrusted.
*Collection* Thomas E. H. Curtis.
*Published* Previously unpublished.
*Remarks* See Remarks Cat. No. 50.

## *Multisided Bottle*
(Cat. No. 52)

### 52 Multisided Bottle with Two Handles

Perhaps Eastern Mediterranean
Probably first century
H 7.1 cm; H body 4.5 cm; max D 3.9-3.4 cm; D rim 2.72-2.56 cm; D base 1.8-1.35 cm
1923.893, Gift of Edward Drummond Libbey

Translucent streaked manganese colored glass. Translucent grayish green handles (10 GY 5/2). Medium thick glass. Many medium-sized and two large, vertically elongated bubbles in neck; several large oval bubbles in body.
Rim tooled. Neck and body blown into a two-part mold of two vertical sections (MCT VIII H). One continuous mold seam around body and base extends onto either side of neck. Handles drawn out thin.
Rim folded outward, upward, inward, and downward. Cylindrical neck. Nine-sided body. Rectangular flat base. Two coil handles applied to shoulder and attached below rim. Handles positioned to one side of mold seam.
On front and back panels, one large circle in raised outline.
*Condition* Intact. Areas of rainbow iridescence on exterior; some surface pitting.
*Collection* Thomas E. H. Curtis.
*Published* Previously unpublished.
*Remarks* Small multisided bottles were made in a number of variants. Many have two handles. Two were excavated in the cemetery of Urbnis, Georgia,[1] and one each at Panticapaeum, Kerch, Ukraine,[2] and Zadar, Croatia,[3] but most are from unknown find spots.[4]

[1] For example:
a) Tblisi, Mus., no. 58:1889, H 5.6 cm, Saginashvili 1970, 50, 95, no. 7;
b) Moscow, Hist. Mus., no. 55738, H 7.1 cm, Saginashvili 1970, 51, 95, no. 8.

[2] St. Petersburg, Hermitage, no. P. 1903.169, Kunina 1973, 124-126, fig. 19, no. 1.

[3] Murano, Mus. Vetrario, a bottle of transparent streaked manganese colored glass, H ca. 7.0 cm, with a design of four raised circles surrounding a central dot on front and back panels.

[4] For example:
a) formerly R. W. Smith Collection, no. 1324, Smith 1957, 65, no. 88, acquired in Lebanon;
b) formerly R. W. Smith Collection, no. 314, Smith 1957, 65, no. 87; *Parke Bernet*, May 2, 1975, no. 186;
c) Murano, Mus. Vetrario, G. Mariacher, "Mostra dei vetri del Museo di Zara," *Bollettino dei Musei Civici Veneziani* 8 (1963) 14, no. 19, fig. 14;
d) Newark, no. 50.1469, Auth 1976, 71, no. 68;
e) RGM, no. 58,33, La Baume 1973, no. C 11, pl. 9.4, with one handle;
f) Loeffler Collection, no. 182, La Baume and Salomonson 1976, 36, no. 65, pl. 7.4;
g, h) Switzerland, private collections, Luzern 1981, 79, nos. 260 and 261;
i) Yale, no. 1955.6.60, Matheson 1980, 51, no. 130;
j) formerly Paris, de Clercq Collection, de Ridder 1909, 150, no. 265, pl. 10;
k) Oppenländer Collection, no. 22205, von Saldern et al. 1974, 147, ill., 154, no. 437;
l) St. Petersburg, Hermitage, no. E. 492, Kunina 1973, 125, fig. 19, no. 2.

## *Bulbous Bottles with a Band of Lozenges*
(Cat. Nos. 53 and 54)

### 53 Bulbous Bottle with Two Handles

Syro-Palestinian
Probably second half of the first century
H 7.4 cm; H body 5.5 cm; max D 4.85-4.75 cm; D rim 2.35-2.2 cm; D base 2.6 cm
1923.681, Gift of Edward Drummond Libbey

Translucent dark purplish blue glass. One translucent streaked similarly colored handle and one translucent manganese colored handle. Medium thin glass. Small spherical bubbles.
Neck free blown. Body blown into a three-part mold of two vertical sections joined to a cup-shaped base section (MCT V D). Edges of vertical mold sections not carefully aligned.
Rim folded outward, upward, and inward. Cylindrical neck. Ovoid body. Flat base. Two coil handles applied to shoulder and attached just below rim. Handles positioned next to mold seams.
Friezes of twenty-one downturned tongues in raised outline on upper body and twenty-three similar upturned tongues on lower body joined by a central band of contiguous lozenges in raised outline (or X's or net pattern) with thickened intersection points. Central band bordered by a groove and a raised ridge above and a raised ridge below.
*Condition* Intact. Incipient surface frosting.
*Collection* Thomas E. H. Curtis.
*Published* Previously unpublished.
*Remarks* The design of this bottle, with a central band of lozenges bordered above and below by tongues, occurs in a number of Series that cannot be easily distinguished from each other on the basis of published descriptions and illustrations. Most vessels have two handles, but a few have only one. The thickened knobs at the intersections of the lozenges on this bottle and its smaller dimensions

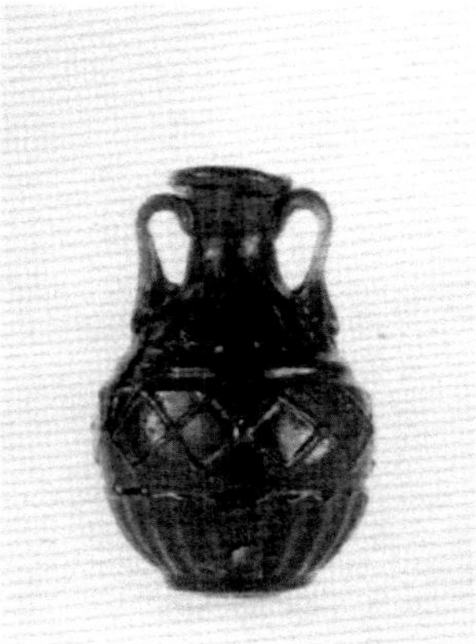

53 (1:2)

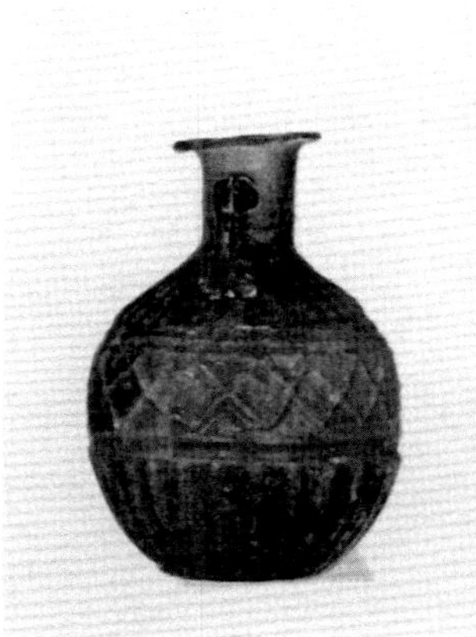

54 (1:2)

differentiate it from Cat. No. 54, which belongs to another Series with a closely related design. Still another Series has three horizontal ridges below the central band.[1] These Series were probably made in different workshops. A few bottles were blown in two-part molds (MCT VIII),[2] indicating that the construction of molds varied from workshop to workshop.

Small bulbous bottles with handles like Cat. Nos. 53 and 54, and presumably also like Cat. Nos. 55-58, functioned as aryballoi, which were filled with oil, used instead of soap in antiquity. Aryballoi, also made of clay and metal, were filled with oil and carried to the gymnasium or palaestra for cleaning after exercise. Several glass examples, such as a bottle in Princeton (Fig. 81)[3] still preserve bronze rings, part of a carrying chain, hanging from the handles. The design may have been inspired by vessels made by Ennion.[4] It is less common than a related design consisting of a central band of horizontal S-shaped scrolls or continuous scrolls (see Cat. No. 55). Find spots indicate that most if not all Series decorated with this tongue and lozenge motif were made and used in the eastern Mediterranean.[5] The undecorated underside suggests a date in the second half or toward the end of the first century.

[1] Saint-Germain-en-Laye, France, Mus. des Antiquités Nationales, from a tomb at Vaison, Lantier 1929, no. 11A.

[2] For example:
- a) Afyon, Turkey, Arch. Mus., no. 2327, from the vicinity of Afyon, Lightfoot 1989, 25, no. 8;
- b) MMA, no. 17.194.221, formerly Gréau Collection, no. 1124, with a small knob on the seam in the center of the underside of the base;
- c) Princeton, N. J., Princeton University, Museum of Art, no. 46-31, said to be from Syria (here Fig. 81).

[3] See above note 2c.

[4] Bands of lozenges occur on jugs and spherical bowls signed by Ennion:

Jugs:
- a) Tel Aviv, no. MH 120058, Harden 1944-45, 89-90, A4iib; Israeli 1964, 34-35, no. 1, figs. 1-3; Lehrer 1979, 7, pl. 1, no. 1, said to have been found in the vicinity of Jerusalem;
- b) CMG, no. 59.1.76, formerly R. W. Smith Collection, Harden 1935, 168, A4iia; Harden 1944-45, 89; Lehrer 1979, 7; *Glass of the Caesars* 1987, 166, no. 87;
- c) MMA, no. 17.194.226, formerly Gréau Collection, Harden 1935, 168, A4ia; Lehrer 1979, 7, pl. 1, no. 4;
- d) Tel Aviv, no. 3974, fragments excavated in the Jewish Quarter of Jerusalem, Israeli 1983, 66-67, figs. 2-3.

Spherical bowls:
- e) Yale, no. 1955.6.66, Harden 1935, 167, Group A3a; Harden 1944-45, 89; Lehrer 1979, 10, pl. 4, no. 1; Matheson 1980, 43-44, no. 118.

[5] Compare, from find spots in the East:
- a, b) formerly Paris, de Clercq collection, de Ridder 1909, 151, nos. 268 and 269 (not ill.), from Tortosa on the Syrian coast;
- c) Adana, Bölge Museum, no. 12.1.72, from Cilicia, southern Turkey, with one handle (unpublished);
- d) See above note 2a;
- e, f) Tblisi, Georgia, Mus., nos. 59:2288 and 57:1551, from a cemetery at Urbnis, Georgia, Saginashvili 1970, 49, 94, nos. 1 and 3;
- g) Yerevan, no. 1964/365, from Garny, tomb 6, with two handles, Arakelian et al. 1969, 56, no. 112;
- h) Yerevan, no. 2100/40, from Garny, tomb 75, possibly local, Arakelian et al. 1969, 57, no. 113;
- i) Yerevan, no. 2100/85, from Garny, tomb 82, possibly local, Arakelian et al. 1969, no. 114;
- j) Yerevan, no. 1836, from Verin Guetashen, possibly local, Arakelian et al. 1969, 57, no. 115.

From find spots in the West:
- k) Murano, Mus. Vetrario, no. unknown, one bottle, transparent yellowish gray olive glass, with two translucent very dark blue handles, from Zadar, Croatia;
- l) See above note 1.

Many more examples have been preserved from unknown find spots, for example:
- m) Carnegie, no. 21868/4, perhaps acquired in Jerusalem, Oliver 1980, 62, no. 56;
- n, o) Oppenländer Collection, nos. 22267 and 22263, von Saldern et al. 1974, 146, nos. 416 and 417;
- p) Athens, Nat. Mus., no. 3032, Weinberg 1992, 126, no. 99.

### 54 Bulbous Bottle with Two Handles

Syro-Palestinian
Probably second half of the first century
H 7.75 cm; H body 5.75 cm; max D 5.15-5.2 cm; D rim 2.49-2.47 cm; D base 2.8 cm
1923.914, Gift of Edward Drummond Libbey

Translucent to transparent streaked manganese colored glass. One similarly colored and one translucent grayish blue handle. Thin glass. Fabric cannot be determined because of weathering.

Neck free blown. Body blown into a three-part mold of two vertical sections joined to a cup-shaped base section (MCT V). Edges of mold carefully aligned. Tips of handles drawn out thin and folded back on top of handle.

Everted rim, barely rounded in flame and partly folded outward, upward, and inward. Cylindrical neck. Ovoid body. Flat base with slight depression in center. Two coil handles applied to shoulder and attached halfway up neck. Handles positioned adjacent to mold seams.

Friezes of twenty-six downturned tongues in raised outline on upper body and twenty-seven upturned flutes on lower body joined by a central band of contiguous lozenges in raised outline (or X's or net pattern) bordered above by a raised ridge.

*Condition* Intact. Iridescent film on exterior; incipient frosting.
*Collection* Thomas E. H. Curtis.
*Published* Previously unpublished.
*Remarks* See Remarks Cat. No. 53.

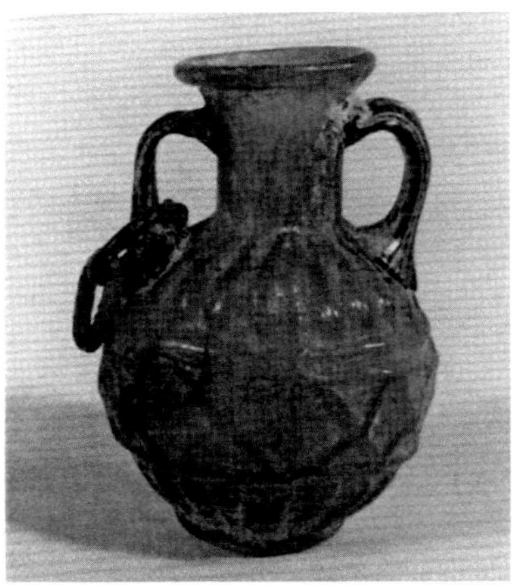

*Fig. 81. Bulbous bottle. This bottle still has two bronze rings, part of a carrying chain, preserved hanging from one handle, suggesting it was used as an aryballos. Princeton, N. J., Princeton University Museum of Art, no. 46-31. Photo: Princeton University Museum of Art.*

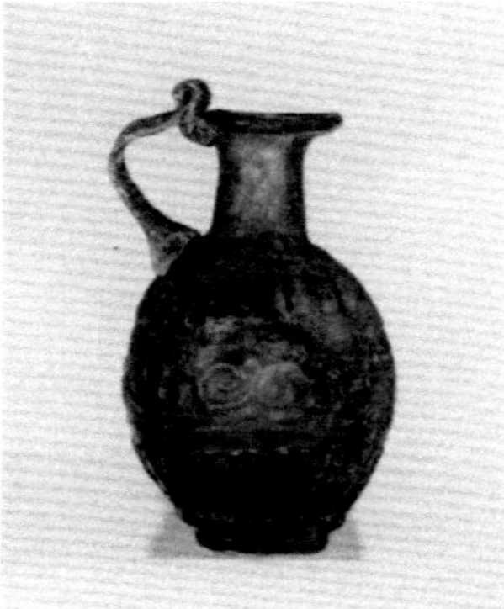

55 (1:2)

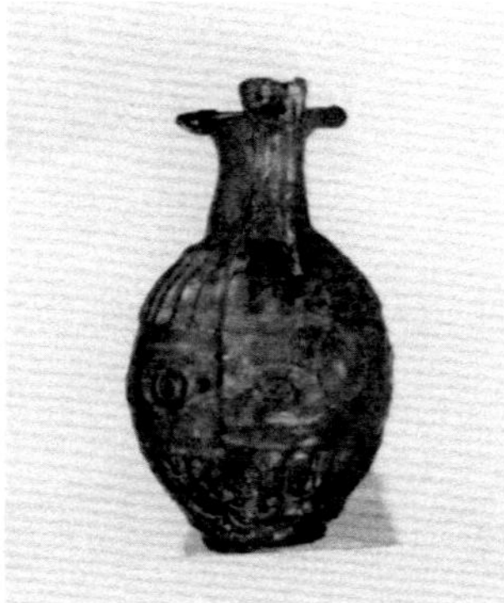

55 (1:2)

### ***Bulbous Bottles with a Band of Scrolls***
(Cat. Nos. 55-58)

**55 Bulbous Bottle with One Handle**

Syro-Palestinian
First century
H 7.3 cm; H w handle 8.0 cm; H body 5.1 cm; max D 4.25 cm; D rim 2.68 cm; D base 1.96-1.93 cm
1923.679, Gift of Edward Drummond Libbey

Translucent streaked manganese colored glass. Translucent streaked grayish blue handle. Medium thin glass. Fabric cannot be determined because of weathering.

Neck free blown. Body blown into a two-part mold of two vertical sections (MCT VIII O). One continuous mold seam extends across underside of base. Edges of mold not carefully aligned; therefore, base is in two planes. Tip of handle drawn out thin and folded back.

Rim folded outward, upward, and inward, except where handle is attached. Cylindrical neck. Ovoid body. Base intended to be flat. Bifurcated handle applied to shoulder and attached to rim with projecting thumbrest above. Handle positioned to one side of mold seam.

Friezes of twenty-four downturned contiguous flutes in raised outline on upper body and twenty-nine upturned flutes on lower body joined by a central band of tendril scrolls bordered above and below by a low narrow ridge and wide shallow groove.

*Condition* Intact. Small areas of iridescence and severe surface pitting.

*Collection* Thomas E. H. Curtis.

*Published* Previously unpublished.

*Remarks* Same Type and possibly same Series as Cat. No. 56, which is finished with two handles. Many Series exist with decoration of a band of scrolls between vertical flutes or tongues; three different Series are represented in the Toledo Museum collection (see also Cat. Nos. 57 and 58).[1] From published descriptions it is generally impossible to decide which Series is meant. Some bottles have only one handle,[2] but most have two[3] (see Cat. No. 56). The Type itself appears to be a miniature version of large mold-blown amphoras, such as one bearing Ennion's signature in the Hermitage Museum, St. Petersburg, decorated with vertical gadroons around neck and bottom, framing three different floral bands in addition to a band of hexagons.[4]

Most if not all Series with this design seem to have been made in two-part MCT VIII molds, with the mold seam extending across the underside of the base. The relationship of one Series to another has not been established, nor has the development of any one Series. A first-century date is suggested by excavated finds in Greece and the Crimea.[5] Find spots point to production in the eastern Mediterranean,[6] with either trade to or a second production center in the western Mediterranean.[7]

[1] Kunina 1973, 117-122, notes four variants in St. Petersburg, Hermitage, from tombs in Panticapaeum, Crimea:
- a, b) first variant: Hermitage, nos. P. 1903-4.19 and P.1871.53, 119, fig. 16, nos. 1 and 2;
- c) second variant: Hermitage, no. P. 1907.94, 119, fig. 16, no. 3;
- d) third variant: Hermitage, no. P. 1909.17, 119, fig. 16, no. 4;
- e, f) fourth variant: Hermitage, nos. P. 1902.148 and P. 1902.102, 119, fig. 16, nos. 5 and 6.

It has not been possible to coordinate Kunina's variants with the three Series represented in the Toledo Museum.

[2] Compare, with one handle:
- a) from a tomb in Samothrace, Greece, located "in an area containing Flavian and somewhat later tombs," i.e. the last quarter of the first century and the early second century, Dusenbery 1967, 40, no. 13, fig. 14;
- b) Oppenländer Collection, no. 2244, von Saldern et al. 1974, 144, ill. 147, no. 412.

[3] A number of bottles with two handles from unknown find spots have been preserved:
- a-c) Newark, nos. 50.1473, 50.1474, and 50.1472, Auth 1976, 71, no. 66; 199, nos. 329 and 330;
- d, e) Leiden, Rijksmuseum van Oudheden, nos. B 1952/8.4 and B 1952/8.5, J. H. Kern, "A Roman Glass Flask in the Leiden Museum," *BABesch* 29 (1954) 38 and 39;
- f, g) formerly Kofler-Truniger Collection, *Kofler-Truniger* 1985, 63, nos. 98 and 99;
- h, i) BM, formerly Felix Slade Collection, Nesbitt 1871, 31, nos. 185 and 186, fig. 43;
- j, k) Berlin, Antikenmuseum, nos. 30219,147 and 1964.1, Platz-Horster 1976, 39, nos. 56 and 57;
- l-n) Oppenländer Collection, nos. 2340, 2211, and 2377, von Saldern et al. 1974, 145 and 146, nos. 413-415;
- o) MFA, no. 15.858, von Saldern 1968, no. 33;
- p) formerly R. W. Smith Collection, no. 389, Smith 1957, 61, no. 75;
- q-s) St. Petersburg, Hermitage, nos. E. 1046, E. 490, and E. 310, Kunina 1973, 121, fig. 17.

[4] St. Petersburg, Hermitage, no. P. 1852.54, Kunina 1973, 101-106, figs. 1-6; Lehrer 1979, 8, pl. 2, nos. 1 and 2.

[5] For example:
- a) see above note 2a;
- b) from Panticapaeum, Kerch, *mogila* 32, in a tomb that also contained unguentaria of the second half of the first century, Kunina and Sorokina 1972, 162, fig. 7, no. 34;

[6] Find spots indicating a Syro-Palestinian origin:
- a) formerly Paris, de Clercq Collection, from Tortosa on the Syrian coast, de Ridder 1909, 150-151, no. 266 (not ill.);
- b) from Amrit, about 14 km south of Tortosa, de Ridder 1909, 151, no. 267, pl. X;
- c) Damascus, Nat. Mus., no. 9002, with two handles, Zouhdi 1964, 52-53, no. 63, fig. 40;
- d) Adana, Bölge Mus., no. 1.10.72, from Kadirli (ancient Flaviopolis), in Cilicia, southern Turkey (unpublished);
- e) see above note 2a;
- f) from a tomb at Thessaloniki, Greece, Syntrivanios Square, *ArchDelt* 26 (1971) B2, 378, pl. 371B;
- g) Athens, Agora, no. G373, Gladys D. Weinberg, pers. com.;
- h-l) from various cemeteries in Armenia, Yerevan, State Hist. Mus., nos. 2241/1, 502, Tblisi, Museum of Georgia, no. $\frac{50-06}{A-1-4}$, and two unknown nos., Arakelian et al. 1969, 55-56 nos. 107-111;
- m, n) from the cemetery at Urbnis, Georgia, Saginashvili 1970, nos. 1 and 2;
- o) Oppenländer Collection, no. 2244, with one handle, formerly Vogell Collection, probably from southern Russia or Ukraine, von Saldern et al. 1974, 144, no. 412.

[7] Find spots possibly indicating an origin in Italy:
- a) from Cuma (ancient Cumae), Campania, E. Gabrici, "Scavi di Cuma: Ritrovamenti dal VI secolo A.C. fino all'età romana," *Monumenti Antichi* 22 (1913) col. 740, pl. 118, no. 9;

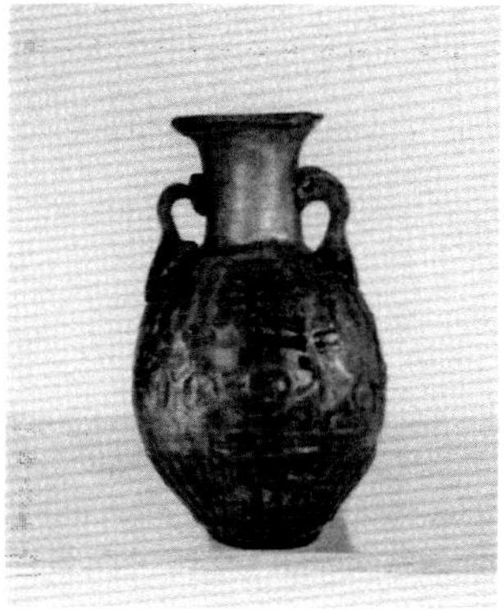
56 (1:2)

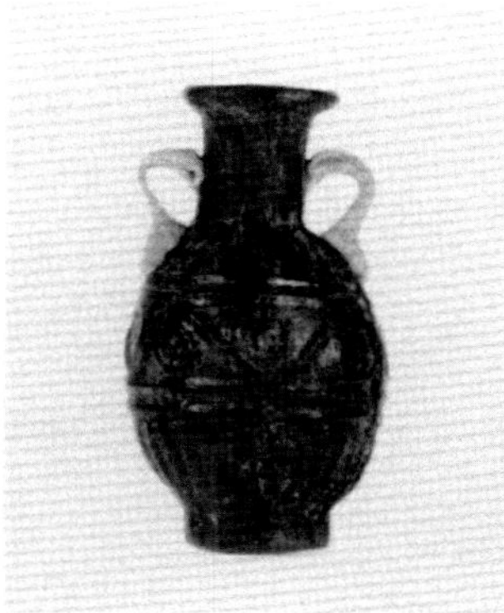
57 (1:2)

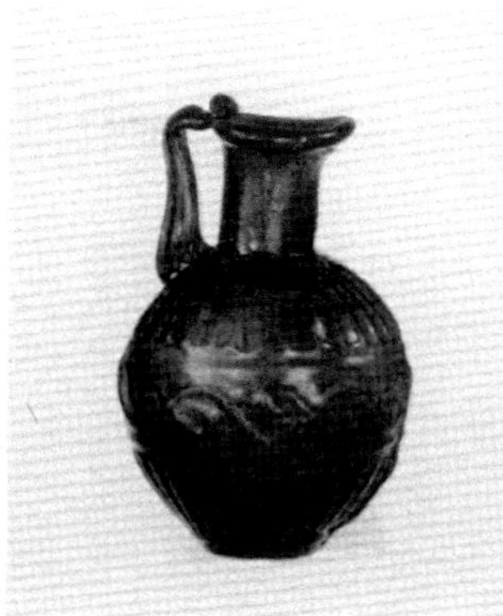
58 (1:2)

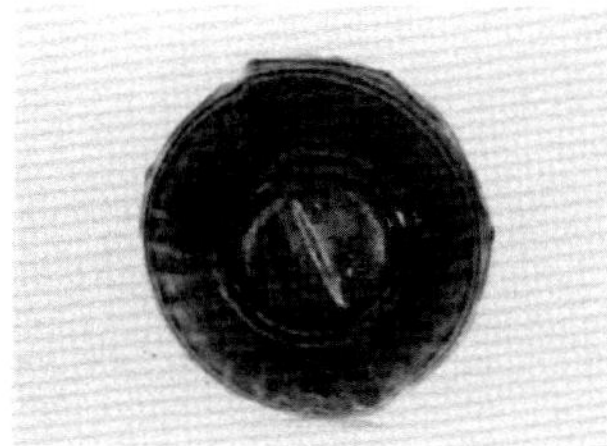
58, underside (1:2)

b) Murano, Mus. Vetrario, no. I.G. VE 353, from the necropolis of Nona, at Zadar, Croatia;
c) Murano, Mus. Vetrario, no. 6514, from San Donato di Zara (Zadar, Croatia), cited by Meconcelli Notarianni 1979, 152, no. 211 (perhaps the same vessel as the previous, note 7b);
d) Obrovac, northern Dalmatia, from a Roman necropolis at Argyruntum (modern Starigrad), in Dalmatia, Yugoslavia; cited by Kern (above note 3e), 39, no. 7;
e, f, g) Belgrade, Serbia, Nat. Mus., nos. 1197/II, 1207/II, and 1205/II, from Budva-Buthoe Dalm, Montenegro, Velickovic 1974, 170-171, figs. 16-18.

### 56 **Bulbous Bottle with Two Handles**

Syro-Palestinian
First century
H 7.3 cm; H body 5.1 cm; max D 4.17 cm; D rim 2.6-2.5 cm; D base 1.8-1.7 cm
1923.680, Gift of Edward Drummond Libbey

Transparent pale olive (near 10 Y 6/2). Similarly colored translucent handles. Medium thin glass. A few pinprick and small spherical bubbles.
Neck free blown. Body blown into a two-part mold of two vertical sections (MCT VIII). One continuous mold seam extends across underside of base. Edges of mold not carefully aligned; therefore, base is in two planes. Tips of handles drawn out thin and folded back to base of handle.
Rim folded outward, upward, and inward for about five-sixths of circumference. Cylindrical neck. Ovoid body. Base intended to be flat. Two coil handles applied to shoulder and attached halfway up neck. Handles positioned about one-third of the distance between seams.
Friezes of twenty-four downturned contiguous flutes in raised outline on upper body and twenty-nine upturned flutes on lower body joined by a central band of tendril scrolls bordered above and below by a raised ridge.
*Condition* Intact. Iridescent film with large areas of thin transparent yellowish brown iridescent layer throughout.
*Collection* Thomas E. H. Curtis.
*Published* Previously unpublished.
*Remarks* Same Type and probably same Series as Cat. No. 55.

### 57 **Bulbous Bottle with Two Handles**

Syro-Palestinian
First century
H 7.78 cm; H body 5.2 cm; max D 4.4-4.3 cm; D rim 2.8-2.6 cm; D base 2.25-2.0 cm
1923.494, Gift of Edward Drummond Libbey

Translucent dark royal blue. Two opaque buff white handles. Medium thin glass. Fabric cannot be determined because of weathering.
Neck free blown. Body blown into a two-part mold of two vertical sections (MCT VIII). One continuous mold seam extends across underside of base. Edges of mold not carefully aligned; therefore, base is in two planes. Tips of handles drawn out thin and folded back.
Everted rim, rounded and thickened in flame; tool mark on interior along about three-quarters of rim. Cylindrical neck. Ovoid body. Base intended to be flat. Two coil handles applied to shoulder and attached halfway up neck. Handles positioned adjacent to seams.
Friezes of about twenty-three downturned raised tongues on upper body, some entirely covered by handle attachment, and twenty-five upturned tongues on lower body joined by a central band of large tendril scrolls bordered above and below by two raised ridges.
*Condition* Intact. Iridescent film and surface frosting.
*Collection* Thomas E. H. Curtis.
*Published* Previously unpublished.
*Remarks* See Remarks Cat. No. 55. Same Type as Cat. Nos. 55 and 56, but a different Series.

### 58 **Bulbous Bottle with One Handle**

Syro-Palestinian
First century
H 7.3 cm; H body 5.1 cm; max D 4.6-4.08 cm; D rim 3.5 cm; D base 1.85-1.77 cm
1923.493, Gift of Edward Drummond Libbey

Translucent dark reddish manganese colored glass. Unintentional bluish green streaks in rim and neck. Medium thin glass. Numerous pinprick and small spherical bubbles in body; linear bubbles in neck, of which three in a row burst at surface.
Neck free blown. Body blown into a two-part mold of two vertical sections (MCT VIII). One continuous mold seam extends across underside of base. Edges of mold not carefully aligned; therefore, base is in two planes. Tip of handle drawn out thin and folded back.
Rim folded outward, upward, and inward. Cylindrical neck tapering slightly downward. Ovoid body. Base intended to be flat. Coil handle applied to shoulder and attached to top of rim with projecting loop above. Handle positioned over and next to seam.
Friezes of twenty-nine downturned contiguous flutes in raised outline on upper body and thirty upturned flutes on lower body joined by a central band of stylized tendril scrolls and dots bordered above and below by a raised ridge.

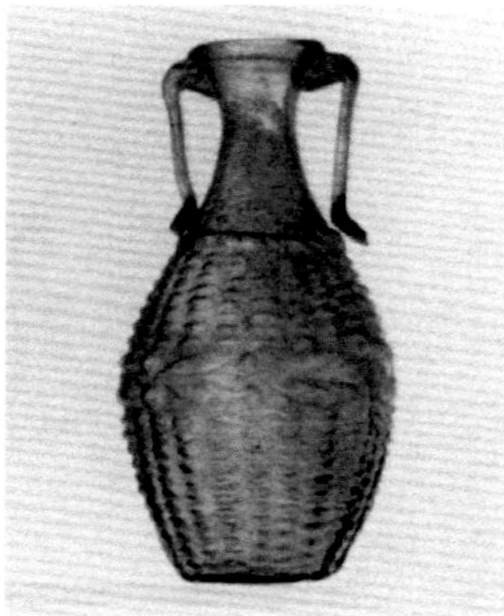

59, side A (1:2)

*Condition* Intact. Incipient surface frosting. Interior incrusted.
*Collection* Thomas E. H. Curtis.
*Published* Previously unpublished.
*Remarks* See Remarks Cat. No. 55. Same Type as Cat. Nos. 55 and 57, but a different Series.

### *Minature Amphora with Floating Handles*

(Cat. No. 59)

**59 Miniature Amphora in Wicker Basket**

Color Plate 11, p. 54; photograph Fig. 58
Syro-Palestinian, perhaps made in Sidon
Workshop of the Floating Handles
First half of the first century
H 8.9 cm; max D 4.5 cm; D rim 2.0 cm; D base 2.4-2.1 cm
1951.375, Purchased with funds from the Libbey Endowment, Gift of Edward Drummond Libbey

Translucent to transparent dusky yellow (5 Y 6/4). Similarly colored handles. Thin glass. Fabric cannot be determined because of weathering.
Rim tooled. Neck and body blown into a two-part mold of two vertical sections (MCT VIII A). One continuous mold seam around body and base extends onto neck. Excess glass at lower tips of handles cut off and pinched to form a small disk.
Rim rounded and thickened in flame; tool mark on interior. Concave neck with curved transition to shoulder. Ovoid body. Flat base. Two coil handles applied to underside of rim and drawn down to shoulder but not attached (see Fig. 58). Handles positioned adjacent to mold seam.
Vessel shaped like a miniature amphora encased in a wicker basket, with a rope around the shoulder and a branch of laurel leaves alternating with laurel berries around the middle of the basket. On side A the leaves point to the right; on side B the leaves point to the left.
*Condition* Intact. Iridescence and pitting on interior.
*Published* Previously unpublished.
*Remarks* Three Series of miniature amphoras have been noted: one with the leaves of the wreath around the middle pointing to the right,[1] one with the leaves pointing to the left,[2] and one without a wreath.[3] Some of the bottles belonging to the two Series with a wreath may have leaves pointing in different directions on either side like this bottle, which could have resulted from combining half a mold from each of the two Series. All Series appear to have been made in the Workshop of the Floating Handles, which was active in the first half of the first century, perhaps in Sidon (see Chapter I: *The Workshop of the Floating Handles*). Ten Types of mold-blown glass bottles have been attributed to this Workshop. The Toledo Museum collection has representatives of four Types: a lenticular bottle with a Medusa face on either side (Cat. No. 62), an acorn-shaped bottle (Cat. No. 113), a head-shaped vessel (Cat. No. 138), and this miniature amphora.
The design of this Type is based on the actual practice of protecting glass transport vessels by wrapping them in wickerwork, as can be seen by a few ancient glass vessels which still retain their wicker baskets.[4]

[1] To judge from publications, numerous examples with the wreath pointing to the right exist, but some may have the wreath on the other side pointing in a different direction like the Toledo bottle:
a) St. Petersburg, Hermitage, no. PAN 586, from Panticapaeum, natural green, with floating handles, Kunina 1973, 111-112, fig. 13; Hermitage Museum, *Khudozhestvennoe Steklo, albom no materialam bistavki v Gosydarstvennom Ermitage ("Art Glass: Guide to Material in the Hermitage Art Museum")* (Leningrad 1967) no. 9, fig. 9;
b) Oppenländer Collection, no. 2354, pale bluish green, with two floating handles, von Saldern et al. 1974, 150, no. 431, 155, ill;
c) MFA, no. 28.226, pale olive green, with one floating handle, von Saldern 1968, no. 34;
d) Yale, no. 1955.6.55, formerly New York, Kouchakji Collection, pale green, with one floating handle, Matheson 1980, 50-51, no. 128;
e) Tel Aviv, no. MH 63058, opaque dull blue, with one floating handle, Israeli 1964, 39, no. 8b, fig. 13 (perhaps also Tel Aviv, no. MH 49358, Israeli 1964, 39, no. 8a, with one handle);
f) Newark, no. 50.1476, blue-green, with one floating handle, Auth 1976, 68, no. 63;
g) formerly Constable-Maxwell Collection, pale green, *Maxwell* 1979, 62, no. 93;
h) Switzerland, private collection, transparent pale blue green, Luzern 1981, 78, no. 253 (this piece and the preceding both have the remains of one handle with a pinched vertical thumbrest and were probably made in the same workshop).

Other vessels belonging to the same Type:
i) BM, no. S 184, formerly Roussel Collection, Nesbitt 1871, 31, no. 184 (not ill.);
j) BM, no. GR 97.1-11.1, said to be from the Greek Islands, cited by Calvi 1968, 100;
k) MMA, no. 17.194.222, said to be from Syria, transparent yellowish olive green, with remains of two floating handles, with a wreath in opposite directions on either side.

[2] Examples with wreath pointing to the left:
a) formerly Kofler-Truniger Collection, opaque white, with two floating handles, *Kofler-Truniger* 1985, 69, no. 112;
b) Aquileia, Mus., no. 2893, perhaps found at Aquileia, handles and area of application not preserved, Calvi 1968, 105, no. 246, pl. 16, no. 4.

[3] Examples without a wreath:
a) Oppenländer Collection, no. 2471, pale bluish green, with two floating handles, von Saldern et al. 1974, 150, no. 432, 155, ill.;
b) Aquileia, Mus., no. RC1032, perhaps found at Aquileia, pale bluish green, Calvi 1968, 105, no. 248, pl. 16, no. 3 (this piece belongs to the same Series as 3a but cannot be assigned to the Workshop of the Floating Handles because the remains of the handle appear to have the excess glass at the top folded back);
c) V & A, no. 1033-1868, cited by Matheson 1980, 50-51, at no. 128.

[4] For example: CMG, no. 77.1.3, from Egypt, fourth century, *JGS* 20 (1978) 120, no. 10. A first- to second-century glass amphora that may have been used as a cremation urn in the Egyptian Fayum was placed in a wicker basket with a removable wicker lid: Mainz, RGZM, no. O.40505, F. J. Hassel, "Glasamphore im Deckelkorb," *JRGZM* 33 (1986) 908-909, fig. 94.

### *Lenticular Bottles*

(Cat. Nos. 60-62)

**60 Lenticular Bottle with Two Handles**

Color Plate 12, p. 54
Probably Eastern Mediterranean
First century
H 8.14 cm; H body 5.7 cm.; max D 6.05-3.74 cm; D rim 3.12-3.14 cm; D base 2.12-1.82 cm
1923.532, Gift of Edward Drummond Libbey

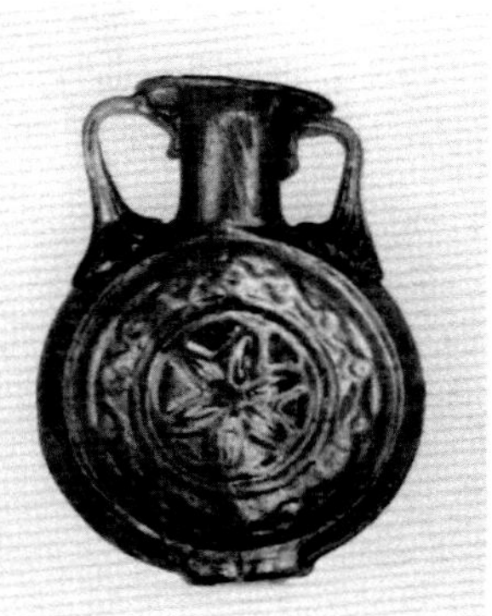

60, side A (1:2)

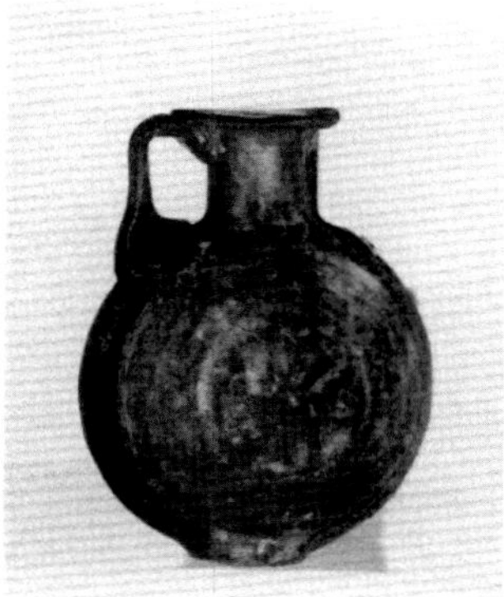

61, side A (1:2)

Translucent brownish yellow. Two translucent blue handles. Thin glass. Large vertically elongated bubbles near rim.
Rim tooled. Neck and body blown into a two-part mold of two circular vertical sections (MCT VIII G). One continuous mold seam along the circumference of body extends across underside of base. Relief moderately crisp on side A, indistinct on side B. Excess glass at tips of handles folded back.
Rim folded outward, upward, and inward, covering nearly half of opening. Cylindrical neck. Lenticular body. Small flat rectangular base. Two coil handles applied to shoulder and attached to underside of rim and top of neck. Handles positioned over mold seams.
On each side of the body, a six-petaled rosette in raised outline, framed on side A by a band of raised linked circles and on side B by a band of raised scrolls, each band bordered by pairs of raised circles.
*Condition* Intact. Exterior covered with purple and golden iridescence; severe surface pitting.
*Collection* Thomas E. H. Curtis.
*Published* Previously unpublished.
*Remarks* Side B from a mold closely related to Cat. No. 61. At least three Series belong to this Type; they have not yet been distinguished from each other in existing publications. A Series of large bottles, with a maximum diameter of 8.2 cm, was carefully executed,[1] while a Series of smaller bottles, to which the Toledo bottles belong,[2] is usually less carefully made. It is not clear whether the variations in diameter (6.1 to 6.5 cm) of this Series are owing to the use of different size molds or to manipulation of the vessels after extraction from the mold. A third Series is even smaller and lacks decoration in the band framing the rosette.[3]
Although a six-petaled rosette is found on a six-sided jug from Yahmour, Syria, which has tentatively been connected with Ennion's style,[4] the motif does not occur on his signed vessels. Ennion did use an elaborate scroll with tendrils on an amphora in St. Petersburg,[5] but scrolls and rosettes are common Hellenistic motifs that were probably used in other workshops as well. Molds for lenticular bottles with a design in shallow relief were already made in the Workshop of the Floating Handles (see Chapter I: *Workshop of the Floating Handles* and Cat. No. 62), which was active during the early first century. However, the handles on this bottle and others of this Type do not float. Bottles of this Type were made in one or more workshops whose other products have not yet been identified.
The workshop that created this Type was probably active in the first century because a fragment of rim, neck, and upper part of body of one bottle was found in a tomb in Jordan that has been dated to the first century.[6] Find spots in Greece, Jordan, and Georgia suggest these bottles were made and used predominantly in the eastern Mediterranean.[7]

[1] For example, a bottle in Berlin, Antikenmuseum, no. 30219, formerly von Gans Collection, with two handles, Platz-Horster 1976, 40, no. 58. The elaborate tendril scroll on this bottle recalls the work of Ennion.

[2] The following examples probably represent more than one Series, although the designs are closely related:
a) from chamber tomb A1 at Petra, Jordan, fragment of upper part of vessel, Horsfield 1941, 140, no. 142, pl. XIX (see also below note 5);
b) Tblisi, Georgia, Mus., no. 1-61:3354, from Urbnis, Georgia, grave 170, transparent purple, Saginashvili 1970, 51, 96, no. 10;
c) BM, no. S 187, formerly Roussel Collection, said to be from the Greek Islands, transparent violet, Nesbitt 1871, 31, no. 187, fig. 44;
d) Newark, no. 50.1463, transparent purple with opaque white handles, Auth 1976, 70, no. 67;
e) Jerusalem, Dobkin Collection, Neuburg 1962, fig. 39;
f) formerly Norton, Mass., Wheaton College, no. G82, transparent blue, Dusenbery 1971, 13, no. 10, fig. 7;
g) Tel Aviv, no. MH 63158, transparent blue, Israeli 1964, 40-41, no. 10, fig. 15;
h) formerly Kofler-Truniger Collection, transparent blue, *Kofler-Truniger* 1985, 63, no. 97;
i) Loeffler Collection, no. 38, transparent yellowish glass with dark blue handles, La Baume and Salomonson 1976, 36, no. 64;
j) MFA, no. 31.132, translucent blue with opaque white handles, von Saldern 1968, no. 32;
k) Oppenländer Collection, no. 2251, transparent natural green, von Saldern et al. 1974, 154-155, no. 436;
l) Los Angeles County Museum of Art, no. M.88.129.49, formerly Hans Cohn Collection, transparent yellowish with blue handles, von Saldern 1980b, 50, no. 43.

[3] For example, MMA, no. 81.10.230, said to be from Greece, transparent dark yellow, H body 5.2 cm, max. D 5.4×3.7.

[4] Harden 1944-45, 82, tomb I, no. 2, pl. V; see also Remarks Cat. No. 63.

[5] St. Petersburg, Hermitage, no. P. 1852.54, Kunina 1973, 102-106, figs. 1-3, cf. Harden 1935, 168, A5a; Lehrer 1979, 6, 7, pl. 2, nos. 1-2.

[6] Petra, Jordan, from chamber tomb A1, Horsfield 1941, 140, no. 142, pl. XIX (according to Barag, transl., 95, the bottle is made of "colorless glass with purple streaks" and the handles are brown; Barag cites Cat. No. 61 as a parallel to this bottle).

[7] A related bottle from the necropolis of Nona, at Zadar, Croatia, now Murano, Italy, Mus. Vetrario, no. I.G. VE 379, shows a spoked star framed by a band of S-shaped scrolls; the mold seam extends across the underside of the base, which is small and rectangular.

### 61 Lenticular Bottle with Two Handles

Probably Eastern Mediterranean
First century
H 7.7 cm; H body 5.8 cm; max D 6.1-3.35 cm; D rim 2.8-2.6 cm; D base 2.05-1.7 cm
1923.533, Gift of Edward Drummond Libbey

Translucent patchy manganese colored glass. Translucent royal blue handle. Thin glass. Fabric cannot be determined because of weathering.
Rim tooled. Neck and body blown into a two-part mold of two circular vertical sections (MCT VIII G). One continuous mold seam along the circumference of the body extends across underside of base. Relief indistinct. Neck irregular in section because of misalignment of mold halves.
Everted rim, rounded in flame. Cylindrical neck. Lenticular body. Small flat rectangular base. Two coil handles applied to shoulder and attached to underside of rim.
On each side of the body, a six-petaled rosette in raised outline, framed by a band of raised scrolls bordered by pairs of raised circles.
*Condition* One handle missing; a hole at point of application.
*Collection* Thomas E. H. Curtis
*Remarks* See Remarks Cat. No. 60. Same Series as Cat. No. 60, side B.

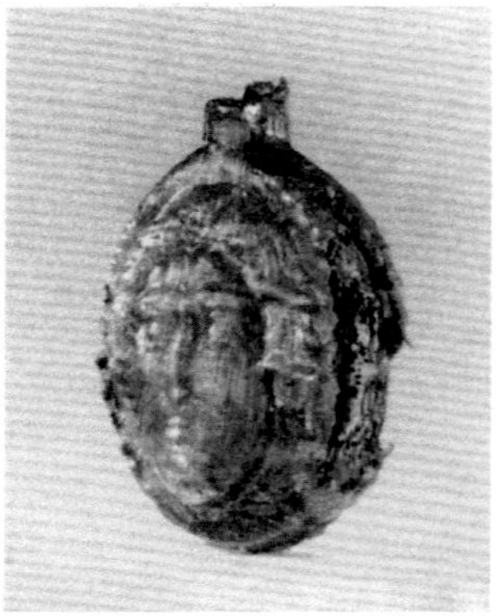

62, side A (1:2)

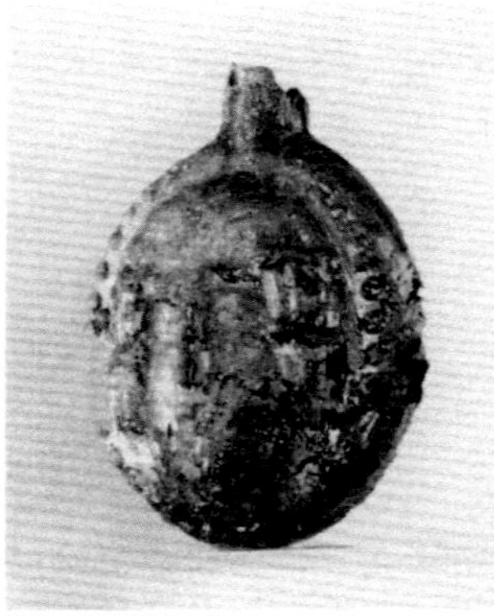

62, side B (1:2)

### 62 **Lenticular Bottle**

Eastern Mediterranean; probably Syro-Palestinian, perhaps made in Sidon
Perhaps Workshop of the Floating Handles
Probably first half of the first century
H (pres). 8.1 cm; max D 6.4 cm; Th (body) 3.3 cm
1923.415, Gift of Edward Drummond Libbey

Translucent natural pale green (5 G 7/2). Medium thin glass. Fabric cannot be determined because of weathering.
Body blown into a two-part mold of two circular vertical sections (MCT VIII F). One continuous mold seam along the circumference of the body. Relief indistinct.
Tubular neck. Circular flat-sided body. Convex base (vessel cannot stand).
On each side of the body, a face of Medusa with similar features: prominent bulging browline, large bulging eyes with recessed pupils, extremely long narrow nose, small mouth with short horizontal lips slightly parted, and heavy jawline wider than the forehead. The hair is parted above the center of the forehead and held back on the sides by narrow horizontal "snakes." Framed on each side by a border of evenly spaced raised knobs.
*Condition* Rim and upper part of neck missing. A photograph of this vessel taken in 1954 and now in the object's documentation file shows a short tubular neck and everted rim still preserved, but no trace of handles. Large holes at either edge of body where handles may have been attached. Cleaned. Original surface lost. Exterior has severe surface pitting, rainbow iridescence, and ice floes. Interior has golden surface pits and areas of dark brown weathering.
*Collection* Thomas E. H. Curtis.
*Published* Previously unpublished.
*Remarks* This vessel belongs to a small group of lenticular bottles showing a frontal Medusa face on either side.[1] The Type was created in the Workshop of the Floating Handles (see Chapter I: *The Workshop of the Floating Handles*), but may have been copied in other workshops. Ten Types of mold-blown glass bottles have been attributed to the Workshop: the Toledo Museum collection has representives of four Types: a miniature amphora cased in wickerwork (Cat. No. 59), an acorn-shaped bottle (Cat. No. 113), a head-shaped vessel (Cat. No. 138), and perhaps this lenticular bottle with a frontal Medusa face on either side. Because the handles are not preserved, it is not possible to determine whether this bottle had floating handles.
The design is related to that of a mold-blown lenticular flask without handles that was excavated at Yahmour, in Syria.[2] Both sides have a curious face that appears clean-shaven and hairy when seen right way up, but bald and bearded when seen upside down. The face is framed by a circle of knobs similar to the frame of dots surrounding the Medusa's face on this piece. The tomb at Yahmour also contained a mold-blown jug like Cat. No. 76 that was made in the first half of the first century. This date corresponds to the time period of the Workshop of the Floating Handles.

[1] See the following examples:
a) Berlin, Staatliche Museen, Antikensammlung, no. M66, Rohde 1980, 161, no. 31, purple, from Miletos, Turkey, grave K 5, first century A.D., handles not preserved;
b) Flushing, N.Y., Queens College, Godwin-Ternbach Museum, no. 60-54, translucent purple, with two floating handles (one missing), Maria Manzari et al., *Ancient and Antique Glass in the Queens College Art Collection* (New York 1976) 9, no. 11;
c) formerly Kofler-Truniger Collection, natural bluish green, with two floating handles (one missing), *Kofler-Truniger* 1985, 59, no. 91; Luzern 1981, 82, no. 277;
d) formerly Rome, Sarti Collection, dark green, with four floating handles, Kisa 1908, III, 752, fig. 288;
e) RGM, no. N300, formerly Niessen Collection, blue green, with two floating handles, Fremersdorf VI 1961, 55, pl. 105 (none of the parallels quoted by Fremersdorf belong to this Type);
f) Tel Aviv, no. MH 72518.

[2] Beirut, Nat. Mus., no. 111, Harden 1944-45, 83, tomb I, no. 3, pl. VI.

## *Hexagonal Bottle*
(Cat. No. 63)

### 63 **Hexagonal Bottle with Two Handles**

Eastern Mediterranean
Probably second half of the first century
H 8.2 cm; H body 6.4 cm; max D 3.85-3.6 cm; D rim 2.3 cm; D base 3.0-2.85 cm
1923.596, Gift of Edward Drummond Libbey

Patchy translucent blue. Two similarly colored handles. Medium thin glass. Fabric cannot be determined because of weathering.
Neck free blown. Body blown into a two-part mold with elongated cross section (MCT VIII J). One continuous mold seam around body and underside of base between scroll panels on front and back. Relief moderately crisp. Handles drawn out thin.
Everted rim, rounded and partially thickened in flame; tool mark on interior. Short tapering neck merges into irregularly convex shoulder. Hexagonal body tapering downward. Flat base. Two coil handles applied to shoulder and attached to underside of rim. Handles positioned over palm frond panels on sides of bottle.
On the shoulder, six lunettes in raised outline; each contains three small circular raised bosses. On the body, six oblong panels framed by four horizontal ribs at top and two horizontal ribs at bottom. Each half of body (as determined by the mold) has a set of three panels. In the panels of each set: 1, a scroll pattern facing right; 2, a palm frond; and 3, a scroll pattern facing left.
*Condition* Complete; broken and repaired. Chip missing from rim; small losses on shoulder and at one corner of bottom. Exterior surface covered with areas of purplish iridescence and light brown weathering.
*Collection* Thomas E. H. Curtis.
*Published* Previously unpublished.
*Remarks* This bottle belongs to one of an unknown number of Series derived from a Type created by Ennion.[1] Each Series appears to be farther removed from the original. This bottle represents the last stage in the development. The first, and also the earliest, Series is represented by a large mold-blown jug excavated in a first-century tomb at Yahmour, Syria.[2] A smaller bottle with two handles, excavated at Constanta (ancient Tomis),[3] represents a second Series. All other bottles are even smaller, varying in height from 9 to 8.2 cm; not one of these can be dated independently.
It is not clear how many Series are represented by four variants in the decorative design of the smallest bottles. The variants could

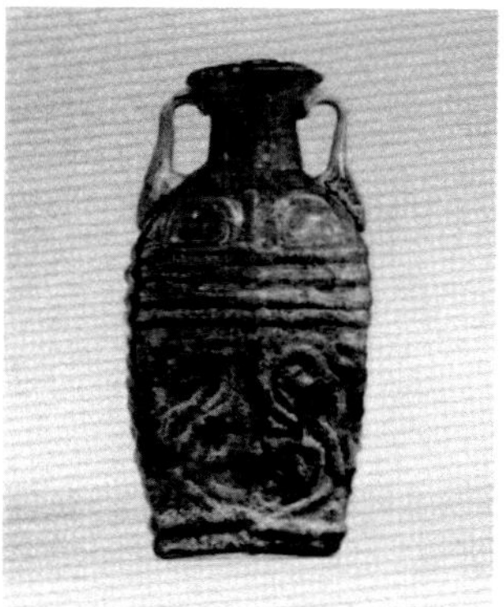

63, panels 3 and 1 (1:2)

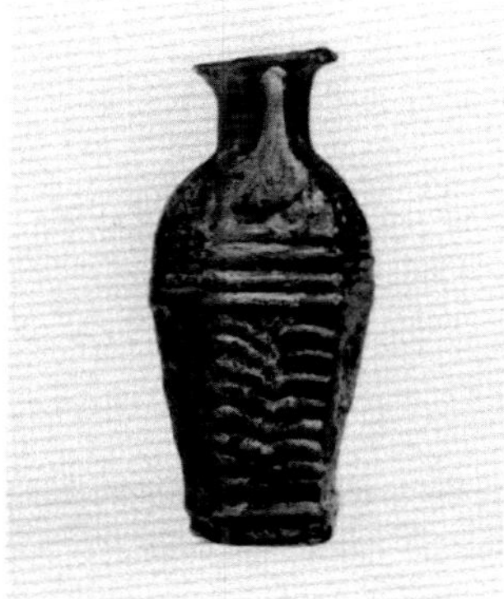

63, panel 2 (1:2)

have resulted either from recutting molds or from depending on different archetypes. The most elaborate, and therefore probably the earliest, variant has stylized leaf tendrils on front and back panels, two heart-shaped ivy leaves in one side panel, and crosshatching in the other side panel; the lunettes on the shoulder appear to contain X-shaped crosses.[4] The second variant has a more slender form with an X-shaped cross in four lunettes and three raised dots in two lunettes; four body panels are decorated with vertical scrolls and two with palm fronds.[5] The decoration of the third variant is similar to this Toledo bottle. This variant is also the most numerous.[6] The fourth variant appears to be a simplified version of the third: the lunettes on the shoulder are empty, and there are only three ribs at the tops of the panels and no ribs below.[7] The position of the handles varies: they are sometimes above the side panels and sometimes over the mold seams. Find spots of all Series point to production in the eastern Mediterranean.[8]

[1] MMA, no. 81.10.224, from Potamia, Cyprus, translucent dark blue, H 14.3 cm, two handles, each with a vertical pinched thumbrest, Lehrer 1979, 9, pl. 3, nos. 1-4, who notes the excavation of a fragment from a similar vessel in the Jewish Quarter of Jerusalem; Israeli 1983, 68-69, fig. 5. The body of the bottle in New York is formed of six rectangular panels tapering downward that show cult objects that have been connected with Dionysiac rites and Jewish festivals (see Remarks Cat. No. 71). The domed shoulder is decorated with lunettes filled alternately with vertical crosses and unidentified circular objects, perhaps libation cups.

[2] Harden 1944-45, 82, tomb I, no. 2, pl. V, H 0.17 cm. The tomb is dated to the middle of the first century by a mold-blown cylindrical jug with floral design similar to Cat. No. 76.

[3] Mihai Bucovala, "Recent Discoveries of Ancient Glass in Romania," *BullAIHV* 8 (1977-80) 225, no. 4, figs. 82, 83, greenish, H 12.6 cm, apparently with a mold seam on the underside of the base.

[4] Newark, no. 50.1451, Auth 1976, 66, no. 60.

[5] Oppenländer Collection, no. 2504, von Saldern et al. 1974, 158, no. 439.

[6] Third variant examples:

a) CMG, no. 59.1.81, formerly R. W. Smith Collection, transparent pale manganese colored, with opaque white handles placed over the mold seams, Smith 1957, 65, no. 83;
b) formerly Moore Collection, with handles over the side panels, Eisen 1927, I, 331, pl. 81a;
c) MMA, no. 12.212.4, with handles over the side panels, Eisen 1927, I, 331, pl. 81b;
d) Yale, no. 1955.6.80, translucent dark blue, with handles over the mold seams, Matheson 1980, 51, no. 129;
e) Switzerland, private collection, translucent blue, with handles over the side panels, Luzern 1981, 79, no. 257.

[7] Murano, Mus. Vetrario, from the necropolis of Nona, Zadar, Croatia, transparent manganese colored, with two dark greenish gray coil handles.

[8] For example:

a) Athens, Arch. Mus., no. 2694, fragmentary bottle found at Tatoi, Attica, Weinberg 1992, 125 and 126, no. 98;
b) MMA, no. 12.212.4, said to be from Beisan (Beth Shean), Israel;
c) bottle that may have come from the Syro-Palestinian coast because it was in the Azeez Khayat Sale, American Art Association, Anderson Galleries, New York, April 1927, 13 and 14, no. 334, is cited by Matheson 1980, 51, at no. 129.

## *Miniature Transport Amphoras*
(Cat. Nos. 64-68)

### 64 Miniature Transport Amphora

Place of manufacture undetermined
Second half of the first century
H 11.0 cm; max D 4.5-4.0 cm; D rim 2.2-2.13 cm
1923.690, Gift of Edward Drummond Libbey

Transparent royal blue. One similarly colored handle and one translucent pale olive (10 Y 6/2) tinted handle. Thin glass. Fabric cannot be determined because of weathering.
Neck free blown. Body blown into a two-part mold of two vertical sections (MCT VIII). One continuous mold seam around body. Edges of mold carefully aligned. Tips of handles drawn out thin and folded back.
Rim folded outward, upward, and inward. Cylindrical neck. Elongated ovoid body with circular cross section. Pointed base. Two coil handles applied to shoulder and attached below rim. Handles positioned adjacent to mold seam.
Body decorated with twenty-seven concentric, horizontal mold-blown ridges from neck to base.
*Condition* Broken and repaired; handles reattached, but most of one handle missing with large loss where it would have been attached to the shoulder. Much of the blue color looks as if it were painted on.
*Collection* Thomas E. H. Curtis.
*Published* Previously unpublished.
*Remarks* The Type imitates large clay transport amphoras. Three glass variants, imitating three distinct classes of clay amphora, exist. This vessel and most of the other glass bottles have a tall slender body in imitation of Graeco-Italic amphoras of the late Hellenistic period (see also Cat. Nos. 65-67).[1] The second variant has an ovoid body tapering abruptly to a point, perhaps in imitation of early first-century amphoras of Dressel Type 19 (see Cat. No. 68). The third variant, not represented in the Toledo Museum collection, has a conical body and a circular flat base.[2]
Numerous variations in size, height of the mold (including or not including part of the neck), number of horizontal ridges cut into the mold, finishing of the base (pointed, rounded, or flat), quality

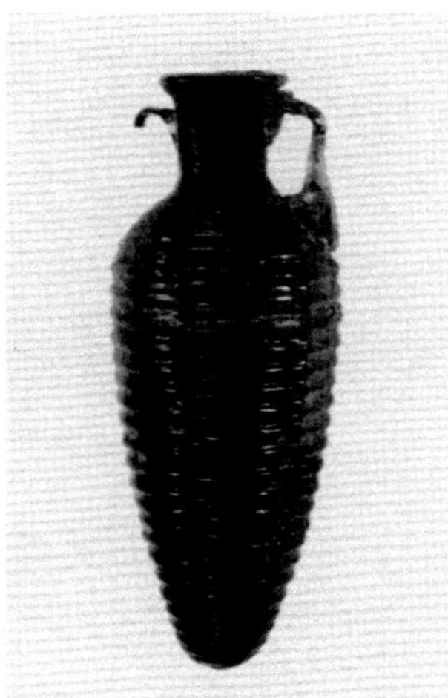

64 (1:2)

of the glass, and care taken in aligning mold parts indicate that a number of workshops produced these popular shapes.[3] The five Toledo examples were all made in different molds and probably represent different Series.

The discovery of two mold-blown amphoras of this Type at Pompeii proves that the Type was created before 79 when the eruption of Mount Vesuvius buried the cities of Campania.[4] A date in the second half of the century is also supported by finds in tombs in Georgia and Armenia.[5]

Although it has been frequently stated that this Type is common in Cyprus and in Syria and that it was made in a Syro-Palestinian workshop,[6] relatively few examples appear to have a Syro-Palestinian or Cypriot provenance.[7] The finds from Georgia, Armenia, and Turkey mentioned above suggest at least one production site was in the eastern Mediterranean. Western Mediterranean finds include two bottles from Pompeii,[8] one from Pavia,[9] one probably found in Aquileia,[10] and several from various sites in Yugoslavia.[11] One bottle from Scupi, in Macedonia, was found in a tomb that also contained two date-shaped bottles imported from the Syro-Palestinian coast (see Remarks Cat. No. 84). It is not clear whether some of these miniature amphoras might have been produced in the West.

[1] Newark, no. 50.1528, dark blue with opaque white handles, H 10.9 cm, Auth 1976, 71, no. 69; the Newark vessel is similar to this piece and to Cat. No. 67. Compare also:
 a) RGM, no. 42,189, H 7.0 cm, La Baume 1973, pl. 45, no. 4;
 b) Loeffler Collection, no. 19, La Baume and Salomonson 1976, 36, no. 63, pl. 7, no. 2;
 c) Spartz 1967, no. 37, pl. 37.

[2] Oppenländer Collection, no. 2345, natural bluish green, with turquoise handles, H 7 cm, von Saldern et al. 1974, 147, no. 425, 151, ill.

[3] Von Saldern et al. 1974, 146-147, nos. 418-425, provides a good sample of the variation in shapes.

[4] Naples, Mus. Naz. Arch., nos. 13676 and 13677, cited by Calvi 1968, 100, n. 187, and by Oliver 1980, 63, at no. 58. The glass amphoras from Pompeii are also mentioned by Ziviello, "Napoli: Museo Archeologico Nazionale," *BullAIHV* 9 (1981-83) 162.

[5] Examples from Georgia and Armenia:
 a) Tblisi, Georgia, no. 58-1910, from Urbnis, tomb 20, ovoid type, H 7.0 cm, Saginashvili 1970, 95-96, no. 9;
 b) Yerevan, Armenia, State History Mus., no. 1964/130, from Garny, tomb 14, yellowish green, tall pointed type, H 13.0 cm, found with two coins of the emperor Augustus (27 B.C.-A.D. 14) and two piriform unguentaria of mid to second-half-of-first-century shape, Arakelian et al. 1969, 58, no. 119;
 c) St. Petersburg, Hermitage, no. K3-6245, from a tomb in Ashnak, reddish violet, ovoid type, H 8.0 cm, Arakelian et al. 1969, 58, no. 120.

[6] Calvi 1968, 100-101; Oliver 1980, at no. 58; Arakelian et al. 1969, 58, nos. 119-120.

[7] Syro-Palestinian examples:
 a) Yale, no. 1934.566e, from Dura Europos, purple fragment, Clairmont 1963, 40-41, no. 154, pl. 21;
 b) formerly Beirut, Kirchner-Schwarz Collection, purple, ovoid type, H 7.3 cm, *Kirchner-Schwarz* 1914, no. 91, pl. 1;
 c) CMG, no. 55.1.73, formerly R. W. Smith Collection, no. 866, acquired in Lebanon, opaque blue, ovoid type, H 8.4 cm, Smith 1957, 65, no. 85;
 d) Damascus, Nat. Mus., Calvi 1968, 100, n. 187;
 e) Beirut, Nat. Mus., Calvi 1968, 100, n. 187.

Oliver 1980, 63, at no. 58, cities dubious Cyprus and Crete examples but the Type does not occur in Vessberg's survey of Cypriot glass: Vessberg 1952:
 f) Nicosia, Cyprus Museum, no. D1858;
 g) Cambridge, Mus. of Archaeology and Ethnology, no. RC.23.793, perhaps from Cyprus, *Fitzwilliam* 1978, 53, no. 57.
 h) Herakleion, Arch. Mus., no. 422;

One example is said to have been found in Turkey:
 i) Istanbul, Kocabaş Collection, from Bursa, Çaliköy, grayish blue, with dark blue rim and handles, H 10.0 cm, *Kocabaş* 1984, 53, no. 77.

[8] See above note 4.

[9] Garlasco (Pavia), Italy, Mus. Arch. Lomellino, from tomb N at Garlasco-Cazzanina, *BullAIHV* 9 (1981-83) 62 (not ill.).

[10] Aquileia, Mus. Arch., transparent blue with blue and white flecks, ovoid type, H 7.5 cm, Calvi 1968, 105, no. 247, pl. 16, no. 5.

[11] Examples from Yugoslavia:
 a) from Scupi, Macedonia, tomb 137, dated to the reigns of Domitian through Hadrian (81-138), Mikulcik 1974, 200-201, 206, pl. 4, no. 382;
 b) from Budva, Montenegro, Cermanović-Kuzmanović 1974, 186, pl. 1, no. 4;
 c) from Doclea, Montenegro, Cermanović-Kuzmanović 1974, 188, pl. 3, no. 16;
 d, e) Murano, Mus. Vetrario, nos. I.G. VE 384 and I.G. VE 385, from the Nona necropolis at Zadar, Croatia, both transparent manganese colored, made in MCT VIII molds.

### 65 Miniature Transport Amphora

Place of manufacture undetermined
Second half of first century
H (pres.) 10.6 cm; max D 4.67-4.57 cm; D rim 2.88-2.86 cm
1923.688, Gift of Edward Drummond Libbey

Transparent to translucent grayish yellow (near 5 GY 7/2). Dark yellow streaks. Opaque blue handles. Thin glass. Small oval bubbles in body.

Neck free blown. Body blown into a two-part mold of two vertical sections (MCT VIII). One continuous mold seam around body. Edges of mold poorly aligned on one side. Tips of handles drawn out thin and folded back.

Rim folded outward, upward, and inward. Tall neck tapering downward. Elongated ovoid body with circular cross section. Pointed base. Two coil handles applied to shoulder and attached to underside of rim. Handles positioned over mold seams.

Body decorated with twenty-three concentric, horizontal mold-blown ridges from neck to base.

*Condition* Broken and restored; small losses in neck, body, and tip of base.

*Collection* Thomas E. H. Curtis.

*Published* Previously unpublished.

*Remarks* See Remarks Cat. No. 64. Fabric and color of handles are very similar to free-blown miniature jar TMA 1923.703.[1]

[1] TMA, *Roman and Early Byzantine Glass* (forthcoming).

### 66 Miniature Transport Amphora

Place of manufacture undetermined
Second half of first century
H 10.6 cm; max D 4.5-4.45 cm; D rim 3.18-3.05 cm
1923.686, Gift of Edward Drummond Libbey

Translucent to transparent pale olive (near 10 Y 6/2). Two translucent pale blue handles. Thin glass. Fabric cannot be determined because of weathering.

Neck free blown. Body blown into a two-part mold of two vertical sections (MCT VIII). One continuous mold seam around body. Edges of mold carefully aligned. Tips of handles drawn out thin and folded back.

Rim folded downward, outward, upward, and inward. Cylindrical neck. Elongated ovoid body with circular cross section. Pointed base. Two coil handles applied to shoulder and attached to underside of rim. Handles positioned adjacent to mold seam.

Body decorated with twenty-four concentric, horizontal mold-blown ridges from neck to just above base.

*Condition* Intact. Interior of neck and entire exterior covered with

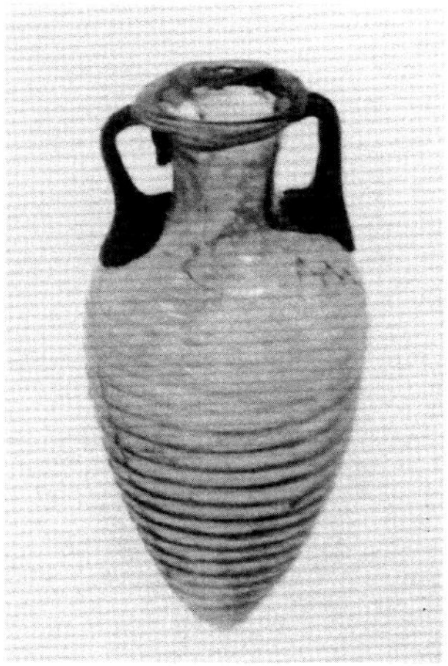

65 (1:2)

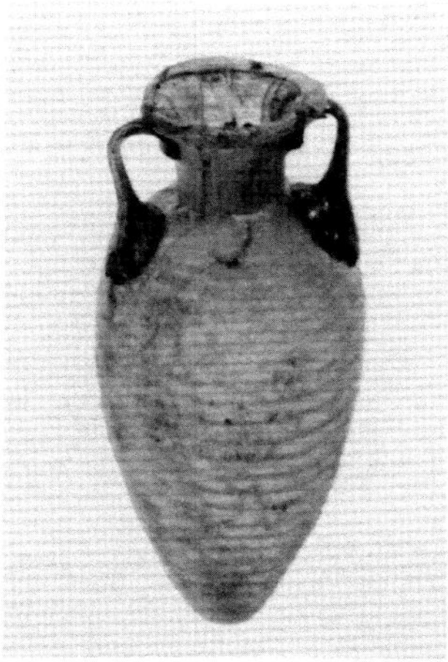

66 (1:2)

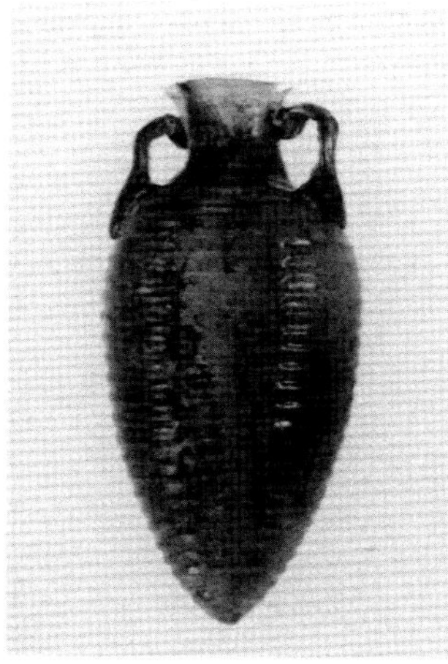

67 (1:2)

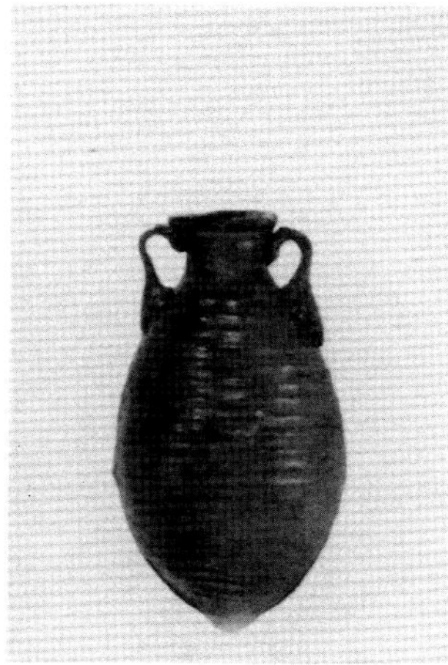

68 (1:2)

rainbow iridescence. Traces of golden iridescence on rim, neck, handles, and upper part of body. Surface pits and frosting.
*Collection* Thomas E. H. Curtis.
*Published* Previously unpublished.
*Remarks* See Remarks Cat. No. 64.

### 67 **Miniature Transport Amphora**

Place of manufacture undetermined
Second half of the first century
H 10.15 cm; max D 4.6 cm; D rim 2.48 cm
1923.687, Gift of Edward Drummond Libbey

Transparent to translucent streaked manganese colored glass. Similarly colored handles. Thin glass. Fabric cannot be determined because of weathering.
Neck free blown. Body blown into a two-part mold of two vertical sections (MCT VIII). One continuous mold seam around body. Edges of mold not carefully aligned. Tips of handles drawn out thin and folded back.
Everted rim, partly folded inward and downward. Cylindrical neck. Elongated ovoid body with circular cross section. Pointed base. Two handles, one bifurcated, the other trifurcated, applied to shoulder and attached halfway up neck. Handles positioned some distance from mold seam.
Body decorated with twenty-six concentric, horizontal mold-blown ridges from neck to base.
*Condition* Small chip missing from rim. Exterior and interior of rim and neck covered with purplish blue iridescence; surface pits and oily frosting.
*Collection* Thomas E. H. Curtis.
*Published* Previously unpublished.
*Remarks* See Remarks Cat. No. 64.

### 68 **Miniature Transport Amphora**

Place of manufacture undetermined
Second half of the first century
H 7.88 cm; max D 4.37-4.15 cm; D rim 2.1 cm
1923.689, Gift of Edward Drummond Libbey

Transparent to translucent manganese colored glass. Translucent pale blue handles. Thin glass. Fabric cannot be determined because of weathering.
Neck free blown. Body blown into a two-part mold of two vertical sections (MCT VIII) with an unintentional diagonal crack on both sides. One continuous mold seam around body. Edges of mold carefully aligned. Tips of handles drawn out thin and folded back.
Everted rim, partly folded inward and downward. Concave neck with curved transition to shoulder. Ovoid body. Pointed base. Two coil handles applied to shoulder and attached below rim. Handles positioned over mold seam.
Body decorated with twenty-two concentric, horizontal mold-blown ridges from neck to just above base.
*Condition* Handles broken off and reattached; surface around attachments altered. Areas of copper-colored crust, especially below one handle.
*Collection* Thomas E. H. Curtis.
*Published* Previously unpublished.
*Remarks* See Remarks Cat. No. 64.[1] The raised diagonal line on the body indicates a crack in the mold through which the glass protruded.

[1] For ovoid shape tapering abruptly to a point, like early first-century clay amphoras of Dressel type 19, compare:
a) St. Petersburg, Hermitage, no. $K^3$-6245, from a tomb in Ashnak, Armenia, Arakelian et al. 1969, 58, no. 120;
b) Aquileia, Mus., probably from Aquileia, Calvi 1968, 105, no. 247;
c) ROM, no. 950.157.177, Hayes 1975, 48, no. 85;
d) Bologna, Mus. Civ. Arch., no. Rom. 57, Meconcelli Notarianni 1979, 152-153, no. 211;
e-h) Oppenländer Collection, nos. 22270, 2343, 22215, and 22333, von Saldern et al. 1974, 146-147, nos. 420-423.

## *Barrel-shaped Bottles*
(Cat. Nos. 69 and 70)

### 69 **Barrel-shaped Bottle with Two Handles**

Probably Syro-Palestinian
Probably second half of the first century
H 8.3 cm; H body 5.2 cm; max D 6.1-5.35 cm; D rim 2.35-2.25 cm
1923.682, Gift of Edward Drummond Libbey

Translucent royal blue. Two similarly colored handles. Thin glass. Small spherical bubbles in neck.
Neck free blown. Body blown into a two-part mold of two vertical sections (MCT VIII K). One continuous mold seam follows central rib around body. Tips of handles drawn out thin.
Rim partly folded outward, upward, and inward. Tall concave neck. Ovoid body. Convex base. Two coil handles applied to shoulder and attached just above midpoint of neck. Handles

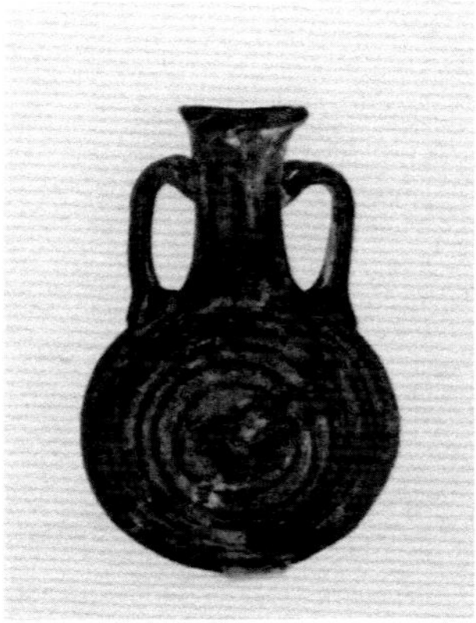

69 (1:2)

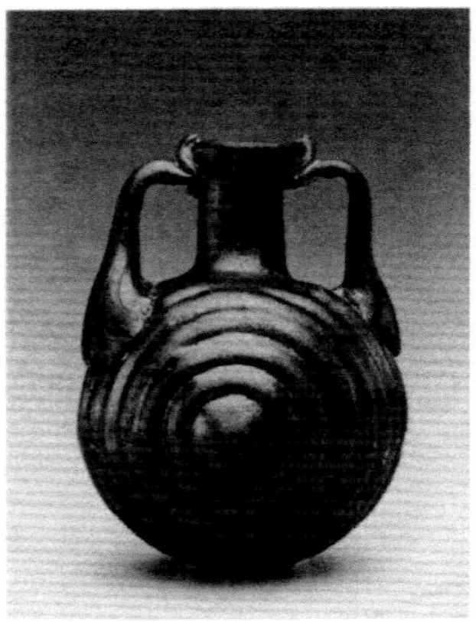

70, end (1:2)

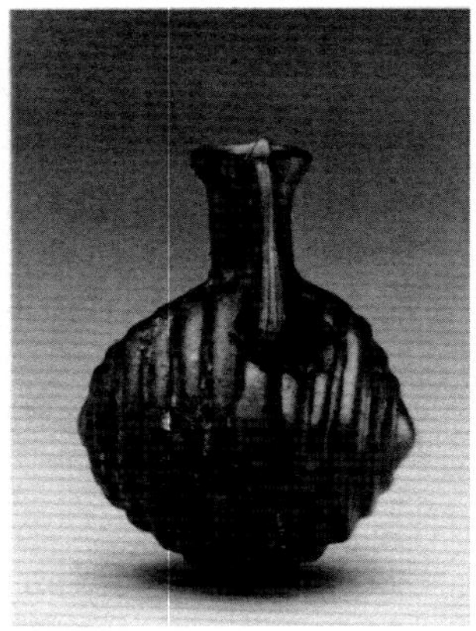

70, side (1:2)

positioned at an oblique angle to mold seam (one handle twisted in an attempt to correct the problem).
On each side of the body, an eight-petaled rosette in raised relief at the center of sixteen concentric circular ribs. Ribs often roughly indicated.
*Condition* Intact. Areas of buff-colored weathering on exterior and interior. Some iridescent film.
*Collection* Thomas E. H. Curtis.
*Published* Previously unpublished.
*Remarks* The mold-blown ribs of this bottle imitate the hoops that hold together the staves of a wooden barrel. Mold-blown bottles shaped like a wooden wine barrel were known in the second half of the first century, as shown by the excavation of a large example in a tomb at Akko on the coast of Israel.[1] The Akko bottle is made of blue glass and has twenty ribs. No mention is made in the description of a rosette on the ends, but published photographs show this detail. Two Toledo bottles, this one and Cat. No. 70, represent two Series that are smaller editions of the design of the Akko bottle. One Series has a rosette on both ends, like this bottle;[2] the other has plain ends, like Cat. No. 70.
Miniature wine barrels are also known in ceramic and bronze.[3] Glass blowers in Gaul and Germany revived making barrel-shaped vessels in the third century.[4]

[1] From Akko, tomb A 61,22, H 15.0, L 14.4 cm, Maria Teresa Fortuna, "I vetri soffiati della necropoli di Akko," *JGS* 7 (1965) 19-20, figs. 6 and 7; Maria Teresa Fortuna, *Campagne di scavo ad Akko 1961-1962 (Memorie dell'Istituto Lombardo — Accademia di scienze e lettere, classe di lettere, scienze morali e storiche* 29, fasc. 4, Milan 1966) 472, figs. 22 and 23. In addition to the barrel-shaped bottle, the tomb contained a glass lamp shaped like a clay lamp (like TMA 1923.542) and numerous glass piriform unguentaria, all of which suggest an early Flavian date, about 69-79: Barag transl. 81-82.

[2] For example:
a) formerly Kofler-Truniger Collection, transparent natural greenish yellow, *Kofler-Truniger* 1985, 65, no. 104;
b) CMG, no. 74.1.21, formerly R. W. Smith Collection, no. 313, opaque pale blue, with two handles applied to the shoulder and attached about halfway up the neck, Smith 1957, 65, no. 84.

[3] Hilgers 1969, 54-56, 164-165, no. 130, s.v. *cupa*; La Baume 1964, 157ff.

[4] Clasina Isings, "Glazen tonnetjes," *Antiek* 5 (1970-71) 137ff.; Morin-Jean 1913, 176-178, Forms 133-135. Compare: Strasbourg, Mus. Arch., Véronique Arveiller-Dulong and Jacques Arveiller, *Le Verre d'époque romaine au musée archéologique de Strasbourg* (Paris 1985) 139, Ch. 67, no. 304, fig. 304 (with parallels); RGM, no. 5378, from Cologne, Fremersdorf VI 1961, 32, pl. 35 (with parallels).

### 70 Barrel-shaped Bottle with Two Handles

Color Plate 12, p. 54
Probably Syro-Palestinian
Probably second half of the first century
H 7.55 cm; H w handle 7.75 cm; H body 5.15 cm; max D 5.67-5.53 cm; D rim 2.15 cm
1923.683, Gift of Edward Drummond Libbey

Translucent dark blue. Two similarly colored handles. Medium thin glass. Fabric cannot be determined because of weathering. Neck free blown. Body blown into a two-part mold of two vertical sections (MCT VIII). One continuous mold seam follows central rib around body. Tips of handles not drawn out; they could have been snapped off when the glass cooled or cut off with shears. Everted rim folded outward, upward, and inward. Cylindrical neck. Ovoid body. Convex base. Two coil handles applied to shoulder and attached to rim. Handles positioned over mold seam. On each side of the body, sixteen concentric circular ribs. Center of each side left undecorated.
*Condition* Intact. Exterior of body covered with green, purple, and rainbow iridescence. Severe surface pitting.
*Collection* Thomas E. H. Curtis.
*Published* Previously unpublished.
*Remarks* See Remarks Cat. No. 69.[1]

[1] Similar examples, with undecorated ends:
a) Newark, no. 50.1484, Auth 1976, 199, no. 331;
b) Carnegie, no. 25139/1, Oliver 1980, 63, no. 57;
c) Oppenländer Collection, no. 2487, without handles, von Saldern et al. 1974, 150-151, no. 435;
d) Tel Aviv, no. MHG 63358.

## *Jugs*
(Cat. Nos. 71-78)

### *Hexagonal Jugs with Dionysiac Symbols*
(Cat. Nos. 71-74)

### 71 Hexagonal Jug with Dionysiac Symbols

Syro-Palestinian, perhaps made in Sidon; or possibly Galilean
Probably second quarter of the first century
H 9.8 cm; H body 6.3 cm; max D 5.1 cm; D rim 3.0 cm; D base 3.1 cm
1923.484, Gift of Edward Drummond Libbey

Translucent to transparent pale olive (10 Y 6/2). Similarly colored handle. Medium thin glass. Fabric cannot be determined because of weathering.

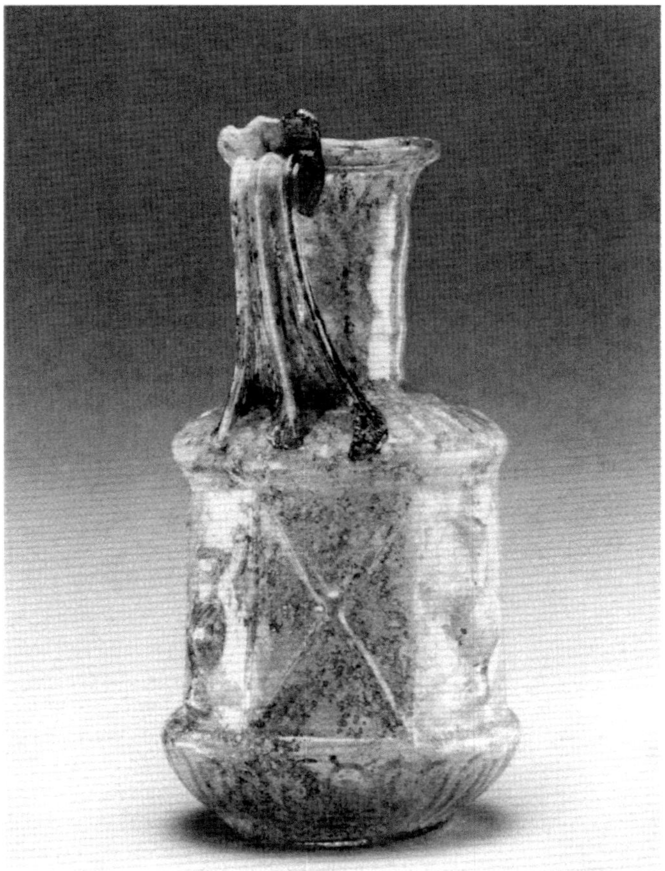

71, panel 1 (1:1)

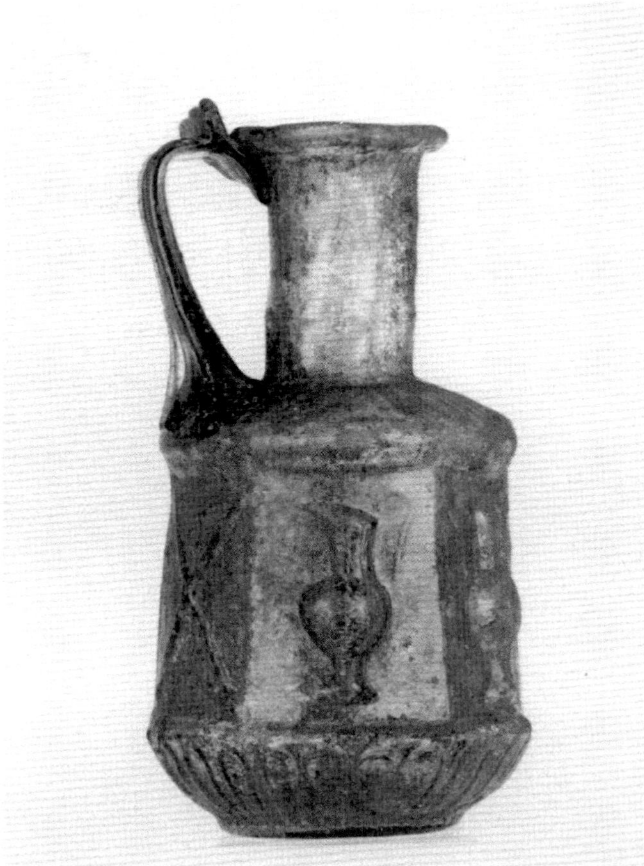

71, panel 2 (1:1)

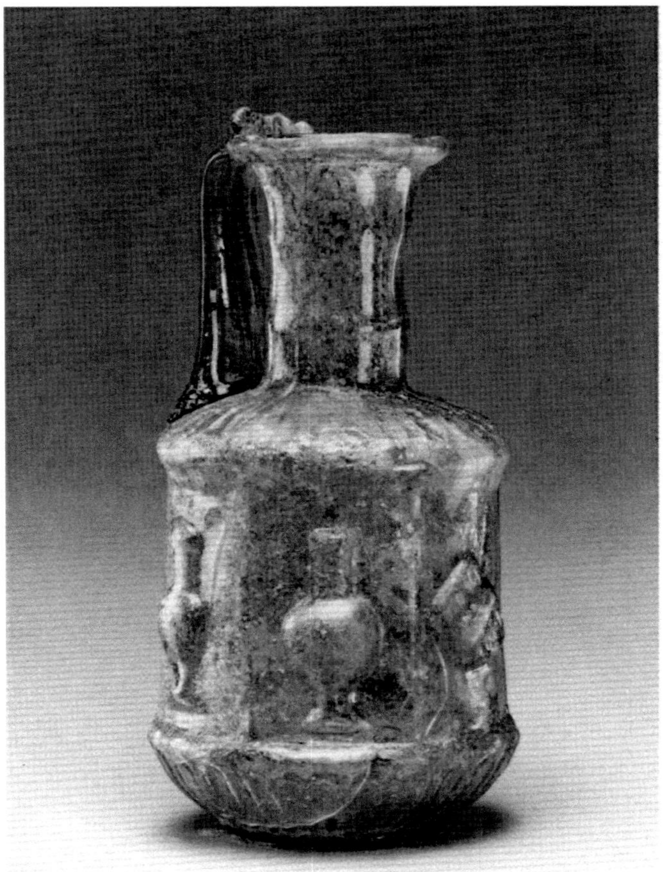

71, panel 3 (1:1)

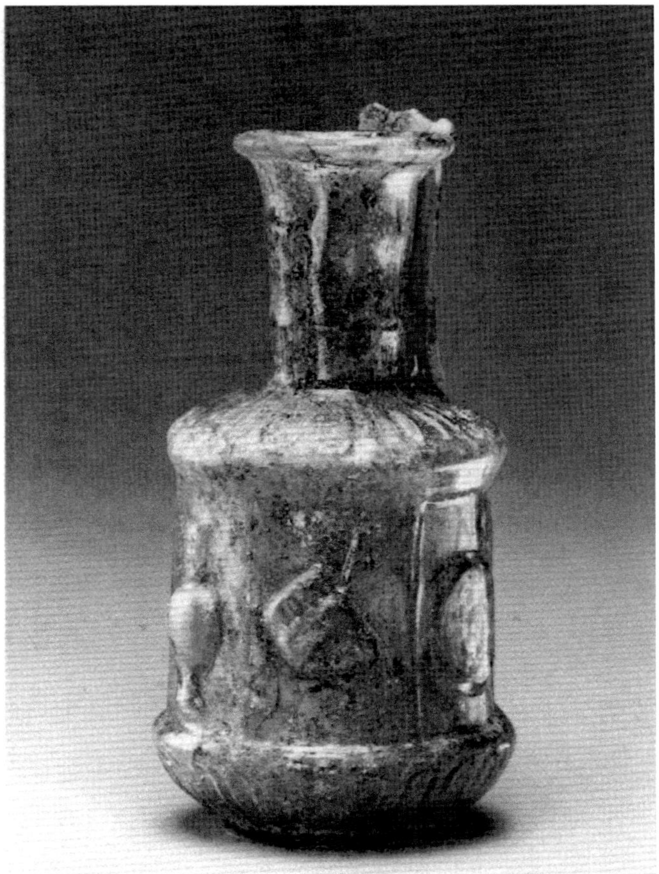

71, panel 4 (1:1)

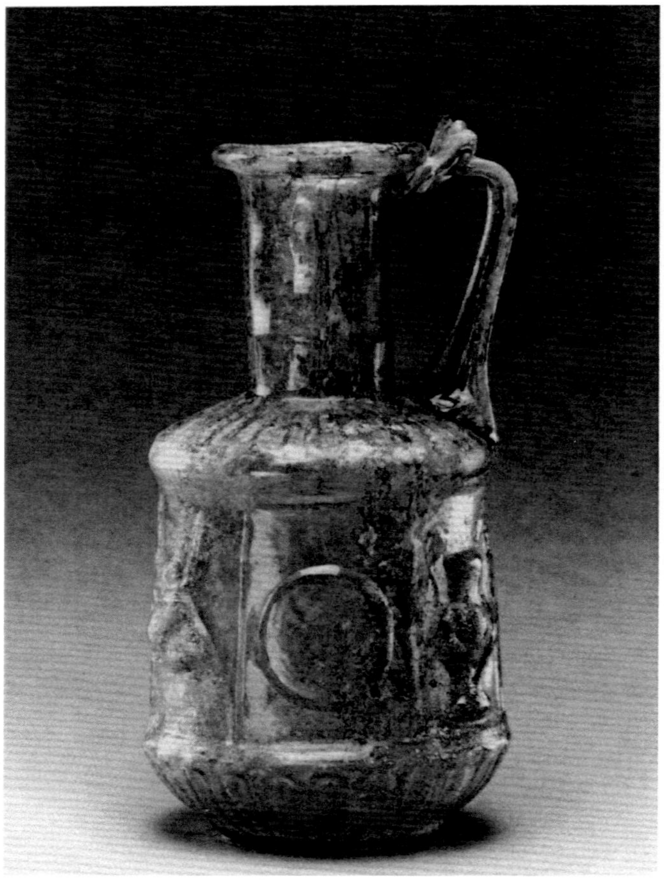

71, panel 5 (1:1)

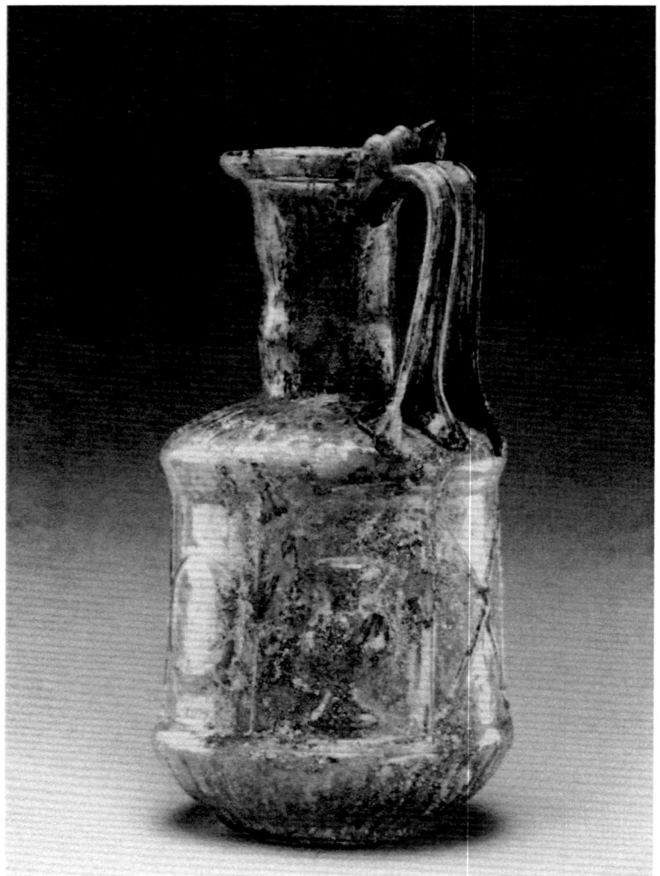

71, panel 6 (1:1)

Neck free blown. Body and lower part of neck blown into a multipart mold of three or two vertical sections joined to a cup-shaped base section which included the base and the fluted area around the bottom (MCT I or MCT V, see Remarks). Mold seams concealed within the decoration. Tip of handle drawn out thin and folded back.

Rim folded outward, upward, and inward; horizontal tool mark around interior of neck, 0.9 cm above base of neck, probably from opening up neck after extraction from the mold. Cylindrical neck; the ridge ca. 0.8 cm above its base marks the upper edge of the mold. Convex shoulder and convex bottom joined by a hexagonal body. Flat base with three raised concentric circles on the underside. Trifurcated handle applied to shoulder and attached to rim with projecting thumbrest above. Handle positioned above panel 1. On the shoulder, a frieze of about twenty-one downturned tongues in raised outline. On the body, six rectangular panels, each containing a Dionysiac symbol. In the panels, from left to right: 1, crossed double-ended thyrsoi; 2, a footed jug with a spout to the left and a high handle to the right; 3, a footed amphora with high handles; 4, a suspended syrinx (pan pipes) with four reeds of increasing lengths and a thick crossbinding; 5, a circular object with smooth surface (probably a phiale);[1] 6, a footed crater with high handles. Around the bottom, a frieze of twenty-four upturned tongues in raised outline.[2]

*Condition* Intact; body cracked.

*Collection* Thomas E. H. Curtis.

*Published* Previously unpublished.

*Remarks* This and the following jugs (Cat. Nos. 72-74) have a hexagonal body divided into six panels, each of which shows an object in high relief. The height of the relief and the fact that the objects represented belong to one class, in this case cult objects, connect these vessels with the large family of small Hexagonal Bottles with High Relief (Cat. Nos. 15-47). Unlike those bottles, which rarely have a handle, the body of this Type is always larger and it is usually finished as a one-handled jug.[3] Five variations in the decorative scheme of the panels have been noted. They affect both the order and details in the actual representation of objects, but the objects themselves appear to be the same on all vessels.[4] It has not been established whether the variants represent individual Series, each derived from a different archetype, or whether the variants are mutations of one or more Series.

### Technique

Mold seams between the panels on the four Toledo jugs are so well concealed that it is not always possible to distinguish whether the molds had three vertical sections (MCT I) or two (MCT V).[5] Whichever of the two constructions was used, they set these jugs apart from the Hexagonal Bottles with High Relief which were blown only in Mold Construction Types II, III, and V A (see *Glossaries* and Cat. Nos. 15-47). The use of a separate cup-shaped section for the bottom and base links these jugs to small cylindrical jugs with low relief blown in MCT I (see Fig. 15 and Cat. No. 76) and to small cylindrical bottles with low relief blown in MCT V C (see Fig. 15 and Cat. No. 48).

The locking devices of the molds must have been quite sophisticated in order to achieve such a perfect fit that it prevented the escape of any glass during blowing. It is not even always possible to distinguish whether the vertical sections included the fluted area around the shoulder of the vessel.[6] This perfection in mold construction suggests a dedication to high standards of quality[7] shared by the workshop(s) blowing these hexagonal jugs and master glass blowers such as Ennion, who is recognized for having taken great care in concealing the mold seams on his vessels.[8]

### Distribution and Origin

Jugs of this Type[9] have been found throughout the Eastern Mediterranean: five in the Syro-Palestinian area,[10] one each in Armenia[11] and the north coast of the Black Sea,[12] and four in Greece.[13] The distribution pattern recalls that of the Hexagonal Bottles with High Relief, Vessels Type, Series A (Cat. Nos. 15-33) and suggests that the jugs were made in or near Sidon. This hypothesis is supported by the iconography of the objects represented in the panels, which points to an origin in Sidon or Beth Shean (see below: *Interpretation*).

### Date

Although a number of jugs were found during excavations (see above), there is no archaeological evidence for dating this Type more narrowly within the first century. A fragment from Knossos, Crete, was found in a Neronian (54-68) or Flavian (69-96) context.[14] The sophisticated construction of the molds used for blowing and the stylistic connections with Hexagonal Bottles with High Relief and with Ennion suggest the jugs were made predominantly in the first half of the first century, perhaps in the second quarter of the century.

### Interpretation

In the past, opinions have been divided as to whether the imagery of these jugs is Dionysiac[15] or Jewish.[16] A Jewish interpretation can be ruled out because the X-shaped design on panel 1, which has previously been interpreted as thyrsoi,[17] torches,[18] scepters,[19] or flutes,[20] can now be identified beyond reasonable doubt as crossed double-ended thyrsoi. The thickened ends of the thyrsoi are barely visible in Cat. No. 71 but very clear in Cat. Nos. 72-74. The double-ended thyrsos, known in antiquity as a *thyrsolongchos*, or spear thyrsos, is first attested in the third century B.C. in Kallixeinos's description of a procession of the Egyptian ruler Ptolemy Philadelphos (285-246 B.C.) in Alexandria (*FGrHist* F 2, 3). One pine cone conceals the spear point, the other conceals the *sauroter*, the spike or ferrule at the lower end of a *longche*, or spear, by which it was stuck into the ground.[21] The *thyrsolongchos* is connected with Dionysos's mythical campaign to India when his followers disguised their spears as thyrsoi, a story that gained prominence after Alexander the Great's Indian campaign in 327 B.C. Depictions of the *thyrsolongchos* are rare in Greece and Italy but common in the Near East, where it appears on coins minted in Sidon, Raphia (Palestine), and Nysa-Skythopolis (Beisan, modern Beth Shean, Israel), as well as on other objects.[22] Crossed *thyrsolonghoi*, as on these glass jugs, are not common. They are found prominently on two panels of a stone altar to Dionysos recently excavated at Beth Shean, Israel, where the pine cones at both ends are executed in detail (see Fig. 56),[23] and on two Types of glass vessels: these Hexagonal Jugs with Dionysiac Symbols (Cat. Nos. 71-74) and an Early Hexagonal Bottle with Mixed Symbols (see Chapter I: *Syro-Palestinian Styles: Hexagonal Bottles with High Relief: Mixed Symbols Type* and Cat. No. 43 *Remarks*). The altar carries an inscription in a rectangular frame saying it was dedicated to Dionysos, "the founder." The epithet refers to a local tradition that Dionysos had founded the city in Galilee known in antiquity as Nysa-Skythopolis and there buried his nurse Nysa. Throughout antiquity Nysa-Skythopolis remained a polytheistic Greek city surrounded by Jewish and Christian communities. Perhaps crossed *thyrsolongchoi* were associated with a local cult of Dionysos at Nysa-Skythopolis.

The function of these jugs has not yet been established, but the archaeological context of a Hexagonal Jug with Dionysiac Symbols from Ayios Stephanos, Greece, illustrates how such jugs may have been associated with Dionysos by their owners.[24] It was found with three tombstones surrounding an altar-like structure. The inscription on one of the tombstones says that the deceased was a young man who died at age 22 after having drunk too much undiluted wine. One cannot help but wonder whether the jug was chosen as a grave gift because the symbols depicted in the panels refer to Dionysos as the god of wine. In addition, a connection with the Eastern Mediterranean is given by the mention, in the inscription, that the young man's father originated from Ephesos in Asia Minor.

1 Eisen 1927, I, 251 identified this object as a circular plate for meat, a suggestion that has not been accepted by other scholars. Oliver 1980, 53, no. 61, identified it as a cymbal, which is unlikely because in antiquity cymbals are usually depicted in profile as, for example, on a silver cup in Toledo (here Fig. 57). Nor is a tambourine or tympanon a likely interpretation, because these instruments are usually shown with small side attachments (see Fig. 57). Although the circular object here has a smooth surface, as is to be expected for a tambourine, it is shown with a series of concentric circles on most other bottles of this Type (see Cat. Nos. 72-74); these circles are a well-known Hellenistic convention for depicting a libation cup or phiale: Lehrer 1979, 9.

2 There are twenty-one individual tongues as well as three contiguous ones which bridge the space between beginning and end of the pattern.

3 For a vessel of the same Type finished with two handles: formerly Kofler-Truniger Collection, translucent cobalt blue, H 9.1 cm, *Kofler-Truniger* 1985, 67, no. 107.

4 In addition to the three variants represented by the Toledo jugs, a fourth variant shows the circular object (with concentric circles) suspended by ties, followed in the next panel by a kantharos suspended by a tie from each handle, for example:

a) Carnegie, no. 24055/6, Oliver 1980, 61, no. 53;
b) MFA, no. 31.130, said to have been found near Tyre, von Saldern 1968, no. 31;
c, d) formerly Kofler-Truniger Collection, *Kofler-Truniger* 1985, 66-67, nos. 106 and 107.

A fifth variant shows objects suspended as in Cat. Nos. 73 and 74 but in an arrangement that follows the order of objects shown in Cat. No. 72 on which the objects are not suspended:

e) Yale, no. 1955.6.272, Matheson 1980, 46-47 no. 123;
f) Philadelphia, The University Mus., no. MS 5016, acquired at Aleppo (unpublished).

5 Cat. No. 72 was blown in an MCT V C mold.

6 The two vertical sections of the MCT V C mold used for Cat. No. 72 probably included one-half of the shoulder.

7 If the molds were not made in glass-blowing workshops, the connection could have been the use of a common expert mold maker.

8 On Ennion's molds: Israeli 1983.

9 In the following discussion of distribution it was not possible to distinguish between variants.

10 For example:

a) MMA, no. 81.10.221, formerly Charvet Collection, said to be from "Syria," translucent purple, with an opaque white handle, Froehner 1879, 64, n. 1, pl. IV, no. 18;
b) Carnegie, no. 26713/5, said to be from Jordan, purple, Oliver 1980, 61, no. 54;
c) see above note 4b;
d) CMG, no. 74.1.23, formerly R. W. Smith Collection, no. 636, acquired in Lebanon, Smith 1957, 63, no. 79;
e) BM, no. GR 1913.5-22.10, cited by Matheson 1980, 47, no. 123.

11 Yerevan, Armenia, State Hist. Mus., no. 1964/121, from Garny, tomb 7, fragmentary bluish juglet, with a greenish handle, H 8.0 cm, Arakelian et al. 1969, 54-55, no. 106, fig. 106; judging from the photograph, the relief of the panels is very indistinct and the shoulder and bottom portions may lack decoration.

12 Sorokina 1967, 70, fig. 2, nos. 20 and 23.

13 For example:

a) from Ayios Stephanos, Chalkis, Boeotia, *ArchDelt* 28 (1973) 300-301, pl. 259c; *AAA* 7, no. 1 (1974) 28, 31, fig. 7;

b) MMA, no. 81.10.218, from Athens, translucent blue, with a blue handle, formerly Charvet Collection, Froehner 1879, 64, n. 1, pl. XXVIII, no. 114;
c) BM, no. S 181, formerly in the Roussel Collection, said to be from the Greek Archipelago, Nesbitt 1871, 30, no. 181;
d) from Knossos, Crete, Price 1988b, 34, fig. 3, no. 28.

14 See above note 13d.

15 Oliver 1980, 61, at no. 53.

16 A. Engle, "The Sidonian Glassmaker's Market," in *Readings in Glass History* 11 (Jerusalem 1980) 23-27, connects the jugs with the Jewish Festival of the Tabernacles, Sukkoth.

17 Oliver 1980, 61, at no. 53; Matheson 1980, 46 and 47, at no. 123.

18 Engle (see above note 16), 24.

19 Eisen 1927, I, 250-251 (with lit.).

20 Kisa 1908, III, 717.

21 Erika Simon, pers. com.

22 *LIMC*, III, s.v. *Dionysos in periferia orientali*, 516, nos. 6, 9c; 521, nos. 71 a, b (coins); 522, no. 81 (Antakya, Turkey, Arch. Mus., no. 958, floor mosaic from Antioch); no. 142 (Paris, Musée du Louvre, no. X4792, tapestry from Antinoopolis [modern Sheikh Abadeh], Egypt); K. Schauenburg, "Pluton und Dionysos," *JdI* 68 (1953) 70, fig. 23 (fragmentary marble funeral relief in Bologna, Museo Civico, said to be from Alexandria, Egypt).

23 Foerster and Tsafrir 1988, 31, photo 18. The other panels of the altar show masks of Dionysos and Pan, a syrinx, and a pedum, all of which are familiar Dionysiac motifs. The syrinx appears also on the Hexagonal Jugs with Dionysiac Symbols (Cat. Nos. 71-74).

24 See above note 13a.

## 72 Hexagonal Jug with Dionysiac Symbols

Syro-Palestinian, perhaps made in Sidon; or possibly Galilean
Probably second quarter of the first century
H 9.5 cm; H w handle 10.0 cm; H body 6.6 cm; max D 5.3 cm; D rim 2.85 cm; D base 2.85 cm
1923.487, Gift of Edward Drummond Libbey

Translucent to transparent light grayish blue. Similarly colored handle. Thin glass. Fabric cannot be determined because of weathering. Neck free blown. Body blown into a multipart mold of two vertical sections, each including one-half of the shoulder, joined to a cup-shaped base section which included the fluted area around the bottom (MCT V). Mold seams concealed within the decoration. Tip of handle drawn out thin.

Rim folded outward, upward, and inward. Tall cylindrical neck. Convex shoulder and convex bottom joined by a hexagonal body. Flat base with two raised concentric circles around a central knob on the underside. Bifurcated handle applied to shoulder and attached to rim with projecting thumbrest above. Handle positioned above panel 3. On the shoulder and the lower part of neck, a frieze of twenty-two thick ribs. On the body, six rectangular panels, each containing a Dionysiac symbol. In the panels, from left to right: 1, crossed double-ended thyrsoi over a raised knob; 2, a footed jug with a spout to the left and a high handle to the right; 3, a footed amphora with high handles; 4, a syrinx (pan pipes) with six reeds of the same length and two narrow crossbindings; 5, a circular object decorated with concentric rings (probably a phiale); and 6, a footed crater with high handles. Around the bottom, a frieze of twenty-four upturned tongues in raised outline.

*Condition* Lengthwise section of handle missing. Golden rainbow iridescence and iridescent film on exterior. Frosting and surface pits.

*Collection* Thomas E. H. Curtis.

*Published* Previously unpublished.

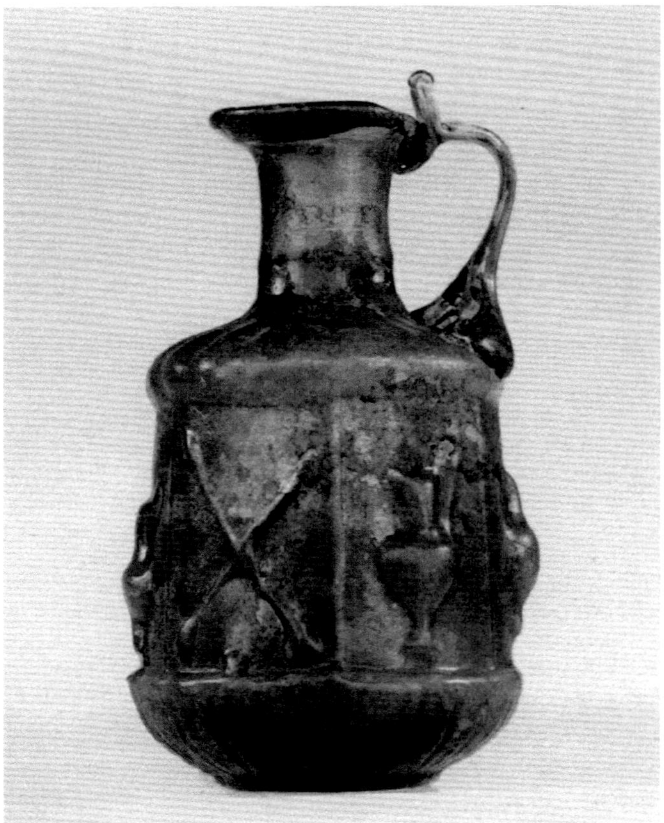

72, panels 1 and 2 (1:1)

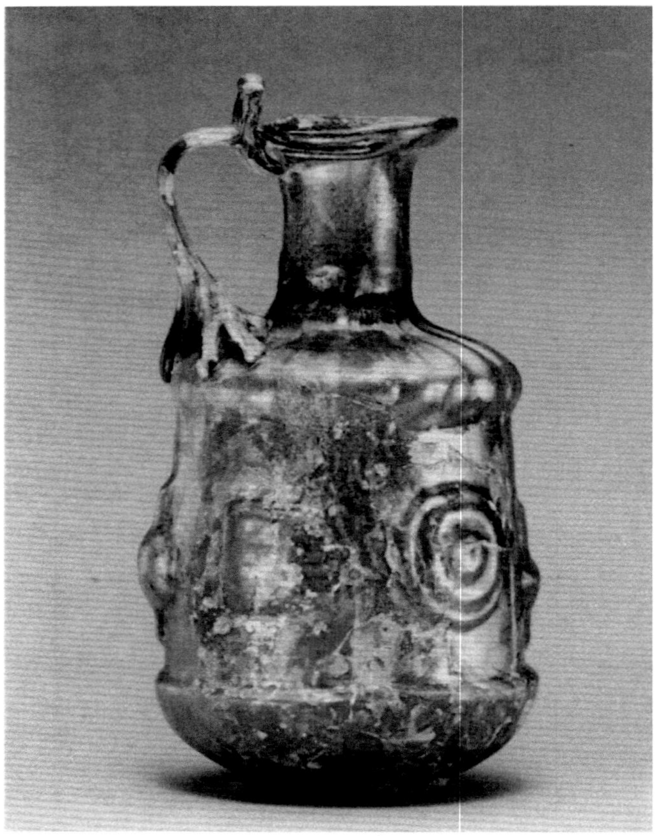

72, panels 4 and 5 (1:1)

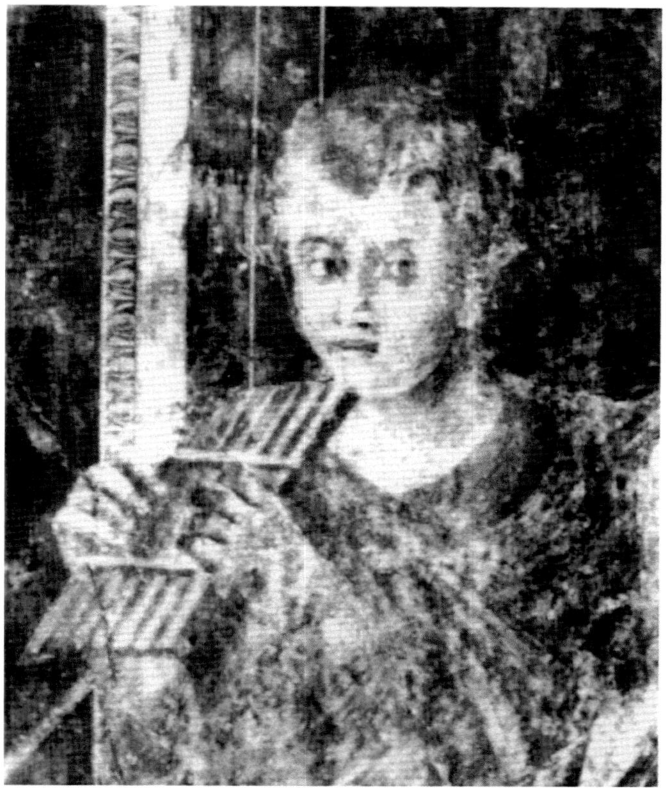

*Fig. 82. Satyr holding a syrinx, detail from a wall painting, Villa of the Mysteries, Pompeii, dating ca. 60-50 B. C. Photo: Alinari/Art Resource, NY.*

*Remarks* See Remarks Cat. No. 71.[1] The object in panel 4 is indistinct. It has been interpreted as a book, gridiron, jewel box, basket, and checkerboard;[2] but it is a syrinx[3] with six reeds of equal length (Fig. 82). This is the common form of the Roman syrinx, or fistula, which was used in an orchestra accompanying pantomime performances. Instead of binding together reeds of different length, the interior of each reed was blocked with wax to produce pipes of varying size.[4]

[1] Possibly same Series as this vessel:
- a) ROM, no. 950.157.87, Hayes 1975, 48, no. 84;
- b) Carnegie, no. 26713/5, Oliver 1980, 61, no. 54;
- c) Oppenländer Collection, no. 2326, von Saldern et al. 1974, 158, no. 443, 144, ill. (the secondary decoration on shoulder and around bottom very indistinct).

[2] Eisen 1927, I, 251, fig. 118, mentions the possibility of identification as a syrinx, but rejects this in favor of a jewel box. Matheson 1980, 46-47, no. 123, suggests a woven basket. Hayes 1975, no. 84, calls it a gridiron motif.

[3] Correctly identified by Oliver 1980, 61, no. 53.

[4] M. Wegner, *Musikleben der Griechen* (Berlin 1949) 58-60, 226f., pl. 2.2, 3a; G. Wille, *Musica Romana* (Amsterdam 1967) 112f.; G. Fleischhauer ed., *Musikgeschichte in Bildern* 2,5 (*Etrurien und Rom*, Leipzig 1963) fig. 12.41. For another type of syrinx, partly of different length reeds and partly of same length reeds: Oliver 1977, 116, no. 76, top left and below right. On pantomime performances: see Remarks Cat. No. 41.

## 73 Hexagonal Jug with Dionysiac Symbols

Syro-Palestinian, perhaps made in Sidon; or possibly Galilean
Probably second quarter of the first century
H 9.6 cm; H body 6.75; max D 5.3 cm; D rim 2.8 cm; D base 2.9 cm
1923.485, Gift of Edward Drummond Libbey

Transparent to translucent grayish blue. Medium thin glass. Bubbles elongated vertically in neck. Remainder of fabric cannot be determined because of weathering.

Neck free blown. Body blown into a multipart mold of three or two vertical sections joined to a cup-shaped base section which included the ribbed area around the bottom (MCT I or MCT V). Mold seams concealed within the decoration. Tip of handle drawn out thin and folded back.

Rim folded outward, upward, and inward. Neck slightly tapered. Convex shoulder and convex bottom joined by a hexagonal body. Flat base with two raised concentric circles around a central knob on the underside. Trifurcated handle applied to shoulder and attached to rim with projecting thumbrest above. Handle positioned above panel 2.

On the shoulder, a frieze of thirty thick ribs. On the body, six rectangular panels, each containing a Dionysiac symbol, all of which except those in panel 1 are suspended by a single line from the top of the panel. In the panels, from left to right: 1, crossed double-ended thyrsoi; 2, a footed amphora with high handles; 3, a circular object decorated with two raised concentric circles and a central knob (probably a phiale); 4, a footed jug with spout to the left and high handle to the right; 5, a syrinx (pan pipes) with six reeds of the same length and two narrow crossbindings; and, 6, a footed amphora with high handles. Around the bottom, a frieze of thirty-seven thick ribs.

*Condition* Intact. On exterior small areas of cream-colored incrustation. Interior incrusted.

*Collection* Thomas E. H. Curtis.

*Published* Previously unpublished.

*Remarks* See Remarks Cat. No. 71. Same Series and possibly same Generation as Cat. No. 74. The quality of the glass and the design on the underside of the base suggest that these two jugs were made in the same workshop as the cylindrical bottle Cat. No. 78, which is decorated with a floral spray around the body.

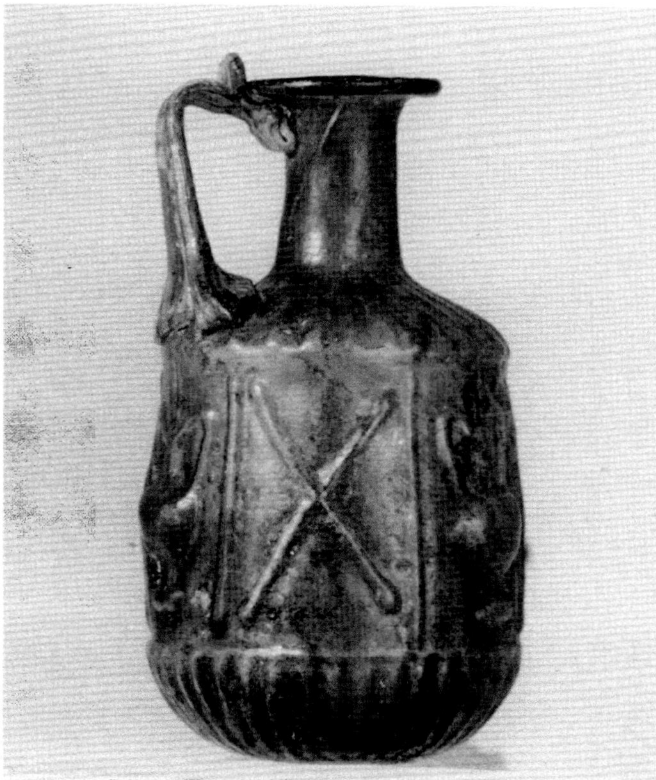

73, panel 1 (1:1)

74 **Hexagonal Jug with Dionysiac Symbols**

Syro-Palestinian, perhaps made in Sidon; or possibly Galilean
Probably second quarter of the first century
H 8.88 cm; H body 6.7 cm; max D 5.3 cm; D rim 2.7 cm; D base 2.9 cm
1923.486, Gift of Edward Drummond Libbey

Translucent streaked manganese colored glass. Translucent pale green handle. Medium thin glass. Fabric cannot be determined because of weathering.
Neck free blown. Body blown into a multipart mold of three or two vertical sections joined to a cup-shaped base section which included the ribbed area around the bottom (MCT I or MCT V). Mold seams concealed within the decoration. Tip of handle drawn out thin and folded back.
Rim folded outward, upward, and inward. Cylindrical neck. Convex shoulder and convex bottom joined by a hexagonal body. Flat base with two raised concentric circles around a central knob on the underside. Bifurcated handle applied to shoulder and attached to rim with projecting thumbrest above. Handle positioned above panel 1.
Decoration like Cat. No. 73.
*Condition* Intact. Exterior covered with remains of buff-colored weathering and iridescence.
*Collection* Thomas E. H. Curtis.
*Published* Previously unpublished.
*Remarks* See Remarks Cat. No. 71. Same Series and possibly same Generation as Cat. No. 73.

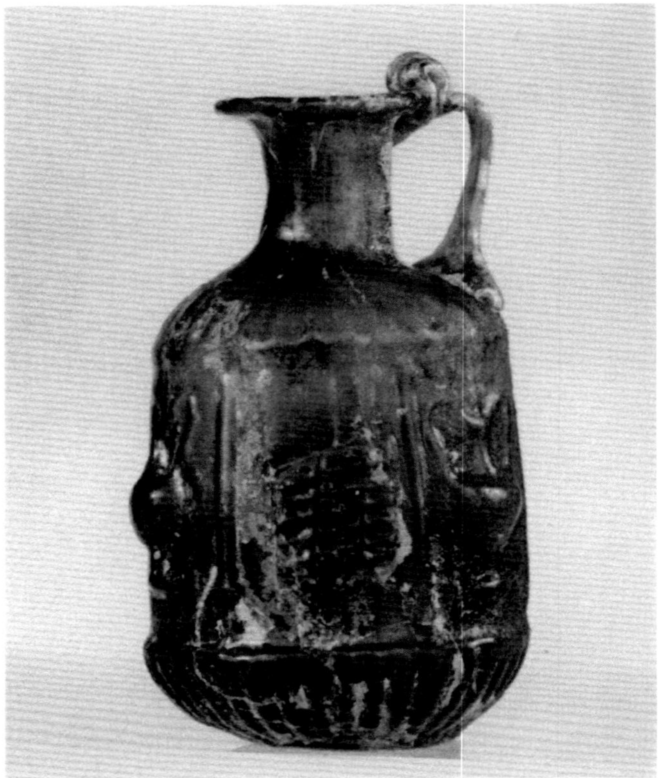

74, panel 5 (1:1)

## *Cylindrical Jugs with Floral Sprays*
(Cat. Nos. 75-78)

75 **Cylindrical Jug with Three Pairs of Floral Sprays**

Syro-Palestinian, perhaps made in Sidon or vicinity
First half of the first century
H 9.33 cm; H body 6.3 cm; max D 5.3 cm; D rim 2.2 cm; D base 3.0 cm
1923.440, Gift of Edward Drummond Libbey

Transparent to translucent grayish yellow green tint (5 GY 7/2). Medium thin glass. Fabric cannot be determined because of weathering.
Neck free blown. Body and lower part of neck blown into a four-part mold of three vertical sections joined to a cup-shaped base section (MCT I). Flashes in mold seams at base of neck; one very large flash under handle. Tip of handle drawn out thin and folded back.
Rim folded outward, upward, and inward. Cylindrical neck tapered slightly downward. Convex shoulder and convex bottom joined by a cylindrical body. On the flat underside of the base, three raised concentric rings surrounding a central dot, the outer ring forming a base ring 3.0 cm in diameter. Bifurcated handle applied to shoulder and attached to rim with projecting thumbrest above. Handle centered over mold seam between mold sections 2 and 3.
On the lower part of the neck and shoulder, in each mold section there are seven or eight downturned tongues in raised outline with raised darts between them. On the body, three pairs of floral sprays arranged horizontally tip to tip. In the mold sections, from left to right: 1, trefoil leaves (ivy); 2, multifoil leaves (grape); and 3, single-pointed leaves alternating with round fruit (not naturalistic, but possibly stylized olive).[1] On the bottom, a frieze of twenty-four upturned tongues in raised outline with raised darts between.
*Condition* Part of rim and most of handle lost. Exterior and interior surfaces covered with buff-colored weathering and areas of purplish and golden iridescence.
*Collection* Thomas E. H. Curtis.
*Published* Perhaps Harden 1944-45, 85, no. e (although the color of this vessel is not the same as described, the position of the handle agrees with the description).
*Remarks* Same Series as Cat. No. 76. Design in all details the same as Cat. No. 76, but this jug was blown in a mold with a slightly larger base section as shown by the difference in the diameters of the base rings (3.0 cm and 2.9 cm respectively). Another peculiarity of this mold is that in section 3 the left branch of pointed leaves with round fruit is much shorter than the right branch, 2.0 and 2.5 cm respectively.
This vessel belongs to a Type that had a wide distribution in antiquity. Known find spots include Yahmour, Syria,[2] Beth Shean, Israel,[3] Limassol, Cyprus,[4] Salzburg[5] and Magdalensberg, Austria,[6] and Mallorca, Spain,[7] while many more examples exist from unrecorded find spots.[8]
One of the three bottles from Beth Shean[9] is said to have been acquired from Azeez Khayat, who also sold to Curtis, to whose collection the Toledo bottles formerly belonged. It is therefore possible, although there is no proof, that the Toledo bottles were also excavated at Beth Shean.
The fragment from Magdalensberg was excavated in a context dated to the year 45 or earlier, providing evidence that this Series must have been created before that date. The eastern Mediterranean find spots indicate that the Series originated in a Syro-Palestinian workshop, perhaps in or near Sidon. The delicate petal-like tongues with darts between recall the work of Ennion.[10] The stylized floral sprays are closely related to the style of Aristeas, a close contemporary of Ennion. A connection of this Series with Ennion was suggested before vessels signed by Aristeas became known.[11] The styles of both masters are closely related, but it is possible to differentiate between them.[12] It is possible, but cannot be proven, that the concept for this design was actually

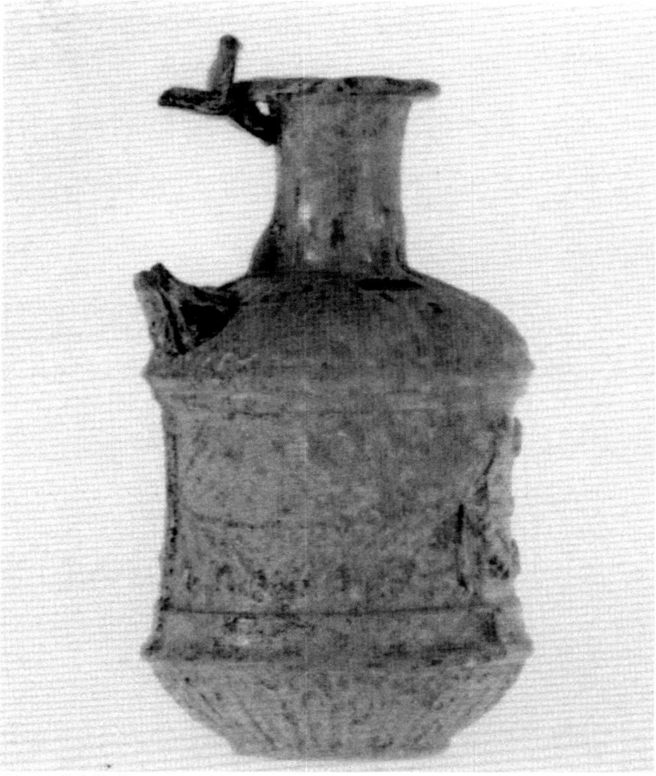

75 panel 3 (1:1)

Aristeas's. A related design, with bows between the floral sprays, occurs on a small group of cylindrical pyxides with pointed lids, the design of which originated in the Syro-Palestinian area, judging from the find spots. [13]
Little research has been done on this Type; its development, therefore, has not yet been clarified. However, comparison with the results of research on the Hexagonal Vessels with High Relief (see Chapter I: *Hexagonal Vessels with High Relief: Vessels Type* and Cat. Nos. 15-47) suggests that it is not likely that all the vessels decorated with floral sprays were made in one workshop.

[1] For identification of the floral sprays, my thanks are owed to Dr. Sam Segal, Amsterdam.

[2] Beirut, Nat. Mus., no. 107, from Yahmour, Syria, tomb II, Harden 1944-45, 84, pl. VII, no. II, 2.

[3] For example:

a, b) MMA, no. 13.198.6, transparent dark royal blue, with two opaque white handles, and no. 37.128.9, transparent royal blue, with one similarly colored handle, formerly Pratt Collection, Harden 1944-45, 84, no. b, 85, no. h, 292;

c) MMA, no. 12.212.3, transparent dark bluish green, with one handle, acquired from Azeez Khayat, Eisen 1927, I, 253, pl. 52.

[4] Limassol, Mus., no. LM 1168, from Kato Palemidhia, about 1 km south of the village, tomb 22, yellowish, H 9.5 cm, found with other glass objects that have not yet been published, Vassos Karageorghis, "Chronique des fouilles et découvertes archéologiques à Chypre en 1987: Musée régional de Limassol," *BCH* 112 (1988), 799, 801, fig. 16.

[5] Salzburg, Stadt-Museum, from a Roman cremation burial, Kisa 1908, III, 718; II, fig. 270; Harden 1944-45, 85, no. d.

[6] Klagenfurt, Austria, Landesmuseum für Kärnten, fragment from Magdelensberg, Czurda-Ruth 1979, 144-145, no. 1055, pls. 9, 22 (ill. upside down).

[7] Barcelona, Provincial Arch. Mus., fragment excavated in the Roman necropolis at "Sa Carroxta," Las Salinas, Campos, Mallorca, Price 1973, 68, no. 2.

[8] To the examples assembled by Harden 1944-45, 84-86, 292, add:

a) V & A, no. Buckley no. 51, Honey 1946, 25-26, pl. 5B;

b) Nicosia, Cyprus Mus., no. 1936-II-19-12, Vessberg 1952, 130, pl. VI, no. 20;

c, d) Tel Aviv, nos. MHG 7058 and 49958, two one-handled bottles, Israeli 1964, 36, nos. 4a, 4b, 37, figs. 6 and 7;

e) Perelada, Gerona, Spain, Mateu Collection, Price 1973, 68, no. 22;

f) Bristol, England, Bomford Collection, Bristol Museum and Art Gallery, *The Bomford Collection of Pre-Roman and Roman Glass on Loan to the City of Bristol Museum and Art Gallery* (Bristol 1976) no. 42;

g) Carnegie, no. 21822/1, one-handled bottle, Oliver 1980, 62, no. 55;

h) Vienna, Kunsthistorisches Mus., no. AS 861, Sunkowski 1956, 18, fig. 31b;

i) Newark, no. 50.1459, formerly Schaefer Collection, with two handles, Auth 1976, 66, no. 59;

j, k) Oppenländer Collection, no. 2365, opaque white, with two translucent blue handles, and no. 2252, natural bluish green, with one handle, von Saldern et al. 1974, 158, nos. 440 and 441;

l) Switzerland, private collection, natural transparent bluish green, with one handle, Luzern 1981, 78, no. 249.

[9] See above note 3c.

[10] See Cat. No. 49 for a discussion of this design.

[11] Harden 1944-45, 84-86.

[12] Calvi 1965a, 14, associated this design with Aristeas.

[13] Pyxides with paired floral sprays joined by bows:

a) Yale, no. 1955.6.58, formerly Moore Collection, said to have been found near Tyre, Matheson 1980, 47-48, no. 128;

b) Paris, Musée du Louvre, no. MNE 107, from Athens, opaque white, de Ridder 1909, 156-157, no. 278, pl. XI;

To these add:

c) Barcelona, Amatller Collection, acquired in Istanbul, opaque white, J. Gudiol y Cunill, *Catálech dels vidres gue integren lo Colecció Amatller* (Barcelona 1925) 17-18, no. 64.

On the shape of pyxides with pointed lids, see Cat. No. 79.

## 76 Cylindrical Jug with Three Pairs of Floral Sprays

Color Plate 13, p. 55
Syro-Palestinian, perhaps made in Sidon or vicinity
First half of the first century

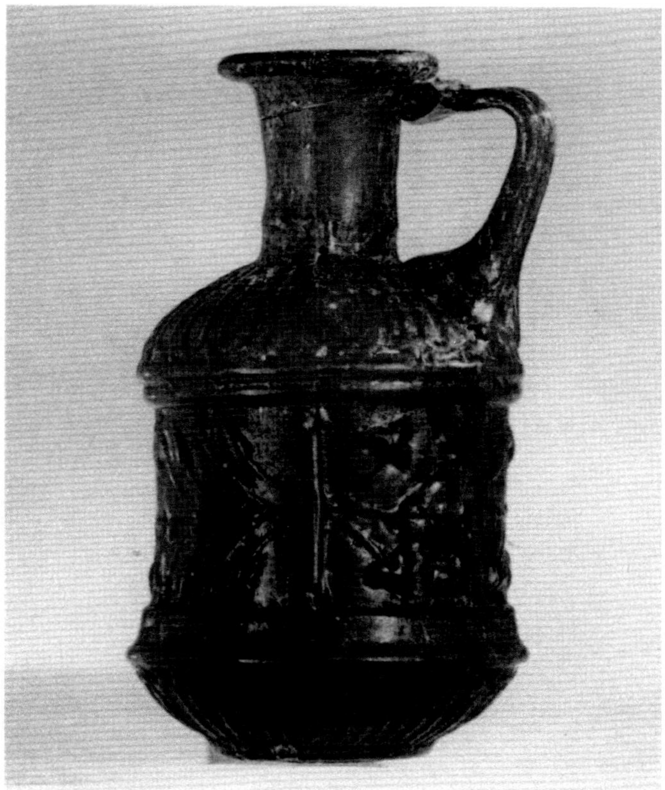

76, panels 3 and 1 (1:1)

H 9.68 cm; H body 6.3 cm; max D 5.3 cm; D rim 2.9 cm; D base 2.9 cm
1923.482, Gift of Edward Drummond Libbey

Translucent royal blue. Brownish green streak at rim. Medium thin glass. One very large elongated bubble in neck. Fabric cannot be determined because of weathering.
Neck free blown. Body and lower part of neck blown into a four-part mold of three vertical sections joined to a cup-shaped base section (MCT I). Tip of handle drawn out thin.
Rim folded outward, upward, and inward. Cylindrical neck tapered slightly downward. Convex shoulder and convex bottom joined by a cylindrical body. Bifurcated handle applied to shoulder and attached to rim with projecting thumbrest above. Handle positioned above trefoil leaves in mold section 1.
Decoration like Cat. No. 75. On the underside of the flat base, three raised concentric rings surrounding a central dot, the outer ring forming a base ring 2.9 cm in diameter.
*Condition* Rim broken and reattached at upper neck; thumbrest missing. Exterior covered with iridescent film; surface pits.
*Collection* Thomas E. H. Curtis.
*Published* Perhaps Harden 1944-45, 85, no. e; it is not clear which of the two Toledo jugs is mentioned; although this one is blue, as described by Harden, the position of the handle does not agree with his description (see Cat. No. 75); Grose 1978, 78, fig. 15, second from left.
*Remarks* See Remarks Cat. No. 75. Same Series as Cat. No. 75, but from a mold with a slightly smaller base section.

### 77 Cylindrical Jug with Three Pairs of Floral Sprays

Perhaps Syro-Palestinian
First half of the first century
H 7.45 cm, H w handle 8.05 cm; H body 5.05 cm; max D 4.2 cm; D rim 2.2 cm; D base 2.6 cm
1923.438, Gift of Edward Drummond Libbey

Opaque pale blue (near 5 B 6/2). Medium thin glass. Fabric cannot be determined because of weathering.
Neck free blown. Body and lower part of neck blown into a four-part mold of three vertical sections joined to a cup-shaped base section (MCT I). Tip of handle drawn out thin.
Rim folded outward, upward, and inward. Cylindrical neck. Convex shoulder and convex bottom joined by a cylindrical body. On the flat underside of the base, a circular channel between base ring (D 2.6 cm) and central disk (D 1.4 cm), with a shallow narrow groove at 0.1 cm from its edge. Bifurcated handle applied to shoulder and attached to rim with projecting thumbrest above; handle has two attachments to shoulder and split to thumbrest. Handle positioned above mold seam between mold sections 1 and 2.
On the lower part of the neck and shoulder, a frieze of twenty-seven downturned flutes. On the body, three pairs of floral sprays arranged horizontally tip to tip. In the mold sections, from left to right: 1, trefoil leaves (ivy) and triple fruit; 2 and 3, trefoil leaves (ivy) alternating with single and triple fruit. On the bottom, a frieze of twenty-nine upturned flutes.
*Condition* Loss on shoulder. The entire vessel covered with a thick opaque yellowish gray weathering crust (5 Y 7/2) which penetrates deeper than the relief of the mold-blown decoration. The crust is peeling off and has been consolidated to prevent further deterioration. The curious disintegration of the glass is evident on the lower body, where the original blue glass can be seen through a hole in the upper crust.
*Collection* Thomas E. H. Curtis.
*Published* Previously unpublished.
*Remarks* See Remarks Cat. No. 75. Closely related to the Series to which Cat. Nos. 75 and 76 belong, but this Series is smaller

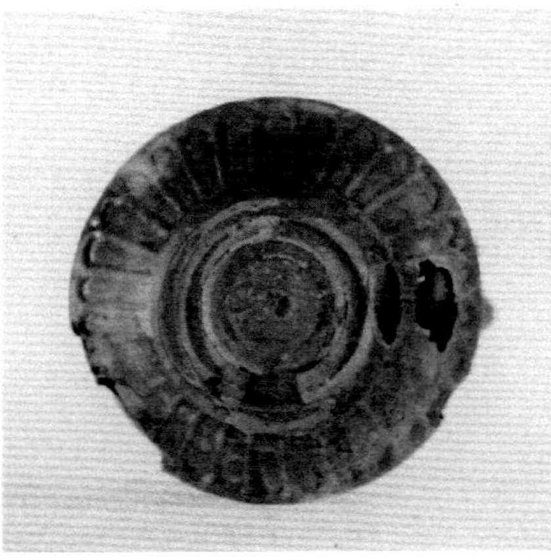
77, underside (1:2)

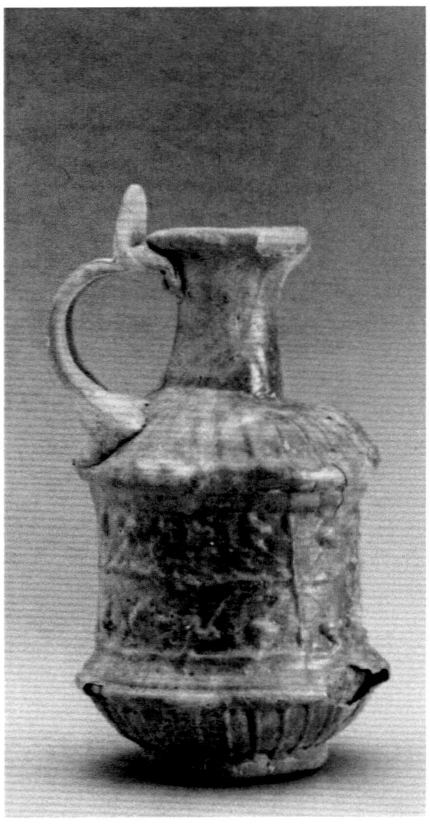
77, panels 2 and 3 (1:2)

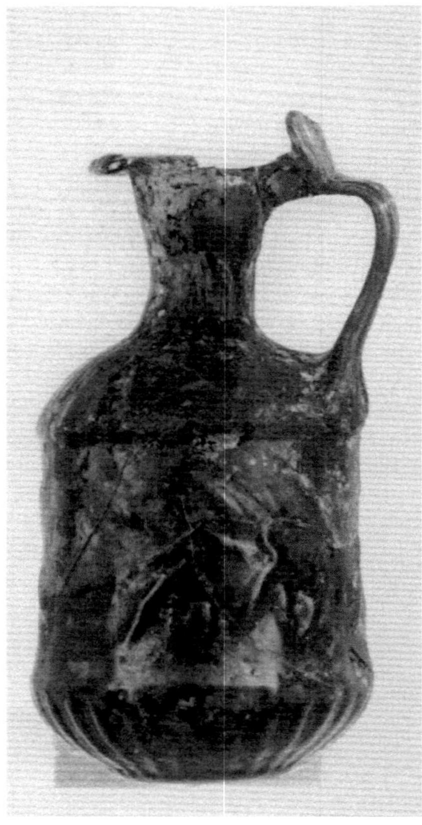
78, side B (1:2)

overall, with less variation in the floral sprays. This Series[1] appears to be less common than the larger one.

[1] Other vessels belonging to this Series:
a) Oppenländer Collection, no. 2576, translucent brown, with one opaque turquoise handle, von Saldern et al. 1974, 158, no. 442, 153 (color pl.);
b) formerly Kofler-Truniger Collection, translucent emerald green, *Kofler-Truniger* 1985, 65, no. 105;
c) Berlin, Staatliche Museen, Antikensammlung, no. 30219, 266, formerly von Gans Collection, Platz-Horster 1976, 40, no. 59;
d) Zurich, Galerie Heidi Vollmoeller, opaque light blue, Luzern 1981, 78, no. 250.

### 78 Cylindrical Bottle with One Pair of Floral Sprays

Perhaps Syro-Palestinian
Probably mid or third quarter of the first century
H 9.12 cm; H body 6.4 cm; max D 4.8 cm; D base 2.5-2.3 cm
1923.483, Gift of Edward Drummond Libbey

Transparent to translucent streaked manganese colored glass. Translucent grayish blue handle. Thin glass. Small oval and spherical bubbles in body.
Neck free blown. Body blown into a three-part(?) mold of two vertical sections joined to a cup-shaped base section which included the ribbed area around the bottom (MCT V). Mold seams between the ends of the floral sprays do not continue onto the shoulder. Relief indistinct. Tip of handle drawn out thin. Vertical crimps at base of neck.
Rim folded outward, upward, and inward. Neck tapers downward. Convex shoulder and convex bottom joined by a cylindrical body. On the flat underside of the base, one broad raised circle around central dimple. Bifurcated handle applied to shoulder and attached to rim with projecting thumbrest above. Handle positioned at a distance from mold seam at back of sprays.
On the shoulder, a frieze of twenty-six fine ribs with thickened lower ends. On the body, two floral sprays arranged horizontally tip to tip, each with single-pointed leaves alternating with round fruit (not naturalistic but possibly stylized olives). Around the bottom, a frieze of thirty fine ribs with thickened upper ends.
*Condition* Broken and repaired; losses in body, shoulder, and rim restored in wax. Some silvery iridescence on body; frosting and surface pits.
*Collection* Thomas E. H. Curtis.
*Published* Eisen 1927, I, 253, pl. 52, below left.
*Remarks* The stylized, stiff floral design of this Type[1] may be derived from designs such as Cat. No. 75. The construction of the mold, with two vertical sections instead of three, the indistinct design on the underside of the base, and the simplification of the outline of this vessel indicate that this jug was not created in the same workshop(s) as Cat. Nos. 75, 76, and 77. The vertical mold seams between the floral sprays do not continue onto the shoulder, so the possibility exists that a fourth, cup-shaped mold section was used to form the shoulder (see Remarks Cat. No. 71). The quality of the glass and the design on the underside of the base suggest that this jug was made in the same workshop as the Hexagonal Jugs with Dionysiac Symbols (Cat. Nos. 73 and 74), but slightly later. Thus, a date around the middle or the third quarter of the first century is probable. The only examples for which a find spot has been recorded are three bottles from the necropolis of Nona, at Zadar, Croatia.[2] They have fine ribs on bottom and shoulder; one is transparent manganese colored with a green handle, two are transparent streaky yellowish amber with a natural bluish green handle.

[1] To the same Type belong:
a) Newark, no. 50.1470, Auth 1976, 198, no. 323;
b) Switzerland, private collection, translucent purple, Luzern 1981, 78, no. 251.

[2] Murano, Mus. Vetrario, on exhibit 1991.

## *Pyxides*
## (Cat. Nos. 79-83)

### *Cylindrical Pyxides*
### (Cat. Nos. 79-83)

### 79 Cylindrical Box (Pyxis) and Conical Lid

Color Plates 14 and 15, p. 55
Campanian or Syro-Palestinian
From Piedimonte d'Alife, Campania, Italy
Second half of the first century
Box: H w lid 7.9 cm; H 5.35 cm; max D 5.95 cm; D rim 5.6 cm; Th rim 0.09 cm; D base 5.82 cm; Lid: H 3.6 cm; max D 6.3 cm; D rim 5.9 cm; Th rim 0.08 cm
1967.5, Purchased with funds from the Libbey Endowment, Gift of Edward Drummond Libbey

Transparent light olive tint (10 Y 5/4). Thin glass. Pinprick bubbles.
Box: blown into a three-part mold of two vertical sections joined to a disk-shaped base section which included the horizontal ridge forming the lower edge of the wall (MCT VII D). Mold seams from rim to ridge partly concealed in the inverted palmettes; on one side the seam protrudes significantly. Lid: blown into a two-part mold of two vertical sections from rim to top of lid (MCT VIII). Relief moderately crisp but difficult to distinguish because of transparency.
Box: rim unworked. Cylindrical body with inset and mold-blown ridge below to support lid. On the flat underside of the base, four raised concentric circles around a small central boss. Lid: rim unworked. Cylindrical lip with conical top.
Box: on the body, between two ridges forming top and bottom borders, a continuous frieze of eight elegant seven-leaf palmettes, alternately upright and inverted. The upright palmettes have thick outturned leaves; the inverted palmettes have thin leaves curling up and inward. Elaborate detailing of the palmette hearts. Lid: sixteen downturned tongues in raised outline bordered by a band of ten three-leaf palmettes, alternately upright and inverted.
*Condition* Box: intact. Incipient pitting. Lid: complete; broken and repaired. Losses in lip and lower part of cover.
*Collections* Giorgio Sangiorgi, Rome (until 1965); Sergio Sangiorgi, Rome (1965-67).
*Published* Sambon 1906, 501, fig. 58; Sangiorgi 1914, no. 102, pl. 18; Harden 1935, 185, app. B, III, no. a; Kern 1954, 33, fig. 6; TMA, *Art in Glass*, 1969, 24, ill.
*Remarks* Same Series as Cat. Nos. 80 and 81, but there is no conclusive evidence that they were made in the same mold.
The decoration is reminiscent of silver made in the first half of the first century A.D.[1] Four variants in mold-blown glass have been distinguished; they probably represent as many Series. The first variant is similar to this pyxis. The second has widely spaced stiff palmettes (Cat. No. 82). The third is slightly narrower than the first two; the palmettes on the body are very widely spaced, and it has a conical lid with gadrooning only.[2] The palmettes of the fourth variant are attached to the ridges bordering the design above and below by short vertical lines.[3] Most preserved pyxides appear to belong to the first variant, represented by this pyxis,[4] but it is not possible to be sure without examining the objects. The design, consisting of a continous frieze of alternately upright and inverted palmettes, also occurs on a few small cylindrical jugs made

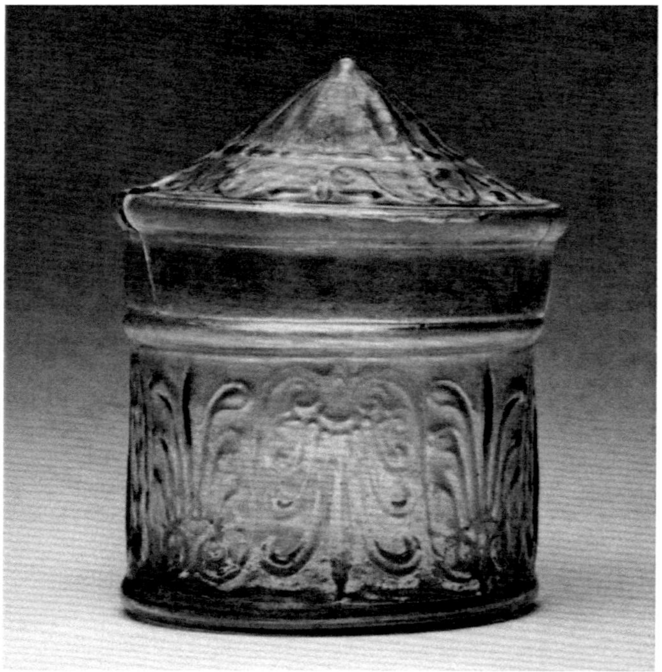

79 (1:1)

made of many different materials, the most common of which were copper, bronze, lead, and wood. Expensive pyxides, however, were made of silver, gold, ivory, and glass. Small pyxides with a cylindrical body and a conical lid were made in silver as early as the third century B.C. [10] Most preserved pyxides are cylindrical, but octagonal ones are also known. [11] Both shapes of pyxis have counterparts that were finished as jugs with closely related mold-blown designs. It is tempting to assume that a jug and a pyxis formed a set, but there are no excavated finds to confirm this hypothesis. The excavation at Pompeii of fragments of a pyxis belonging to the same Series as this one proves that this Type was created before the destruction of the city by the eruption of Mount Vesuvius in 79. [12] This pyxis is one of two reported to have been found in Campania (see note 4a). It was found at Piedimonte d'Alife (ancient Allifae). Fragments of one or two similar pyxides were excavated at Pompeii, but other find spots point to the eastern Mediterranean. One example is said to be from Homs, Syria, while others were acquired in Lebanon. [13] The related jugs (see note 5) have a similar distribution: a pair of identical jugs is reported as having been found at Tel el Hassan, north of the Sea of Galilee, while another is said to be from the Rhineland. Future research may show that this Type of pyxis was made both in the East and in the West. It is tempting to assume that the Type originated in one of the workshops on the Syro-Palestinian coast, perhaps in Sidon, and that one or more Series were subsequently produced in a Campanian glass workshop.

in closely related molds. [5] There is no evidence for assigning the creation of this design to Ennion. The frieze of alternately upright and inverted palmettes has no counterpart among the designs known from Ennion's signed work.

The Toledo pyxides of this type were all made in a variation of an MCT VII mold that had a disk-shaped base section with a low upright rim forming the top of a ridge at the lower edge of the wall. According to ancient authors, [6] pyxides were used to hold various substances such as cosmetics, medicines, magical ingredients, poison, and honey. Silver pyxides with an interior glass cup were used as inkwells in the Augustan period (27 B.C.-A.D. 14) in north Italy. [7] A number of fourth-century B.C. Cypriot male statues are represented holding pyxides and a libation cup or phiale in their hands. [8] The shape of a pyxis could vary, but it always had a lid. They are said to have looked like a *capsella* [9] (a diminutive form of Latin *capsa*, a satchel for books), which suggests that the original form of a pyxis was a small cylindrical container because ancient books were usually in the form of a papyrus roll. Pyxides were

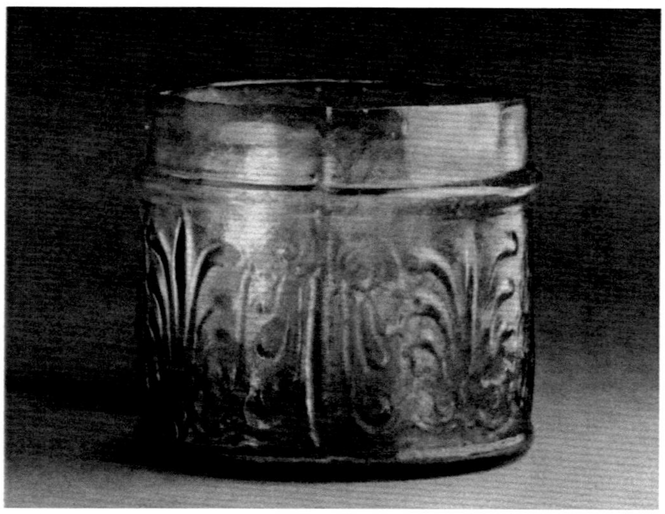

79 (1:1)

[1] For example:

a) Saint Louis Art Mus., no. 11:1928, silver-plated bronze dish from Trinquetaille, near Arles, France, Oliver 1977, 132-133, no. 84;
b) Paris, Louvre, no. BJ 1997-1998 (MNC 1976-1977), four salts from Boscoreale, Italy, Oliver 1977, 136, 137, nos. 87-88.

[2] Examples of third variant:

a) Newark, no. 50.1475, translucent deep blue, with three concentric circles on the underside of the base, H 5.5 cm, D rim 4.8 cm, max D 5.5 cm, Auth 1976, 67, no. 61;

Possibly the same variant:

b) formerly Constable-Maxwell Collection, transparent yellowish green, lid not preserved, H 5.1 cm, max D 5.5 cm, *Maxwell* 1979, 97, no. 155.

[3] Los Angeles County Museum of Art, no. M.88.129.48a, b, formerly Hans Cohn Collection, von Saldern 1980b, 50, no. 42, pale olive green, lid not preserved.

[4] Examples of first variant:

a) Leiden, Rijksmuseum van Oudheden, no. P.P. 22, from Pompeii, Kern 1954c, 34-35, fig. 7;
b) Yale, no. 1955.6.57, said to be from Homs, Syria, Matheson 1980, 48, no. 125;
c, d) CMG, nos. 55.1.69 (opaque white) and 55.1.68, (transparent blue), formerly R. W. Smith Collection, nos. 695 and 862, both acquired in Lebanon, both with lids, Smith 1957, 67, nos. 90 and 91;
e) Harden 1935, 185, app. B, III, no. 1.

[5] For example:

a, b) Newark, nos. 50.1462, and 50.1461, two identical jugs made of blue green glass, with deep blue handles, said to have been found at Tel el Hassan, north of the Sea of Galilee, Auth 1976, 68, no. 62 and 197, no. 321;
c) formerly R. W. Smith Collection, no. 361, pale green, with a lustrous blue handle, Smith 1957, 65, no. 82. The absence of a ridge forming the lower border of the palmette frieze may indicate that these jugs were blown in molds consisting of the two vertical sections of a pyxis mold, without the base section.

Other jugs appear to have been blown in a pyxis mold that included the base section, for they have a ridge along the lower border of the design:

d) Wroclaw (Breslau), Poland, Mus. of Silesia, no. S II.138, said to be from the Rhineland, Kisa 1908, II, 637, fig. 269; Kern 1954c, 36, fig. 8;
e) Sidon, Lebanon, private collection, Neuberg 1962, fig. 35.

[6] For a discussion of literary sources: Hilgers 1969, 265-267, no. 308.

[7] C. Calvi, "Calamai d'argento di Altino," *Aquileia Nostra* 57 (1986) 494-513.

[8] John L. Myres, *The Metropolitan Museum of Art. Handbook of the Cesnola Collection of Antiquities from Cyprus* (New York 1914) 185, 186, fig. 1201.

[9] Hilgers 1969, 266: *pyxides vasa modica... in similitudinem capsellae.*

[10] For example, MMA, no. 03.24.6, silver pyxis from the Bolsena hoard, G. M. A. Richter, *Greek, Roman, and Etruscan Bronzes* (New York 1915) 180ff., ill. facing 180; Strong 1966, 103f., pl. 28B; Oliver 1977, 55, no. 22, dated second half of the third century.

[11] A distinct group of octagonal glass pyxides is discussed at Cat. No. 48.

[12] Leiden, Rijksmuseum van Oudheden, no. P.P. 22, Kern 1954c, 34-35, fig. 7. A second pyxis, now in the Antiquarium at Pompeii, no. 14092, has been published in a drawing only; it is not possible to judge whether it was blown in a related mold. The pyxis is blue and was excavated at Pompeii in the House of Fabius Rufus, Scatozza Höricht 1991, 79, fig. 13, b.

[13] See above note 4b, c, and d.

### 80 Cylindrical Box (Pyxis)

Color Plate 15, p. 55
Campanian or Syro-Palestinian
Second half of the first century
H 5.3 cm; max D 5.84 cm; D rim 5.4 cm; Th rim 0.06 cm; D base 5.8 cm
1923.442, Gift of Edward Drummond Libbey

Transparent patchy dark blue. Very thin glass. Fabric cannot be determined because of weathering.

Blown into a three-part mold of two vertical sections joined to a disk-shaped base section which included the horizontal ridge forming the lower edge of the wall (MCT VII). Mold seams from rim to ridge partly concealed in the inverted palmettes. Relief moderately crisp.

Rim unworked. Cylindrical body with inset and mold-blown ridge below to support lid. On the flat underside of the base, four raised concentric circles around a small central boss.

On the body, between two ridges forming top and bottom borders, a continuous frieze of eight elegant seven-leaf palmettes, alternately upright and inverted. The upright palmettes have thick outturned leaves; the inverted palmettes have thin leaves curling up and inward. Elaborate detailing of the palmette hearts.

*Condition* Complete; chips from rim and a hole in the ridge below. Glass has become paper-thin in places due to weathering. Exterior severely frosted. Interior incrusted. Lid missing.

*Collection* Thomas E. H. Curtis.

*Published* Perhaps Richter 1914, 85, fig. 15 (it is not clear which of the two pyxides in the Curtis Collection was illustrated by Richter, Cat. No. 80 or Cat. No. 81); Harden 1935, 185, app. B, III, no. e.

*Remarks* See Remarks Cat. No. 79. Same Series as Cat. Nos. 79 and 81, but there is no conclusive evidence that they were made in the same mold.

### 81 Cylindrical Box (Pyxis)

Color Plate 15, p. 55
Campanian or Syro-Palestinian
Second half of the first century
H 5.4 cm; max D 6.0 cm; D rim 5.55 cm; Th rim 0.11 cm; D base 5.98 cm
1923.488, Gift of Edward Drummond Libbey

Opaque grayish (purplish) blue. Thin glass.

Blown into a three-part mold of two vertical sections joined to a disk-shaped base section which included the horizontal ridge forming the lower edge of the wall (MCT VII). Mold seams from rim to ridge partly concealed in the inverted palmettes. Relief crisp. Rim unworked. Cylindrical body with inset and mold-blown ridge below to support lid. On the flat underside of the base, four raised concentric circles around a small central boss.

On the body, between two ridges forming top and bottom borders, a continuous frieze of eight elegant seven-leaf palmettes, alternately upright and inverted. The upright palmettes have thick outturned leaves. The inverted palmettes have thin leaves curling up and inward. Elaborate detailing of the palmette hearts.

*Condition* Fragmentary; most of rim, approximately half of body, and entire edge of base restored. Severe surface pitting. Lid missing.

*Collection* Thomas E. H. Curtis.

*Published* Perhaps Richter 1914, 85, fig. 15 (it is not clear which of the two pyxides in the Curtis Collection was illustrated by Richter, Cat. No. 80 or Cat. No. 81); Harden 1935, 185, app. B, III, no. d.

*Remarks* See Remarks Cat. No. 79. Same Series as Cat. Nos. 79 and 80, but there is no conclusive evidence that they were made in the same mold.

### 82 Cylindrical Box with Three Handles (Pyxis)

Color Plate 15, p. 55
Probably Syro-Palestinian
Reported to be from near Hebron, Israel
Probably second half of the first century
H 5.4 cm; max D 6.0 cm; D rim 5.74 cm; Th rim 0.16 cm; D base 5.7 cm

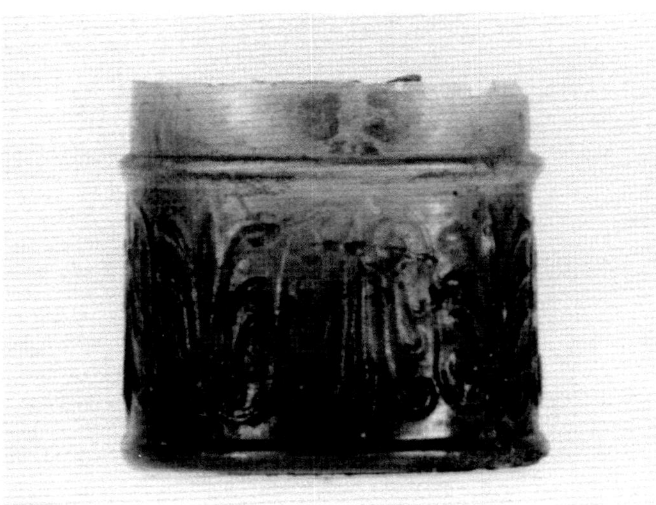

80 (1:1)

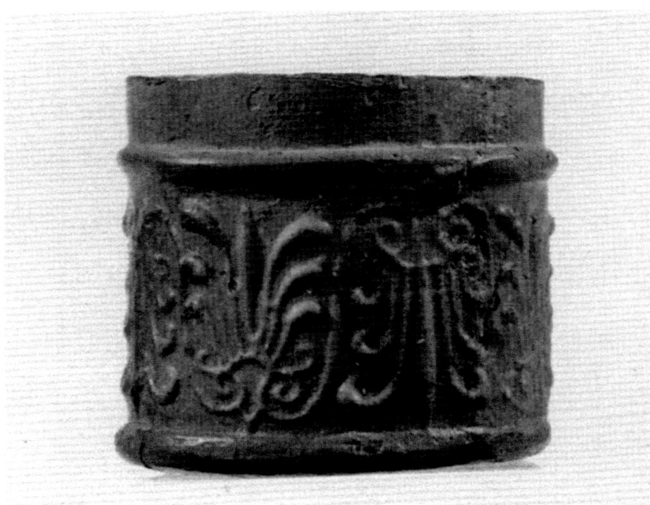

81 (1:1)

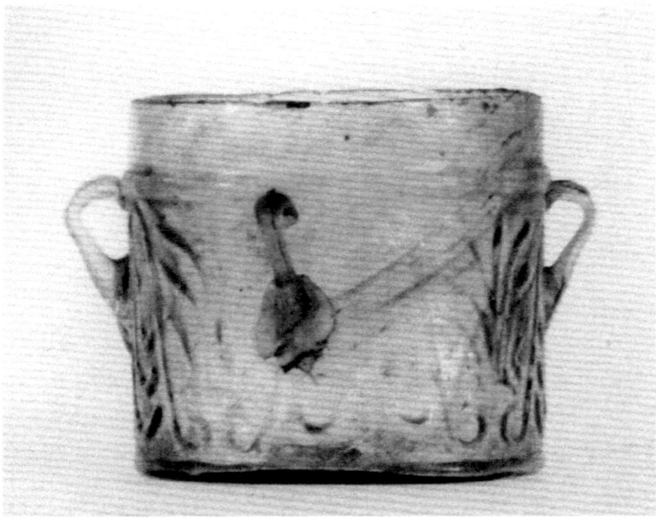

82 (1:1)

1956.3, Purchased with funds from the Libbey Endowment, Gift of Edward Drummond Libbey

Transparent decolorized glass. Pale green and streaked manganese colored areas. Similarly colored handles. Medium thick glass. Pinprick and small spherical bubbles.
Blown into a three-part mold of two vertical sections joined to a disk-shaped base section which included the horizontal ridge forming the lower edge of the wall (MCT VII). Mold seams from rim to ridge partly concealed in the inverted palmettes. Relief moderately crisp but difficult to distinguish because of transparency. Rim unworked. Cylindrical body with inset and mold-blown ridge below to support lid. On the flat underside of the base, four raised concentric circles around a small central depression. Three coil handles attached to middle of body and applied to lower edge of ridge for lid. Handles positioned adjacent to mold seams on half of body and between seams on other half.
On the body, between two ridges forming top and bottom borders, a continuous frieze of eight stiff seven-leaf palmettes, alternately upright and inverted. Little variation in thickness of leaves. Schematic rendering of the palmette hearts.
*Condition* Intact. Lid missing.
*Collection* Azeez Khayat, New York.
*Published* Riefstahl 1961, 37, ill.; Labino 1968, 21, fig. 8.
*Remarks* The design [1] lacks the elegance of the workshop of the previous three pyxides (Cat. Nos. 79-81). No parallels are known for handles on this type of pyxis. For a discussion of the Type, see Remarks Cat. No. 79.

[1] Compare: Oppenländer Collection, no. 2366, von Saldern et al. 1974, 158, no. 444.

### 83 Cylindrical Box (Pyxis)

Probably Syro-Palestinian
First half of the first century
H 3.6 cm; max D 6.6 cm; D rim 6.6 cm; Th rim 0.12 cm; D base 3.7 cm
1923.489, Gift of Edward Drummond Libbey

Transparent colorless (?) glass. Thin glass. Fabric cannot be determined because of weathering.
Blown into a three-part mold of two vertical sections joined to a cup-shaped base section (MCT V F). Mold seams concealed in stems of plants. Relief moderately crisp.
Straight unworked rim. Short cylindrical body with mold-blown ridges. Stepped bottom with hollow base ring. On the flat underside of the base, three raised circles.
On the body, between two ridges forming top and bottom borders, four stylized symmetrical plants with stylized ring-shaped flowers alternating with ring-shaped circlets.
*Condition* Complete, except for a few chips from rim; cracked. Exterior and interior covered with golden iridescence; severe surface pitting.
*Collection* Thomas E. H. Curtis.
*Published* Previously unpublished.
*Remarks* This pyxis belongs to an uncommon Type, one example of which is said to have been found in Aleppo, Syria. [1] The linear stylization of these flowers and plants recalls the stylization of wreaths on a group of cylindrical beakers with mold-blown victory inscriptions, most of which were found in Cyprus and Sidon (Cat. Nos. 2-4). [2] It is tempting to conclude that the beakers and this Series of pyxides were originally created in the same workshop, which must have been located in the eastern Mediterranean, perhaps in Sidon or one of the other glass centers on the Syro-Palestinian coast.

[1] Said to have been found in Aleppo:
a) formerly Munich, Helbing Gallery, thick dark blue, Helbing 1913, 43, no. 659, pl. 33;

Other examples:
b) MFA, no. 50.2277, transparent to translucent blue, von Saldern 1968, no. 29;
c) Oppenländer Collection, no. 22608, dark blue, von Saldern et al. 1974, 159, no. 445.

A distantly related pyxis, decorated with a more naturalistic spray of flowers but blown into a mold of three vertical sections, was excavated near Agrigento, Sicily: G. Caputo, "Pisside di vetro decorato a rilievo da Palma di Montechiaro," *Festschrift T. Dohrn* (1982) 29-31, pl. 1.

[2] Harden 1935, 176-179, class K1iii. Compare also the linear style of a cup excavated at Corinth, Greece: *Corinth* XII, 96, no. 610, and a beaker of unknown provenance: formerly Constable-Maxwell Collection, *Maxwell* 1979, 168-169, no. 301.

## *Fruit-shaped Bottles*
(Cat. Nos. 84-113)

### *Date-shaped Bottles* [1]
(Cat. Nos. 84-107)

[1] For a full discussion of this Type, see Chapter I: *Syro-Palestinian Styles: Date-shaped Bottles.*

### 84 Date-shaped Bottle

Color Plate 16, p. 56

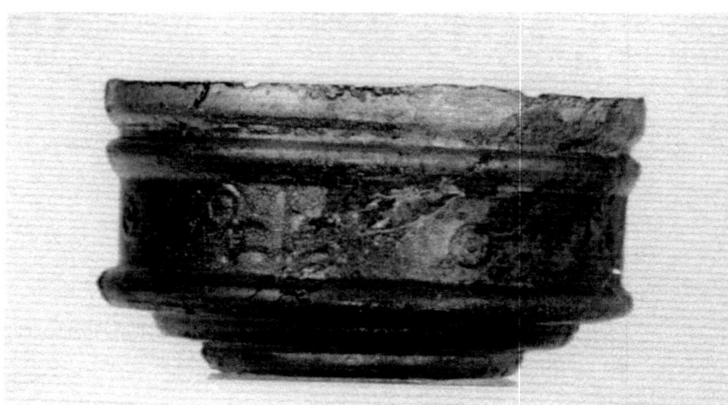

83 (1:1)

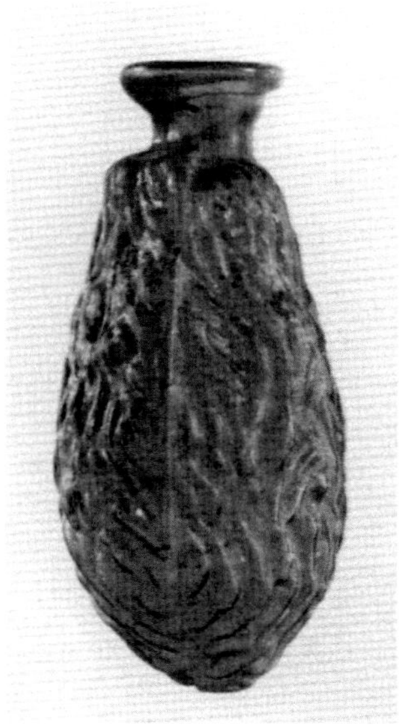
84 (1:1)

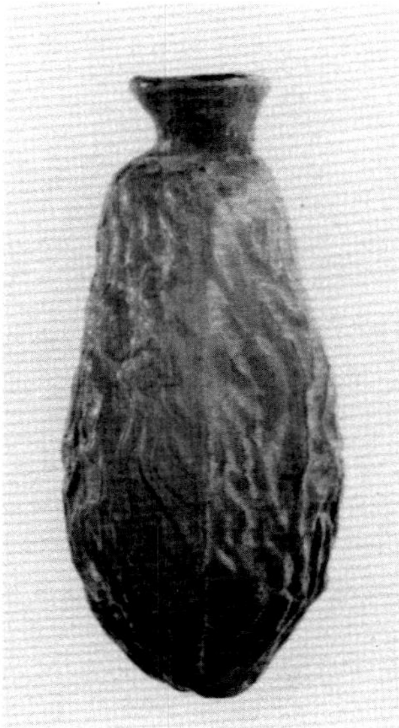
85 (1:1)

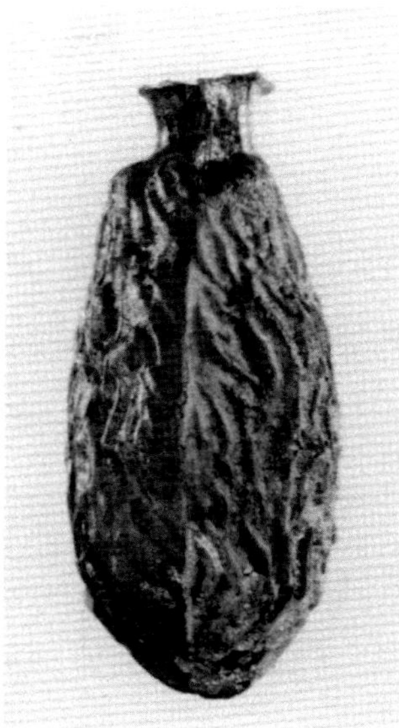
86 (1:1)

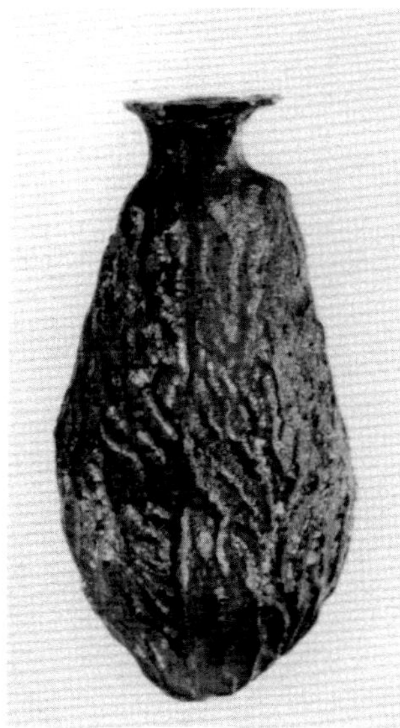
87 (1:1)

Phoenician
Mid first to early second century
H 6.63 cm; H body 5.7 cm; max D 3.1-3.1 cm; D rim 1.78-1.7 cm
1923.578, Gift of Edward Drummond Libbey

Translucent yellowish brown. Medium thin glass. Fabric cannot be determined because of weathering.
Rim tooled. Body blown into a two-part mold of two vertical sections. One continuous mold seam around body, extends onto one side of neck (MCT VIII).
Everted rim, folded inward and downward for two-thirds of circumference. Flaring neck. Body shaped like a fat date.
Relief pattern of short, fat, wavy ridges and hooks, imitating the wrinkles in the skin of a ripe date.
*Condition* Small chip from rim. Exterior surface slightly dulled with areas of rainbow iridescence.
*Collection* Thomas E. H. Curtis.
*Published* Previously unpublished.
*Remarks* Isings 1957, Form 78 D. One of five date-shaped bottles in the Toledo Museum collection made in closely related molds (Cat. Nos. 84-88).

85 **Date-shaped Bottle**

Phoenician
Mid first to early second century
H 6.5 cm; H body 5.7 cm; max D 3.1-3.0 cm; D rim 1.5-1.48 cm
1923.585, Gift of Edward Drummond Libbey

Translucent yellowish brown. Medium thin glass. Fabric cannot be determined because of weathering.
Rim tooled. Body blown into a two-part mold of two vertical sections. One continuous mold seam around body, extends onto both sides of neck (MCT VIII). Mold sections not well aligned.
Everted rim, folded inward and downward. Flaring neck. Body shaped like a fat date with a square cross section.
Relief pattern of short wavy ridges and hooks, imitating the wrinkles in the skin of a ripe date.
*Condition* Intact. Exterior covered with a layer of milky weathering which has created a network of fine surface cracks.
*Collection* Thomas E. H. Curtis.
*Published* Previously unpublished.
*Remarks* One of five date-shaped bottles in the Toledo Museum collection made in closely related molds (Cat. Nos. 84-88).

86 **Date-shaped Bottle**

Phoenician
Mid first to early second century
H (pres.) 6.55 cm; H body 5.7 cm; max D 3.1-3.0 cm
1923.575, Gift of Edward Drummond Libbey

Translucent yellowish brown. Medium thin glass. Fabric cannot be determined because of weathering.
Body blown into a two-part mold of two vertical sections. One continuous mold seam around the body, extends onto both sides of neck (MCT VIII). Mold sections not well aligned.
Flaring neck. Body shaped like a fat date, pinched irregularly lengthwise, with an irregular cross section.
Relief pattern of short wavy ridges and hooks, imitating the wrinkles in the skin of a ripe date.
*Condition* Rim missing; small losses in body; cracked. Exterior covered with areas of rainbow iridescence and milky weathering; severe surface pitting.
*Collection* Thomas E. H. Curtis.
*Published* Previously unpublished.
*Remarks* One of five date-shaped bottles in the Toledo Museum collection made in closely related molds (Cat. Nos. 84-88).

87 **Date-shaped Bottle**

Phoenician
Mid first to early second century
H 6.4 cm; H body 5.7 cm; max D 3.1-3.1 cm; D rim 1.83-1.66 cm
1923.590, Gift of Edward Drummond Libbey

Translucent yellowish brown. Medium thin glass. Fabric cannot be determined because of weathering.
Rim tooled. Body blown into a two-part mold of two vertical sections. One continuous mold seam around body, extends onto both sides of neck (MCT VIII).
Everted rim, folded upward, inward, and flattened. Flaring neck. Body shaped like a fat date with a circular cross section.
Relief pattern of short wavy ridges and hooks, imitating the wrinkles in the skin of a ripe date.
*Condition* Section of rim missing. Exterior covered with rosy golden and green iridescence; rim covered with milky weathering.
*Collection* Thomas E. H. Curtis.
*Published* Previously unpublished.

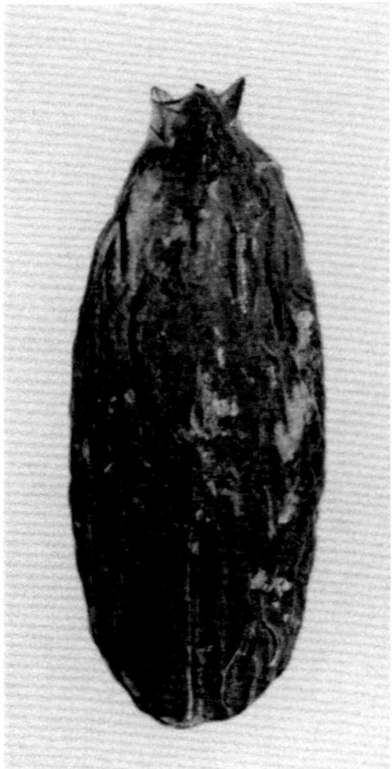

88 (1:1)

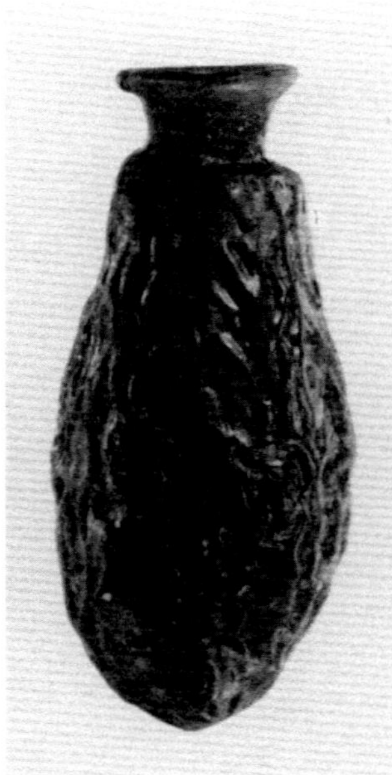

89 (1:1)

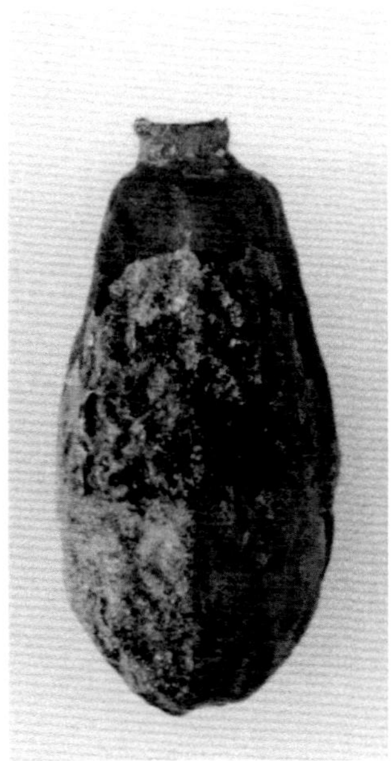

90 (1:1)

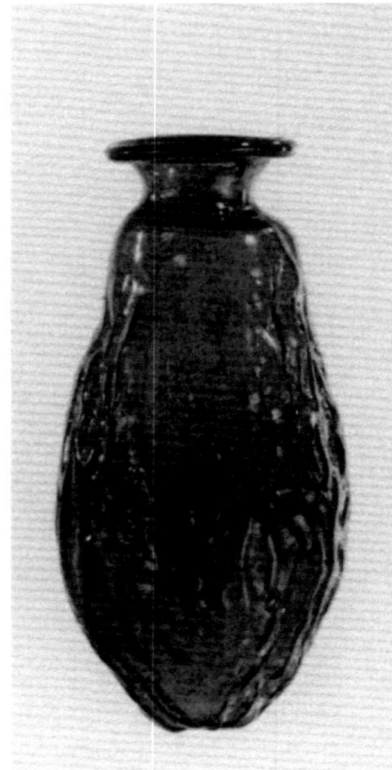

91 (1:1)

*Remarks* One of five date-shaped bottles in the Toledo Museum collection made in closely related molds (Cat. Nos. 84-88).

88 **Date-shaped Bottle**

Phoenician
Mid first to early second century
H (pres.) 6.98 cm; H body 6.1 cm; max D 2.7-2.7 cm
1923.583, Gift of Edward Drummond Libbey

Translucent yellowish brown. Medium thin glass. Fabric cannot be determined because of weathering.
Body blown into a two-part mold of two vertical sections. One continuous mold seam around body, extends onto both sides of neck (MCT VIII). Body elongated.
Flaring neck. Body shaped like a date with an irregular cross section. Relief pattern of short wavy ridges and hooks, imitating the wrinkles in the skin of a ripe date.
*Condition* Rim and part of upper body missing. Exterior covered with rosy golden and green iridescence; severe surface pitting.
*Collection* Thomas E. H. Curtis.
*Published* Previously unpublished.
*Remarks* One of five date-shaped bottles in the Toledo Museum collection made in closely related molds (Cat. Nos. 84-88). This bottle was elongated after extraction from the mold, so that it no longer resembles a fat date like others belonging to the same Series.

89 **Date-shaped Bottle**

Color Plate 16, p. 56
Phoenician
Mid first to early second century
H 7.0 cm; H body 6.1 cm; max D 3.2-3.2 cm; D rim 1.93-1.91 cm
1923.593, Gift of Edward Drummond Libbey

Translucent dark royal blue. Medium thin glass. Small bubbles. Rim tooled. Body blown into a two-part mold of two vertical sections. One continuous mold seam around body, concealed in the pattern of wrinkles on one side, extends onto both sides of neck (MCT VIII).
Everted rim, rolled inward. Short flaring neck. Body shaped like a date with a nearly circular cross section.
Relief pattern of short wavy ridges and hooks, imitating the wrinkles in the skin of a ripe date.
*Condition* Intact. Exterior covered with numerous areas of cream-colored weathering and iridescence; surface pitting. Half of body weathered dull black.
*Collection* Thomas E. H. Curtis.
*Published* Previously unpublished.
*Remarks* One of two date-shaped bottles in the Toledo Museum collection made in closely related molds (Cat. Nos. 89 and 90).

90 **Date-shaped Bottle**

Phoenician
Mid first to early second century
H (pres.) 6.4 cm; H body 6.0 cm; max D 3.0-3.0 cm
1923.572, Gift of Edward Drummond Libbey

Translucent yellowish brown. Medium thin glass. Fabric cannot be determined because of weathering.
Body blown into a two-part mold of two vertical sections. One continuous mold seam around body, concealed in the pattern of wrinkles on one side, extends onto both sides of neck (MCT VIII).
Short flaring neck. Body shaped like a date with a nearly circular cross section.
Relief pattern of short wavy ridges and hooks, imitating the wrinkles in the skin of a ripe date.
*Condition* Rim missing. Exterior covered with thick creamy golden iridescence; where the iridescence has flaked off, deep purple blue iridescence; severe surface pitting.
*Collection* Thomas E. H. Curtis.
*Published* Previously unpublished.
*Remarks* One of two date-shaped bottles in the Toledo Museum collection made in closely related molds (Cat. Nos. 89 and 90).

91 **Date-shaped Bottle**

Phoenician
Mid first to early second century
H 6.4 cm; H body 5.7 cm; max D 3.1-3.1 cm; D rim 2.25-2.12 cm
1923.2413, Gift of Edward Drummond Libbey

Translucent yellowish brown. Medium thin glass. A cluster of small, vertically elongated bubbles in body.

Rim tooled. Body blown into a two-part mold of two vertical sections. One continuous but inconspicuous mold seam around body, concealed in the pattern of wrinkles, extends onto one side of neck (MCT VIII).
Everted rim, folded upward, inward, downward, and slightly flattened. Short flaring neck. Body shaped like a fat date with a circular cross section.
Relief pattern of short wavy ridges and hooks, imitating the wrinkles in the skin of a ripe date.
*Condition* Intact. Exterior surface glossy with areas of incipient pitting; traces of golden iridescence in furrows of wrinkles. Interior with a few patches of cream-colored incrustation.
*Collection* Thomas E. H. Curtis.
*Published* Previously unpublished.

92 **Date-shaped Bottle**

Phoenician
Mid first to early second century
H 6.6 cm; H body 5.8 cm; max D 3.25-2.6 cm; D rim 2.06-2.1 cm
1923.591, Gift of Edward Drummond Libbey

Translucent light manganese colored. Medium thin glass. Small bubbles, vertically elongated in body and neck, some spherical in body.
Rim tooled. Body blown into a two-part mold of two vertical sections. One continuous mold seam around body, partly concealed in the pattern of wrinkles (MCT VIII).
Everted rim, irregularly rolled inward. Flaring neck. Body shaped like a fat date with an irregular cross section.
Relief pattern of short wavy ridges and hooks, imitating the wrinkles in the skin of a ripe date.
*Condition* Intact; surface scratched. Exterior covered with light iridescent film.
*Collection* Thomas E. H. Curtis.
*Published* Previously unpublished.

93 **Date-shaped Bottle**

Color Plate 16, p. 56

Phoenician
Mid first to early second century
H 7.4 cm; H body 6.43 cm; max D 3.1-3.1 cm; D rim 2.14-2.1 cm
1923.577, Gift of Edward Drummond Libbey

Translucent brownish yellow. Medium thin glass. Small bubbles, vertically elongated in body and neck, some spherical in body.
Rim tooled. Body blown into a two-part mold of two vertical sections. Exceptionally pronounced continuous mold seam around body, extends onto both sides of neck (MCT VIII).
Everted rim, folded inward and downward. Flaring neck. Body shaped like a date with an irregular cross section.
Relief pattern of wavy ridges and hooks, imitating the wrinkles in the skin of a ripe date.
*Condition* Intact. Exterior covered with areas of milky weathering and traces of rainbow iridescence.
*Collection* Thomas E. H. Curtis.
*Published* Previously unpublished.
*Remarks* One of three date-shaped bottles in the Toledo Museum collection made in closely related molds (Cat. Nos. 93-95).

94 **Date-shaped Bottle**

Phoenician
Mid first to early second century
H 7.5 cm; H body 6.4 cm; max D 3.1-3.1 cm; D rim 2.0-1.75 cm
1923.586, Gift of Edward Drummond Libbey

Translucent brownish yellow. Medium thin glass. Fabric cannot be determined because of weathering.
Rim tooled. Body blown into a two-part mold of two vertical sections. Exceptionally pronounced continuous mold seam around body, extends onto one side of neck (MCT VIII).
Everted rim, folded inward and downward. Flaring neck. Body shaped like a date with an irregular cross section.
Relief pattern of wavy ridges and hooks, imitating the wrinkles in the body of a ripe date.
*Condition* Small loss along one mold seam. Exterior covered with golden iridescence; severe surface pitting.
*Collection* Thomas E. H. Curtis.
*Published* Previously unpublished.

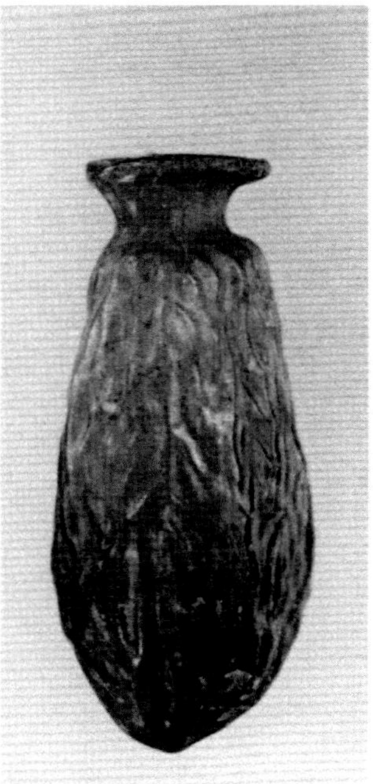

92 (1:1)

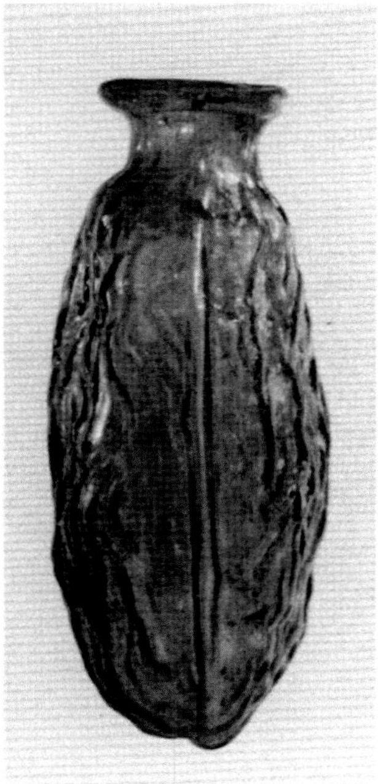

93 (1:1)

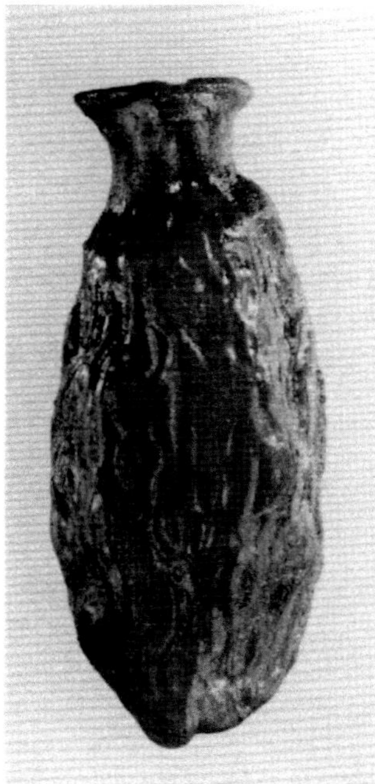

94 (1:1)

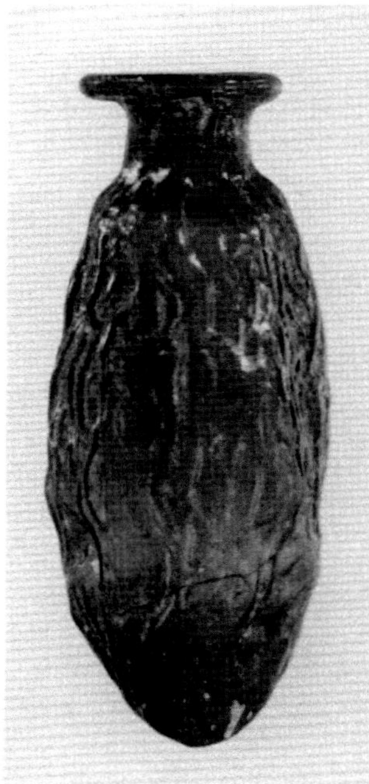

95 (1:1)

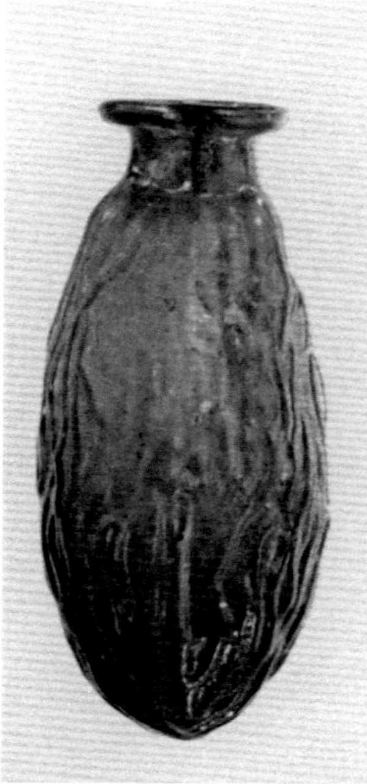

96 (1:1)

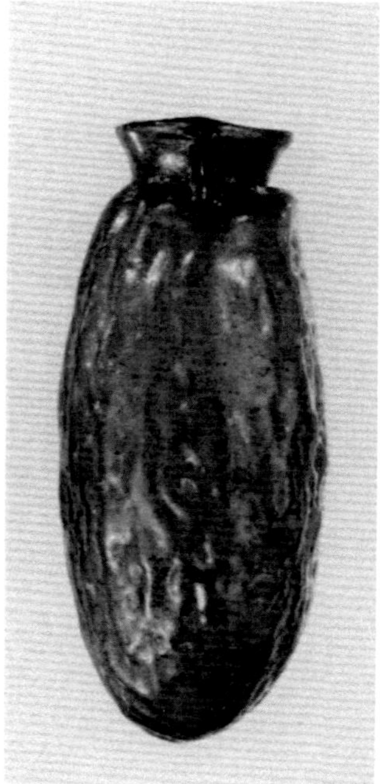

97 (1:1)

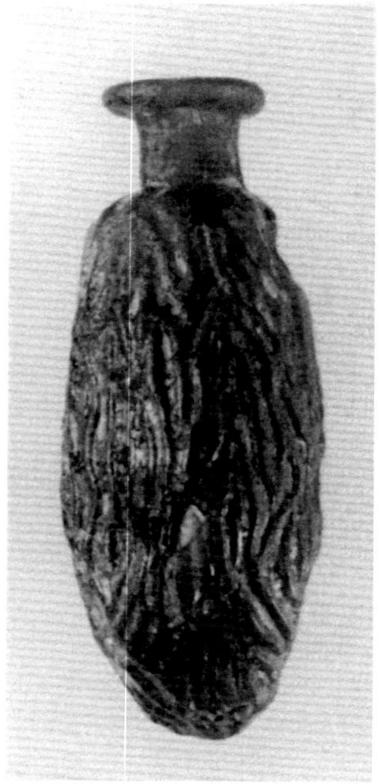

98 (1:1)

*Remarks* One of three date-shaped bottles in the Toledo Museum collection made in closely related molds (Cat. Nos. 93-95).

### 95 **Date-shaped Bottle**

Phoenician
Mid first to early second century
H (rest.) 7.5 cm; H body (rest.) 6.4 cm; max D 3.1-3.1 cm; D rim 2.27-1.94 cm
1923.589, Gift of Edward Drummond Libbey

Translucent yellowish brown. Medium thin glass. Numerous small spherical and vertically elongated bubbles.
Rim tooled. Body blown into a two-part mold of two vertical sections. One continuous mold seam around body, extends onto both sides of neck (MCT VIII).
Everted rim, folded inward and downward. Flaring neck. Body shaped like a date with an irregular cross section.
Relief pattern of wavy ridges and hooks, imitating the wrinkles in the skin of a ripe date.
*Condition* Bottom broken and restored. Exterior covered with small areas of rainbow iridescence.
*Collection* Thomas E. H. Curtis.
*Published* Previously unpublished.
*Remarks* One of three date-shaped bottles in the Toledo Museum collection made in closely related molds (Cat. Nos. 93-95).

### 96 **Date-shaped Bottle**

Phoenician
Mid first to early second century
H 7.1 cm; H body 6.27 cm; max D 3.1-3.1 cm; D rim 2.23-2.06 cm
1923.582, Gift of Edward Drummond Libbey

Translucent dark yellowish brown. Medium thin glass. Small bubbles, vertically elongated in body and neck, some spherical in body.
Rim tooled. Body blown into a two-part mold of two vertical sections. One continuous mold seam around body, very pronounced on one side of neck (MCT VIII).
Everted rim, folded inward and downward. Flaring neck. Body shaped like a date with a nearly circular cross section.
Relief pattern of long wavy ridges and hooks, imitating the wrinkles in the skin of a ripe date.
*Condition* Intact. Large area of dull weathering on upper body.
*Collection* Thomas E. H. Curtis.
*Published* Previously unpublished.

### 97 **Date-shaped Bottle**

Phoenician
Mid first to early second century
H 6.96 cm; H body 6.3 cm; max D 3.0-2.9 cm; D rim 2.25-2.1 cm
1923.2418, Gift of Edward Drummond Libbey

Translucent to transparent manganese colored glass. Thin glass. Fabric cannot be determined because of weathering.
Rim tooled. Body blown into a two-part mold of two vertical sections. One continuous mold seam around body, very pronounced extends onto both sides of neck (MCT VIII).
Everted rim, rounded in flame. Short flaring neck. Body shaped like a date with a nearly circular cross section.
Indistinct relief pattern of wavy ridges and hooks, imitating the wrinkles in the skin of a ripe date.
*Condition* Intact. Exterior covered with small areas of golden iridescence, most of which has flaked off, leaving a glossy surface with incipient pitting in the valleys between the wrinkles.
*Collection* Thomas E. H. Curtis.
*Published* Previously unpublished.
*Remarks* The indistinct pattern may indicate the bottle was slightly expanded after being removed from the mold.

### 98 **Date-shaped Bottle**

Phoenician
Mid first to early second century
H 7.38 cm; H body 6.26 cm; max D 2.7-2.7 cm; D rim 1.9-1.84 cm
1923.581, Gift of Edward Drummond Libbey

Translucent streaked brownish yellow. Medium thin glass. Small bubbles, vertically elongated in neck; one burst bubble on surface of body. A few black specks.
Rim tooled. Body blown into a two-part mold of two vertical

sections. One continuous mold seam around body, partly concealed in the pattern of wrinkles on one side (MCT VIII).
Everted rim, rolled inward and downward. Short cylindrical neck. Body shaped like a date with an irregular cross section.
Relief pattern of long wavy ridges and hooks, imitating the wrinkles in the skin of a ripe date.
*Condition* Intact; deep scratch on body. Exterior covered with a thin layer of milky weathering and areas of iridescence. Interior with incrustation.
*Collection* Thomas E. H. Curtis.
*Published* Previously unpublished.

99 **Date-shaped Bottle**

Phoenician
Mid first to early second century
H 7.4 cm; H body 6.3 cm; max D 3.5-2.83 cm; D rim 1.77-1.75 cm
1923.588, Gift of Edward Drummond Libbey

Translucent yellowish brown. Medium thin glass. Fabric cannot be determined because of weathering.
Rim tooled. Body blown into a two-part mold of two vertical sections. One continuous mold seam around body (MCT VIII).
Everted rim, rolled inward. Cylindrical neck. Body shaped like a date with a nearly circular cross section.
Relief pattern of short wavy ridges and hooks, imitating the wrinkles in the skin of a ripe date.
*Condition* Intact. Exterior covered with golden iridescence over white weathering; severe surface pitting. On interior of rim and neck, grayish white mother-of-pearl weathering.
*Collection* Thomas E. H. Curtis.
*Published* Previously unpublished.

100 **Date-shaped Bottle**

Phoenician
Mid first to early second century
H 6.96 cm; H body 5.6 cm; max D 2.85-2.8 cm; D rim 1.87 cm
1923.576, Gift of Edward Drummond Libbey

Translucent dark yellow. Thin glass. Fabric cannot be determined because of weathering.
Rim tooled. Body blown into a two-part mold of two vertical sections. One continuous mold seam around body, partly concealed in the pattern of wrinkles, extends very pronounced onto one side of neck (MCT VIII).
Everted rim, rolled inward. Cylindrical neck. Body shaped like a date with a nearly square cross section.
Relief pattern of short wavy ridges and hooks, imitating the wrinkles in the skin of a ripe date.
*Condition* Broken and repaired; two large losses on upper body, rim chipped. Exterior covered with iridescent film.
*Collection* Thomas E. H. Curtis.
*Published* Previously unpublished.

101 **Date-shaped Bottle**

Phoenician
Mid first to early second century
H 7.15 cm; H body 6.9 cm; max D 3.2-2.8 cm; D rim 1.96-1.83 cm
1923.580, Gift of Edward Drummond Libbey

Streaked translucent to transparent dark yellow. Medium thin glass. Small spherical and vertically elongated bubbles.
Rim tooled. Body blown into a two-part mold of two vertical sections. One continuous mold seam around body, extends onto both sides of neck (MCT VIII).
Everted rim, rolled inward. Flaring neck. Body shaped like a date with an irregular cross section.
Relief pattern of long wavy ridges and hooks, imitating the wrinkles in the skin of a ripe date.
*Condition* Intact. Exterior covered with iridescent film; surface pitting.
*Collection* Thomas E. H. Curtis.
*Published* Previously unpublished.

102 **Date-shaped Bottle**

Phoenician
Mid first to early second century
H (pres.) 7.3 cm; H body (pres.) 5.5 cm; max D 3.1-3.1 cm; D rim 2.53-2.5 cm
1923.592, Gift of Edward Drummond Libbey

Translucent to transparent manganese colored glass. Thin glass. Fabric cannot be determined because of weathering.

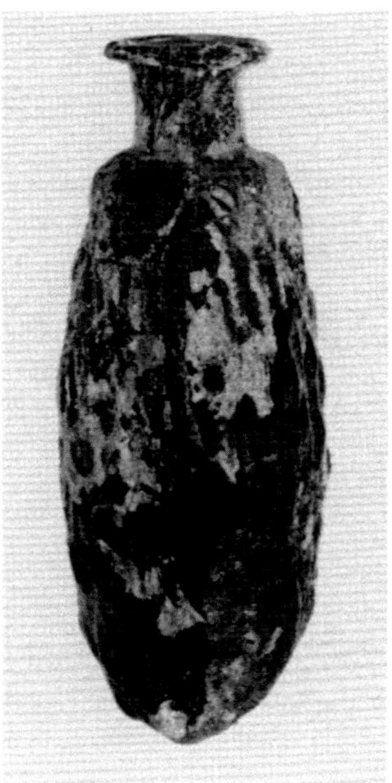

99 (1:1)

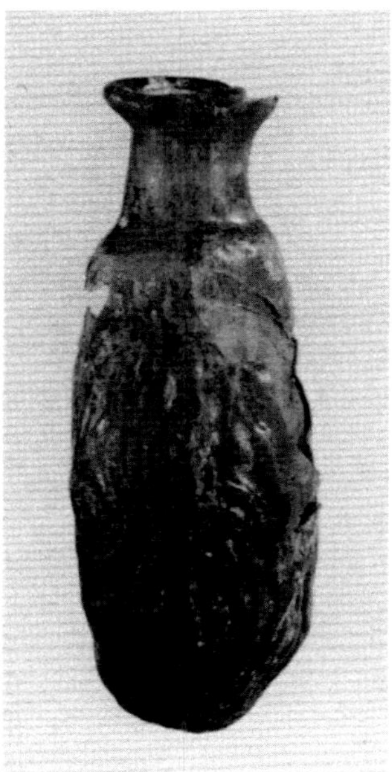

100 (1:1)

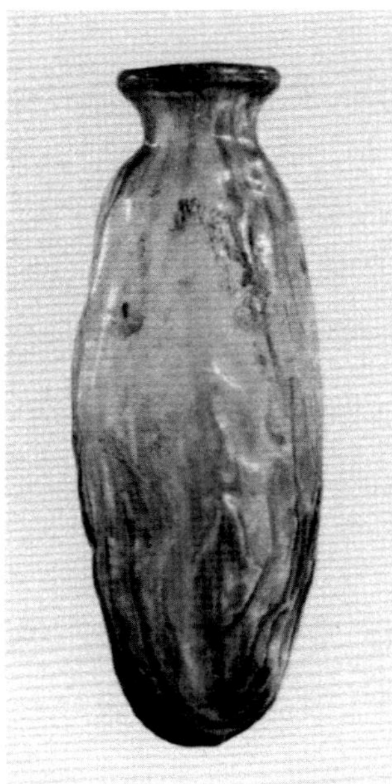

101 (1:1)

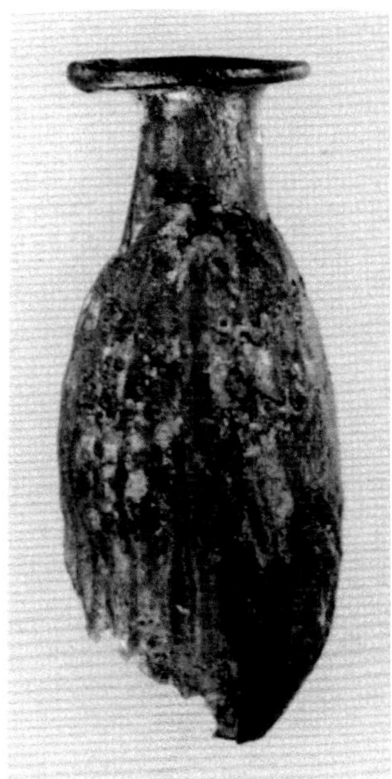

102 (1:1)

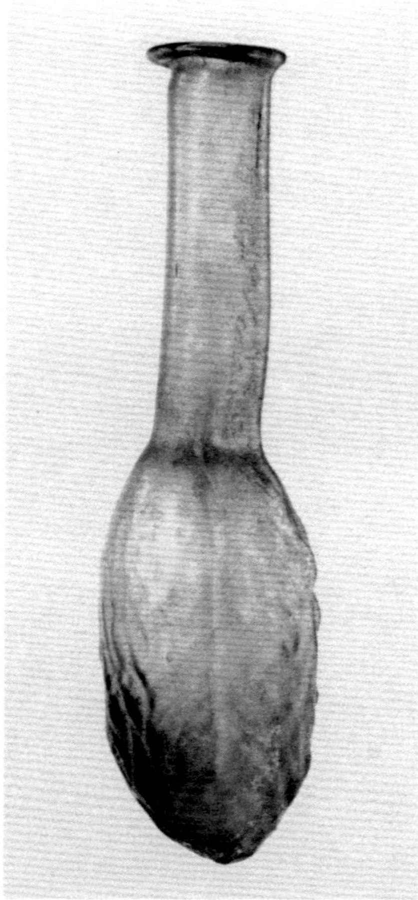

103 (1:1)

Rim tooled. Body blown into a two-part mold of two vertical sections. One continuous mold seam around body (MCT VIII).
Everted rim, folded upward and inward. Tall cylindrical neck. Body shaped like a date with a nearly circular cross section.
Relief pattern of long wavy ridges and hooks, imitating the wrinkles in the skin of a ripe date.
*Condition* Lower part of body missing. Exterior covered with purple iridescence; severe surface pitting.
*Collection* Thomas E. H. Curtis.
*Published* Previously unpublished.
*Remarks* A date-shaped bottle with a similar long neck was excavated in Aquileia, Italy.[1]

[1] Aquileia, Mus. Arch., no. 50305, from a tomb at Aquileia, Calvi 1968, 106, no. 251, pl. 17, no. 3.

### 103 **Date-shaped Bottle**

Color Plate 16, p. 56
Probably Phoenician
Mid first to early second century
H 11.29 cm; H body 6.0 cm; max D 3.26-2.94 cm; D rim 1.97-1.9 cm
1923.595, Gift of Edward Drummond Libbey

Translucent grayish green (near 10 GY 5/2). Medium thin glass. Small bubbles, vertically elongated in neck, spherical in body.
Rim tooled. Body blown into a two-part mold of two vertical sections. One continuous mold seam around body (MCT VIII).
Everted rim, folded inward and downward. Very tall (about half of total height) cylindrical neck. Body shaped like a date with a nearly circular cross section.
Relief pattern of knobs and short straight ridges, imitating the wrinkles in the skin of a ripe date and forming a herringbone pattern along the edges of the mold seam.
*Condition* Intact. Exterior of body and lower part of neck covered with iridescent film. Interior with incrustation.
*Collection* Thomas E. H. Curtis.
*Published* Previously unpublished.
*Remarks* The neck of this date is unusually long. The pattern on the body is so stylized that the mold may have been modeled instead of cast directly from a date.

### 104 **Date-shaped Bottle**

Color Plate 16, p. 56
Phoenician
Mid first to early second century
H 7.33 cm; H body 6.22 cm; max D 3.0-2.6 cm; D rim 2.0-1.86 cm
1923.571, Gift of Edward Drummond Libbey

Translucent dark brownish yellow. Medium thin glass. Numerous pinprick and small elongated bubbles, vertical in neck, horizontal in body. Black streaks in rim and neck.
Rim tooled. Body blown into a three-part mold of three vertical sections. One continuous mold seam around body, extends onto both sides of neck, and a central mold seam on one side; mold seams concealed in the pattern of wrinkles (MCT IV B).
Everted rim, folded inward. Flaring neck. Body shaped like a date with a diamond cross section.
Relief pattern of short straight ridges, imitating the wrinkles in the skin of a ripe date.
*Condition* Intact. Interior with incrustation.
*Collection* Thomas E. H. Curtis.
*Published* Previously unpublished.
*Remarks* One of three date-shaped bottles in the Toledo Museum collection made in closely related molds (Cat. Nos. 104-106).

### 105 **Date-shaped Bottle**

Phoenician
Mid first to early second century

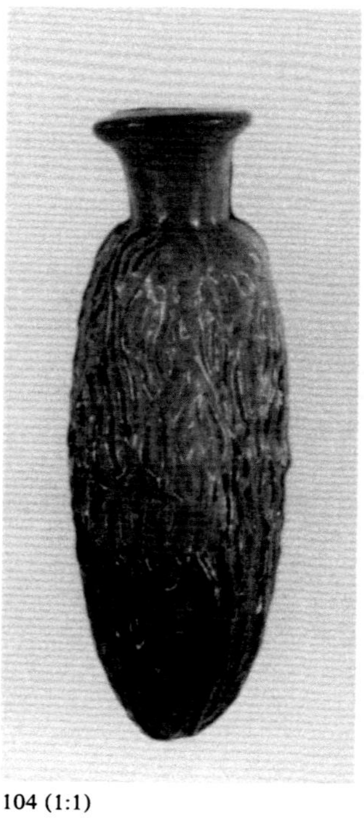

104 (1:1)

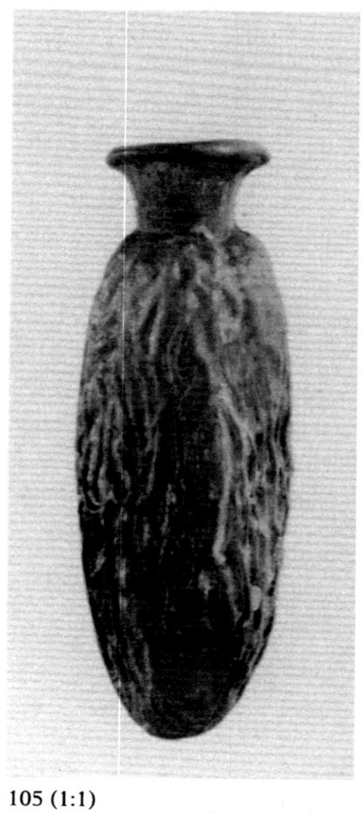

105 (1:1)

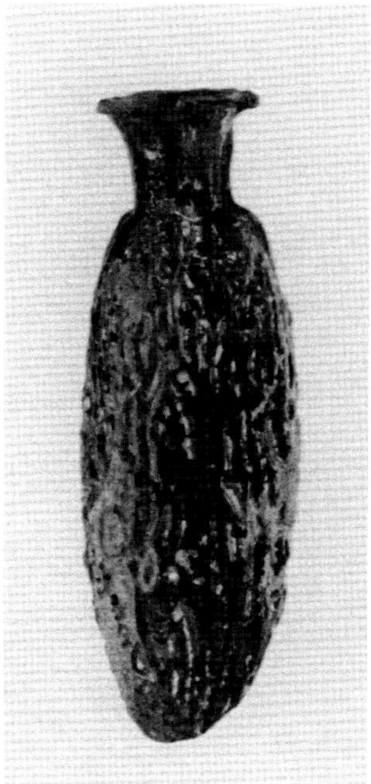

106 (1:1)

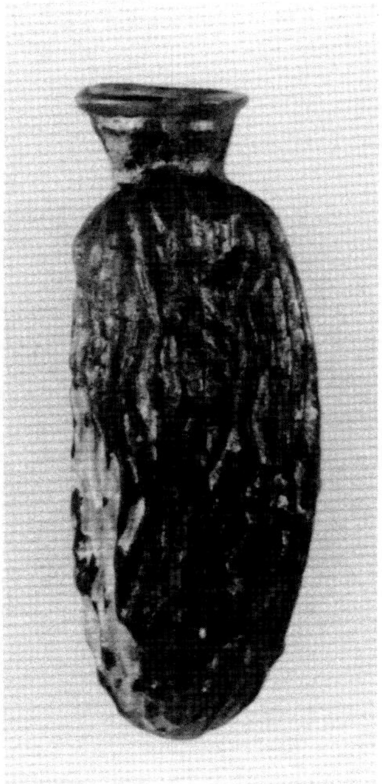

107 (1:1)

H 7.0 cm; H body 5.93 cm; max D 2.9-2.58 cm; D rim 1.93-1.9 cm
1923.579, Gift of Edward Drummond Libbey

Translucent dark yellow. Thin glass. Fabric cannot be determined because of weathering.
Rim tooled. Body blown into a three-part mold of three vertical sections. One continuous mold seam around body, extends onto both sides of neck, and a central mold seam on one side; mold seams concealed in the pattern of wrinkles (MCT IV).
Everted rim, folded inward. Flaring neck. Body shaped like a date with a nearly diamond cross section.
Relief pattern of short straight ridges, imitating the wrinkles in the skin of a ripe date.
*Condition* Intact. Exterior covered with purple and golden iridescence; slight surface pitting. Interior with cream-colored weathering.
*Collection* Thomas E. H. Curtis.
*Published* Previously unpublished.
*Remarks* One of three date-shaped bottles in the Toledo Museum collection made in closely related molds (Cat. Nos. 104-106).

### 106 **Date-shaped Bottle**

Phoenician
Mid first to early second century
H 7.33 cm; H body 6.0 cm; max D 2.9-2.5 cm; D rim 1.85-1.83 cm
1923.573, Gift of Edward Drummond Libbey

Translucent streaked brownish yellow. Medium thin glass. Fabric cannot be determined because of weathering.
Rim tooled. Body blown into a three-part mold of three vertical sections. One continuous mold seam around body and a central mold seam; mold seams concealed in the pattern of wrinkles (MCT IV).
Everted rim, folded inward. Flaring neck. Body shaped like a date with a nearly diamond cross section.
Relief pattern of short straight ridges, imitating the wrinkles in the skin of a ripe date.
*Condition* Intact. Exterior covered with a thick layer of golden and silvery iridescence; severe surface pitting. Interior with incrustation.
*Collection* Thomas E. H. Curtis.
*Published* Previously unpublished.
*Remarks* One of three date-shaped bottles in the Toledo Museum collection made in closely related molds (Cat. Nos. 104-106).

### 107 **Date-shaped Bottle**

Phoenician
Mid first to early second century
H 7.25 cm; H body 6.35 cm; max D 2.7-2.55 cm; D rim 1.93-1.88 cm
1923.570, Gift of Edward Drummond Libbey

Translucent brownish yellow. Medium thin glass. Fabric cannot be determined because of weathering.
Rim tooled. Body blown into a three-part mold of three vertical sections. One continuous mold seam around body and a central mold seam; mold seams concealed in the pattern of wrinkles (MCT IV).
Everted rim, rounded and thickened in flame; tool mark on interior. Short flaring neck. Body shaped like a date with a nearly triangular cross section.
Relief pattern of long wavy ridges, imitating the wrinkles in the skin of a ripe date.
*Condition* Intact. Exterior almost entirely covered with purple and silvery iridescence. Interior with cream-colored incrustation.
*Collection* Thomas E. H. Curtis.
*Published* Previously unpublished.

## *Almond-shaped Bottle*
(Cat. No. 108)

### 108 **Almond-shaped Bottle**

Color Plate 17, p. 56
Probably Phoenician
Mid first to early second century
H 8.9 cm; H body 7.4 cm; max D 4.25-3.16 cm; D rim 2.05-1.9 cm
1923.594, Gift of Edward Drummond Libbey

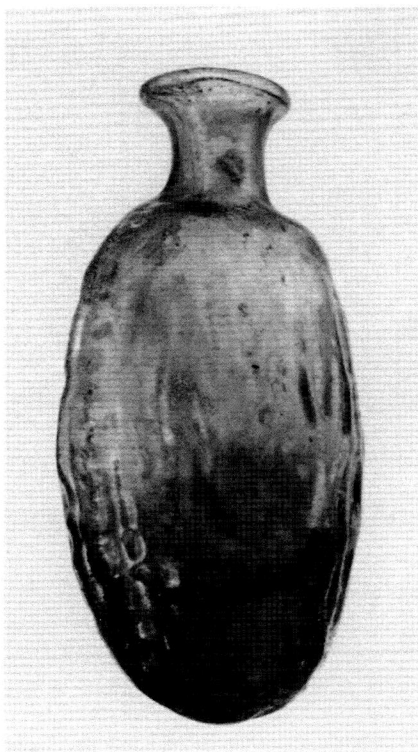

108 (1:1)

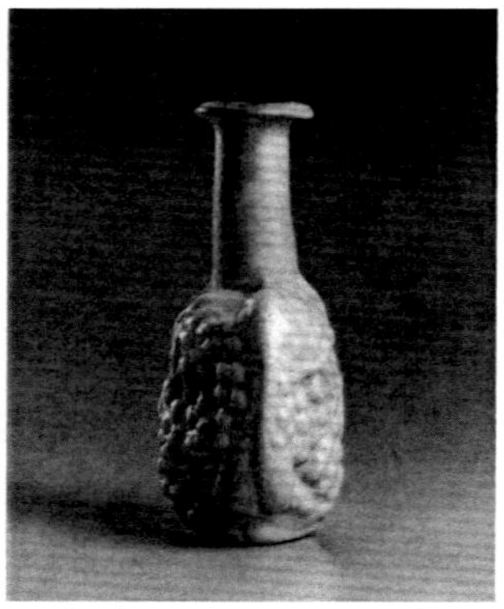
109 (1:2)

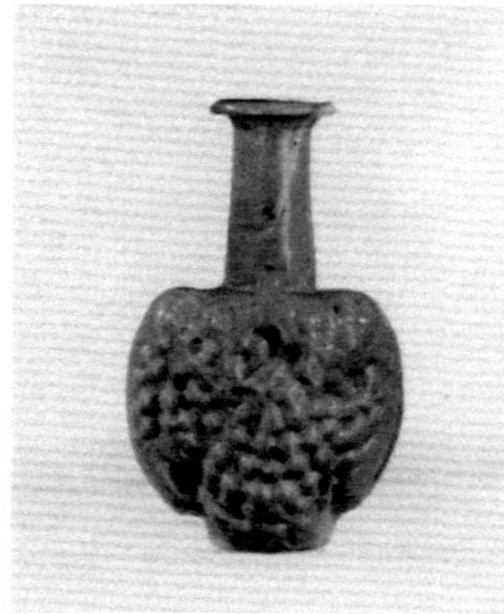
109 (1:2)

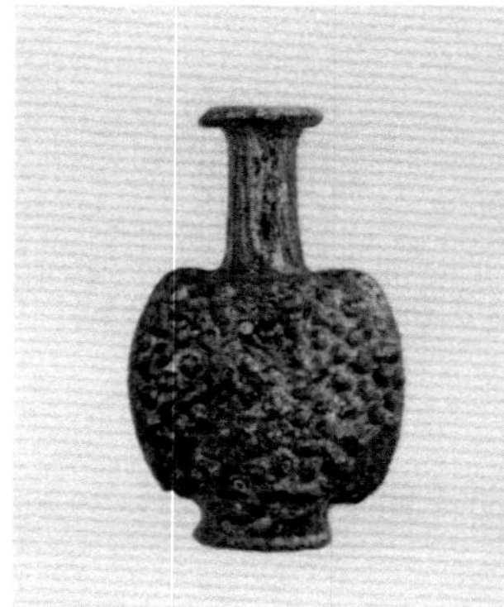
110 (1:2)

Translucent to transparent brownish yellow. Medium thin to thick glass. Fabric cannot be determined because of weathering.
Neck free blown. Body blown into a two-part mold of two vertical sections.
One continuous mold seam around body (MCT VIII). Vertical crimps on the shoulder above the mold seams, probably made by the tool used to hold the vessel while finishing neck and rim. Everted rim, rounded in flame; tool mark on interior. Cylindrical neck. Body shaped like an almond. Lower part of body solid. Relief pattern of long, straight, stylized ridges on body and of short vertical crimps along the seam, probably imitating the flat edge of an almond shell.
*Condition* Intact. Exterior covered with iridescent film.
*Collection* Thomas E. H. Curtis.
*Published* Previously unpublished.

### *Grape Bunch-shaped Bottles*
(Cat. Nos. 109 and 110)

### 109 **Grape Bunch-shaped Bottle**

Color Plate 17, p. 56
Place of manufacture undetermined
Second half of the first century
H 7.55 cm; H body 4.4 cm; max D 4.5-3.1 cm; D rim 2.05-2.0 cm; D base 1.85 cm
1923.886, Gift of Edward Drummond Libbey

Opaque streaked pale purple blue; black streaks on interior of neck and white streak on exterior of neck. Medium thin glass.
Neck free blown. Body blown into a two-part mold of two vertical sections. One continuous mold seam around body, extends onto underside of base (MCT VIII L). Relief crisp.
Rim folded outward, upward, and inward. Tall cylindrical neck. Body shaped like a bunch of grapes. Concave underside of base. Body shaped like a three-lobed bunch of small round grapes with a vine leaf hanging from the shoulder on either side. Slender profile tapering toward top.
*Condition* Complete; broken and repaired. Exterior covered with areas of coffee-colored weathering on neck and body.
*Collection* Thomas E. H. Curtis.
*Published* Previously unpublished.
*Remarks* Same Type and probably same Series as Cat. No. 110. Bottles in the shape of a bunch of grapes were blown in many different glass centers and during many different periods in the history of Roman glass.[1] This bottle and Cat. No. 110 represent one of the earliest Series.[2] Excavated finds demonstrate that glass bottles belonging to this Series were known in Greece and in Italy by the third quarter of the first century,[3] but the location of the workshop or workshops producing the Series has not been determined. Representations of three-lobed grape bunches were common throughout the Mediterranean.[4]

[1] Isings 1957, 94, Form 78 E.
[2] For other Series of grape-shaped bottles, see Cat. No. 119.
[3] From Greece:
a) from Monasteriaki Kephala, Knossos, Crete, Roman chamber tomb with a burial dated third quarter of first century; pres. H 7.0 cm, pres. D 3.7 cm, greenish blue, made in a two-part mold, Carington Smith 1982, 280-281, no. 69, fig. 6, pl. 39j;

From Italy:
b) Naples, Mus. Naz. Arch., no. 13651, from Pompeii and thus before 79, cited by Calvi 1968, 103, n. 195 (perhaps the same bottle as cited by Isings 1957, 94, at Form 78 E);
c) Aquileia, Italy, Mus. Arch., no. 12676, from Aquileia, opaque white, with exceptionally tall neck, H 8.3 cm, Calvi 1968, 106, no. 254, pl. no. 2;

From unknown find spots:
d) formerly R. W. Smith Collection, no. 644, acquired in Lebanon, opaque white, H 7.4 cm, Smith 1957, 133, no. 257;
e) ROM, no. 950.157.261, yellowish green, H 7.4 cm, Hayes 1975, 49, no. 90;
f) Oppenländer Collection, no. 2461, color not mentioned, H 7.4 cm, von Saldern et al. 1974, 174, no. 475.

[4] During the first century, the motif occurs on Jewish tomb sculpture and Jewish coins: Goodenough III, figs. 21-24, 677, 694-697. Compare also hexagonal bottles with high relief of the Fruit Type, Cat. Nos. 36-40.

### 110 **Grape Bunch-shaped Bottle**

Place of manufacture undetermined
Second half of the first century
H 7.35 cm; H body 4.58 cm; max D 4.4-3.45 cm; D rim 2.0-1.9 cm; D base 2.4-2.2 cm
1923.885, Gift of Edward Drummond Libbey

Transparent blue? (color cannot be clearly determined because of weathering). Medium thin glass. Fabric cannot be determined because of weathering.
Neck free blown. Body blown into a two-part mold of two vertical sections. One continuous mold seam around body (MCT VIII). Relief crisp.
Rim folded outward, upward, and inward. Tall cylindrical neck. Body shaped like a bunch of grapes. Circular flat underside of base. Body shaped like a three-lobed bunch of small round grapes with a vine leaf hanging from the shoulder on either side.
*Condition* Complete; small losses on some of bulges that form individual grapes.
*Collection* Thomas E. H. Curtis.
*Published* Previously unpublished.
*Remarks* Same Type and probably same Series as Cat. No. 109.

### *Cone-shaped Bottles*
(Cat. Nos. 111 and 112)

111 **Pine Cone-shaped Bottle**

Color Plate 17, p. 56
Place of manufacture undetermined
Mid to second half of the first century
H 9.2 cm; H body 6.6 cm; max D 5.85-5.65 cm; D rim 2.8 cm
1923.678, Gift of Edward Drummond Libbey

Translucent grayish yellow green (5 GY 7/2). Thin glass. Numerous small spherical bubbles in neck and body; some vertically elongated bubbles in neck.
Rim tooled. Neck and body blown into a two-part mold of two vertical sections. One continuous mold seam around body, extends onto both sides of neck (MCT VIII E).
Rim folded outward, upward, and inward. Cylindrical neck. Body shaped like a hanging pine cone.
On the body, a pattern of oval scales with central dimples, imitating the scales of a pine cone.
*Condition* Intact. Exterior covered with small areas of iridescent film. Interior with traces of rusty brown material (perhaps remains of contents?).
*Collection* Thomas E. H. Curtis.
*Published* Grose 1978, 83, fig. 20; Grose 1984, 37, ill.
*Remarks* Various Types of bottles shaped like a cone are known. All are characterized by scales with a central dimple, but this is the only Type with a pointed base.[1] It represents a hanging pine cone. Most pine cone-shaped bottles with pointed bases are transparent to translucent yellow or yellowish brown or green; the dimensions are similar to those of the Toledo bottle.
The shape suggests the original contents may have been oil with a pine scent.[2] There appear to be two variants of pointed pine cone-shaped bottles, one with a cylindrical neck and fire-polished or folded rim, like this bottle,[3] and the other usually broken off below the rim, a circumstance which suggests that these bottles were originally filled on the spot with powder or scents and that the mouth was then fire-closed,[4] as was common with colored glass bottles shaped like a sphere[5] or a bird.[6] Like the spheres and bird-shaped bottles, pine cone-shaped bottles which were fire-closed are found only in the western part of the Roman empire. The largest concentration of the Type appears to occur in southern and western Switzerland and in northern Italy. Six bottles were excavated in the Locarno area;[7] numerous fragments were excavated at Avenches (ancient Aventicum), the administration center of the Helvetii on the Swiss plateau;[8] one was found near Pavia,[9] and three were found in Piedmont.[10]
The fragments from Avenches were found at the site of a glass workshop active between 40 and 70. The large quantity of fragments and the circumstance that the colors are the same as those of chunks of raw glass excavated on the site show that these bottles belonged to the production of the workshop. Since none of the pine cone-shaped bottles that preserve their rim appears to have been found in northern Italy or Switzerland, it is not likely that they were made, or even used, in that area.
The eastern Mediterranean was famous for its mold-blown bottles imitating fruit, but unlike date-shaped bottles, which are found throughout the Roman empire and predominantly in Syria (see Cat. Nos. 84–107 and Chapter I: *Date-shaped Bottles*), pointed pine cone-shaped bottles are rare in the Near East. The sole known find spot of pine cone-shaped bottles with a rim is a tomb at Knossos, Crete. These bottles were found with burials dated to the second half of the first century and the third quarter of the first century.[11] This date indicates that the bottles with a rim were contemporary with those that were fire-closed, but it does not give a clue to their origin.[12]

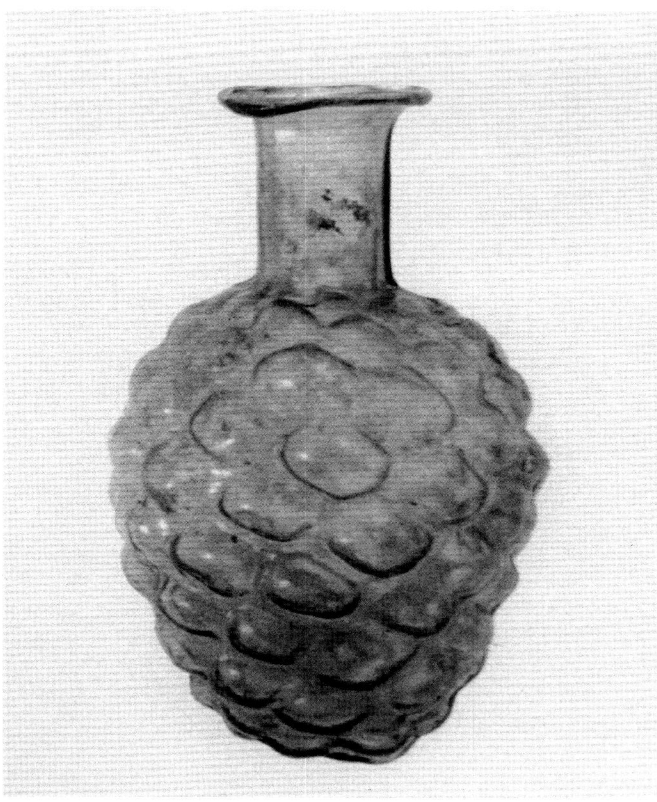

111 (1:1)

[1] For cone-shaped bottles with an upturned point, see Cat. No. 112.

[2] See Chapter I: *Syro Palestinian Styles: Date-shaped Bottles*, where a similar connection between contents and shape of container is suggested.

[3] Published examples with a rim include:
- a, b) from Monasteriaki Kephala, Knossos, Crete, two fragments, one rose mauve colored, the other faint greenish blue, Carington Smith 1982, 280, nos. 63 and 64;
- c) Newark, no. 50.1447, amber, H 10.5 cm, Auth 1976, 73, no. 73;
- d) V & A, no. 1016-1868, Honey 1946, 26, pl. 5c;
- e) Switzerland, private collection, Neuburg 1962, color pl. IXc;
- f) Oppenländer Collection, no. 2250, von Saldern et al. 1974, 172, no. 471.

[4] Biaggio Simona 1991, I, 157; Morel et al. 1992, 9.

[5] Isings 1957, 25-26, Form 10.

[6] Isings 1957, 27, Form 11.

[7] For examples without a rim from Locarno and vicinity, southern Switzerland:
- a) Bellinzona, Switzerland, Ufficio Cantonale dei Monumenti Storici, no. 1936:0598, from Minusio Cadra, tomb 6, Biaggio Simona 1991, no. 163.2.041;
- b) Locarno, Museo Civico, no. 1931:0070, from Muralto, Grand Hotel (no context); Biaggio Simona 1991, no. 176.1.262;
- c) Locarno, Museo Civico, no. 1931:0368, from Tenero, tomb 2, Biaggio Simona 1991, no. 236.1.001;
- d) Locarno, Museo Civico, no. 1931:0368, from Tenero, tomb 5, Biaggio Simona 1991, no. 236.1.003;
- e) Locarno, Museo Civico, no. 1931:0680, from Locarno, find spot not recorded, Aldo Crivelli, "I vetri romani di Locarno," *Verbanus* 1 (1979) pl. 11; Biaggio Simona 1991, no. 000.1.025; Carazzetti and Biaggio Simona 1988, 38-40, no. 7, fig. 7;
- f) Locarno, Museo Civico, no. 1931.1875b, from Locarno, find spot not recorded, Biaggio Simona 1991, no. 000.1.034.

[8] Morel et al. 1992, 9, figs. 17.6, and 20.

[9] Pavia, Italy, Musei Civici, no. A682, Maccabruni 1983, 35, no. 9, fig. 9.

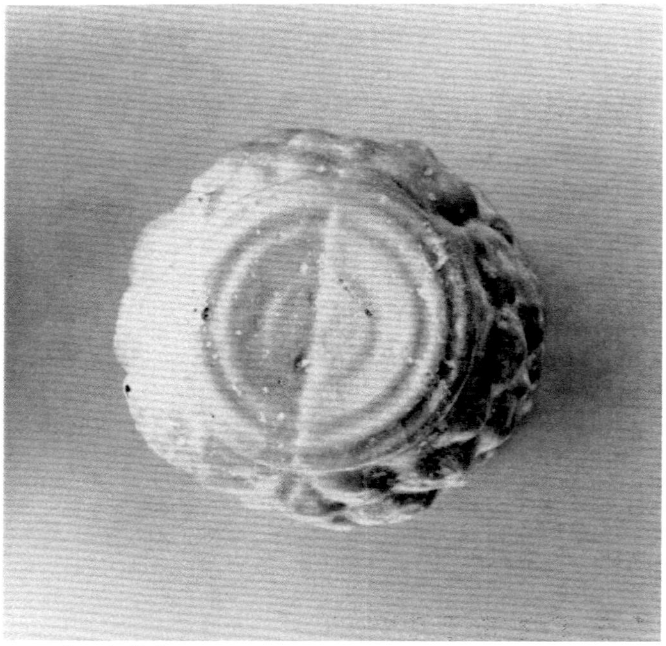

112, underside (1:1)

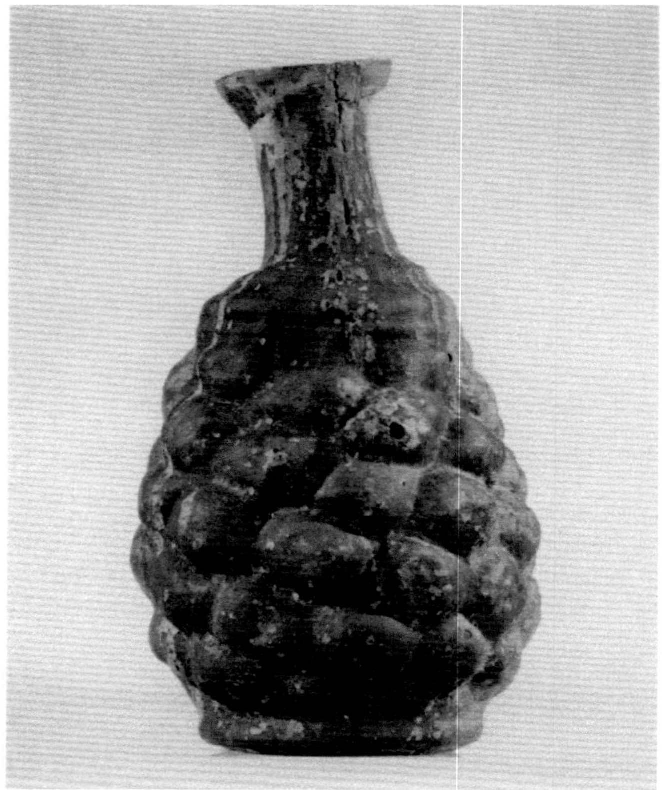

112 (1:1)

[10] Bottles from Masera, Gravellone Toce, and Alba, cited by Biaggio Simona 1991, I, 157, n. 82.

[11] See above note 3a, b.

[12] An Italian origin was suggested by Carington Smith 1982, 280, on the basis of finds from northern Italy and Switzerland, but at the time of her publication it had not yet been observed that all bottles from that area lack a rim.

### 112 **Cedar Cone-shaped Bottle**

Perhaps Western Mediterranean
Probably second half of the first century
H 9.5 cm; H body 6.9 cm; max D 5.45-5.2 cm; D rim 2.5-2.33 cm; D base 3.24 cm
1923.887, Gift of Edward Drummond Libbey

Opaque pale blue. Medium thin glass.
Neck free blown. Body blown into a two-part mold of two vertical sections. One continuous mold seam around body, extends onto underside of base (MCT VIII D).
Everted rim, rounded in flame. Concave neck. Piriform body shaped like a cedar cone. Circular flat underside of base.
On the shoulder, a band of three horizontal ridges. On the body, a close-set pattern of diamond-shaped raised scales with central dimples, imitating the scales of a cedar cone. On the underside, three slightly raised concentric ridges.
*Condition* Complete; several small losses on body. Exterior covered with large areas of buff-colored weathering and greenish and purplish iridescence. On interior of neck, thick white weathering.
*Collection* Thomas E. H. Curtis.
*Published* Previously unpublished.
*Remarks* Three Series of cedar cone-shaped bottles have been recognized. All have flat bases and appear to represent an upright cone,[1] presumably in imitation of a female cedar cone (See Remarks Cat. No. 36 and Fig. 74).
The opaque pale blue color of this bottle is common in the first century. The construction of the mold, with two vertical sections bisecting the design of concentric ridges on the underside of the bottle, is unusual; it may indicate that this bottle was made in the West.

[1] No exact parallel for the Toledo bottle exists. The second Series appears to be a degenerated version of the Toledo bottle; it has a flat underside without concentric ridges. For example:
- a) RGM, no. N5905, blue green, H 7.4 cm, La Baume 1973, no. C 15, pl. 11, no. 3;
- b) Carnegie, no. 222557/1, natural green, H 7.5 cm, Oliver 1980, 72, no. 68;

Perhaps also:
- c) Tel Aviv, no. MH 9058.

The third Series has a funnel neck, which suggests a date in the second century or later. For example:
- a) Trier, Germany, Rheinisches Landesmuseum, no. G.733, from Trier, with two low concentric rings on the underside of the base, Goethert-Polaschek 1977, 138, Form 83, no. 775, pl. 56;
- b) Carnegie, no. 25136/8, Oliver 1980, 72, no. 69.

## *Acorn-shaped Bottle*
(Cat. No. 113)

### 113 **Acorn-shaped Bottle**

Color Plate 11, p. 54
Syro-Palestinian, perhaps made in Sidon
Workshop of the Floating Handles
First half of the first century
H 9.0 cm; H body 7.15 cm; max D 4.2-4.2 cm; D rim 2.1-2.0 cm
1923.441, Gift of Edward Drummond Libbey

Opaque white. Medium thin glass.
Rim tooled. Neck and body blown into a two-part mold of two vertical sections. One continuous mold seam around body and

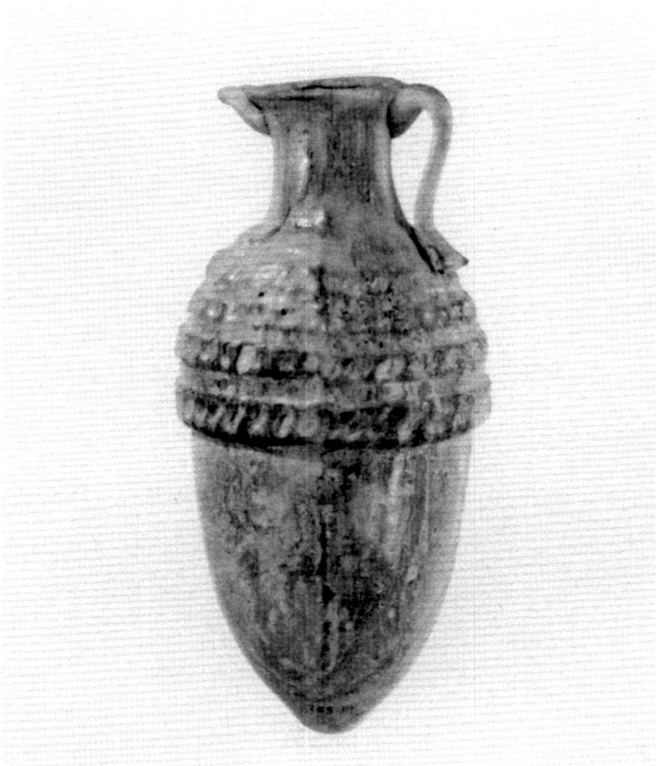

113 (1:1)

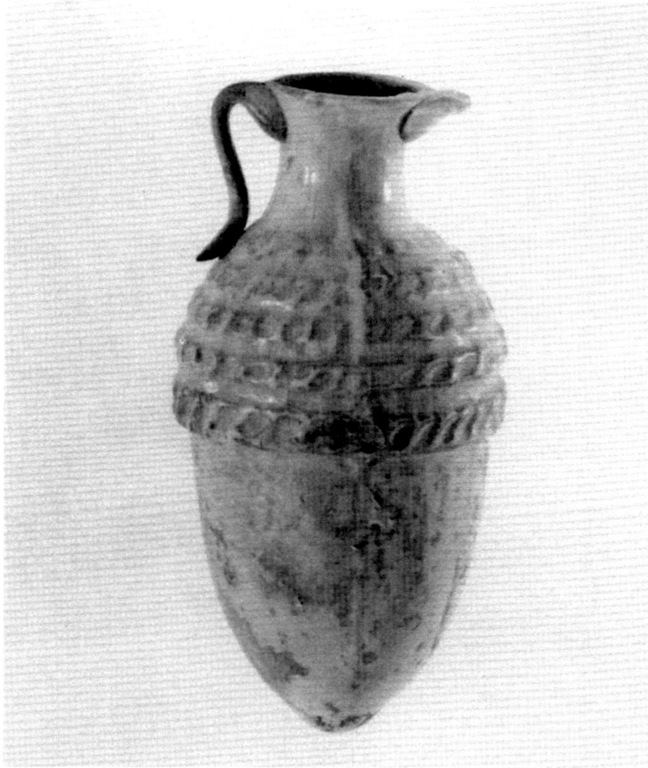

113 (1:1)

knobbed pointed base (MCT VIII). Excess glass at lower tips of handles cut off and pinched to form a small disk.
Rim irregularly pulled outward and inward. Concave neck. Body shaped like an acorn tapering to a slight knob at base. Two coil handles applied to underside of rim and drawn down to shoulder but not attached. Handles positioned halfway between mold seams. On the shoulder, a pattern of five horizontal bands of linked boxes, imitating the hard woody cupule surrounding the nut.
*Condition* One handle missing below point of application. Small loss between lower two decorated bands. Area of severe weathering on one side where surface has split away from underlying glass; this area has purplish brown weathering.
*Collection* Thomas E. H. Curtis.
*Published* Richter 1914, 85, fig. 12.
*Remarks* This bottle belongs to the Acorn Type, small bottles shaped like an acorn,[1] that was made in the Workshop of the Floating Handles (see Chapter I: *The Workshop of the Floating Handles*). Ten Types of mold-blown glass vessels have been attributed to this Workshop. The Toledo Museum collection has representatives of four Types: a miniature amphora encased in wickerwork (Cat. No. 59), a lenticular bottle with a Medusa face on either side (Cat. No. 62), a head-shaped vessel (Cat. No. 138), and this acorn-shaped bottle.

Paper inside has : L 1265 / 14 written on it. This number is also painted in red on the underside of the base.

[1] Other examples belonging to this Type:

a) CMG, no. 55.1.72, formerly R. W. Smith Collection, no. 860, acquired in Lebanon, opaque white, with remains of two floating handles, Smith 1957, 65, no. 86;
b) formerly Kofler-Truniger Collection, said to be from Syria, translucent green, with two floating handles (one broken off near the top), *Kofler-Truniger* 1985, 62, no. 95.

## *Jar*
(Cat. No. 114)

### *Spherical Jar with Floral Design*
(Cat. No. 114)

**114 Miniature Spherical Jar with Basket Handle**

Italian or Eastern Mediterranean
First century
H 3.55 cm; H w handle (pres.) 4.7 cm; max D 4.35-4.2 cm; D rim 3.45 cm; D base 2.0 cm
1923.416, Gift of Edward Drummond Libbey

Color cannot be determined because of weathering. Thin glass. Fabric cannot be determined because of weathering.
Rim tooled. Body blown into a two-part mold of two vertical sections. One continuous mold seam around body, extends onto underside of base (MCT VIII).
Rim rounded and thickened in flame; tool mark on interior and constriction below. Spherical body. Flat base. Handle positioned above tips of branches midway between mold seams.
On either side of body, four branches with alternating leaves and berries, suspended from rim so that tips meet in center. At top, trefoil leaves (ivy) with bunches of three round berries; below, single pointed leaves (laurel) alternating with round fruit. Below each ivy branch, a thick fillet; between the tips of the laurel branches, a small ox skull (bucranium).
*Condition* Broken and repaired. Most of handle missing. Three small holes in body. The glass has devitrified to a chalky buff substance with numerous surface pits on exterior. Interior covered with dusky yellowish brown layer (10 YR 2/2).
*Collection* Thomas E. H. Curtis.
*Published* Previously unpublished.
*Remarks* Suspended branches were common decorative motifs in

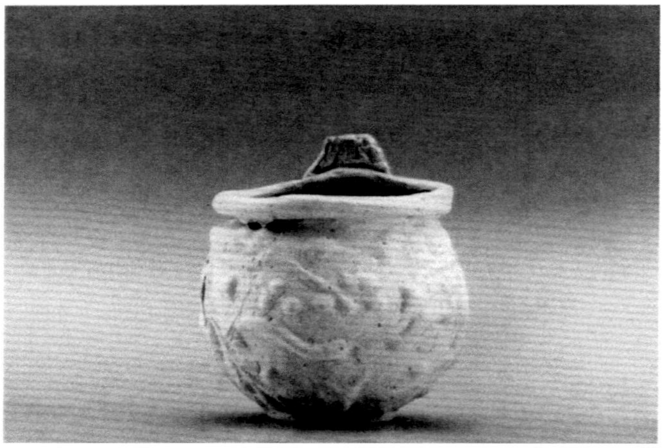

114 (1:1)

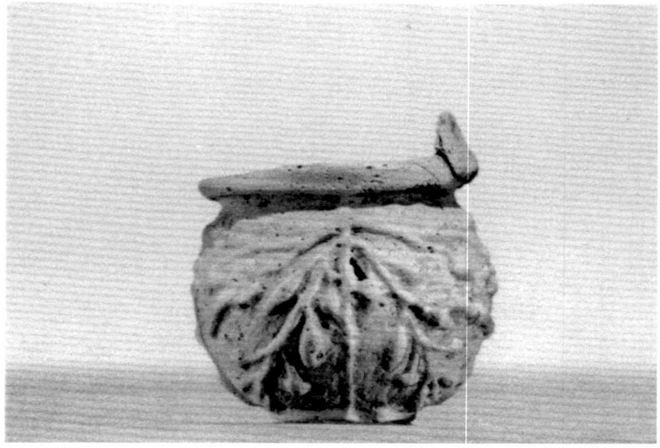

114 (1:1)

the early Roman empire,[1] but their combination with fillets and bucrania, skulls of sacrificial animals, which refer to religious ceremonies, is unusual on vessels. These motifs are far more common in the sculpted architectural ornamentation of temples and altars, for example the Ara Pacis Augustae in Rome (erected 13-9 B.C.).[2] The naturalistic style of the branches on this vessel with their succulent leaves and fruit is unparalleled on vessels by Ennion or Aristeas. The design represents the work of a craftsman whose style appears to be close to that of craftsmen creating molds for glazed ceramic wares.

[1] For example, silver cups from the treasure of Hildesheim Berlin-Charlottenburg, Staatliche Museen Preussischer Kulturbesitz: U. Gehrig, *Hildesheimer Silberschatz im Antikenmuseum* (1980) 13f., color pl. II, 19, no. 22; E. Simon, *Augustus: Kunst und Leben in Rom um die Zeitenwende* (Munich 1986) color pl. 10. For imports of glazed pottery into Italy: Claudia Maccabruni, "Tipologia della ceramica invetriata di età romana nell'area del Ticino. Considerazioni preliminari," *Giornata di Studio sulla ceramica invetriata — Como 1981,* in *Archeologia dell'Italia Settentrionale* 2: *La ceramica invetriata tardo-romana e alto-medioevale* (Como 1985). For glass examples:

a) St. Petersburg, Hermitage, P. 1852.54, amphora by Ennion, Lehrer 1979, 8, pl. 2, nos. 1 and 2.

For a small group of mold-blown cylindrical pyxides:

b) Yale, no. 1955.6.58, Matheson 1980, 47-48, no. 124 (with parallels).

On garlands in Roman art: M. Honroth, *Stadtrömische Girlanden* (Vienna 1971).

[2] Simon 1967, 13-14, pls. 7 and 8. Compare a marble base showing bucrania between white poplar branches, H. 73 m, from Rome, a sanctuary of Hercules, Rome, Museo Nazionale delle Terme, no. 417, Helbig[4], III, no. 2297; Rita Paris, in Antonio Giuliano ed. *Museo Nazionale Romano, Le Sculture* I. 1 (Rome 1979) 264-265, no. 162.

# II
# The Second to Fourth Centuries

During the second to fourth centuries the technique of mold blowing was used almost exclusively to make bottles and small containers. The production of fine mold-blown tablewares ceased shortly before or around the end of the first century. Although mold-blown tablewares were still in common use under the first two Flavian emperors, Vespasian (69-79) and Titus (79-81), as can be seen from dated finds from Jerusalem[1] and Masada,[2] Israel, and the cities of Campania buried by Mount Vesuvius,[3] production in the western part of the empire seems to have come to an end soon afterward.[4] The latest deposits in the West containing substantial numbers of mold-blown fragments are pits filled around the year 100 at Nijmegen (ancient Noviomagus), Netherlands.[5]

In the East mold-blown tablewares of relatively good quality were still produced throughout the second century, but the quantity was very small (see Cat. Nos. 115-119 and head-shaped bottles Cat. Nos. 142, 145, and 146). One of the more attractive types of tableware may have been made in Asia Minor:[6] mold-blown beakers decorated with mythological figures in high relief.[7] Almost one half of the beakers is said to be from Turkey. Three were excavated in northwest Turkey;[8] the find spots of the others are not named. Finds from controlled excavations at Masada, Israel,[9] Cyzicus, Turkey,[10] and Vindonissa, Switzerland,[11] date these beakers to the late first and early second centuries. If some were made in northwest Turkey, as the distribution pattern suggests, the art of mold blowing must have been introduced there by the fourth quarter of the first century. How long mold blowing continued to be practiced in this area has not been determined. Six fragments of mythological beakers from Dura Europos, destroyed and abandoned in 256, may date considerably later than the first century.[12] These fragments show figures comparable to those on first-century beakers, but one of the base fragments bears a mold-blown drinking exhortation in Greek, ΠΙΕ ΖΗΣΑΙΣ (*pie zesais*), "drink and live [long]," surrounded by a wreath, as on a bottle in Yale which has been tentatively dated to the third century.[13] The relationship between the Dura

[1] Avigad 1980, 107-108, 117, figs. 95-96, 112; Israeli 1983, 65-69.

[2] Barag 1991, 139, notes various classes of mold-blown wares from Masada deposits not later than about 73-74, but only a few fragments from the period 74-115.

[3] Scatozza Höricht 1991, 76-85.

[4] Price 1991, 74, notes that mold-blown tablewares are not common on western sites founded after about 75-80.

[5] Isings 1980.

[6] On glass from Asia Minor: G. F. Bass and F. H. van Doorninck "A Fourth Century Shipwreck at Yassi Ada," *AJA* 75 (1971) 27-37; Emel Erten Yagcı, "Hatay Müzesindeki bir grup cam eser," in *Uluslararası Anadolu Cam Sanatı Sempozyumu (1st International Anatolian Glass Symposium*, April 26-27, 1988, published Istanbul 1990) 30-36; Lightfoot 1989; Lightfoot 1990; Lightfoot and Arslan 1992; von Saldern 1980a; Stern 1984; Stern 1985a; Stern 1989a; Stern 1989b. Lightfoot 1989, 11, n. 8 and n. 10 lists Turkish publications and reports for sites in Turkey where glass has been excavated.

[7] Weinberg 1972, 26-47; Wight 1988, 71-76.

[8] Of a total of twenty-four beakers listed by Weinberg 1972, 26-47, ten are reported to have been found in Asia Minor, five from the Near East (Syria or Palestine), three from Switzerland, and one each from Kerch and Yugoslavia. Another example was found at Cosa (David F. Grose, pers. com.). This distribution is even more telling when confined to the beakers of Weinberg's Group I: out of eight examples listed, six have an Asia Minor provenance, three from excavations in northwest Turkey:

a) BM, no. 78.10-20.1, from Cyzicus, Weinberg 1972, 30, no. 3;
b) BM, number unknown, from Gallipoli, Weinberg 1972, 47;
c) Istanbul, Arch. Mus., no. 1637, from Balikesir (village of Çekirdeki), Weinberg 1972, 30-31.

[9] Weinberg 1972, 46-47, describes unpublished fragments of two beakers from Masada; these have now been assigned to the period 74-115 on the basis of coins: Barag 1991, 139.

[10] See above note 8a, found in a tomb dated to the first half of the second century on the basis of a lamp, D. M. Bailey, *Catalogue of Lamps in the British Museum* III (London 1988) 414, no. Q 3297, pl. 146.

[11] Vindonissa (modern Windisch), Switzerland, Mus., nos. 1941: 15 and 5215, from Vindonissa, dated after 60/75, Berger 1960, 51-52, nos. 126 and 127.

[12] Clairmont 1963, 14, no. 27, pl. XVIII; Matheson 1980, 105-106, fig. 1.

[13] Yale, no. 1955.6.51, Matheson 1980, 105-107, no. 280.

fragments and the bottle in Yale is not clear, but the possibility cannot be excluded that some mold-blown vessels with high relief were still being blown in the late second and third centuries in Asia Minor (for example, head-shaped flasks such as Cat. No. 148).

During the third century, mold blowing was a little more common than during the second century (see bottles Cat. Nos. 120-133 and head-shaped vessels Cat. Nos. 147-157), but it was apparently no longer a prestigious technique valued by consumers for display at dinner parties. Mold blowing had degenerated to mass production of inexpensive but attractive small bottles in transparent to translucent natural tints of glass.

During the fourth century mold blowing experienced a limited revival. It was still used almost exclusively for mass-produced wares (Cat. Nos. 134-136), but the technique was also adapted by one or more workshops to produce fine large figural objects such as a shell-shaped dish (Cat. No. 137), head-shaped vessels (Cat. Nos. 158-164), and vessels imitating statuettes.[14]

## CUT GLASS AND THE DECLINE OF MOLD-BLOWN TABLEWARES

The decline of mold-blown tablewares probably resulted from a change of fashion that took place in the third quarter of the first century. According to Pliny (*Natural History* 36.198), in his time, between about 50 and 77, the most highly valued glass was colorless and transparent, resembling rock crystal as closely as possible.[15] Pliny goes on to say that for drinking vessels the use of glass had ousted metals such as silver and gold. The new preference for transparent, decolorized vessels over translucent and opaque colored glass was probably the main reason fine mold-blown tablewares went out of fashion, since mold-blown designs are much less legible in clear glass (compare Cat. No. 33 with Cat. Nos. 26-32).

The new fashion was for colorless drinking cups decorated with figural motifs cut in high relief, such as a fragmentary cage cup excavated at Nijmegen, Netherlands, in a tomb dated about 80-100.[16] Early cut designs[17] were not very different in style and subject from the scrolls and vegetal motifs popular on mold-blown glass[18] and metal[19] vessels with high relief. Some of these early cut glass patterns were in turn copied on mold-blown wares (Cat. Nos. 8-11). Decoration cut in high relief did not remain popular for very long, perhaps in part because the high edges of designs made such vessels unpleasant to handle, but also because vessels decorated in this manner required a great deal of labor to carve and could not be turned out in sufficient quantity to satisfy growing demand, as had been possible with mold-blown tablewares.

The need to increase production to meet consumer demand probably explains why, by the early second century, facet cutting became the standard method of decorating colorless tablewares fashionable in both East and West. Facet-cut designs were relatively easy to sketch on the surface of a vessel and could be executed relatively quickly. They did not demand the high technical skills needed to design or carve figural designs in high relief. Moreover, facet cutting had the added advantage that it accentuated the rock-crystal-like shimmer of a vessel because each facet reflected light differently. On the other hand, mold-blown imitation facets do not read well because all surfaces of mold-blown glass are equally smooth. There is no differentiation in the light being reflected, as can be seen, for example, in the rows of sunken dots imitating cut facets above and below the main decoration of Cat. Nos. 115-118. Numerous simple patterns of facets were used to decorate drinking cups[20] and other fine tablewares.[21] As far as fine tablewares were concerned, mold-blowing appears to have been relegated to a support function. Vessels to be decorated with facets were blown into molds to impart the layout of the design.[22]

[14] For example:

a) RGZM, no. 292, from Cologne, Magusstrasse, vessel in the shape of a monkey holding a syrinx, Fremersdorf VI 1961, pls. 177-179, with parallels from northwestern Europe;

b, c) Yale, nos. 1955.6.81, provenance unknown, and 1955.6.274, said to have been found near Homs, Syria, vessels in the shape of seated Tyches, probably from a Syro-Palestinian workshop, Matheson 1980, 102-103, nos. 276 and 277.

[15] *Maximus tamen honos in candido tralucentibus, quam proxima crystalli similitudine*, translation by D. E. Eicholz, *The Loeb Classical Library*, no. 419 (London 1962, reprinted 1971, 1989).

[16] Annelies Koster, "Ein Diatretglas des 1. Jahrhunderts aus Nijmegen," *KölnJb* 22 (1989) 69-72, figs. 3-5; Annelies Koster and David Whitehouse, "Early Roman Cage Cups," *JGS* 31 (1989) 25-33, figs. 1-3.

[17] Axel von Saldern, "Roman Glass with Decoration Cut in High-Relief," in Newby and Painter 1991, 112-117, pls. 28 and 29.

[18] For example, Naples, Nat. Mus., no. 110489, from Pompeii, Spinazzola 1928, pl. 228; *Collezioni Napoli* 1989, 220-221, no. 25.

[19] U. Gehrig, *Hildesheimer Silberschatz im Antikenmuseum* (Berlin 1980); F. Baratte, *Le trésor d'orfevrerie romaine de Boscoreale* (Paris 1986); L. Pirzio Biroli Stefanelli et al., *L'Argento dei Romani. Vasellame da tavola e d'apparato* (Rome 1991).

[20] On facet-cut beakers: Andrew Oliver, Jr., "Early Roman Faceted Glass," *JGS* 26 (1984) 35-48 (with lit. and list of examples). Additional finds from the West:

a-c) Frankfurt, Germany, Mus. für Vor- und Frühgeschichte, nos. α23883, α24085, and α20169, from Nida-Heddernheim, Welker 1985, 25-26, pl. 7, nos. 82-84;

d) Frankfurt, Germany, A. Schröder Collection, from Nida-Heddernheim, Welker 1985, 27, no. 85;

e) fragments from Vitudurum, Switzerland, Rütti 1988, 55-57;

f) fragments from Augst, Switzerland, Rütti 1991, I, 44-45, Form AR 45, II, 58-59, nos. 1196-1210, pls. 51-52.

[21] Jennifer Price, "Late Hellenistic and Early Imperial Cast Vessel Glass in Spain," *AnnAIHV* 10 (1985, publ. 1987) 72-80, figs. 4-5; Grose 1991, 12-18. A few exceptional pieces of glass tableware, including a magnificent circular plate decorated with cut facets, were excavated in Israel where they were part of hoards hidden in caves during the revolt of Bar Kokhba against the Roman emperor Hadrian in 135, but there were very few mold-blown vessels among these finds: Barag 1963, 101-109; *Holy Land* 1986, 260, figs. 113-114. On the fragments of one mold-blown globular bottle or bowl from the Cave of Horror: Dan P. Barag, "Glass Vessels from the Cave of Horror," *IEJ* 12 (1962) 213, nos. 15-16, figs. 14-15, pl. 33A.

[22] Clay molds for bowls to be decorated with cut facets have been excavated at Tanais, in southern Russia, Sorokina 1977c, 118-119, fig. 4, no. 2.

## MOLD CONSTRUCTION TYPES (MCTs III, VII, VIII, and IX)

All the mold construction types invented in the first century included the bottoms of the vessels (MCTs I-VIII). Some of these mold types continued to be used during the second and third centuries: MCT III (Cat. Nos. 115-118), MCT VII (Cat. Nos. 120-128), and MCT VIII (Cat. Nos. 119 and 129-136). In the third century, however, the construction of molds seems to have been simplified: some multipart molds were open at the bottom (MCT IX). In the Toledo collection, MCT IX is documented by only a few head-shaped vessels (Cat. Nos. 149-156, 158, 160-164). Cat. No. 137 was made in a one-part cup-shaped mold.

## SHAPES

The forms made by mold blowing after the first century were most frequently small globular bottles and head-shaped vessels. A few second- and third-century flasks and "bottles" may have been used as decanters (Cat. Nos. 115-118, 120-128). Shell-shaped dishes made in the fourth century may also have been tablewares (Cat. No. 137).

A new shape, developed before the middle of the third century,[23] was the sprinkler, a bottle with an internal diaphragm at the base of the neck formed by a cut-out with a tiny aperture of about 0.3-0.5 cm. Sprinklers are probably the type of flask referred to in Jewish sources as a "flask whose collar (*ogen*) is squeezed inside it."[24] The narrowing of the aperture evidently had a function, but precisely what this was remains a matter for conjecture.[25] Sprinklers were made in a large variety of forms and sizes, for example, jars, tubes, amphorisks, animal-shaped bottles, and head-shaped flasks. Glass blowers made sprinklers using a large variety of techniques and decorated them in every known fashion, but the most common type of decoration was simple geometric mold-blown surface patterns. Fourth-century sprinklers are often decorated with expanded mold-blown designs;[26] but during the third century sprinklers were usually small, compact globular bottles decorated with mold-blown patterns that were not expanded (Cat. Nos. 129-133). Sprinklers are found frequently in Syria and in inland (eastern) Palestine, but they are not common in coastal (western) Palestine. Two production areas have been identified: one in Syria, the other in Mesopotamia.[27]

## DECORATION

Elaborate mold-blown figurative decoration in high relief, such as had been popular in the first half of the first century, was rarely produced after the end of the first century. Designs tended to be linear, usually forming simple geometric patterns like circles, lozenges, squares, and dots. These were relatively easy to arrange on a bulbous or spherical mold.

During the second century, at least one glass house experimented with designs in intaglio or sunken relief instead of in raised relief. Sunken designs were easy to execute because the lines could be scooped or melted out of a wax model or even hammered directly into a metal mold (see Cat. No. 115).[28] Decoration in sunken relief is known from a small family of flasks and jugs that may have been used as tablewares (Cat. Nos. 115-118).

A few mold-blown bottles were made in the shape of fruit (Cat. No. 119), but most were no longer shaped naturalistically. In the eastern Mediterranean a distinctive Type of Bottle with Stylized Grape Pattern was produced during the third century (see Cat. Nos. 120-128).

The decrease in production of mold-blown wares in the second to fourth centuries is reflected in the Toledo collection. Only twenty-three vessels can be assigned to this period, in addition to perhaps twenty-four of the head-shaped vessels discussed in Chapter III.

23 The earliest datable sprinklers are those from Dura Europos, abandoned in 256.

24 Tosephta, Mikwaoth 6, cited by *Function and Design in the Talmudic Period*, exh. cat., Haaretz Museum (Tel Aviv 1978, 2nd ed. 1979) 53*.

25 Eisen 1927, I, 333-334, suggested the vessels were used for sprinkling rose water. It seems unlikely that they were lamps, as suggested by Bremen 1964, 198, no. 17e.

26 Large mold-blown sprinklers were made by expanding the pattern after extraction from the mold and will therefore be included in a future catalogue of the Toledo Collection: TMA, *Roman and Early Byzantine Glass* (forthcoming).

27 For the most recent discussion: Stern 1977a, 95-100, which treats not only Syrian-type sprinklers but also Mesopotamian ones. On Mesopotamian sprinklers from Coche, Iraq: Negro Ponzi 1984, 33-40.

28 Metal molds with raised elements designed to create vessels with sunken relief are known to have been used for Jewish and Christian pilgrim flasks during the sixth and seventh centuries: see also below Chapter IV and Cat. Nos. 169-177.

# CATALOGUE

## *Flasks with Mold-blown Designs in Sunken Relief* (Cat. Nos. 115-118)

### 115 **Spherical Flask with Collar at Shoulder**

Color plate 18, p. 57
Eastern Mediterranean, perhaps Syro-Palestinian
Probably second century
H 13.1 cm; H body 8.0 cm; max D 8.0 cm; D rim 4.6 cm; D base 3.9×3.8 cm
1923.1084, Gift of Edward Drummond Libbey

Transparent yellowish olive brown (not in rock color chart). Medium thin glass. Pinprick and small bubbles.
Neck and mouth free blown. Body blown into a four-part mold of three vertical sections joined to a disk-shaped base section (MCT III). Relief crisp except just above the base. No pontil mark.
Rim outsplayed at a 45-degree angle, rounded and thickened in flame. Tall tubular neck, with slight bulge above constriction at its base. Concave shoulder above shoulder collar. Spherical body with walls curving down to vertical edge above concave base.
On the body, sunken relief design of nine interlocking circles with a dot in the center of each, bordered above and below by a double row of dots. On underside of base, a raised circle around a small central knob.
*Condition* Intact. Some cloudiness on exterior.
*Collection* Thomas E. H. Curtis.
*Published* Previously unpublished.
*Remarks* Same Type as Cat. No. 116, from a closely related mold, and perhaps also Cat. No. 117. Mold-blown vessels decorated with this design are usually yellowish brown (this bottle), manganese-colored purple (Cat. Nos. 116 and 117), natural bluish green, or decolorized glass.
The sunken relief design of this Type was probably melted or scooped out of a model made of wax or clay. Chill marks on the lower part of the body and the underside of the base suggest that the mold may have been made of metal.
Two variants have been identified, perhaps belonging to the same Series. In the first variant the upper edge of the mold had a short upright rim that formed a collar above the shoulder of the vessel (Cat. Nos. 115-117); the second variant does not have a shoulder collar (Cat. No. 118). Most vessels with collars are spherical flasks without handles, such as this;[1] a few are finished as jugs with a circular mouth.[2] However, most jugs were blown in molds that lacked the upright rim (see Cat. No. 118). Other shapes without a shoulder collar include jugs with a trefoil mouth,[3] aryballoi,[4] and jars.[5] The bodies of these vessels appear to be slightly smaller than those of vessels made in molds with an upright rim, and the relief is slightly less crisp. These features may therefore indicate that the second variant was copied from a vessel belonging to the first variant and that both variants belong to the same Series. The finishing of mouth and vessel rim varies considerably within each vessel shape, indicating that they were perhaps executed by different glass blowers.
The dates assigned to this Type in recent publications fluctuate among second century,[6] first half of third century,[7] and third to fourth centuries,[8] but excavated parallels and stylistic evidence point strongly to a date in the second century. A trefoil mouth jug from Georgia was found with a late first-century to early second-century horizontally ribbed amphora like Cat. No. 64.[9] The most important stylistic evidence for dating is provided by a tall-necked flask or bottle in Cologne.[10] It has a double-folded collar rim (outward, downward, outward, upward, and outward) of a type common in the last quarter of the first and the first half of the second centuries.[11] The use of a complicated four-part MCT III mold and the lack of a pontil mark also support a relatively early date for this Type.
Finds have been reported from Israel,[12] Syria,[13] Cyprus,[14] Georgia,[15] and, perhaps, Croatia.[16] These provenances suggest a Near Eastern origin for the Type, but give no clue to the location of the workshop. A Near Eastern origin is supported by the double-folded rim of the bottle in Cologne, because this type of rim was especially popular in Cyprus[17] and Asia Minor.[18] It has been suggested that aryballoi with this type of rim were made in Pergamon, in Asia Minor.[19]

1 For example:
a) MMA, no. 81.10.217, formerly Charvet-Marquand Collection, formerly Cesnola Collection, from Idalion, Cyprus, with shoulder collar, H 11.4 cm, max D 7.9 cm, D rim 4.9 cm, Cesnola 1903, pl. 78, no. 4; John L. Myres, *The Metropolitan Museum of Art. Handbook oj the Cesnola Collection of Antiquities from Cyprus* (New York 1914) 567;
b) Murano, Mus. Vetrario, reported to be from the necropolis at Nona, Zadar, Croatia, decolorized glass (unpublished);
c) Philadelphia, University Museum, no. MS 4953, acquired at Aleppo (unpublished);
d) RGM, no. N1055, olive green, with a very tall neck, H 16.5 cm, La Baume 1973, no. C 23, pl. 13, no. 3;
e) RGZM, Behn 1913-14, 10, fig. 9.

2 Damascus, Nat. Mus., Abdul Hak 1951, pl. XLVII, no. 2a (cannot be identified in text).

3 Examples of jugs with a trefoil mouth:
a) Tblisi, Georgia, no. 1-58: 1901, from Urbnis, tomb 20, Saginashvili 1970, 96, no. 12, fig. 12;
b) Newark, no. 50.1489, Auth 1976, 79, no. 83;
c) ROM, no. 950.157.71, Hayes 1975, 50, no. 95, pl. 8;
d) Carnegie, no. 22554/5, Oliver 1980, 72, no. 70;
e) BM, no. S209, Nesbitt 1871, 35, no. 209, pl. VII, no. 2;
f) Oppenländer Collection, no. 2346, von Saldern et al. 1974, 176, no. 488, ill. p. 175.

4 Examples of aryballoi:
a) South Germany, private collection, Luzern 1981, 87, no. 308;
b) formerly R. W. Smith Collection, no. 1369, Smith 1957, 133, no. 254.

5 Sorokina 1967, 70, fig. 2, no. 24.

6 Sorokina 1984, 258, pl. LXXI, no. 102.

7 Hayes 1975, 50, at no. 95.

8 La Baume 1973, at no. C 23; Auth 1976, 79, at no. 83; von Saldern et al. 1974, 176, at no. 488.

9 See above note 3a.

10 See above note 1d.

11 This rim type occurs predominantly on utilitarian wares such as cylindrical or square bottles and aryballoi, for example, a cylindrical bottle from a cave in the Judaean Desert used in the Bar Kochba period (before 135): Barag 1963, 102, no. 1-5.7, fig. 38, pl. 30. A spherical bottle with two handles was found in a tomb at Palmyra, dated about 150: K. Michalowski, *Palmyra 1959* (Warsaw 1960) 199, no. 35, fig. 222. Vessels with this type of rim must have been made some time before 79, because they have also been excavated at Herculaneum: Ercolano, Antiquarium, no. E 157, spherical jug, Scatozza Höricht 1986, 49, no. 97, pl. XXXI, and a globular bottle without handles cited by Isings 1957, 90, Form 70. On cylindrical bottles with this type of rim: D. Charlesworth, "Roman Square Bottles," *JGS* 8 (1966) 27, fig. 3a (Isings rim type 2a). On aryballoi: Sorokina 1987, 43, figs. 2, no. 5, and 3 (a variation of Isings rim type B).

12 Jerusalem, Rockefeller Mus., no. 1656.42.97, possibly from near Nahariya, cited by Auth 1976, 79, at no. 83.

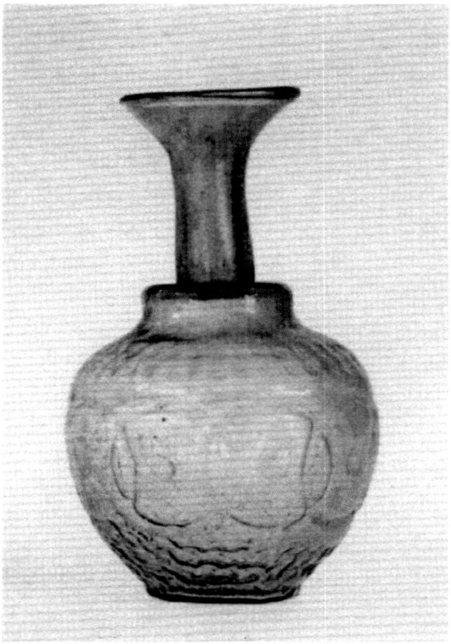
115 (1:2)

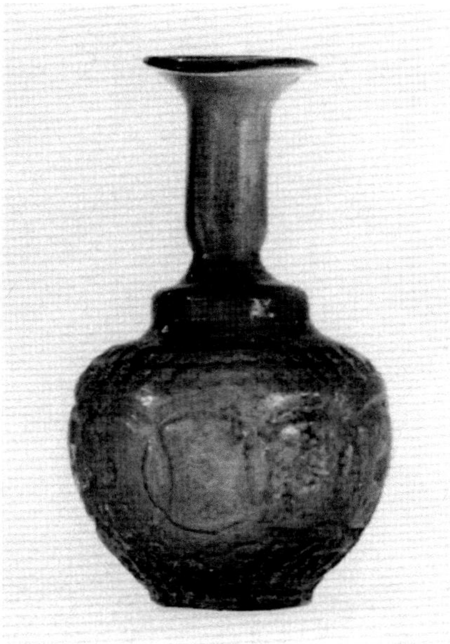
116 (1:2)

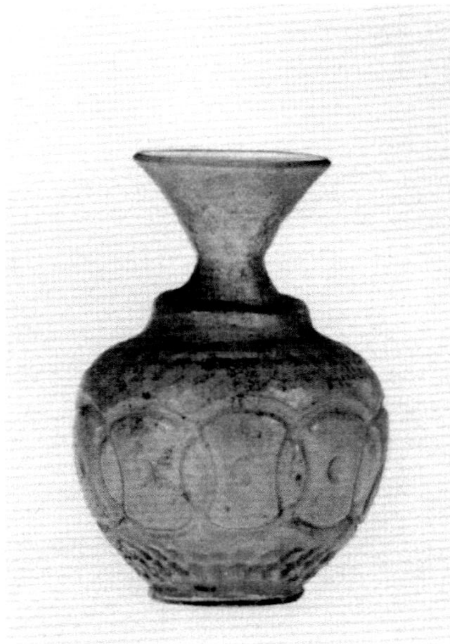
117 (1:2)

[13] See above note 1c and note 2.
[14] See above note 1a and BM, no. 98.7-21.1, cited by Auth 1976, 79, at no. 83.
[15] See above note 3a.
[16] See above note 1b.
[17] Vessberg 1952, pls. V, nos. 8 and 18, and VII, nos. 10, 18, 21-22.
[18] Sorokina 1984, 257, pl. 70.
[19] Sorokina 1987, 42.

### 116 **Spherical Flask with Collar at Shoulder**

Color plate 18, p. 57
Eastern Mediterranean, perhaps Syro-Palestinian
Probably second century
H 13.8 cm; H body 8.0 cm; max D 8.05 cm; D rim 4.5-4.4 cm; D base 4.0 cm
1923.1085, Gift of Edward Drummond Libbey

Transparent manganese colored streaked grayish red (10 R 4/2). Medium thin glass. Pinprick bubbles elongated vertically in neck.
Neck and mouth free blown. Body blown into a four-part mold of three vertical sections joined to a disk-shaped base section (MCT III). Relief crisp except just above the base. No pontil mark.
Everted rim, rounded and partially thickened in flame, with tool mark along interior. Tall tubular neck, with slight bulge above constriction at its base. Shoulder sloping to horizontal edge above shoulder collar. Spherical body with walls curving down to vertical edge above concave base.
On the body, sunken relief design of nine interlocking circles with a dot in the center of each, bordered above and below by a double row of dots. On underside of base, a raised circle around a small central knob.
*Condition* Intact. Exterior partly cloudy and covered with iridescence.
*Collection* Thomas E. H. Curtis.
*Published* Grose 1978, fig. 20; Grose 1984b, 37, ill.
*Remarks* See Remarks Cat. No. 115. Same Type as Cat. No. 115, from a closely related mold.

### 117 **Spherical Flask with Collar at Shoulder**

Color plate 18, p. 57
Eastern Mediterranean, perhaps Syro-Palestinian
Probably second century
H 11.3 cm; H body 8.0 cm; max D 8.0 cm; D rim 5.0 cm; D base 3.9 cm
1923.889, Gift of Edward Drummond Libbey

Transparent very pale light brownish gray (5 YR 6/1) with darker manganese colored swirls (glass not well mixed). Medium thin glass. Small diagonally elongated bubbles in neck.
Neck and mouth free blown. Body blown into a four-part mold of three vertical sections joined to a disk-shaped base section (MCT III). Relief crisp. No pontil mark.
Rim rounded in flame. Funnel neck, with constriction at its base. Shoulder sloping to horizontal edge above shoulder collar. Spherical body with walls curving down to vertical edge above slightly concave base.
On the body, sunken relief design of nine interlocking circles with a dot in the center of each, bordered above and below by a double row of dots. On underside of base, a raised circle around a small central knob.
*Condition* Intact. Exterior with purple iridescence. Interior cloudy and pitted.
*Collection* Thomas E. H. Curtis.
*Published* Previously unpublished.
*Remarks* Shoulder collapsed on one side during blowing or annealing.
See Cat. No. 115. Same Type as Cat. Nos. 115 and 116, from a related mold. A flask with identical dimensions and a similar funnel-shaped neck was excavated at Idalion, Cyprus.[1]

[1] MMA, no. 81.10.217, formerly Charvet-Marquand Collection, formerly Cesnola Collection, from Idalion, Cyprus, with shoulder collar, H 11.4 cm, max D 7.9 cm, D rim 4.9 cm, Cesnola 1903, pl. 78, no. 4; John L. Myres, *The Metropolitan Museum of Art. Handbook of the Cesnola Collection of Antiquities from Cyprus* (New York 1914) 567.

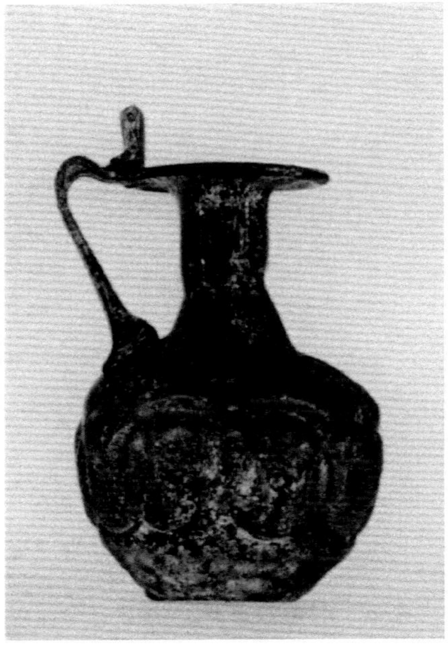

118 (1:2)

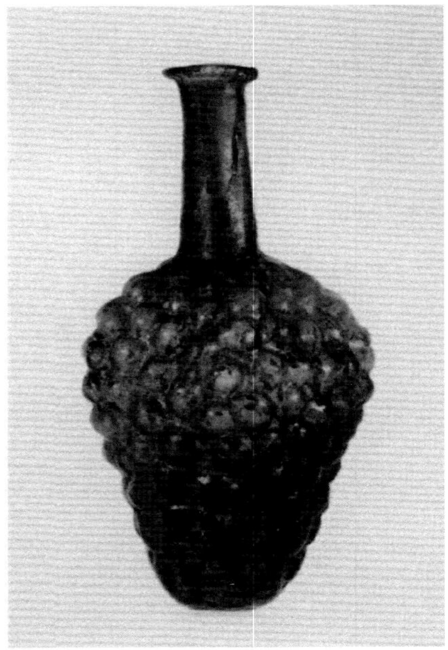

119 (1:2)

118 **Spherical Jug without Collar at Shoulder**

Eastern Mediterranean, perhaps Syro-Palestinian
Probably second century
H 11.0 cm; H with handle 12.85 cm; H body 6.6 cm; max D 7.9 cm; D rim 5.2 cm; D base 4.0-3.8 cm
1923.888, Gift of Edward Drummond Libbey

Transparent to translucent manganese colored dusky red (near 5 R 3/4). Similarly colored handle. Medium thin glass. Fabric cannot be determined because of weathering.
Neck and mouth free blown. Body blown into a four-part mold of three vertical sections joined to a disk-shaped base section (MCT III). Relief moderately crisp. No pontil mark.
Rim folded outward, upward, inward, and flattened. Circular mouth. Cylindrical neck, with constriction at its base. Strongly sloped shoulder. Spherical body. Slightly concave base. Angular strap handle applied to shoulder and attached to rim where it is folded upward, backward, downward, and pinched to form a vertical projection.
On the body, sunken relief design of nine interlocking circles with a dot in the center of each; double row of dots below; faint remains of a double row of dots above on one-third of the body corresponding to one vertical mold section. Underside of base undecorated.
*Condition* Complete. Vertical projection of handle reattached. Exterior with purple and green iridescence and severe surface pitting.
*Collection* Thomas E. H. Curtis.
*Published* Previously unpublished.
*Remarks* See Remarks Cat. No. 115. Same Type as Cat. Nos. 115-117, but without a shoulder collar. Jugs such as this, blown in molds without an upright rim, usually have a trefoil mouth.[1] A jug with a circular mouth, like this one, presumably found in Syria,[2] has a shoulder collar and was therefore blown in a mold with an upright rim.

[1] For examples of jugs trefoil mouth, see Cat. No. 115, note 3a-f.
[2] Damascus, Nat. Mus., Abdul Hak 1951, pl. XLVII, no. 2a (cannot be identified in text).

## *Bottle Shaped like a Naturalistic Cluster of Grapes* (Cat. No. 119)

119 **Grape Cluster-shaped Bottle**

Probably Syro-Palestinian
Probably second half of second century
H 14.0 cm; H body 9.05 cm; D 7.4-7.15 cm; D rim 2.4 cm
1951.373, Purchased with funds from the Libbey Endowment, Gift of Edward Drummond Libbey

Translucent to transparent patchy manganese colored glass. Medium thin glass. Fabric cannot be determined because of weathering.
Neck free blown. Body blown into a two-part mold of two vertical sections with a roughly square circumference like head-flask Cat. No. 141 (MCT VIII). Relief crisp except near base where it has lost definition and the walls are smooth.[1] No pontil mark.
Rim folded outward, upward, inward, and downward. Tall cylindrical neck. Irregular inverted piriform body. Poorly flattened base (vessel cannot stand).
Body shaped like a cluster of grapes; four smaller clusters of grapes around shoulder.
*Condition* Intact. Exterior covered with oily iridescent film. Incipient surface pitting and dark specks.
*Published* Previously unpublished.
*Remarks* Bottles in the shape of a naturalistic grape cluster were made in both the eastern and western parts of the Roman empire. Grape-cluster bottles made in the second century and later are larger than those made in the first century (compare Cat. Nos. 109 and 110). A fired clay half-mold (MCT VIII) for a grape-cluster bottle was excavated at the site of a glass-working factory at Macquenoise, in southern Belgium, showing that some bottles were made there.[2] Bottles found in the West usually have two handles,[3] while those from eastern find spots have no handles.[4] This bottle belongs to the group made in the East.
Fragments excavated at Dura Europos[5] indicate that in the East production of these large, realistic grape cluster-shaped bottles began before 256, when the city was abandoned, but the eastern vessels may date much earlier if they are contemporary with western finds, which usually appear in contexts dated to the end of the second century.

[1] The mold seam is not visible on the underside of the base because the vessel was reheated after extraction from the mold. A circular chill mark on the underside of the base is the result of flattening the base, probably with a wet wooden paddle.

[2] Chambon 1955, 31, and frontispiece; *Glass of the Caesars* 1987, 151-152.

[3] Isings 1957, 108-109, Form 91A, with parallels from finds dated to the end of the second century.

[4] Examples of the same Type:

a) Hama, Syria, Mus., no. 942, purple, from the region of Hama, Abdul-Rahim Masri, "Collection de verres du Musée de Hama," *BullAIHV* 3 (1964) 72, no. 4, fig. 70;
b) Oppenländer Collection, no. 2402, formerly Mavrogordato Collection, purple, von Saldern et al. 1974, 173, no. 472;
c) formerly Beirut, Kirchner-Schwarz Collection, purple, *Kirchner-Schwarz* 1914, 4, no. 40, pl. 2;
d) Loeffler Collection, no. 22, reddish purple, La Baume and Salomonson 1976, 38, no. 76, pl. 8, no. 4;
e) CMG, no. 64.1.8, formerly R. W. Smith Collection, no. 390, deep blue, Smith 1957, 135, no. 259;
f) Tel Aviv, no. MHG 9158.

[5] Yale, no. 1934.556, amber, and no. 1931.606b, natural bluish green, Clairmont 1963, 38-39, nos. 148 and 149.

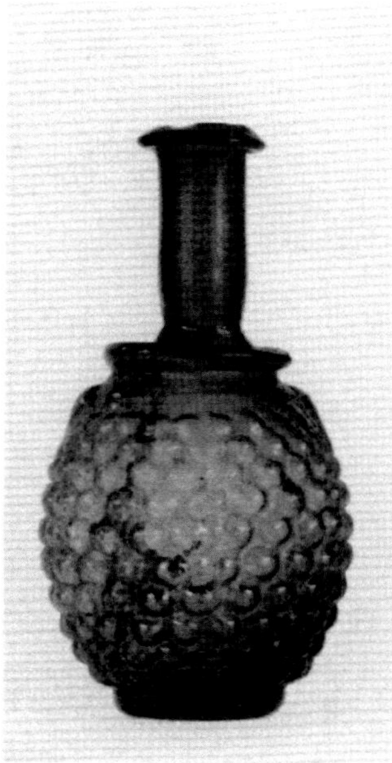
120 (1:2)

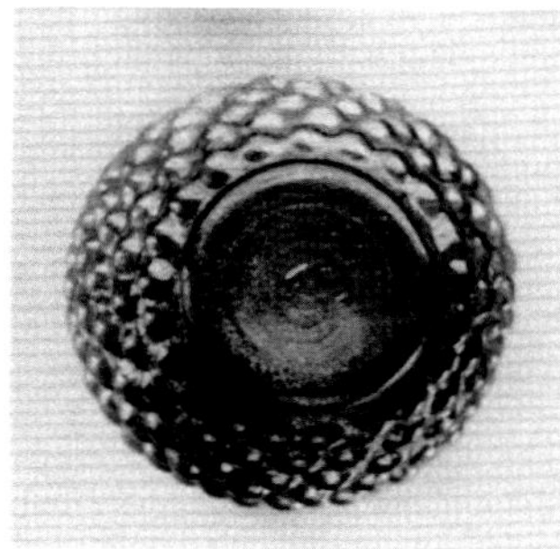
120, underside (1:2)

## *Bulbous Bottles with Stylized Grape Pattern* (Cat. Nos. 120-128)

### 120 **Ovoid Bottle with Stylized Grape Pattern**

Color plate 19, p. 57
Probably Syrian or inland Palestinian
Probably third century
H 12.9 cm; H body 8.35 cm; D 6.7-6.65 cm; D rim 2.8-2.6 cm; D base 3.5 cm
1923.1077, Gift of Edward Drummond Libbey

Translucent light olive brown (near 5 Y 5/6). Medium thin glass. Numerous small to medium vertically elongated linear bubbles in neck.

Neck and projecting roll free blown. Body and collar above shoulder blown into a three-part mold of two vertical sections joined to a disk-shaped base section (MCT VII). Vertical mold seams halfway between the leaves. Relief crisp. Pontil mark ca. 1.0 cm.

Triangular hollow rim, folded downward and outward, then upward and inward. Tall cylindrical neck, with constriction at its base. Hollow projecting roll above shoulder collar. Ovoid body. Slightly concave base.

On the body, stylized pattern of grapes consisting of ten interlocking rows of twenty-one contiguous hemispherical knobs. At the shoulder, opposite each other and centered between the mold seams, small triangular leaves with a central vein. On underside of base, three narrow raised concentric circles (see detail photograph).

*Classification* Stylized Grape Bottle, Series A.

*Condition* Intact. Exterior with iridescent film. Interior with brownish weathering.

*Collection* Thomas E. H. Curtis.

*Published* Previously unpublished.

*Remarks* Same Type as Cat. Nos. 121-128. Same Series and probably same Generation as Cat. No. 121. The hemispherical knobs of this bottle represent a stylized cluster of grapes, as can be determined from the two small triangular leaves at the shoulder of the vessel. Most bottles belonging to this Type have a hollow projecting roll, or overblow, just above the shoulder collar formed by the upright rim of the mold, which may have been inspired by the shoulder collars of bottles such as Cat. Nos. 115-117. The projecting roll is lacking on Cat. Nos. 123 and 126. Although the finishing of neck and rim varies greatly from bottle to bottle, the height of the mold-blown body is usually about 8.5-8.0 cm. The glass is commonly an intense yellowish brown or natural bluish green, presumably in imitation of grapes. A few bottles have been provided with handles, for example Cat. No. 121, with two small loop handles from shoulder to base of neck, and Cat. Nos. 125 and 126, each with one handle from shoulder to rim. One bottle is finished as a sprinkler with an interior diaphragm at the base of the neck (Cat. No. 123).

Three Series are represented among the nine Toledo bottles. Series A,[1] to which this bottle belongs, has ten rows of large, contiguous knobs and a single leaf in the center of each mold half. The underside of the base of the earliest bottles is decorated with narrow concentric rings in relief (in addition to this bottle, see Cat. Nos. 121 and 123). Later generations have less crisp relief and an undecorated base (e.g. Cat. Nos. 124-126). On the Toledo bottles with undecorated bases the position of the knobs along the mold seams differs, indicating that they were made in different sets of molds.

Series B has eleven interlocking rows of small, relatively widely spaced knobs and two small vestigial leaves, one adjacent to and the other on a vertical mold seam (Cat. No. 127). Series C may have been recut from Series B: the knobs are unevenly spaced and no leaves appear at the shoulder (Cat. No. 128). The undersides of the bases of Series B and C are not decorated. The fact that all three Series have very similar small solid pontil marks, 0.8-1.2 cm, may indicate that they were blown in one workshop.

Provenances suggest that stylized grape bottles of this Type were used in the Syro-Palestinian area[2] and, perhaps, in Egypt.[3] None has been excavated in an independently datable context. The third-century date is based on stylistic evidence. The earliest datable occurrence of a tall neck rising from an overblow or projecting roll is a head flask excavated in a second-century tomb at Constanta, Romania.[4] The latest stylized grape flasks have no projecting roll but are finished like jugs with an added thread around the exterior of the mouth and at the base of the neck (see Cat. No. 126). This type of decoration is characteristic of fourth-century jugs in Palestine.[5]

[1] Numerous examples exist of Series A, but none from a controlled excavation. For example:

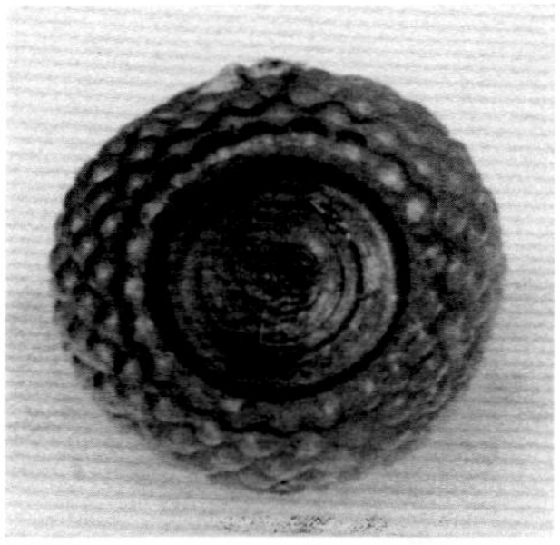
121, underside (1:2)

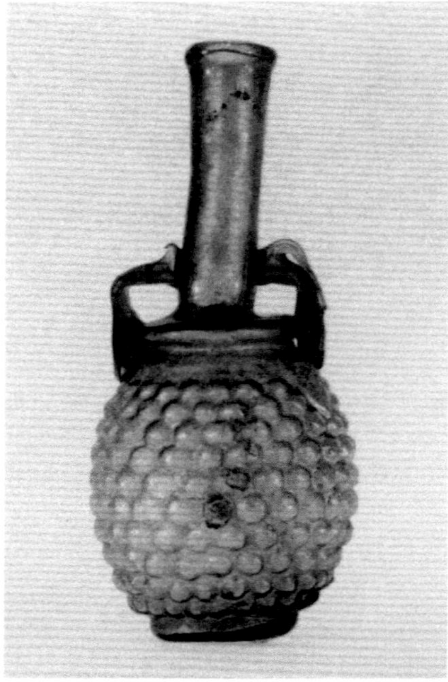
121 (1:2)

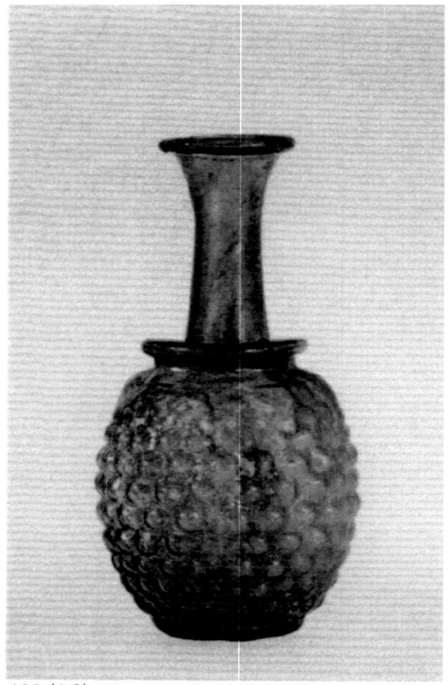
122 (1:2)

a) Newark, no. 50.1465, amber, Auth 1976, 73, no. 72;
b) ROM, no. 950.157.85, Hayes 1975, 49, no. 91, pl. 7;
c) Boston, MFA, no. 99.78, von Saldern 1968, no. 38;
d) Oppenländer Collection, no. 2221, amber, von Saldern et al. 1974, 171-172, no. 470.

[2] Examples from Syro-Palestinian area:

a) Nazareth, Israel, Mus. Francescanum, from Beisan (Beth Shean), Bagatti 1967a, 229, no. 77, figs. 3 and 7;
b) Damascus, Nat. Mus., no. 10.675, from Khisfine, tomb II, brown, Abdul Hak 1951, 108, pl. XLI, no. 2b;
c) Damascus, Nat. Mus., no. 14925, Zouhdi 1964, 49, no. 52, fig. 35;
d) formerly J. A. Durighello Collection, probably from Palestine, Durighello 1911, 9, no. 51, pl. 1;
e) formerly Azeez Khayat Collection, Khayat 1913, no. 183, ill.;
f) formerly Azeez Khayat Collection, Khayat 1919, 24, no. 245, ill.;
g) formerly Azeez Khayat Collection, Khayat 1917, no. 251, ill.;
h) Philadelphia, University Museum, no. MS 5114, acquired at Yebna (unpublished).

[3] The provenances of two bottles in the Museum at Al Minya, in Middle Egypt, are unknown: one is made of natural bluish gray glass, H ca. 12 cm; the other is light purple, H ca. 9 cm. Fragments of mold-blown bulbous bottles decorated with knobs on the body have been excavated at Wadi Qitna and other Egyptian sites: Eugen Strouhal, *Wadi Qitna and Kalabsha South*, I: *Archaeology* (Prague 1984) 210, no. P2918, from Wadi Qitna, citing parallels from Kalabsha-North and Karanog. However, it is not clear whether these Egyptian finds belong to the Type discussed here. The Egyptian finds do not have a projecting roll; they may be more closely related to a flask from Karanis, now in Cairo, Egyptian Museum, no. J51406, Harden 1936, 213, no. 628, pls. 8 and 18, which appears to have been blown in a two-part mold of Type VIII and is somewhat smaller than most of the bottles from the Syro-Palestinian area.

[4] Constanta, Arch. Mus., no. 5041, Bucovala 1968, 115-116, no. 237. Compare Cat. No. 147. See also Chapter III: *Survey: Second Century*, Type E, Series 2 (here Fig. 91).

[5] TMA, *Roman and Early Byzantine Glass* (forthcoming).

### 121 Ovoid Bottle with Stylized Grape Pattern

Probably Syrian or inland Palestinian
Probably third century
H 15.4 cm; H body 8.5 cm; max D 6.6 cm; D rim 2.3 cm; D base 3.47 cm
1923.1225, Gift of Edward Drummond Libbey

Translucent grayish yellow green (near 5 GY 7/2) to dusky yellow green (5 GY 5/2). Similarly colored handles. Medium thick glass. Pinprick to large vertically elongated bubbles in neck.
Neck and projecting roll free blown. Body and collar above shoulder blown into a three-part mold of two vertical sections joined to a disk-shaped base section (MCT VII). Vertical mold seams halfway between the leaves. Relief crisp. Pontil mark ca. 1.1 cm. Tips of handles drawn out thin and folded back.
Rim rounded and thickened in flame; tool mark on the interior. Tall cylindrical neck, with constriction at its base. Hollow projecting roll above shoulder collar. Ovoid body. Slightly concave base. Two coil handles attached to shoulder and applied to lower part of neck, positioned over leaves.
Decoration like Cat. No. 120 (see detail photograph of underside of base).
*Classification* Stylized Grape Bottle, Series A.
*Condition* Two small losses in body restored. Backfold of one handle missing.
*Collection* Thomas E. H. Curtis.
*Published* Previously unpublished.
*Remarks* Same Type as Cat. No. 120. Same Series and probably same Generation as Cat. No. 120. Handles are rare on this Type of vessel.

### 122 Ovoid Bottle with Stylized Grape Pattern

Probably Syrian or inland Palestinian
Probably third century
H 13.0 cm; H body 7.7 cm; D 6.7-6.3 cm; D rim 3.5-3.4 cm; D base 3.7×3.3 cm
1923.1080, Gift of Edward Drummond Libbey

Translucent to transparent dark grayish green (10 Y 6/6) to light olive brown (5 Y 5/6). Medium thick glass. Small vertically elongated oval and linear bubbles in neck and small spherical bubbles in body.
Neck and projecting roll free blown. Body and collar above shoulder blown into a three-part mold of two vertical sections joined to a disk-shaped base section (MCT VII). Vertical mold seams

halfway between the leaves. Relief crisp. Pontil mark ca. 1.1 cm. Flaring rim, folded outward, upward, inward, and downward. Tall cylindrical neck. Hollow projecting roll above shoulder collar. Ovoid body. Flat base.
On the body, stylized pattern of grapes consisting of ten interlocking rows of twenty-one contiguous hemispherical knobs. At the shoulder, opposite each other and centered between the mold seams, two small vestigial leaves without central vein. Underside of base undecorated.
*Classification* Stylized Grape Bottle, Series A.
*Condition* Intact. Exterior with iridescent film.
*Collection* Thomas E. H. Curtis.
*Published* Previously unpublished.
*Remarks* Same Type as Cat. No. 120. Same Series as Cat. No. 120 but a younger Generation.

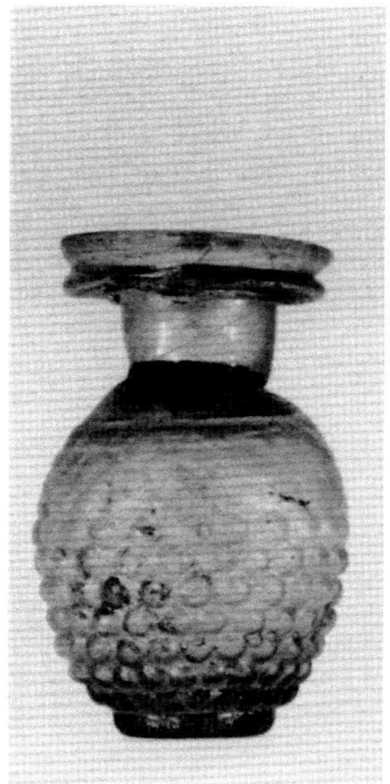

123 (1:2)

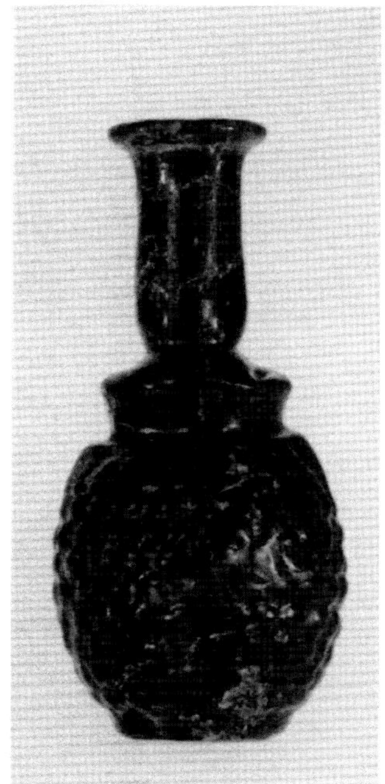

124 (1:2)

### 123 **Ovoid Bottle with Stylized Grape Pattern (Sprinkler)**

Probably Syrian or inland Palestinian
Probably third century
H 11.4 cm; H body 8.1 cm; D 7 cm; D rim 6.2-5.0 cm, D base 3.6-3.3 cm
1923.1078, Gift of Edward Drummond Libbey

Translucent to transparent grayish yellow green (near 5 GY 7/2). Similarly colored thread. Medium thin glass. Numerous small spherical bubbles, elongated horizontally in rim.
Neck and mouth free blown. Body blown into a three-part mold of two vertical sections joined to a disk-shaped base section (MCT VII). Vertical mold seams halfway between the leaves. Relief moderately crisp but indistinct in upper part of body. Pontil mark ca. 0.8 cm. Added thread.
Circular flaring mouth with hollow collar formed by a pushed-out projecting roll. Short cylindrical neck, with an interior diaphragm at its base. Ovoid body. Flat base.
At top of neck, an applied thread. On the body, stylized pattern of grapes consisting of ten interlocking rows of twenty-one contiguous hemispherical knobs. At the shoulder, opposite each other and centered between the mold seams, two small leaves with a central vein. On underside of base, three narrow raised concentric circles.
*Classification* Stylized Grape Bottle, Series A.
*Condition* Intact. Exterior with iridescent film.
*Collection* Thomas E. H. Curtis.
*Published* Previously unpublished.
*Remarks* Same Type as Cat. No. 120. The relief on the shoulder and upper part of body lost its definition when the paraison was reheated to form the interior diaphragm. On sprinklers, see *Glossaries* and Chapter II: *The Second to Fourth Centuries: Shapes.*

### 124 **Ovoid Bottle with Stylized Grape Pattern**

Probably Syrian or inland Palestinian
Probably third century
H 13.9 cm; H body 8.65 cm; D 6.4-6.3 cm; D rim 3.5-3.46 cm; D base 3.5 cm
1923.1083, Gift of Edward Drummond Libbey

Translucent dark grayish green (near 10 GY 4/2). Thick glass. Numerous pinprick and small vertically elongated bubbles in neck.
Neck free blown. Body and collar above shoulder blown into a three-part mold of two vertical sections joined to a disk-shaped base section (MCT VII). Vertical mold seams halfway between the leaves. Relief indistinct. Pontil mark ca. 1.22 cm.
Rim folded outward, upward, and inward. Tall cylindrical neck, with constriction at its base. Collar above shoulder. Ovoid body. Slightly concave base.
On the body, stylized pattern of grapes consisting of ten interlocking rows of twenty-one contiguous hemispherical knobs. At the shoulder, opposite each other and centered between the mold seams, two small vestigial leaves. Underside of base undecorated.
*Classification* Stylized Grape Bottle, Series A.
*Condition* Complete; one small chip from outer layer of glass at edge of base. Exterior with areas of iridescence.
*Collection* Thomas E. H. Curtis.
*Published* Previously unpublished.
*Remarks* Same Type as Cat. No. 120. Same pattern as Cat. No. 120, but a younger Generation made in a different set of molds, as can be seen by the arrangement of the knobs along the mold seams. Probably same Series and same Generation as Cat. No. 125.

### 125 **Ovoid Jug with Stylized Grape Pattern**

Probably Syrian or inland Palestinian
Probably third century
H 15.0 cm; H with handle 15.5 cm; H body 8.9 cm; D 6.5 cm; D rim 6.1 cm; D base 3.5 cm
1923.1076, Gift of Edward Drummond Libbey

Translucent manganese colored glass. Translucent pale green handle. Thick glass. Fabric cannot be determined because of weathering.
Neck and projecting roll free blown. Body and collar above shoulder blown into a three-part mold of two vertical sections joined to a disk-shaped base section (MCT VII). Vertical mold seams halfway between the leaves. Relief moderately crisp. Pontil mark ca. 1.25 cm.
Flaring mouth, with rim rounded and thickened in flame. Tall slightly tapering neck with curved transition to shoulder. Indistinct projecting roll above shoulder collar. Ovoid body. Flat base. Coil handle applied to shoulder and attached to rim, positioned over leaf.
On the body, stylized pattern of grapes consisting of ten interlocking rows of twenty-one contiguous hemispherical knobs. At the shoulder, opposite each other and centered between the mold

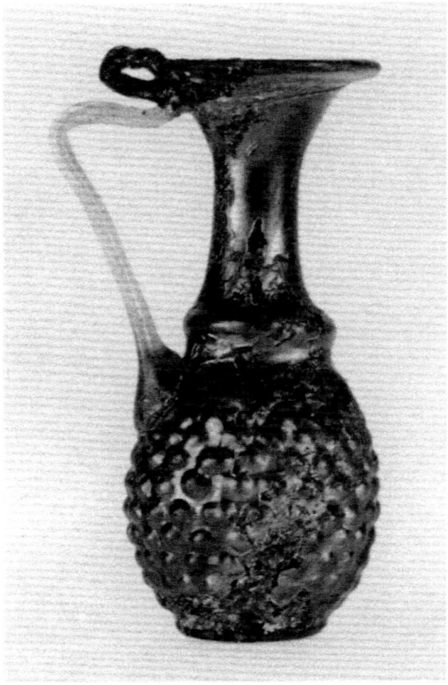
125 (1:2)

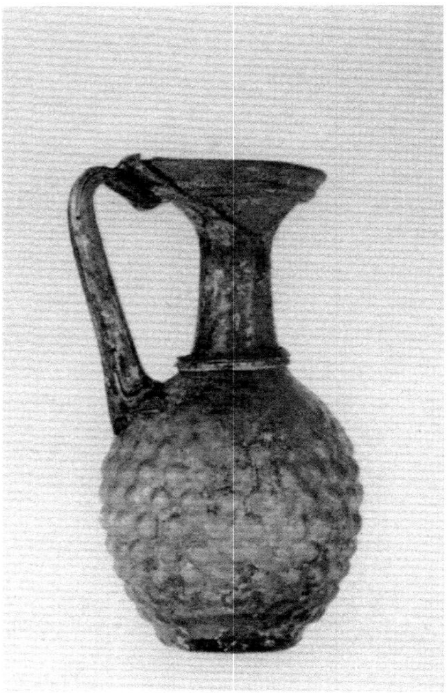
126 (1:2)

seams, two small vestigial leaves. Underside of base undecorated.
*Classification* Stylized Grape Bottle, Series A.
*Condition* Nearly complete; top of handle missing. Exterior and interior covered with thick green, blue, and brown iridescence; severe surface pitting.
*Collection* Thomas E. H. Curtis.
*Published* Previously unpublished.
*Remarks* Same Type as Cat. No. 120. Same Series and probably same Generation as Cat. No. 124.

126 **Ovoid Jug with Stylized Grape Pattern**

Probably Syrian or inland Palestinian
Late third or early fourth century
H 12.5 cm; H body 7.2 cm; D 6.5-6.4 cm; D rim 4.8 cm; D base 3.4×3.2 cm
1923.1075, Gift of Edward Drummond Libbey

Translucent grayish green (10 GY 5/2). Similarly colored handle and coils. Medium thick glass. Fabric cannot be determined because of weathering.
Neck and collar free blown. Body blown into a three-part mold of two vertical sections joined to a disk-shaped base section (MCT VII). Vertical mold seams halfway between the leaves. Relief indistinct. Pontil mark ca. 1.05 cm. Tip of handle cut off. Added coil.
Flaring mouth, with rim rounded in flame. Tall, slightly tapering neck with curved transition to shoulder. Ovoid body. Flat base. Coil handle applied to shoulder and attached to rim, positioned over leaf.
Double spiral coil below rim and at the base of the neck. On the body, stylized pattern of grapes consisting of ten interlocking rows of twenty-one contiguous hemispherical knobs. At the shoulder, opposite each other and centered between the mold seams, two small vestigial leaves. Underside of base undecorated.
*Classification* Stylized Grape Bottle, Series A.
*Condition* About three-fifths of mouth restored. Strain crack from lower neck to middle of body. Exterior iridescent; areas of severe surface pitting.
*Collection* Thomas E. H. Curtis.
*Published* Previously unpublished.
*Remarks* Same Type as Cat. No. 120. This jug lacks the characteristic shoulder collar and pushed-out projecting roll. Instead, certain features such as the rim coil and the coil around the base of the neck may be reminiscent of fourth-century jugs made in Palestine.[1]

[1] For example, TMA, no. 1923.980; see TMA, *Roman and Early Byzantine Glass* (forthcoming).

127 **Ovoid Bottle with Stylized Grape Pattern**

Color plate 19, p. 57
Probably Syrian or inland Palestinian
Probably third century
H 13.0 cm; H body 8.1 cm; D 6.5-6.3 cm; D rim 3.95-3.8 cm; D base 3.55×3.4 cm
1923.1082, Gift of Edward Drummond Libbey

Translucent moderate brown. Medium thick glass. Numerous medium to large linear bubbles in neck; some burst at surface of interior.
Neck and projecting roll free blown. Body and collar above shoulder blown into a three-part mold of two vertical sections joined to a disk-shaped base section (MCT VII). Vertical mold seams adjacent to and under the leaves. Relief crisp. Pontil mark ca. 1.0 cm in diameter.
Triangular hollow rim, folded downward-outward, then upward-inward. Tall cylindrical neck, with constriction at its base. Hollow projecting roll above shoulder collar. Ovoid body. Flat base. On the body, stylized pattern of grapes consisting of eleven interlocking rows of twenty-one small, widely spaced hemispherical knobs. At the shoulder next to mold seams, two small vestigial leaves. Underside of base undecorated.
*Classification* Stylized Grape Bottle, Series B.
*Condition* Intact.
*Collection* Thomas E. H. Curtis.
*Published* Previously unpublished.
*Remarks* Same Type as Cat. No. 120. Series B is characterized by smaller, noncontiguous knobs arranged in eleven instead of ten

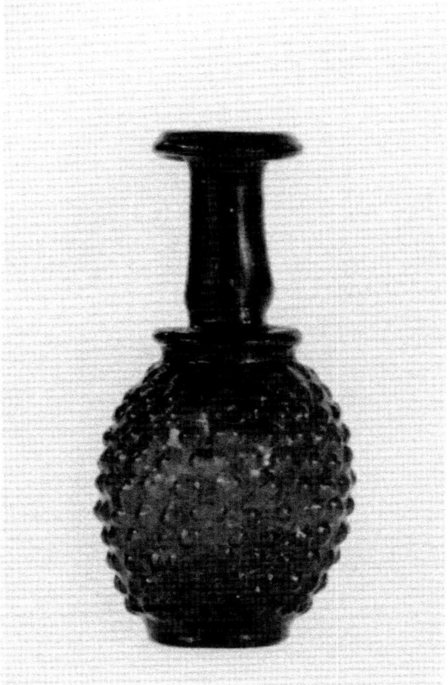

127 (1:2)

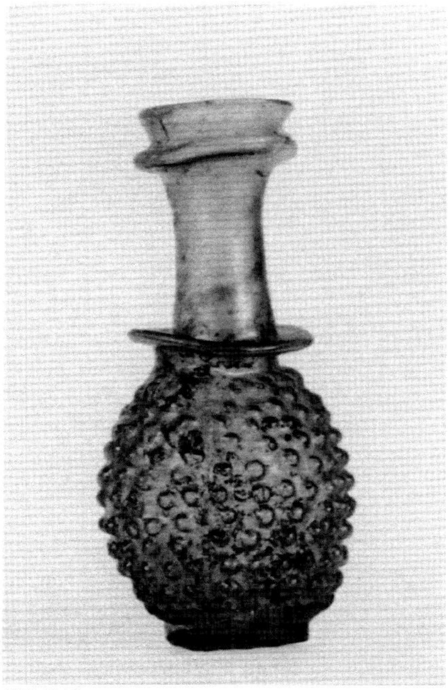

128 (1:2)

alternating rows.[1] The leaves lack a central vein and are adjacent to the mold seams.

[1] Additional examples:
a) Yale, no. 1955.6.79, Matheson 1980, 104-105, no. 279;
b) formerly R. W. Smith Collection, no. 138, Smith 1957, 133, no. 258.

128 **Ovoid Bottle with Stylized Grape Pattern**

Color plate 19, p. 57
Probably Syrian or inland Palestinian
Probably third century
H 14.2 cm; H body 8.4 cm; D 6.5-6.1 cm; D rim 4.1-3.95 cm; D base 3.55 cm
1923.1079, Gift of Edward Drummond Libbey

Translucent pale olive (10 Y 6/20). Medium thick glass. Small vertical linear bubbles in neck.
Neck and collar free blown. Body blown into a three-part mold with two vertical sections joined to a disk-shaped base section (MCT VII). Relief crisp. Pontil mark ca. 1.4 cm in diameter.
Flaring rim, rounded in flame, with hollow pushed-out collar below. Tall cylindrical neck. Hollow projecting roll above shoulder collar. Ovoid body. Base slightly concave.
On the body, stylized pattern of grapes consisting of eleven irregular rows of small, widely spaced hemispherical knobs. Underside of base undecorated.
*Classification* Stylized Grape Bottle, Series C.
*Condition* Complete. Strain crack in half of rim collar.
*Collection* Thomas E. H. Curtis.
*Published* Previously unpublished.
*Remarks* Same Type as Cat. No. 120. Series C is characterized by a lack of leaves.[1] The mold(s) may have been recut from those used for Series B (see Cat. No. 127). The irregular spacing of the knobs resembles the stylized knobby hair of double head flask Cat. No. 157.

[1] Additional examples:
a) Damascus, Nat. Mus., no. 14925, from Syria, Zouhdi 1964, 49, no. 52, fig. 35;
b) Berlin, Staatliche Museen, Antikensammlung, no. 30219, 143, Platz-Horster 1976, 42, no. 64.

## *Bulbous Bottles with Geometric Designs* (Cat. Nos. 129-136)

129 **Spherical Bottle with Lozenge Pattern (Sprinkler)**

Syrian or inland Palestinian
First half of third century
H 6.9 cm; max D 5.55 cm; D rim 4.5 cm; D base 2.4 cm
1923.1334, Gift of Edward Drummond Libbey

Color cannot be determined because of weathering. Medium thin glass. Blowing spirals. Fabric cannot be determined because of weathering.
Rim and neck free blown. Body blown into a two-part mold with two vertical sections (MCT VIII). One continuous mold seam extends around body and across underside of base. Relief crisp on one side and indistinct on the other. No pontil mark.
Collar rim outsplayed horizontally and very thin at edge (unworked?),[1] with a cut-out below edge. Tubular neck, with a constriction at its base where an interior cut-out forms a diaphragm with an aperture of ca. 0.3 cm. Spherical body. Flattened base.
On the body, twelve rows of ten lozenges in raised outline, the bottom row consisting of tongues with pointed tips in raised outline. The top two rows are barely visible.
*Condition* Complete, except for a small hole in the body where the glass has weathered through. Exterior and interior covered with a thick iridescent, creamy-white layer of weathering.
*Collection* Thomas E. H. Curtis.
*Published* Grose 1978, 82, fig. 20 (acc. no. mistakenly cited as 1923.1134); Grose 1984, 37, ill.
*Remarks* Same Type as Cat. No. 130. A pattern of lozenges in raised outline is not only the most common but also the earliest documented design on mold-blown sprinkler bottles. Fragments of three bottles decorated with this design were found at Dura Europos, where they predate the destruction of the city in 256.[2] The pattern

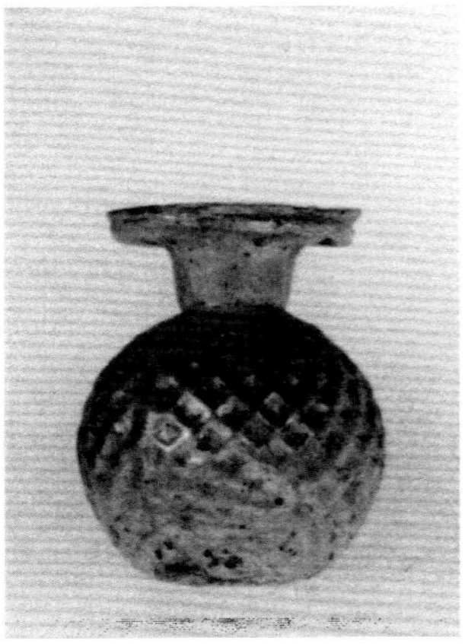

129 (1:2)

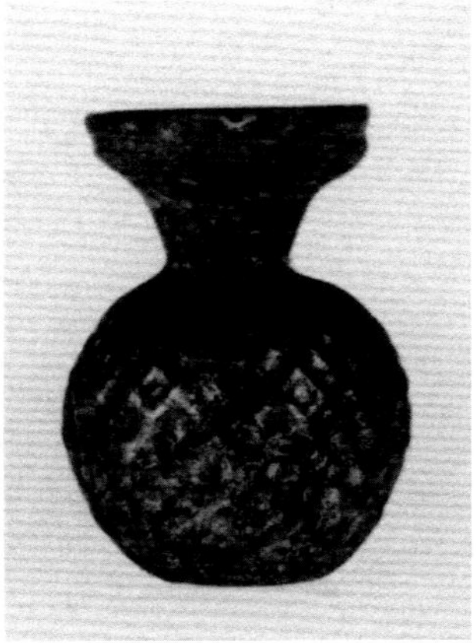

130 (1:2)

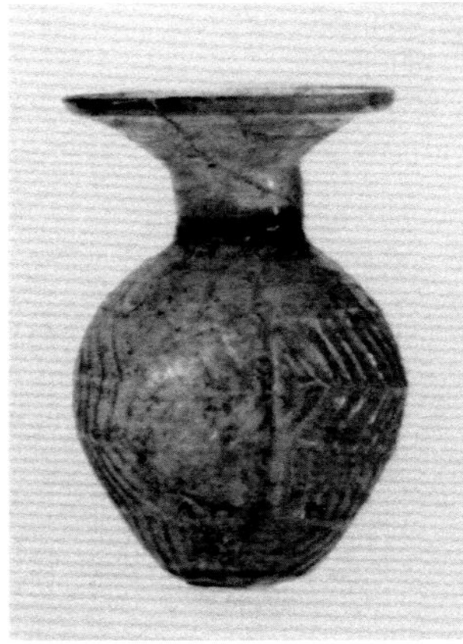

131 (1:2)

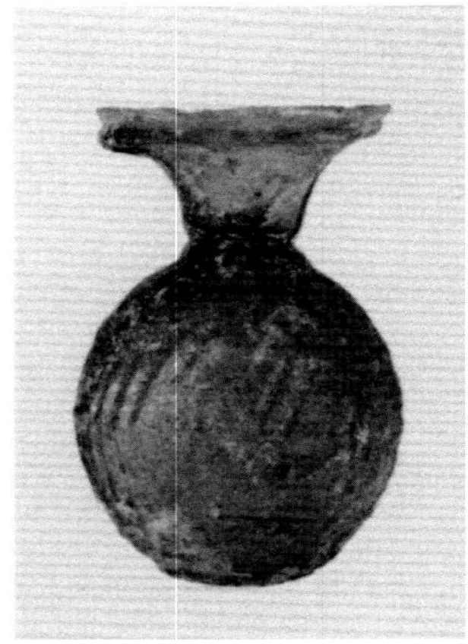

132 (1:2)

occurs frequently on bottles found in the Syro-Palestinian area.[3] Related patterns, but expanded, on slightly larger sprinklers, have been found in northern Mesopotamian tombs dated from the last quarter of the third to the beginning of the fourth centuries.[4] On sprinklers, see *Glossaries: Descriptive Terms for Vessel Shapes* and Chapter II: *The Second to Fourth Centuries: Shapes*.

1 A similar thin, unworked edge has been noted on other sprinklers with this design, for example:
  a) Newark, no. 50.1491, Auth 1976, 78, no. 81;
  b) Berlin, Staatliche Museen, Antikensammlung, no. 30291, 245, Platz-Horster 1976, 42, no. 66;
  c) see Cat. No. 132.

2 Yale, no. 1934.565a, and collections unknown, Clairmont 1963, 14, 41, nos. 28, 29, and 157; nos. 28 and 29 are classified as first-century vessels, but the large pontil mark makes this very unlikely, because pontil marks do not occur this early. The design, with pointed tongues in the bottom row, is identical to that of Toledo bottles Cat. Nos. 129 and 130; the diameters of the bases indicate that the bottles from Dura Europos were of a similar or slightly larger size.

3 For example:
  a) Damascus, Nat. Mus., no. 6924, from Syria, Zouhdi 1964, 49, no. 51, fig. 34;
  b) formerly J. A. Durighello Collection, probably from Palestine, Durighello 1911, 14, no. 102, pl. 6;
  c) ROM, no. 950.157.64, probably from Palestine, Hayes 1975, 78, no. 280, pl. 7;

4 M. Negro Ponzi, "Sasanian Glassware from Tell Mahuz (Northern Mesopotamia)," *Mesopotamia* 3-4 (1968-69) nos. 41-42, fig. 155 (from Tell Mahuz), 309 (date).

### 130 **Spherical Bottle with Lozenge Pattern (Sprinkler)**

Syrian or inland Palestinian
First half of third century
H 8.5 cm; max D 6.3 cm; D rim 4.9 cm; D base 2.8 cm
1923.1335, Gift of Edward Drummond Libbey

Color cannot be determined because of weathering. Medium thick glass. Blowing spirals. Fabric cannot be determined because of weathering.
Rim and neck free blown. Body blown into a two-part mold with two vertical sections (MCT VIII). One continuous mold seam extends around body and across underside of base. Relief indistinct. No pontil mark.
Collar rim outsplayed at an angle and rounded in flame, with an open cut-out below edge. Neck tapers downward to a constriction at its base where an interior cut-out forms a diaphragm with an aperture of ca. 0.3 cm. Spherical body. Flattened base.
On the body, eleven rows of twelve lozenges in raised outline, the bottom row consisting of tongues with pointed tips in raised outline. The top two rows are barely visible.
*Condition* Complete. Exterior and interior covered with iridescent brown on black layer of weathering (see Remarks).
*Collection* Thomas E. H. Curtis.
*Published* Previously unpublished.
*Remarks* Same Type as Cat. No. 129. The curious brown on black layer of weathering suggests that this vessel may have been originally acquired from Azeez Khayat, for it also occurs on a vessel formerly in the Azeez Khayat Collection now in Toronto.[1]

1 ROM, no. 950.157.64, Hayes 1975, 78, no. 280, pl. 7.

### 131 **Bulbous Bottle with Herringbone Pattern (Sprinkler)**

Syrian or inland Palestinian
Third century
H 9.0 cm; max D 6.0-5.7 cm; D rim 5.8 cm
1923.1340, Gift of Edward Drummond Libbey

Transparent natural pale green (10 G 6/2). Medium thin glass. Numerous pinprick and small bubbles. Black specks. Stone.
Neck and mouth free blown. Body blown into a two-part mold with two vertical sections (MCT VIII). One continuous mold seam extends around body and across underside of base. Edges of mold not carefully aligned. Relief crisp. Pontil mark ca. 1.8 cm in diameter. Tooled.
Rim outsplayed at a 45-degree angle, rounded and thickened in flame. Short tubular neck, with a constriction at its base where an interior cut-out forms a diaphragm with an aperture of ca. 0.3 cm. Ovoid body with greatest diameter near middle. Convex base (vessel cannot stand).
On the body, two horizontal herringbone motifs to right, bordered at shoulder by a row of eleven squares in raised outline.
*Condition* Complete, but broken and repaired. Slightly cloudy on interior of rim.
*Collection* Thomas E. H. Curtis.
*Published* Previously unpublished.
*Remarks* Judging from published parallels, this herringbone pattern[1] appears to have been less common than the lozenge pattern. No examples of this Type, with a herringbone pattern, have been published from controlled excavations. However, the small compact size and the finishing of the rim indicate that these vessels date from the same period as small spherical bottles with lozenges (Cat. Nos. 129 and 130) and may have been made in the same workshops. This particular bottle is unusual in that it was finished on a pontil.[2] A bottle with a related pattern in the Toledo Museum

collection has no pontil mark and may therefore have been made in a different workshop (see Cat. No. 132).
On base, oval white sticker with double red border and in pencil "ZD" or "BZ."

[1] For example:
a) Newark, no. 50.1485, Auth 1976, 78, no. 80;
b) Carnegie, no. 24602/1, Oliver 1980, 120, no. 208;
c) Oppenländer Collection, no. 22281, von Saldern et al. 1974, 176, no. 485, 168, color pl.;
d) Tel Aviv, no. MH 39758;
e) RGM, no. 67,499, herringbone combined with a row of raised dots, La Baume 1973, no. C 24, pl. 13, no. 4.

[2] For other sprinklers with a herringbone pattern finished on a pontil: see above note 1a and b.

### 132 Spherical Bottle with Herringbone Pattern (Sprinkler)

Syrian or inland Palestinian
Third century
H 8.65 cm; max D 5.8 cm; D rim 5.3 cm
1923.1341, Gift of Edward Drummond Libbey

Transparent natural pale green (5 G 7/2). Medium thin glass. Fabric cannot be determined because of weathering.
Neck and mouth free blown. Body blown into a two-part mold with two vertical sections (MCT VIII). One continuous mold seam extends around body and across underside of base. Edges of mold carefully aligned. Relief indistinct. No pontil mark. Tooled.
Rim outsplayed at an angle and very thin at edge (unworked?),[1] with a slight cut-out near the edge. Flaring neck, with a constriction at its base where an interior cut-out forms a diaphragm with an aperture of ca. 0.5 cm. Spherical body. Convex base (vessel cannot stand).
On the body, two horizontal herringbone motifs to right.
*Condition* Complete, except for a hole in middle of body where the glass has weathered through.
*Collection* Thomas E. H. Curtis.
*Published* Previously unpublished.
*Remarks* See Remarks Cat. No. 131. Stern 1977a, 99, rim type C.

[1] See Cat. No. 129.

### 133 Spherical Bottle with Raised Dot Pattern (Sprinkler)

Syrian or inland Palestinian
Probably third century
H 10.5 cm; max D 7.0 cm; D rim 5.4 cm
1923.1339, Gift of Edward Drummond Libbey

Translucent to transparent manganese colored dusky red (5 R 3/4). Medium thin glass. Fabric cannot be determined because of weathering.
Neck and mouth free blown. Body blown into a two-part mold with two vertical sections (MCT VIII). Underside of base smoothed, mold seam not visible. Edges of mold carefully aligned. Relief indistinct. No pontil mark. Tooled.
Rim outsplayed at a 45-degree angle, folded outward, upward, and inward. Tubular slightly bulging neck, with a constriction at its base where an interior cut-out forms a diaphragm with an aperture of ca. 0.6 cm. Spherical body. Convex base (vessel cannot stand).
On the body, six rows of ten raised dots, bordered above by horizontal raised ridges. A seventh row of dots near the bottom borders a row of small empty cavities instead of dots.
*Condition* Intact. Entire exterior and interior of mouth covered

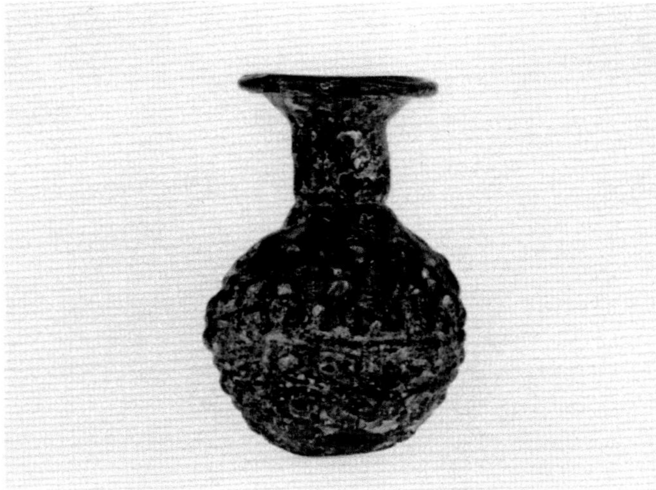
133 (1:2)

with flaking golden and cream-colored iridescence, ice floes, and surface pits.
*Collection* Thomas E. H. Curtis.
*Published* Previously unpublished.
*Remarks* Raised dots occur in combinations with various other patterns, for example with a herringbone motif[1] or set in squares[2] instead of rows. The pattern, small compact size, and finishing of the rim associate bottles of this Type with the small bulbous bottles with lozenge pattern (Cat. Nos. 129 and 130) and herringbone pattern (Cat. Nos. 131 and 132).

[1] RGM, no. 67,499, La Baume 1973, no. C 24, pl. 13, no. 4.
[2] For example:
a) ROM, no. 950.157.185, Hayes 1975, 78, no. 281, pl. 7;
b) Oppenländer Collection, no. 22262, von Saldern et al. 1974, 175, no. 483, 168, color pl.;
c) Tel Aviv, no. MHG 39658.

### 134 Spherical Bottle with Concentric Circles Pattern

Inland Palestinian or Syrian
Fourth century
H 13.4 cm; max D 8.7-8.5 cm; D rim 6.4-6.0 cm
1923.1332, Gift of Edward Drummond Libbey

Transparent manganese colored patchy grayish red (near 10 R 4/2). Translucent natural green coil, the exact color of which cannot be determined because of weathering. Medium thin glass. Blowing spirals. Fabric cannot be determined because of weathering.
Neck and mouth free blown. Body blown into a two-part mold with two vertical sections (MCT VIII). One continuous mold seam extends around body and across underside of base. Edges of mold carefully aligned, but the pattern is not perfectly continuous. No pontil mark. Added coil.
Rim outsplayed at a 45-degree angle, folded outward, upward, and inward. Short tubular neck, with bulge above constriction at its base (no inner diaphragm). Strongly sloped shoulder. Spherical body. Convex base (vessel cannot stand).
Rim coil. On the body, concentric circles with irregular hatched forms in the spaces between the circles.
*Condition* Complete, except for a small hole in the body. Original surface not preserved. Exterior and interior rough with purple and golden iridescence.
*Collection* Thomas E. H. Curtis.
*Published* Previously unpublished.

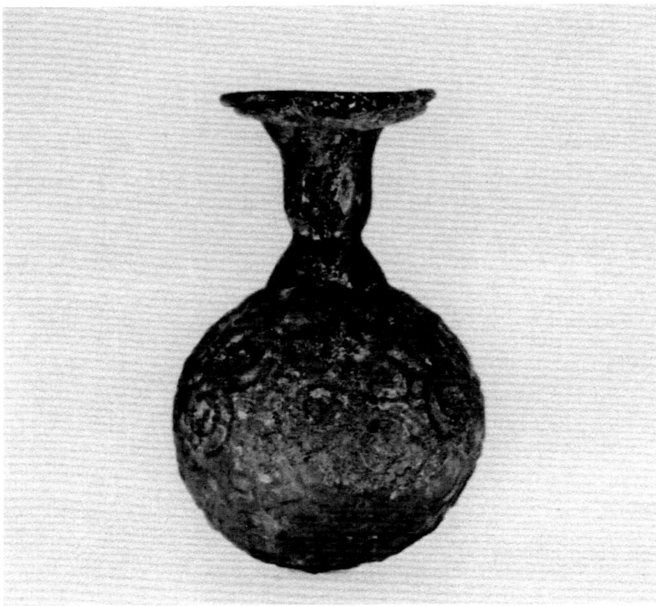

134 (1:2)

*Remarks* Vessels with this and related patterns, often expanded, are usually provided with a handle and finished as jugs.[1] A closely related pattern occurs on expanded mold-blown sprinklers, some of which may actually have been blown in the same molds as unexpanded vessels.[2] Bottles and jugs usually have a rim coil, a decorative element typical of fourth-century Syro-Palestinian vessels.

[1] For example:
- a) Damascus, Nat. Mus., no. 6924, from Homs, Syria, Abdul Hak 1965, 31-32, fig. 17;
- b) RGZM, no. O.6585, said to be from Aleppo, Behn 1913-14, 10, pl. 2, no. 21;
- c) RGM, no. N5985, La Baume 1973, no. C 22, pl. 13, no. 2;
- d) Oxford, Ashmolean Mus., number unknown, Harden 1964b, 23, fig. 5, top right;
- e) Kassel, no. G.139, Spartz 1967, no. 96, pl. 22;
- f) Jerusalem, Dobkin Collection, Neuburg 1962, pl. 37.

[2] For example, TMA 1953.150, TMA, *Roman and Early Byzantine Glass* (forthcoming).

### 135 **Bulbous Bottle with Lozenge Pattern**

Probably Syro-Palestinian
Probably late third century
H 7.4 cm; H body 6.0 cm; max D 7.3 cm; D rim 6.3 cm; D base 4.3 cm
1923.1333, Gift of Edward Drummond Libbey

Transparent natural green. Exact color cannot be determined because of weathering. Thin glass. Fabric cannot be determined because of weathering.
Neck and mouth free blown. Body blown into a two-part mold with two vertical sections (MCT VIII). One continuous mold seam extends around body and across underside of base. Edges of mold carefully aligned. No pontil mark.
Wide horizontal brim, folded outward, upward, inward, and flattened. Short cylindrical neck. Sloping shoulder. Bulbous body with greatest diameter at middle. Concave base, edges not clearly defined.
On the body and underside of the base, thirteen rows of lozenges with a central dot, nine columns across on one side of the vessel and eight on the other. Pattern widens slightly toward greatest diameter.
*Condition* Complete. Severe surface pitting; four small holes on one side of body where weathering has consumed the glass. Exterior and interior covered with a buff-colored layer of iridescent weathering.
*Collection* Thomas E. H. Curtis.
*Published* Previously unpublished.
*Remarks* Pattern and technique are closely related to the geometric patterns decorating small third-century sprinkler bottles such as Cat. Nos. 129-133. The shape may also be compared to free-blown jars made in the Syro-Palestinian area during the late third and fourth centuries.[1]

[1] For example, TMA 1923.1036, TMA, *Roman and Byzantine Glass* (forthcoming).

### 136 **Bulbous Bottle-Jar with Lozenge Pattern**

Probably Syro-Palestinian
Probably third or early fourth century
H 7.7 cm; max D 6.7 cm; D rim 4.7 cm; D base 3.2 cm
1923.1338, Gift of Edward Drummond Libbey

Transparent natural green. Exact color cannot be determined because of weathering. Medium thin glass. Small spherical bubbles. Blowing spirals.
Neck and mouth free blown. Body blown into a two-part mold with two vertical sections (MCT VIII). Mold seams concealed in design; not visible on bottom of body and underside of base.[1] Edges of mold carefully aligned. Relief very shallow, blurred in upper part of body. No pontil mark.
Everted rim, folded outward, upward, and inward. Flaring straight-walled neck. Horizontal shoulder. Bulbous body with greatest diameter at middle. Concave base, edges not clearly defined.
On the body, four panels of lozenges in raised outline alternate with four columns of six vertical chevrons. Panels containing seven rows of five lozenges alternate with seven rows of four lozenges. Pattern widens slightly toward greatest diameter.
*Condition* Intact. Exterior and interior covered with golden, green, and purple iridescence. Incipient pitting on interior.
*Collection* Thomas E. H. Curtis.
*Published* Previously unpublished.
*Remarks* The shape may be compared to third- to fourth-century free-blown spherical jars made in Palestine.[2] Mold-blown examples are rare.[3]

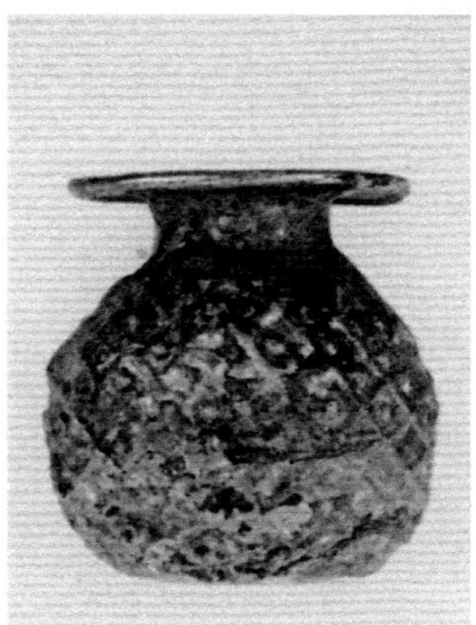

135 (1:2)

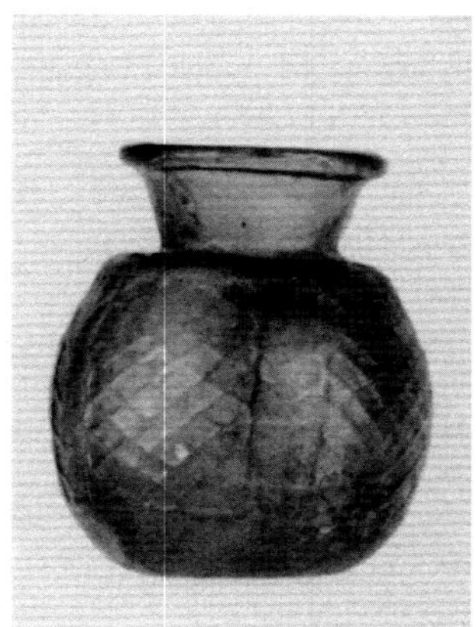

136 (1:2)

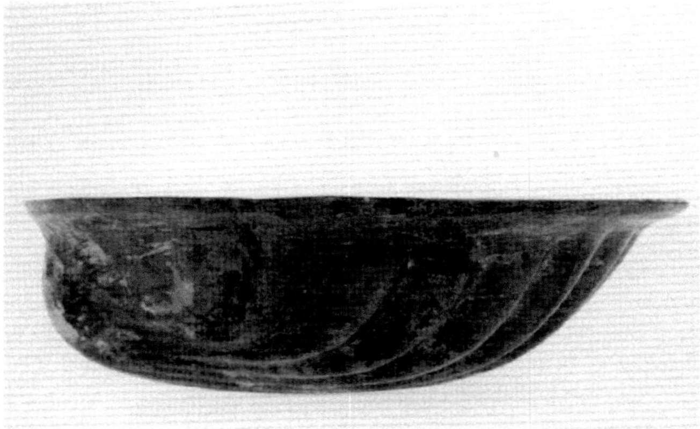

137 (1:2)

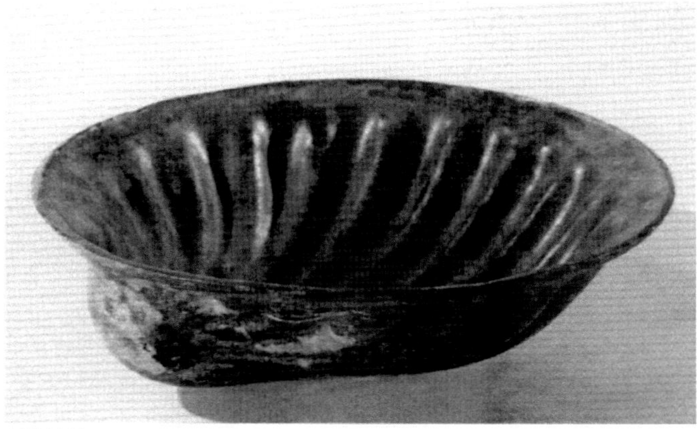

137 (1:2)

[1] The mold seams were probably smoothed out when the bottom of the vessel was reheated in order to collapse the base to make it concave.

[2] For example, TMA 1923.809, 1923.810, 1923.819, and 1923.1898, TMA, *Roman and Early Byzantine Glass* (forthcoming).

[3] See, for example, Oppenländer Collection, no. 22251, von Saldern et al. 1974, 175, no. 481, 178, ill.

## *Dish in the Shape of a Shell* (Cat. No. 137)

### 137 **Dish in the Shape of a Shell**

Color plate 20, p. 58
Perhaps made in Cologne
Probably first half of fourth century
H 5.0 cm; D rim 15.9-15.8 cm; Th rim 0.24 cm
1981.93, Purchased with funds from the Libbey Endowment, Gift of Edward Drummond Libbey

Translucent to opaque dark cobalt blue. Thin glass. Blowing spirals. Fabric cannot be determined because of weathering.
Blown into an open one-part, cup-shaped mold. No pontil mark. Ground.
Flaring concave rim, lightly ground; edge partly horizontal, partly sloping outward. Shallow body with circular circumference in the form of a stylized scallop shell.
Nineteen tapering ribs radiate from a well-developed ocellus in a fan pattern imitating the structure of a bivalve.
*Condition* Broken and repaired from many pieces; some missing fragments near rim reconstructed with plastic. Exterior pitted and covered with areas of blue and green iridescence and reddish brown weathering. Interior pitted and covered with reddish brown iridescence.
*Collections* Margarita Haagen, Aachen, Germany; Elizabeth Haagen Mueller, near Zurich, Switzerland.
*Published* K. T. Luckner, "Recent Acquisitions of Ancient Glass at the Toledo Museum of Art," *Decorative Arts Society Newsletter* 8, nos. 2 and 3 (1982) 10-11; "Recent Important Acquisitions," *JGS* 24 (1988) 88, no. 6, ill.; Grose 1984b, 33, ill.
*Remarks* Although there is an undercut between the rim of the dish and the ocellus, the scallops opposite the ocellus slant so gently outward that the paraison could have been extracted from the mold by tilting the blowpipe at an angle away from the ocellus and sliding or pulling the paraison out of the mold. The vessel was probably annealed with the moil still attached. When the vessel was cold, the moil was cracked off[1] and the rim lightly ground.
Dishes shaped like shells became popular in the Hellenistic period, particularly in silver and semiprecious stones. In glass, two distinct Types of shell-shaped dishes are known: an early Type, dating from before 79, which was made by sagging,[2] and a fourth-century Type which was blown in an open one-part mold. The earlier Type is more realistic; it includes the shell's mantle edges and has an undulated rim. Find spots appear to suggest an Italian origin. The shape of this shell-shaped dish is characteristic of the stylized, fourth-century Type.[3] There is no indication of mantle edges, and the rim is unnaturalistically regular. The dark blue color of this bowl is exceptional.[4] Most, if not all, other published glass dishes of this Type are made of decolorized glass. They appear to fall into three distinct size groups, the first with a diameter of about 15.5-16 cm, the second with a diameter of about 13.5 cm, and the third with a diameter of about 9.5-10.5 cm. Find spots are concentrated in the area around Cologne (ancient Colonia Agrippinensis)[5] and at Intercisa (ancient Alba Regia), Hungary.[6]

[1] The moil could have been cracked off by scoring the vessel at the desired point of fracture and wrapping a trail of hot glass around it: Weinberg 1988, 89.

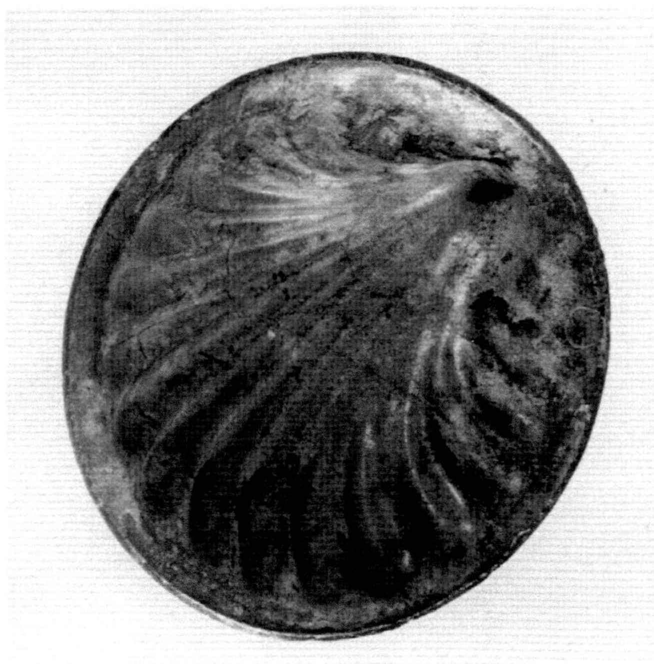

137, exterior (1:2)

[2] For example:

a) from Pompeii, A. Maiuri, *La Casa del Menandro* (Naples 1932) pl. 62;
b) said to be from southern Russia, CMG, no. 66.1.29, formerly Sangiorgi Collection, Goldstein 1979, 139-140, no. 288, pl. 17;
c) from Urbino, Italy, fragment of an emerald green shell, Mercando 1982, 229, fig. 99.

On the technique: Goldstein 1979, 139-140, at no. 288.

[3] The shape without mantle edges was already made in silver before 79: Strong 1966, 152 and 153, pl. 42A, a large silver shell-shaped bowl from the Casa del Menandro, Pompeii. On shell-shaped dishes of various materials: P. La Baume, *Römisches Kunstgewerbe* (Braunschweig 1964) 175-178. A silver bowl with a diameter of about 23.6 cm was excavated in a fourth-century sarcophagus at Zülpich-Enzen, near Bonn, Germany, Follmann-Schulz, 1989, 57-58, no. 24, fig. 4, with citation of a second example from Cologne, now in Bonn, Rheinisches Landesmus., no. 6433.

[4] A similar blue glass characterizes a group of fourth-century vessels made in an as yet unlocalized workshop in the eastern Mediterranean: TMA, *Roman and Early Byzantine Glass* (forthcoming), for example TMA 1923.736. The exact relationship between this shell-shaped bowl and the eastern Mediterranean group has not yet been established.

[5] RGM, nos. 1024, N310, N311, 29.117, N312, and N313, six examples from tombs in Cologne, three large ones with diameters of about 15.5-16 cm and three smaller ones with diameters of about 9.5-10.5 cm, Fremersdorf VI 1961, 73-74, pls. 154-158.

[6] Dunaújváros, Hungary, Intercisa Mus., diameter 13.6 cm, from Intercisa, grave 38, Barkóczi 1988, 212, no. 541, pls. 63 and 114. Two more shell-shaped bowls from Intercisa are mentioned by E. B. Vago, *Ausgrabungen in Intercisa (1957-1969). Alba Regia II* (Budapest 1971) 189, n. 85.

# III
# Head-shaped Vessels: The First to Fourth Centuries

Mold-blown vessels in the shape of a head in the round, or two or more heads back-to-back, are called head-shaped vessels.[1] The Toledo Museum's collection of thirty-one glass head-shaped vessels is exceptionally large and diverse. Reflecting its principal source, the Curtis Collection,[2] it is representative of eastern Mediterranean production. The following remarks and observations on the development of glass head-shaped vessels are based, first, on an analysis of technique and Mold Construction Types (MCTs) and, second, on a chronological survey of head-shaped vessels from dated finds.[3]

## TYPOLOGY OF FORM AND FORMAT

**Form** Most head-shaped vessels of glass are bottles or flasks[4] with a narrow neck rising from the top of the head; some are finished with one or two handles. A few glass head-shaped vessels are drinking cups with a short upright rim around a large opening on the top of the head; these are occasionally finished with one handle. The Forms — jugs, bottles, aryballoi, flasks, and cups — are revivals of older head-shaped vessels in other media, perhaps in keeping with the antiquarian interests prevailing in the Augustan period (27 B.C.–A.D. 14).

It is not clear of what material head-shaped vessels were first made in the Roman period. Jugs and one-handled flasks in the shape of a single head are the earliest datable Form of head-shaped vessels in glass. This Form probably originated in the eastern Mediterranean, where it has been found in glass (see Cat. No. 138) and in clay.[5] The glass jugs may have been made before the death of Augustus, or slightly later; clay examples have been excavated in tombs of the Augustan period.

The problem is more complicated when trying to establish where and in which material head-shaped drinking cups were first created. Most clay head-shaped cups are finished as skyphoi or kantharoi with two handles,[6] forms that were not made in mold-blown glass. It has been suggested that the clay cups replaced head-shaped jugs and flasks in Asia Minor during the early first century,[7] but the dating of both jugs and cups is based on stylistic grounds and has not been confirmed by datable finds from excavations. The earliest

[1] Vessels with mold-blown decoration in low relief with the motif of a face are classified in this catalogue by the shape of the vessel, e.g. Cat. No. 62 with a Medusa face on either side, classified as a lenticular bottle.

[2] For Thomas E. H. Curtis, see below note 22 and Grose 1989, 19-21.

[3] Clay head-shaped vessels have recently been discussed by Mandel 1988, 199-206, but no attempt has been made to discuss glass head-shaped vessels as a group since Kisa 1908, III, 751-759.

[4] In this catalogue, head-shaped vessels that appear to have functioned as storage containers that were closed are called bottles, while those that were probably not closed and might have functioned as tablewares are called flasks: see *Glossaries: Descriptive Terms for Vessel Shapes.*

[5] Clay jugs in the shape of a single head, of a Form very similar to those of glass, were excavated on the Greek island of Samothrace in tombs dated from the late first century B.C. to the early first century A.D. About ten head-shaped jugs were excavated, some from the same mold. Three Types have been illustrated:

a) Negroid head, first thought to be fourth or third century B.C., but subsequently recognized as late first century B.C. or early first century A.D. after a head flask from the same mold appeared in an undisturbed tomb, Dusenbery 1959, 168, fig. 9; Dusenbery 1964, 186;
b) female head with Isis coiffure, feather, fillets, and corumbs, Dusenbery 1964, 186-187, fig. 4;
c) horned head, discovered in an undisturbed tile tomb that also contained a glass gold-band alabastron; the horned head recalls side B of Cat. Nos. 143 and 144, G. Daux, *BCH* 84 (1960) 805-806, fig. 1.

A fourth Type was excavated in mainland Greece, at Echinos:

d) flask shaped like a female head, with a roll of hair or a fillet across the forehead, *ArchDelt* 24 (1969) 184, pl. 194.

[6] For example, Mandel 1988, 228-229, nos. P 187-194, pls. 21-22. The Form was well known in archaic and classical Greece, but there is no continuous tradition into the Roman period.

[7] Mandel 1988, 205.

head-shaped cups of any material from an independently datable context are glass cups without a handle from Pompeii, which must date from before 79 when the town was destroyed by the eruption of Mount Vesuvius (see below: First Century, Type A, Single Ethnic Head, and Cat. No. 139). Glass cups with one handle at the back of the head are dated to the second half of the second and early third centuries (see below: Second Century, Type E, Single Dionysos-Antinous, Series 1).[8]

**Format** Glass head-shaped vessels occur in three Formats. The first is a single head, e.g. Cat. Nos. 138-140, 147, and 148. The second is janiform, two heads joined back to back with the faces looking in opposite directions. The third has multiple heads.[9] Bottles and flasks were made in all three Formats, cups only as single heads.

Glass bottles and flasks shaped like a single head are usually somewhat larger than those shaped as double or multiple heads, which are generally about 7.0-8.0 cm high, except for a fourth-century Type made in the West (see below: Fourth Century, Type A, Double Clean-shaven Male). Cat. No. 140, a single head of opaque white glass, is exceptionally small (H 7.35 cm). Both single and double head-shaped vessels occur occasionally with handles, e.g. Cat. No. 138 (single head) and Cat. Nos. 161-164 (double heads). The cups are relatively large vessels, about 10 cm high, which occur both without a handle (Cat. No. 139) and with a handle (see below: Second Century, Type E, Single Dionysos-Antinous, Series 1).

From the end of the first to the end of the fourth centuries, glass was the most common material used for janiform vessels. It is difficult to pinpoint the date and origin of the janiform Format. There is no evidence for the production of janiform glass vessels in the first half of the first century, which is somewhat surprising because the Workshop of the Floating Handles was already producing mold-blown lenticular bottles with a face in low relief on either side of the body (Cat. No. 62), a shape that is thought to precede typologically that of a vessel fully shaped like two heads in janiform fashion.[10]

The idea itself had already been explored long before the invention of glass blowing. Ionic potters in the eastern Mediterranean combined two heads in janiform fashion in the seventh and sixth centuries B.C., and the Format remained popular in Attic and later in South Italian pottery.[11] The heads were either similar or deliberate contrasts. There is insufficient datable evidence from excavations to answer the question whether there was a continuing unbroken tradition of janiform vessels in clay or bronze in the Hellenistic period between the end of the fourth century B.C. and the first century A.D. Head-shaped vessels in materials other than glass are usually single heads.[12]

A few janiform vessels in bronze[13] and clay[14] have been assigned to the Hellenistic period on stylistic evidence, but not one was found in an independently datable context.

The creation of janiform glass vessels has recently been dated to the second century.[15] Archaeological evidence now suggests this Format was first made in glass in the second half of the first century (see below First Century, Type C, Double Medusa, see Fig. 86). It is an attractive hypothesis that Roman glass blowers (as opposed to other artisans) reintroduced the fashion,[16] but there are not sufficient datable objects to prove this.

Nor is there sufficient evidence to support the suggestion put forward here that the janiform format might have been

[8] Compare two head-shaped clay cups with one handle at the back of the head, Bergama, Museum, no. 544 (Dionysos) and no. 816 (maenad), Mandel 1988, 229-230, nos. P 197 and P 198, pl. 20, attributed to a third- or fourth-century workshop, active in Pergamon, Asia Minor (ibid., p. 25).

[9] For example, bottles with four faces looking in opposing directions, which are usually finished as sprinklers with an inner diaphragm at the base of the neck:

a) CMG, no. 59.1.151, formerly R. W. Smith Collection, no. 715, Smith 1957, 143-145, no. 287;
b) Newark, no. 50.1436, Auth 1976, 75, no. 76;
c) Yale, no. 1955.6.74, Matheson 1980, 73, no. 190.

[10] Richter 1967, 51; Kisa 1908, III, 751.

[11] Compare, for example:

a) clay aryballos with two Gorgon heads, Berlin, Staatliche Museen, Antikensammlung, no. 1961.4, A. Greifenhagen, "Einige figürliche Vasen," *BerlMus* n.s. 8.1 (1963) 7, figs. 1-4;
b) faïence aryballos, from Cyprus, formerly Berlin, Richter 1967, fig. 8;
c) BM, no. A 1233, E. Buschor, "Das Krokodil des Sotades," *MüJb* 11 (1921) 34, fig. 50.

On archaic Greek head vases: M. I. Maximova, *Les vases plastiques dans l'antiquité. Epoque archaique* (Paris 1927) vol. 1, 148-164; vol. 2, pls. 30-31; J. D. Beazley, "Charinos," *JHS* 49 (1929) 38-78. On later Greek and South Italian head vases: H. Sichtermann, *Griechische Vasen in Unteritalien aus der Sammlung Jatta in Ruvo* (Tübingen 1966) 63, nos. 120-127, pls. 161-167.

[12] Richter 1967, 48-53, 114-118, nos. 118-161, lists head-shaped vessels of various materials made in the Roman period. Examples of clay head-shaped vessels:

a) RGM, no. K151, cup-shaped satyr head, La Baume and Salomonson 1976, 78, no. 272, pl. 40, no. 4;
b) RGM, no. K68, cup-shaped satyr head, said to have been found near Pergamon, dated second to first centuries B.C., La Baume and Salomonson 1976, 78, no. 273, pl. 41, no. 1;
c) RGM, no. K150, child's head, dated first century A.D., La Baume and Salomonson 1976, 82, no. 297, pl. 44, no. 1;
d) RGM, no. K97, wreathed bearded head, dated second century A.D., La Baume and Salomonson 1976, 84, no. 308, pl. 46, no. 1;
e-k) RGM, nos. N98 to N104, North Africa, dated late third to early fourth centuries, La Baume and Salomonson 1976, 169-172, nos. 624-630, 633.

On clay head-shaped vessels from excavations in Greece: see above note 5. On bronze head-shaped vessels: Peter Goessler, "Antike Büstengefässchen aus Metall," in *Antike Plastik. Walther Amelung zum sechzigsten Geburtstag* (Berlin 1928) 75-86; Cristo M. Danoff, "Grabfund aus Warbovka in Nordbulgarien," *AA* (1937) 339-347, dated first half of third century; Antonio Garcia y Bellido, *Esculturas romanas de España y Portugal* (Madrid 1949) 453-458, nos. 484-490, pls. 338-341; F. Coarelli, "Nuovi elementi per la cronologia di Begram: cinque recipienti bronzei in forma di busto," *ArchCl* 13 (1961) 168-179, pls. 75 and 76; L. Pressouyre, "A propos d'un 'balsamaire' trouvé à Lamaurelle (L. et G.)," *RA* 1 (1962) 165-181; Frances F. Jones, "A Bronze Head-Vase," *Record of the Art Museum Princeton University* 46, no. 2 (1987) 17-23.

[13] S. Haynes, "Etruskische Bronzekopfgefässe aus hellenistischer Zeit," *JRGZM* 6 (1959) 122-123, nos. 1-9, pls. 55-57; D. G. Mitten and S. F. Doeringer eds., *Master Bronzes of the Classical World* (Mainz 1968) 222, no. 225.

[14] La Baume and Salomonson 1976, 86, no. 321, pl. 46, no. 4, reddish clay with yellowish red slip, said to have been found in Asia Minor, dated second to first century B.C. by Salomonson.

[15] Matheson 1980, 72, at no. 188; Barag 1985a, 93, at no. 118.

[16] First suggested by Matheson 1980, 72, at no. 188.

reinvented in Italy. Most janiform Double Medusa Type bottles were made in the eastern Mediterranean. However, the earliest datable glass janiform head flask of the Roman period is a double Medusa head-shaped vessel found at Vigorovea, in northern Italy (see Fig. 86), in a tomb containing numerous glass vessels, all of which are forms known to have been made in Italy (see also below: Survey: First Century, Type C). Although this hints that the janiform head flask may also have been made in Italy, it does not prove it. The Double Medusa Type remained popular in Italy in second-century tombs (see below Second Century, Type B, Double Medusa, see Fig. 88).

## SUBJECTS

The subjects of head-shaped vessels must have been dictated by consumer demand. Deities and mythological creatures are most common, starting with Medusa, who was appreciated for her apotropaic qualities (First Century, Type C; Second Century, Type B; Fourth Century, Type C; see also Cat. Nos. 142, 152 and 153, and 158) and followed closely by Dionysos (First Century, Type B; Second Century, Type E; see also Cat. Nos. 140, 143, 144, 147, and 148) and probably most of the Chubby Childlike Types discussed below. A second group consists of unusual faces, mostly foreigners with interesting ethnic features — especially Negroid[17] (First Century, Type A; Second Century, Type F; see also Cat. No. 139) — and a limited range of grotesques[18] (see Cat. Nos. 157 and 160). Representations of ordinary people are rare in glass;[19] they have not yet been identified among eastern Mediterranean head-shaped vessels, although several Types of large head flasks showing male heads were made in northwestern Europe in the third to fourth centuries: see below in the Survey, Third Century, Type B; Fourth Century, Type A; Fourth Century, Type B.[20]

Some subjects have not yet been identified, including the many janiform chubby childlike faces, sometimes in the combination of one smiling and one pouting face (Third Century, Type A; see also Cat. Nos. 149-151, 154, 155, and 161-164). While it has been proposed that these juxtapose male and female faces, there is no evidence for this. They may represent Dionysos or Eros, both of whom were frequently represented as a child.[21] Representations of Dionysos as a child were especially appropriate in Sidon and its vicinity where most of the bottles depicting chubby childlike faces were probably made (see Cat. No. 149).

The subjects popular in the East and the West differed. Ethnic types, caricatures, and ordinary people occur in greater numbers in Italy and northwestern Europe, while head flasks showing Dionysiac subjects and childlike faces are most common in the eastern Mediterranean. Medusa faces appear to have been common in both East and West.

## DISTRIBUTION

During the first century, glass head flasks seem to have been made both in Syro-Palestinian workshops and in Italy, but during the second and third centuries, eastern Mediterranean glass blowers produced by far the largest output. The Syro-Palestinian coast appears to have been one of the main production areas,[22] but production in the eastern Mediterranean seems to have ceased around the middle of the fourth century. From the third century onward glass head flasks were common in northwestern Europe, and Cologne appears to have been a major production center. Germany and Gaul were the main producers of glass head flasks during the fourth century.[23] Head flasks have been excavated in fifth-century contexts in Rome, but their date and places of manufacture have not yet been established; see below: Fifth Century, Type A, Two Similar Faces (see Fig. 99).

## TECHNIQUE

Despite their high relief, which increased the risk of undercuts making it impossible to extract the vessel after blowing, most head-shaped vessels were blown in molds with only two vertical sections (MCT VIII). The mold seams are

[17] On the typology of Blacks in Hellenistic art: Hausmann 1962, 255-281, pls. 74-80; see also: F. M. Snowden, Jr., *Blacks in Antiquity: Ethiopians in the Greco-Roman Experience* (Cambridge, Mass. 1970); Ladislas Bunger ed., *The Image of the Black in Western Art*, I: *From the Pharaohs to the Fall of the Roman Empire* (Cambridge, Mass and London 1976, reprinted 1991). Most Hellenistic vessels and figurines shaped like Blacks show the entire figure, not just the head.

[18] On vessels and lamps of glass, bronze, and clay in the shape of grotesques: P. La Baume, *Römisches Kunstgewerbe zwischen Christi Geburt und 400* (Braunschweig 1964) 221ff. On the origin of the Grotesque Type and its possible relation to mime masks: J. P. Cèbe, *La caricature et la parodie dans le monde romain antique* (Paris 1966) 354-355.

[19] Apparently, they are common in clay head-shaped vessels: Mandel 1988, 204.

[20] On large "Constantinian"-style glass head-shaped vessels made in northwestern Europe: RGM, no. N299, Fremersdorf VI 1961, 78, pls. 174-175 (with lit.). A unique blue glass head flask, a single head wearing a helmet, is in Hamburg, MKG, no. 1978.25, H (pres.) 15.1 cm, Axel von Saldern, "Museum für Kunst und Gewerbe: Erwerbungen der Antikenabteilung im Jahre 1978, Kopfflasche," *JbHambKuSamml* 24 (1979) 199-200, ill.

[21] For the fanciful interpretation as "marriage vessels," implying an identification as male and female faces, see below note 22. An interpretation as Dionysos or Eros is more likely: Dionysos: *LIMC* III.1, 478-482, nos. 664-705; III.2, 376-381, figs. 664-705; Eros: *LIMC* III.1, 859, nos. 48-62, 861, nos. 65-70; III.2, 612-613, figs. 48d-68b.

[22] Many objects entered the Toledo Museum collection through the Curtis Collection; Curtis's principal source was the New York dealer Azeez Khayat, whose sources were in Syria and Palestine: Grose 1989, 19-21. Khayat 1906, unpaginated, no. 198, ill.; Khayat 1913, no. 27, two examples, one with two, one with three-and-a-half heads (ill. opposite text no. 59), no. 57 (ill. opposite text no. 59); nos. 127, 269, 301 (not ill.). The double head flasks are described as consisting of one male and one female head and are referred to as "marriage vessels."

[23] It is interesting to note that during the late third and fourth centuries various North African factories made not only clay head flasks but also clay oil lamps in the shape of human heads, e.g. the workshop of Navigius: J. W. Salomonson, "Spätrömische rote Tonware mit Reliefverzierung aus nordafrikanischen Werkstätten," *BABesch* 44 (1969) 4-109 (*Archeologische Studien Ned. Hist. Inst. Rome, Scripta Minora II*); *MEFRA* 82 (1970) 769ff. Double head flasks made in this area differ from those made at all other production sites by the fact that the heads are not placed opposite each other, janiform fashion, but above each other. Janiform heads occur only on the necks of lagynoi. For an overview of different types of North African head flasks: J. W. Salomonson, in La Baume and Salomonson 1976, 159-175, nos. 613-635, pls. 63-68.

concealed in the hair, on vessels shaped like a single head among the locks and strands, on double head-shaped vessels along the boundary between the two heads. The frontal views of the faces were thus considered the most important. In a few instances the mold seam runs down the nose through the middle of the face, indicating that the profile is the principal view (e.g. Cat. No. 138).[24]

A pontil mark is unusual (for examples, see Cat. Nos. 145-147, 160, and 164). The rims of most vessels must have been finished while the vessel was still in the mold or perhaps with the aid of a clamping device. A few bottles have squashed faces (e.g. Cat. No. 145), which may have resulted from the use of such a clamping device or from reheating.

## MOLD CONSTRUCTION TYPES

Four types of mold construction have been noted among the head-shaped vessels of the Toledo Museum collection: MCTs VI, VII, VIII, and IX (see *Glossaries: Mold Construction Types* and Fig. 15).

MCT VI is rare. It has been noted for one head flask only (Cat. No. 140). The disk-shaped base section of the mold was semipermanently attached to one of the two vertical sections of the mold, a feature that recalls MCT II, which has been dated to the second and third quarters of the first century. The circumference of MCT VI combines one half of a circular section with one half of a square section. This recalls the circumference of MCT VIII B, which was used in the Workshop of the Floating Handles (see Chapter I: *Syro-Palestinian Styles: The Workshop of the Floating Handles* and Cat. No. 138). This workshop was probably located in or near Sidon, which is where MCT II also appears to have been used. This suggests an origin for MCT VI in the Syro-Palestinian area, as does the one known find spot of a parallel for Cat. No. 140. None of these parallels can be dated independently, but the similarities with MCTs II and VIII B suggest that MCT VI was approximately contemporary, probably created around the middle or in the third quarter of the first century.

MCT VII molds had a separate flat disk-shaped base section that sometimes bears decoration (Cat. Nos. 139, 142, and 148). This mold construction appears to have been the rule for cups with ethnic features made during the first century (see below: First Century, Type A; see also Cat. No. 139). The undersides of the bases of First Century, Type A, Series 1 and 3, are decorated with raised concentric rings, while Series 2 shows a relief head in profile. In most other Types a decorated base is rare. Cat. No. 142 has a geometric design of concentric lozenges on the underside of the base, and Cat. No. 148 has a head in intaglio.

MCT VIII molds included about half of the base with each of the two vertical sections. This was the most versatile type of mold construction for all shapes of bottles. It was used throughout the first century and into the second century to make both single and double head flasks (e.g. Cat. Nos. 138 [single] and 141 and 146 [double]).

[24] Head flasks with a mold seam through the face were also made in the northwestern part of the Roman empire: RGM, no. 976, Fremersdorf VI 1961, 56, pl. 108 (with parallels); they probably date from the second century, judging by the form of the spout which connects them with Isings 1957, Form 88.

MCT IX represents an important simplification of mold construction because these molds had no base section but were open at the underside. All datable head-shaped vessels made in MCT IX molds date from the early third to the late fourth centuries (Cat. Nos. 149-156, 158, and 160-163). This suggests that the use of molds open at the bottom was common by the early third century.

## SURVEY: HEAD-SHAPED VESSELS FROM DATED FINDS

The following survey, which does not pretend to be complete, assembles published head-shaped vessels found during controlled excavations. The vessels are arranged in chronological order by century according to the date of the earliest context in which they occur.

### *The First Century*

Four Types of head-shaped vessels have been excavated in first-century contexts. First Century Types A, B, and D are single heads, while Type C is a double head. Of these, Type C is the most important because it proves that double head-shaped vessels were blown in the first century, perhaps already in the third quarter of the century. The molds used for cups of Type A appear to be related to flasks showing a Single Ethnic Head made in the second century (see below, Second Century, Type F).

TYPE A. SINGLE ETHNIC HEAD (see Cat. No. 139). A cup shaped like a single male head with strong Negroid features. [25] The latest example of a Single Ethnic Head vessel is not a cup but a flask, see below: Second Century, Type F. Colors noted for this type are brilliant green, natural bluish green, and dark manganese-colored purple, the last perhaps in imitation of dark skin.

The occurrence of the Single Ethnic Head Type at Herculaneum and Pompeii is significant because it demonstrates that the Type was created before these cities were destroyed by the eruption of Mount Vesuvius in 79.[26] A cup from Mangalia Nord (ancient Neptun), Romania, was found in a tomb dated late first to early second century.[27] An example from Caerleon, Wales, was found with Flavian rubbish, i.e. fourth quarter of the first century, "below the metalling of the legionary parade-ground datable to about 140."[28]

[25] On the subject: Hausmann 1962; Snowden 1970 (above note 17); Mariette de Vos, *L'Egittomania in pitture e mosaici romano-campani della prima età imperiale* (Leiden 1980).

[26] For example:
a) from Pompeii, Isings 1957, Form 78a;
b) from Herculaneum, with two concentric rings on the underside of the base, Ercolano, Antiquarium, no. E2372, Scatozza Höricht 1986, Form 33, no. 103, pls. XII, and XXXIII.

According to Scatozza Höricht, the head flasks from Pompeii and Herculaneum are not from the same mold.

[27] Constanta, Romania, Arch. Mus., no. 14851, from a necropolis north of Callatis, Mangalia, Bucovala 1968, 53, no. 58; Bucovala 1979, 93-94, no. 3, fig. 3.

[28] Boon 1967, 95-97, fig. 2.

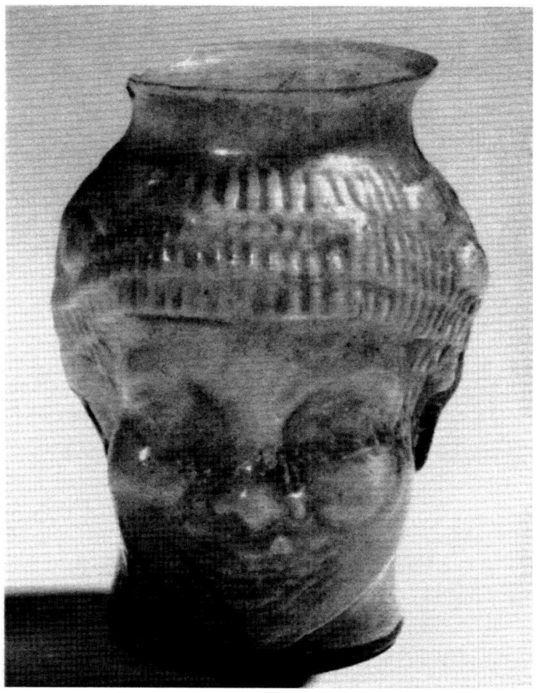

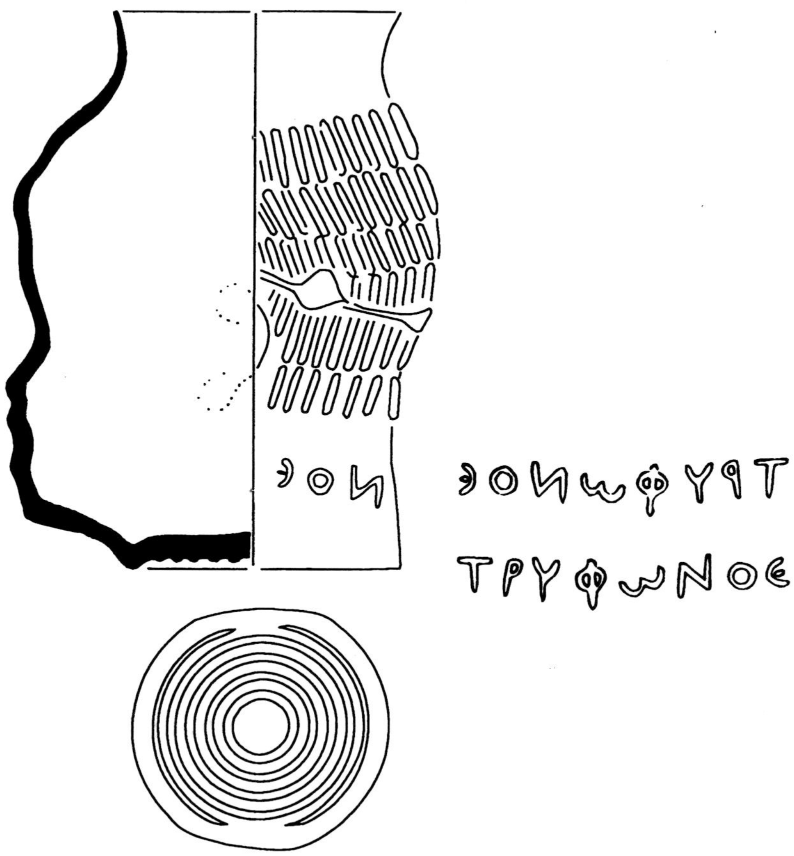

*Fig. 83. First Century, Type A, Series 1. Head flask in the shape of a single ethnic head, signed in Greek. The signature is retrograde on the nape:* ΤΡΥΦΩΝΟϹ, Tryphonos *("by Tryphon"). On the underside of the base, five concentric mold-blown circles. Said to be from Phoenicia, Ville de Paris, Musée du Petit Palais, no. Dutuit 247. Photo: Cliché photothéque des musées de la Ville de Paris, copyright by SPADEM 1993. Profile drawing (right): Courtesy Christian Landes.*

It is not clear where head flasks with ethnic features were first made. All examples from controlled excavations come from Italy and the western provinces,[29] possibly suggesting that most if not all Series were created in the western part of the Roman empire, perhaps in Italy. The find spots in Britain indicate a western origin for Series 2, signed in Latin by Bugaddi. The origin of Series 1, signed by Tryphon is more problematic, because the sole vessel representing this Series is said to have been found in Phoenicia, but the statement cannot be verified. The mold-blown name in Greek may indicate that this Series was made in the eastern part of the Roman empire, where Greek was the official language, but the overwhelmingly western find spots for all other examples of the Single Ethnic Head[30] indicate that an origin in the West must be considered as a possibility. This Type of glass head-shaped vessel is the only one that has close counterparts in bronze (see Cat. No. 139).

**Series 1 and 2** include a maker's name in the mold, Series 1 in Greek: ΤΡΥΦΩΝΟϹ (*Tryphonos*, genitive of the proper name Tryphon) (Fig. 83),[31] and Series 2 in Latin: *C. Caesi Bugaddi* (Fig. 84). The Bugaddi Series is also noteworthy because it carries on the underside of the base, in relief, a radiate male bust in profile that has been identified as Sol Helios.[32]

[29] See the following examples:

a) see above note 26a;
b) see above note 26b;
c) see above note 27;
d) Lattes (Hérault), France, Mus. Arch., no. 982.208, fragment of yellow glass preserving one eye, one cheek, and one brow, from Lattes, Landes 1984, 339, fig. 1;
e) MMA, no. 81.10.226, from Cologne, formerly Charvet Collection, light blue green, H 10.48 cm; Froehner 1879, inset on frontispiece; Kisa 1908, III, 756, fig. 301; Boon 1967, 97, n. 1; Snowden 1970 (above note 17) 28, fig. 59;
f) see above note 28;
g) BM, no. P. 1973.7-2.386, H 10.5 cm, bluish green, from London, Jennifer Price, "A Roman Mould-Blown Negro-head Glass Beaker from London," *AntJ* 54 (1974) 291-292; here Fig. 84;
h) Newcastle upon Tyne, Black Gate Museum, said to be from the fort at South Shields, North Britain, deep blue, D. Charlesworth, "Roman Glass in Northern Britain," *Archaeologia Aeliana*, 4th ser., 37 (1959) 54-55, pl. V, no. 4.

[30] A single example, CMG, no. 55.1.92, formerly R. W. Smith Collection, no. 886, Smith 1957, 145, no. 288, was acquired in Egypt, but the find spot is unknown.

[31] Formerly Paris, Dutuit Collection, no. 247, Sambon 1906, 500, fig. 3; Sambon 1928, 19, fig. 35, no. 24. According to Kisa 1908, III, 756, this vessel was found in Phoenicia. The signature ΤΡΥΦΩΝΟϹ, *Tryphonos* ("by Tryphon") occurs also in a mold-blown inscription on the base of a square bottle excavated in the Athenian Agora, no. G 131 (Gladys D. Weinberg, pers. com.), but there is no way of determining whether the Agora bottle and the head-shaped cup were made in the same workshop.

[32] Although similar designs on the undersides of vessels have been identified as coin impressions (Taborelli 1982a), the Sol-Helios head was probably not made directly from a coin impression because it is shown in relief instead of in intaglio. It also lacks an inscription surrounding the head; instead, a whip is shown in front of the head.

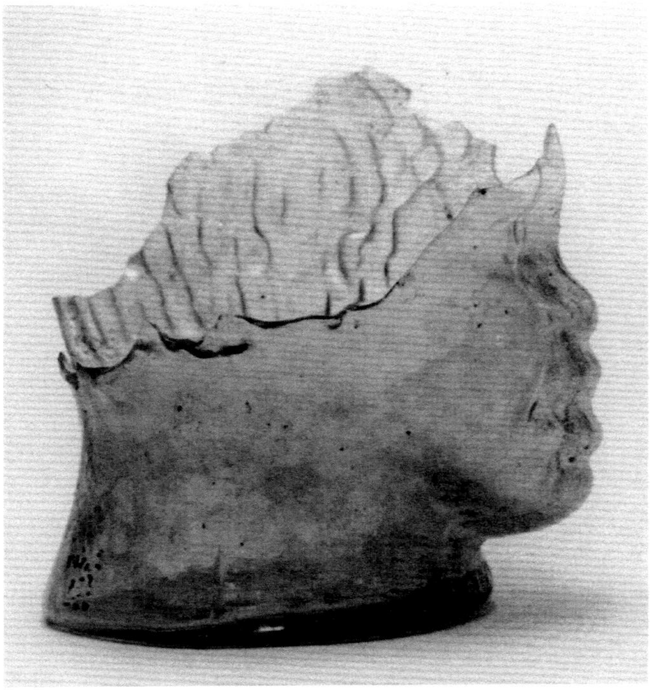 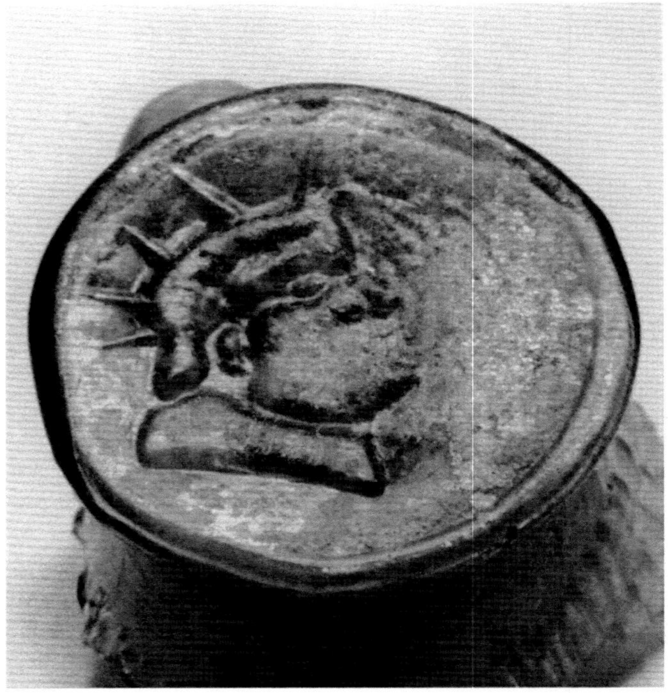

*Fig. 84. First Century, Type A, Series 2. Head flask in the shape of a single ethnic head, signed in Latin. The signature reads from left to right:* C. CAESI, *on the nape, and* BUGADDI, *under the chin. On the underside of the base, a radiate male bust in profile. From London, H 10.5 cm, bluish green, London, British Museum, no. P. 1973.7-2.386. Photo: Copyright British Museum.*

**Series 3** (here Cat. No. 139) probably lacks a maker's name.[33]

**Series 4** is represented by a fragmentary cup excavated at Herculaneum.[34] It has three concentric circles on the underside of the base like Cat. No. 139, but the profile shows a protruding mouth and lower face.

More Series may be represented among unpublished finds from Pompeii.[35]

TYPE B. CHILD IN BUNCH OF GRAPES. A fragment of a head-shaped vessel showing a childlike face, tentatively identified as the infant Dionysos, emerging from the center of a bunch of knob-shaped grapes was found in a tomb in Crete, dated second half of first century (Fig. 85).[36] The base of the vessel appears to have been finished in an irregular point so it cannot stand, like some first-century bottles shaped like a bunch of grapes which it resembles. Too little of the vessel is preserved to determine whether there was a face on both sides. The best parallel is a head-shaped bottle which shows a face on one side only.[37] The Cretan fragment is transparent yellow, similar to the glass of a bottle shaped like a bunch of grapes from the same tomb, thought to have been made in northern Italy.[38]

TYPE C. DOUBLE MEDUSA. A double head-shaped bottle of transparent royal blue glass (Fig. 86) was found in a tomb at Vigorovea, northern Italy, containing coins of the Roman emperors Tiberius (minted between 34 and 36), Caligula (minted in 37), and Claudius (minted in 41).[39] The bottle was blown in an MCT VIII mold. The rim was folded outward, upward, and inward, but not flattened. The Medusa faces on sides A and B are very similar and have features somewhat resembling the Medusa type known as the Medusa Rondanini,[40] but the expression is forbidding: frowning eyes with hollow dots for pupils and ornamentally carved brows. The execution of eyes, lips, brows, and knotted snakes below the chin suggests the mold was made from a clay model on which these elements were applied in the form of thin

Two examples are known:
a) see above note 28;
b) see above note 29g.

33 Irregular mold-blown ridges on the front of the neck of Cat. No. 139 may be the remains of a maker's inscription, but this is not certain.

34 See above note 26b.

35 Scatozza Höricht 1986, 52, mentions the existence in Pompeii of other glass cups, similar and vaguely similar ("confronti vaghi ed esemplari simili") but not identical to the Herculaneum cup and not blown in the same mold. According to *BullAIHV* 9 (1981-83) 162, there are two Negroid head-shaped vessels from Pompeii in Naples, Mus. Naz. Arch., no. 129404, prune-colored, *Collezioni Napoli* 1989, 224 and 225, no. 46, ill.

36 Carington Smith 1982, 281, no. 72, fig. 6, pl. 39, nos. k and l, from a first-century tomb at Monasteriaki Kephala, where it was associated with burial V, of a woman.

37 Loeffler Collection, no. 9, acquired in Istanbul, H 10.6 cm, transparent amber, La Baume and Salomonson 1976, 38, no. 77, pl. 9, no. 1.

38 Carington Smith 1982, 281.

39 Padua, Musei Civici, no. 2482, from a tomb at Vigorovea, excavated in 1954, H 8.0 cm, D base 2.6×2.5 cm, max D 4.6×4.0 cm, Calvi 1959, 3-9, figs. 1 and 2; Zampieri 1984, 113-122, esp. p. 119, no. 14.10. My thanks are owed to Mr. Zampieri for permission to study the vessel.

40 E. Buschor, *Medusa Rondanini* (Stuttgart 1958).

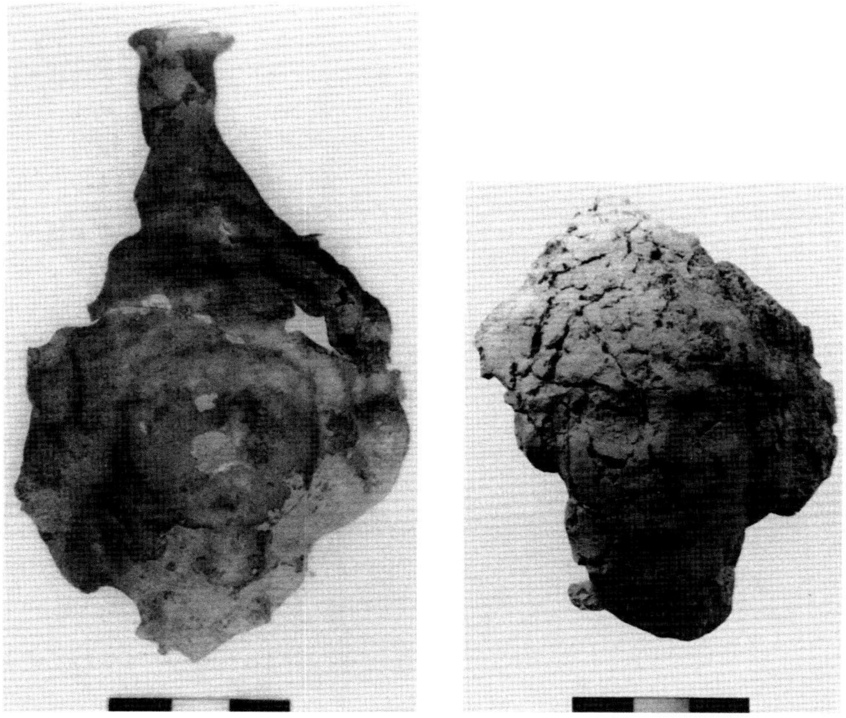

*Fig. 85. First Century, Type B. Fragment of a head-shaped vessel with a child-like face in the center of a bunch of grapes* (left) *and a mud impression of the interior excavated with it* (right). *From Monasteriaki Kephala, H (pres.) 10.2 cm, translucent yellow-gold, Heraklion, Crete, Archaeological Museum. Photo: Courtesy J. Carington Smith.*

coils. The slender profile of the faces, tapering upward, recalls bottles shaped like a grape bunch such as Cat. Nos. 109 and 110. The bottle from Vigorovea is similar, but not identical, to Second Century, Type B, Double Medusa.

In addition to the head-shaped bottle, the tomb at Vigorovea included fragments of eleven other glass vessels:

1) mosaic-glass bowl;
2) deep ribbed bowl (Isings Form 3);[41]
3) footed kantharos with handles formed from twisted rods;
4) squat square bottle (Isings Form 50a);
5) large jar with a lid (Isings Form 67a);
6-11) five tubular unguentaria.

[41] Isings 1957, 19-20.

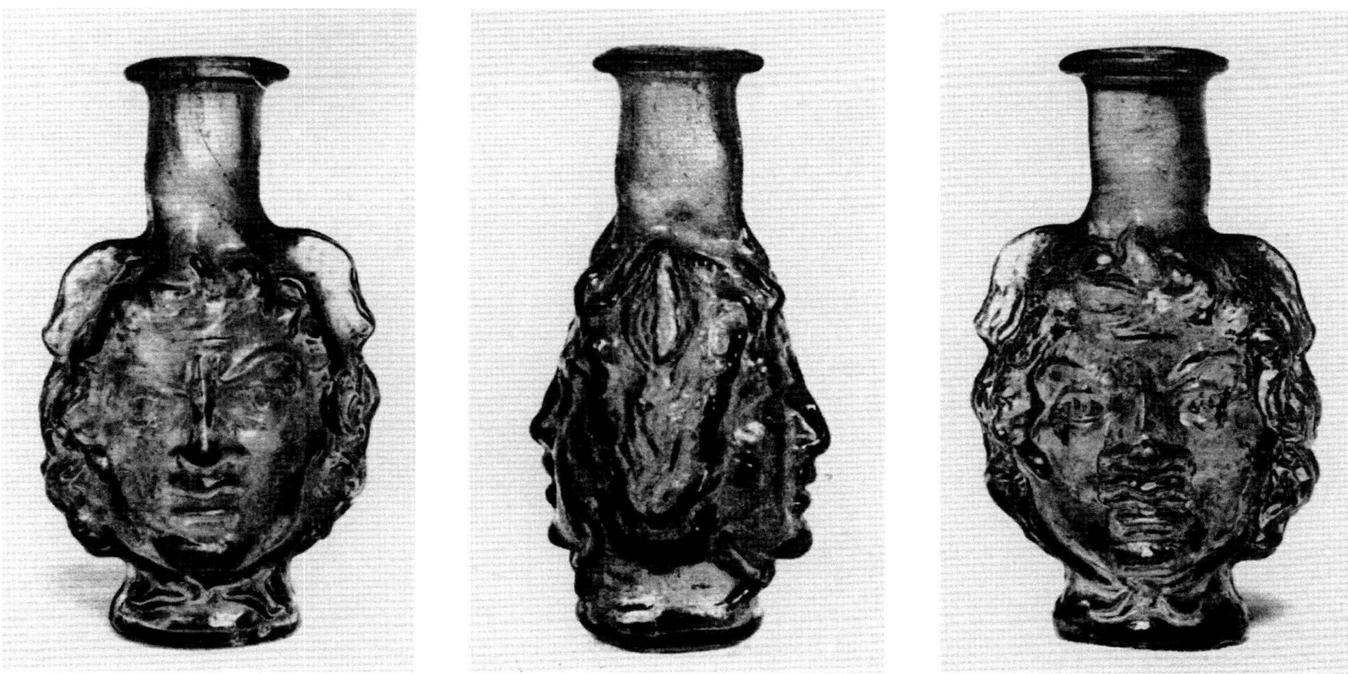

*Fig. 86. First Century, Type C. Double head flask with two Medusas. This is the earliest known double head flask of the Roman period. From Vigorovea, H 8.3 cm, transparent blue, Padua, Musei Civici, no. 2482. Photo: Musei Civici, Padua.*

All these vessels are forms that could have been made in Italy. This suggests, but does not prove, that the head flask was also made in the West. The tomb is dated to the second half of the first century. The glass vessels are types that occur in contexts dated before 50, but many were still in use in Pompeii at the time that city was destroyed in 79.

TYPE D. SINGLE CHILD. A tall-necked head-shaped bottle in the form of a single child's head was excavated in a child's tomb at Esvres, near Tours, in central France (Fig. 87).[42] The context suggests a late first- or early second-century date. No other examples of this Type have been published.

### *The Second Century*

Most head-shaped vessels excavated in second-century contexts are double heads shaped like bottles without a handle, ranging in height from about 8.5 to 9.5 cm. Types A, D, and E are from the eastern Mediterranean, while Type B is possibly Italian and Type C is of unknown origin. Type F is a slightly larger single head-shaped bottle. Types E and F are interesting because they suggest the use of the same molds for blowing cups and flasks.

TYPE A. ONE SERIOUS FACE, ONE SMILING FACE (here Cat. No. 141). A double head-shaped bottle excavated at Korakou, Cyprus,[43] represents this Type, which must have been very popular in the eastern Mediterranean, judging from the numerous preserved parallels. The bottle from Cyprus is translucent dark blue, has a slender profile, and shows a serious face with a fillet over the forehead back to back with a broad, smiling face with wavy hair. A date in the Hadrianic period (117-138) has been suggested.[44] About half of the vessels assigned to this Type are made of translucent glass, usually dark blue; the others are opaque white.

TYPE B. DOUBLE MEDUSA, TWO SIMILAR SEDUCTIVE MEDUSAS (Fig. 88). This Type shows two similar Medusa faces, a subject already created in the first century (see above: First Century, Type C). It is represented by an opaque white bottle from a tomb at Portorecanati, Italy.[45] The tomb has been dated to the second century on the basis of a small clay plate of gray ware and this head flask. The motif below the chin is described by the excavator as acanthus, but in the photographs it looks like the knotted snakes of Medusa's hair. The profile view of the vessel is slender, like that of Second Century, Type A. A few parallels exist in other collections.[46]

TYPE C. DOUBLE KNOBBY-HAIR. A head-shaped bottle of blue glass, excavated at Ravenna, represents this Type (Fig. 89).[47] It has two heads, presumably female, rounded knobby cheeks, very large eyes with iris and pupil, and rounded knobby curls. The faces have a pointed profile resulting from the fact that the forehead and lower half of the face recede. As a pointed profile has not been noted on flasks made in the eastern Mediterranean,[48] it may be a western characteristic, in which case an origin in northern Italy seems most probable. The Ravenna bottle has been published as dated to the second century, but no mention is made of accompanying finds to support this date.[49] Knobby hair also occurs on head-shaped vessels made in the eastern Mediterranean in the third and fourth centuries. We may therefore assume

[42] H ca. 7.4 cm, Octave Bobeau, "Un cimetière gallo-romain à Esvres (Indre-et-Loire)," *Bulletin archéologique du comité des travaux historiques et scientifiques* 1909, 224-225, fig. 4. The tomb also contained two glass bottles shaped like a bunch of grapes (like Cat. No. 110), two glass spheres with a lead coating on the interior, and two terra-cotta toy figurines. On glass vessels with a lead coating: Morel et al. 1992, 6, 10-11, fig. 20, which describes a mid first-century glass workshop at Avenches, Switzerland, with remains suggesting that some vessels made there were coated with lead on the spot; Rütti 1988, 78-79, fig. 39.

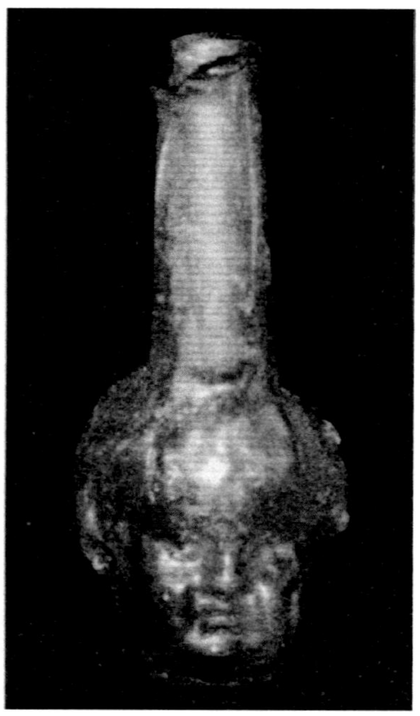

*Fig. 87. First Century, Type D. Head flask in the shape of a single child's head. From Esvres, colorless, H ca. 7.4 cm, Commune d'Esvres, France, no number. Photo: Jacques Dubois, Musée Archéologique, Tours.*

[43] Cyprus, Nat. Arch. Mus., no. 1963/XI-22/9, H 8.5 cm, from a tomb at Korakou, Cyprus, which contained as the only other find a pair of earrings difficult to date, Karageorghis 1964, 310-311, fig. 32a-c; cf. Kyriakos Nicolaou, "Découvertes archéologiques de verres antiques faites à Chypre dans la décade 1962-1971," *BullAIHV* 6 (1971-72) 128-130, no. 11.

[44] Barag 1985a, 93, at no. 118.

[45] Mercando 1974, 320-322, 325, fig. 247, from Portorecanati, tomb 167.

[46] Compare examples probably of the same Type:
a) formerly R. W. Smith Collection, no. 863, opaque white, acquired in Lebanon, Smith 1957, 143, no. 286;
b) RGM, no. N1070, opaque white, find spot unknown, La Baume 1973, no. C13, pl. 10;
c) Moscow, Nat. Hist. Mus., no. 54791, Sorokina 1968, 187, fig. 3, nos. 1-2, Faces A and B ill.

A similar bronze head flask, Moscow, Nat. Hist. Mus., no. 28737, was excavated in Kerch: Sorokina 1968, 187, fig. 3, no. 3.

[47] Ravenna, Mus. Nat., no. 10020, from Ravenna, cemetery at "Le Palazette," tomb 13 (excavated in 1966), Montanari 1967, 31, no. 43, fig. 43; Maioli 1974, 17, 24, fig. 13.

[48] The knobby hair and circle eyes with recessed pupils are similar to those of Cat. No. 149, made in the eastern Mediterranean, but the triangular pointed profile of the face is different, showing that the Ravenna bottle belongs to a different Series.

[49] Maioli 1974, 24. The cemetery itself contained material from the first through the third or early fourth centuries.

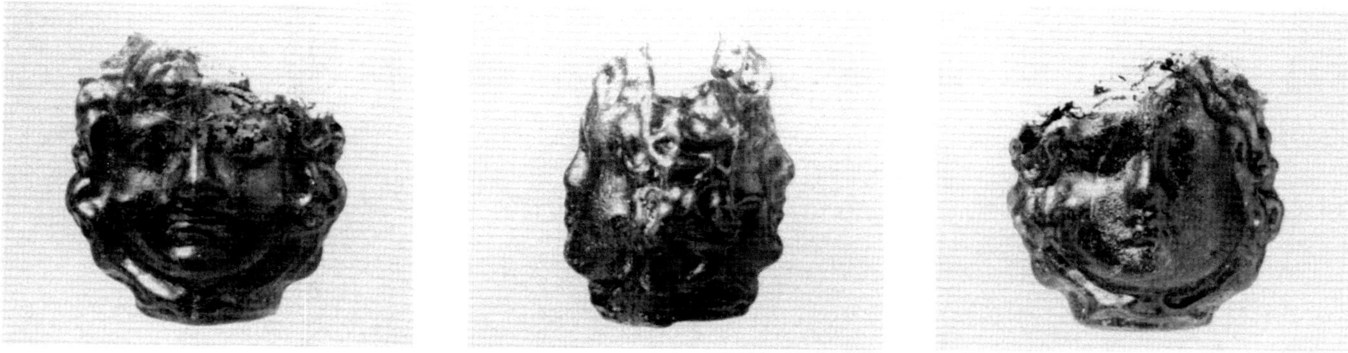

*Fig. 88. Second Century, Type B. Double head flask with two Medusas. From Portorecanati, H (pres.) 4.2 cm, opaque white, Ancona, Italy, Museo Archeologico Nazionale delle Marche, no. 26509. Photo: Museo Archeologico, Ancona.*

that the Ravenna bottle was made not very long before that time, perhaps in the second half of the second century or in the early third century.

TYPE D. TWO SIMILAR FACES OF UNDETERMINED SEX (compare here Cat. Nos. 145 and 146). A bottle or flask excavated at Garny, Armenia,[50] is natural green and shows two similar faces of undetermined sex. The excavator assigns a second-century date to the tomb in which the vessel was found but does not give the basis for the dating. If the vessel from Garny belongs to the same Type as Cat. Nos. 145 and 146, it was made in an MCT VIII mold and the hair does not consist of round knobs around the face but only above the forehead, with rounded tufts along the sides of the face and irregular horizontal ridges behind. The stylization of the hair and the mold construction support a second-century date. The vessel from Garny is thought to have been imported from the Syro-Palestinian coast, like much of the glass excavated in Armenia. An inland Palestinian or Syrian origin seems probable for this Type, especially since the Toledo Museum flasks are made of the translucent manganese colored purple glass popular in that area.[51]

TYPE E. SINGLE DIONYSOS-ANTINOUS. Two forms of head-shaped vessels, each showing a single head wearing a wreath of ivy, have been assigned a second-century date by their excavators. One is a cup with a handle at the back of the head (Type E, Series 1), while the other is a tall-necked flask (Type E, Series 2). (The relationship between cups and flasks is similar to the relationship between cups of Single Ethnic Head, First Century, Type A, and flasks of the Single Ethnic Head, Second Century, Type F.) The molds for Type E may have been inspired by an image of the Roman emperor Hadrian's friend Antinous in the guise of the Greek god Dionysos (see Cat. No. 148). The excavated examples are here assigned to two Series because the flask has a design in intaglio on the underside of the base, while the base of the cup is plain.

[50] Yerevan, Armenia, Nat. Mus., no. 1965/29, from Garny, slab-stone tomb 27, described as showing two female heads, H 9.5 cm, Arakelian et al. 1969, no. 133, fig. 133.

[51] Stern 1977a, 156.

***Series 1*** cups are represented by a cup with a wide upright rim (Fig. 90), excavated at Kep on the northern coast of the Black Sea in a tomb containing objects dated to the second half of the second and early third centuries. The cup, however, has been dated on stylistic grounds to the reigns of the Roman emperors Hadrian (117-138) or Antoninus Pius (138-161).[52]

A clay head-shaped cup with two handles on either side of the face, made in a stylistically related mold showing an idealized face with a lion-like mane of hair reminiscent of portraits of Alexander the Great,[53] has been assigned a late Antonine date, i.e. second half of the second century.[54]

[52] Sorokina 1977a, 133-135, fig. 5, from Kep on the Taman peninsula, tomb 133. For a discussion of the tomb finds: Sorokina 1968.

[53] Karlsruhe, Badisches Landesmus., no. 68/48, Mandel 1988, 229, no. P 194, pl. 22. The cup is inscribed ΤΡΩΙΛΟC (*Troilos*) in Greek along the rim above the forehead. It is not clear whether the inscription refers to the maker of the vessel or to the legendary Trojan prince who was slain by Achilles.

[54] Mandel 1988, 25.

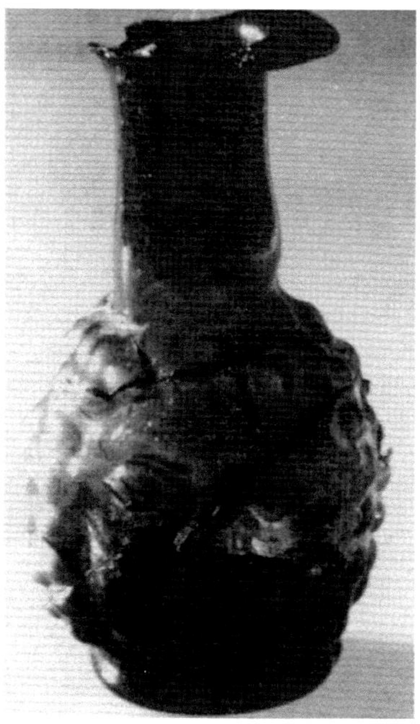

*Fig. 89. Second Century, Type C. Double head flask with knobby hair. The pointed profiles of the faces are not apparent in this three-quarter view. From Ravenna, H 8.5 cm, blue, Ravenna, Museo Nazionale di Ravenna, no. 10020. Photo: Museo Nazionale, Ravenna.*

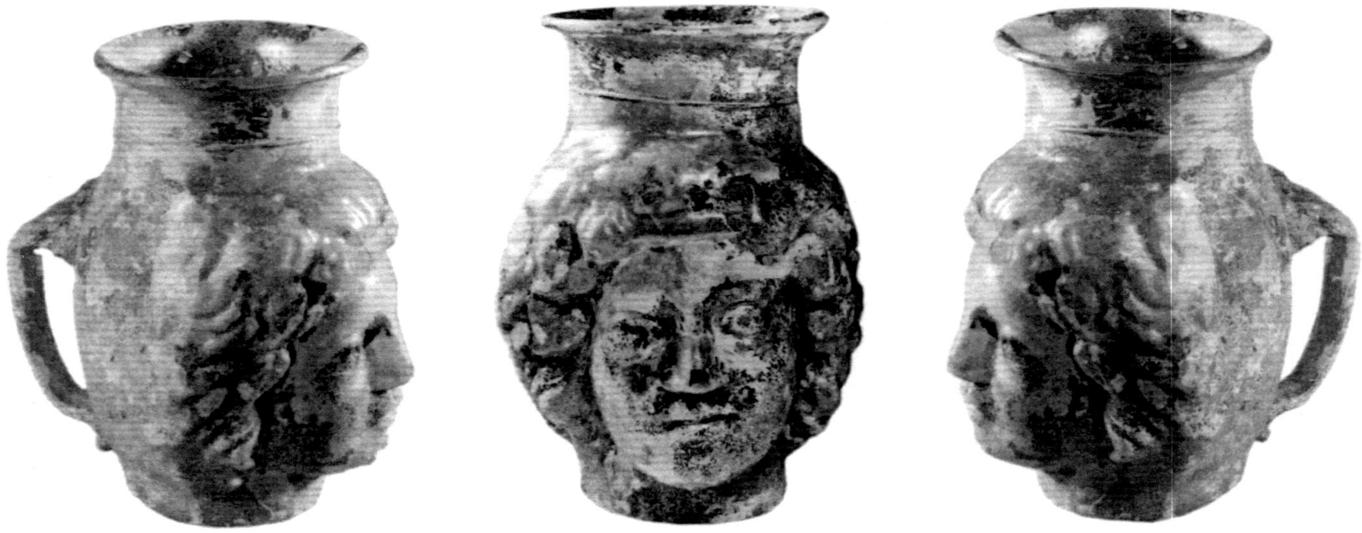

*Fig. 90. Second Century, Type E, Series 1. Cup in the shape of a single head of Dionysos-Antinous. From Kep on the Taman peninsula, tomb 133, Moscow, Historical Museum, no. 99549, N23. Photos: Courtesy Nina Sorokina.*

***Series 2*** flasks (here Cat. No. 148) are represented by a tall-necked flask excavated in a tomb at Constanta (ancient Tomis) Romania, in 1959.[55] A closely related flask with a relief on the underside of the base, probably belonging to the same Series, was excavated at Karantinny Cove (Fig. 91).[56] Both flasks have a mask in relief on the underside of the base, a rare design that connects them with Cat. No. 148. Very little of the face of the flask from Constanta is preserved, but the coiffure as seen from behind is so similar to that of Cat. No. 148 that the flasks may have been made in closely related molds. According to the excavator, the flask from Constanta was found in a second-century context.[57] If that date is correct and the flask was made in the second century — a date supported by the use of MCT VII[58] — then this flask is a very early example of a tall neck rising from an overblow or projecting roll, a neck shape that became common in the third century in the Syro-Palestinian area. The free-blown overblow marked the transition between the upright rim of the mold and the free-blown neck.

[55] Constanta, Arch. Mus., no. 5041, Bucovala 1968, 115-116, no. 237, H 21.0 cm, D 9.0 cm.

[56] St. Petersburg, Hermitage, no. X 1902.14, excavated in 1902, H 11 cm, ill. by G. Sokolov, *Antique Art on the Northern Black Sea Coast* (Leningrad 1974) 137, no. 146, from grave 1386, rock-cut tomb near Karantinny Cove, Tauric Chersonesos. The flask from Karantinny Cove does not have a handle and the rim appears to be broken; it was probably finished with a tall neck rising from a wide upright rim like the flask from Constanta.

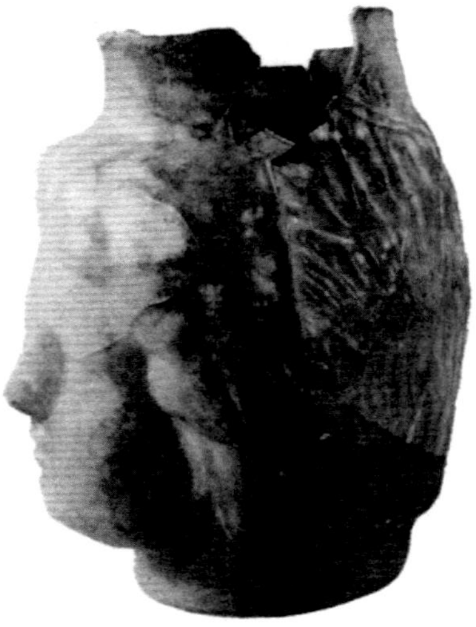

*Fig. 91. Second Century, Type E, Series 2. Flask, originally probably with a tall neck, in the shape of a single head of Dionysos-Antinous. From Karantinny Cove, H (pres.) 11 cm, transparent green, St. Petersburg, Hermitage, no. X 1902.14. Photo: from G. Sokolov,* Antique Art on the Northern Black Sea Coast *(Leningrad 1974) fig. 146.*

Type F. Single Ethnic Head. A curious flask with a tall neck and an overblow between body and neck was excavated at Alzey, Germany (Fig. 92).[59] The flask looks as though it might have been made in a mold for a cup like those discussed under First Century, Type A. The overblow functioned as a transition from the mold-blown body to the tall neck, a relationship similar to that between cups and flasks of the Second Century, Type E, Single Dionysos, Series 1 and 2.

## *The Third Century*

The excavated head-shaped vessels found in third-century contexts are all double heads. Those found in the eastern Mediterranean show youthful faces with full cheeks and knobby hair (Type A), while those made in the West lack the characteristic knobby hair (Type B).

Type A. Double Knobby-Hair (here Cat. Nos. 149-156). The two faces are usually similar, both youthful. The most

[57] Bucovala 1968, 116.

[58] Comparison with Cat. No. 148 suggests the Constanta flask was made in an MCT VII mold.

[59] RGZM, no. 017611, H. Klumbach, "Alzey, Vicus und Kastell," *Alzeyer Gesichtsblätter* 2 (1965) 21-22, fig. 16 (cited by Landes 1984, 342, fig. 5).

characteristic feature is the stylization of the hair in a uniform pattern of rounded knobs. This may have been intended to recall the full, curly hair of Roman portraits of the second half of the second century,[60] but the technique used to carve the molds — the knobs were probably hollowed out directly in the clay mold — appears to be closely related to Bottles with Stylized Grape Pattern such as Cat. Nos. 120-128. These were blown in Syro-Palestinian workshops, and the same origin is suggested for most of the head-shaped vessels with knobby hair, with the possible exception of the late second-century bottles, Type C, discussed above. Four Series belonging to the Double Knobby-Hair Type have been excavated in third-century contexts. Of these, Series 3 is exceptional because it carries a mold-blown *chi-rho* monogram on the underside of the base. Fragments of nine double head-shaped vessels and one single head flask, of assorted sizes and colors, are recorded as having been excavated at Dura, but neither photographs nor detailed descriptions are published so it is not known whether they belong to Type A. All must predate 256, the year the city was abandoned.[61]

***Series 1*** is represented by a double head with knobby hair found with a coin of Commodus (177-192) in a tomb at Dura Europos (Fig. 93).[62]

***Series 2*** is represented by a double head dated to the end of the second to early third century, excavated at Csongrad, Hungary.[63]

***Series 3*** is represented by a double head with knobby hair and a mold-blown *chi-rho* monogram on the underside of the base (Fig. 94). It was excavated in a period C house at Karanis, in the Egyptian Fayum,[64] dated early second to mid third century by the excavator. The heads have characteristic knobby hair The *chi-rho* monogram cannot be used to date the flask. In Christian contexts this form of monogram first appears in the Near East in the mid fourth century, but it is also known on Armenian and Parthian coins of the first century B.C. and on Jewish coins in A.D. 38.[65]

***Series 4*** is represented by a double head excavated at Gurob in the Fayum (Fig. 95). It has not only knobby hair but also knobby cheeks and a knob-shaped chin. The tomb has been dated early fourth century, "about the time of Constantine" (ruled 306-337).[66] Judging from the rest of the glass found

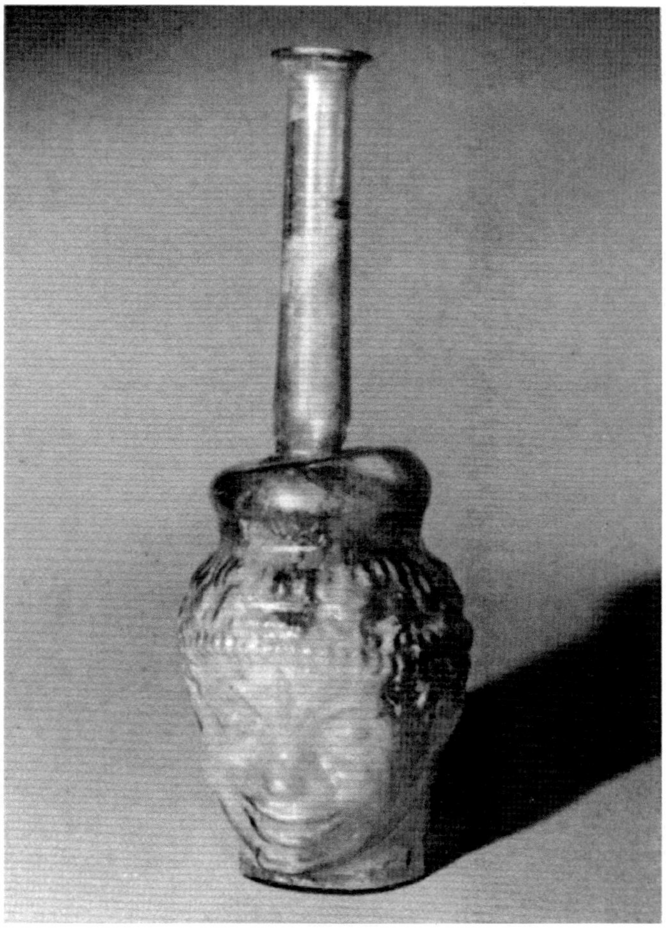

*Fig. 92. Second Century, Type F. Head flask in the shape of a single ethnic head. From Alzey, H 22.6 cm, green with little iridescence, Mainz, Germany, Römisch-Germanisches Zentralmuseum, no. 017611. Photo: Römisch-Germanisches Zentralmuseum.*

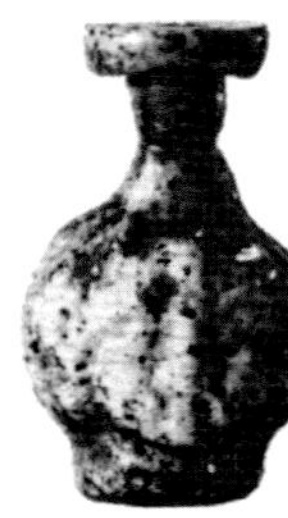

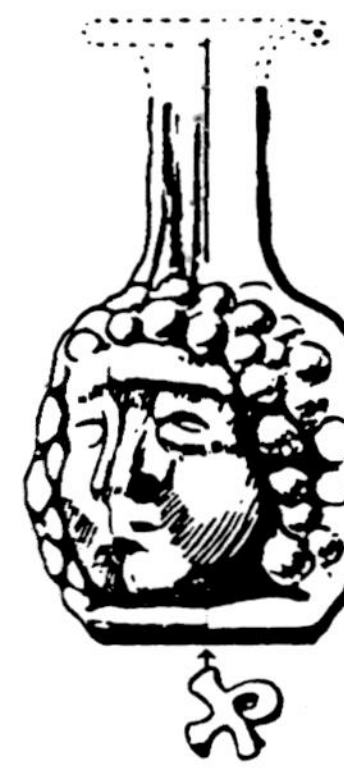

*Fig. 93. Third Century, Type A, Series 1. Double head flask with knobby hair. From Dura Europos, H 6.5 cm. Damascus, National Museum, Dura Excavations, no. I 363. Photo: Yale University Art Gallery, Dura Europos Collection.*

*Fig. 94. Third Century, Type A, Series 3. Double head flask with knobby hair, with ch=rho monogram on the underside of the base. From Karanis, H 8.5 cm, green, Ann Arbor, The University of Michigan, Kelsey Museum of Archaeology, no. 6168. Drawing: Reprinted with permission from Harden 1936, pl. 18, no. 629.*

60 Compare the hair styles of the Roman emperors Lucius Verus (161-169) and Commodus (177-192).

61 Clairmont 1963, 37-38, nos. 132-145.

62 Damascus, Nat. Mus., from tomb 31, Clairmont 1963, 37, no. 132, pl. XXI; *The Excavations at Dura Europos, Preliminary Report* IX,2 (New Haven 1946) 64, no. 4, pl. XLIX. Perhaps similar to an undatable head-shaped vessel of opaque yellow and green glass, H 6.0 cm, excavated at Homs, Syria, Abdul Hak 1965, 28-29, fig. 7.

63 M. Parducz, *Romaikoi leletek Csongradon, Dolgozatok a M. Kir. Ferenc Jozsef Tudomany* (*Egyetem Archaeologiai Intezetebolz* 12, Budapest 1936) 63ff., pl. XL, no. 1 a-b (cited by Barkóczi 1988, 213-214, at no. 544).

64 Ann Arbor, Mich., Kelsey Museum of Archaeology, no. 6168, from Karanis, field inv. no. 27/C 45J/B1, Harden 1936, 214, no. 629, pl. XVIII.

65 *EAA* II, 950-951, s.v. *croce.*

66 W. M. Flinders Petrie, *Illahun, Kahun and Gurob 1889-90* (London 1891) 29, pl. XXXIII, no. 7.

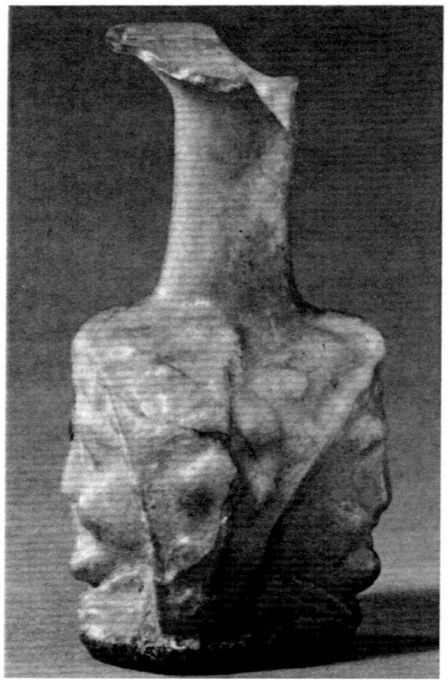

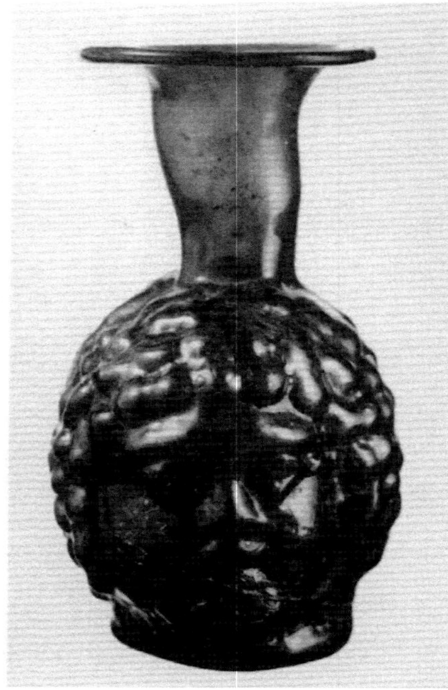

*Fig. 95. Third Century, Type A, Series 4. Double head flask with knobby hair and knobby cheeks. From Gurob, H 8.4 cm, opaque white, London, British Museum, no. GR 1890. 11-15.2. Photo: Copyright British Museum.*

*Fig. 96. Third Century, Type A, perhaps Series 4. Double head flask with knobby hair and knobby cheeks. From Gebel Adda, H 8.0 cm, blue, Toronto, Royal Ontario Museum, no. 973.24.453. Photograph: The Royal Ontario Museum.*

in the tomb, the tomb was used in the second half of the third and in the early fourth centuries. A vessel of blue glass excavated at Gebel Adda in Egyptian Nubia (Fig. 96) is said to be similar to the vessel from Gurob.[67] The vessel from Gebel Adda is thought to have belonged to a Meroitic period burial in Cemetery Three, dated before the middle of the fourth century, but how long before is not known.

TYPE B. ONE BEARDED, ONE CLEAN-SHAVEN MALE. A double head shaped like an aryballos with two small handles (Fig. 97) was excavated in a tomb at Esch, in the Netherlands, and has been firmly dated to the first half of the third century on the basis of the tomb contents.[68] The aryballos has two relatively large male faces, one bearded with a mustache, the other clean-shaven and young. Two vessels belonging to the same Type and probably to the same Series were excavated at Cologne,[69] but the aryballos from Esch is the only one to have been excavated in an independently dated context. The Series must have been created before the middle of the third century.[70] Find spots indicate a northwestern European origin for this Type.

### *The Fourth Century*

From the Egyptian finds discussed above, Third Century, Type A, Series 3 and 4, it seems that double head-shaped vessels with stylized knobby hair were still in use (if not actually still being produced) during the first half of the fourth century.[71] But most head-shaped vessels found in fourth-century contexts are western types that are not represented in the Toledo Museum collection (Types A and B). They illustrate a trend toward larger size and a single head format

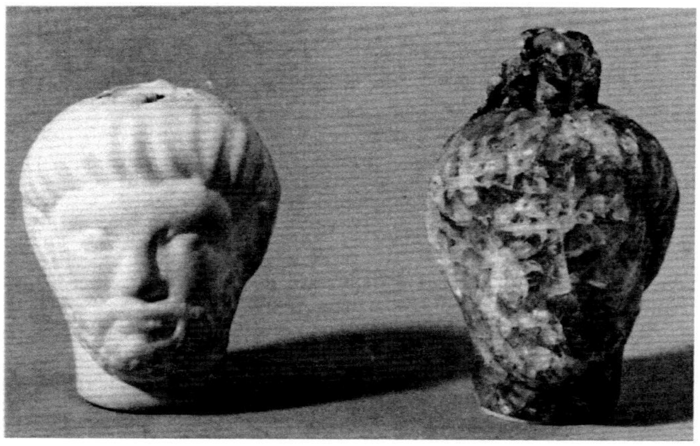

*Fig. 97. Third Century, Type B. Double head flask with bearded and clean-shaven male heads, plus cast (left) which gives a clearer idea of the shape. From Esch, H (pres.) 6.0 cm, colorless, 's-Hertogenbosch, Noordbrabants Museum, no. 9906. Photo: Noordbrabants Museum.*

[67] Nicolas B. Millet "Gebel Adda — Preliminary Report for 1963," *JARCE* 2 (1963) 158, fig. 10, 159.

[68] Decolorized glass, H (head) 6.0 cm, L. J.A.M. van den Hurk, "The Tumuli from the Roman Period of Esch, Province of North Brabant, III" *BerRob* 27 (1977), 104-105, fig. 22; 125, no. 19; pl. III, no. 1, from grave V.

[69] For example:

- a, b) RGM, nos. N5897, N297, from Cologne, Fritz Fremersdorf, *Das Naturfarbene Sogenannte Blaugrüne Glass in Köln* (*Die Denkmälern des römischer Köln* 4, Cologne 1958) 21, pl. 3;
- c, d) V&A, nos. 993-1868 and 999-1868, find spots unknown, Honey 1946, 26, pl. 5 D;
- e) formerly R. W. Smith Collection, Smith 1957, 145, no. 290.

[70] Vessels with aryballos handles were made from the first through the fourth centuries.

[71] A double head-shaped vessel with knobby hair was excavated in the hypogaeum at Homs, Syria, green, H 7.8 cm, Abdul Hak 1965, 30, fig. 11. The finds from Homs cannot be dated independently because the tomb contains glass from the third through the fifth centuries.

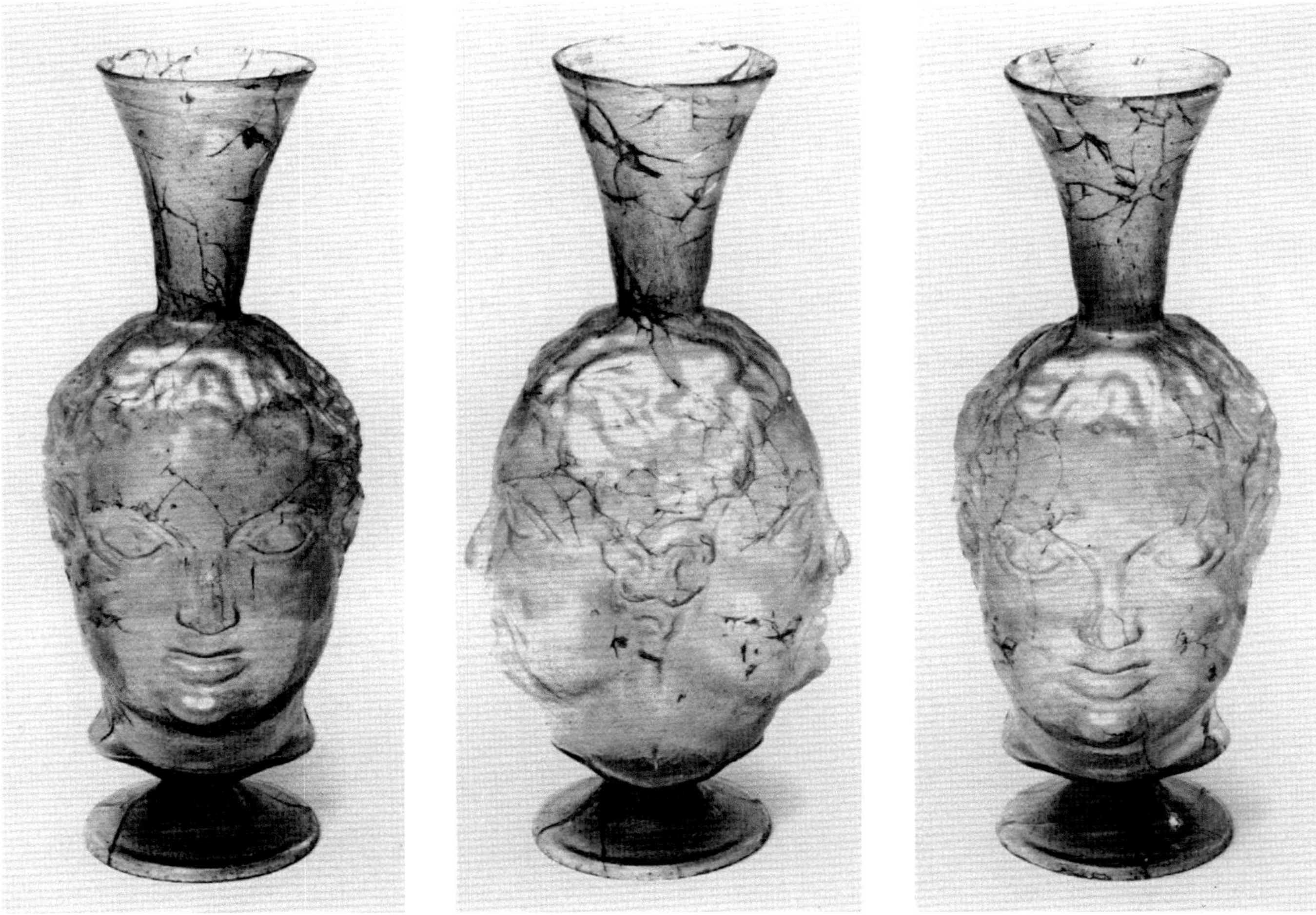

*Fig. 98. Fourth Century, Type A. Double head flask with two clean-shaven male heads. From Worms, H ca. 24.0 cm, Worms, Museum der Stadt, no. WM 267. Photo: Worms City Archives.*

that took place throughout the Roman empire [72] The only head-shaped vessel from a datable fourth-century context in the Syro-Palestinian area, Type C, is problematic because its stylistic features fit best with second-century vessels.

TYPE A. DOUBLE CLEAN-SHAVEN MALE. A large flask with a tall funnel neck and two clean-shaven, mature male heads was found with other glasses at Worms, Germany, in a sarcophagus dated to the first half of the fourth century (Fig. 98). [73] The beardless faces suggest the type was created after the Roman emperor Constantine reintroduced the fashion of shaving in 307. Find spots suggest that head flasks of this Type were made in Cologne and perhaps in Gaul. [74]

TYPE B. SINGLE CLEAN-SHAVEN MALE. A single head flask with one handle at the back of the head was excavated at Steinfort, Luxemburg, in a cemetery used during the late

[72] Compare a group of large, single head-shaped vessels of translucent dark blue glass made in an unidentified eastern Mediterranean workshop during the fourth century:

a) CMG, no. 59.1.150, formerly R. W. Smith Collection, Smith 1957, 141, no. 279; *Glass of the Caesars* 1987, 175, no. 96;
b) Berlin, Antikenmuseum, no. 30219, 208, Platz-Horster 1976, 45, no. 70;
c) Oppenländer Collection, no. 22351, *JGS* 19 (1977) 170, no. 7; then *Kofler-Truniger* 1985, 57, no. 86 (now Malibu, The J. Paul Getty Museum, acc. no. 85.AF.320, *GettyMusJ* 14 (1986) 196, no. 70, ill.).

Not one of the head flasks was found in a datable context; the group can nevertheless be dated because they were made in the same workshop as a characteristic group of free-blown jugs. On the group: Spartz 1967, at no. 98; TMA 1923.736, in TMA, *Roman and Early Byzantine Glass* (forthcoming).

[73] Worms, Museum der Stadt, no. WM 267, H ca. 24.0 cm, from Worms, Maria Münster cemetery, Helmut Bernhard *Römische Gläser in Worms*, exh. cat., Museum der Stadt Worms (Worms 1979) 13-15, figs. 8 and 23, no. 80.

[74] For example:

a) RGM, no. 946, from Cologne, Kisa 1908, III, 745, fig. 299; Fremersdorf VI 1961, 77, pls. 170-171;
b) RGM, no. 56.408, from Cologne, Fremersdorf VI 1961, 77-78, pls. 172-173;
c) RGM, no. 34.292, acquired at Gondorf but possibly from Cologne, Fremersdorf VI 1961, 78, pl. 176;
d) Trier, Landesmuseum, no. 03,342, from Trier, Goethert-Polaschek 1977, 257-258, Form 159, no. 1532, pl. 80.

Other find spots mentioned are Mainz-Finthen and Boulogne-sur-Mer: Sambon 1928, 49, fig. 96, no. 50.

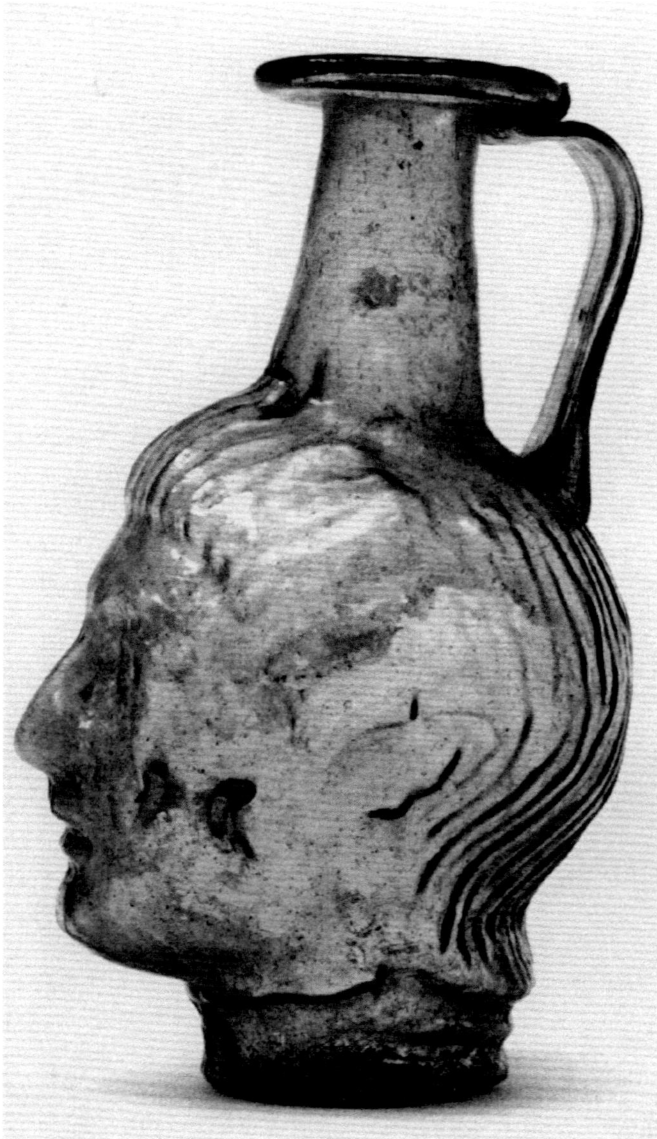

*Fig. 99. Fourth Century, Type B. Head flask in the shape of a single clean-shaven male. From Steinfort, H 14.4 cm, transparent greenish, Luxembourg, Musée National d'Histoire et d'Art, no. 26984. Photo: Musée National d'Histoire et d'Art, Luxembourg (Gr.-D. de Luxembourg).*

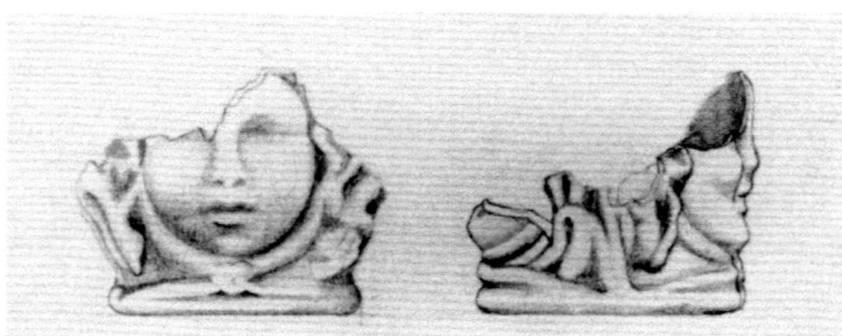

*Fig. 100. Fourth Century, Type C. Double head flask with two Medusas. From Pella of the Decapolis, H (pres.) 5.1 cm. Drawing: Courtesy Basil Hennessy, The University of Sydney.*

fourth and early fifth centuries (Fig. 99).[75] This head flask also has a clean-shaven face. Find spots suggest this Type was not made in Cologne,[76] but the location of the workshop has not yet been identified.

TYPE C. DOUBLE MEDUSA. A fragmentary vessel featuring two Medusa heads back to back, with knotted snakes beneath the chins was found in a tomb at Pella, Jordan, with other glass vessels that appear to date from the late third to the end of the fourth centuries (Fig. 100).[77] Since it is the sole representative of this Type found in an independently dated context, it cannot be used to date the Type; the only conclusion that can be drawn from this find is that the head flask was created before the end of the fourth century. From a stylistic point of view the small size of this vessel, the absence of knobby hair, and the subject represented — a double head of Medusa — suggest a second-century date (compare above: Second Century, Type B). The possibility of its having been an heirloom or reused in the fourth century cannot be excluded.[78]

### *The Fifth Century*

A few fragments of head-shaped vessels have been found in Rome in contexts dated to the late fourth to mid fifth centuries, but fragments of only one vessel have been published.

TYPE A. TWO SIMILAR FACES. Fragments of a double head-shaped vessel of green glass were found in a building on the Lungotevere Testaccio, Rome (Fig. 101).[79] The vessel has two similar faces characterized by a nose with flaring nostrils, a straight mouth, and a knob-shaped chin. Front and profile views of the head, as well as the wavy bands of hair slanting down to a meeting point at the seam between the two heads, are similar to Cat. Nos. 145 and 146. More important than these similarities, however, are the differences: the vessel from Rome, with a base diameter of 4.9 cm, is almost twice the size of Cat. Nos. 145 and 146, while its hair does not consist of a row of knobs around the face but is rendered as wavy bands of curls drawn back horizontally. The discovery of this vessel in a room filled with glass fragments that included waste from glass working has led to the hypothesis that this shape was made in Rome.[80] But the fact that there was only one such vessel among numerous other fragments is reason for caution. Not all fragments found

[75] Wilhelm 1969, 31, no. 129, 79, ill. Another example was excavated in the Grand Hôle cemetery, at Chassemy (Aisnes), in a similar fourth- to fifth-century context, Lantier 1929, 15, no. 32, pl. 32.

[76] An example in the Museum at Kreuznach on the Rhine is mentioned by Wilhelm 1969, 31, no. 129, who records no finds from Cologne.

[77] From Pella, tomb 39A, H (pres.) ca. 5.1 cm, McNicoll and Hennessy 1982, 536, pl. CXVI, no. 11.

[78] Foy 1977-80, 212, fig. 76, no. 4, notes the discovery of a third-century double head-shaped vessel in a fifth-century sarcophagus excavated at Marseilles.

[79] Sternini 1989, 40-42, pl. 9, no. 54a-c. The author also mentions fragments of a head-shaped vessel, otherwise undescribed, from the *tabernae* near the Temple of Magna Mater on the Palatine, which she is preparing for publication.

[80] Sternini 1989, 55.

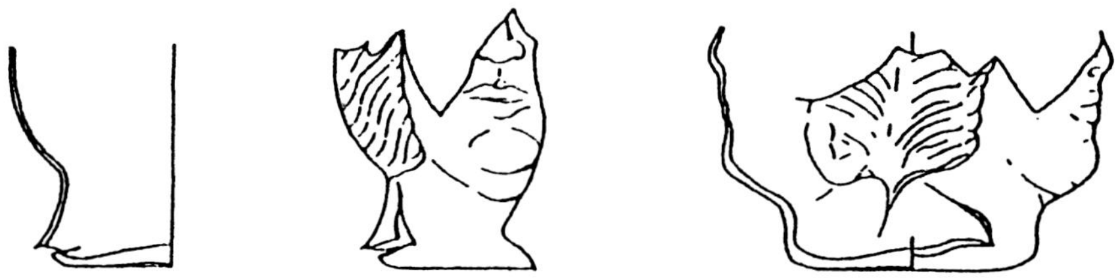

*Fig. 101. Fifth Century, Type A. Double head flask with two similar faces. From Rome, Lungotevere Testaccio, H ca. 5.5 cm, green. Drawing: Reprinted with permission from Sternini 1989, fig. 54.*

in a glass-worker's dump can be interpreted as waste from production; glass collected for recycling can also be expected.[81]

[81] Gladys Davidson Weinberg, "Specialized Production in a Late Roman Glass Factory," *ErIsr* 19 (1987) 62-70; Weinberg 1988, on vessels excavated at the site of a glass-worker's furnace at Jalame, Israel.

## NOTE: ORDER OF CATALOGUE ENTRIES

Despite uncertainties in dating, the catalogue of head-shaped vessels is arranged in a loosely chronological order. The dates proposed in the catalogue refer to the Toledo Museum vessels, but because the development of the Types to which they belong has not been researched, the dates are given a wide range.

# CATALOGUE

*Vessels Shaped Like a Single Head, First Century*
(Cat. Nos. 138-140)

138 **Jug in the Shape of a Female Head, possibly Livia as Hera**[1]

Color plate 22, p. 60
Syro-Palestinian, perhaps made in Sidon
Workshop of the Floating Handles
First half of the first century, possibly as early as A.D. 14
H 11.0 cm; H without thumbrest 10.0 cm; H body 7.2 cm; max D 4.2 cm; D rim 2.07 cm; D base 3.3 cm (along the seam)-2.85 cm (ear to ear)
1967.8, Purchased with funds from the Libbey Endowment, Gift of Edward Drummond Libbey

Transparent pale blue green (5 BG 7/2). Medium thin glass. Fabric cannot be determined because of weathering.
Rim tooled. Neck and body blown into a two-part mold of two vertical sections, each including one half of the base (MCT VIII B). One continuous mold seam, through center of face and back of head, extends across underside of base. Relief crisp. No pontil mark.
Rim folded upward and inward. Cylindrical neck. Body in the shape of a head on a tall cylindrical neck. Oval slightly concave base. Handle applied to rim and pulled downward; at point of application folded upward, outward, downward, and pinched to form a vertical thumbrest.
Body in the shape of a female head with deep-set almond-shaped eyes under brows that form diagonal lines descending from the bridge of the nose, a narrow projecting nose, short horizontal lips slightly parted, a very heavy protruding chin, and a tall cylindrical neck with three Venus rings.[2] Teardrop-shaped earrings hang from visible ear lobes. The hair is worn in a very severe style, parted in the middle with a nodus over the center of the forehead, drawn down close to the head in wavy locks from the part to a roll which frames the face, and tied in a chignon at the nape. A corkscrew curl emerges from the corner of the chignon on either side and is drawn diagonally forward down the neck. A short lock curls forward in front of each ear. Underside of base undecorated.
*Condition* Nodus, chignon, and nose partly restored. Holes in hair roll near back on left side of head. Only point of application and vertical thumbrest preserved from handle. Exterior covered with a very thick flaking layer of silver iridescence. Interior with areas of beige and dark brown weathering.
*Collections* Giorgio Sangiorgi (1886-1965), Rome (by 1914-65); Sergio Sangiorgi, Rome (1965-67).
*Published* Sangiorgi 1914, no. 107, pl. XXII; *TMA Mus News* 12.4 (1969) ill.; TMA, *Art in Glass* 1969, 24, ill.; Grose 1978, 82, fig. 19.
*Remarks* This jug is important because it is the earliest example of

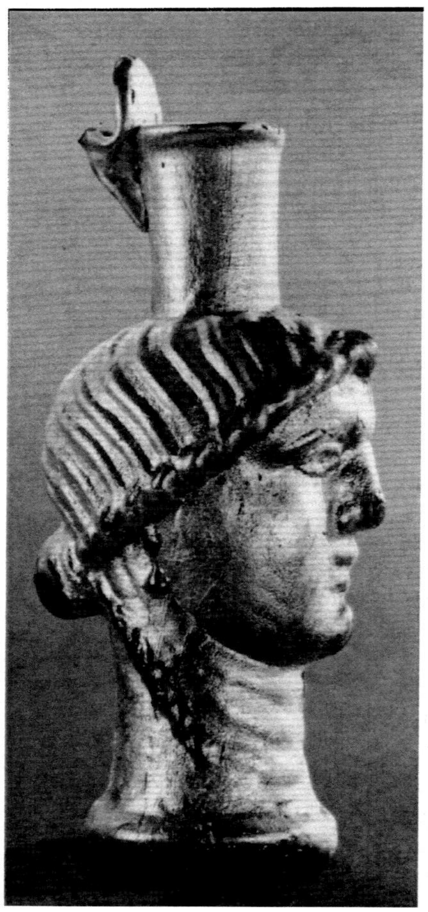

138 (1:1)

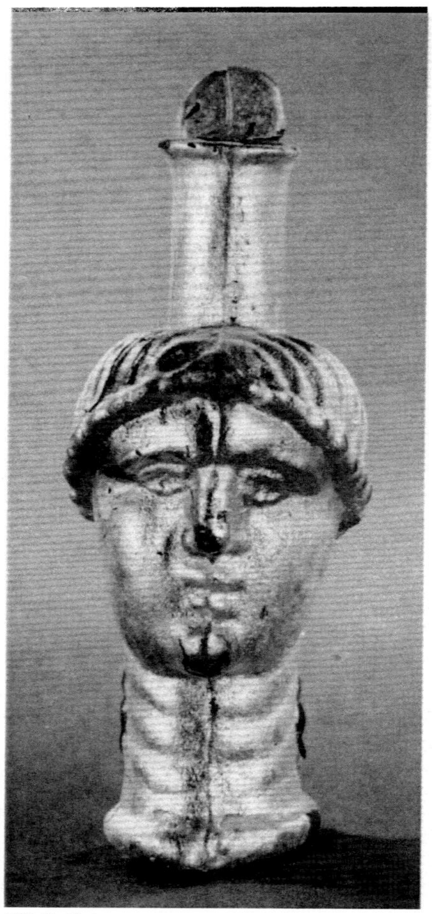

138 (1:1)

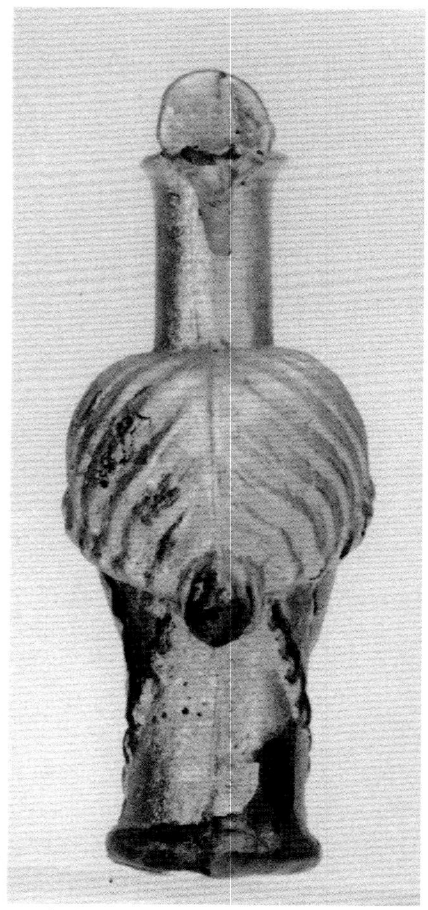

138 (1:1)

a mold-blown glass head-shaped vessel, a shape that was to have a long life in the history of glass. Some ten examples of this Type are known.[3] The jugs vary in height from about 10.0 to 11.0 cm. All reported find spots are in the Near East. The handle of this jug is missing, but several examples belonging to the same Type have the distinctive handle that characterizes them as products of the Workshop of the Floating Handles (Fig. 102), which was active in the first quarter of the first century (see Chapter I: *Syro-Palestinian Styles: The Workshop of the Floating Handles*). The workshop may therefore be credited with the invention of this distinctive shape.

Ten Types of mold-blown glass vessels have been attributed to the Workshop of the Floating Handles. The Toledo Museum collection has representatives of four Types: a miniature amphora encased in wickerwork (Cat. No. 59), a lenticular bottle with a Medusa face on either side (Cat. No. 62), an acorn-shaped bottle (Cat. No. 113), and this head-shaped vessel. The workshop was probably the first to make glass head-shaped vessels.

### Technique

Some of the roundness of the features of this head may be the result of the technique of mold blowing. When a glass bubble, or paraison, is blown, it can be difficult to make it fill a mold evenly because air trapped in hollows prevents the expanding glass from touching the wall of the mold. Modern practice is to pierce a hole in the wall of the mold through which air in the mold can escape. In the case of these ancient head-shaped jugs, filling the nose may have presented a problem, because in spite of the fact that the two mold halves met along the top of the nose, which would have allowed some of the air to escape, the noses of some jugs are more successful than those of others. The nose of the jug in Toledo is well filled out compared with the flat nose of the jug at Stanford (Fig. 103). However, problems with expanding the paraison cannot account for the fact that forehead, nose, and chin of all examples appear to be in one plane, and this feature must have been part of the mold's design.

### Interpretation

These head-shaped vessels are intriguing because they were probably made on the Syro-Palestinian coast but show a typical hairstyle from the city of Rome with a small nodus, or knot, over the forehead and a chignon at the nape of the neck. The Latin poet Ovid, writing around the beginning of the first century A.D., remarks (*Ars Amatoria*, III, 137-140):

> *A simple parting suits an oval face*
> ...
> *A round one needs a dainty cluster (nodus) posed*
> *Above the brow, that leaves the ears disclosed.*[4]

In imperial iconography the nodus coiffure was associated predominantly with the women of the emperor Augustus's household. This is the reason it has been proposed that this head flask represents Augustus's wife, Livia, whom he married in 38 B.C. and who survived her husband until she died, aged 86, in A.D. 29.[5] The nodus hairstyle was fashionable in Rome in the third quarter of the first century B.C.,[6] but Livia is known to have worn it into the first century. In Asia Minor the nodus hairstyle is first attested slightly later than in Rome.[7] Interpretation of the jugs as portraits of Livia is an attractive explanation for the hairstyle, but it faces two major problems. First, the physiognomy of Livia is well known and does not resemble these vessels. She had a slender face with

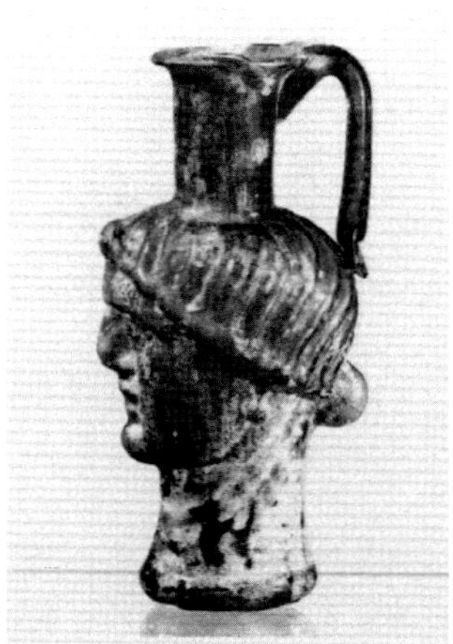

*Fig. 102. Jug in the shape of a female head, Workshop of the Floating Handles. H 10.2 cm, transparent brown, New Haven, Conn., Yale University Art Gallery, no. 1953.28.14. Photo: Yale University Art Gallery.*

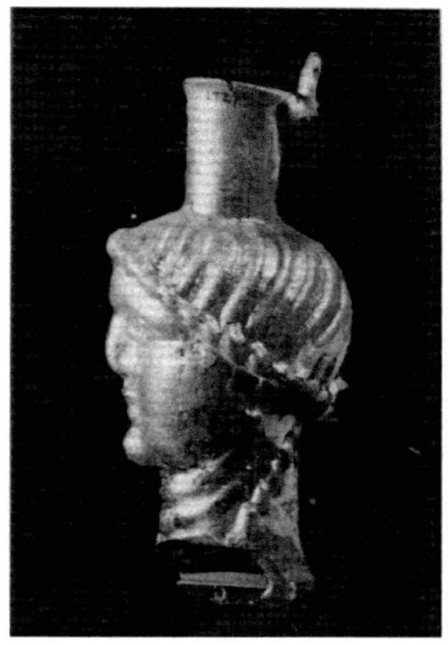

*Fig. 103. Jug in the shape of a female head, Workshop of the Floating Handles. H 10.0 cm, light green, Palo Alto, Stanford University Museum, no. 17277. Photo: Stanford University Museum.*

*Fig. 104.* Aes *of Pergamon, probably minted 16-13 B. C., with heads of* (left) *Livia as Hera (obv.)* and (right) *Julia as Aphrodite (rev.). Although the physiognomy and the hairstyle of Livia differ markedly from those in most sculptural portraits of her, the coin portrait is identified by an inscription. Paris, Cabinet des Médailles, from a cast in New York, The Metropolitan Museum of Art. Photo: Courtesy Maxwell Anderson.*

full cheeks, a rather straight forehead from which a long pointed nose projected slightly, a small mouth, a small determined chin, and a long slender neck.[8] Livia's features do not resemble the flat cheeks, short snub nose, and protruding rounded chin of the lady portrayed by these jugs. Second, there are no other imperial portrait head-shaped vessels in glass among the many head-shaped vessel Formats preserved from antiquity.

In the Hellenistic Near East variations of the nodus coiffure are also known from coin representations of goddesses such as Diana and Victory.[9] Gods and mythological beings are the most common subjects of glass head-shaped vessels (see Chapter III: *Head-shaped Vessels: Subjects*). One of the characteristics of the Hellenistic period was the worship of rulers as gods. Representations which have recognizable features of rulers, yet are not strictly portraits of those rulers, were a concomitant development of the period.[10] Augustus and members of his family were frequently worshipped in the guise of a traditional god or goddess. A number of coin emissions showing Livia as the personification of a goddess were minted in the Near East during Livia's lifetime. Coins with an image of Salus Augusta (A.D. 22) show how idealized features of the goddess or personification and portrait features of Livia could be mingled.[11] Various cities in Asia Minor minted coins showing Livia as Hera. In Pergamon, where a cult of Livia as Hera existed,[12] an issue probably minted in the years 16-13 B.C. shows Livia on the obverse (Fig. 104, left) and her stepdaughter Julia on the reverse (Fig. 104, right).[13] The coins are of special interest because the image of Livia has a crude, rigid profile and a square chin bearing

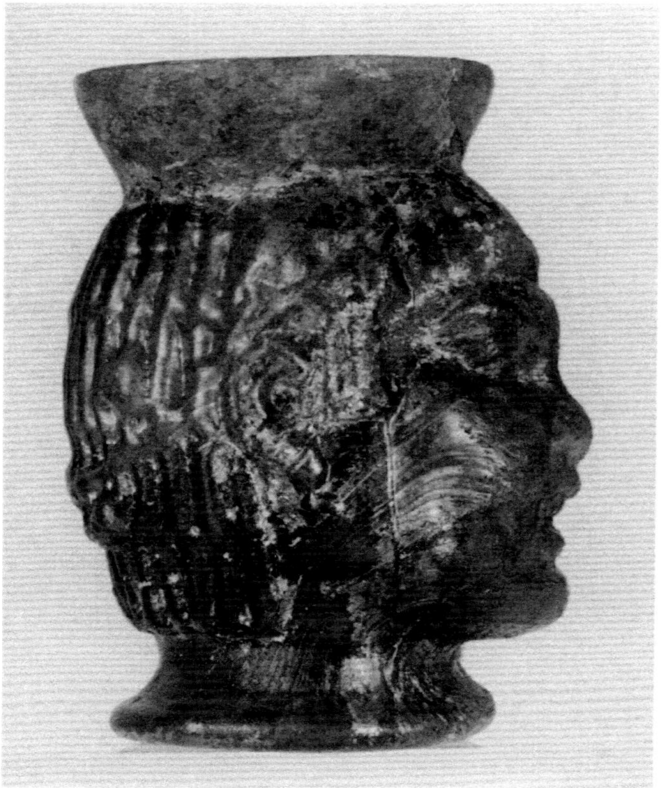

139 (1:1)

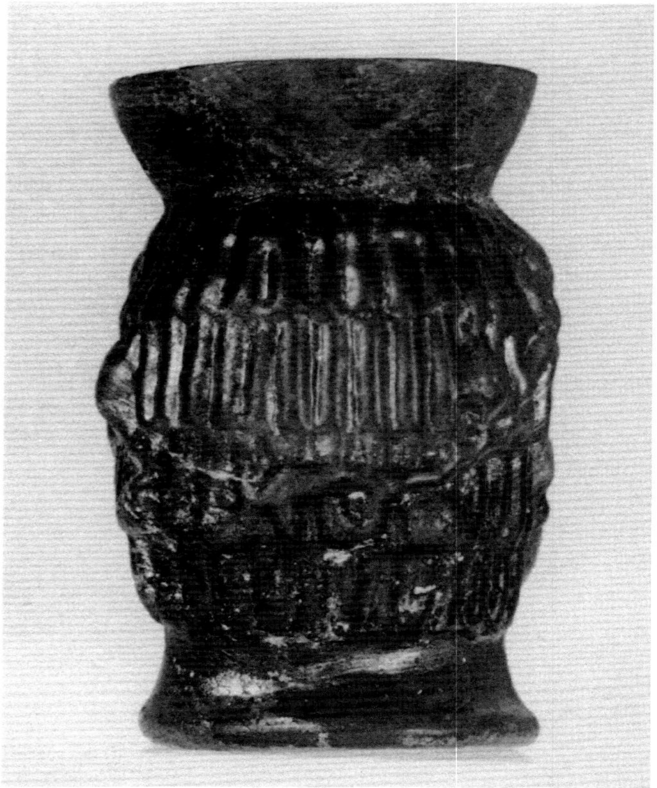

139 (1:1)

little resemblance to Livia's physiognomy but recalling the profile of the glass head flasks. A Near Eastern coin such as this one from Pergamon could have provided the model for the mold in which these glass jugs were blown. The use of a coin as a model would also explain the curious location of the seam passing though the center of the face of the vessels. Head-shaped vessels are usually meant to be seen full face, but in this Type the placement of the seam indicates that the profile is the principal view.

The addition of shoulder locks to the nodus coiffure may have been an idealizing trait. The inscription on the Pergamene coin identifies its image as Livia-Hera. Representations of Livia as Hera were the counterpart of Augustus in the guise of Jupiter. But representations of Livia as Hera may have had an independent significance as well, for Hera was the Greek counterpart of Juno, who in Rome was associated with marriage and progeny.[14] This aspect of Juno would have been of topical interest in A.D. 4, when Augustus designated Livia's son Tiberius as his successor, and again in 14, when Tiberius succeeded Augustus.[15] Such an event could have inspired the creation of this head-shaped vessel.

[1] On this vessel and the Type: Stern 1992b.

[2] Venus rings were a common attribute of beauty in the Near East, as can be seen, for example, on a marble portrait of Livia from the Fayum (see below note 8) and on many private portraits of the time, for example two busts from the so-called Licinian Tomb, Rome, now in Copenhagen, Ny Carlsberg Glyptotek, cat. nos. 603, 604: V. Poulsen, *Glyptothèque Ny Carlsberg. Les portraits romains* I: *République et dynastie julienne* (Copenhagen 1962) nos. 69 and 70, pls. 118-121. The convention may have originated in Egypt, where it is found on glass inlays as early as the late Eighteenth to early Nineteenth Dynasties (end of the fourteenth to early thirteenth centuries B.C.): see Stern 1994, 146, no. 13. On Venus rings in connection with the *gens Julia*: Erika Simon, *Augustus. Kunst und Leben in Rom um die Zeitenwende* (Munich 1986) 63.

[3] Known examples:

a) the Toledo example, transparent bluish green;

b) formerly Paris, de Clercq Collection, said to be from Tortosa on the Syrian coast, bluish, presence of handle cannot be determined from illustration, de Ridder 1909, 157, no. 279, pl. XI;

c) Damascus, Nat. Mus., handle not visible on photograph, Zouhdi 1960, pl. 6, third from right;

d) Palo Alto, Stanford University Mus., no. 17277, formerly Leland Stanford Jr. Collection, said to have been found in Rhodes, H 10.0 cm, light green, with a vertically pinched thumbrest, handle missing, here Fig. 103;

e) CMG, no. 55.1.91, formerly R. W. Smith Collection, no. 670, acquired in Lebanon, amber, without handle, Smith 1957, 143, no. 281;

f) formerly R. W. Smith Collection, no. 651, acquired in Lebanon, pale bluish green, with one floating handle and a vertically pinched thumbrest, Smith 1957, 143, no. 282; auc. cat., Sotheby Parke Bernet, New York, May 2, 1975, lot 183;

g) Yale, no. 1953.28.14, formerly Rockefeller Collection, formerly Duveen Collection, transparent brown, with a floating handle, Matheson 1980, 59-60, no. 138, here Fig. 102;

h) Berlin, Staatliche Museen, Antikensammlung, no. 30219, 135, formerly von Gans Collection, transparent blue green, with one floating handle and a vertically pinched thumbrest, Platz-Horster 1976, 45, no. 71;

i) Hartford, Wadsworth Athenaeum, no. 1930.55, colorless with a pale green tinge, cited by Smith 1957,143, at no. 281;

j) Washington, D.C., Smithsonian Institution, formerly R. Guggenheim Collection, cited by Smith 1957, 143, at no. 281.

[4] *Longa probat facies capitis discrimina puri... . Exiguum summa nodum sibi fronte relinqui, Ut pateant aures, ora rotunda volunt.* Ovid. *The Love Poems*, translated by A. D. Melville (Oxford 1990).

[5] Matheson 1980, 59-60.

[6] The nodus hairstyle was first attested in Rome shortly before 40 B.C.: L. Furnée-van Zwet, "Fashion in Women's Hairdress in the First Century of the Roman Empire," *BABesch* 31 (1956) 3-7, survey nos. 1-17; Karin Polaschek, "Studien zu einem Frauenkopf im Landesmuseum Trier," *TrZ* 35 (1972) 145-155, figs. 5-6. The exact date of the disappearance of the nodus hairstyle in Rome is disputed, but it was some time between 20 B.C. and the end of the first century B.C.: Furnée-van Zwet 1956, 7; Polaschek 1972, 145-146.

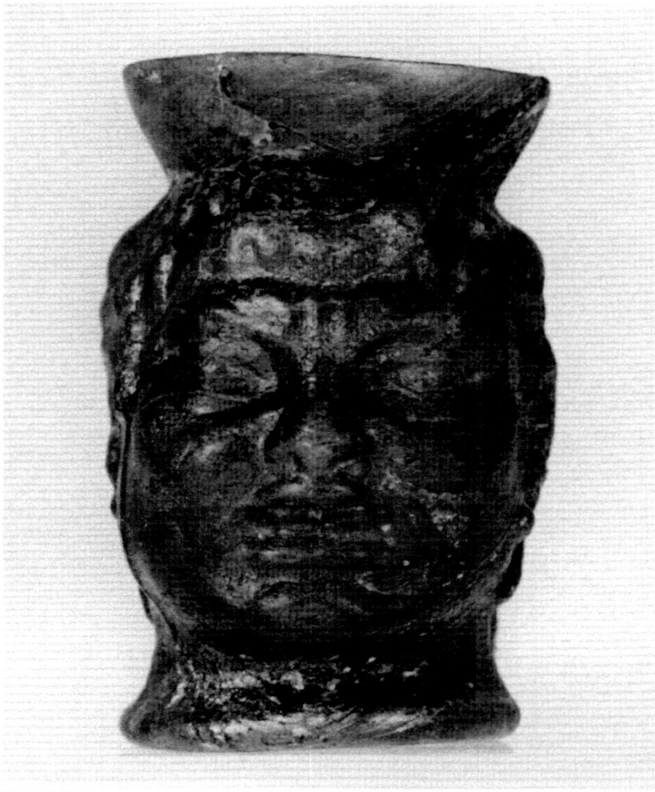

139 (1:1)

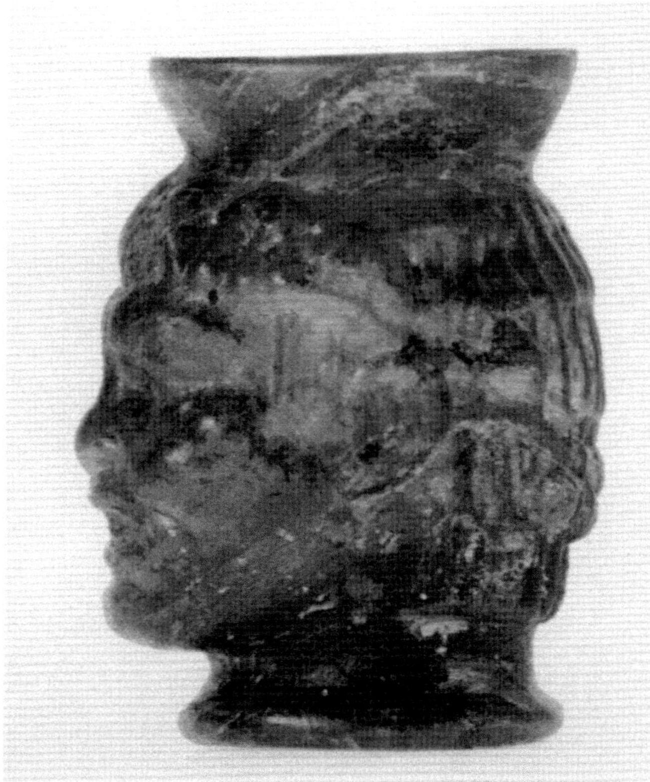

139 (1:1)

[7] Furnée-van Zwet 1956 (see above note 6), 4, no. 6. Compare two heads of terra-cotta figurines found in undatable contexts at Pergamon: E. Töpperwein, *Terrakotten von Pergamon* (*Pergamenische Forschungen*, 3, Berlin 1976) 81, nos. 371-372, pl. 52. Also compare a curious portrait head of Livia of white marble in Marmaris, Ilk Okulu, from an unknown find spot: J. Inan and E. Rosenbaum, *Roman and Early Byzantine Portrait Sculpture in Asia Minor* (London 1966) 60-61, no. 11, pl. 7.

[8] For example, a portrait of Livia from the Fayum, now Copenhagen, Ny Carlsberg Glyptotek, no. 615, sculpted between 4 and 14, Poulsen 1962 (see above note 2), 65-71, no. 34, pls. 52-54. On Livia's iconography: Walter Hatto Gross, *Iulia Augusta. Untersuchungen zur Grundlegung einer Livia-Ikonographie* (*Abhandlungen der Akademie der Wissenschaften in Göttingen, Philologisch-Historisch Klasse, Dritte Folge*, no. 52, Göttingen 1962); H. Bartels, *Studien zum Frauenporträt der augusteischen Zeit* (Munich 1962); Klaus Fittschen and Paul Zanker, *Katalog der römischen Porträts in den Capitolinischen Museen und den anderen kommunalen Sammlungen der Stadt Rom* III: *Kaiserinnen- und Prinzessinnenbildnisse Frauenporträts* (Mainz 1983) 1-2.

[9] Walter Trillmich, *Das Torlonia-Mädchen, Zu Herkunft und Entstehung des kaiserzeitlichen Frauenporträts, Abh. Akad. Wissenschaften Göttingen, Phil.-Hist. Klasse*, 3. Folge, no. 99 (Göttingen 1976) 44-45.

[10] On representations of Hellenistic rulers as gods: Hans Peter Laubscher, "Triptolemos und die Ptolemäer," *JbHambKuSamml* 6/7 (1988) 11-40, esp. 35, nn. 1-2 (with lit.).

[11] Gross 1962 (see above note 8) 19ff.

[12] On the worship of Livia as Hera at Pergamon: ibid., 11; on Livia as Juno: ibid., 49, n. 26, citing Ward, *Studi e materiali della storia delle religioni* (1933) 24 ff. See also Willem den Boer ed., *Le culte des souverains dans l'empire romain, Fondation Hardt, Entretiens* 19 (Geneva 1973); S. R. F. Price, *Rituals and Power. The Roman Imperial Cult in Asia Minor* (Cambridge 1984) 69-70.

[13] Paris, Bibliothèque Nationale, no. 1195, Gross 1962 (see above note 8), 28-30, 71-72, pl. 4, nos. 6-8.

[14] Simon 1990, 94.

[15] Two later occasions are less likely: Livia's death in 29 and, as suggested by Matheson 1980 Livia's deification by the emperor Claudius in 41: Stern 1992b, 398.

## 139 **Cup in the Shape of a Male Negroid Head** [1]

Color plate 21, p. 59
Probably made in Italy
Second half of the first century
H 9.5 cm; H body 7.3 cm; max D 7.0 cm; D rim 5.7 cm; D base 5.3 cm
1923.457, Gift of Edward Drummond Libbey

Transparent manganese colored grayish red purple (5 RP 4/2).

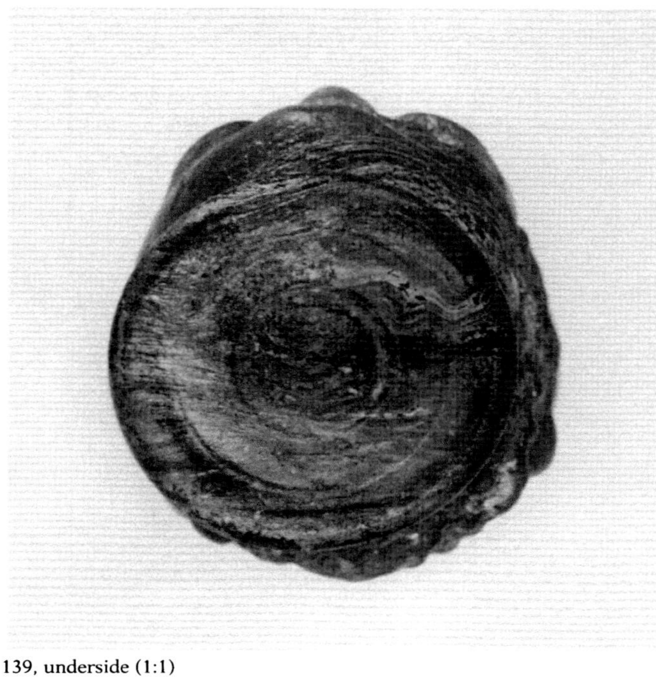

139, underside (1:1)

Medium thin glass. Blowing spirals. Fabric cannot be determined because of weathering.
Rim free blown. Body blown into a three-part mold of two vertical sections and a disk-shaped base section (MCT VII A). Mold seams on either side of the face along the two locks in front of the ear. Relief crisp. No pontil mark.
Flaring rim. Rim ground. Body in the shape of a head on a sturdy neck. Flaring, nearly circular flat base, with decoration on the underside.
Body in the shape of a male head with Negroid features wearing a wreath. The childlike profile contrasts surprisingly with the frontal view, which shows a broad square face with a grim expression: knit brow, eyebrows arched high over narrow eyes with small round pupils, eyelids modeled with deep pouches under the eyes, fleshy dimpled cheeks with the underlying bone structure strongly represented, wide flat nose, half-open mouth with a pronounced upper lip and visible teeth, dimpled chin. Decorative crescent-shaped ears with a groove along the center. Spherical earrings dangle from the lobes on both sides. The thick coils of hair are carefully aligned vertically to make a decorative pattern. There are approximately eleven short curls over the forehead and two in front of each ear. There are four tiers of curls on the back of the head. The wreath, tied at the back of the head, consists of leaves and berries. On underside of base, mold-blown design of three raised concentric circles with a raised dot at the center. A series of indistinct raised ridges on the front of the neck may be the remains of a mold-blown inscription.
*Condition* Body broken into many pieces and repaired. Half of rim restored. Exterior and interior covered with rainbow iridescence and severe surface pitting.
*Collection* Thomas E. H. Curtis.
*Exhibitions* New York, The Metropolitan Museum of Art (while in Curtis Collection, per Richter 1914).
*Published* Richter 1914, 77, fig. 6; Grose 1978, 82, fig. 19.
*Remarks* Cups shaped like a head with Negroid features were one of the most popular shapes of glass head-shaped vessels during the first century. At least four different Series are known. Two bear mold-blown maker's inscriptions (see Chapter III: *Survey: First Century*, Type A, Single Ethnic Head). This vessel belongs to Series 3.[2] A cup excavated in Romania may belong to the same Series, but it cannot have been blown in the same mold because it has only one concentric circle on the underside of the base.[3]
Finds in Herculaneum[4] and Pompeii[5] date the beginning of production of this Type firmly before the eruption of Mount Vesuvius in 79.
In spite of the Negroid features, there is no evidence to support the common hypothesis that these cups were made in Alexandria.[6] It is possible that Series 1 (here Fig. 83), signed in Greek by Tryphon,[7] was made in the eastern Mediterranean, but all other find spots are in the western half of the Roman empire. This distribution pattern is comparable to that of a group of bronze bust-shaped vessels with Negroid features that have also been assigned to Alexandria solely because of their iconography.[8] For the glass vessels an origin in Alexandria is out of the question, not only because all known find spots are in the West but also because Alexandrian glass workers were slow to take up the new technique of glass blowing.[9] Sophisticated mold-blowing techniques were not practiced in Egypt during the first century A.D. During the first century B.C., and especially after the annexation of Egypt in 30 B.C., Egyptianizing motifs became a fashion in Rome and remained popular in Italy.[10] Moreover, during the first century A.D. Roman legionaries from Africa were stationed at western sites, so that artists creating molds for glass blowers and bronze casters in these areas could also have had first-hand knowledge of a variety of ethnic Negroid features.

1 See Chapter III: *Survey: First Century, Type A.*

2 Other examples that may belong to this Series:
a) CMG, no. 79.1.63, formerly Strauss Collection, *CMG Annual Report 1978*, 4, pl. 2; *JGS* 22 (1980) 103, no. 4, ill.;
b) Sotheby's, London, auc. cat., July 13-14, 1981, no. 452, color ill.;
c) Eisen 1927, I, 330, pl. 75, formerly Mrs. H. O. Havemeyer Collection, New York, said to come from Tel el Hosan, north of the Sea of Galilee;
d) BM, no. GR 1868.5-1.201, formerly Raoul Rochette Collection, formerly Slade Collection, no. S201, Nesbitt 1871, no. 201, pl. VI, no. 1;
e) perhaps Yüksel Erimtan Collection, no. 144, from Asia Minor, light blue green, H max 9.7 cm, D rim 5.6 cm, D base 5.0 cm, Lightfoot and Arslan 1992, 60-61, no. 22, ill., color cover (described as having a pontil mark).

3 Constanta, Romania, Arch. Mus., no. 14851, from a tomb at Neptun, north of Callatis (ancient Mangalia), greenish, H 9.3 cm, dated second half of the first century or beginning of the second, Bucovala 1968, no. 58; Bucovala 1979, 93, no. 3, fig. 3.

4 Ercolano, Antiquarium, no. E2372, excavated in 1961, Scatozza Höricht 1986, 52, no. 103, pls. 12 and 33.

5 Scatozza Höricht 1986, 52, mentions the existence in Pompeii of other glass cups, similar and vaguely similar ("confronti vaghi ed esemplari simili") but not identical to the Herculaneum cup and not blown in the same mold. According to *BullAIHV* 9 (1981-83) 162, there are two Negroid head-shaped vessels from Pompeii in Naples, Mus. Naz. Arch., no. 129404, prune-colored, *Collezioni Napoli* 1989, 224 and 225, no. 46, ill.

6 Kisa 1908, III, 756; Landes 1984, 343. Scatozza Höricht 1991, 85, while supporting the hypothesis of an eastern Mediterranean origin for the Type, correctly notes that some were made along the Campanian coast.

7 Formerly Paris, Dutuit Collection, no. 247, Sambon 1906, 500, fig. 35; Sambon 1928, 19, fig. 35 and frontispiece, reported to have been found in Phoenicia by Kisa 1908, III, 756. CMG, no. 55.1.92, formerly R. W. Smith Collection, no. 886, Smith 1957, 145, no. 288, acquired in Cairo, is said to be similar to the Dutuit cup, but no mention is made of an inscription.

8 Peter Goessler, "Antike Büstengefässchen aus Metall," *Antike Plastik: Walther Amelung zum Geburtstag* (Berlin 1928) 83-84, figs. 9-12, mentions numerous finds from Germany (Siegburg, Cologne), Austria (Graz), Hungary (Budapest), and Italy (Ostia); L. Pressouyre, "A propos d'un 'balsamaire' trouvé à Lamaurelle (L.-et-G.)," *RA* 1 (1962) 165-181, cites additional finds from North Africa, including Volubilis, p. 166, and Djemila, p. 170; F. Coarelli, "Nuovi elementi per la cronologia di Begram: cinque recipienti bronzei in forma di busto," *ArchCl* 13 (1961) 168-179. However, A. Radnoti suggests they were made in France, in the Saône-Rhône region. The bronze vessels differ from the glass cups in that they were not drinking vessels but rather closed containers shaped like a bust; they have a large opening with a lid at the top of the head and rings in which a bail handle is inserted.

9 Stern 1993, 26.

10 Mariette de Vos, *L'Egittomania in pitture e mosaici romano-campani della prima età imperiale* (Leiden 1980).

### 140 Bottle in the Shape of a Male Head, probably Dionysos

Eastern Mediterranean, perhaps Syro-Palestinian
Second half of the first century
H 7.35 cm; H body (front) 4.5 cm; H body (back) 4.1 cm; max D 3.8 cm; D rim 2.6 cm; D base 2.1 cm
1923.537, Gift of Edward Drummond Libbey

Opaque white. Medium thick glass. Some vertically elongated bubbles weathered through on surface of neck.
Rim and neck free blown. Body blown into a three-part mold of two vertical sections and a disk-shaped base section permanently attached to the vertical section with the back of the head (MCT

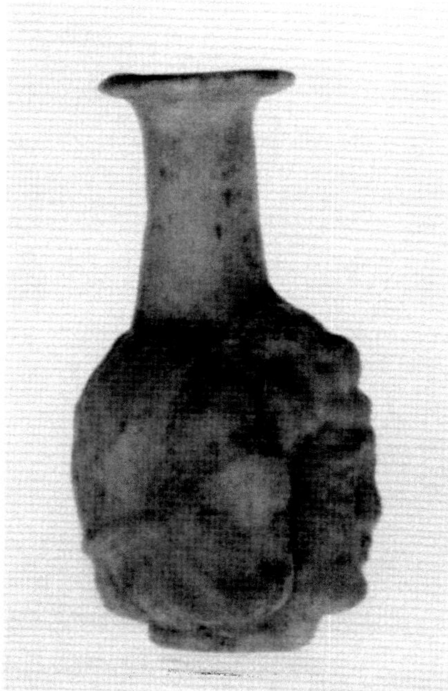

140 (1:1)

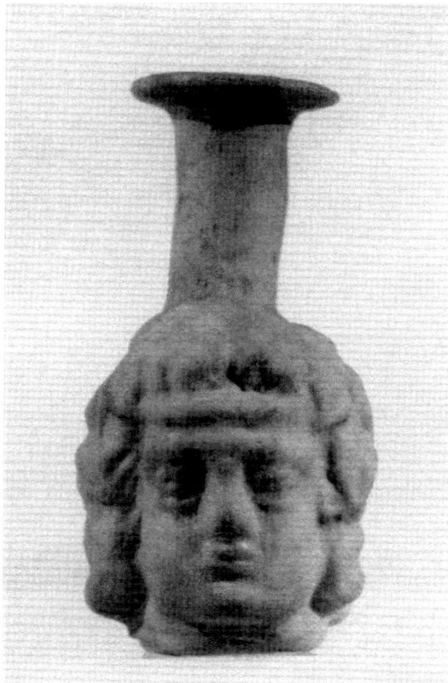

140 (1:1)

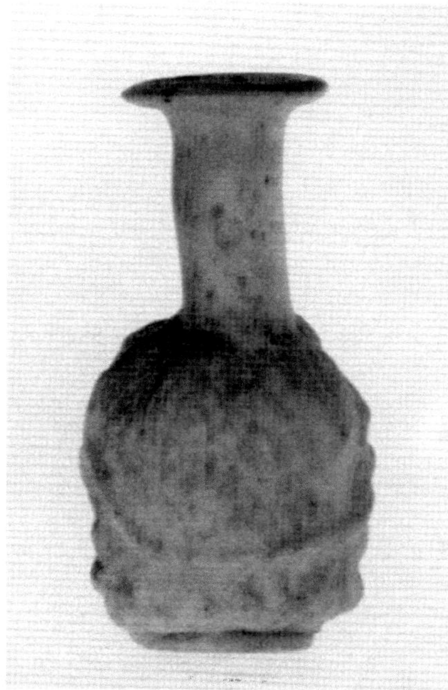

140 (1:1)

VI). Mold seams concealed in hair behind ears. Relief moderately crisp. No pontil mark.

Everted rim, folded outward, upward, inward, and downward. Tubular neck, slightly crooked. Body in the shape of a head. Circular flat base, with depressed flat circular area on the underside.

Body in the shape of a youthful head wearing a fillet and wreath. The face is long and rectangular, with large protruding eyes with recessed pupils, a broad nose, small straight lips, and a round chin with heavy jaw. The face is slightly wider at the jaw than at the forehead. The hair surrounds the face in tufts and thick curls that extend down to the level of the chin. The hair on the back of the head is flatter, with irregular vertical ridges defining the locks. A fillet stretches across the forehead, while the tips of an ivy wreath meet above the fillet over the center of the forehead, with two groups of berries, one on each tip. On the front of the head the wreath consists of large heart-shaped leaves, four on either side of the face. The wreath, without adornment, dips low around the back of the head.

*Condition* Intact. Cleaned. Original surface not preserved. Exterior rough with areas of light brown weathering. Interior has areas of brown encrustation.

*Collection* Thomas E. H. Curtis.

*Published* Previously unpublished.

*Remarks* A bottle of the same Type, but somewhat more expanded, was found at Tel Nabi Mind, Syria.[1] Although there are no parallels from a dated context, the first-century date seems certain because the rare MCT VI mold closely resembles MCT II, which was used mostly in the second and third quarters of the first century (see Remarks Cat. No. 26).

The identification as Dionysos is based on the presence of the fillet.[2] Compare here Cat. No. 148, where an ivy wreath and fillet suffice to indicate a personification of Dionysos.

[1] Damascus, Nat. Mus., no. 5741, opaque white, H 8.0 cm, Zouhdi 1964, 44, no. 14, fig. 26; Al-'Ush et al. 1969, fig. 39, second from left.

[2] Stern 1994, 316, no. 126.

## *Double Head-shaped Vessels, Late First to End of Second Century* (Cat. Nos. 141–146)

### 141 **Double Head-shaped Bottle: One Serious Face, One Smiling Face**[1]

Color plate 23, p. 60
Eastern Mediterranean, perhaps Syro-Palestinian
Late first to early second century
H 6.8 cm; H body 4.8 cm; max D 3.7 cm; D rim 2.0 cm; D base 2.3 cm
1923.536, Gift of Edward Drummond Libbey

Opaque white. Medium thick glass. A few black specks. Blowing spirals.

Rim and neck free blown. Body blown into a two-part mold of two vertical sections, each including one-half of the base (MCT VIII M). One continuous mold seam concealed in hair at junction of heads, extends across underside of base. Relief moderately crisp. No pontil mark.

Everted rim, folded outward, upward, inward, and downward. Tubular neck. Body in the shape of a double head. Circular flat base, with circular groove on the underside.

Body in the shape of two heads back to back: a serious face, A, and a smiling face, B. The serious face A is more idealized than the smiling face. The eyes have recessed pupils, the nose is straight and narrow, and the lips are horizontal and slightly parted. A curved fillet stretches across the forehead at the temples; curls and hook-shaped locks of hair fall over it. Beside the cheeks straight coils of hair are held in place by two diagonal wing-shaped locks on either side. The smiling face B is more round, broad, and full than face A. It has protruding eyes with recessed pupils, a wide flat nose, and full cheeks. The hair is distinctive for two locks, parted at the center of the forehead, that curve horizontally across the top of the forehead, then curl downward and inward above the temples like hooks. The other locks of hair around the face are also hook-shaped.

*Condition* Intact. Cleaned, leaving soapy texture on exterior. Exterior has streaks of brown weathering.

*Collection* Thomas E. H. Curtis.

*Published* Previously unpublished.

*Remarks* Double head-shaped vessels appear to have been reinvented by glass blowers around the middle of the first century (see Chapter III: *Typology of Form* and *Format*). This particular Type is one of the earliest double head-shaped vessels made in the eastern Mediterranean.[2] About half of the known examples are opaque white; most of the rest are transparent blue, one transparent purplish red. A curious opaque white bottle, reported to have been found in the vicinity of Olbia on the northern coast of the Black Sea, is encased in a metal cage; a metal pincers with an adjustable ring around the shaft is said to have been found upright in the vessel.[3]

The only datable example of a vessel belonging to this Type was excavated in a second-century tomb at Korakou, Cyprus (see note 2c). Stylistic comparisons suggest that the Type was created at the end of the first or the beginning of the second century.[4] Recessed pupils, as in both faces of this flask, began to appear in marble portraits of the Hadrianic period (117-138), but they occur earlier in other materials (see Cat. No. 140). The coiffure of face A, with two diagonal wing-shaped locks holding back the hair on either side of the face, is reminiscent of the coiffure of Medusa on either side of lenticular flask Cat. No. 62, which probably dates from the first half of the first century because it belongs to a Type created in the Workshop of the Floating Handles. The proportions of face B recall double head-shaped vessels with two similar Medusa heads with knotted snakes below the chin; the earliest of these known was excavated in a tomb in northern Italy dated to the second half of first century (see Chapter III: *Survey: First Century, Type C* and Fig. 86),[5] while others were excavated in second-century tombs in central Italy (see Chapter III: *Survey: Second Century, Type B* and Fig. 88).[6]

[1] See Chapter III: *Survey: Second Century, Type A.*

[2] See the following examples from known find spots:

a) formerly Paris, de Clercq Collection, from Sidon, opaque white, de Ridder 1909, 158, no. 280, Face A ill. pl. XII;
b) from Amrit, Syria, opaque white, de Ridder 1909, 159, no. 281, Face A ill. pl. XII;
c) Cyprus, Nat. Arch. Mus., no. 1963/XI-22/9, from Korakou, a second-century tomb, translucent dark blue, Karageorghis 1964, 310-311, fig. 32a-c; Kyriakos Nicolaou, "Découvertes archéologiques de verre antiques faites à Chypre dans la décade 1962-1971," *BullAIHV* 6 (1971-72) 128-130, no. 11, fig. 100, profiles of Faces A and B, ill.;
d) Adana, Turkey, Bölge Museum, no. 24.52.72, from Cilicia, unpublished;
e) BM, no. 91528, S. 2454 from the palace at Nineveh (modern Kuyunjik), translucent dark blue, Barag 1985a, 93, no. 118, pl. 14 and color pl. D, faces A, B, and profile ill.;
f) Damascus, Nat. Mus., find spot and color not mentioned, Zouhdi 1960, pl. 6, left, face B ill.;
g) from Tauric Chersonesos, in the Crimea, Sorokina 1968, 186, 188, fig. 3, nos. 8-9, faces A and B ill.

For examples from unknown find spots, see:

h) Aquileia, Italy, Arch. Mus., no. 45874, perhaps from the vicinity of Aquileia, northern Italy, but provenance not mentioned, translucent dark blue, Calvi 1968, 106, no. 256, pl. 16, no. 6, profile ill.;
i) formerly Kofler-Truniger Collection, opaque white, *Kofler-Truniger* 1985, 69, no. 113; Luzern 1981, 82, no. 278, face A ill.;
j) formerly Kofler-Truniger Collection, translucent deep blue, *Kofler-Truniger* 1985, 69, no. 110, face A ill.; Luzern 1981, 82, no. 279, profile ill.;
k, l) formerly Constable-Maxwell Collection, *Maxwell* 1979, 64-65, nos. 98 and 100, opaque white, face A ill.;
m) Oppenländer Collection, no. 2494, opaque white, von Saldern et al. 1974, 169, no. 460, 171, face A ill.;
n) Los Angeles, County Museum of Art, no. M.88.129.53, formerly Hans Cohn Collection, von Saldern 1980b, 54, no. 47, color pl. 5, face A ill.;
o) formerly R. W. Smith Collection, no. 129, deep purplish red, Smith 1957, 142-143, no. 283, profile ill.;
p) Carnegie, no. 25456/2, opaque white, Oliver 1980, 69, no. 62, Face B ill.;
q) Tel Aviv, no. MHG 9958, unpublished.

Closely related:

r) Dusenbery 1971, 16, figs. 13-14, opaque white, face A and profile ill.

[3] Formerly R. W. Smith Collection, no. 228, formerly Berlin, Schiller Collection, opaque white, Smith 1957, 140-143, no. 280, face A ill.

[4] Barag 1985a, 93, at no. 118, suggests the faces were inspired by portraits of the Roman emperor Hadrian's friend Antinous and dates this Type to the Hadrianic period, 117-138. On Antinous's portraits, see Remarks Cat. No. 148.

[5] Padua, Musei Civici, no. 2482, from a tomb at Vigorovea, excavated in 1954, H 8.0 cm, Calvi 1959, 3-9, figs. 1-2; Zampieri 1984, 113, 122, esp. 119, no. 14.10.

[6] Ancona, Italy, Mus. Naz. no. 26509, from Portorecanati, tomb 167, opaque white, Mercando 1974, 320-326, fig. 247, faces A, B, and profile ill.

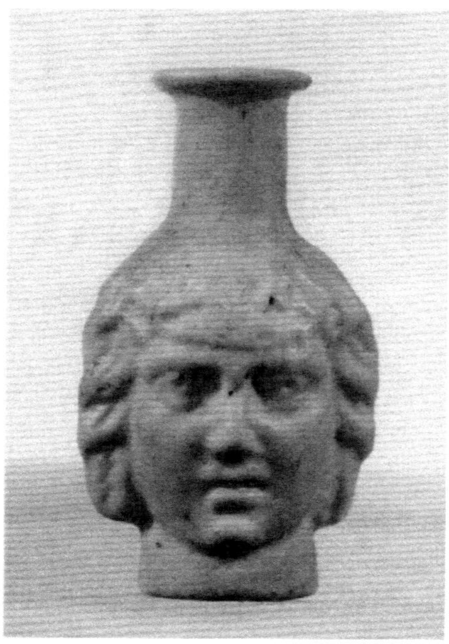
141, side A (1:1)

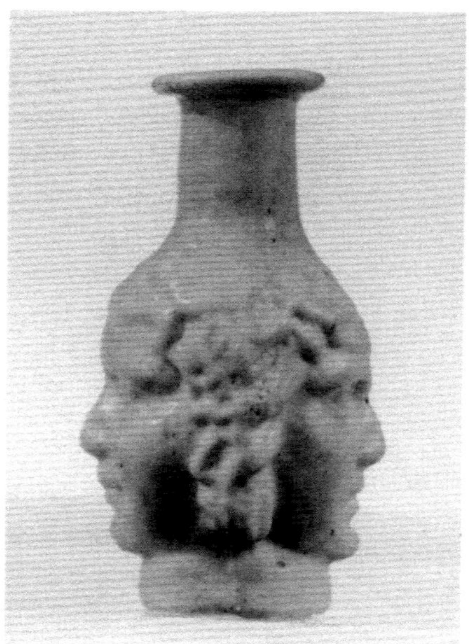
141, profile B:A (1:1)

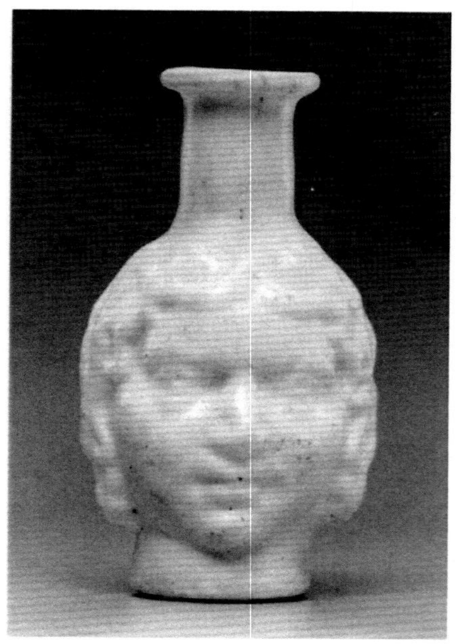
141, side B (1:1)

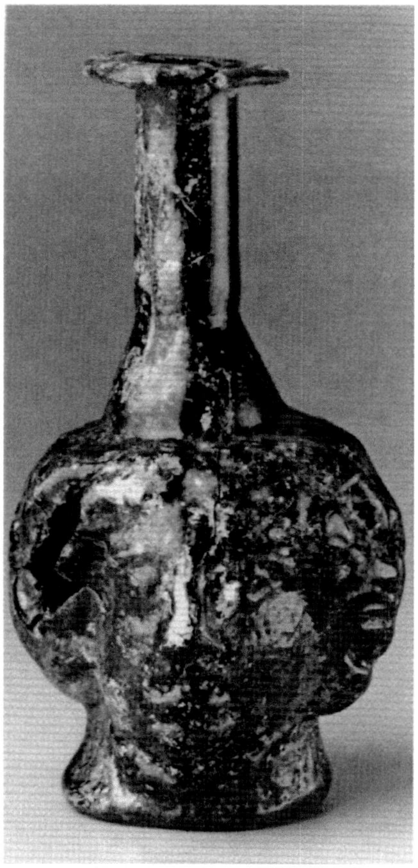

142, side A (1:1)

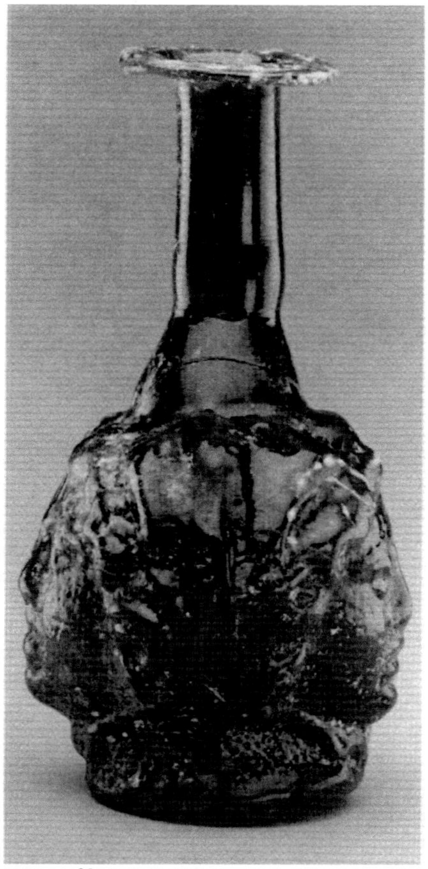

142, profile B:A (1:1)

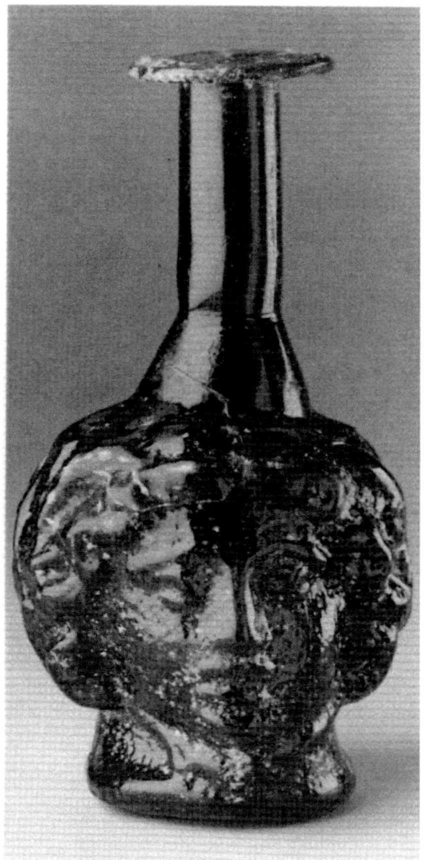

142, side B (1:1)

142 **Double Head-shaped Bottle: Two Similar Medusas**

Color plate 23, p. 60
Place of manufacture undetermined
Perhaps late first or second century
H 10.5 cm; H body (side A) 5.6 cm; H body (side B) 5.5 cm; max D 5.4 cm; D rim 2.8 cm; D base 3.7 cm
1923.458, Gift of Edward Drummond Libbey

Translucent manganese colored moderate red (near 5 R 5/4). Medium thin glass. Fabric cannot be determined because of weathering. Rim and neck free blown. Body blown into a three-part mold of two vertical sections and a separate disk-shaped base section (MCT VII). Mold seams concealed in hair at junction of heads. Relief moderately crisp. No pontil mark.

Horizontal rim, folded outward, upward, and inward. Tubular neck, with constriction below middle; neck widens significantly below constriction. Body in the shape of a double head. Circular slightly concave base, with decoration on the underside.

Body in the shape of two heads of Medusa back to back with similar features: arching browline, large almond-shaped eyes with recessed pupils, narrow nose, straight lips slightly parted, and a narrow prominent chin. Face A is narrower than face B. On side B, wings can be distinguished in the hair above the forehead. Beneath the chins of both faces is a thick roll with loops to left and right which represents the knotted snakes of Medusa. The snaky hair is represented by curving locks to chin level, with a central part over the forehead. On underside of base, two diamonds with concave sides in raised outline surrounding a sunken umbilicus.

*Condition* Broken and repaired from many pieces; several fragments of the hair of face A are missing. Exterior covered with areas of thick dark encrustation and areas of light beige weathering and golden iridescence. Interior has areas of dark weathering.

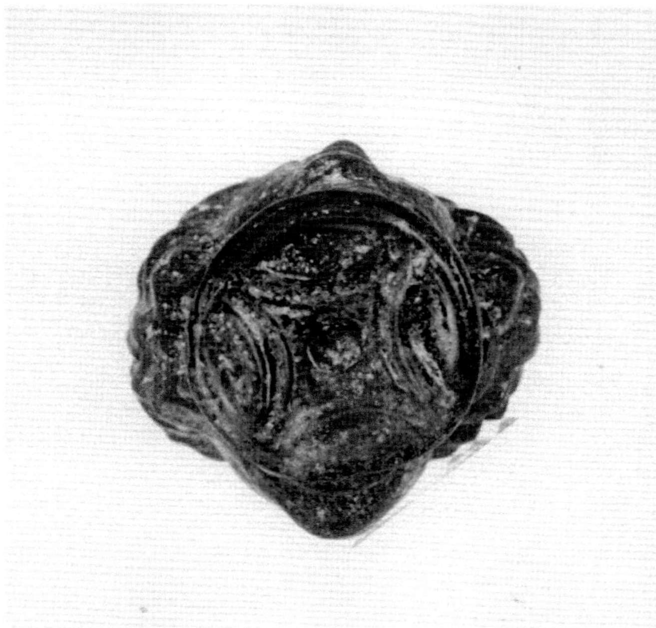

142, underside (1:1)

*Collection* Thomas E. H. Curtis.

*Published* Previously unpublished.

*Remarks* Same subject as Cat. No. 158, but a different Type, modeled

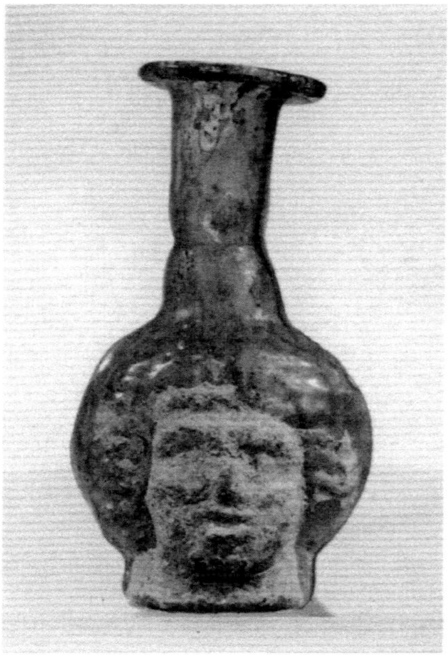

143, side A (1:1)

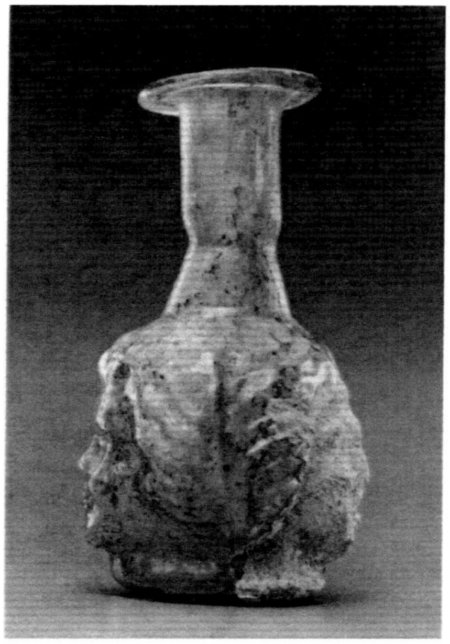

143, profile B:A (1:1)

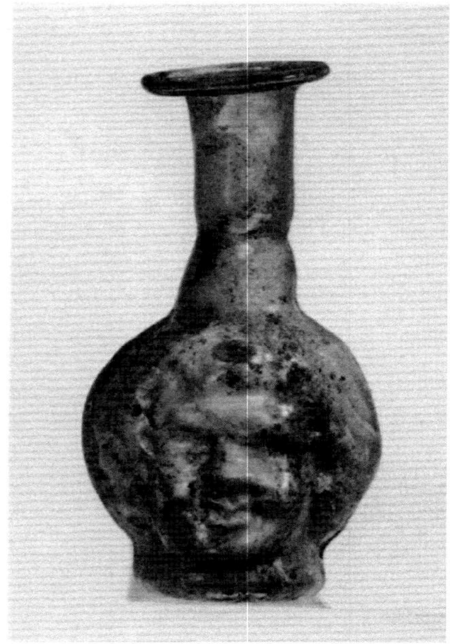

143, side B (1:1)

on the Hellenistic representation of Medusa as a beautiful and seductive woman known as the Rondanini Medusa. [1] In addition to the Series represented by the Toledo bottle, three Series of double head-shaped bottles modeled on this particular appearance of Medusa can be identified, [2] but the heads are all slightly smaller than this bottle and they have neither the characteristic constriction in the neck nor a base with mold-blown decoration. A bottle with a similar constriction near the base of the neck, excavated at Budva, Montenegro, shows two Medusa heads and may belong to the same Series as this bottle. [3] However, it is not known whether the bottle from Budva has a design on the underside. Head-shaped vessels with a decorated base are rare. They are blown in MCT VII molds and have been dated to the first century and to the late second and early third centuries, [4] but it is not known whether this mold construction type remained in continuous use during the second century or was reinvented at the later date. [5] The bottle from Budva is dated by the excavator to the first century and is thought to be an import from Syria. However, no eastern Mediterranean parallels are known.

[1] On the Rondanini Medusa: E. Buschor, *Medusa Rondanini* (Stuttgart 1958).

[2] The relationship between the Toledo Series and the following three Series of Medusas has not yet been established.

*Series a*, with both faces broadest at the jawline:

a) Ancona, Mus. Naz., no. 26509, from a second-century burial at Portorecanati, central Italy, opaque white, H (pres.) 4.2 cm, Mercando 1974, 320-322, fig. 247 (here Fig. 88); see also Chapter III: *Survey: Second Century, Type B.*

*Series b*, with knob-shaped earrings:

a) CMG, formerly R. W. Smith Collection, no. 863, acquired in Lebanon, opaque white, H 6.1 cm, Smith 1957, 143-144, no. 286;
b) Cologne, private collection, opaque white, Kisa 1908, III, 752, fig. 291 (perhaps the same as RGM, no. N1070, formerly Niessen Collection, opaque white, H 7.2 cm, La Baume 1973, no. C13, pl. 10);

*Series c*, with both faces elongated:

a) Trier, Rheinisches Landesmuseum, no. 6416, excavated at Trier, Paulin 1882, translucent yellow brown, Goethert-Polaschek 1977, 257, no. 1531, pl. 79;
b) Berlin, Staatliche Museen, Antikensammlung, no. 30219, 152, formerly von Gans Collection, translucent purple, H 8.0 cm, Platz-Horster 1976, 47, no. 77;
c) Oppenländer Collection, no. 2208, translucent dark blue with an opaque yellow streak, H 7.2 cm, von Saldern et al. 1974, 170, no. 461.

[3] Cermanović-Kuzmanović 1974, 176, 186, pl. I, no. 5. A similar bottle of dark manganese-purple glass with a concave-sided diamond on the underside of the base was sold at Sotheby's, London, auc. cat. July 13-14, 1981, no. 428, color ill.

[4] See Chapter III: *Survey: First Century, Type A, Series 2*; *Second Century, Type E, Series 2*; see also Remarks Cat. Nos. 139 and 148.

[5] For example, a head-shaped vessel from Karanis with a *chi-rho* monogram on the underside of the base, see Chapter III: *Survey: Third Century, Type A, Series 3.*

### 143 Double Head-shaped Bottle: Possibly Dionysos and Satyr

Place of manufacture undetermined
Perhaps second century
H 7.0 cm; H body 3.8 cm; max D 3.8 cm; D rim 2.4 cm, D base 2.9 cm (nose to nose)-2.4 cm (along the seam)
1923.517, Gift of Edward Drummond Libbey

Transparent pale olive (10 Y 6/2) with green streak in neck. Medium thin glass. Blowing spirals. Vertically elongated bubbles in neck; oval bubbles in body.

Rim and neck free blown. Body blown into a two-part mold of two vertical sections, each including one half of the base (MCT VIII). Continuous mold seam concealed in hair at junction of heads, extends across underside of base. Mold edges not well aligned. Relief moderately indistinct. No pontil mark.

Horizontal rim, folded outward, upward, inward, and flattened. Tapering tubular neck, with constriction below middle. Body in the shape of a double head. Oval flat base.

Body in the shape of two heads back to back: an idealized face, A, and a laughing male face, B. Face A is oval, broadest at the cheekbones, with hollow eyes, prominent cheekbones, narrow nose, straight lips slightly parted, and round chin. The hair sweeps back from the face horizontally; a symposiast's fillet, with two horseshoe-shaped loops rising from it, stretches across the forehead. Face B has small, deep-set hollow eyes, broad nose, pronounced cheeks, and a short V-shaped mouth fixed in a broad smile. The hair sweeps downward away from the face at a diagonal, except at the forehead where there are several full tufts of hair and two small bumps at the corners that probably represent horns.

*Condition* Intact. Partly cleaned. Original surface not preserved. Exterior of face A and interior of face B have thick beige encrustation. Exterior and interior have iridescent film and pitting.
*Collection* Thomas E. H. Curtis.
*Published* Previously unpublished.
*Remarks* Same Type as Cat. No. 144 where the fillet of Face A and the horns of Face B are more distinct. Face A, with the slightly opened mouth and the symposiast's fillet, probably represents Dionysos, the god of wine.[1] Face B, with horns, is one of Dionysos's companions, a satyr or Pan.[2] Cast or stamped glass relief decorations comparable in design and size to face A occur as handle attachments on glass jugs[3] and other vessels[4] produced between about 50 and 125 and in use until about 150.[5] They also appear on fourth-century vessels found in the East[6] and the West.[7] However, most attachments have been preserved without their vessels and are difficult to date.[8] The use of an MCT VIII mold and the small size of this bottle suggest a second-century rather than a fourth-century date.

1 Related to face A:
- a) Loeffler Collection, no. 144, reported to have been found near Aachen, Germany, pale green, La Baume and Salomonson 1976, 37, pl. 8, no. 70;
- b) Berlin, Staatliche Museen, Antikensammlung, no. 30219, 161, formerly von Gans Collection, colorless, Platz-Horster 1976, 47, no. 76, p. 47;
- c) Newark, no. 50.1434, emerald green, Auth 1976, 200, no. 340;
- d) Switzerland, private collection, translucent turquoise, Luzern 1981, 83, no. 280;
- e) BM, no. WAA 100659: 1905.5-13.8, olive green, *Masterpieces* 1968, 61, no. 76; not included in Barag 1985a.

2 For face B compare: formerly R. W. Smith Collection, acquired in Lebanon, manganese colored with white streaks, Smith 1957, 143, no. 285. A clay single head-shaped bottle excavated in an Augustan tomb on the Greek island of Samothrace shows a related face: G. Daux, *BCH* 84 (1960) 804-805, fig. 1.

3 For example:
- a-k) RGM, nos. 912, N6233, N6234, 27.2912, N160b, N6163, N160c, N6164, N160d, N160a, and 24.376, Fremersdorf VI 1961, 43, 50-52, pls. 73, 100-102;
- l) Sangiorgi 1914, pl. II, no. 82.

For the shape of these jugs: Isings 1957, 72-74, Form 55. For glass attachments very similar to this face:
- m-n) Split, Croatia, Arch. Mus., no. G1047, from Salona, and no. G1514, from Dalmatia, Rnjak 1979, 177-178, nos. 266 and 267; ibid. for theater masks in other materials.
- o) Switzerland, private collection, mask with a wide open mouth, Luzern 1981, 93, no. 343a.

4 For example:
- a) BM, no. GR 1865.12-26.1122, from the vicinity of Atella, southern Italy, translucent wine red urn, with two opaque white theater masks below the handles, *Masterpieces* 1968, 60, no. 75; Tait 1991, 68, figs. 83 and 84;
- b) CMG, no. 66.1.41, formerly Sangiorgi Collection, transparent yellow jug, with an opaque pale blue maenad mask below the handle, *Glass of the Caesars* 1987, 118, no. 50;
- c) RGM, no. 912 (see above note 3a), from a tomb in Cologne, natural bluish green, with a translucent dark green Medusa mask, *Glass of the Caesars* 1987, 119, no. 51.

5 On the date: Donald B. Harden, "The Glass Jug," in Martin Biddle, "Two Flavian Burials from Grange Road, Winchester," *AntJ* 47 (1967) 238-240.

6 For example: Jerusalem, Dobkin Collection, jug with a comic theater mask at the attachment of the handle, Israeli 1980, 111-114.

7 For example:
- a) RGM, no. 24.311, from Cologne, shallow bowl-shaped glass lamp with cut design, the masks have a horizontal forehead band and horizontally combed hair like the Toledo bottle, excavated from a late fourth-century tomb with a coin of the Roman emperor Maximianus (r. 286-305), Fremersdorf VI 1961, 69 and 70, pl. 139;
- b) Leiden, Netherlands, Mus. of Antiquities, no. K1914/2.45, found at Cuyk, Netherlands, glass jar with applied lion masks on the wall, Isings 1964, 59-63.

On applied theater masks: Isings 1964, 62 n. 21.

8 For example, theater masks: Cairo, Egyptian Museum, from Egypt, Edgar 1905, 77-78, nos. 32762-32767, pl. X; Athens, Benaki Museum, nos. 3195, 3197, and 14486, Clairmont 1977, 12, nos. 38-40, pls. II-III. For a survey of theater masks as decorative elements in finds from Yugoslavia: Rnjak 1979 passim, and pp. 58-65, 176-181, nos. 261-280 for glass attachments. On theater masks in general: T. B. L. Webster, *Monuments Illustrating Tragedy and Satyrplay* (*BICS, Supplement no.* 20, London 1967); T. B. L. Webster, *Monuments Illustrating New Comedy* (*BICS, Supplement no.* 24, London 1969).

### 144 Double Head-shaped Bottle: Possibly Dionysos and Satyr

Photograph, Fig. 2
Place of manufacture undetermined
Perhaps second century
H 6.5 cm; H body 3.75 cm; max D 3.6 cm; D rim 2.0 cm; D base 2.25 cm (nose to nose)-1.8 (along the seam)
1923.524, Gift of Edward Drummond Libbey

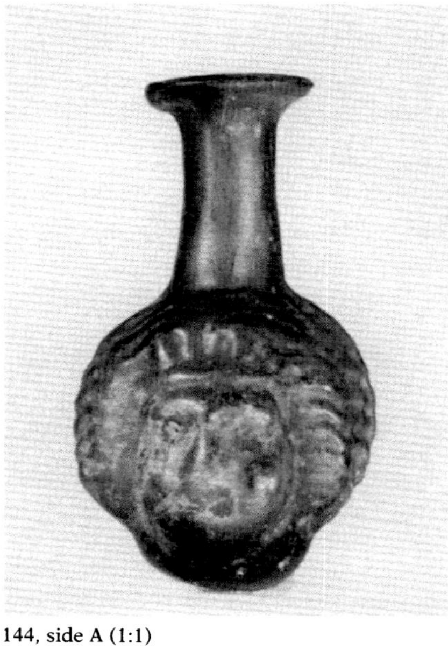

144, side A (1:1)

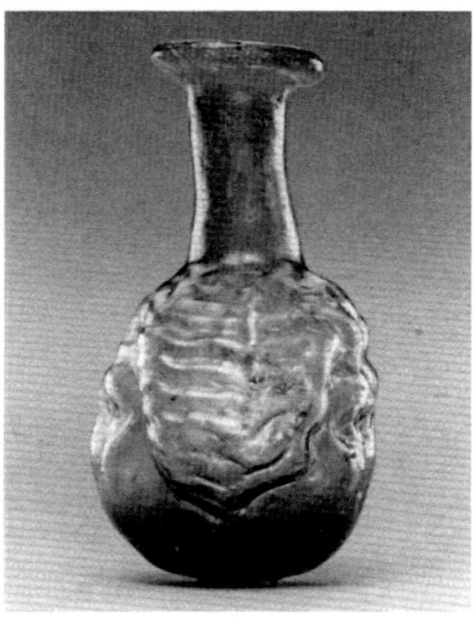

144, profile A:B (1:1)

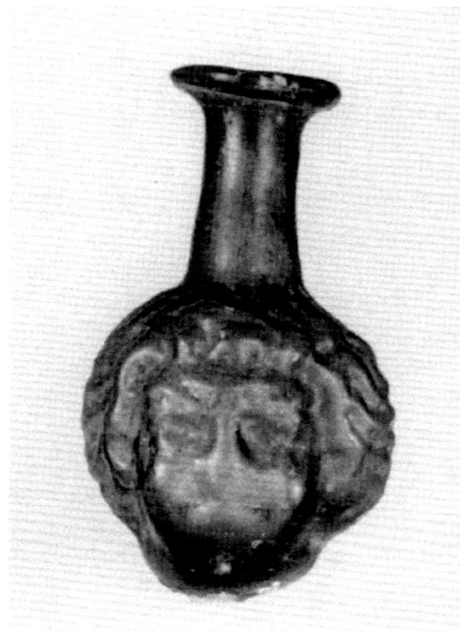

144, side B (1:1)

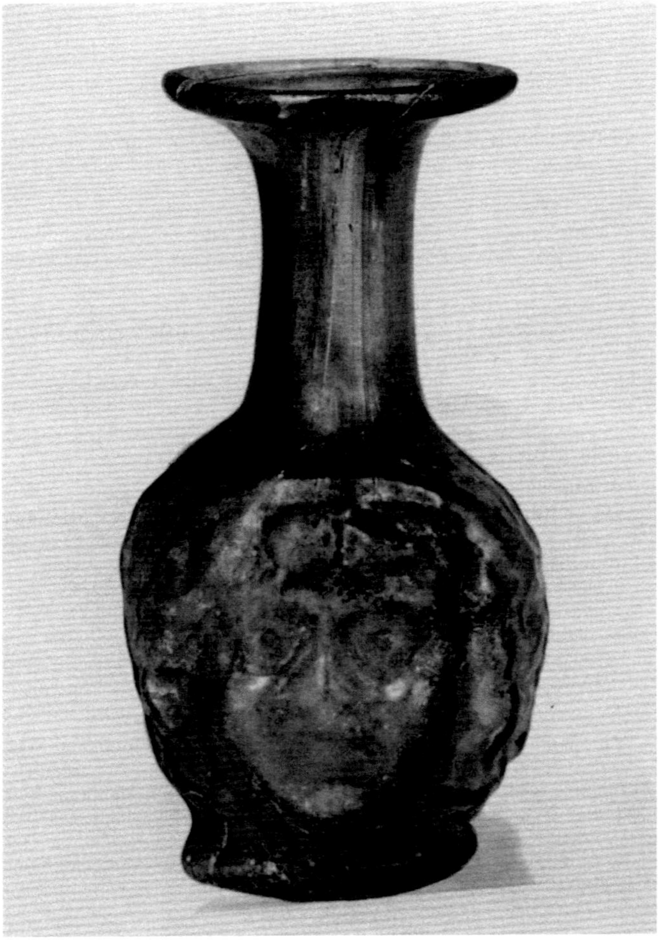

145, side A (1:1)

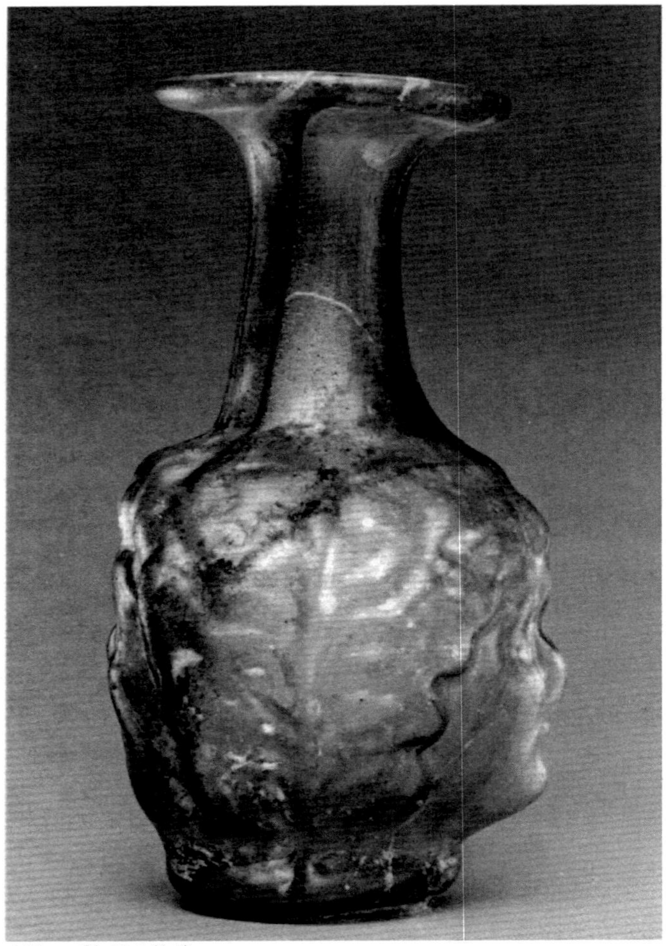

145, profile B:A (1:1)

Transparent grayish green with red streaks near rim. Medium thin glass. Numerous small and medium-sized bubbles, vertically elongated in neck, ovoid in body (see Fig. 2, left and right).
Rim and neck free blown. Body blown into a two-part mold of two vertical sections, each including one half of the base (MCT VIII). Continuous mold seam concealed in hair at junction of heads. No mold seam visible on underside of base; the bottom of this vessel was expanded after extraction from the mold, which caused the seam to blow out. Relief indistinct on lower half of vessel. No pontil mark.
Flaring rim, rounded in flame, with tool mark along part of interior. Tubular neck; upper part of neck slightly crooked down to top of head. Body in the shape of a double head. Oval convex base (vessel cannot stand).
Body in the shape of two heads back to back: a serious face, A, and a laughing male face, B. Face A is narrow at the forehead, widening toward cheeks. The eyes are small with recessed pupils. The lower half of the face is indistinct. A symposiast's fillet, with two horseshoe-shaped loops rising from it, stretches across the forehead. The hair forms a symmetrical pattern of wavy horizontal lines above and on either side of the face. Face B has full features with deep-set eyes, a broad forehead with a ridge, a broad nose, fleshy cheeks, and wide mouth with dimples at the corners. Sweeping back from the forehead are several tufts of hair and two small curved horns at the corners. The first layer of locks at the side of the face forms a frame; behind it wavy locks form a horizontal pattern.
*Condition* Intact. Deep scratch on underside. Exterior has areas of beige weathering. Interior slightly pitted.
*Collection* Thomas E. H. Curtis.
*Published* Previously unpublished.
*Remarks* Same Type as Cat. No. 143.[1]

[1] Face A related to the following examples:
a) BM, no. S173, formerly Roussell Collection, purple, Nesbitt 1871, 29, no. 173, fig. 40;
b) Oppenländer Collection, no. 2367, purple, with handle, von Saldern et al. 1974, 170, no. 462.

### 145 **Double Head-shaped Flask: Two Similar Female Faces**

Eastern Mediterranean, perhaps inland Palestinian or Syrian
Probably late second century
H 11.5 cm; H body 6.7 cm; max D 6.3-5.7 cm; D rim 5.0 cm; D base 4.0 (along the seam)-3.9 cm (nose to nose)
1923.525, Gift of Edward Drummond Libbey

Transparent streaked manganese colored pale red purple (5 RP 6/2). Medium thin glass. A few small spherical and vertically elongated bubbles in neck and body; several large vertically elongated bubbles in neck. Blowing spirals.
Rim and neck free blown. Body blown into a two-part mold of two vertical sections, each including one part of the base (MCT VIII). Continuous mold seam concealed in hair at junction of heads, extends across underside of base. Relief moderately indistinct. Ring pontil about 1.8 cm in diameter (see below Remarks).
Hollow flaring rim, folded outward, upward, and inward. Slightly concave tubular neck. Body in the shape of a double head. Nearly circular flat base.
Body in the shape of two heads back to back with similar features: large round ring-shaped eyes, high knobby cheek bones, small nose

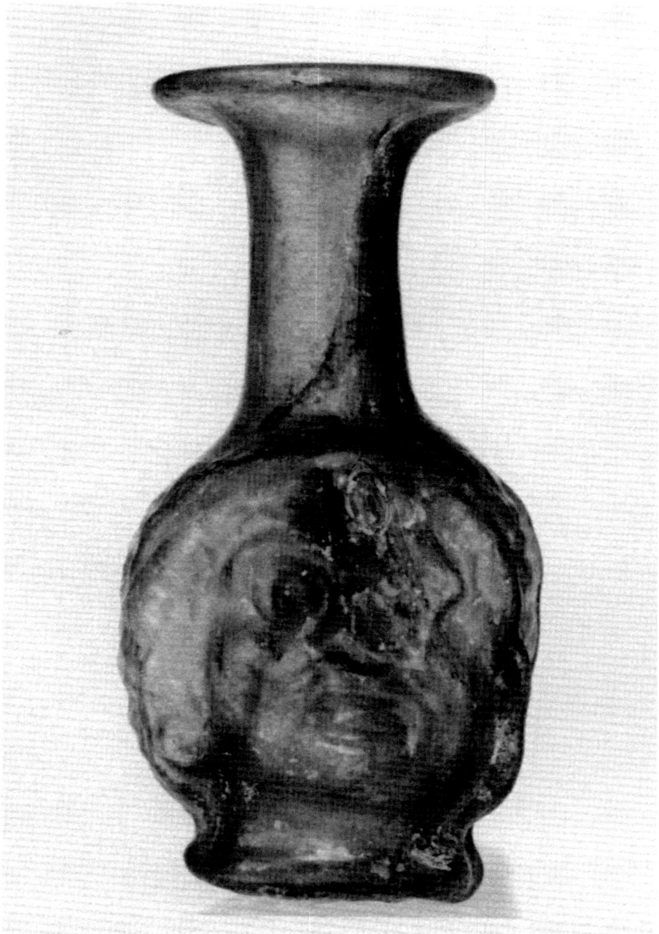

145, side B (1:1)

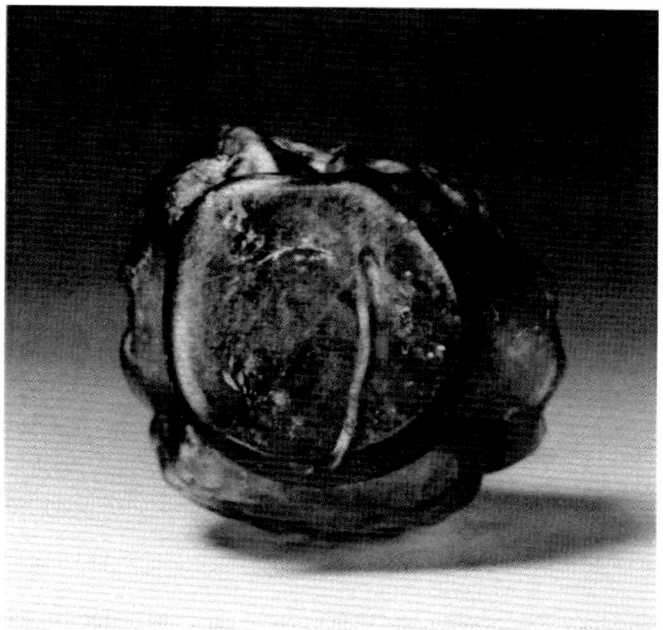

145, underside (1:1)

with short bridge, small mouth with small straight lips, and prominent knobby chin. The features of face B were flattened and distorted, probably during removal from the mold. The hair around both faces has four characteristic knobs above the forehead and rounded tufts on either side of the face, with irregular ridges aligned horizontally to the mold seam.
*Condition* Intact. Cleaned. Exterior covered with areas of surface pitting, green and purple iridescence, and brown discoloration. Interior has areas of golden iridescence and brown discoloration.
*Collection* Thomas E. H. Curtis.
*Published* Previously unpublished.
*Remarks* Same Type and same Series as Cat. No. 146, from a related mold.[1] The mold halves included unequal portions of the base (see detail photograph and Fig. 15, Mold VIII N), the portion attached to side A being about one third of the base, while the portion attached to B is about two thirds. The mold seam between the two parts of the underside of the base is straight, but the portion attached to B is recessed about 0.1 cm from portion A. A circle of small gashes marks the place where the rim of the pontil was attached to the vessel; when the pontil was pulled away, the glass on the underside of the base was broken out. (On ring-shaped pontil scars, see *Advances in Roman Glass Technology: The Pontil Technique*).
A flask or bottle of translucent purple glass belonging to the same Type and probably the same Series was excavated in Armenia in a slab stone tomb dated to the second century by the excavator.[2] This date is supported by the use of an MCT VIII mold. The stylization of the hair also suggests a late second-century date. The purple color of this flask suggests an inland Palestinian or Syrian origin.

[1] Same Type, probably same Series: formerly Constable-Maxwell Collection, *Maxwell* 1979, 71, no. 112.

[2] Yerevan, Armenia, Mus., no. 1965/29, from Garny, tomb 27, H 10.5 cm, Arakelian et al. 1969, 62, no. 133, fig. 133. See Chapter III: *Survey: Second Century, Type D.*

### 146 Double Head-shaped Flask: Two Similar Female Faces

Eastern Mediterranean, perhaps inland Palestinian or Syrian
Probably second century
H (pres.) 9.5 cm; H body 6.5 cm; max D 6.6-5.6 cm; D base 4.0 cm (nose to nose)-3.6 cm (along the seam)
1923.677, Gift of Edward Drummond Libbey

Translucent manganese colored grayish red purple (5 RP 4/2). Thick glass. Fabric cannot be determined.
Neck free blown. Body blown into a two-part mold of two vertical sections, each including one part of the base (MCT VIII N). Continuous mold seam concealed in hair at junction of heads, extends across underside of base. Relief moderately crisp. Traces of ring pontil that cannot be measured.
Tubular neck. Body in the shape of a double head. Oval flat base. Body in the shape of two heads back to back with similar features: large round ring-shaped eyes, high knobby cheek bones, small nose with short bridge, mouth with small straight lips. Face A is slightly crooked and wider at the cheekbones than face B. The hair around both faces has four characteristic knobs above the forehead and rounded tufts on either side of the face, with irregular ridges aligned horizontally to the mold seam.

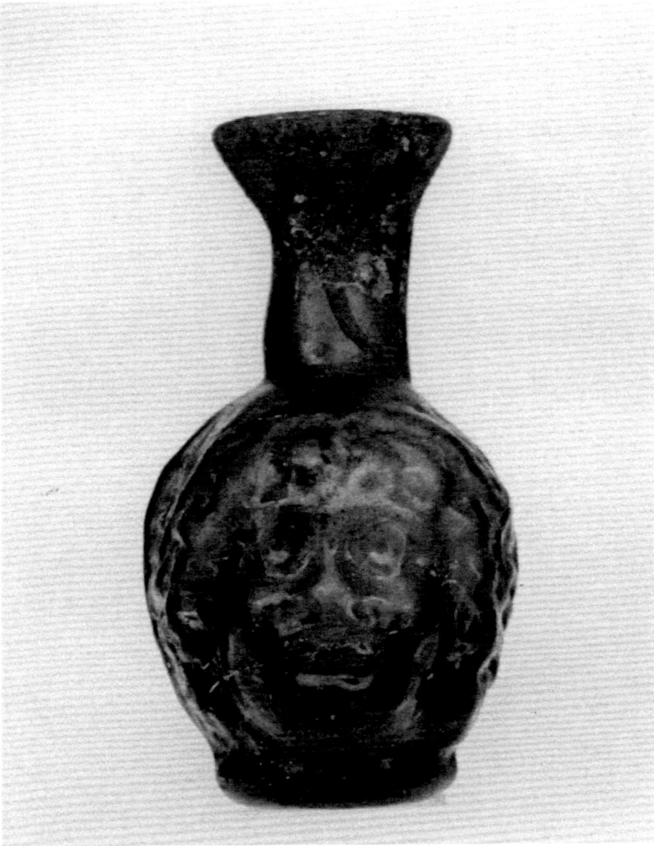

146, side A (1:1)

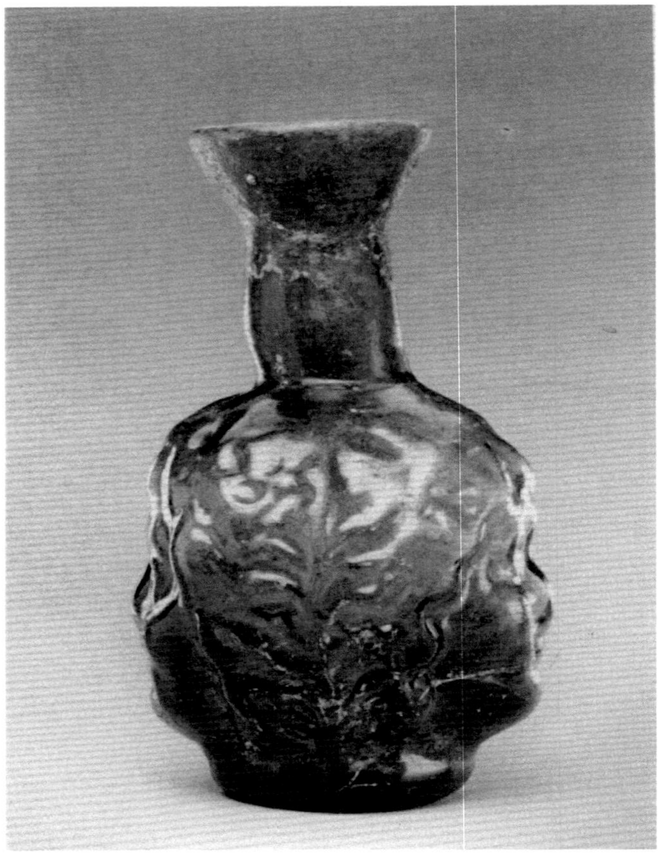

146, profile B:A (1:1)

*Condition* Rim, mouth, and upper part of neck not preserved, but restored. Cleaned.
*Collection* Thomas E. H. Curtis.
*Published* Previously unpublished.
*Remarks* Same Type and same Series as Cat. No. 145, from a related mold.

## *Vessels Shaped Like a Single Head, Late Second to First Half of Third Century* (Cat. Nos. 147 and 148)

### 147 **Flask in the Shape of an Idealized Male Head, Perhaps Dionysos**

Color plate 24, p. 61
Eastern Mediterranean, probably Syro-Palestinian, or perhaps Cypriot
Late second to early third century
H 20.4 cm; H body 8.9 cm; max D 7.6 cm; D rim 2.7 cm; D base 5.5 cm (ear to ear)-4.6 cm (along the seam)
1923.414, Gift of Edward Drummond Libbey

Transparent natural moderate greenish yellow tint. Medium thin glass.
Neck and projecting roll free blown. Body and shoulder collar blown into a two-part mold of two vertical sections, each including one half of the base (MCT VIII). Continuous mold seam, through hair, extends across underside of base. Relief crisp. Pontil mark ca. 1.3 cm.
Rim rounded and thickened in flame. Tall tubular neck tapering downward, with constriction near its base. Hollow projecting roll above shoulder collar. Body in the shape of a head. Oval slightly concave base.
Body in the shape of a beardless youthful head wearing an ivy wreath. The face is narrow and oval with a high forehead; it has wide open, almond-shaped eyes with recessed pupils, a narrow well-proportioned nose, and short fleshy lips slightly parted. The face is widest at the jaw. Around the face is an ivy wreath with clusters of berries at forehead and ears and with a heart-shaped leaf at each temple. The hair is parted in the middle, and the locks of hair framing the face are thick and swept back into a roll. The hair on the back of the head is flat with sinuous vertical strands. Below the chin is a mold-blown inscription in Greek: ΥΠΕΡΕΧΕΙ (*hyperekhei*), "he/she/it surpasses, excels"; a second, now illegible, inscription can be made out just below the projecting roll on the back of the vessel.
*Condition* Complete, except for tip of nose. Cleaned. Original surface not preserved in some areas. Scratches on neck. Exterior cloudy and rough with reddish brown weathering and areas of purple iridescence. Interior has dark brown weathering and pitting.
*Collection* Thomas E. H. Curtis.
*Published* Richter 1914, 83, fig. 11; Harden 1935, 183, pl. XXIII; Labino 1968, 21, fig. 9, 22; Grose 1978, 78, 82, fig. 19.
*Remarks* The delicate features and feminine hairstyle suggest that this flask shows a head of the Greek god Dionysos, who was often portrayed with a woman's coiffure. The hairstyle, with a roll around the face, recalls fourth-century B.C. Greek sculptures. The meaning of the inscription is ambiguous. It may refer to the contents (wine), to the elegant shape of the flask, or, possibly, to the power of Dionysos, who surpasses all.
The projecting roll above the shoulder collar formed by the rim of the mold was formed by an overblow and marks the transition between mold-blown body and free-blown neck. The earliest occurrence of a tall neck rising from such an overblow appears to be a head flask excavated in a second-century tomb at Constanta

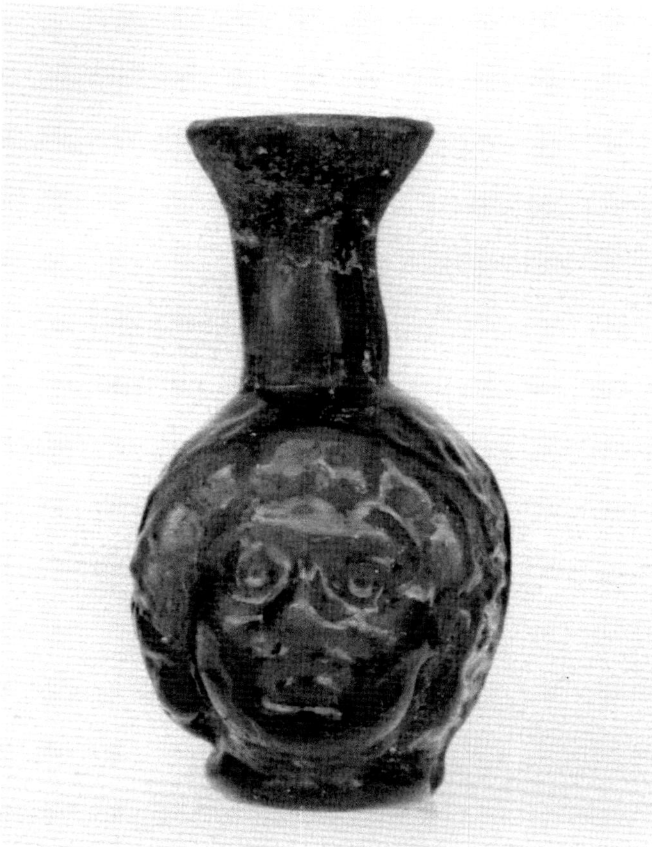

146, side B (1:1)

(ancient Tomis), Romania,[1] but this neck shape was not common in the Syro-Palestinian area until the third century (see Cat. No. 120). A number of parallels showing the same facial features and the same inscription are mentioned in older literature, but it is not clear whether they all represent different vessels. One head flask, found at Idalion, Cyprus, must be a different vessel, because it does not have a tall neck.[2] Two other references to a head flask of this Type may actually be early mentions of the Toledo head flask, which cannot be traced farther back than 1914 when it was part of the Curtis Collection,[3] or of a second head flask now in a private collection, New York, in all respects similar to the Toledo flask.[4] The first reference is to a flask said in 1906 to have been found recently;[5] the second is illustrated in a watercolor in the Museum at Kiev.[6] A flask similar in shape and technique, but showing a young man's head and bearing a different inscription, was found in a tomb at Potamia, near Idalion.[7] Finding two such exceptional head flasks at Idalion probably indicates that these Types of head flasks were made in the area, either on the Syro-Palestinian coast, or, perhaps, Cyprus.

[1] Constanta, Arch. Mus., no. 5041, H 21.0 cm, Bucovala 1968, 115-116, no. 237. See Chapter III: *Survey: Second Century, Type E, Series 2.*

[2] M. Beaudoin and E. Pottier, "Inscriptions de l'île de Chypre (1)," *BCH* 3 (1879) 163, no. 2 (not ill.), a head flask inscribed ΥΠΕΡΕΧΕΙ, *hyperekhei*, then in the collection of Dr. Tsepis, Larnaka. According to the description, it was a "vase de verre légèrement irisé, représentant une tête de femme. L'ouverture du vase est sur le sommet du crâne. Autour du cou on lit l'inscription en lettres en relief." The height, 9.0 cm, corresponds to that of the Toledo head flask from the base to the top of the shoulder collar, suggesting that the flask from Idalion originally had a tall neck rising from an overblow but that this had not been preserved.

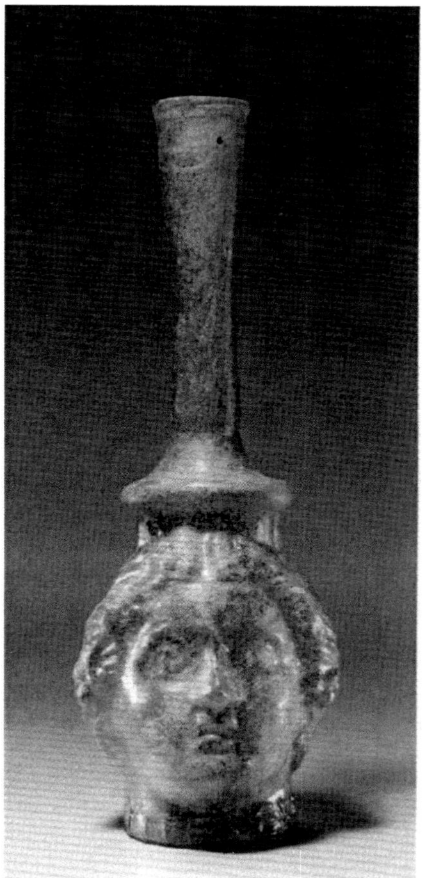

147 (1:2)

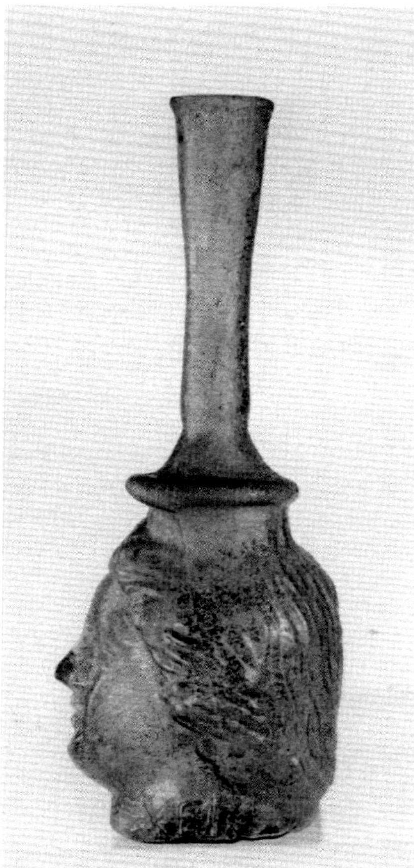

147 (1:2)

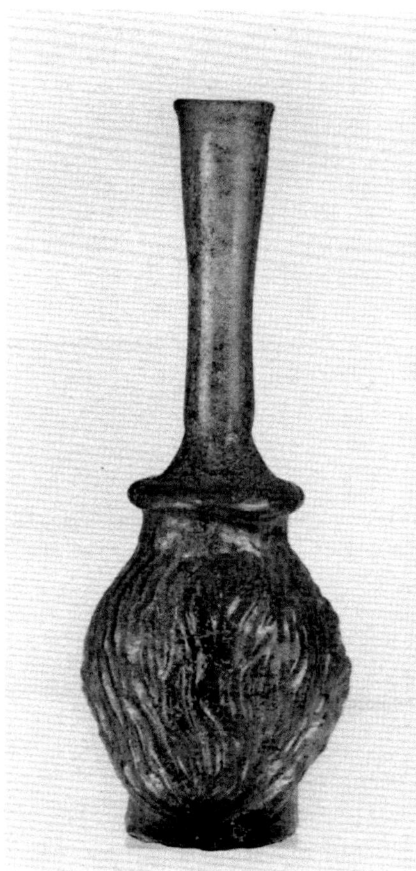

147, back (1:2)

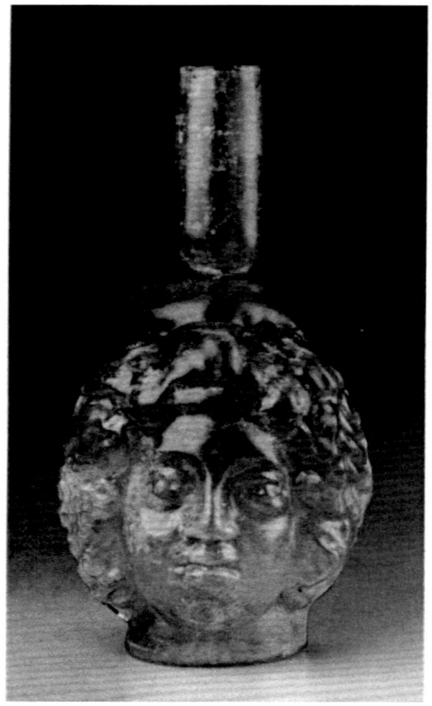

148 (1:2)

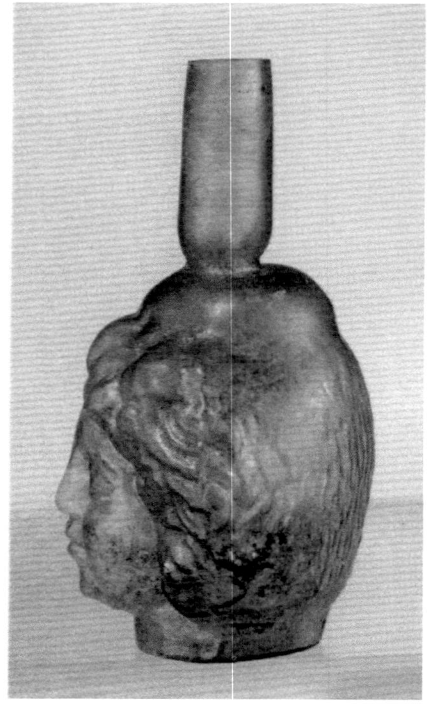

148 (1:2)

[3] Richter 1914, 83, fig. 11.

[4] New York, private collection, acquired in London.

[5] Sambon 1906, 499-450, fig. 52; Sambon 1928, 27, fig. 52, present whereabouts unknown; cf. Harden 1935, 183. In Sambon's drawing the outline of the overblow is shorter and the projecting roll less pronounced than in the Toledo flask, while the hair swept back at the nape is longer.

[6] Sorokina 1968, 184, fig. 3, no. 6. The watercolor, made by B. Pharmakowski, a Russian archaeologist and correspondent of the German Archaeological Institute active in the early twentieth century, shows a flask very similar to the Sambon flask and the flasks in Toledo and New York.

[7] BM, no. GR 1876.11-14.3, dated first century A.D., *Masterpieces* 1968, 54, no. 63; *Glass of the Caesars* 1987, 160, no. 82; correctly dated third century by V. Tatton Brown, in Tait 1991, 83, fig. 102 (color).

### 148 Flask in the Shape of a Male Head, Possibly Antinous as Dionysos [1]

Color plate 25, p. 61
Eastern Mediterranean, perhaps northwestern Asia Minor
First half of third century
H 16.6 cm; H body 10.7 cm; max D 8.5 cm; D rim 2.2 cm; D base 5.0 cm
1979.53, Purchased with funds from the Libbey Endowment, Gift of Edward Drummond Libbey

Translucent natural very pale green (10 G 8/2). Medium thin glass. A few vertically elongated bubbles in neck, some small spherical bubbles in body.

Neck free blown. Body blown into a three-part mold of two vertical sections and a separate disk-shaped base section (MCT VII). Mold seams concealed in hair. Relief crisp. No pontil mark.

Ground rim slanting inward. Tall cylindrical neck, with constriction at its base. Horizontal shoulder formed by overblow. Body in the shape of a head. Circular concave base, with intaglio mold-blown design on underside.

Around neck at regular intervals, four horizontal bands of wheel-cut incisions, a fifth band on the overblow. Body in the shape of a beardless youthful head wearing an ivy wreath. The oval face has idealized fleshy features: heavy brows, large almond-shaped eyes with heavy upper and lower lids and recessed pupils, narrow well-proportioned nose, carefully rendered, slightly parted full lips, round chin with a large central dimple and a slight roll beneath the chin, and heavy jaw wider than the forehead. Around the face is an ivy wreath with a fillet across the forehead, clusters of three round berries at the temples, and three heart-shaped leaves on each side of the face. The hair is rendered as large tufts around the face and very flat, irregular vertical ridges on the back of the head. On the underside of the base is a mold-blown mask of Medusa slightly turned to one side with grotesque features: large symmetrical face with broad semicircular chin, low straight forehead, small triangular bulging eyes, and small mouth with short full lips. Stiff locks of hair pointing outward surround the face on three sides; a half-round ornamental collar below the chin represents the knotted snakes. Below the mask is a bow-shaped ornament.

*Condition* Intact. Exterior and interior cloudy with areas of milky weathering, iridescent film, and incipient pitting.

*Collections* S. Moussaief, London; Mr. and Mrs. Andrew Constable-Maxwell (until 1979).

*Published* *Maxwell* 1979, 71, no. 113; *TMA Mus News* 21 (1979) 79; *JGS* 22 (1980) 89.

*Remarks* Mold-blown designs on the undersides of the bases of head-shaped vessels are rare. They have been noted on two Types: on first-century glass cups with Negroid features (see Chapter III: *Survey: First Century, Type A, Series* 2) and on this Type of flask representing Antinous as Dionysos. Both Types are related by the circumstance that they occur in two Forms: cups and tall-necked flasks.

Only one Form of each Type is decorated on the underside of the base: the cups of the Negroid Type and the flasks of the Dionysos Type. The significance of the decoration on the underside is unknown.

Two parallels for the Toledo flask are known. Both have the same facial features and coiffure and a design on the underside of the base, but it has not been possible to determine whether the design on the underside is the same on all three vessels. The first parallel

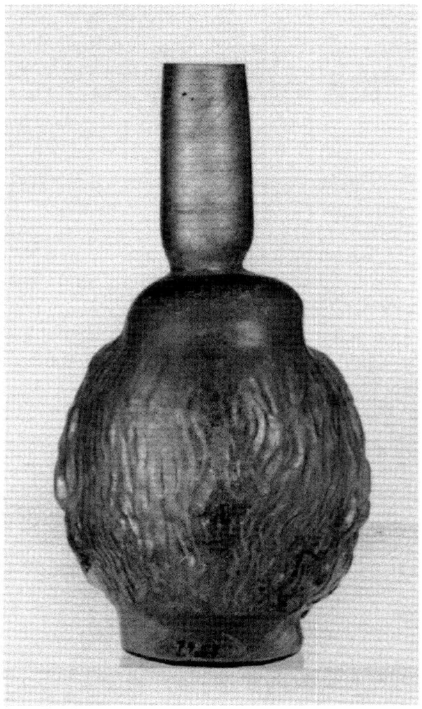

148, back (1:2)

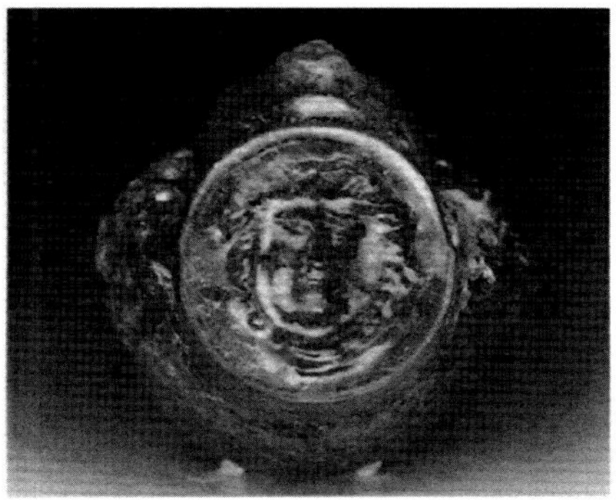

148, underside (1:2)

was excavated in Tauric Chersonesos in the Crimea (see Fig. 91)[2] and the second at Constanta (ancient Tomis), Romania.[3] The design on the underside of the vessel from the Crimea is described as "a female head in relief." The Crimean vessel now looks like a cup, but the edge of the short upright rim appears to be broken, and it might be the remains of an upright collar formed by the edge of a mold, an indication that it had a tall neck rising from an overblow like the flask from Constanta. Most of the face of the Romanian flask is lost, but the back of the head is preserved, and the coiffure appears to be the same as that on this Toledo flask. The flask from Constanta was finished with a projecting roll or overblow above a shoulder collar and a tall neck like Cat. No. 147.

## Date

A second-century date for the creation of this Type is indicated by the flask from Constanta, which was excavated in a tomb dated to the second century. Iconographical and stylistic evidence also points to a second-century date for creation of molds featuring Antinous as Dionysos. A cup with similar facial features and one handle at the back of the head (see Chapter III: *Survey: Second Century, Type E, Series* 1) was excavated at Kep, in the Taman peninsula opposite the Crimea, in a tomb containing clay lamps and vessels used during the second half of the second and first half of the third centuries.[4] The resemblance of the facial features to the Roman emperor Hadrian's friend Antinous is reason to date the cup to the reign of Hadrian (117-138) or that of his successor Antoninus Pius (138-161).[5] A ceramic head-shaped cup, with similar facial features and a lionlike mane[6] like that commonly used to render the hair of Alexander the Great, has been dated independently for stylistic reasons to the late Antonine period (about 161-192).[7]

Flasks of this Type are probably somewhat later than cups, because the flasks have a characteristic tall neck that is not common during the second century. Two neck shapes are usually associated with this Type: a tall cylindrical neck with a ground rim, such as that on this Toledo flask, and an even taller, tubular neck with its rim rounded in flame, rising from a pronounced overblow, as on the flask from Constanta.[8] The cylindrical neck with a ground rim and fine, wheel-cut incisions was common in the eastern Mediterranean during the first half of the third century[9] on spherical bottles and flasks, as shown by fragments of many examples from Dura Europos which date before the destruction of that city in 256.[10] In the East, spherical bottles of this shape were succeeded in the second half of the third century by a debased neck shape with a thickened rim rounded in flame that remained common into the fourth century.[11] Judging from the finishing of the neck and rim, the Toledo bottle was blown before the middle of the third century, presumably in the early third or very late second century. An early third-century date is supported by the similarity of the Medusa on the underside of the bottle to a Medusa carved on the breastplate of an early third-century marble bust of a member of the imperial family.[12]

## Distribution and Origin

Find spots in the Crimea,[13] Turkey,[14] and Romania[15] indicate an eastern Mediterranean origin for this Type, even though an intaglio design on the underside of the base appears to have been unusual in this part of the Roman empire. Designs in intaglio on the underside of the base are associated primarily with glass vessels made in the West, such as Single Ethnic Type cups signed by C. Caesi Bugaddi (see Chapter III: *Survey: First Century, Type A, Series* 2 and Fig. 84) and first- and second-century glass unguentaria excavated in the western part of the Roman empire bearing impressions made from coins.[16] However, the technique of using a coin to make an impression on the underside of a glass bottle may also have been practiced in Asia Minor, as hinted by a recent find from Ficarolo, in northern Italy, with the impression of a coin minted in the year 128/129 for the Koinon of Bithynia, in northwestern Asia Minor.[17]

Very little is known about the glass industry in northwestern Asia Minor, but a concentration of finds of mold-blown beakers with

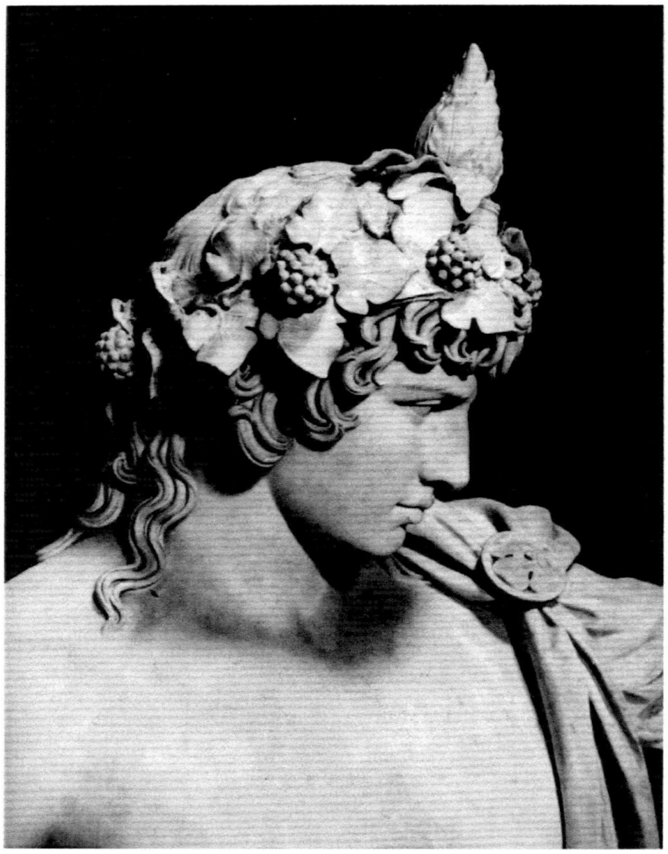

*Fig. 105. Detail, statue of Antinous as Dionysos. From Palestrina, H 3.26 m, marble, Rome, Musei Vaticani, Sala Rotondo, no. 256. Photo: Musei Vaticani, Archivio Fotographico.*

mythological scenes from that area [18] suggests that during the late first and early second centuries the technique of mold blowing was practiced there. An origin in northwestern Asia Minor would be appropriate for head-shaped vessels of the Antinous-Dionysos Type, because Antinous came from that area. The bow-shaped ornament below the Medusa mask on the underside of the flask also suggests an origin in Asia Minor because similar ornaments are carved below the Medusa masks on the Gorgon frieze of Temple III at Didyma, Turkey. [19]

#### Interpretation

Antinous was a native of Bithynia in northwestern Asia Minor. His portraits are characterized by fleshy sensual features, heavy eyebrows, and full lips. After his mysterious death in the Nile in the year 130, Antinous was deified and statues dedicated to him in many parts of the empire. The glass vessels showing Antinous with a Dionysiac ivy wreath may have been based on a statue type showing Antinous in the guise of Dionysos, such as Fig. 105. [20]

[1] See Chapter III: *Survey: Second Century, Type E, Series 2.*

[2] St. Petersburg, Hermitage, no. X 1902.14, excavated in 1902, H 11 cm, from grave 1386, in a rock-cut tomb near Karantinny Cove, G. Sokolov, *Antique Art on the Northern Black Sea Coast* (Leningrad 1974) 137, no. 146, ill. (color), here Fig. 91.

[3] Constanta, Arch. Mus., no. 5041, H 21.0 cm, D 9.0 cm, Bucovala 1968, 115-116, no. 237, figs. 237a, b, back and underside of base ill.

[4] From Kep, tomb 133, Sorokina 1968, 185-186, figs. 1 and 2, here Fig. 90.

[5] Sorokina 1977a, 133-135, fig. 5.

[6] Karlsruhe, Badisches Landesmuseum, no. 68/48, thought to have been made in Pergamon, Asia Minor, clay cup with two vertical handles, inscribed ΤΡΩΙΛΟC, *Troilos*, in Greek on the rim, Mandel 1988, 229, no. P 194, pl. 22.

[7] Mandel 1988, 25.

[8] Compare, for example, two flasks of which the underside has not been described:

a) Oppenländer Collection, no. 22225, decolorized, H 19.4 cm., von Saldern et al. 1974, 171, no. 467, color pl. p. 161;
b) Leiden, Rijksmuseum van Oudheden, no. B 1975/9.3, said to have been found in southwestern Turkey.

[9] In the West this type of neck was not common before the second half of the third century; it was especially common in the fourth century: Isings 1957, 121-122, Form 103.

[10] Clairmont 1963, 112-115, nos. 540-541, 543-561.

[11] For example, a bottle from Hanita, Israel, tomb 15, used ca. 220-330, Dan P. Barag, *Hanita, Tomb XV. A Tomb of the Third and Early Fourth Century C. E.* (*'Atiqot* [English Series] 13, Jerusalem 1978) 23, 26, no. 45, fig. 12, pl. IV.

[12] Vatican, Museo Gregoriano Profano, formerly Lateran Collection, no. 10135, Helbig[4], I, 798; E. Buschor, *Medusa Rondanini* (Stuttgart 1958) 25, no. 4, pl. 33, no. 4.

[13] See above notes 2 and 4.

[14] See above note 8b.

[15] See above note 3.

[16] Taborelli 1982a, 315-340, esp. pp. 325-326, listing intaglio designs made from impressions of coins of the emperors Nero (54-68), Domitian (81-96), Hadrian (117-138), and Marcus Aurelius (161-176). He notes unguentaria from Italy (Rome, Sentinum, Reggio Emilia, Nesazio, Cremona, and Modena), Croatia (Pula), and England (London). On the origin and function of these stamps: L. Taborelli, "Nuovi esemplari di bolli già noti su contenitori vitrei dall'area centro-italica," *Picus, Studi e ricerche sulle marche nell'antichità* 3 (1983) 23-69; L. Taborelli, "A proposito della genesi del bollo sui contenitori vitrei (Note sul commercio delle sostanze medicinali e aromatiche tra l'età ellenistica e quella imperiale)," *Athenaeum, Studi periodici di letteratura e storia dell'antichità*, n.s. 63 (1985) 198-217. On stamped unguentaria from Spain: J. Price, "Roman Unguent Bottles from Rio Tinto (Huelva) in Spain," *JGS* 19 (1977) 30-39.

[17] H. Büsing, "Der Münzabdruck im Boden einer Glasflasche aus Ficarolo (I)," *AntW* 22 (1991) 21. On the date of the coin: *Sylloge Nummorum Graecorum, Copenhagen. Nationalmuseet* 18: *Bosporus. Bithynia* (Copenhagen 1942-48) no. 324; D. R. Sear, *Greek Imperial Coinage* (London 1982) no. 1143.

[18] On beakers with mythological scenes: Weinberg 1972, 26-47; Wight 1988, 71-76. See also Chapter II: *The Second to Fourth Centuries.*

[19] Buschor 1958 (see above note 12), 39, figs. 2-5. On the Gorgon frieze, which was added after the end of the second century: *PECS* 273, s.v. *Didyma* (K. Tuchelt).

[20] Vatican Museums, Sala Rotonda, no. 256, from Palestrina (ancient Praeneste), S. Aurigemma, *Villa Adriana* (Rome 1961) fig. 3.

## *Double Head-shaped Flasks with Knobby Hair, Third Century* (Cat. Nos. 149-157)

### 149 Double Head-shaped Flask: Two Similar Chubby Child-like Faces

Color plate 26, p. 62
Eastern Mediterranean, possibly Syro-Palestinian
Probably early third century
H 8.6 cm; H body 5.25 cm; max D 4.6 cm; D rim 3.6 cm; D base 3.3 cm (nose to nose)-3.0 cm (seam to seam)
1923.515, Gift of Edward Drummond Libbey

Transparent dusky yellow (5 Y 6/4). Medium thin glass. A few vertically elongated bubbles in neck.
Rim and upper portion of neck free blown. Lower part of neck and body blown into a two-part mold of two vertical sections, probably open at the base (MCT IX). Mold seams concealed in hair

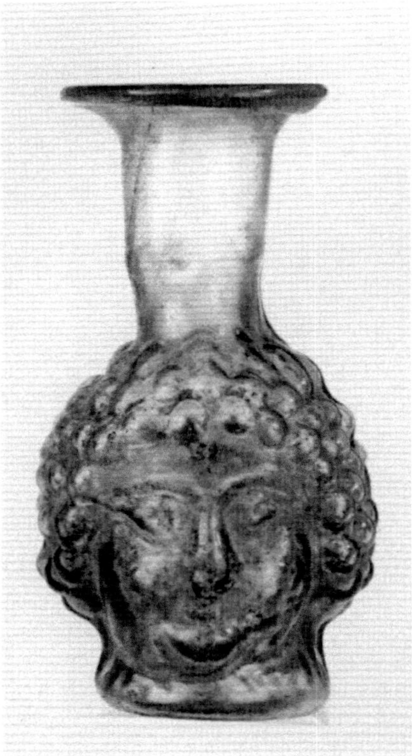
149, side A (1:1)

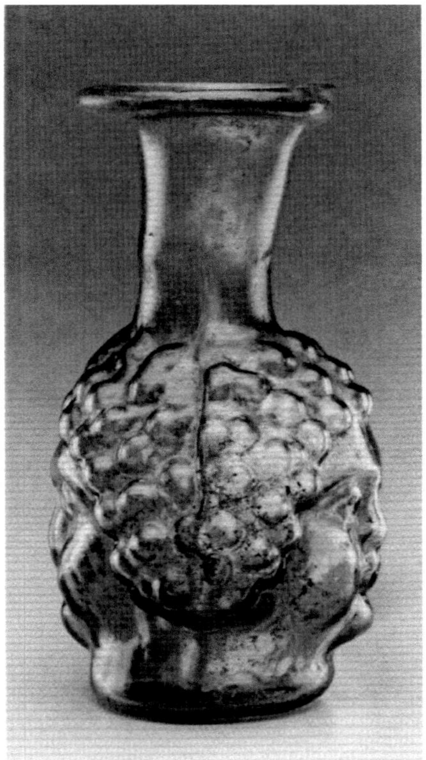
149, profile A:B (1:1)

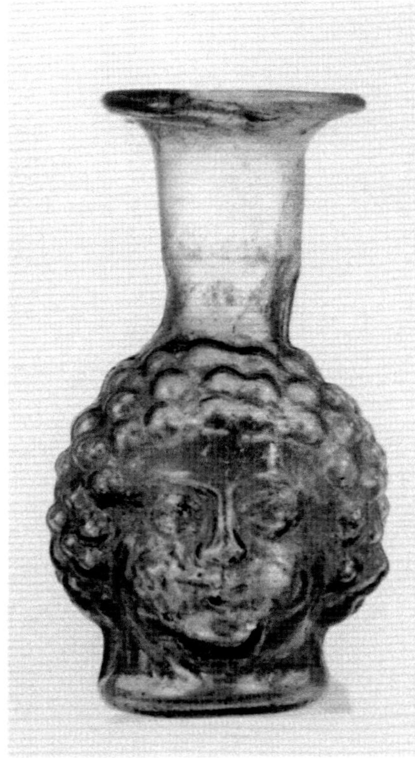
149, side B (1:1)

at junction of heads. No mold seam on underside of base. Relief crisp. No pontil mark.

Irregularly outsplayed rim, folded outward, upward, and inward. Tubular neck tapering downward; upper part of neck crooked down to constriction marking top of mold (about two thirds of the way down neck; H from base to top of mold 6.2 cm). Body in the shape of a double head. Oval flat base.

Body in the shape of two heads back to back with similar childlike features: prominent arched browline, large wide almond-shaped eyes with recessed pupils, and protruding knobby chin. Only the noses are significantly different, with a large broad nose on face A and a narrower nose on face B. Face B is wider at the jaw, giving the face a piriform shape, while face A is more rectangular. The hair around both faces is rendered as distinct, evenly spaced knobs arranged in three regular tiers framing the face to chin level; the hair of face B recedes slightly at the corners of the forehead.

*Condition* Intact. Cleaned. Exterior slightly rough with incipient pitting and small areas of iridescent film. Interior has areas of reddish brown weathering.

*Collection* Thomas E. H. Curtis.

*Published* Previously unpublished.

*Remarks* Same Type as Cat. Nos. 150 and 151.[1] Knobby hair appears to be characteristic of third-century head-shaped bottles and flasks. A double head flask with stylized knobby hair was excavated at Dura Europos[2] in a tomb which also contained a coin of Commodus (ruled 177-192). See Chapter III: *Survey: Third Century, Type A*. The use of a mold open at the base (MCT IX) was common in the third and fourth centuries.

The two childlike faces of this flask may represent Eros, the Greek god of love, or Dionysos, the Greek god of wine, both of whom were frequently depicted as a child and both of whom were worshipped throughout the Syro-Palestinian area during the second and third centuries. Interpreting the vessels as representations of Dionysos as a child is particularly attractive because Dionysos was traditionally thought to have been raised in the Syro-Palestinian area. According to an early Greek tradition of the seventh century B.C., Dionysos buried his nurse Nysa in Nysa-Skythopolis (modern Beth Shean, Israel);[3] while a fifth-century poet associated Dionysos's childhood with a Sidonian nurse named Mystis who taught him the Dionysiac mysteries.[4]

[1] Same Type or related:

a) Warsaw, Muzeum Narodowe, no. 42074, formerly Wl. Semerau-Siemianowskiego Collection (close to face B), B. Filarska, *Szkła Starożytne, Katalog naczyn*, Muzeum Narodowe w Warszawie (Warsaw 1952) 112-113, no. 87, pl. XVIII, no. 3;
b) RGM, no. N1070, La Baume 1973, no. C 14, pl. 10, nos. 2 and 3;
c) Hanover, Kestner-Museum, no. 1929, 253, Liepmann 1982, 43, no. 17;
d) Oppenländer Collection, no. 2342, von Saldern et al. 1974, 170-171, no. 466.

Compare also:

e) Newark, no. 50.1437, said to have been found in Palestine, H 8.3 cm, blue-green, Auth 1976, 74, no. 74;
f) from a cemetery with Roman and Christian tombs at Saloniki, northern Greece, *ArchDelt* 21, B2 (1966) 336 (no description), pl. 351;
g) Yale, no. 1930.413, said to have been found at Tel-el-Horus, Syria, H 8.7 cm, pale green, with an illegible inscription stamped on the underside of the base, Matheson 1980, 73, no. 191;
h) Murano, Mus. Vetrario, from the necropolis of Nona, at Zadar, Croatia, transparent grayish olive glass.

[2] Clairmont 1963, 37, no. 132, pl. XXI. For fragments of double head-shaped vessels, also excavated at Dura, but not ill.: Clairmont 1963, 37-38, nos. 133-144.

[3] Terpandros, Fr. 8; see also Chapter I: *Hexagonal Bottles with High Relief: Interpretation* and Cat. No. 71.

[4] Nonnos, *Dionysiaka* IX, 98-131.

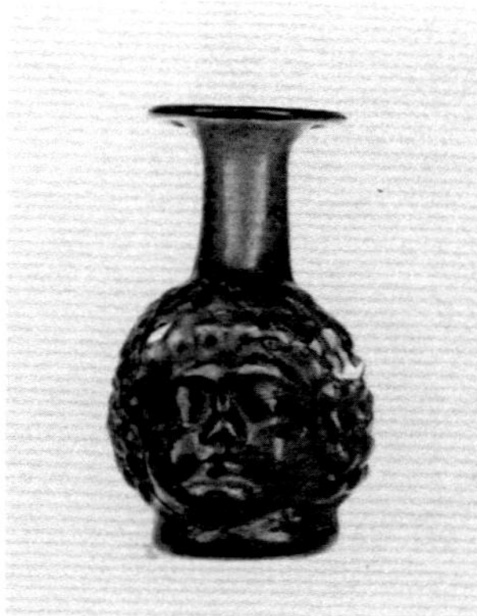
150, side A (1:2)

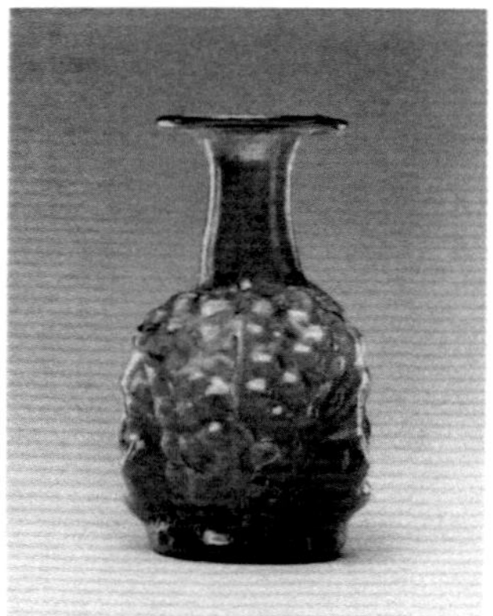
150, profile B:A (1:2)

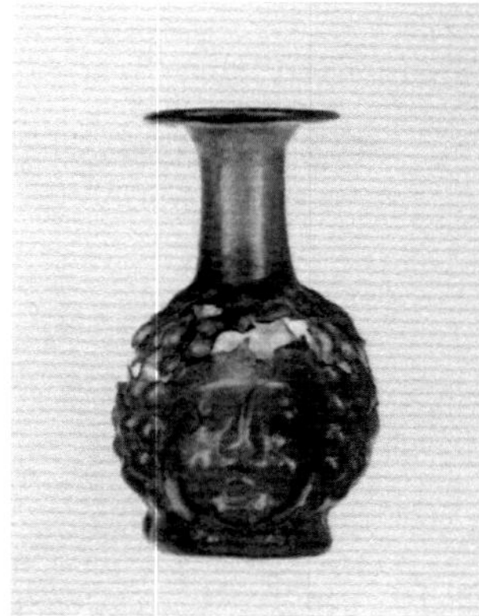
150, side B (1:2)

150 **Double Head-shaped Flask: Two Similar Chubby Child-like Faces**

Eastern Mediterranean, possibly Syro-Palestinian
Probably early third century
H 7.8 cm; H body 4.85 cm; max D 4.5 cm; D rim 3.4 cm; D base 3.55 cm (nose to nose)-3.1 cm (seam to seam)
1923.521, Gift of Edward Drummond Libbey

Transparent deep blue (not in rock color chart). Thin glass. Blowing spirals. Vertically elongated bubbles in neck.
Rim and neck free blown. Body blown into a two-part mold of two vertical sections, probably open at the base (MCT IX). Mold seams concealed in hair at junction of heads. No mold seam on underside of base. Relief moderately crisp. No pontil mark.
Flaring rim, folded outward, upward, inward, and downward. Tubular neck tapering upward. Body in the shape of a double head. Oval base, with slight depression in center of underside.
Body in the shape of two heads back to back with similar childlike features: prominent arched browline, large round eyes, fat cheeks, large nose, and small mouth with dimples at the corners. Face A is more exaggerated than face B, with larger eyes, fleshier features, and a very broad jaw. The hair around both faces is rendered as distinct, evenly spaced knobs arranged in three regular tiers (only two tiers on the right side of face A) framing the face to chin level.
*Condition* Body broken into small pieces and repaired; many fragments from hair of both sides missing. Cleaned. Exterior and interior have small areas of beige weathering and pitting.
*Collection* Thomas E. H. Curtis.
*Published* Previously unpublished.
*Remarks* For the Type, see Remarks Cat. No. 149. Same Type as Cat. Nos. 149 and 151; perhaps same Series as Cat. No. 151. [1]

[1] Related faces:
a) Carnegie, no. 21868/2, Oliver 1980, 64, no. 61;
b) Oppenländer Collection, no. 2342, von Saldern et al. 1974, 170-171, no. 466.

151 **Double Head-shaped Flask: Two Similar Chubby Child-like Faces**

Eastern Mediterranean, possibly Syro-Palestinian
Probably early third century
H 8.5 cm; H body 4.7 cm; max D 4.4 cm; D rim 3.5-3.3 cm; D base 3.43 cm (nose to nose)-2.9 cm (seam to seam)
1923.518, Gift of Edward Drummond Libbey

Transparent pale olive (10 Y 6/2). Medium thin glass. Numerous vertically elongated bubbles in neck. Black specks.
Rim and upper portion of neck free blown. Body blown into a two-part mold of two vertical sections, probably open at the base (MCT IX). Mold seams concealed in hair at junction of heads. No mold seam on underside of base. Relief moderately indistinct. No pontil mark.
Trefoil mouth, aligned approximately with mold seams. Hollow rim, folded outward, upward, and inward. Tubular neck, with constriction just below center of neck. Body in the shape of a double head. Oval flat base.
For description, see Cat. No. 150.
*Condition* Intact. Cracks in neck. Exterior covered with very thin green and purple iridescent film. Interior has areas of dark brown and beige weathering.
*Collection* Thomas E. H. Curtis.
*Published* Previously unpublished.

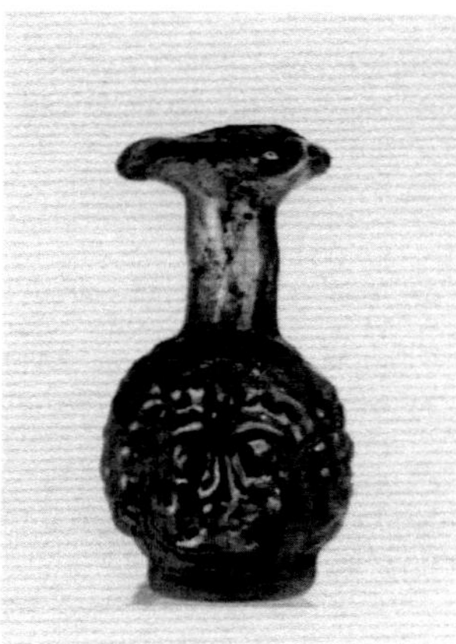
151, side A (1:2)

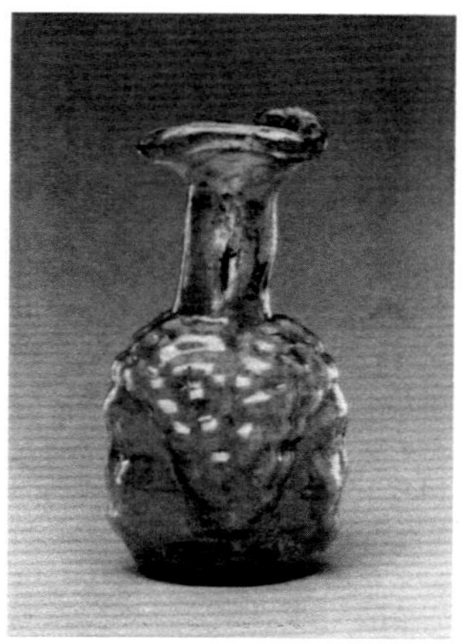
151, profile A:B (1:2)

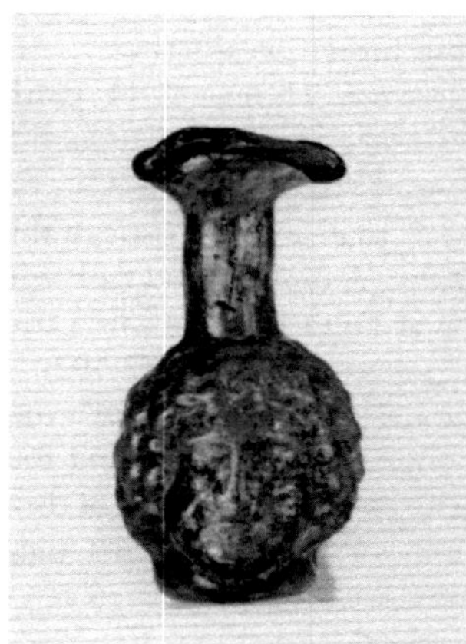
151, side B (1:2)

*Remarks* For the Type, see Remarks Cat. No. 149. Same Type as Cat. Nos. 149 and 150; perhaps same Series as Cat. No. 150.[1]

[1] Related faces:

a) from a Roman or Christian tomb at Thessalonike, northern Greece, *ArchDelt* 21, B2 (1966) 336 (no description), pl. 351;
b) from Tarsus, grave 3, yellowish brown, Hetty Goldman ed., *Excavations at Gözlu Kule, Tarsus*, I: *The Hellenistic and Roman Periods* (Princeton 1950) 402, no. 26, fig. 274, no. 26.

### 152 Double Head-shaped Flask: Two Similar Idealized Faces, Perhaps Medusas

Eastern Mediterranean, possibly Syro-Palestinian
Probably early third century
H (pres.) 7.0 cm; H body 5.2 cm; max D 4.8 cm; D base 3.5 cm (nose to nose)-3.2 cm (seam to seam)
1923.523, Gift of Edward Drummond Libbey

Transparent moderate greenish yellow (10 Y 7/4). Medium thin glass. Fabric cannot be determined because of weathering.
Preserved portion of neck and body blown into a two-part mold of two vertical sections, probably open at the base (MCT IX). Mold seams concealed in hair at junction of heads. No mold seam on underside of base. Relief crisp. No pontil mark.
Tubular neck. Body in the shape of a double head. Flat diamond-shaped base with rounded corners.
Body in the shape of two heads back to back with similar features: prominent arched browline, almond-shaped eyes with recessed pupils, narrow nose with flaring nostrils, small mouth with straight lips slightly parted, and fat cheeks. Face A is slightly crooked at jaw level; the mouth of face B is indented more deeply. Both faces have a collar-shaped band below the chin. The hair around both faces is rendered as distinct, evenly spaced knobs arranged in three regular tiers framing the faces to below chin level.

*Condition* Rim and upper portion of neck missing. Exterior surface covered with beige frosting. Interior scratched.
*Collection* Thomas E. H. Curtis.
*Published* Previously unpublished.
*Remarks* Same Type as Cat. No. 153 but from a different Series. The collar-shaped bands below the chins may be a degenerated form of the knotted snakes of Medusa,[1] or they may represent torques or necklaces such as appear on bronze head flasks representing maenads,[2] Dionysos,[3] and Antinous.[4]
The knobby stylization of the hair suggests a date similar to that of the head-shaped flasks with childlike faces (Cat. Nos. 149-151). The hair, with longer locks falling on either side of the head, recalls the coiffures of the ladies of the imperial Severan dynasty (ruled 193-235), many of whom came from Syria.[5]

[1] Cf. Cat. No. 142.
[2] S. Haynes, "Etruskische Bronzekopfgefässe aus hellenistischer Zeit," *JRGZM* 6 (1959) 122-123, Group VI, nos. 2-7, pls. 56 and 57.
[3] F. Jones, "A Bronze Head Vase," *Record of the Art Museum. Princeton University* 46.2 (1987) 22-23, Type E, cat. nos. 1-7, figs. 10-11.
[4] Jones 1987 (see above note 3) 21, Type B, cat. nos. 1-5, fig. 5.
[5] Compare the portraits of Julia Domna, e.g. Munich, Glyptothek, no. W354, Bernard Andreae, *The Art of Rome* (New York 1979) pl. 114.

### 153 Double Head-shaped Flask: Two Similar Idealized Faces Perhaps Medusas

Eastern Mediterranean, possibly Syro-Palestinian
Third century
H 8.8 cm; H body 4.9 cm; max D 4.9 cm; D rim 3.1 cm; D base 3.2 cm
1923.459, Gift of Edward Drummond Libbey

Transparent manganese colored grayish red purple (near 5 RP 4/2,

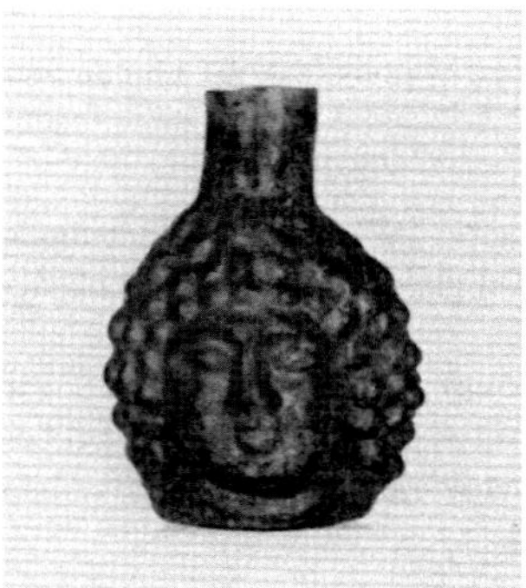

152, side A (1:2)

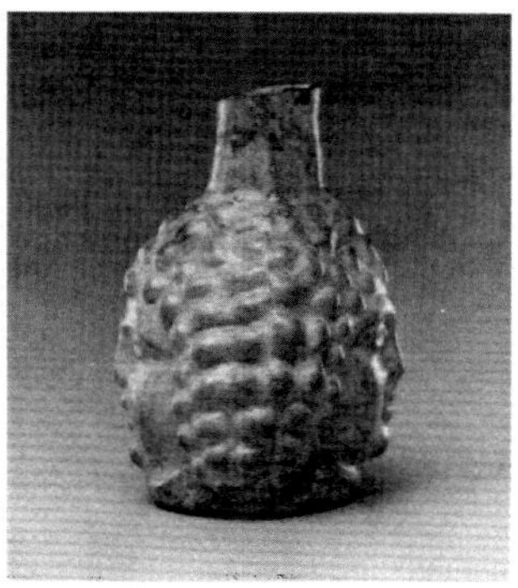

152, profile B:A (1:2)

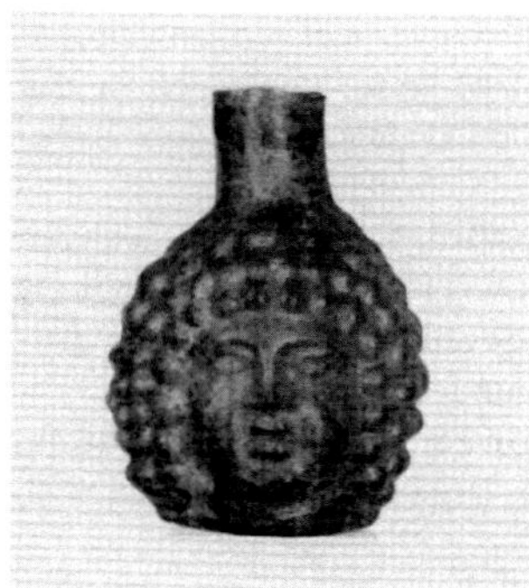

152, side B (1:2)

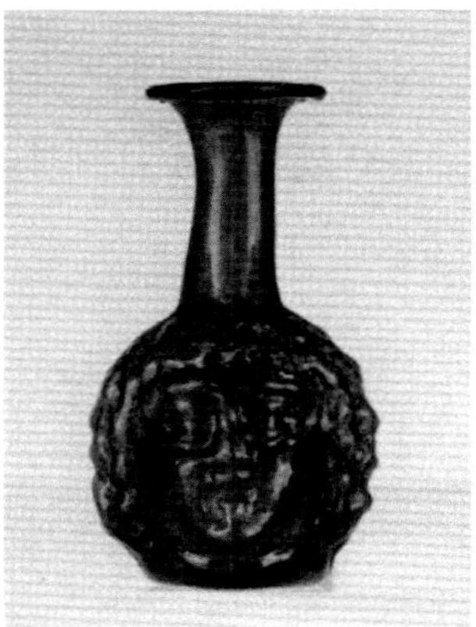

153, side A (1:2)

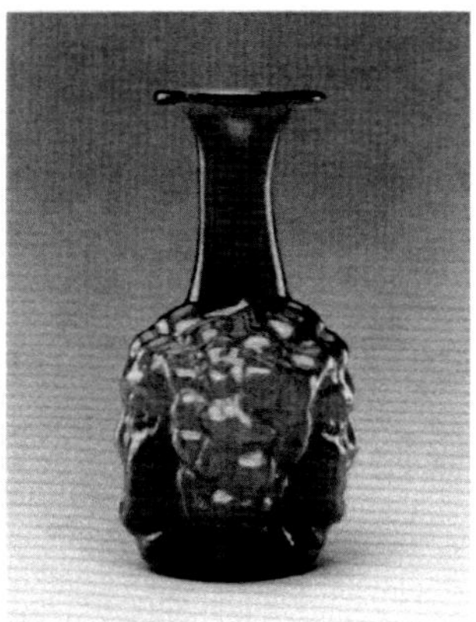

153, profile B:A (1:2)

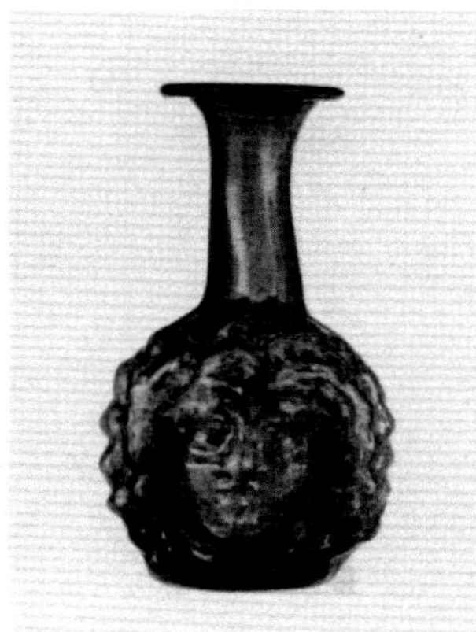

153, side B (1:2)

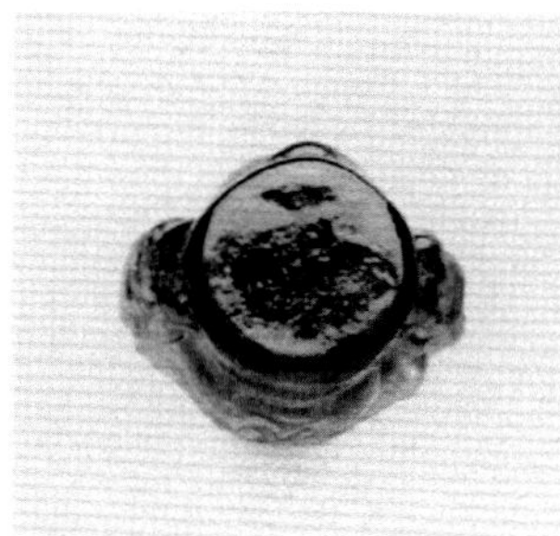

153, underside (1:2)

but more red). Medium thin glass. Numerous vertically elongated bubbles in neck.
Rim and neck free blown. Body blown into a two-part mold of two vertical sections, probably open at the base (MCT IX). Mold seams concealed in hair at junction of heads. No mold seam on underside of base. Relief moderately crisp. No pontil mark.
Flaring hollow rim, folded outward, upward, and inward. Tubular neck tapering upward. Body in the shape of a double head. Circular flat base.
Body in the shape of two heads back to back with similar features: prominent arched browline, large hollow eyes, small indistinct mouth, and a collar-shaped band below the chin. The hair of side A is predominantly knobby and arranged in two tiers; the hair of side B is rendered as irregular tufts and ridges arranged in two tiers.
*Condition* Intact. Exterior has area of milky weathering; interior has traces of brown weathering.
*Collection* Thomas E. H. Curtis.
*Published* Previously unpublished.
*Remarks* Same Type as Cat. No. 152, but from a different Series. The underside of the base shows small pits and inclusions of dust characteristic of molds that did not include a base section (MCT IX). Dust and debris from the floor of the workshop adhered to the glass during expansion in the mold (see detail photograph).

### 154 Double Head-shaped Flask: Two Similar Youthful Faces

Eastern Mediterranean, possibly Syro-Palestinian
Third century
H 8.2 cm; H body 5.1 cm; max D 4.7 cm; D rim 3.3 cm; D base 3.85 cm (seam to seam)-3.4 cm (nose to nose)
1923.526, Gift of Edward Drummond Libbey

Transparent natural yellow green; exact color cannot be determined because of weathering. Medium thin glass. Fabric cannot be determined because of weathering.
Rim and upper part of neck free blown. Lower part of neck and body blown into a two-part mold of two vertical sections, probably open at the base (MCT IX). Mold seams concealed in hair at junction of heads. No mold seam on underside of base. Relief indistinct. No pontil mark.
Horizontal rim, folded outward, upward, and inward. Tubular neck, upper part slightly crooked down to constriction marking top of mold. Body in the shape of a double head. Irregularly flattened diamond-shaped base with rounded corners.
Body in the shape of two heads back to back with similar youthful features: prominent arched browline, large ring-shaped eyes with depressed centers, large nose, small mouth with straight lips slightly parted, and fat cheeks. The hair around both faces is rendered as circular knobs, sometimes joined to form wide ridges, arranged in three tiers framing the faces to below chin level.
*Condition* Exterior covered with a thick layer of black encrustation, areas of rainbow iridescence, and severe surface pitting. Interior covered with dark weathering.
*Collection* Thomas E. H. Curtis.
*Published* Previously unpublished.
*Remarks* Face A is distorted by a vertical ridge of glass between the tip of the nose and the chin, resulting from molten glass protruding through a crack in the mold. The indistinct features are due in part to the mold not being completely filled during blowing. The resemblance of the hair to a wig, with locks falling to the shoulders, and the outline of the forehead recall Cat. No. 152, but this Type does not have a collar-shaped band below the chin. Both faces are related to Cat. No. 155, face A, and Cat. No. 156, face A. The Type was common in the eastern Mediterranean.[1] At least one example found its way in antiquity to southern France, where it was excavated in a fifth-century sarcophagus at Marseilles, St. Victor.[2]

[1] From a *hypogeum* at Homs, Syria, Abdul Hak 1965, 30, fig. 11.
From unknown find spots:
a) ROM, no. 950.157.84, Hayes 1975, 50, no. 94;
b) Berlin, Staatliche Museen, Antikensammlung, no. 30219, 148, Platz-Horster 1976, 47, no. 75.
Related: Vienna, Kunsthistorisches Mus., no. AS 1584, Sunkowsky 1956, fig. 31a.
[2] Foy 1977-80, 212, fig. 76, no. 4. On the date of the sarcophagus: Danièle Foy, "Verreries paléochrétiennes découvertes récemment à Marseille," *RANarb* 10 (1977) 274, n. 1, and *Gallia* 32 (1974) 517, fig. 23.

### 155 Double Head-shaped Flask: Two Similar Youthful Faces

Eastern Mediterranean, possibly Syro-Palestinian
Third century
H 8.65 cm; H body (side A) 5.2 cm; H body (side B) 4.9 cm; max D 4.6 cm; D rim 3.3 cm; D base 3.6 cm
1923.460, Gift of Edward Drummond Libbey

Transparent natural yellow green; exact color cannot be determined because of weathering. Medium thin glass. Fabric cannot be determined because of weathering.
Rim and upper part of neck free blown. Lower part of neck and body blown into a two-part mold of two vertical sections, probably open at the base (MCT IX). Mold seams concealed in hair

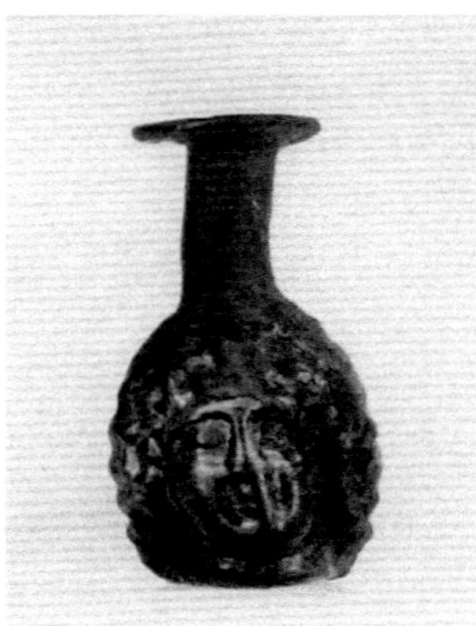
154, side A (1:2)

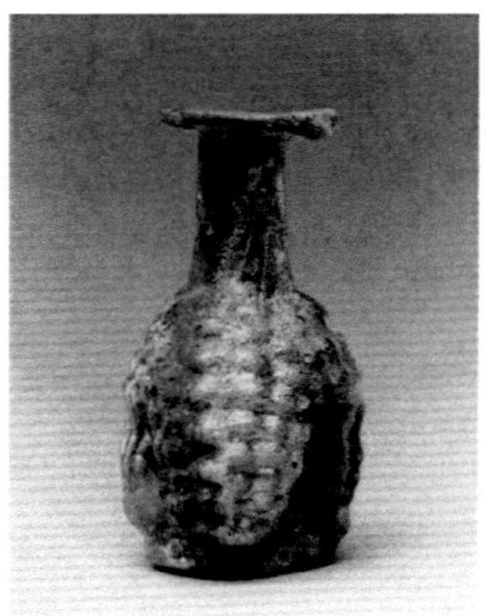
154, profile A:B (1:2)

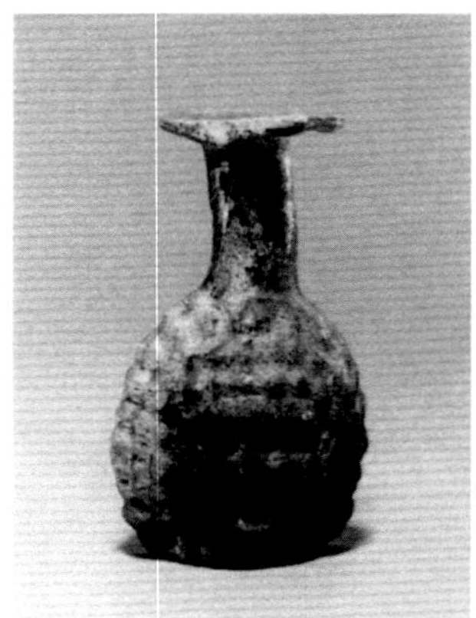
154, side B (1:2)

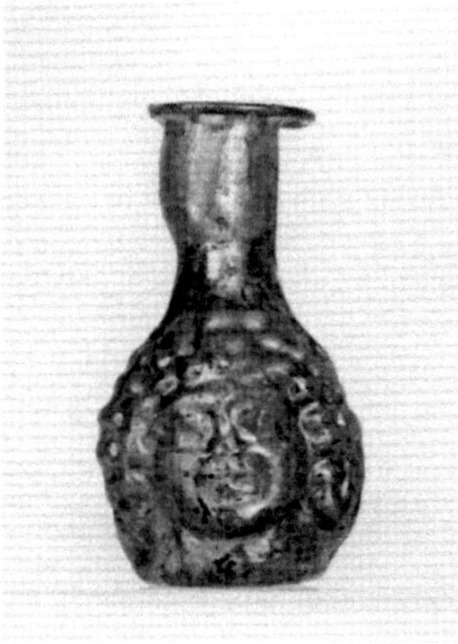

155, side A (1:2)

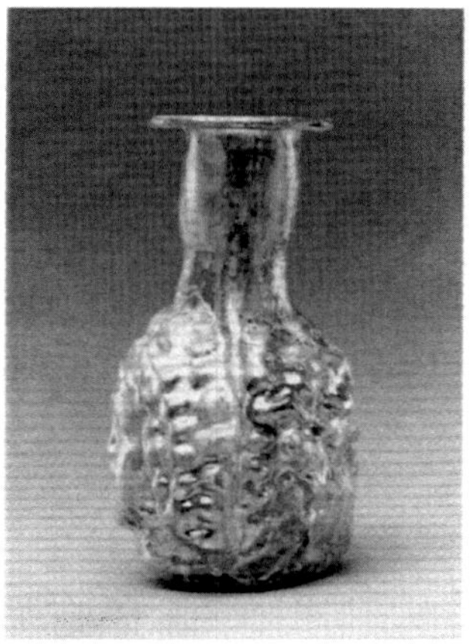

155, profile A:B (1:2)

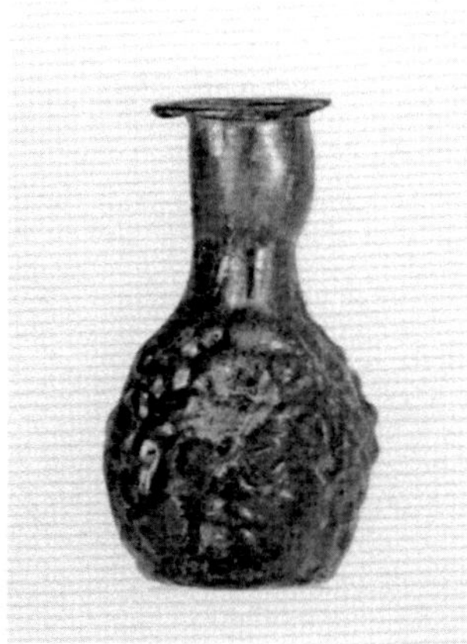

155, side B (1:2)

at junction of heads. No mold seam on underside of base. Relief moderately crisp. No pontil mark.

Horizontal rim, folded outward, upward, inward, and flattened. Tubular neck, with bulge on one side above constriction marking top of mold (about two thirds of the way down neck). Body in the shape of a double head. Nearly circular flat base.

Body in the shape of two heads back to back with similar serious features. Face A has a prominent browline, large round bulging eyes, narrow nose, and slightly parted straight lips. Face B appears squashed: it has heavy planar features consisting of a prominent browline, protruding eyes, wide flat nose, massive cheeks, and wide chin with hardly any tuck under chin to throat. The hair around both faces is rendered predominantly as circular knobs, with some ridges of connected knobs above the forehead, framing the faces to below chin level.

*Condition* Approximately one third of rim missing. Long weathering cracks on exterior of side A. Exterior has heavy purple and green iridescence and severe surface pitting. Interior has thick blackish brown weathering layer and severe pitting on interior.

*Collection* Thomas E. H. Curtis.

*Published* Previously unpublished.

*Remarks* Face A is distorted by a crack in the glass, extending from the base up to the nose, which accentuates a failure to fill the mold evenly, so that the photograph gives the erroneous impression of a flash resulting from a crack in the mold.

Face A is related to Cat. Nos. 154 and 156. Face B is related to a head flask in the Oppenländer Collection. [1]

[1] Oppenländer Collection, no. 22245, von Saldern et al. 1974, 170, no. 465.

### 156 Double Head-shaped Flask: One Smiling Face, One Serious Face

Eastern Mediterranean, possibly Syro-Palestinian
Third century
H 8.0 cm; H body 4.9 cm; max D 4.8 cm; D rim 4.0 cm; D base 3.7 cm (nose to nose)-3.5 cm (seam to seam)
1923.514, Gift of Edward Drummond Libbey

Transparent dusky yellow green (near 5 GY 5/2). Medium thin glass. Numerous vertically elongated bubbles in neck; some oval bubbles in body; one large bubble in base. Black specks. Blowing spirals.

Rim and upper part of neck free blown. Lower part of neck and body blown into a two-part mold of two vertical sections, probably open at the base (MCT IX). Mold seams concealed in hair at junction of heads. No mold seam on underside of base. Relief relatively indistinct; lower part of face B deformed during blowing. No pontil mark.

Flaring hollow rim, folded outward, upward, and inward. Tubular neck tapering downward, with constriction marking top of mold below middle of neck. Body in the shape of a double head. Nearly square flat base.

Body in the shape of two heads back to back: one smiling face, A, and one serious face, B. Face A has rounded, childlike features: prominent browline, wide ring-shaped eyes, fat cheeks, and dimples at the corners of the mouth. The hair around the face is rendered as three tiers of small regular knobs framing the face to below chin level. Face B is elongated, with an indistinct droll expression, large protruding eyes, and a long triangular nose. Mouth and chin are indistinct. The hair around the face is rendered as irregular ridges and knobs framing the face to below chin level.

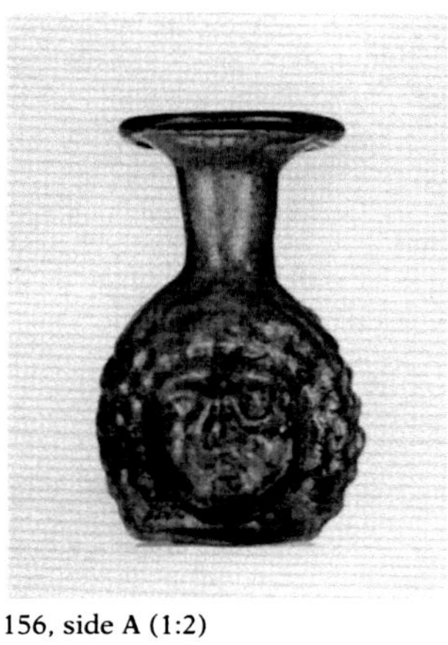

156, side A (1:2)

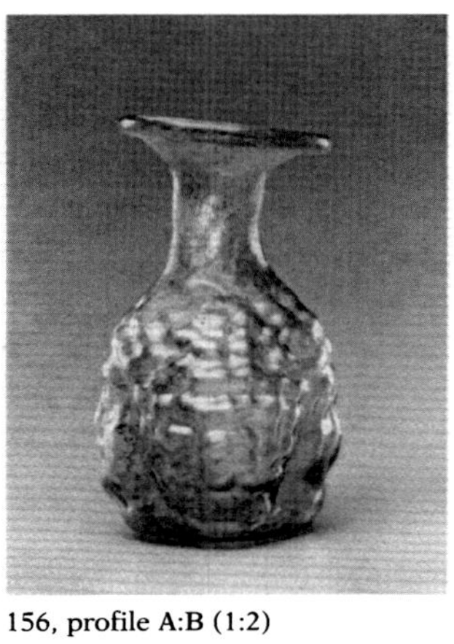

156, profile A:B (1:2)

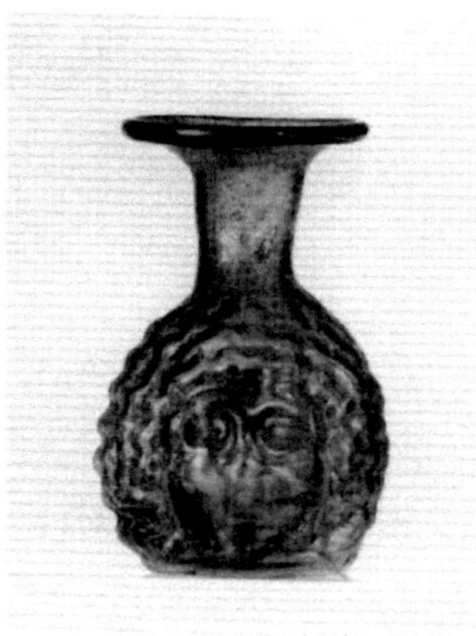

156, side B (1:2)

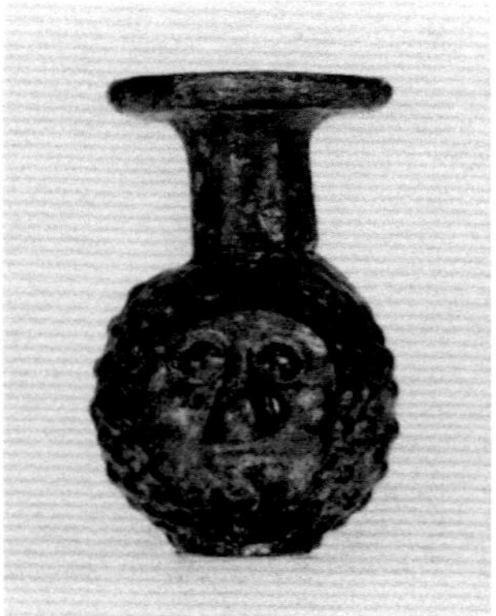

157, side A (1:2)

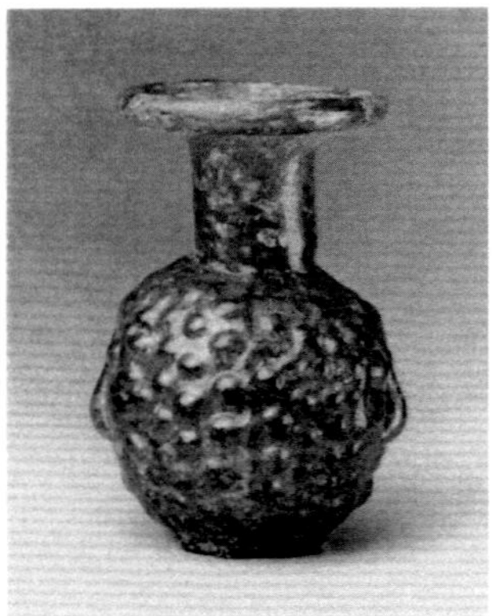

157, profile B:A (1:2)

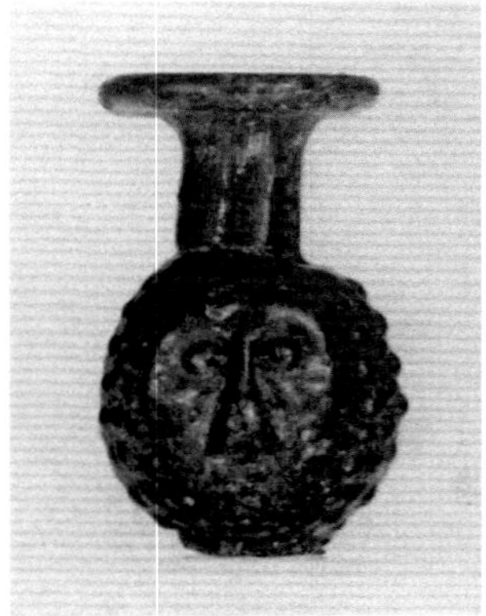

157, side B (1:2)

*Condition* Intact. Exterior covered with iridescent film, pitting, and areas of brown discoloration. Interior pitted, with areas of brown weathering.
*Collection* Thomas E. H. Curtis.
*Published* Previously unpublished.
*Remarks* Face A is related to Cat. Nos. 154 and 155.

### 157 **Double Head-shaped Flask with Internal Diaphragm (Sprinkler): Two Similar Stylized Droll Faces**

Color plate 26, p. 62
Eastern Mediterranean, probably Syro-Palestinian
Probably early third century
H 8.3 cm; H body (side A) 5.0 cm; H body (side B) 5.2 cm; max D 5.1 cm; D rim 4.8 cm; D base 2.7 cm (nose to nose)-2.2 cm (seam to seam)
1923.1224, Gift of Edward Drummond Libbey

Transparent natural pale green (10 G 6/2). Medium thick glass. Numerous small spherical and vertically elongated bubbles in body and neck. Black streaks. Stone.
Rim and neck free blown. Body blown into a two-part mold of two vertical sections, each including one half of the base (MCT VIII). One continuous mold seam concealed in hair at junction of heads, extends across underside of base. Tooled diaphragm. Relief crisp. No pontil mark.
Thick outsplayed hollow rim, folded outward, upward, and inward. Tubular neck, with constriction at its base where an interior cut-out forms a diaphragm with an aperture of 0.8 cm. Spherical body in the shape of a double head. Oval flat base.
Body in the shape of two heads back to back with similar mask-like features, giving each face a droll expression: flat oval face, protruding knobs for eyes and chin, long large triangular nose, and small straight mouth with lips slightly parted. The hair around both faces is schematic, rendered as distinct, evenly spaced small knobs arranged in four interlocking tiers framing the face, except on the left side of face B where there are only three interlocking rows.
*Condition* Intact.
*Collection* Thomas E. H. Curtis.
*Published* Previously unpublished.
*Remarks* Very similar to a sprinkler formerly in the Kofler-Truniger Collection.[1] The use of an MCT VIII mold suggests a date in the second or early third century. Stylized knobby hair was fashionable in the third century; the small, widely spaced knobs of this flask are very similar to those of Bulbous Bottles with Stylized Grape Pattern, such as Cat. No. 128.
An inner diaphragm at the base of the neck is common in glass made in Mesopotamia and the Syro-Palestinian area,[2] but not in Italy or the West. The droll faces of this head flask may represent comic actors in theater performances.[3]

[1] Formerly Kofler-Truniger Collection, translucent yellowish green, H 8.4 cm, *Kofler-Truniger* 1985, 58, no. 89; Luzern 1981, 83, no. 287. See also Raleigh, N.C., North Carolina Museum of Art, no. G.50.8.42, unpublished, Janet D. Jones, pers. com.

[2] Stern 1977a, 95-100.

[3] J. P. Cèbe, *La caricature et la parodie dans le monde romain antique* (Paris 1966) 37-57 (with lit.).

## *Double Head-shaped Flasks, Late Third to Fourth Century* (Cat. Nos. 158-160)

### 158 **Double Head-shaped Flask: Two Similar Medusas with Wavy Hair**

Color plate 27, p. 62
Probably Near Eastern
Probably late third or fourth century
H 8.4 cm; H body 4.65 cm; max D 4.7 cm; D rim 3.2 cm; D base 3.65 cm (nose to nose)-3.4 cm (seam to seam)
1923.516, Gift of Edward Drummond Libbey

Transparent pale blue (near 5 PB 5/2), with darker grayish blue areas (near 5 PB 5/2). Medium thin glass.
Rim and neck free blown. Body blown into a two-part mold of two vertical sections, probably open at the base (MCT IX). Mold seams concealed in hair at junction of heads. No mold seam on underside of base. Relief moderately crisp. No pontil mark.
Flaring hollow rim, folded outward, upward, and inward. Tubular neck tapering downward, with constriction near its base marking top of mold. Body in the shape of a double head. Slightly oval flat base.
Body in the shape of two heads of Medusa back to back with similar features: arched browline, narrow nose, knobby cheeks, thin lips, broad round chin, and an indistinct roll under the jawline representing the knotted snakes of Medusa. The hair around both faces is carefully arranged in irregular locks rendered as wavy lines, generally in a vertical orientation, framing the faces to below chin level. Two locks are arranged like pincers over the center of each forehead. Along the right cheek of both faces a few locks curve down to the cheek at an angle. Face A is narrow and more square, with stronger differentiation between chin and neck than face B.
*Condition* Body cracked throughout. Exterior scratched and covered with silvery iridescence and beige weathering. Interior has areas of beige weathering.
*Collection* Thomas E. H. Curtis.
*Published* Previously unpublished.

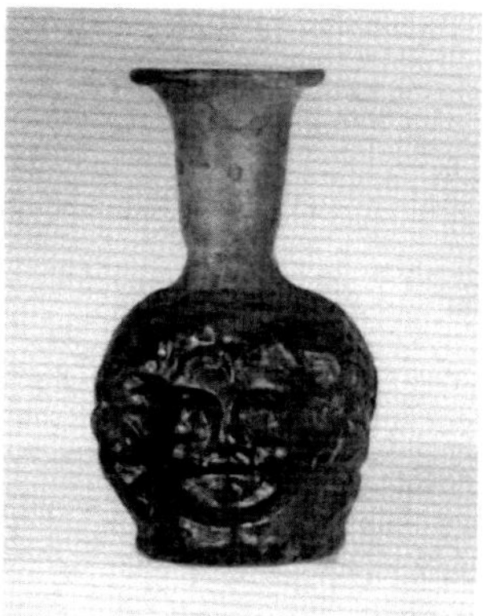
158, side A (1:2)

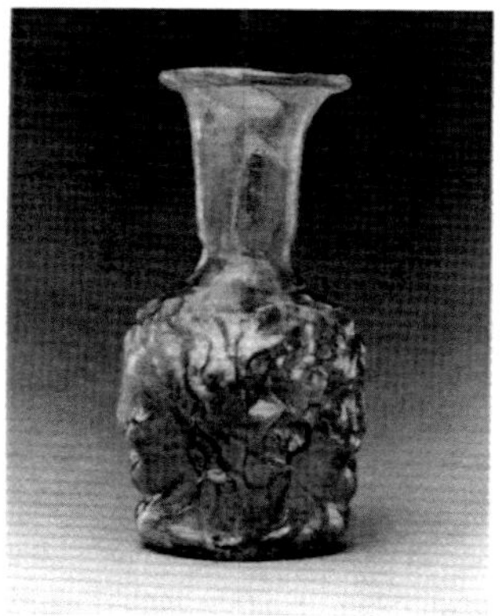
158, profile B:A (1:2)

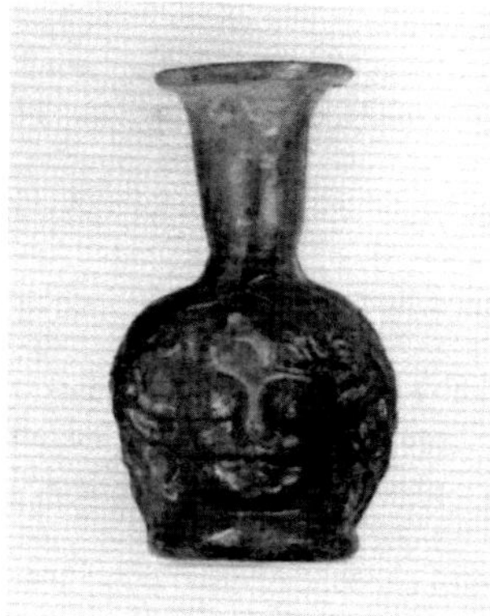
158, side B (1:2)

*Remarks* Other flasks belonging to the same Type [1] but with more distinct features reveal that the indistinct roll under the jaw was originally a pair of knotted snakes, indicating that the faces represent Medusa. [2] Recorded find spots point to a Near Eastern origin. [3] A second-century date has been proposed for this Type because it does not have the stylized knobby hair characteristic of many third-century head flasks, [4] but the only parallel excavated in a datable context is a fragmentary vessel from Pella of the Decapolis, Jordan (see Fig. 100). [5] It was found in a tomb that appears to have been used from the late third to the early fifth centuries, judging from the rest of the glass found with it (see Chapter III: *Survey: Fourth Century, Type C*). Although the reuse of vessels in antiquity has been documented a number of times, [6] the late date of the Pella tomb must be taken seriously in the absence of any other evidence for dating this Type. The apparent use of MCT IX also suggests a date in the third or fourth centuries.

[1] For example:
a) Yale, no. 1955.6.76, formerly Moore Collection, H 8.0 cm, pale brown, Matheson 1980, 72, no. 188;
b) formerly Constable-Maxwell Collection, *Maxwell* 1979, 72, no. 115, almost colorless;
c) Berlin, Staatliche Museen, Antikensammlung, no. 30219,133, formerly von Gans Collection, natural green, Platz-Horster 1976, 48, no. 78;
d) Oppenländer Collection, no. 22303, light green, von Saldern et al. 1974, 170, no. 464.

[2] On Medusa: see Remarks Cat. No. 142.

[3] Examples belonging to the same Type:
a) formerly Paris, de Clercq Collection, from Tyre, bluish, de Ridder 1909, 160, no. 285, pl. XII;
b) Nicosia, Cyprus Mus., no. D 1686, honey colored, Vessberg 1952, 136, pl. VIII, no. 10;
c) Stockholm, Nat. Mus., Cyprus Coll. no. 811, acquired in Cairo, amethyst, Vessberg 1952, 136, pl. VIII, no. 11, pl. XVI, no. 3.

[4] Matheson 1980, 72, at no. 188.

[5] From tomb 39A, H (pres.) 5.1 cm, McNicoll and Hennessey 1982, 536, pl. CXVI, no. 11.

[6] For the reuse of a third-century glass double head-shaped vessel in a fifth-century sarcophagus at St. Victor, Marseilles: Foy 1977-80, 212, fig. 76, no. 4. On the date of the sarcophagus: *Gallia* 32 (1974) 517, fig. 23.

### 159 Double Head-shaped Flask: Two Similar Male Heads, Probably Hercules Wearing a Lion's-Head Helmet

Place of manufacture undetermined
Third or fourth century
H 9.9 cm; H body 6.9 cm; max D 5.4 cm; D rim ca. 5 cm; D base 4.2 cm (nose to nose)-3.3 cm (seam to seam)
1923.741, Gift of Edward Drummond Libbey

Transparent manganese colored pale red purple (near 5 RP 6/2). Thick glass. Fabric cannot be determined because of weathering. Rim and upper portion of neck free blown. Body and perhaps lower part of neck blown into a two-part mold of two vertical sections, probably open at the base (MCT IX). Mold seams through helmets at junction of heads. No mold seam on underside of base. Relief indistinct. No pontil mark.

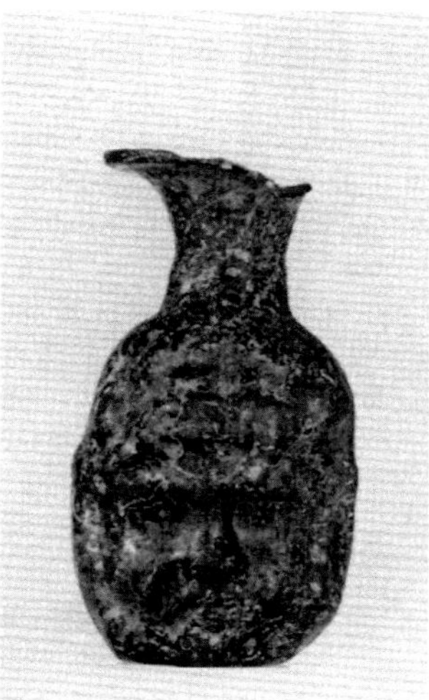
159, side A (1:2)

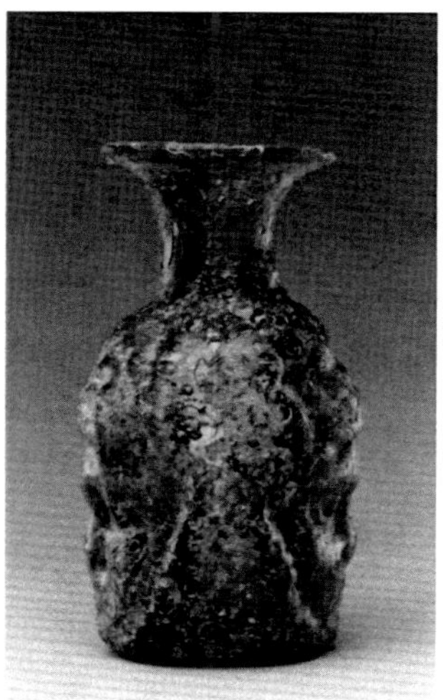
159, profile B:A (1:2)

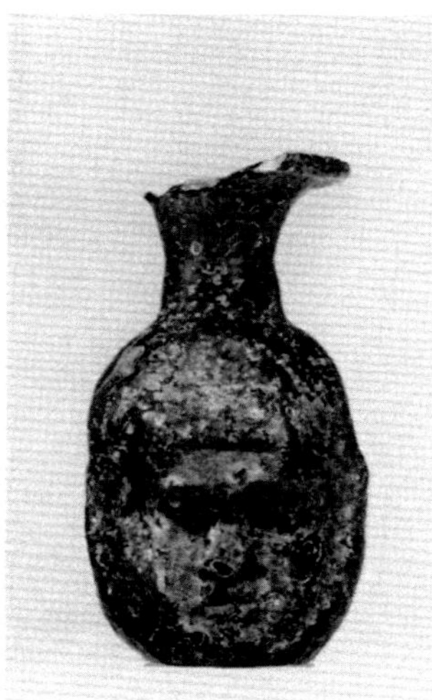
159, side B (1:2)

Flaring rim, irregularly folded outward, upward, and inward. Tubular neck, with constriction at middle probably marking top of mold. Body in the shape of a double head. Oval concave base.
Body in the shape of two male heads back to back with similar, although indistinct, features: rectangular face, deep sunken eyes, and broad flat nose. Close-fitting bands with small circular depressions above the forehead suggest stylized lion's teeth; large circular bulges at either end of the row of teeth suggest lion's fangs. Irregular oval bulges above the band of teeth may represent lion's nostrils and cheeks; they are more pronounced above face A. Jagged lines framing the lower part of the face may represent a curly beard or a lion's lower jaw.
*Condition* Two thirds of rim and mouth missing. Exterior severely pitted with ice floes and areas of green and golden iridescence. Interior has areas of brown weathering.
*Collection* Thomas E. H. Curtis.
*Published* Previously unpublished.
*Remarks* The closest parallel for this Type is a double head flask in a Swiss private collection,[1] where the beard, or lion's lower jaw, has been identified as a triangular acanthus leaf with sawed edges, even though the photograph shows the lion's head helmet more clearly than the Toledo Museum flask. If the headgear of this Type has correctly been interpreted here as a lion's-head helmet,[2] this Type represents the Greek hero Herakles, known to the Romans as Hercules. The lion's skin was a standard attribute of Herakles, a trophy of his first labor, the killing of the Nemean lion. Herakles is often shown wearing it around his shoulders with the lion's head on his head as a helmet.[3] A third- or fourth-century date is suggested for this flask because it appears to have been blown in an MCT IX mold. An early third-century date is possible in view of the general resemblance of this Type to an over life-size, granite portrait of the emperor Caracalla (ruled 211-217) from Egypt, now in Philadelphia.[4]

[1] Switzerland, private collection, translucent purple, H 10.9 cm, Luzern 1981, 83, no. 284.

[2] Identification suggested by Timothy A. Motz during drawing of the vessel.

[3] Compare, for example, *LIMC* IV, 742-743, nos. 211-243.

[4] Philadelphia, University Mus., University of Pennsylvania, no. E976, *Romans and Barbarians*, exh. cat., Museum of Fine Arts, Boston (1977) 25, no. 31, ill.

160 **Double Head-shaped Flask: Two Similar Grotesque Heads**

Color plate 27, p. 62
Probably Eastern Mediterranean
Probably late third or fourth century
H 13.6 cm; H body (without overblow) 7.6 cm; max D 6.5 cm; D rim 3.4 cm; D base 3.5 cm
1923.428, Gift of Edward Drummond Libbey

Transparent streaked manganese colored. Medium thin glass. Blowing spirals. Fabric cannot be determined because of weathering. Rim and neck free blown. Body blown into a two-part mold of two vertical sections, probably open at the base (MCT IX). Mold seams through leaf at junction of heads. No mold seam on underside of base. Relief moderately crisp. Pontil scar ca. 1.2 cm in diameter.
Flaring rim, rounded and irregularly thickened in flame, with a tool mark on the interior. Tubular neck, bulging above constriction at its base. Projecting roll formed by overblow marks shoulder. Ovoid body in the shape of a double head. Circular flat base.
Body in the shape of two heads back to back with similar features: knitted brow, eyebrows arching over bulging knobby eyes, short nose with carefully rendered nostrils, large frowning mouth with a long downturned upper lip and a short straight lower lip, small protruding chin with a cleft, and large puffed cheeks. The face is framed by a horizontal ridge above the forehead with a row of about eight small, circular knobs above it and by a double roll below the chin with a central circular element that may represent the knotted snakes of Medusa. There is no hair. A large acanthus leaf disguises the mold seam on either side; there are five lobes on each side of one leaf and four lobes on each side of the other leaf.
*Condition* Face B broken and restored, but nose and fragments from right cheek missing. Leaf to right of face B restored. Cleaned. Original surface not preserved. Exterior, especially of body, severely pitted and made thin by weathering with areas of iridescent film and beige weathering. Interior pitted with areas of beige and dark brown weathering.
*Collections* It is not clear whether a head flask formerly in the Sambon Collection is the same as the Toledo flask;[1] Thomas E. H. Curtis.
*Published* Grose 1978, 78, 82, fig. 19.

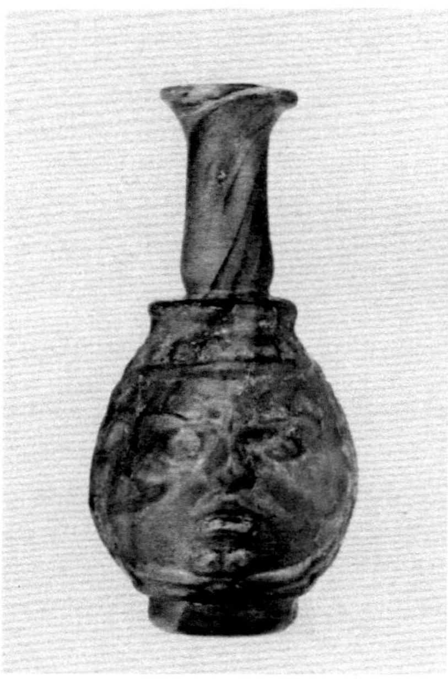
160, side A (1:2)

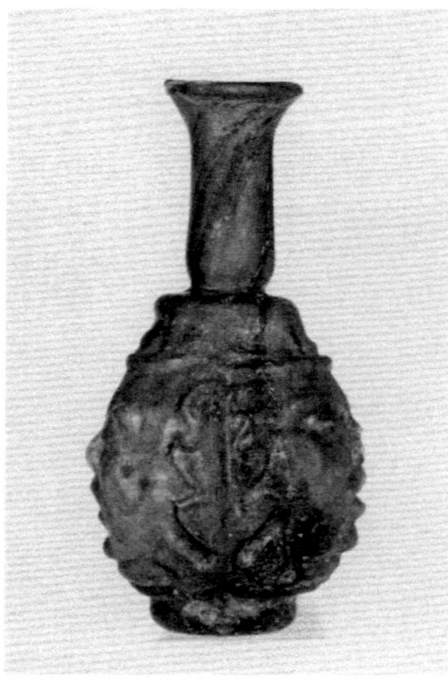
160, profile A:B (1:2)

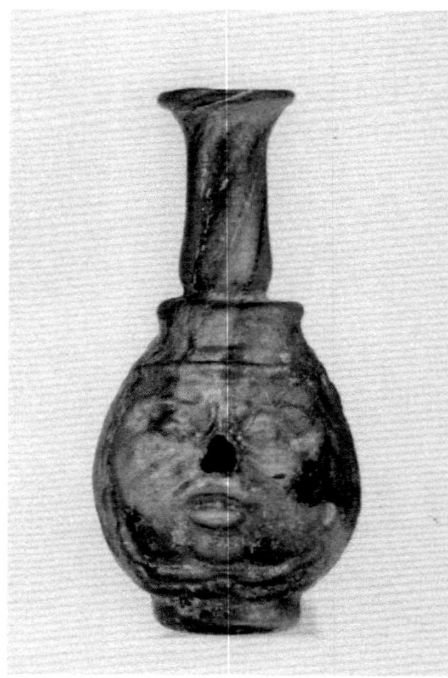
160, side B (1:2)

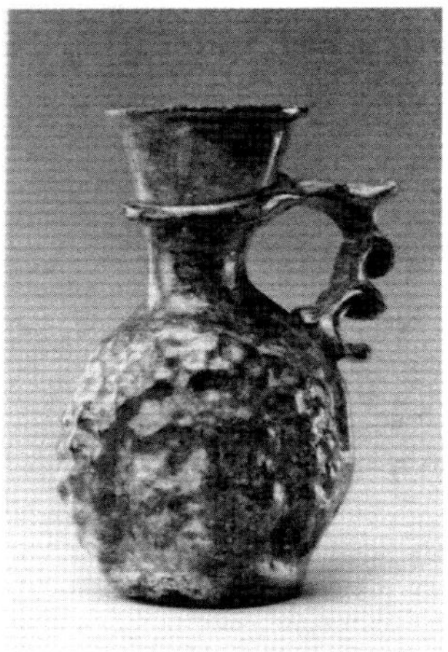

161, side A (1:2)

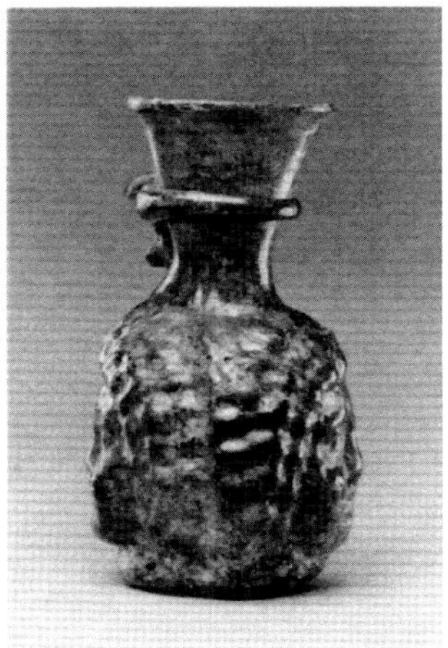

161, profile B:A (1:2)

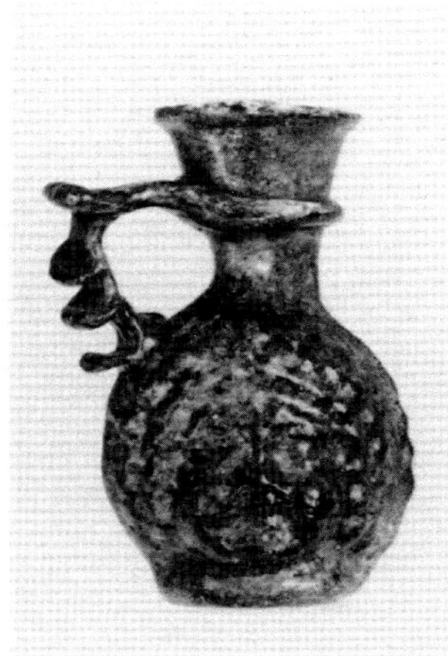

161, side B (1:2)

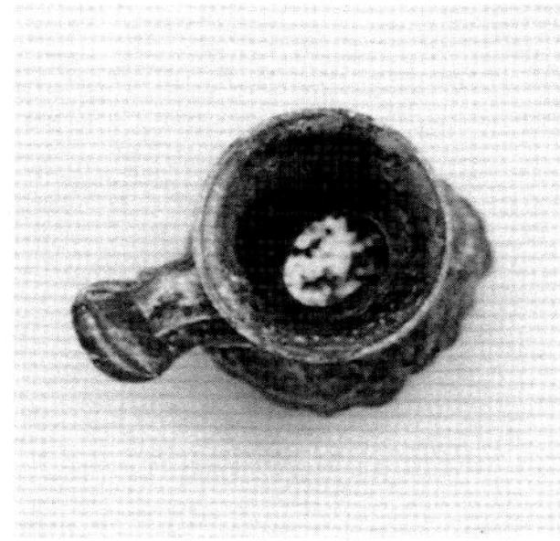

161, mouth and neck (1:2)

*Remarks* The projecting roll from which the tall neck of this flask rises was common in third-century mold-blown bottles made in the Syro-Palestinian area (see Remarks Cat. No. 147). Although grotesque images seem to have originated in Egypt and Asia Minor,[2] glass head flasks showing grotesque faces were much less common in the eastern Mediterranean than in western Europe. This flask was probably an eastern Mediterranean counterpart[3] of grotesque head flasks made in the West.[4]

[1] Sambon 1928, 30, 33, fig. 61.

[2] On the origin of the grotesque type and its possible connection with Roman mime masks: J. P. Cèbe, *La caricature et la parodie dans le monde romain antique* (Paris 1966) 354-355 (with lit.).

[3] Same Type:

a) formerly in the Sambon Collection, perhaps the same as this TMA vessel, see above note 1;
b) Newark, no. 50.1433, transparent yellow green, Auth 1976, 75, no. 77;
c) ROM, no. 950.157.83, green, Hayes 1975, 49, no. 92, pl. 7.

[4] For example, RGM, nos. 56.1, 761, N296, Fremersdorf VI 1961, 76-77, pls. 166-167, 169 (with parallels). One of the factories may have been in Cologne.

## *Double Head-shaped Jugs with Knobby Hair and One Handle, Fourth Century* (Cat. Nos. 161-164)

### 161 Double Head-shaped Jug with Crimped Handle: Two Childlike Faces, One Serious, One Smiling

Eastern Mediterranean, possibly Syro-Palestinian
Fourth century
H 9.5 cm; H body 5.7 cm; max D 5.8 cm; D rim 3.8 cm; D base 3.7 cm
1923.742, Gift of Edward Drummond Libbey

Translucent to transparent manganese colored grayish red purple (near 5 RP 4/2, but more red). Similarly colored neck coil and handle. Medium thin glass. Fabric cannot be determined because of weathering.

Rim and neck free blown. Body blown into a two-part mold of two vertical sections, probably open at the base (MCT IX). Mold seams concealed in hair at junction of heads. Edges of mold not carefully aligned. No mold seam on underside of base. Relief moderately crisp. No pontil mark. Neck squashed at base (see detail photograph) where it was held with the left hand between the round legs of a clamping tool (tongs?).

Unworked everted rim, with slight bulge below. Funnel neck. Body in the shape of a double head. Nearly circular, irregularly flattened base. Ribbon handle with four crimps, originating from neck coil and attached to shoulder.

Neck coil around middle of neck from right to left. Body in the shape of two heads back to back: one serious face, A, and one smiling face, B. Face A has arched browlines, well-proportioned nose and mouth, and chubby childlike cheeks. The chin-length hair is rendered as distinct, evenly spaced knobs arranged in three (in some places four) tiers. Face B has wide round eyes, turned-up nose with recessed nostrils, and a smile which is reflected in the bulging cheek muscles. The chin-length hair is rendered as evenly spaced knobs arranged in three tiers.

*Condition* Intact. Exterior covered with dark brown weathering. Interior has ice floe texture and rainbow iridescence.

*Collection* Thomas E. H. Curtis.

*Published* Previously unpublished.

*Remarks* Same Type as Cat. Nos. 162 and 163,[1] but larger.[2] Although no examples of this Type have been excavated in datable contexts, technical and stylistic features date it firmly to the fourth century: the funnel-shaped neck with its unworked rim and bulging edge may be compared to a fourth-century free-blown glass jug in the Toledo Museum collection.[3] Handles with pinched protrusions can also be found on fourth-century vessels.[4] The two youthful faces are probably derived from third-century head flasks such as Cat. No. 154.

Judging from Cat. Nos. 161-163, numerous tiny circular indentations on both faces may be a technical feature characteristic of this Type. The indentations outline eyes, mouth, and the first row of knobs representing hair (and in one case also between the knobs of the hair, see Cat. No. 162). They could have been caused by dirt or dust in the mold or by the use of a pointed instrument on the model before making a mold from it.

[1] Same Type:

a) Damascus, Nat. Mus. no. 5521, from Djebel Druze, south Syria, Zouhdi 1964, 49, no. 47, fig. 33;
b) Newark, no. 50.1438, Auth 1976, 201, no. 342;
c) Yale, no. 1955.6.75, Matheson 1980, 123-124, no. 332;
d) formerly Sambon Collection, Sambon 1928, 30, fig. 59.

Compare also:

e) formerly A. Rudebeck Collection, exceptionally large head flask, with pinched handle, standing on a wound coil base, Sotheby's, London, auc. cat., November 20, 1987, 40, lot 82;
f) Hanover, Germany, Kestner-Museum, no. 1929, 253, without handle, Liepmann 1982, 43, no. 17;
g) formerly Kofler-Truniger Collection, with handle but without pinched protrusions, *Kofler-Truniger* 1985, 56, no. 85; Luzern 1981, 83, no. 286.

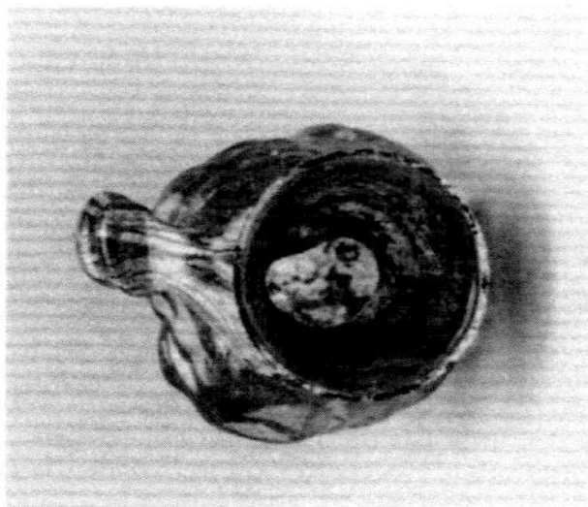
162, mouth and neck (1:2)

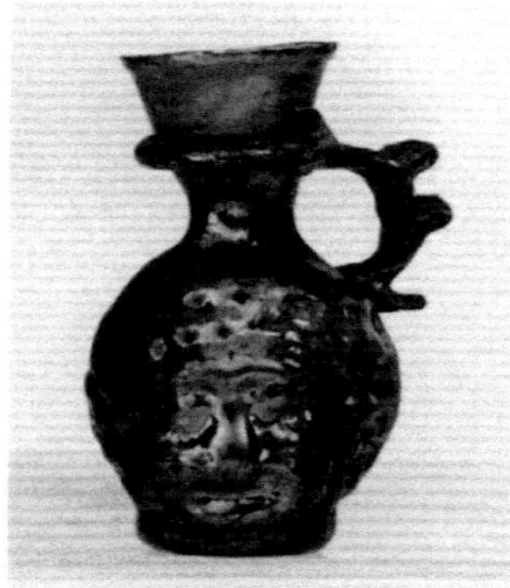
162, side A (1:2)

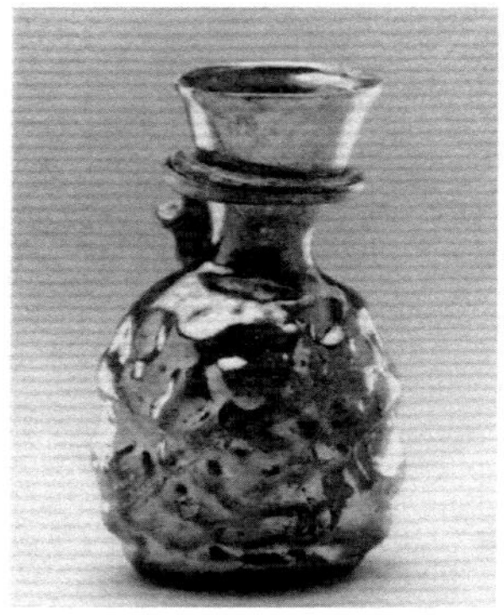
162, profile B:A (1:2)

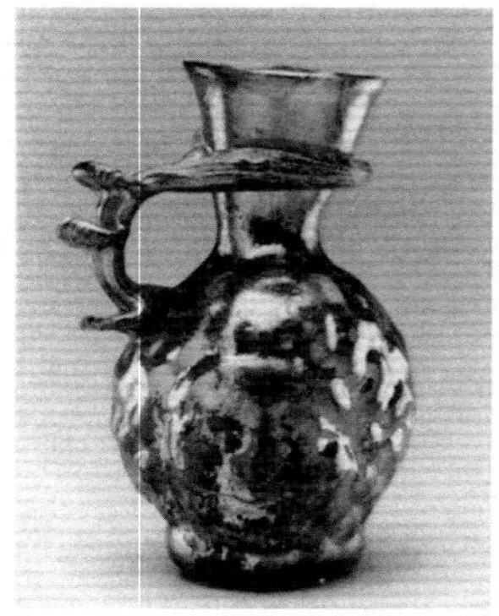
162, side B (1:2)

[2] Carnegie, no. 21342/3, similar large example, translucent purple, Oliver 1980, 119, no. 205.

[3] TMA 1923.736, TMA, *Roman and Early Byzantine Glass* (forthcoming).

[4] Compare:

a) formerly Kofler-Truniger Collection, jug with a pinched handle and a wound coil base of fourth-century type, *Kofler-Truniger* 1985, 54-55, no. 84;

b) CMG, no. 59.1.152, formerly R. W. Smith Collection, no. 1328, Smith 1957, 137, no. 267.

## 162 Double Head-shaped Jug with Crimped Handle: Two Childlike Faces, One Serious, One Smiling

Color plate 27, p. 62
Eastern Mediterranean, possibly Syro-Palestinian
Fourth century
H 8.4 cm; H body 5.1 cm; max D 4.9 cm; D rim 3.4 cm; D base 3.9 cm (nose to nose)-3.2 cm (seam to seam)
1923.520, Gift of Edward Drummond Libbey

Transparent pale blue; exact color cannot be determined because of weathering. Similarly colored neck coil and handle. Medium thin glass. Fabric cannot be determined because of weathering. Rim and neck free blown. Body blown into a two-part mold of two vertical sections, probably open at the base (MCT IX). Mold seams concealed in hair at junction of heads. No mold seam on underside of base. Relief indistinct. No pontil mark. Neck squashed at base (see detail photograph) where it was held with the left hand between the round legs of a clamping tool (tongs?). Unworked everted rim, with slight bulge below. Funnel neck. Body in the shape of a double head. Oval flat base, with sunken area in middle. Ribbon handle with three crimps, originating from neck coil and attached to shoulder.

Neck coil around middle of neck from right to left. Body in the shape of two heads back to back: one serious face, A, and one smiling face, B. Face A has prominent arched browlines, well-proportioned nose, indistinct mouth, fat cheeks, and massive jaw wider than the forehead. The chin-length hair is rendered as irregular ridges formed from connected knobs. Face B has prominent arched browlines, wide eyes, well-proportioned nose with recessed nostrils, indistinct smiling mouth, knob-like chin, and fat cheeks. The chin-length hair is similar to that of face A.

*Condition* Intact. Rim cracked; neck cracked near rim coil. Exterior covered with glossy rainbow iridescence, some large pits, and areas of brown discoloration on neck. Interior pitted with areas of beige weathering.

*Collection* Thomas E. H. Curtis.

*Published* Previously unpublished.

*Remarks* Same Type as Cat. Nos. 161 and 163, but blown in a mold that was smaller than that used for either. On the tiny circular indentations on cheeks and around eyes, mouth, and edge of face, as well as between the knobs forming the hair of this jug, see Remarks Cat. No. 161.

## 163 Double Head-shaped Jug with Crimped Handle: Two Childlike Faces, One Serious, One Smiling

Eastern Mediterranean, possibly Syro-Palestinian
Fourth century
H 7.0 cm; H body 5.7 cm; max D (pres.) 5.1 cm; D rim 3.4 cm; D base 3.9 cm (nose to nose)-3.6 cm (seam to seam)
1923.522, Gift of Edward Drummond Libbey

Transparent natural decolorized glass. Similarly colored neck coil and handle. Very thin glass. Blowing spirals. Fabric cannot be determined because of weathering.

Rim and neck free blown. Body blown into a two-part mold of two vertical sections, probably open at the base (MCT IX). Mold seams concealed in hair at junction of heads. No mold seam on underside of base. Relief moderately indistinct. No pontil mark. No deformation at base of neck.

Unworked everted rim, with slight bulge below. Funnel neck. Body in the shape of a double head. Nearly circular, irregularly

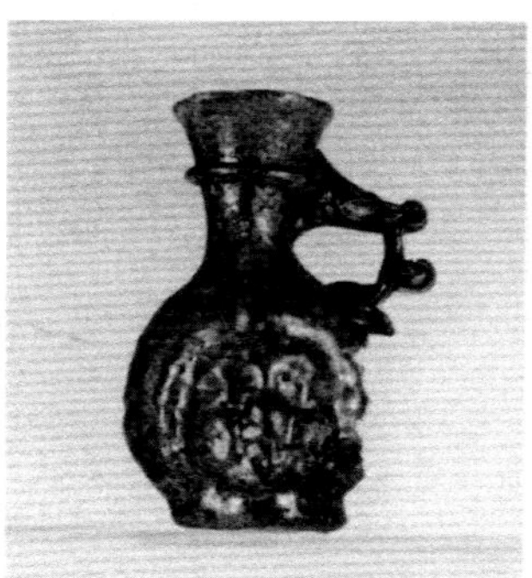
163, side A (1:2)

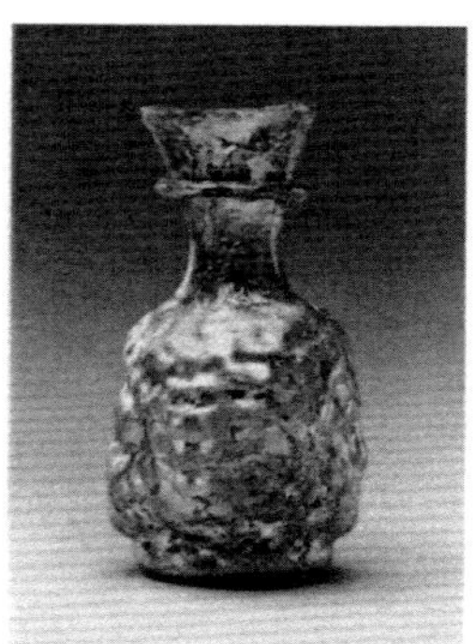
163, profile B:A (1:2)

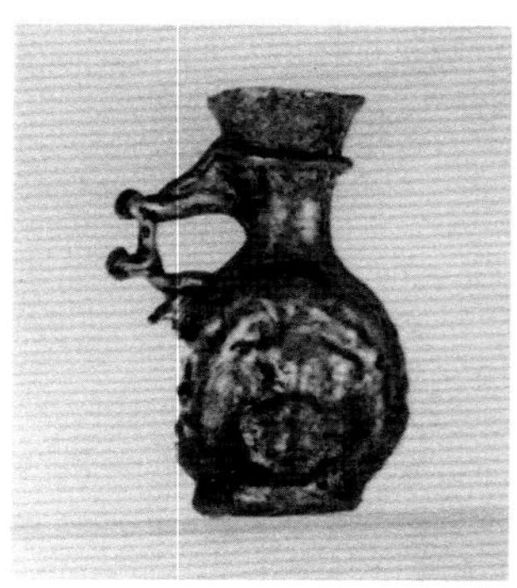
163, side B (1:2)

flattened base. Ribbon handle with three crimps, originating from neck coil and attached to shoulder.
Neck coil around middle of neck from right to left. Body in the shape of two heads back to back: one serious face, A, and one smiling face, B. Face A has prominent arched browlines, well-proportioned nose, indistinct mouth, fat cheeks, and massive jaw wider than the forehead. The chin-length hair is rendered as irregular ridges formed from connected knobs. Face B has prominent arched browlines, wide eyes, well-proportioned nose with recessed nostrils, indistinct smiling mouth, knob-like chin, and fat cheeks. The chin-length hair is similar to that of side A.
*Condition* Large hole in body below handle and small hole in left cheek of face A. Exterior covered with rainbow iridescence, severe surface pitting, and ice floes. Interior has areas of dark weathering.
*Collection* Thomas E. H. Curtis.
*Published* Previously unpublished.
*Remarks* Same Type as Cat. Nos. 161 and 162, but blown in a mold that was smaller than that used for Cat. No. 161 and larger than that used for Cat. No. 162. On the tiny circular indentations on cheeks and around eyes, mouth, and edge of face, see Remarks Cat. No. 161.

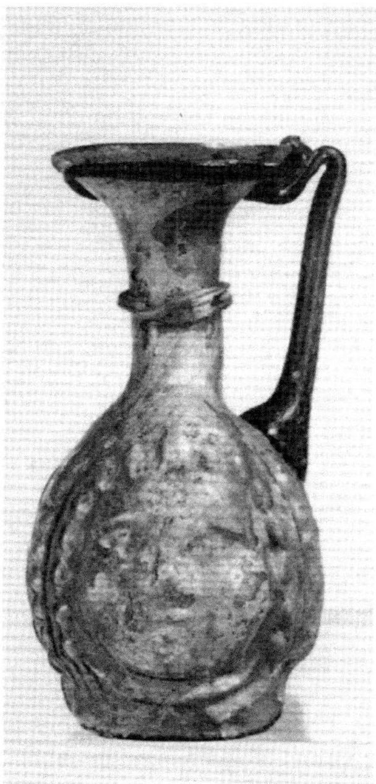
164, side A (1:2)

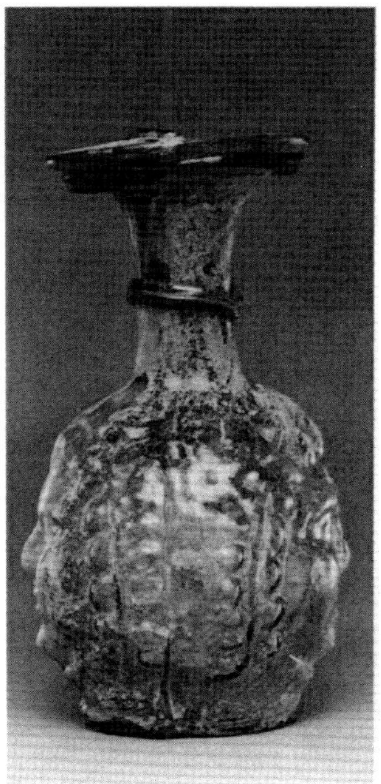
164, profile B:A (1:2)

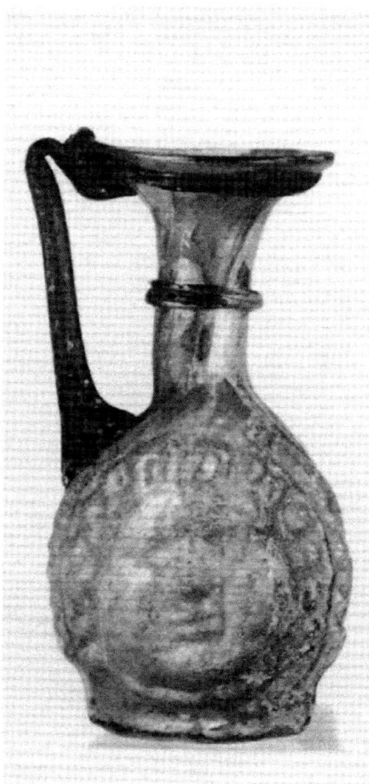
164, side B (1:2)

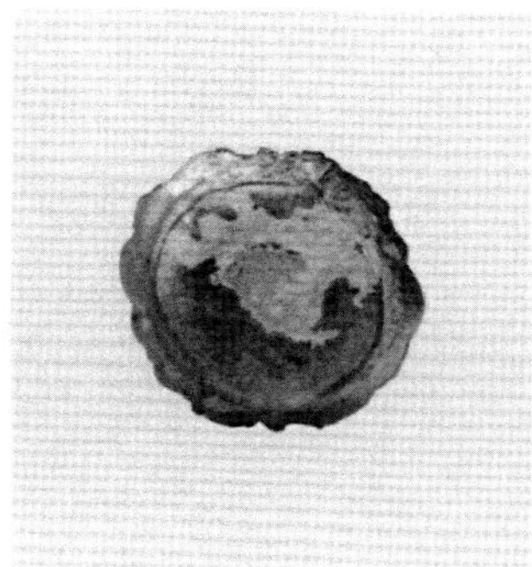
164, underside (1:2)

164 **Double Head-shaped Jug with Coil Handle: Two Chubby Faces, One Smiling, One Serious**

Probably inland Palestinian or Syrian
Fourth century
H 13.2 cm; H body 7.5 cm; max D 6.8 cm; D rim 5.5-5.0 cm; D base 4.7 cm (nose to nose)-4.3 cm (seam to seam)
1923.512, Gift of Edward Drummond Libbey

Transparent light dusky yellow (5 Y 6/4). Translucent grayish green (10 G 4/2) coils and handle. Medium thin glass. Blowing spirals. Fabric cannot be determined because of weathering.
Rim, mouth, and neck free blown. Body blown into a two-part mold of two vertical sections, probably open at the base (MCT IX). Mold seams concealed in hair at junction of heads. Mold edges not carefully aligned. No mold seam on underside of base. Relief moderately crisp. Remains of ring pontil 2.0 cm in diameter (see below Remarks). Excess glass at tip of handle clipped off.
Circular mouth with rim rounded in flame. Tubular neck. Body in the shape of a double head. Oval flat base. Coil handle applied to shoulder and attached to rim where it is folded downward, inward, and upward to form a closed loop.
Rim coil from right to left. Neck coil around middle of neck from right to left. Body in the shape of two heads back to back: one smiling face, A, and one serious face, B. Both have high foreheads, small round eyes, large noses, small mouths with lips slightly parted, fleshy cheeks, and torques(?) below the chins. Face A has a very short lower lip and flattened chin, while face B has lips of equal length, knobby chin, and nose slanting to the right. The hair around both faces is rendered as a row of sixteen distinct, evenly spaced knobs with one knob centered over the forehead, then a continuous ridge, followed by another row of at least twenty-one evenly spaced knobs.
*Condition* Complete. Mouth and neck cracked. Exterior has areas of cloudiness, purple iridescence, and beige weathering. Interior has purple iridescent weathering and numerous surface pits.
*Collection* Thomas E. H. Curtis.
*Published* Previously unpublished.
*Remarks* The poor alignment of the mold halves is clearly visible on the underside of the base where the vertical mold seams end: side A is shifted in relation to side B. The ring pontil mark consists here of a low, narrow ridge of glass that was pulled off the pontil and remained stuck to the underside of the vessel (see detail photograph). The diameters of ring pontil and vessel neck are the same, strongly suggesting that the blowpipe was used as the pontil. This Type is usually finished as a jug with one handle and decorated with a rim or neck coil.[1] With its chubby childlike faces, the Type recalls smaller vessels also finished with one handle such as Cat. No. 161. The addition of rim and neck coils was common in the fourth century,[2] while the color combination, dusky yellow body with grayish green coils and handle, suggests an inland Palestinian or Syrian origin.[3]
For a modern imitation, see Cat. No. 165.

[1] Same Type:

a) formerly Constable-Maxwell Collection, *Maxwell* 1979, no. 111, face A ill.;

b) Berlin, Staatliche Museen, Antikensammlung, no. 30219, 247, Platz-Horster 1976, 46, no. 74, face B ill.

[2] See TMA 1923.975, TMA, *Roman and Early Byzantine Glass* (forthcoming).

[3] Stern 1977a, 156.

## *Later Imitations of Roman Head-shaped Vessels* (Cat. Nos. 165–168)

### 165 **Double Head-shaped Jug with Coil Handle: Two Chubby Faces, One Smiling, One Serious**

Place of manufacture undetermined
Probably nineteenth or early twentieth century
H 14.2 cm; H body 7.5 cm; max D 6.7 cm; D rim 6.1–5.7 cm; D base 4.6 cm (nose to nose)-4.1 cm (seam to seam)
1923.527, Gift of Edward Drummond Libbey

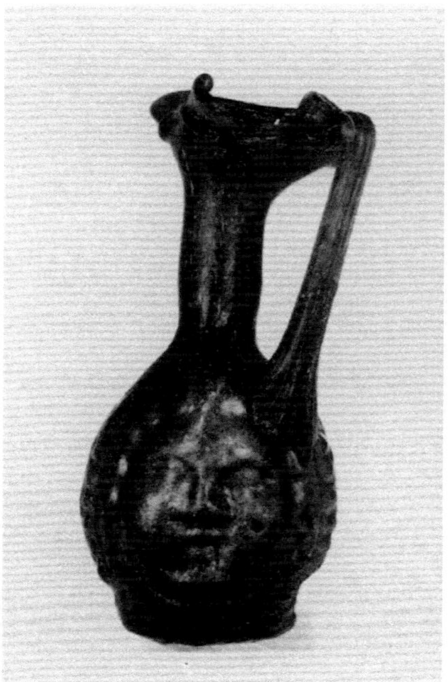

165, side A (1:2)

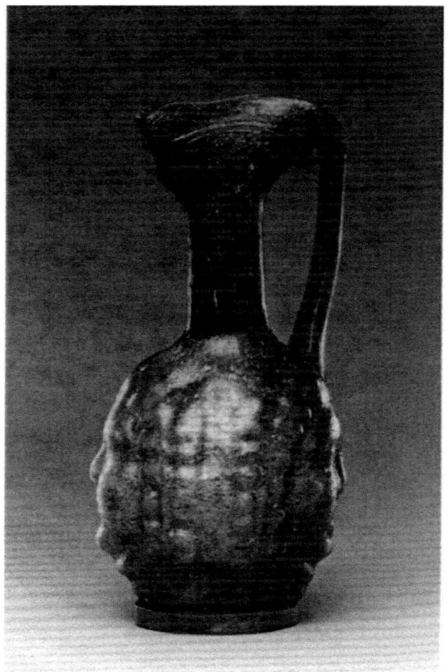

165, profile B:A (1:2)

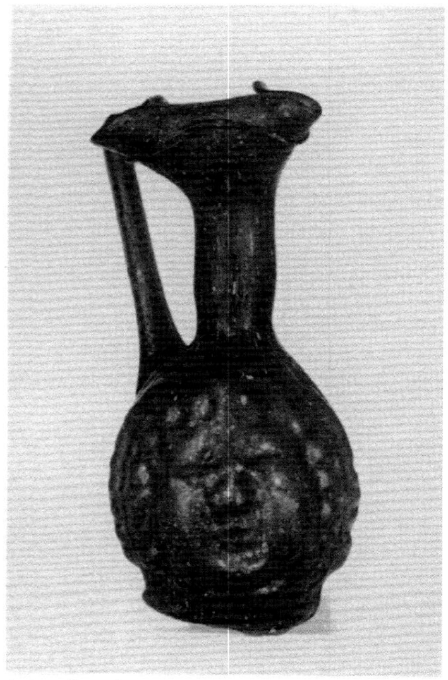

165, side B (1:2)

Translucent grayish blue green (5 BG 5/2). Medium thin glass. Very bubbly glass with black specks; many bubbles burst at surface. Mouth and neck free blown. Body blown into a two-part mold of two vertical sections, probably open at the base (MCT IX). Mold seams concealed in hair at junction of heads. No mold seam on underside of base. Relief indistinct. Ring pontil ca. 1.9 cm in diameter. Excess glass at tip of handle folded back against top of handle.

Spouted mouth (not aligned with mold seam), with rim rounded in flame. Tubular neck, with slight constriction at its base marking top of mold. Body in the shape of a double head. Oval flattened base, slightly convex (vessel cannot stand). Coil handle applied to forehead of face A and attached to rim where it is folded downward, inward, and upward to form a closed loop.

For description, see Cat. No. 164.

*Condition* Intact. Interior of neck has thick dark brown stain.

*Collection* Thomas E. H. Curtis.

*Published* Previously unpublished.

*Remarks* The bright turquoise color and extreme seediness of the glass of this vessel are unknown in ancient Roman glass. The ridge marking the mold seam between the two heads appears to be part of the design rather than a true mold seam. The vessel is perhaps a modern imitation, made in a mold taken from a vessel like Cat. No. 164.

### 166 **Double Head-shaped Flask (Sprinkler): Two Similar Smiling Faces**

Perhaps United States
Nineteenth century
H 9.9 cm; H body 7.1 cm; max D 5.9 cm; D rim 4.9 cm; D base 3.5 cm (nose to nose)-3.1 (seam to seam)
1923.528, Gift of Edward Drummond Libbey

Transparent natural pale green (10 G 6/2). Medium thin glass. Blowing spirals. Numerous medium and small bubbles, vertically elongated in neck, spherical in body. Black specks in neck and rim. Rim and neck free blown. Body blown into a two-part mold of two vertical sections, each including part of the base. One continuous mold seam concealed in hair at junction of heads, extends across underside of base. U-shaped lock in mold seam in center of underside of base (see detail photograph).[1] Relief crisp. No pontil mark.

Horizontal hollow rim, folded outward, upward, inward, and flattened. Short tubular neck, with constriction at its base where an interior cut-out forms a diaphragm with an aperture of 0.8 cm. Body in the shape of a double head. Oval flat base.

Body in the shape of two heads back to back with similar features: large eyes with recessed pupils, short upturned nose, knob-shaped cheeks, lips slightly parted in a smile, and pronounced dimples in the cheeks and chin. The hair around both faces is parted in the middle above the center of the forehead, with a long thick lock (or wing?) over each half of forehead and distinct evenly spaced knobs framing the face to below chin level. Behind the first tier of knobs the hair is rendered as irregular shallow vertical ridges.

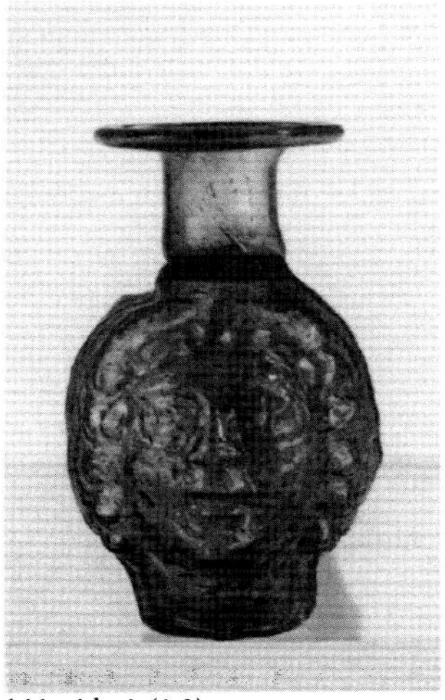
166, side A (1:2)

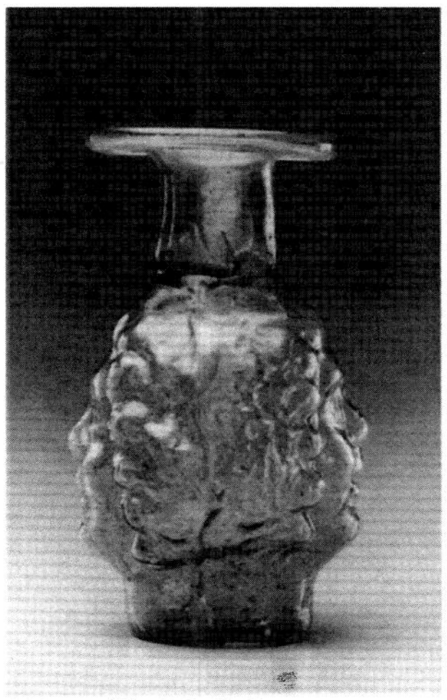
166, profile A:B (1:2)

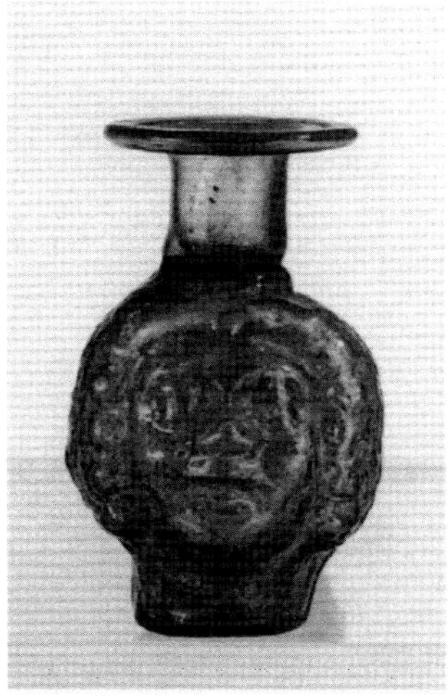
166, side B (1:2)

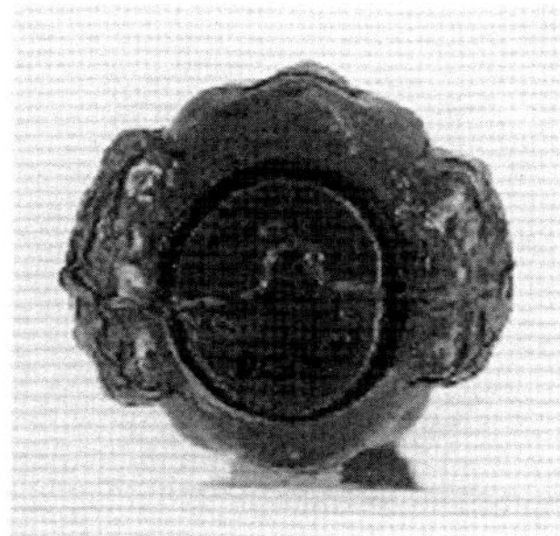
166, underside (1:2)

On face B each long lock seems to be subdivided into two locks; the dimple in the left cheek is missing. The dimple in the chin of face A is deeper and more defined.
*Condition* Intact. On exterior and interior, areas of brown discoloration and incipient pitting.
*Collection* Thomas E. H. Curtis.
*Published* Previously unpublished.
*Remarks* From the same mold as Cat. No. 167.[2] The large round eyes which dominate both faces and the regular classicistic facial features are characteristic of the Constantinian period (306-361), suggesting a date in the first half of the fourth century for the creation of this Type.
However, the U-shaped lock in the center of the mold seam on the underside of the base is not known in ancient glass. Metal molds consisting of two mold halves joined by a U-shaped lock on the base were used for bottles and figured flasks made in the United States between about 1822 and 1875.[3] The head flask sprinklers made in this type of mold may have been made as historicizing vessels rather than as deliberate deceptions.

[1] The underside of the U points toward face A of the vessel.
[2] For related vessels, possibly belonging to the same Type:
a) formerly Kofler-Truniger Collection, natural bluish green, *Kofler-Truniger* 1985, no. 88; Luzern 1981, no. 285;
b) Switzerland, private collection, Neuburg 1962, fig. 45;
c) Sotheby's, London, auc. cat., July 13-14, 1981, 191, no. 427, color ill.

[3] Helen McKearin and Kenneth M. Wilson, *American Bottles and Flasks and Their Ancestry* (New York 1978) 517-518, fig. 36, Type 5. This type of mold occurs both with and without a pontil mark: K. M. Wilson, *The Toledo Museum of Art. American Glass, 1760-1930* (New York 1994) nos. 39, 41, 57-59, 64, 69, 74, 86, 92.

### 167 Double Head-shaped Flask (Sprinkler): Two Similar Smiling Faces

Perhaps United States
Nineteenth century
H 9.7 cm; H body 7.1 cm; max D 6.0 cm; D rim 4.7 cm; D base 3.5 cm (nose to nose)-3.1 cm (along seam)
1923.513, Gift of Edward Drummond Libbey

Transparent natural pale green (10 G 6/2). Medium thin glass. Blowing spirals. Numerous medium and small bubbles, vertically elongated in neck, spherical in body. Black specks in neck and rim. Rim and neck free blown. Body blown into a two-part mold of two vertical sections, each including part of the base. One continuous mold seam concealed in hair at junction of heads, extends across underside of base. U-shaped lock in mold seam in center of underside of base (see detail photograph at Cat. No. 166). Relief crisp. No pontil mark.
Horizontal hollow rim, folded outward, upward, inward, and flattened. Short tubular neck, with constriction at its base where an interior cut-out forms a diaphragm with an aperture of 0.8 cm. Body in the shape of a double head. Oval flat base.

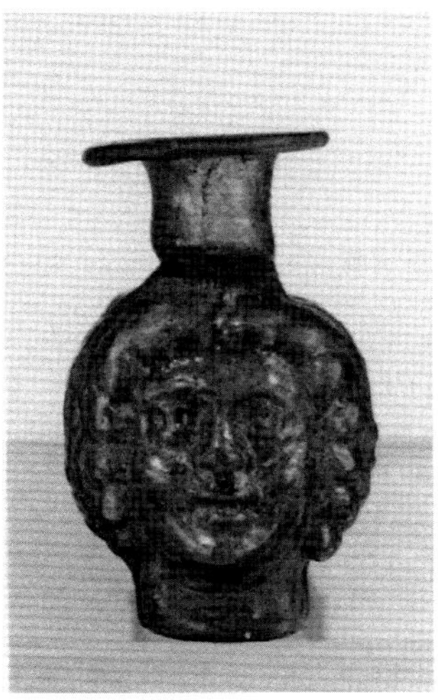
167, side A (1:2)

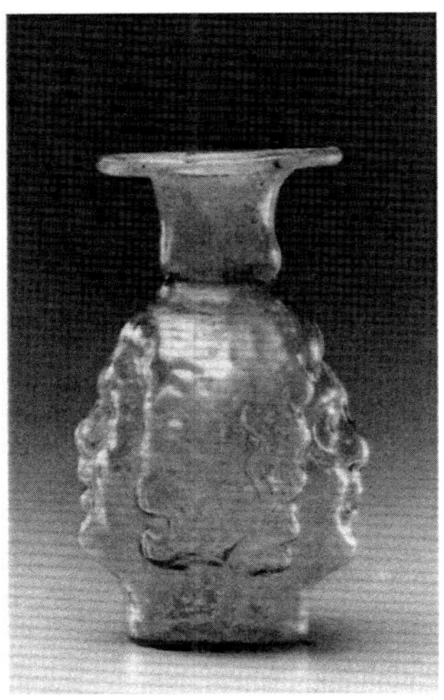
167, profile B:A (1:2)

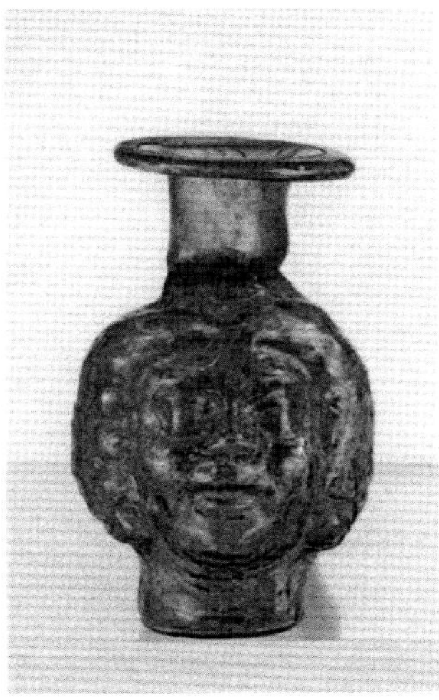
167, side B (1:2)

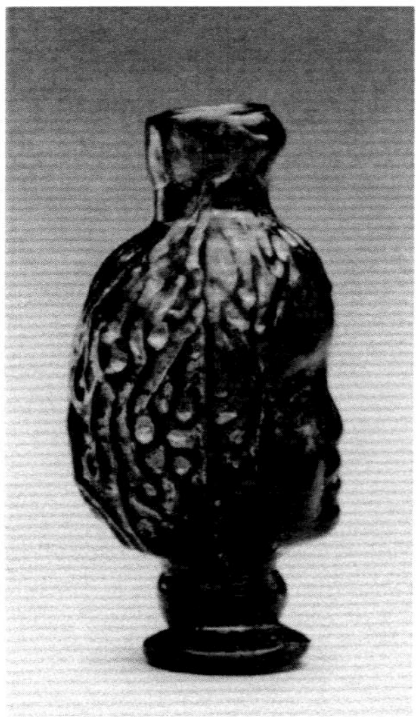

168 (1:1)

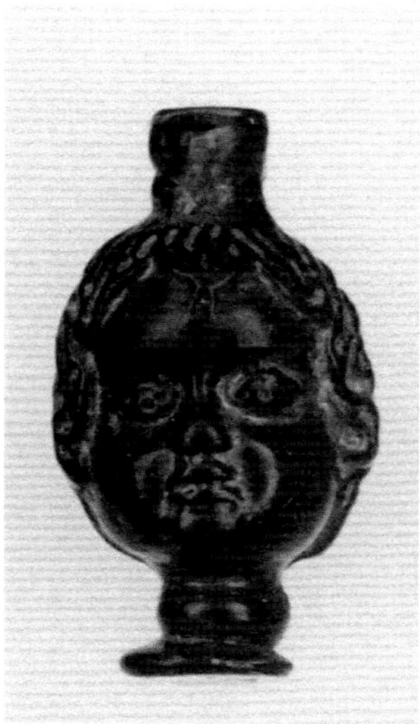

168 (1:1)

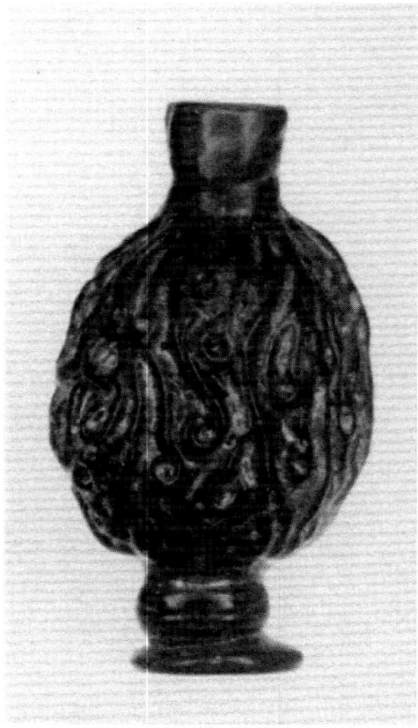

168, back (1:1)

For description, see Cat. No. 166.
*Condition* Intact. On exterior and interior, areas of brown discoloration and incipient pitting.
*Collections* Thomas E. H. Curtis.
*Published* Previously unpublished.
*Remarks* From the same mold as Cat. No. 166.

### 168 **Head-shaped Bottle on Stemmed Foot**

Place of manufacture undetermined
Probably nineteenth century
H 5.7 cm; H body 3.55 cm; max D 3.1 cm; D rim 1.3-1.2 cm; D base 1.7 cm
1923.519, Gift of Edward Drummond Libbey

Translucent light bluish gray glass with manganese colored streak appearing black in reflected light. Medium thick glass. A few small spherical bubbles in body.
Neck and body blown into a two-part mold of two vertical sections, open at the base. Mold seams through hair. Stem free blown. Applied base. Relief crisp. No pontil mark.
Straight rim rounded in flame. Short bulging tubular neck. Body in the shape of a head, below which the stem is in the form of a knob. Thin picked-up disk base.
Body in the shape of a head with childlike features: large almond-shaped eyes with recessed pupils, triangular nose with sharp angles, carefully rendered vertical ridges of lip below the nose, small mouth with lips slightly parted, heavy cheeks and chin. Face widest at the jaw. Forehead, tip of nose, and chin in one plane. Stylized hair with individual locks rendered as irregular wavy ridges ending in circular hooks, especially on back of head. Hair falls over center of forehead in a triangular lock with a loop suggesting a jewel at the tip.
*Condition* Intact. Exterior covered with iridescent film and areas of silvery weathering in low areas.
*Collection* Thomas E. H. Curtis.
*Published* Previously unpublished.
*Remarks* Although the iridescence on this curious small bottle appears authentic, an analysis of technical details, material, and style shows that this Type is not ancient.[1] The finishing of the rim and the combination of a hollow knob-shaped stem[2] with a thin picked-up disk as base have no parallels in Roman or Byzantine glass. The use of light bluish gray glass to simulate black glass is also unusual in Roman and Byzantine glass.[3] The stylization of the hair is not paralleled in antiquity. The shape of the neck of the flask recalls Islamic bottles, while the sentimental girlish appearance of the face is reminiscent of nineteenth-century dolls. Taken together, these observations indicate that this bottle is not ancient.

[1] A similar head-shaped bottle in New York, MMA, no. 17.194.254, formerly Gréau Collection, Froehner 1903, 159, no. 1147, III, pl. 208, nos. 4 and 4 bis, was made in the same workshop but has an opening in the stem only.

[2] The stem knob was probably formed by an overblow through the opening at the bottom of the mold.

[3] No ancient glass is black. Ancient glass which appears to be black is usually translucent dark purple (manganese), dark blue, or dark green in transmitted light.

# IV
# The Fourth to Early Seventh Centuries*

Blowing glass into molds continued as a glass-working technique in the provinces of the eastern Roman empire through the early seventh century, influencing the famous glass workshops of the Sasanian and eventually the early Islamic kingdoms. Most products were made by free blowing or expanded mold blowing, blowing the paraison into a mold to impose an overall pattern, removing the mold, and then further inflating the paraison to the desired size and shape.[1] Some full-size mold-blown vessels continued to take advantage of the technique's ability to create a sharp, precise design. In addition to the late antique head-shaped flasks discussed in Chapter III, the Toledo Museum has twenty-three examples of late Roman and early Byzantine full-size mold-blown glass made in Syria and Palestine during the fourth through early seventh centuries (Cat. Nos. 169-192).[2] All these vessels are jugs or jars. They are distinctive for their decoration, which includes many symbols clearly linked to Judaism or Christianity, and for their mold-making technique, which produced designs in low relief. The Toledo Museum's vessels are here divided into three groups based on their decoration, both the surface treatment — sunken (intaglio) relief or low relief — and the motifs, which need to be interpreted in the context of the special role of the Syro-Palestinian region in Christianity and Judaism from the reign of Constantine (306-337) onward. The first group, Vessels with Jewish and Christian Symbols in Sunken Relief (Cat. Nos. 169-177), often called "pilgrim vessels" or sometimes the "Jerusalem group," has already been well researched and published by Barag.[3] The second group is Vessels with Designs in Sunken Relief (Cat. Nos. 178-189). A small third group is Vessels with Christian Symbols in Low Relief (Cat. Nos. 189-192), which bears designs in raised outline.

·These Syro-Palestinian glasswares are best viewed in the context of the religious politics of the fourth through sixth centuries. In 324 the emperor Constantine defeated his last rival for control of the eastern half of the empire and became sole ruler, attributing his success to his conversion to Christianity. The focus of imperial power thus abruptly turned toward the Christian Church and the Greek-speaking lands of the eastern Mediterranean, and was symbolized by the dedication of a new imperial capital, Constantinople. Founded on the site of the ancient city of Byzantium, the city was the source of the term Byzantine now used to describe the later Roman empire from the fourth century onward. Dramatic shifts in the religious, political, and economic life of the Syro-Palestinian region occurred with regularity after the onset of Roman rule in the first century B.C. Oppressive measures had favored the pagan cities of the coast, the Decapolis, and Samaria over the Jewish population following the destruction of the Second Temple in 70 and suppression of the Bar Kokhba Revolt of 132-135. Before the fourth century Christianity was not as common as in the rest of the empire, but it became widespread as Constantine began a tradition of ostentatious imperial patronage to erect and embellish churches in the many Christian holy places. By the end of the century, within three generations of Constantine's conversion, the emperor Gratian had outlawed all public pagan ceremonies. Roman power also gradually became limited to the eastern Mediterranean; in 395 the rule of the empire was divided into East and West, in 410

* This chapter, as well as the Remarks for Cat. Nos. 189-192, was composed in the Museum's Publications Office after the author left the Museum's employ. The section below on *Vessels with Jewish and Christian Symbols in Sunken Relief* is based on her preliminary draft of 1989.

[1] The Toledo Museum's extensive collection of free-blown and expanded mold-blown glass will be published in TMA, *Roman and Early Byzantine Glass* (forthcoming).

[2] On the large collection of ancient and Islamic glass, mostly from the Near East, assembled by Thomas E. H. Curtis and purchased in 1917 after his death by Edward Drummond Libbey for The Toledo Museum of Art, see Grose 1989, 17-20. On Azeez Khayat, the New York dealer who was Curtis's principal source, see also Bergman 1974, 238-244.

[3] Barag 1970c; Barag 1971a; Barag 1971c.

and again in 455 the city of Rome was sacked, and in 476 rule of the West, including North Africa, passed permanently into the hands of non-Romans. The fourth to sixth centuries in the Near East thus witnessed a many-faceted shift away from the cultural life of hellenized cities like Skythopolis, Caesarea, and Samaria toward the new Christian centers. Protected by distance from the northern frontier and by a line of forts staving off Sasanian ambitions, Syria and Palestine enjoyed centuries of prosperity thanks to imperial patronage, the influx of wealthy Roman aristocrats fleeing western estates lost to Goth and Vandal invaders, extensive Christian monastic settlements in the Judaean desert, and a steady stream of pilgrims from all around the empire.[4] The tranquillity ended in the early seventh century, with the Arab capture of Damascus (634) and Jerusalem (636) foreshadowing the decisive defeat of the Byzantine emperor Heraclius in 638, marking the end of Roman culture in Syria and Palestine as well as in Egypt and North Africa.

Polytheistic, Jewish, and Christian artists of these centuries employed the same technologies and the same classical motifs to communicate meanings. Even on objects of daily life, such as lamps and tablewares, the use of religious symbols — Jewish (menorah, shofar, etrog) and Christian (cross, saint, peacock, fish) — increased, perhaps reflecting the increasing polarization of the times. Representations of pagan gods and heroes continued to be made, perhaps mostly as nostalgic tokens of classical education, into the seventh century and later.[5] And what seem at this distance to be common, overlapping symbols — lulav/palm branch, tree of life, peacock, ivy, ark — may have held different meanings for different religions. Religious turmoil continued as the government dealt, with varying degrees of tolerance, with Christian orthodox and heretical sects, dozens of surviving pagan cults, and the Jewish congregations of the day.

## VESSELS WITH JEWISH AND CHRISTIAN SYMBOLS IN SUNKEN RELIEF ("Pilgrim Vessels")
(Cat. Nos. 169-177)

The nine vessels catalogued here belong to a group of about a hundred surviving examples decorated with Jewish and Christian religious symbols in intaglio, or sunken relief, that have been identified as pilgrim vessels.[6] Most are in remarkably good condition, having been found relatively intact in tombs. They are most often made of brown glass, in shades varying from pale brown to nearly black, occasionally of natural greenish or blue-green glass.

There are no differences between the shapes of the Christian and Jewish vessels, which occur in only two forms: jugs and jars. The bodies of both forms are usually squat and hexagonal in section, but a few of the jars have octagonal bodies. The molds for both shapes were easy to design within a circular circumference. However, since none of the demonstrably Christian vessels is octagonal, the hexagonal shape of the Christian vessels may have had a symbolic meaning.[7]

The jugs have tall cylindrical necks with wide spouted mouths. The handles are applied at the shoulders and attached at the rims; the most salient characteristic is that the handles are made from hollow tubes.[8] The jars usually have short, wide necks and wide funnel mouths, with the rims folded far down into the interior. They are most common without handles but may have from two to six handles.

The glass blower blew the body of the vessel into a mold with designs in low relief on its inner surface. These produced a sunken design in intaglio on the outside of the glass. The designs were probably hammered into a thin, rectangular metal sheet. Harden mentions the existence of a hexagonal object of thin bronze or copper in the Rockefeller Museum, Jerusalem, which was probably a mold for the body and base of vessels like these,[9] but at least four molds must have been open at top and bottom because the entire group of designs is upside down on some vessels.[10] For these vessels the glass blower was able to insert the blowpipe with its gather of glass into either end of the mold.

Barag divides these pilgrim vessels into three Classes according to the symbols appearing on them.[11] This catalogue follows the order of Barag's classification:

| | |
|---|---|
| **A**: | Vessels with Christian symbols |
| **B and BB**: | Vessels with Jewish symbols |
| **C and CC**: | Vessels with symbols found in both Class A and Class B, but without clear-cut Christian or Jewish symbols such as a cross or menorah |

### *Class A: Vessels with Christian Symbols*
(Cat. Nos. 169 and 170)

Three distinct forms of cross appear on alternating panels of the body. Barag postulates that each form emphasized one or more aspects of a monumental gem-encrusted cross, with knobs at the ends of the bars or short cross-bars, erected on the rock of Mount Golgotha: 1, mounted on three steps, 2, mounted on a globe or omphalos, and 3, standing on a tree trunk flanked by stylized leaves.[12] On the vessels of Classes

[4] *Age of Spirituality* 1977; Timothy D. Barnes, *Constantine and Eusebius* (Cambridge, Mass. 1981) 245-260; *Holy Land* 1986, 195-199, 241-249; Robin Lane Fox, *Pagans and Christians* (San Francisco 1986) 663-681; E. D. Hunt, *Holy Land Pilgrimage in the Later Roman Empire, A. D. 312-460* (Oxford 1982).

[5] R. Brilliant, in *Age of Spirituality* 1977, 126-131; U. Avida, in *Holy Land* 1986, 198-199.

[6] See above note 3; *Glass of the Caesars* 1987, 177, no. 98.

[7] Barag 1971a, 51-63, concludes that Early Christian baptisteries, baptismal fonts, and related minor arts were represented as hexagons because the ciborium above the Holy Sepulcher and the ciborium over the cross at Golgotha were hexagonal. For a six-sided altar of Dionysos found at Beth Shean, Israel (see Fig. 56) and the possible meaning of the shape in an earlier context: see Chapter I: *Syro-Palestinian Styles: Hexagonal Bottles in High Relief: Masks Type* and *Mixed Symbols Type*.

[8] Barag 1970a, 46 and n. 62, details evidence for tubular handles being of Byzantine date.

[9] *Glass of the Caesars* 1987, 152.

[10] Barag 1970c, 37-38, mentions one Jewish vessel of Class B I and six Christian vessels belonging to Classes A V, A VII, and C IV.

[11] Barag 1970c, 38.

[12] Barag 1970c, 44. On the emphasis on worship of the cross in Palestinian iconography, see Kurt Weitzmann, "Loca Sancta and the Representational Arts of Palestine," *Dumbarton Oaks Papers* 28 (1974) 31-55.

A I-IV the crosses occur in a set order (see Cat. Nos. 169 and 170, panels 1, 3, and 5), which probably had a symbolic meaning, alternating with lozenges in no particular order (panels 2, 4, and 6).

### *Classes B and BB: Vessels with Jewish Symbols* (Cat. Nos. 171-175)

The hexagonal vessels are named Class B, the octagonal ones BB. The objects depicted on the Jewish vessels (menorah, shofar, incense shovel, stylized tree, aedicula) are more varied than those on the Christian vessels, but their interpretation is less certain. The seven-branched menorah may be flanked by ritual objects such as a shofar (ram's horn) or an incense shovel (see Cat. Nos. 171 and 172 [both Class B II]) or may be alone (see Cat. No. 175 [Class BB I] and Cat. Nos. 173 and 174 [Class B V]). The menorah is a well-known Jewish symbol,[13] but the meaning of an empty aedicula (see Remarks Cat. Nos. 171) or an aedicula framing a stylized tree (see Remarks Cat. No. 175) is not clear. Nor have all the objects depicted been identified (e.g. Cat. Nos. 173 and 174, panel 2). Some symbols, such as concentric lozenges and vessels, appear both in the Christian and in the Jewish Series, but they probably had different meanings.

### *Classes C and CC: Vessels with Unassigned Symbols* (Cat. Nos. 176 and 177)

This group of approximately twenty-five surviving examples consists of hexagonal (Class C) and octagonal (Class CC) vessels. The grouping is artificial, indicating that the symbols do not refer to recognizably Christian or Jewish objects such as a cross or a menorah.[14] The vessels themselves, however, were probably intended for use by Christians or Jews. Vessels of Class C IV carry a variety of symbols that resemble those on contemporary Christian vessels made in other glass workshops. The recent excavation of two identical jars of Class C III in a Christian tomb at Ma'in[15] supports a Christian interpretation both for that Class and for Class C II with its closely related symbols. Barag recently suggested that "with the exception of the octagonal Class CC I, which seems to be Jewish, most or perhaps all of the vessels of Group C are Christian vessels,"[16] but there is no specific evidence as yet for assigning Class C I (see Cat. No. 176) to the Christian or the Jewish Series.

The Christian tomb at Ma'in in which four pilgrim vessels were excavated was contemporary with a local church dated by an inscription to 595-604.[17] This accords well with the dates Barag suggests for Classes A I-IV: 578 through 614 or 636/638. The stepped cross at Golgotha[18] which figures so prominently in Classes A I-IV first appears on coins struck by Tiberius II Constantine in 578. It was torn down by the Sasanian Persian army in 614 during the attack on Jerusalem. The cross on the vessels which Barag assigns to the period 615-636 is of a different type and may represent the replacement cross erected after the Persians were expelled.[19] There are no independent criteria for dating the various Jewish vessels. Production of the Christian vessels probably came to an end soon after the Arab conquest of Jerusalem in 636, but Jewish vessels must have ceased to be made somewhat earlier, in 629, because in that year the Byzantine emperor Heraclius forbade Jews to come within three miles of Jerusalem.

A few pilgrim jugs have been found during controlled excavations in Jordan and Israel, and others are said to be from the Damascus area, Galilee, Tyre, Cyprus, and Egypt. Similarities in shape, color, technique, and style of decoration suggest that the Jewish vessels are not only contemporary with the Christian ones but also produced in the same workshop. The cross of Golgotha depicted on the Christian vessels points to a close connection with Jerusalem. As in the rest of the Roman empire, glass blowers in Palestine were not allowed to pursue their craft within cities because of the fear of fire; therefore the workshop producing these vessels was probably not located in Jerusalem itself, but somewhere in the immediate vicinity.[20]

A curious aspect of this group of vessels is the fact that although they were probably made in one workshop, the religious motifs on the sides of the body designate them for use by Christians or Jews. These religious groups did not normally mingle in ancient Palestine. The existence of Jewish glass makers in the eastern Mediterranean is attested by two stories popular during the sixth century, in both of which the protagonist is a Jewish glass maker. One story comes from Homs (ancient Emesa) in Syria; the other is from Constantinople.[21] We do not know whether the craftsmen who made these pilgrim vessels were Christians or Jews, but they clearly had no problems in producing objects for sale to customers of both creeds. This economic flexibility has also been recognized earlier, during the fourth century, when workshops making luxury wares such as gold glass and cut glass decorated their products for different groups of customers with Christian, Jewish, and polytheistic motifs. The fact that

[13] On the meaning of the menorah: Goodenough 1953, IV, 71-98; XII, 79-83; Nahman Avigad ed., *Beth Shearim* III: *The Excavations 1953-1958* (Jerusalem and New Brunswick, N. J. 1976), 268-274; D. Sperber, "The History of the Menorah," *Journal of Jewish Studies* 16 (1965) 135-159; L. Yarden, *The Tree of Light* (Uppsala 1972); C. L. Meyers, *The Tabernacle Menorah: A Synthetic Study of a Symbol from the Biblical Cult* (Missoula, Montana 1976); Dan P. Barag, "The Menorah in the Roman and Byzantine Periods: A Messianic Symbol," *Bulletin of the Anglo-Israel Archaeological Society* (1985-86) 44-47.

[14] Barag 1971a, 45-50.

[15] Barag 1985b, 366, 368-369; pl. 68, photo 21; pl. 70, fig. 70; pl. 71, photo 22.

[16] Barag 1985b, 368 and n. 7.

[17] M. Piccirillo, "Le antichità bizantine di Ma'in e dintorni," *Liber Annuus* 35 (1985) 350-351.

[18] The precise date, occasion, meaning, and appearance of the cross erected at Golgotha are still matters of dispute: Barag 1970c, 38-42. See also Cat. No. 169, note 4.

[19] Barag 1970c, 44-46.

[20] Barag 1970c, 62.

[21] Barag 1971c, col. 611. In the late seventh century the caliph 'Abd al-Malik (ruled 685-705) hired a group of Jewish glass makers to make the lamps and vessels for the new mosque in Jerusalem.

these pilgrim vessels do not include a group with polytheistic symbols may be a result of their function, but it was also true that by the sixth century pagan religious practices were illegal throughout the Roman empire.

Christian pilgrims visiting the Holy Land during the sixth century are known to have taken home souvenirs, or *eulogiai*, blessings, including tesserae, or tokens, and vessels of clay and metal as well as glass.[22] Decorated with various Christian symbols, the vessels held holy oil from sacred places they visited, hence the common name for the group, "pilgrim vessels." The vessels with images of the cross on Golgotha probably held oil from the Church of the Holy Sepulcher in Jerusalem. A contemporary anonymous author describes how Antoninus of Piacenza, who made a pilgrimage to the Holy Land around the year 570, received oil from a bronze lamp burning in the tomb shown to pilgrims as Christ's burial place.[23]

Jewish pilgrims visiting Jerusalem may also have followed the custom of taking home oil, water, or soil from holy places or famous synagogues on Mount Zion or near the Western Wall.[24] It is also possible that they used the vessels to transport oil with which to anoint the Holy Rock (*Even Hashtiya*) once a year when they ascended the Temple Mount to mourn the destruction of the Second Temple (by the Roman emperor Titus in A.D. 70).[25] None of these activities would have been possible after 629 when the Byzantine emperor Heraclius forbade Jews to enter Jerusalem.

Christian vessels have been excavated in Near Eastern tombs,[26] suggesting that Christian pilgrims were sometimes buried with the holy oil they carried back from Jerusalem. Jewish visitors to Jerusalem may have followed a similar custom. The trees that frequently appear on Jewish vessels (e.g. Cat. Nos. 171-175) are ancient symbols of life and immortality and therefore appropriate for funerary offerings, but, as Barag points out, grave gifts were probably not the primary function of these vessels.

Nothing is known about the find circumstances of the nine vessels in The Toledo Museum of Art, except that all but one (Cat. No. 170) entered the Museum as part of the Curtis Collection.[27] However, at least four of the Museum's Jewish vessels appear to form pairs consisting of a jug and a jar of the same class (Cat. Nos. 171 and 172 [Class B II]; Cat. Nos. 173 and 174 [Class B V]).[28] This is akin to a phenomenon recently noted by Barag in connection with the finds made in Ma'in: the jugs and jars from this tomb form two pairs, each made in one mold.[29] In contrast to the Jewish pairs in the Toledo Museum, the Christian pairs from Ma'in not only belong to the same class but are also the same shape, i.e. the two jugs form one pair and the two jars another.

## VESSELS WITH DESIGNS IN SUNKEN RELIEF (Cat. Nos. 178-188)

Contemporary with the Vessels with Jewish and Christian Symbols in Sunken Relief (Cat. Nos. 169-177), the Toledo Museum has eleven mostly rectangular, hexagonal, and octagonal mold-blown vessels. Most are decorated with symbols similar to those on the pilgrim vessels, but with no symbols incontestably identified as Jewish or Christian. The glass blower used multipart molds, probably with a base plate, to impart similar intaglio, or sunken, relief. Mold seams cannot be distinguished, but the edges of the panels are clear and sharp (except for Cat. No. 184, which was slightly expanded), so the mold sections must have fit tightly together. All were blown of natural yellowish, greenish, or blue-green glass. Many were finished with folded and fire-polished rims and further embellished with thread trailed on the neck.

The eleven Toledo Museum vessels can be arranged into two major groups. The Museum has only a few examples of each, and it has not been possible to verify the exact appearance of individual bottles from published descriptions of other bottles. In order to avoid confusion with the terminology used elsewhere in this volume, the term Class is used here in the same sense that Barag introduced it for the Jerusalem group of pilgrim vessels, for subgroups that appear to be related to each other.

### *Vessels with Symbols in Sunken Relief, perhaps Christian* (Cat. Nos. 178-186)[30]

The first group has most commonly six, but sometimes five or eight, rectangular panels on the body. Each panel is carefully impressed with sunken relief of lozenge (sometimes called diamond), palm frond (sometimes called herringbone), and lattice patterns. The designs occur in several variations within each Class, and the differences may be associated with individual workshops or with the development of each form and its decoration over time. Eisen suggested the existence of two subgroups, one of natural colored glass, the other of

[22] M. E. Frazer, in *Age of Spirituality* 1977, 564-568, 576-578, nos. 515-516, 585-588, nos. 524-527.

[23] Barag 1970c, 46-48.

[24] Barag 1970c, 62.

[25] Avi Yonah 1984, 164.

[26] For example:
- a) from Catacomb 20 at Beth Shearim, Barag 1976, 207f, no. 50, fig. 99b;
- b) from a tomb at Ma'in, Barag 1985b, 365-369, nos. 1-4;
- c) from a mausoleum at Kusr es-Sabah, about 1.5 km south of Amman, N. I. Khairy, "Al Quweismeh Family Tomb," *PEQ* 112 (1980) 59, fig. 7B.

[27] On Thomas E. H. Curtis, see above note 2. A Jewish jar in Yale acquired from Azeez Khayat in 1902 is said to have been found at Beth Shearim (Yale, no. 1930.439, H 7.9 cm, Matheson 1980, 131, no. 350). Khayat was one of the principal dealers who helped Curtis form his collection. It is tempting to think that some of the pilgrim vessels in the Toledo Museum may have come from the same source.

[28] Two jugs belonging to Class C IV, Cat. Nos. 177 and 193, form a dubious pair because the antiquity of the latter is extremely doubtful.

[29] Barag 1985b, 365-369, nos. 1-4. This phenomenon is also seen in two jugs of Class A I acquired at the same time by the Freer Gallery, Washington, D.C.: Richard Ettinghausen, *Ancient Glass in the Freer Gallery of Art* (Washington, D.C. 1962) 22, fig. 64.

[30] Eisen 1927, II, 461, 473-474, first identified the group and calls these flasks with symbols of Bible and cross. Barag 1970c, 35 n. 1, projected short surveys of these vessels that have not yet been published.

brownish glass. A significant technical distinction from the pilgrim vessels is that all of these vessels except Cat. No. 184 were blown into molds that included a separate base plate, perhaps attached to one side of the mold, bearing a design that was impressed on the underside. Hexagonal jugs Cat. Nos. 178 and 179 bear a mold-blown sunken outline design of twelve petals; Cat. Nos. 180-182 have a star with six forked arms; Cat. No. 185 has six radiating double lines with dots in the intervals; and the pentagonal jug Cat. No. 183 has a rosette with five petals. Even the octagonal jug Cat. No. 186 has a design, although it cannot be interpreted. All lack pontil marks, indicating that the vessels were held with a clamping tool while the neck, mouth, and handles were finished. The consistency of the technique suggests that a single workshop may have been responsible for these vessels.

Cat. Nos. 178-182 form a subgroup closely related in fabric, size, and execution of the mold designs. They seem to be closely related to several vessels in other collections, most notably:

***Class 1***, the plain type (here Cat. Nos. 178 and 179): two groups of three panels with 1 and 4, three lozenges arranged vertically with central circles; 2 and 5, lattice (also called cross-hatch or network) pattern; 3 and 6, palm branch with varying numbers of leaves on either side of the central stem.[31]

***Class 2***, the large mask type: two groups of three panels with, left to right, 1 and 4, three lozenges arranged vertically with central circles; 2 and 5, lattice pattern; 3 and 6, palm branch with large masks in the lower half of the panel.[32]

***Class 3***, the small mask type: similar to Class 2, but with a small mask in the top part of panels 3 and 6.[33]

It has been noticed that examples of Classes 1 and 2 have come into at least three collections in pairs, making it likely that they were produced in the same workshop.[34] Cat. Nos. 183-186 are closely related to Cat. Nos. 178-192 in conception but seem to represent independent Classes.

*Inverted Conical Jugs with Overall Geometric Pattern*
(Cat. Nos. 187 and 188)

The second group is characterized by inverted conical form and a body decorated with an overall geometric pattern of sunken hexagons. Both Toledo jugs were slightly inflated after removal from the one-part molds. The rounded shape, smaller size, and colors — blue and blue-green — distinguish these jugs from the first group. Both jugs have a design of nine radial indentations on the undersides of their bases. But the find spots of similar jugs suggest manufacture in the same inland Palestine-Syria area as the first group. Related, somewhat larger jugs in the Toledo Museum collection may be expanded products from similar molds.[35] Such designs do not appear on the related expanded mold-blown examples, suggesting that it is possible the design was imparted by a part-size pattern mold open at both ends. So that the finished jugs could stand, the glass blower flattened the base against a surface bearing the pattern as the paraison was expanded.

## JUGS WITH CHRISTIAN SYMBOLS IN LOW RELIEF
(Cat. Nos. 189-192)

A group of carefully made, tall hexagonal and rectangular jugs forms a group characterized by form, low raised relief, and subject matter. All were blown in natural yellowish, greenish, or blue-green glass. The decoration combines some panels with Christian subject matter, notably standing human figures, some on columns who probably represent Stylite saints (see Cat. No. 189), with other panels bearing geometric decoration similar to both the Vessels with Symbols in Sunken Relief (Cat. Nos. 178-188) and the Vessels with Jewish and Christian Symbols in Sunken Relief (Cat. No. 169-177), such as concentric lozenges, palm branches (herringbone pattern), and lattice network.

There has been no formal survey of this group,[36] but some of the recognized Classes include:

***Class 1***: four-sided jug; on the panels, from left to right: 1, cross *fourchée* with two dots above and below each arm; 2, human figure on a column with ladder to left (Stylite saint?);

[31] Eisen 1927, II, 461, 472-473, identifies the palm fronds as trees, symbolic of the Tree of Life. Here only designs such as panels 3 and 4 on Cat. No. 190 are considered likely trees. Examples from known find spots:
a) from a tomb of sixth-century date at El-Lejjun (Ajlun), Jordan, Harden 1971, 79, fig. 1;
b) from Syria, Damascus Museum, no. 636, Zouhdi 1964, no. 53, fig. 36.

[32] Eisen 1927, II, 549, interprets these masks as "Christ faces." Matheson 1980, 137, at no. 360, notes that similar masks occur at the base of vine-decorated columns in contemporary churches, but they seem to have no specifically Christian significance. Additional examples from known find spots suggest manufacture in northern Palestine or Syria:
a) from Bacbuck near Tyre, BM, no. 1900.12-17.1, Harden 1971, 81, pl. V, no. C;
b) probably from Syria, Yale, no. 1955.51.142, formerly Kouchakji Collection, formerly Moore Collection, pale greenish blue, H 13.9 cm, Eisen 1927, II, 472, 549, fig. 234; Matheson 1980, 136-137, no. 360;
c) presumably from Syria, Brooklyn Museum, number unknown, Eisen 1927, II, pl. 116 right.

[33] Eisen 1927, II, 461. Additional examples:
a) Carnegie, inv. no. 24055/3, with five-petaled rosette on underside of base, blue-green, H 18.0 cm, Oliver 1980, 130, no. 227, color ill. 127;
b) Amman Museum, Amman, Jordan, no. 5160, Harden 1971, 79, fig. 1, no. E.

[34] Matheson 1980, 137, at no. 361.

[35] For example a jug with twelve rows of dots in sunken relief, with thumbrest and spouted mouth, concave base, TMA acc. no. 1923.1342, amber, H 12.5 cm. Less closely related is a group of small, expanded mold-blown jugs with vertical ribs, TMA acc. nos. 1923.959-.962; these may well date after the fourth century: only three fragments were found at the glass-working site at Jalame, closely dated to 351-383, and Weinberg concludes that they probably post-date the factory: Weinberg 1988, 79-80, fig. 438, no. 343. Both types will be published in TMA, *Roman and Early Byzantine Glass* (forthcoming).

[36] Eisen 1927, II, 473-474, pls. 118-120, identified this as a group of Christian flasks with images of the cross and human figures. Barag 1970c, 35, n. 1, projected short surveys of these vessels that have not yet been published. *Glass of the Caesars* 1987, 176, no. 97.

3, lattice with raised dots at the intersections; 4, bird, probably a peacock.[37]

***Class 2*** (here Cat. No. 189): hexagonal jug; on the panels, from left to right: 1, cross *fourchée*; 2, stylized palm branch; 3, lattice; 4, standing human figure with quartered wheel (Stylite saint?); 5, same as panel 3; 6, stylized palm branch.

***Class 3*** (here Cat. No. 190): hexagonal jug, two-part mold; on the panels, from left to right: 1, cross *fourchée* below dot enclosed by a circle; 2, *ampulla* out of which a snake or vine rises; 3, stylized tree; 4, stylized tree; 5, saint in shroud with two crosses; 6, lattice.

***Class 4***: hexagonal jug, two-part mold; on the panels, from left to right: 1, standing stylized saint; 2, vertical row of four circles; 3, lattice; 4, standing human figure in short tunic; 5, same as panel 2; 6, same as panel 3.[38]

The subject matter of this group of jugs with low relief is clearly Christian, since it seems to be related to the veneration of famous saints who retreated from the world by living on top of large columns where they dedicated their lives to prayer and meditation. The first and most popular of the Stylite saints, Saint Symeon the Elder (ca. 389-459) was the most famous of these ascetics. The ruins of his column and shrine at Qal'at Saman, near Antioch, still attracted pilgrims in the nineteenth century.[39] This and the few known find spots may argue for manufacture in Syria.

During the late Roman and early Byzantine era of the fourth through early seventh centuries, distinctive symbols and iconography distinguished the art and artifacts of the many contemporary religious groups — polytheistic, Jewish, and Christian. But they understood the same artistic language and employed the same materials. For Jews and Christians alike, Palestine was the Holy Land, and for both, the most sacred goal of travel and pilgrimage was the city of Jerusalem. Both groups, and the craftsmen who served them, catered to a large population of people with the same religion living outside Palestine. Glass vessels such as these jugs and jars are vivid relics of the cultural life of Syria and Palestine.

[37] Eisen 1927, II, 483-484, identified the group. Additional examples:
a) Yale, no. 1955.6.149, formerly Moore Collection, light green with dark green handle, H 14.2 cm, Eisen 1927, II, 483; Matheson 1980, 132-133, no. 353 (with parallels);
b) Louvre, inv. OA 6417, from Egypt, cited by Eisen 1927, II, 484.

[38] For example, Yale, no. 1955.6.148, formerly Moore Collection, color, H 20.6 cm, two-part mold, Matheson 1980, 133-134, no. 355 (with parallels).

[39] See Remarks Cat. No. 189. For Symeon and other Stylite saints, see Hippolyte Delehaye, *Les saints stylites* (Paris 1923).

# CATALOGUE

## *Vessels with Jewish and Christian Symbols in Sunken Relief* (Cat. Nos. 169-177)

### ***Hexagonal Vessels with Christian Symbols (Barag Class A)*** (Cat. Nos. 169 and 170)

169 **Hexagonal Jug with Christian Symbols**

Palestinian, probably made in the vicinity of Jerusalem
About 578-614
H 15.4 cm; D rim 6.5 cm; max W body 7.9 cm; W base 7.2-6.3 cm
1923.1361, Gift of Edward Drummond Libbey

Transparent to translucent dark brownish yellow. Similarly colored handle. Medium thin glass. Fabric cannot be determined because of weathering.

Rim and neck free blown. Body blown into a mold with designs in low relief on the interior. Mold seams cannot be detected. Relief moderately crisp. Ring pontil mark ca. 1.4 cm in diameter. Excess glass at tip of handle folded back against top of handle.

Spouted mouth, with rim folded outward, upward, inward, and downward. Cylindrical neck. Horizontal shoulder. Hexagonal body. Flat base. Hollow tubular handle applied to edge of shoulder above upper left corner of panel 2 and attached to edge of rim. On the body, six decorated rectangular panels with designs in intaglio; each panel bordered by recessed dots. In the panels, from left to right: 1, cross *fourchée* above three steps, with the base of the cross forming a fourth step; 2, two concentric lozenges with a cross with equal arms in the center and a recessed dot in each corner of the panel; 3, cross *fourchée* similar to that in panel 1, but with a bulbous base rising from the center of two concentric circles rendered in perspective so that the interior circle appears as a crescent; 4, two concentric lozenges with four small recessed dots in the center and a recessed dot in each corner of the panel; 5, cross with arms of equal length, with a short perpendicular element at the end of each arm standing on an indistinct element that may be a tree trunk with four large fleshy leaves extending upward, and four recessed dots in corners of panel; 6, two concentric lozenges with a single recessed dot in the center and a recessed dot in each corner of the panel.

*Classification* Barag A I 3.

*Condition* Parts of mouth and handle restored. Small holes in panels 4 and 5; a large hole in the handle. Cleaned. Original surface not preserved. Exterior and interior covered with rainbow iridescence, severe surface pitting, ice floes, and beige weathering.

*Collection* Thomas E. H. Curtis.

*Published* Barag 1970c, 55 (A I 3), 58, fig. 1.

*Remarks* This vessel belongs to Barag's Class A I with Christian symbols.[1] With at least sixteen examples, this is the largest Class of vessels with Christian symbols. All known examples are jugs.[2] In this Class three distinct crosses appear, in a specific order, each depicting one or more aspects of a monumental cross erected in the second half of the fourth century or later on the rock of Golgotha, the site of Jesus's crucifixion. The crosses in panels 1, 3, and 5 alternate with concentric lozenges in panels 2, 4, and 6 that probably have a symbolic meaning. An attractive theory suggests they represent bound copies of the holy scriptures.[3]

The stepped cross in panel 1 is the most significant for dating the vessels (see also below) because it has been identified with the elaborate, gem-studded cross on Golgotha itself.[4] Pilgrims visiting Jerusalem during the sixth century mention steps leading to the cross which stood on Golgotha until the destruction of the Church of the Holy Sepulcher in 614. The second cross, in panel 3, is similar to the first, but its lower end is slightly bulbous, and it rises from the center of two concentric circles that have been interpreted as the omphalos, or navel, of the Earth.[5] If this interpretation is

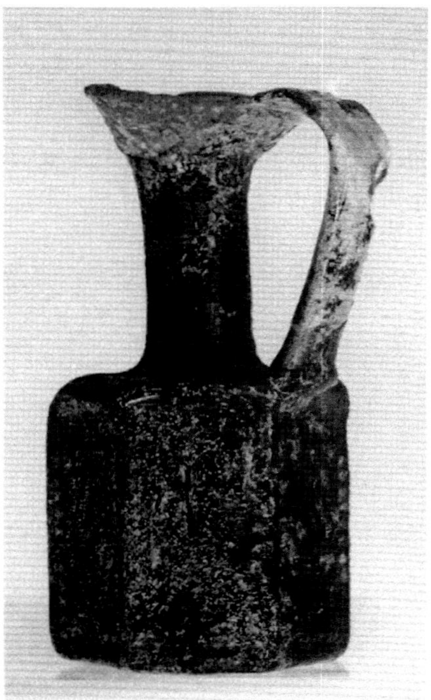

169, panels 6, 1, and 2 (1:2)

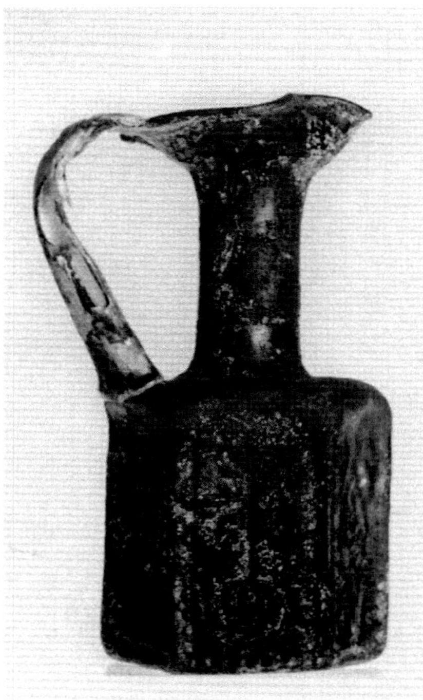

169, panels 2, 3, and 4 (1:2)

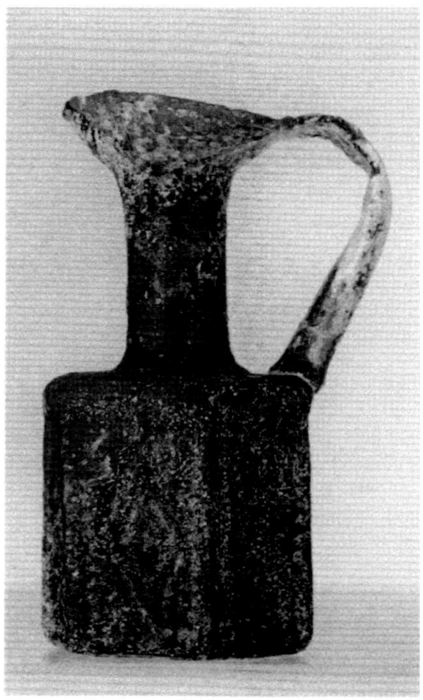

169, panels 5, 6, and 1 (1:2)

correct, the cross on the omphalos would refer to a Christian literary tradition since the middle of the fourth century that linked the site of Golgotha with the center of the earth. The third cross, on panel 5, has equally long arms and top and rises from four large curved leaves. The Leafed Cross connected ancient oriental traditions of the Tree of Life with the early Christian concept of the Life-Giving Cross.[6] Christian belief drew parallels between Christ's Cross and the Tree of Life on the one hand and between Golgotha and Paradise on the other.

Barag has established an approximate date of 578 to 614 for Classes A I-IV, which is supported by the discovery of four pilgrim vessels, including two of Class A I, in a tomb at Ma'in that can be dated to the decade 595 to 604.[7] The prominent place of the cross on Golgotha on vessels belonging to Classes A I-IV connects them with Jerusalem where they were sold. Barag locates the workshop that produced them in the vicinity of Jerusalem. These vessels were probably filled with oil from the Holy Sepulcher that pilgrims took home from Jerusalem.

[1] Barag 1970c, 38. See also Chapter IV: *The Fourth to Early Seventh Centuries: Vessels with Jewish and Christian Designs in Sunken Relief.*

[2] Barag 1970c, 38-44, 55. To Barag's parallels add:
- a-b) Collection Hamarneh and Madaba Museum, excavated at Ma'in, Israel, Barag 1985b, 365-366, nos. 1-2, fig. 6, photo 18-25;
- c) formerly Constable Maxwell Collection, *Maxwell* 1979, 182, no. 321;
- d) Carnegie, no. 24037/1, Oliver 1980, 129, no. 225;
- e) Oppenländer Collection, no. 22319, von Saldern et al. 1974a, 182, no. 505;
- f) Los Angeles County Museum of Art, no. M.88.129.75, formerly Hans Cohn Collection, von Saldern 1980b, 75, no. 70, color pl. 6;
- g) formerly Breitbart Collection, jug, brown, Sotheby's, New York, auc. cat., June 20, 1990, no. 111, ill., Sotheby Parke Bernet, New York, September 26-28, 1972, no. 136, ill.

Yale, no. 1956.6.152 (Barag 1970c, 55, no. A I 4) has since been published by Matheson 1980, 129-130, no. 349. Formerly Kofler-Truniger Collection, no. K 2848 M (Barag 1970c, 55, no. A I 10) has since been published in Luzern 1981, 91, no. 333; *Kofler-Truniger* 1985, 12, no. 8.

[3] Eisen 1927, II, 505, 507, suggests the lozenges with one depression in the center represent the Old Testament, while lozenges with four depressions in the center represent the New Testament with its four Gospels.

[4] It may be the gemmed cross on a starry globe represented on a sixth-century silver paten made in Constantinople, found before 1867 in western Siberia, Hermitage, no. V 209, D 18.6 cm, *Age of Spirituality* 1977, 537-538, no. 482. In the exergue are four "streams" representing the four rivers of Paradise.

[5] Barag 1970c, 41-42.

[6] D. Talbot Rice, "The Leaved Cross," *Byzantinoslavica* 11 (1950) 72-81.

[7] See Chapter IV: *The Fourth to Early Seventh Centuries: Vessels with Jewish and Christian Symbols in Sunken Relief*; Barag 1985b, 365-369, nos. 1-4.

## 170 **Hexagonal Jar with Christian Symbols**

Palestinian, probably made in the vicinity of Jerusalem
About 578-614
H 9.2 cm; D rim 5.0 cm; max W body 7.1 cm; W base 6.3-5.7 cm
1936.12, Gift of Edward Drummond Libbey

Transparent to translucent dark brownish yellow. Medium thin glass. Fabric cannot be determined because of weathering.

Rim and neck free blown. Body blown into a mold with designs in low relief on the interior. Mold seams cannot be detected. Relief moderately crisp. Off-center pontil mark ca. 1.2 cm in diameter. Funnel mouth, with rim folded outward, upward, inward, and downward to narrowest point of opening. Shoulder gently sloped. Hexagonal body. Base slightly concave.

On the body, six decorated rectangular panels with designs in intaglio; each panel bordered by recessed dots. In the panels, from left to right: 1, cross *fourchée* above three steps, with the base of the cross forming a fourth step; 2, two concentric lozenges within which is a short horizontal line with a dot above and below and a recessed dot in each corner of the panel; 3, cross *fourchée* similar to that in panel 1, but with a bulbous base rising from the center of two concentric circles rendered in perspective so that the interior circle appears as a crescent; 4, two concentric lozenges with a single recessed dot in the center and a recessed dot in each corner of the panel; 5, cross with arms of equal length, with a short perpendicular element at the end of each arm standing on an indistinct base which may be a tree trunk with leaves extending upward; 6, two concentric lozenges with four recessed dots in the center and a recessed dot in each corner of the panel.

*Classification* Barag A II 2.

*Condition* Intact. Exterior glossy in areas, scratched, and pitted in other areas with blue and silver iridescence. Interior pitted in areas with blue iridescence.

*Published* Barag 1970c, 55 (A II 2) 58, fig. 1.

*Remarks* This jar belongs to Barag's Class A II with Christian symbols, a small Class consisting of one jug and one jar.[1] The crosses are the same as in Class A I (see Cat. No. 169), the differences being that the leaves at the foot of the third cross, in panel 5, are less lush and that there are no dots in the corners of this panel. In the center of the lozenge in panel 2 a horizontal bar between dots (possibly a debased or worn-out cross) replaces the cross in this position in Class A I. Panels 4 and 6 are the same as in Class A I but are interchanged. It would appear that the mold for Class A I became worn out, was taken apart, and panel 5 replaced by a newly hammered rectangle; panels 4 and 6 could have been interchanged when the mold was reassembled.

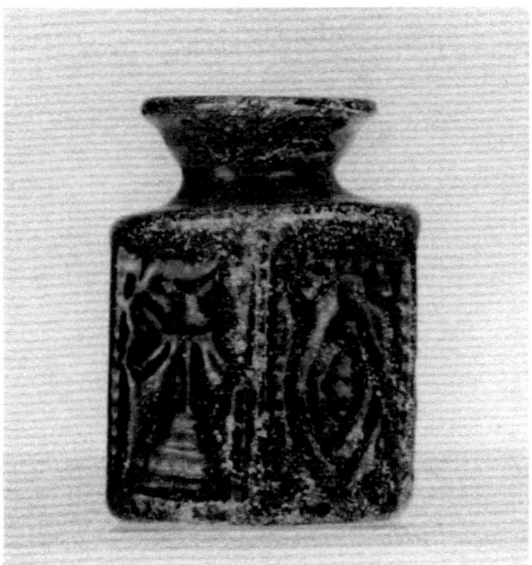

170, panels 1 and 2 (1:2)

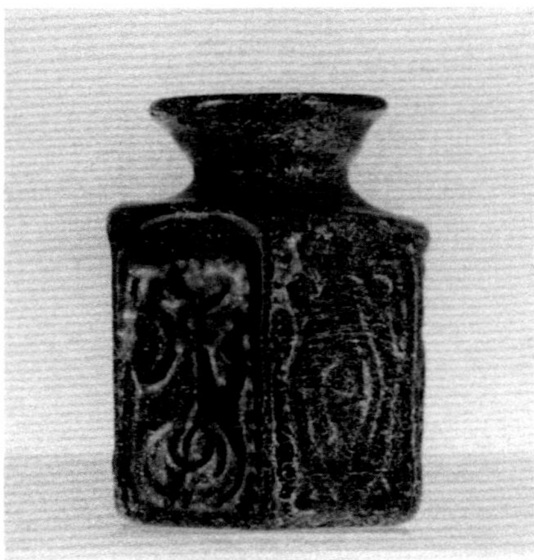

170, panels 3 and 4 (1:2)

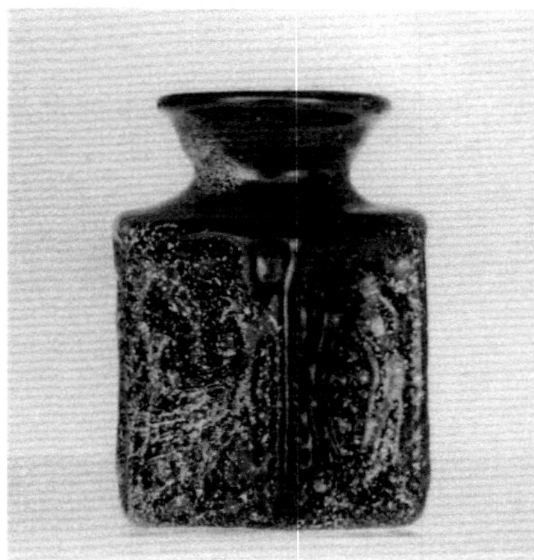

170, panels 5 and 6 (1:2)

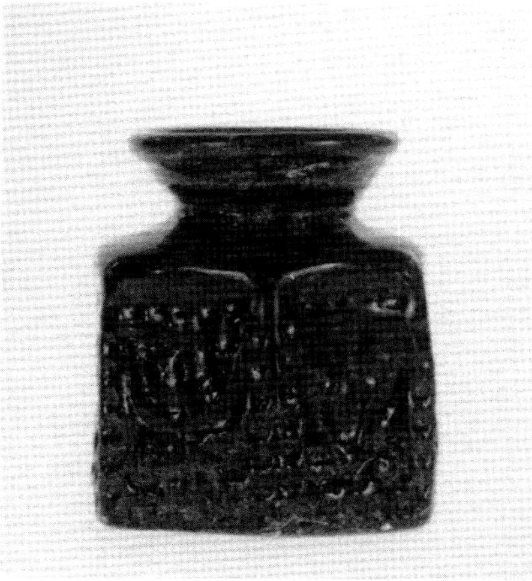

171, panels 1 and 2 (1:2)

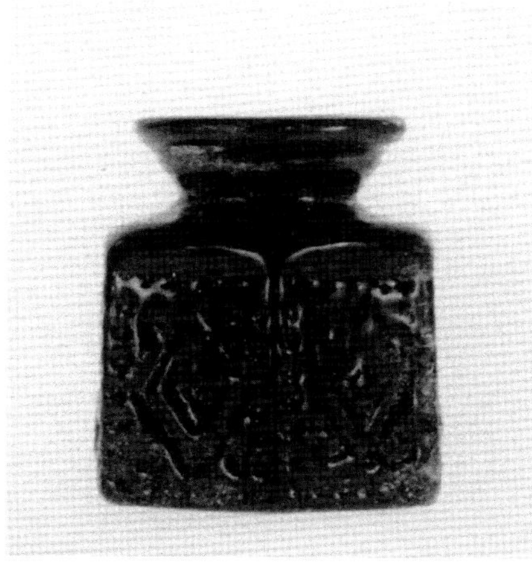

171, panels 3 and 4 (1:2)

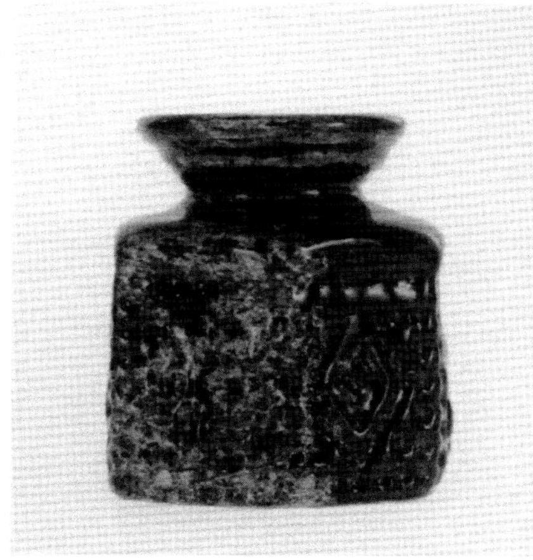

171, panels 5 and 6 (1:2)

[1] Barag 1970c, 55, for parallels.

## *Vessels with Jewish Symbols (Barag Class B)*

(Cat. Nos. 171-175)

### 171 **Hexagonal Jar with Jewish Symbols**

Palestinian, probably made in the vicinity of Jerusalem
About 578-629
H 8.5 cm; D rim 5.8 cm; max W body 8.1 cm; W base 7.5-6.8 cm
1923.1359, Gift of Edward Drummond Libbey

Transparent to translucent dark brownish orange. Medium thin glass. Fabric cannot be determined because of weathering.

Rim and neck free blown. Body blown into a mold with designs in low relief on the interior. Mold seams cannot be detected. Relief moderately distinct. Pontil mark 1.7 cm in diameter.

Funnel mouth, with rim folded outward, upward, inward, and downward to narrowest point of opening, with tool marks on interior. Shoulder gently sloped. Hexagonal body. Flat base.

On the body, six decorated rectangular panels with designs in intaglio; each panel bordered by recessed dots. In the panels, from left to right: 1, menorah on a three-legged stand; to the right of the menorah an incense shovel, to the left, a shofar (ram's horn); 2, stylized amphora with crescents in the lower corners of the panel; 3, two concentric lozenges with a single recessed dot in the center and a recessed dot in each corner of the panel; 4, indistinct, perhaps a jar on curving legs or a plant; 5, stylized date palm, with a recessed dot in each of the lower corners of the panel; 6, two concentric lozenges as in panel 3.

*Classification* Barag B II 1.

*Condition* Intact. Exterior glossy. Exterior has one panel and adjacent areas of base pitted with beige weathering; small areas of dull green and blue iridescence appear to have been repaired. Part of interior of rim pitted with beige weathering.

*Collection* Thomas E. H. Curtis.

*Published* Goodenough 1953, I, 170; III, figs. 408-413; Neuburg 1962, fig. 55; Rudolf M. Riefstahl, "The Complexities of Ancient Glass," *Apollo* 86 (1967), 434, fig. 16; Labino 1968, 21, 24, fig. 11; *TMA, Art in Glass* 1969, 32; Barag 1970c, 49, 56 (B II 1), 60, fig. 15; Charles F. Gunther, "How Glass is Made," *TMA Mus News* 15 (1972) 3-22; A. Engle, *Ancient Glass in Its Context* (*Readings in Glass History* 10, Jerusalem 1978) 84, fig. 69 (right); Grose 1978, 90, fig. 33; Vose 1984, 52, color pl.

*Remarks* This jar belongs to Barag's Class B II with Jewish symbols.[1] Although nothing is known about its find circumstances, it may have formed a pair with the jug Cat. No. 172.

The motifs on vessels of the Jewish series are more varied than those on the Christian series (see Remarks Cat. No. 169). The only certain Jewish symbol is the menorah on panel 1. Accompanied by sacred implements such as an incense shovel (referring to holy rituals) and a shofar (referring to Jewish festivals), the menorah evokes the Menorah of the Temple in Jerusalem, taken to Rome after the destruction of the Second Temple in 70 and depicted in relief on the Arch of Titus in Rome. Recessed dots representing flames appear to rise from a horizontal crossbar above the arms of the menorah.[2] The flames would actually have come from lighted oil lamps set on the crossbar, as can be seen on a sixth-century mosaic excavated in a synagogue in Beth Shean where conical glass oil lamps are depicted.[3] The amphora in panel 2 may also be a sacred temple object.[4] The concentric lozenges in panels 3 and 6 may perhaps represent bookbindings as in the Christian series, the single dot in the center of the lozenges possibly indicating the Old Testament (see Cat. No. 169, panel 6). The motif on panel 4 is not clear. To judge from the vessels in Toledo it could be a plant or tree.[5] The stylized date palm in panel 5 is borrowed from polytheistic art where it symbolizes life, hope, victory, and mortality. In Jewish art it has the same significance.[6]

Similarities in shape, color, technique, and style of decoration suggest that the vessels with Jewish motifs were made in the same workshop as the Christian series (see Cat. Nos. 169 and 170). The workshop was probably located in the vicinity of Jerusalem where the vessels were sold to pilgrims visiting sacred places. The Jewish vessels may have contained holy oil or earth from the Temple Mount. Barag established an approximate date for the Christian series of about 578 to 614 or 636/638 on the basis of the motifs represented and historic events. The vessels with Jewish symbols are contemporary but were probably not made after the year 629 when the Byzantine emperor Heraclius forbade Jews to come within three miles of Jerusalem.

[1] For parallels see Barag 1970c, 56. To these add Los Angeles County Museum of Art, no. M.88.129.77, formerly Hans Cohn Collection, von Saldern 1980b, 76, no. 72, color pl. 8. ROM, no. L952.10 (Barag BII 2) has since been published by Hayes 1975, 50-51, no. 97, pl. 8.

[2] Cf. Jerusalem, Schloessinger Collection, Institute of Archaeology, Hebrew University, the menorah forming the handle of a sixth-century bronze oil lamp, *Holy Land* 1986, 241, no. 126.

[3] In addition to conical lamp TMA 1983.45, TMA, *Roman and Early Byzantine Glass* (forthcoming), surviving lamps and menorahs include:

a) Israel Department of Antiquities and Museums, no. 63-962, *Holy Land* 1986, 267-268, no. 148 (with lit.);

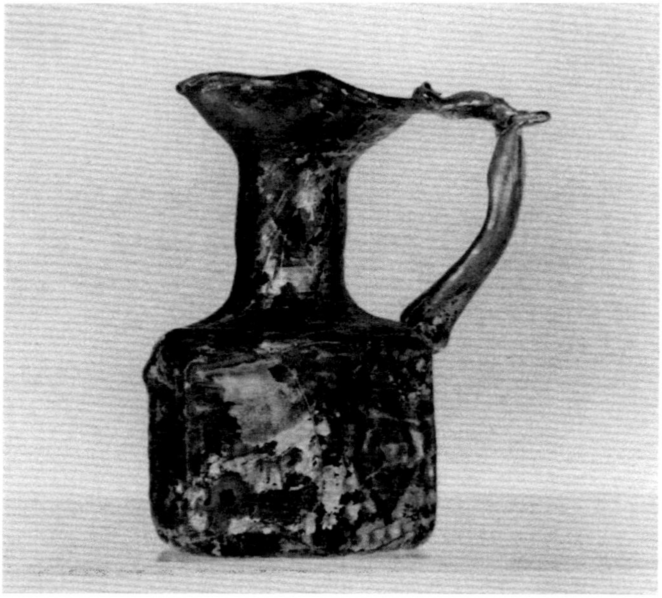

172, panels 5 and 6 (1:2)

b) solid-cast bronze menorah with crossbar for oil lamps, found during excavations in the synagogue at En-Gedi, Israel Department of Antiquities and Museums, no. 70-612, *Holy Land* 1986, 246-247, no. 130 (with lit.);

c) large Byzantine candelabrum on exhibit in the Antikenmuseum, Basel, in 1988 may also be a crossbar from a menorah.

[4] On sacred vessels used in the Temple in Jerusalem: Chapter I: *The First Century: Syro-Palestinian Styles: Hexagonal Bottles with High Relief: Vessels Type.*

[5] Barag 1970c, 49, suggests it is a vessel with curving legs. Goodenough 1953, I, 170, sees in it a tree.

[6] Goodenough 1953, I, passim.

## 172 **Hexagonal Jug with Jewish Symbols**

Palestinian, probably made in the vicinity of Jerusalem
About 578-629
H 13.3 cm; max W body 7.9 cm; D rim 6.7 cm; W base 7.1-6.3 cm
1923.1363, Gift of Edward Drummond Libbey

Transparent to translucent brownish yellow. Similarly colored handle. Medium thin glass. Fabric cannot be determined because of weathering.
Mouth and neck free blown. Body blown into a mold with designs in low relief on the interior. Mold seams cannot be detected. Relief indistinct. Pontil mark 1.4 cm in diameter. Excess glass at tip of handle drawn out thin and pressed flat against top of handle.
Spouted mouth, with rim folded outward, upward, inward, and downward. Tubular neck, with constriction at its base. Shoulder slightly sloped. Hexagonal body. Concave base. Hollow tubular handle applied to shoulder above upper left corner of panel 1 and attached to edge of rim, with horizontal thumbrest ca. 2.5 cm from rim.
Decoration like Cat. No. 171.
*Classification* Barag B II 3.
*Condition* Broken and repaired; several fragments missing from shoulder under application point of handle in panel 1 and at panel 4. Exterior and interior severely pitted and covered with rainbow iridescence.
*Collection* Thomas E. H. Curtis.
*Published* Goodenough 1953, I, 170; III, figs. 394-399; Barag 1970c, 56 (B II 3), 60, fig. 16.
*Remarks* This jug belongs to Barag's Class B II (see Remarks Cat. No. 171). Although both vessels belong to one Series, there is no conclusive evidence that the two vessels were actually blown into the same mold. The flames above the menorah are too damaged to distinguish the presence of a cross bar in panel 1.

## 173 **Hexagonal Jar with Jewish Symbols**

Color plate 28, p. 63
Palestinian, probably made in the vicinity of Jerusalem
About 578-629
H 8.0 cm; max W body 7.7 cm; D rim 5.8 cm; W base 7.2-6.8 cm
1923.1358, Gift of Edward Drummond Libbey

Transparent to translucent dark brownish orange. Medium thin glass. Fabric cannot be determined because of dark color.
Mouth and neck free blown. Body blown into a mold with designs in low relief on the interior. Mold seams cannot be detected. Relief crisp. Ring pontil mark ca. 1.5 cm in diameter.
Funnel mouth, with partly hollow rim folded outward, upward, inward, and downward to narrowest point of opening. Horizontal shoulder. Hexagonal body. Flat base.
On the body, six decorated rectangular panels with designs in intaglio; each panel bordered by recessed dots. In the panels, from left to right: 1, menorah with flames on a three-legged stand; 2, X-shaped element with a stylized heart-shaped leaf at the end of each arm and at the intersection a horizontal cross bar with a loop at each end; 3, two concentric lozenges with a recessed dot at each corner of the panel; 4, two concentric lozenges bordered by small recessed dots and with a crescent in each corner of the panel; 5, empty aedicula consisting of columns with double-tiered bases and cushion-shaped capitals supporting a half-round arch decorated with recessed dots; 6, stylized palm tree. Underside of base undecorated.
*Classification* Barag B V 1.

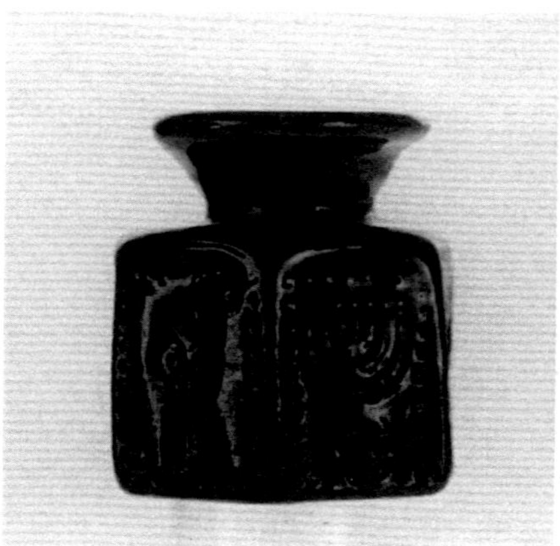

173, panels 6 and 1 (1:2)

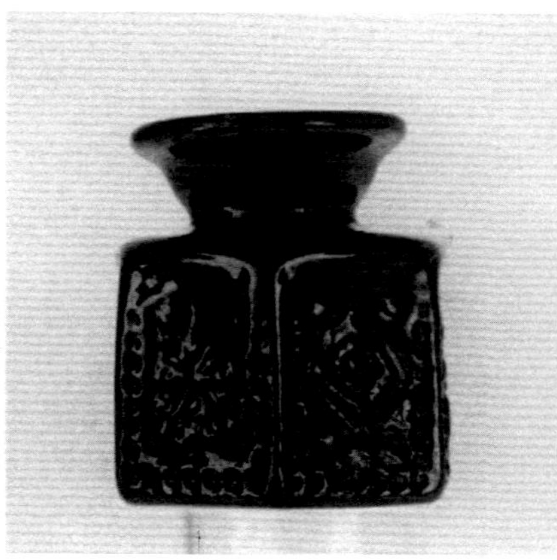

173, panels 2 and 3 (1:2)

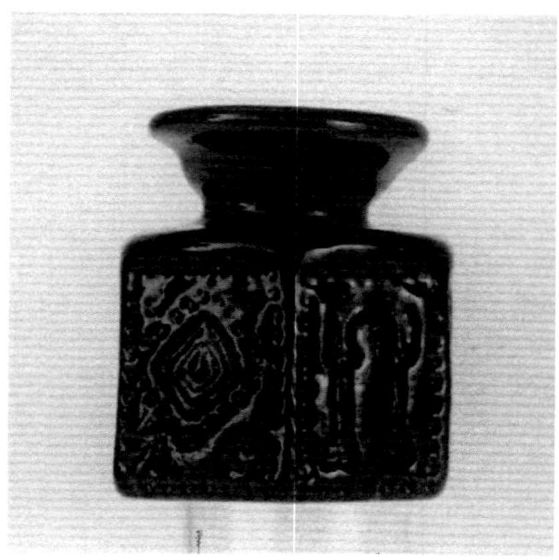

173, panels 4 and 5 (1:2)

*Condition* Intact. Very little weathering. Exterior and interior have areas of white weathering and some scratches and pitting.
*Collection* Thomas E. H. Curtis.
*Published* Reifenberg 1950, ill. 53; Goodenough 1953, I, 171; III, figs. 420-425; Cecil Roth ed., *Jewish ArtAn Illustrated History* (New York 1961), ill. 243; Barag 1970c, 56 (B V 1), 60, fig. 19; Renov 1974, 19, 21.
*Remarks* This jar belongs to Barag's Class B V with Jewish motifs.[1] It seems to form a pair with Cat. No. 174.
The menorah in panel 1 lacks the characteristic cross bar on which to set conical glass oil lamps[2] and is shown without the sacred utensils appearing in Cat. Nos. 171 and 172.
No explanation for the enigmatic object or objects represented in panel 2 has found universal acceptance. The design occurs only in this one Class of Jewish pilgrim vessels. Goodenough calls it a "Solar Cross" and suggests it may have been an ancient symbol of light,[3] but Barag notes that it is a common decorative motif of the period.[4]
The empty aedicula in panel 5 occurs only in the Jewish series. It resembles the Torah shrine flanked by menorahs and other sacred objects represented on many objects and buildings during the Byzantine period, especially in mosaic pavements in synagogues, for example at Beth Shean.[5] It is tempting to think the smaller representations on glass vessels were intended to evoke the shrine in the Temple in Jerusalem in the same way as the menorah in panel 1 of Cat. No. 171. In Barag's closely related series B I, an arched aedicula actually encloses menorah, shofar, and incense shovel. For the lozenges in panels 3 and 4 and the tree in panel 6, see Cat. No. 171.

[1] See Barag 1970c, 56-57, for a list of vessels belonging to this series. Yale, no. 1930.439 (Barag B V 4) has since been published by Matheson 1980, 131, no. 350. Berlin, Staatliche Museen, Antikensammlung, no. 30219,168 (Barag B V 10) has since been published by Platz-Horster 1976, 43, no. 67.

[2] A large menorah with lit conical lamps is represented on the mosaic pavement excavated at Hammath Tiberias, *Holy Land* 1986, 199, fig. 81.

[3] Goodenough 1953, I, 171.

[4] Barag 1970c, 51, n. 81.

[5] The Beth Shean mosaic is reproduced in *Holy Land* 1986, 267-268, no. 148, color.

### 174 Hexagonal Jug with Jewish Symbols

Color plate 28, p. 63
Palestinian, probably made in the vicinity of Jerusalem
About 578-629
H 13.0 cm; max W body 7.7 cm; D rim 7.0-3.6 cm; W base 7.2-6.2 cm
1923.1360, Gift of Edward Drummond Libbey

Transparent to translucent brownish yellow with dark streaks. Similarly colored handle. Medium thin glass. Blowing spirals. Small spherical bubbles in mouth and body, medium-sized vertically elongated bubbles in neck.
Mouth and neck free blown. Body blown into a mold with designs in low relief on the interior. Mold seams cannot be detected. Relief moderately crisp. Ring pontil mark ca. 1.45 cm in diameter. Excess glass at tip of handle drawn out thin and pressed flat against top of handle.
Spouted mouth, with rim folded inward and downward. Cylindrical neck. Horizontal shoulder. Hexagonal body. Flat base. Hollow tubular handle applied to shoulder above junction of panels 4 and 5 and attached to edge of rim with horizontal thumbrest ca. 2.7 cm from rim.
Decoration like Cat. No. 173.

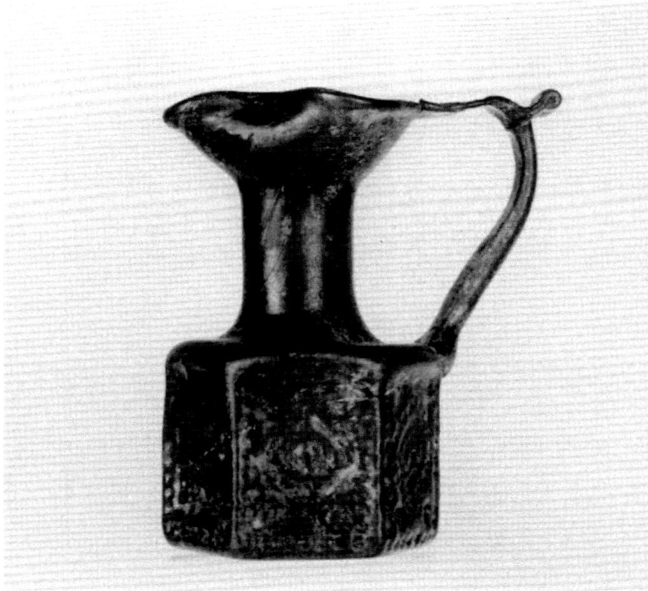

174, panels 2, 3, and 4 (1:2)

*Classification* Barag B V 9.
*Condition* Complete; cracks in mouth adjacent to attachment point of handle. Exterior surface cloudy, pitted, and covered with iridescent film. Interior has iridescent film and incipient pitting.
*Collection* Thomas E. H. Curtis.
*Published* Reifenberg 1937, 60; Reifenberg 1950, ill. 153; Goodenough 1953, I, 169-170; III, figs. 388-393; Barag 1970c, 57 (B V 9), 61, fig. 22; Renov 1974, 19, fig. 2; Bezalel Narkiss, in *Age of Spirituality* 1977, 386-388, no. 354, ill.
*Remarks* See Remarks Cat. No. 173.

### 175 Octagonal Jar with Jewish Symbols

Palestinian, probably made in the vicinity of Jerusalem
About 578-629
H 10.0-9.7 cm; max W body 10.6 cm; D rim 6.7 cm; W base 10.0-9.2 cm
1923.1362, Gift of Edward Drummond Libbey

Translucent dark yellowish brown. Medium thick glass. Blowing spirals. Fabric cannot be determined because of weathering.
Mouth and neck free blown. Body blown into a mold with designs in low relief on the interior. Mold seams cannot be detected. Relief indistinct. Pontil mark ca. 1.8 cm in diameter.
Funnel mouth, with hollow rim folded outward, upward, inward, and downward to narrowest point of opening. Horizontal shoulder. Octagonal body. Flat base.
On the body, eight decorated rectangular panels with designs in sunken relief; each panel bordered by recessed dots. In the panels, from left to right: 1, menorah with flames on a three-legged stand; 2, aedicula consisting of columns with thickened bases and capitals supporting a half round arch framing an indistinct figure (tree?); 3, stylized palm(?) tree; 4, stylized tree; 5, two concentric lozenges with a recessed dot in their centers and a recessed dot in each of the lower corners of the panel; 6, unidentified motif; 7, stylized amphora; 8, stylized tree with a recessed dot in each corner of the panel. Underside of base undecorated.
*Classification* Barag BB I 1.
*Condition* Reconstructed from numerous fragments; several small losses remain as small holes. Exterior covered with rainbow iridescence and severe surface pitting. Interior severely pitted with areas of beige weathering and rainbow iridescence.

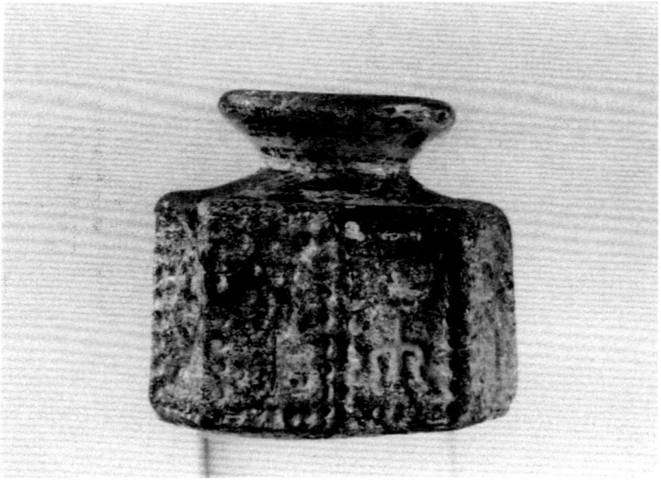
175, panels 8 and 1 (1:2)

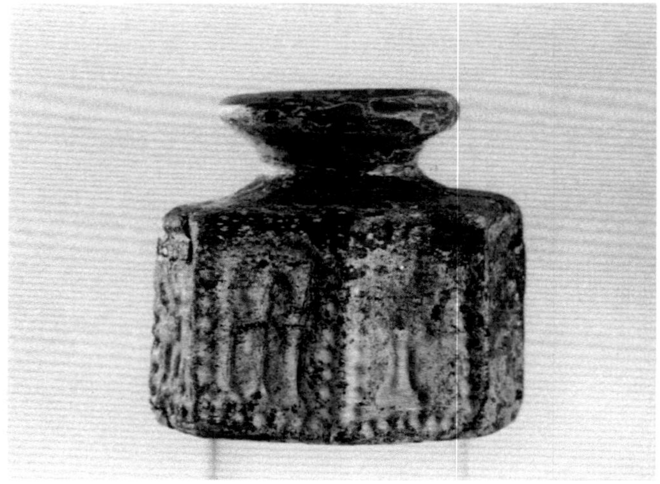
175, panels 2 and 3 (1:2)

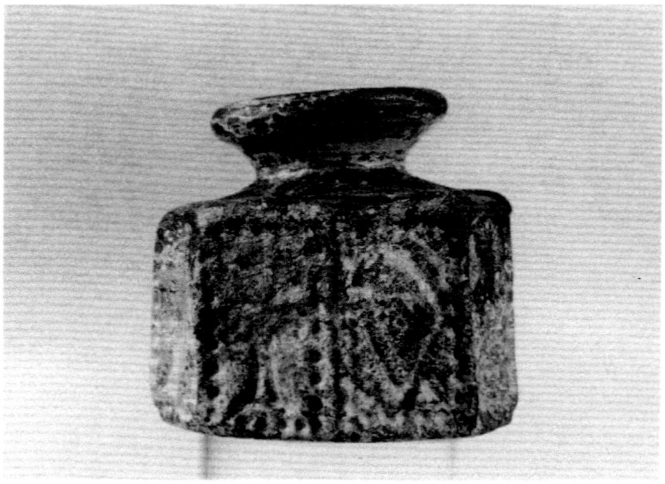
175, panels 4 and 5 (1:2)

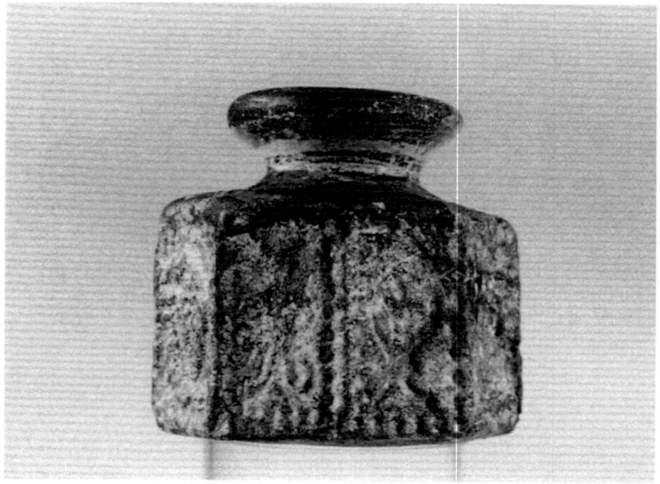
175, panels 6 and 7 (1:2)

*Collection* Thomas E. H. Curtis.
*Published* Goodenough 1953, I, 170-171; III, figs. 414-419; Barag 1970c, 57 (BB I 1), 61, fig. 2; *Glass of the Caesars* 1987, 177, cited at no. 98.
*Remarks* This jar belongs to Barag's Class BB I. See Remarks Cat. Nos. 171 and 173. According to Barag, a shofar (ram's horn) is usually shown below the menorah in panel 1 of this class, but that object cannot be distinguished on this jar. The octagonal shape is known only from jars. Barag suggests it occurred only in Jewish vessels.[1]

[1] See Barag 1970c, 50, 57, for a list of examples belonging to Class BB I, including CMG, no. 50.1.34, formerly J. P. Morgan Collection, H 10.8 cm, *Glass of the Caesars* 1987, 177, no. 98.

## *Hexagonal Jugs with Unassigned Symbols (Barag Class C)*
(Cat. Nos. 176 and 177)

### 176 Hexagonal Jug with Unassigned Symbols

Palestinian, probably made in the vicinity of Jerusalem
Late sixth to early seventh centuries
H 9.5 cm; max W body 7.2 cm; D rim 3.0 cm; W base 6.6-6.0 cm
1923.1357, Gift of Edward Drummond Libbey

Transparent to translucent dark brownish yellow. Medium thin glass. Fabric cannot be determined because of weathering.
Neck free blown. Body blown into a mold with designs in low relief on the interior. Mold seams cannot be detected. Relief crisp. Pontil mark ca. 1.6 cm in diameter.
Rim ground (see below, *Condition*). Short, slightly convex tubular neck, with constriction at its base. Horizontal shoulder. Hexagonal body. Slightly concave base.
On the body, six decorated rectangular panels with designs in intaglio; each panel bordered by recessed dots. In the panels, from left to right: 1, two concentric lozenges with a single recessed dot in their center and recessed dots in each corner of the panel; 2, stylized palm tree; 3, stylized tree; 4, two concentric lozenges without a central recessed dot but with a recessed dot in each corner of the panel; 5, stylized tree; 6, very stylized palm tree.
*Classification* Barag C I 3.
*Condition* Upper portion of neck and entire mouth missing; it is not clear whether the rim was ground in antiquity after breakage or in modern times. A large fragment missing from shoulder at border of panels 2 and 3. Cleaned. Original surface not preserved. Exterior rough, covered with rainbow iridescence and surface pitting. Interior has areas of beige weathering.
*Collection* Thomas E. H. Curtis.
*Published* Barag 1971a, 49 (C I 3).
*Remarks* Barag's Class C with unassigned symbols is an artificial grouping because although the motifs cannot be identified as

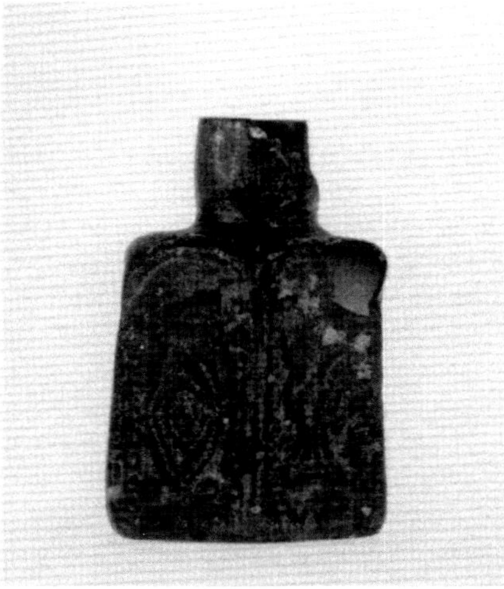

176, panels 1 and 2 (1:2)

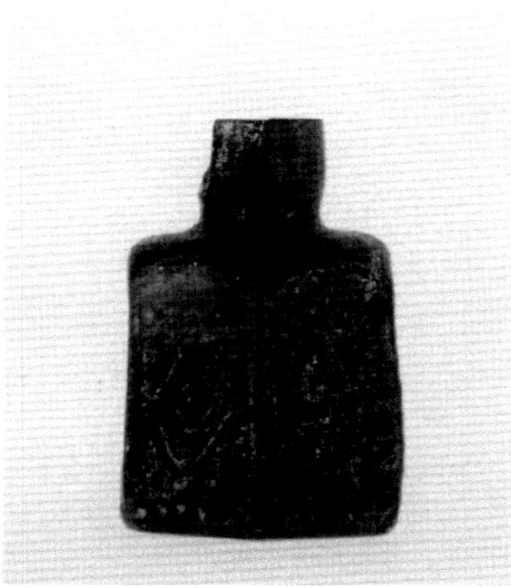

176, panels 4 and 5 (1:2)

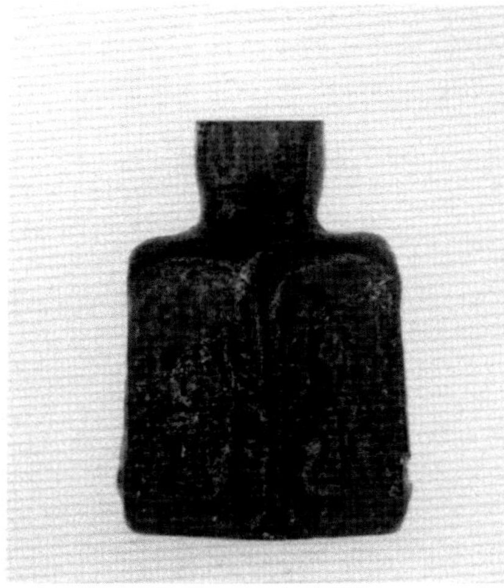

176, panels 5 and 6 (1:2)

specifically Christian or Jewish, the customers who bought the vessels in antiquity doubtless had no difficulty in recognizing what they represented.

This jug belongs to Barag's Class C I with concentric lozenges and various kinds of trees.[1] The lozenges, which have been interpreted as bookbindings, occur in both the Christian and the Jewish Classes. None of the lozenges in Class C I has the four dots that are thought to represent the New Testament (compare Cat. No. 169, panel 4). This may indicate that they are Jewish, but there is no conclusive evidence. Trees do not appear on the Christian vessels of Classes A I-VIII, but they do appear on Classes C II and C III that have recently been identified as Christian.[2] On the other hand, trees are a frequent motif on all the identifiably Jewish Classes (B I-IX and BB I-II). Comparison of the trees in Class C I with those on the Jewish and Christian vessels shows that the tree in panel 2 resembles the palm branches in panels 3 and 5 of the Jewish vessels of Barag's Class B IV, while the tree in panel 3 is best paralleled by the tree in panel 2 of Barag's Christian Class C II. There is therefore as yet no specific evidence for identifying Class C I as Jewish or Christian.

[1] See Barag 1971a, 49, for a list of vessels belonging to Class C I. To these add a jug, *Maxwell* 1979, 183, no. 322. Yale, no. 1955.6.154 (Barag no. C I 1) has since been published by Matheson 1980, 131-132, no. 351.

[2] See Chapter IV: *The Fourth to Early Seventh Centuries: Vessels with Jewish and Christian Designs in Sunken Relief* and Remarks Cat. No. 177.

### 177 **Hexagonal Jug with Christian Symbols**

Palestinian, probably made in the vicinity of Jerusalem
About 578-636 or 638
H 15.9 cm; max W body 6.9 cm; D rim 6.9 cm; W base 5.8-5.1 cm
1923.1354, Gift of Edward Drummond Libbey

Transparent to translucent dusky yellow (near 5 Y 6/4). Similarly colored handle. Medium thin glass. Blowing spirals. Some small spherical bubbles in mouth and body, vertically elongated bubbles in neck.

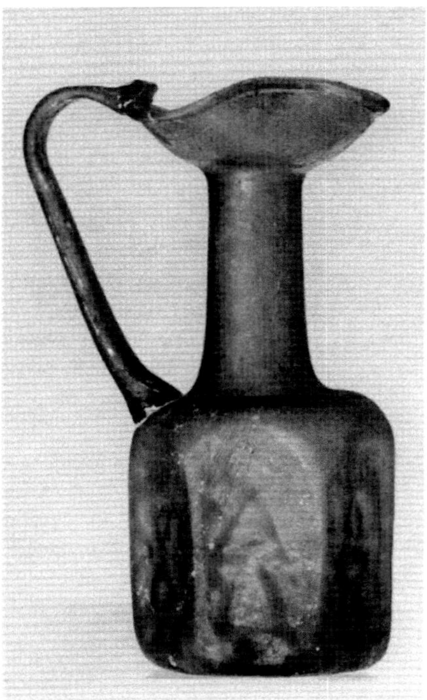

177, panels 1, 2, and 3 (1:2)

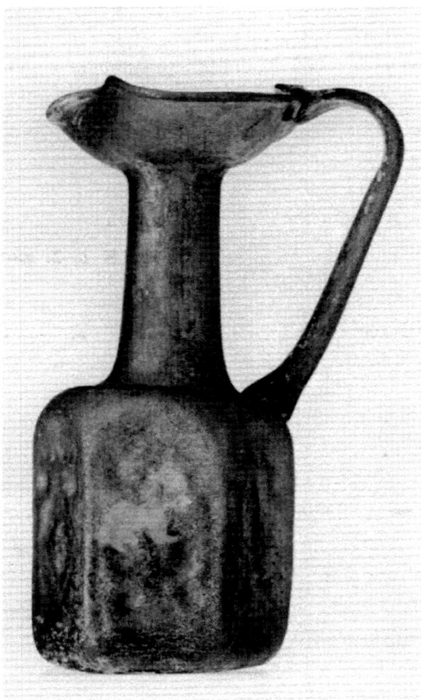

177, panels 4, 5, and 6 (1:2)

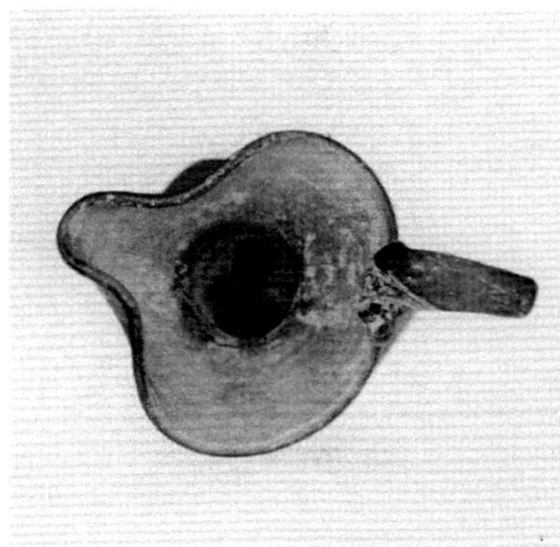

177, mouth and neck (1:2)

Mouth and neck free blown. Body blown into a mold with designs in low relief on the interior. Mold seams cannot be detected. Relief indistinct. Excess glass at tip of handle drawn out thin and folded back against top of handle. Ring pontil mark ca. 1.6 cm in diameter.
Spouted mouth, with rim folded outward, upward, inward, and downward. Tubular neck with slight upward taper. Horizontal shoulder. Hexagonal body. Slightly concave base. Hollow tubular handle applied to shoulder above junction of panels 6 and 1 and attached to rim.
On the body, six decorated rectangular panels with designs in intaglio; each panel bordered by recessed dots. In the panels, from left to right: 1, stylized human (?) figure;[1] 2, unidentified object consisting of two superimposed triangular elements and a recessed dot below, possibly a stylized human (?) figure upside down; 3, stylized human (?) figure; 4, two concentric lozenges with a recessed dot in each corner of the panel; 5, stylized human (?) figure with halo (?); and 6, stylized human (?) figure with wings (?).
*Classification* Barag C IV 3.
*Condition* Complete; crack in upper body adjacent to handle application. Cleaned. Original surface not preserved. Exterior and interior cloudy and rough with surface pitting, and areas of rosy iridescence, and beige weathering.
*Collection* Thomas E. H. Curtis.
*Published* Barag 1971a, 49 (C IV 3), 53, fig. 39.
*Remarks* This jug belongs to Barag's Class C IV with unassigned symbols.[2] The stylized human figures, however, characterize it as Christian since Jewish custom in Palestine did not permit the representation of animals or human forms. The stylized figures resemble the stylized figures identified as saints appearing on another group of vessels with Christian motifs (see Chapter IV: *Jugs with Christian Symbols in Low Relief*). The forms in panels 1 and 3 resemble carelessly rendered versions of the saint on panel 4 of Cat. No. 189, while the figure in panel 5 resembles the saint wrapped in burial bandages in Cat. No. 190, panel 5. For the concentric lozenges in panel 4, see Cat. No. 169.

[1] Barag 1971a, 46, interprets this form as an amphora, but this cannot be recognized on this vessel.

[2] See Barag 1971a, 49, for a list of vessels belonging to this class.

## *Vessels with Designs in Sunken Relief*
(Cat. Nos. 178-188)

### *Vessels with Symbols in Sunken Relief, perhaps Christian*
(Cat. Nos. 178-186)

#### 178 **Hexagonal Jug**

Eastern Mediterranean, probably Syrian
Sixth to early seventh centuries
H 19.1-18.6 cm; W body 8.5-8.3 cm; D rim 3.2 cm; W base 6.4-6.0 cm
1923.1349, Gift of Edward Drummond Libbey

Transparent natural pale green (10 G 6/2). Translucent similarly colored coil and handle. Thin glass. Blowing spirals. Medium-size vertically elongated bubbles in neck. Black specks in handle.
Mouth and neck free blown. Body probably blown into a multipart mold with a base plate. Relief crisp. Pontil mark ca. 1.4 cm in diameter. Excess glass at tip of handle continues into rim coil.
Circular mouth, with unworked rim edge with rim coil. Tapering neck. Shoulder slightly sloped. Hexagonal body with downward taper. Kicked base. Curved coil handle applied to shoulder above panel 1 and attached to upper neck where it continues as rim coil.
Around top edge of rim, rim coil from left to right. On the body, six decorated rectangular panels with designs in sunken relief. In the panels, from left to right: 1, vertical column of four lozenges with a circle in the center of each; 2, lattice; 3, stylized palm branch with thirteen leaves on each side; 4, same as panel 1; 5, same as panel 2; 6, stylized palm branch with sixteen leaves on each side. On the underside of base, in sunken relief, a rosette with twelve petals.
*Condition* Intact. Cleaned. Original surface not preserved. Exterior cloudy and rough with iridescent film, and areas of beige weathering. Interior has areas of beige weathering and pitting.
*Collection* Thomas E. H. Curtis.
*Published* Previously unpublished.
*Remarks* The decoration is similar to Cat. Nos. 179-182. Many other examples are known, and several variations of the panel designs have been recognized.[1] Similarities include the same sequence of three panel designs repeated twice, the presence of a rosette or star on the underside of the base, and similar natural pale green fabric. Parallels suggest manufacture in Syria.[2] However, caution is needed when attempting to assign individual vessels to one workshop. Research for this catalogue has shown that even vessels previously thought to have come from one workshop were actually made in different workshops, see Chapter I: *Hexagonal Vessels with High Relief.*

[1] Eisen 1927, II, 461, 472-473, first recognized the group. See also Harden 1971 and Chapter IV: *Vessels with Designs in Sunken Relief.*

[2] For example, mostly from find spots in Syria or northern Palestine:
a) from a sixth-century tomb at el-Lejjun (Ajlun), Jordan, Donald B. Harden, "Some Tomb Groups of Late Roman Date in the Amman Museum," *AnnAIHV* 3 (1964a, published 1966) fig. 12; Harden 1971, 79, fig. I, no. E;
b) from Syria, Damascus Museum, no. 636, Zouhdi 1964, no. 53, fig. 36;
c) from Anatolia, Bremen 1964, 197, no. 17d;
d) from Kirbet al Karak, Israel, Chicago, University of Chicago, Oriental Institute Museum, no. A30432, unpublished;
e) perhaps from Syria, *BullAIHV* 3 (1964) 50, fig. 36;
f) from Samaria-Sebaste, Crowfoot 1957, 416, fig. 96, no. 8;
g) fragment of lower part of a vessel, Barag transl. 63, nn. 153 and 154;
h) Carnegie, inv. 24055/3, Oliver 1980, 130, no. 227, color ill. 127;
i) Newark, no. 50.1468, yellow-green, H 13.5 cm, Auth 1976, 84, at no. 89, 201, no. 345;
j) Yale, no. 1955.6.143, pale bluish green, H 13.1 cm, formerly Moore Collection, Matheson 1980, no. 361, not ill.;
h) bronze mold with similar motifs, from Sebaste, Samaria, Crowfoot 1957, 450, fig. 109, no. 9.

#### 179 **Hexagonal Jar with Basket Handle**

Eastern Mediterranean, probably Syrian
Sixth to early seventh centuries
H 12.3 cm; W handle 20.1 cm; W body 8.5-7.9 cm; D rim 7.5 cm; W base 5.7-5.0 cm
1923.1364, Gift of Edward Drummond Libbey

Transparent natural pale green (near 10 G 6/2). Translucent similarly colored handle. Thin to medium thin glass. Blowing spirals. Black specks in handles. Fabric cannot be determined because of weathering.
Mouth and neck free blown. Body probably blown into a multipart mold with a base plate. Relief crisp. Pontil mark ca. 1.4 cm in diameter. Excess glass at tip of handle drawn out thin against handle.
Outsplayed rim folded outward, upward, and inward. Deeply concave neck. Shoulder slightly sloped. Hexagonal body with downward taper. Concave base with kick. Two curved coil handles

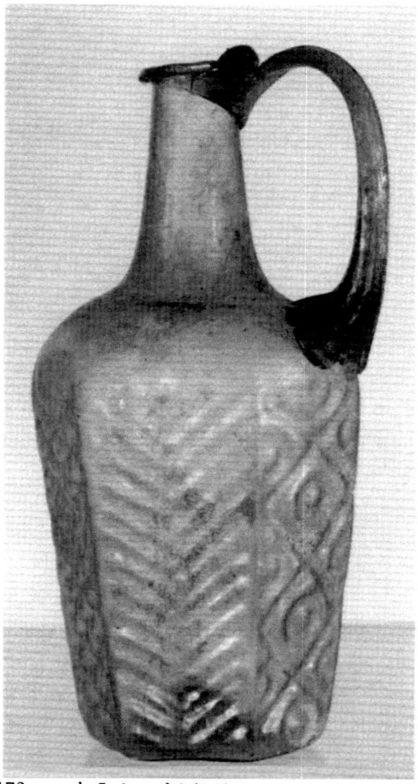

178, panels 5, 6, and 1 (1:2)

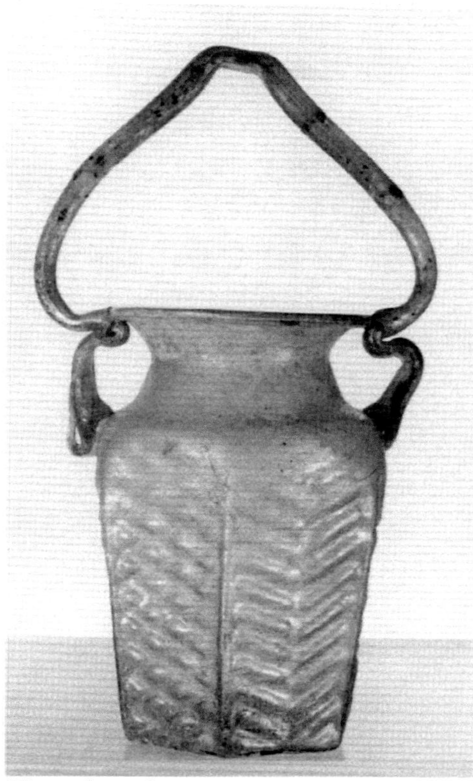

179, panels 2 and 3 (1:2)

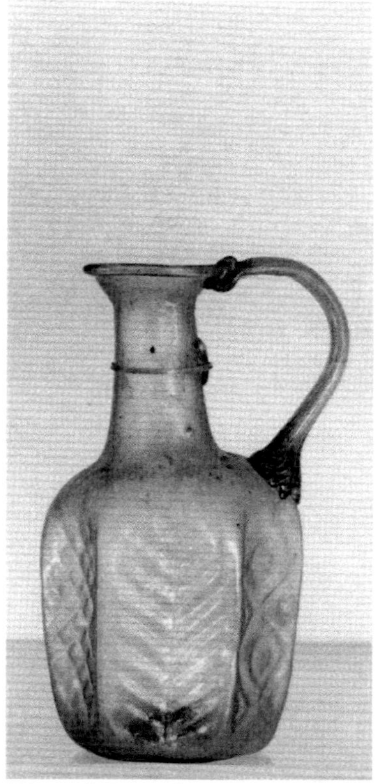

180, panels 5, 6, and 1 (1:2)

applied to shoulder and attached to edge of rim above panels 1 and 4, where they are folded inward, upward, and outward to form a closed loop; separate basket handle made from a coil with circular section applied to the top of one side handle and attached to the opposite side handle; tool mark at top of basket handle.
On the body, six decorated rectangular panels with designs in sunken relief. In the panels, from left to right: 1, vertical column of three lozenges with a circle in the center of each; 2, lattice; 3, stylized palm branch with eleven leaves on each side; 4, same as panel 1; 5, same as panel 2; 6, stylized palm branch with twelve leaves on each side. On underside of base, in sunken outline, a rosette with twelve petals.
*Condition* Complete; basket handle broken off and repaired. Cracks and small loss in upper body on panels 5 and 6. Cleaned. Original surface not preserved. Exterior and interior cloudy and rough with iridescent film, remains of beige weathering, pitting, and ice floes.
*Collection* Thomas E. H. Curtis.
*Published* Previously unpublished.
*Remarks* See Remarks Cat. No. 178.[1] The high, angular basket handle recalls the more extravagant multitiered handles attached to kohl tubes made in the fifth and sixth centuries in inland Palestine.[2]

[1] Compare:
a) Newark, no. 50.1480, pale blue green, H 8.7 cm, Auth 1976, 81, no. 86, with a related but squat body;
b) formerly Constable-Maxwell Collection, with top of basket handle twisted four times to form a vertical loop, pale bluish green, H 18.1 cm, *Maxwell* 1979, 178-179, no. 318.

[2] TMA, *Roman and Early Byzantine Glass* (forthcoming). See also E. Marianne Stern "Regionalism in Roman and Byzantine Palestine," abstract of paper presented at the 91st general meeting of the Archaeological Institute of America, Boston, December 27-30, 1989, published *AJA* 94 (1990) 331. Strong regionalism in shapes has also been noted for Byzantine period pottery in Palestine.

### 180 **Hexagonal Jug**

Color plate 29, p. 63
Eastern Mediterranean, probably Syrian
Sixth to early seventh centuries
H 13.5 cm; max W body 6.9 cm; D rim 3.8 cm; W base 5.5-5.2 cm
1923.1350, Gift of Edward Drummond Libbey

Transparent natural pale green (near 10 GY 6/2). Translucent similarly colored handle and coil. Thin glass. Numerous vertically elongated bubbles in neck; a few spherical bubbles in body. Black specks in neck and shoulder.
Mouth and neck free blown. Body probably blown into a multipart mold with a base plate. Relief crisp. Pontil mark ca. 1.1 cm in diameter. Excess glass at tip of handle folded back against top of handle.
Circular mouth, with rim folded outward, upward, and inward. Tubular neck with slight upward taper. Shoulder slightly sloped. Hexagonal body with slight downward taper. Kicked base. Curved coil handle applied to shoulder between panels 1 and 2 and attached to edge of rim where it is folded downward/inward and upward/outward to form a closed loop.
Around middle of neck, neck coil from left to right, with buckle at point of attachment. On the body, six decorated rectangular panels with designs in sunken relief. In the panels, from left to right: 1, two and a quarter lozenges with a circle in the center of each; 2, lattice; 3, stylized palm branch with eleven or twelve leaves on each side; 4, same as panel 1; 5, same as panel 2; 6, stylized palm branch with nine leaves on each side. On underside of base, in sunken outline, a star with six forked arms.
*Condition* Intact. Cleaned. Original surface not preserved. Exterior cloudy and rough with iridescent film. Interior pitted with areas of cream-colored weathering.
*Collection* Thomas E. H. Curtis.
*Published* Previously unpublished.
*Remarks* See Remarks Cat. No. 178.

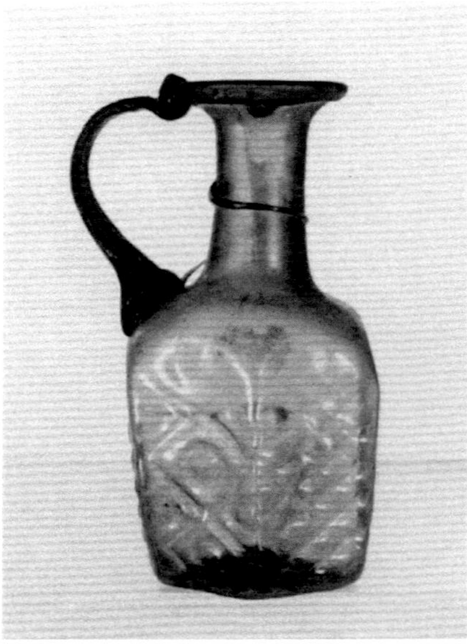

181, panels 4 and 5 (1:2)

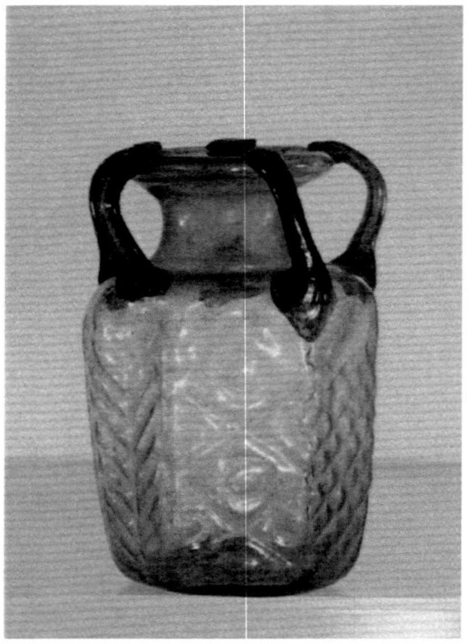

182, panels 6, 1, and 2 (1:2)

181 **Hexagonal Jug**

Color plate 29, p. 63
Eastern Mediterranean, probably Syrian
Sixth to early seventh centuries
H 12.7 cm; W body 6.7-6.3 cm; D rim 4.3 cm; W base 5.5-4.8 cm
1923.1365, Gift of Edward Drummond Libbey

Transparent dusky yellow (near 5 Y 6/4). Translucent brown coil and handle. Thin glass. A few vertically elongated bubbles in neck. Black specks in handle.
Mouth and neck free blown. Body probably blown into a multipart mold with a base plate. Relief crisp. Pontil mark ca. 1.1 cm in diameter. Excess glass at tip of handle folded back against top of handle.
Circular mouth, with rim folded outward, upward, and inward. Tubular neck. Shoulder slightly sloped. Hexagonal body with slight downward taper. Kicked base. Curved coil handle applied to shoulder above panel 3 and attached to edge of rim where it is folded downward/inward and upward/outward to form a closed loop. Around middle of neck, neck coil from left to right. On the body, six decorated rectangular panels with designs in sunken relief. In the panels, from left to right: 1, two and a half lozenges with a circle in the center of each; 2, lattice; 3, stylized palm branch with eight leaves on each side; 4, same as panel 1; 5, same as panel 2; 6, stylized palm branch with eleven leaves on each side. On underside of base, in sunken outline, a star with six forked arms.
*Condition* Intact. Cleaned. Original surface not preserved. Exterior cloudy and rough with iridescent film and surface pitting. Interior has areas of cream-colored weathering and ice floes.
*Collection* Thomas E. H. Curtis.
*Published* Previously unpublished.
*Remarks* See Remarks Cat. No. 178.

182 **Hexagonal Jar with Multiple Handles**

Eastern Mediterranean, probably Syrian
Reported to be from Hebron, Israel
Sixth to early seventh centuries
H 11.1 cm; W body 7.0-6.4 cm; D rim 5.1 cm; W base 5.8-5.3 cm
1953.138, Purchased with funds from the Libbey Endowment, gift of Edward Drummond Libbey

Transparent natural pale green (near 5 G 7/2). Five translucent streaked grayish olive green handles. Medium thin glass. Numerous small spherical bubbles in body, some vertically elongated in neck.
Rim and neck free blown. Body probably blown into a multipart mold with a base plate. Relief crisp. Pontil mark ca. 1.1 cm in diameter. Excess glass at tips of handles folded back against tops of handles.
Hollow rim, folded outward, upward, inward, and downward. Deeply concave neck. Horizontal shoulder. Hexagonal body with slight downward taper. Kicked base. Five curved coil handles applied to shoulder and attached to top of rim.
On the body, six decorated rectangular panels with designs in sunken relief. On the panels, from left to right: 1, two and a half lozenges with a circle in center of each; 2, lattice; 3, stylized palm branch with and nine or ten leaves on each side; 4, same as panel 1; 5, same as panel 2; 6, stylized palm branch with nine leaves on each side. On underside of base, in sunken outline, a star with six forked arms.
*Condition* Complete; stress cracks on bottom.
*Collection* Acquired from Susette Khayat, New York; probably from the collection of her father, Azeez Khayat.
*Published* Previously unpublished.
*Remarks* See Remarks Cat. No. 178.

183 **Pentagonal Jug**

Color plate 29, p. 63
Eastern Mediterranean, probably Syrian
Sixth to early seventh centuries
H 21.2 cm; W body 8.8 cm; D rim 6.5 cm; W base 6.8 cm
1923.1351, Gift of Edward Drummond Libbey

Transparent natural pale green (near 5 G 7/2). Similarly colored thread. Translucent yellowish green handle (not in rock color chart). Medium thin glass. Blowing spirals. A few small and medium vertically elongated bubbles in neck.
Mouth and neck free blown. Body probably blown into a multipart mold with a base plate. Relief crisp. Pontil mark ca. 1.6 cm in diameter. Excess glass at tip of handle folded back against top of handle (before it was folded); tool mark at fold of handle.
Circular mouth, with rim folded outward, upward, and inward. Tubular neck. Shoulder slightly sloped. Pentagonal body with

downward taper. Kicked base. Curved coil handle applied to shoulder over end of neck coil above panel 3 and attached to edge of rim where it is folded downward/inward and upward/outward to form a closed loop.
Around lower half of neck, four revolutions of thread trailed on downward from left to right. On the body, five decorated rectangular panels with designs in sunken relief. In the panels, from left to right: 1, two and a half lozenges with a circle in the center of each; 2, lattice; 3, stylized palm branch with twelve curved leaves on each side; 4, same as panel 2; 5, stylized palm branch with twelve curved leaves on the left side and eleven on the right side. On underside of base, in sunken outline, a rosette with five petals.
*Condition* Complete; cracked mouth from attachment of handle to rim on opposite side. Cleaned. Original surface not preserved. Exterior cloudy and rough, covered with milky weathering and iridescent film. Interior pitted and covered with cream-colored weathering and iridescent film.
*Collection* Thomas E. H. Curtis.
*Published* Previously unpublished.
*Remarks* The designs and sequence of panels closely resemble those on the sides of hexagonal jugs and jars Cat. Nos. 178-182, but this jug has only five sides.

184 **Hexagonal Jug**

Eastern Mediterranean, perhaps Syrian
Probably sixth to early seventh centuries
H 10.4 cm; max W body 5.6 cm; D rim 3.0 cm; W base 4.3-4.0 cm
1923.1355, Gift of Edward Drummond Libbey

Transparent to translucent dark royal blue. Translucent similarly colored handle and coil. Thin glass. A few small spherical bubbles.

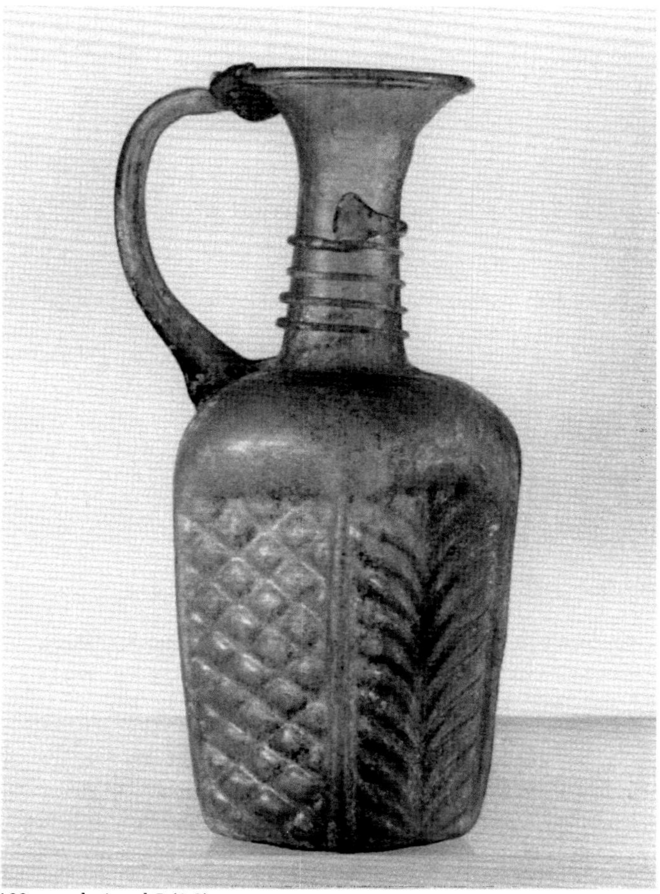

183, panels 4 and 5 (1:2)

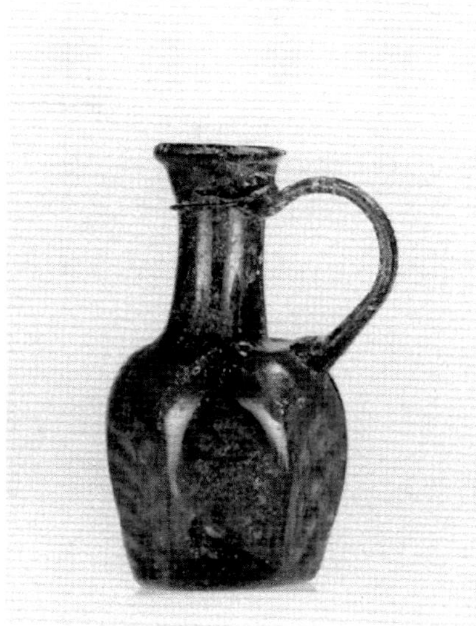

184, panels 6, 1, and 2 (1:2)

Rim and neck free blown. Body probably blown into a multipart mold, possibly without a base plate. Relief moderately indistinct, perhaps due to being slightly expanded after extraction from the mold. Pontil mark ca. 0.7 cm in diameter. Excess glass at tip of handle continues into neck coil.
Rim folded outward, upward, and inward. Tubular neck. Shoulder slightly sloped with concavity on one side. Hexagonal body with a slight downward taper. Concave base. Solid coil handle applied to shoulder above panel 2 and attached to upper part of neck.
Around upper part of neck, neck coil continuing from attachment point of handle, from left to right. On the body, six decorated panels with designs in sunken relief. In the panels, from left to right: 1, two lozenges with a circle in the center of each; 2, lattice; 3, dots; 4, same as panel 1; 5, lattice; 6, stylized palm branch with six leaves on each side. Underside of base undecorated.
*Condition* Intact. Exterior glossy in areas, pitted and cloudy in others with areas of beige weathering and a rust colored spot on base. Interior has areas of beige weathering.
*Collection* Thomas E. H. Curtis.
*Remarks* The similarities to Cat. Nos. 178-183 suggest manufacture in the same general area.[1] The underside of the base bears no design but has many small irregularities that may have been produced by placing the open mold on a slightly rough surface.

[1] Compare: Oppenländer Collection, no. 2425, dark blue, H 10.4 cm, von Saldern et al. 1974, 181-182, no. 503.

185 **Hexagonal Jug**

Eastern Mediterranean, perhaps Syrian
Probably sixth century
H 10.8 cm; W body 4.9-4.4 cm; D rim 5.1-4.1 cm; W base 4.2-3.9 cm
1923.1343, Gift of Edward Drummond Libbey

Transparent brownish yellow (near 5 Y 5/6). Translucent similarly colored hollow(?) tubular(?) handle. Thin glass. Blowing spirals. A few small bubbles, vertically elongated in neck, spherical in body. Mouth and neck free blown. Body probably blown into a multipart mold with a base plate. Relief moderately crisp. Pontil mark ca. 1.1 cm in diameter. Excess glass at tip of handle drawn out thin against handle.

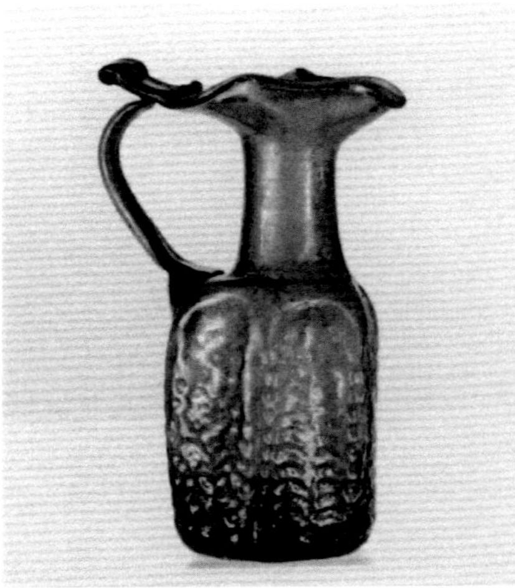

185, panels 2 and 3 (1:2)

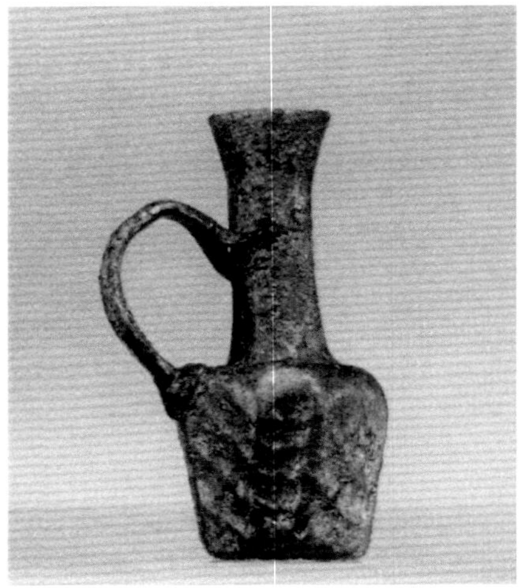

186 (1:2)

Spouted mouth, with rim folded outward, upward, and inward. Tubular neck. Shoulder slightly sloped. Hexagonal body. Kicked base. Curved coil handle applied to shoulder above panel 1 and applied to edge of rim; pinched thumbrest 2.0 cm from rim.
On the body, six decorated panels with designs in sunken relief, each with a border of dots. In the panels, from left to right: 1, lattice; 2, floral pattern composed of dots; 3, stylized palm branch with eleven curved leaves on each side; 4, stylized palm branch with ten straight leaves on each side; 5, irregular pattern of nine dots, perhaps a floral pattern like panel 2; 6, chevron pattern composed of five vertical columns of nineteen dots. On underside of base, in sunken outline, six radiating double lines surrounded by dots in the spaces.
*Condition* Complete, but lower body cracked. Exterior and interior slightly cloudy with areas of cream-colored weathering.
*Collection* Thomas E. H. Curtis.
*Published* Previously unpublished.
*Remarks* The similarities to Cat. Nos. 178–183 suggest manufacture in the same general area.

186 **Octagonal Jug**

Eastern Mediterranean, perhaps Syrian
Probably sixth to seventh centuries
H 10.0 cm; max W body 4.65 cm; D rim 2.7 cm; W base 3.5 cm
1923.1356, Gift of Edward Drummond Libbey

Transparent natural green. Exact color cannot be determined because of weathering. Thin glass. Fabric cannot be determined because of weathering.
Mouth and neck free blown. Body probably blown into a multipart mold with a base plate. Relief indistinct. Pontil mark ca. 1.25 cm in diameter. Excess glass at tip of handle continues into neck coil.
Circular mouth, with unworked rim. Tubular neck. Horizontal shoulder. Octagonal body with downward taper. Kicked base. Curved coil handle applied to shoulder above a narrow panel and attached to middle of neck where it continues into neck coil.
Neck coil continuing from handle around middle of neck from left to right. On the body, eight decorated panels with designs in sunken relief. On each of the four wide panels, lozenges; on each of the four narrow panels, a vertical column of four or five rectangles. On underside of base, an indistinct mold-blown (?) design.
*Condition* Most of neck coil missing. Exterior severely pitted and covered with thick, flaking rainbow iridescence. Interior severely pitted with areas of dark brown weathering.
*Collection* Thomas E. H. Curtis.
*Published* Previously unpublished.
*Remarks* The similarities to Cat. Nos. 178–185 suggest manufacture in the same area. The four narrow panels are deeply indented.

### *Inverted Conical Jugs with Overall Geometric Patterns in Sunken Relief*
(Cat. Nos. 187 and 188) [1]

187 **Inverted Conical Jug**

Probably Palestinian
Probably sixth to early seventh centuries
H 9.2 cm; max D body 4.7 cm; D rim 3.9 cm; D base 3.2–2.9 cm
1923.1337, Gift of Edward Drummond Libbey

Transparent blue (darker than 5 B 6/2). Translucent similarly colored coil and handle. Thin glass. Blowing spirals. Very bubbly with black specks.
Rim and neck free blown. Body probably blown into a one-part patterned mold and slightly expanded. Pontil mark ca. 1.2 cm in diameter. Excess glass at tip of handle folded back against top of handle.
Circular mouth, with rim folded outward, upward, and inward. Tubular neck tapering upward. Inverted conical body. Kicked base. Curved coil handle applied to shoulder and attached to rim where it is folded down to form a closed loop against the side of the mouth.
Neck coil, from left to right around middle of neck. On the body, mold-blown pattern of seven alternating rows of sunken hexagons. On underside of base, nine radial indentations.
*Condition* Intact. Exterior and interior have areas of brown weathering and incipient pitting.
*Collection* Thomas E. H. Curtis.
*Published* Auth 1976, 86, cited at no. 91.
*Remarks* Same Type as Cat. No. 188, but the neck of this jug is noticeably off center. These two jugs represent a small group of vessels blown into a one-part mold that provided the pattern as well as the inverted, slightly conical shape of the vessel. At least two different molds can be identified: the first with seven rows of sunken dots on the body and a star with nine rays on the base (as used for both Toledo Museum jugs) [2] and the second with five to six rows of dots on the body and a star with eight rays on the underside of the base. [3] Most of the jugs were blown in tints of blue or natural blue-green glass. The sunken design and strongly curved handle suggest a sixth- to seventh-century date for this group,

but a fourth-century date cannot be excluded. Find spots suggest that these vessels were used and might have been made in inland Palestine.[4]

1 See also Chapter IV: *Inverted Conical Jugs with Overall Geometric Pattern.*

2 For example, Newark no. 50.1467, yellow-green, H 14.5 cm, Auth 1976, 86, no. 91.

3 For example, ROM, no. 950.157.75, deep turquoise green, H 8.5 cm, Hayes 1975, 112, no. 425, pl. 27.

4 Examples from known find spots:
- a) excavated at Kefer Kenna, a few km north of Nazareth, Bagatti 1967a, 231, no. 112, fig. 5, no. 112;
- b) Krefeld, Germany, Bremen Collection, from an unspecified tomb in Palestine, Bremen 1964, 188, no. 8;
- c) formerly Azeez Khayat collection, probably from northern Palestine, Khayat 1905, ill.

### 188 **Inverted Conical Jug**

Probably Palestinian
Probably sixth to early seventh centuries
H 8.8 cm; max D body 4.7 cm; D rim 3.3 cm; D base 3.3 cm
1923.1336, Gift of Edward Drummond Libbey

Transparent moderate blue green (near 5 BG 4/6). Translucent similarly colored coil and handle. Thin glass. Very bubbly. Blowing spirals.
Mouth and neck free blown. Body probably blown into a one-part patterned mold and slightly expanded. Pontil mark ca. 1.2 cm in diameter. Excess glass at tip of handle folded back against top of handle.
Rim folded outward, upward, and inward. Circular mouth. Tubular neck tapering upward. Inverted conical body. Base with low tooled kick. Curved coil handle applied to shoulder and attached to rim where it is folded down to form a closed loop against the side of the mouth.
Neck coil, from left to right around middle of neck. On the body, mold-blown pattern of seven alternating rows of sunken hexagons. On underside of base, nine radial indentations.
*Condition* Complete, except for small hole in upper body. Exterior and interior covered with light brown weathering.
*Collection* Thomas E. H. Curtis.
*Published* Auth 1976, 86, cited at no. 91.
*Remarks* See Remarks Cat. No. 187.

## *Jugs with Christian Symbols in Low Relief* (Cat. Nos. 189-192)

### 189 **Hexagonal Jug with Christian Symbols**

Eastern Mediterranean, probably Syrian
Fifth to seventh centuries
H 19.1 cm; W body 6.75-6.0 cm; D rim 6.1-6.0 cm; W base 6.2-5.6 cm
1923.1352, Gift of Edward Drummond Libbey

Transparent decolorized glass. Exact color cannot be determined because of weathering. Thin glass. Blowing spirals. Numerous vertically elongated bubbles in neck and body.
Mouth and neck free blown. Body probably blown into a multipart mold. Mold seams cannot be detected. Relief crisp. Pontil mark ca. 1.4 cm in diameter. Excess glass at tip of handle drawn back against top of handle.
Spouted mouth, with rim rounded and thickened in flame. Tapering neck. Horizontal shoulder. Hexagonal body. Kicked base. Curved coil handle applied to shoulder above panel 3 and attached

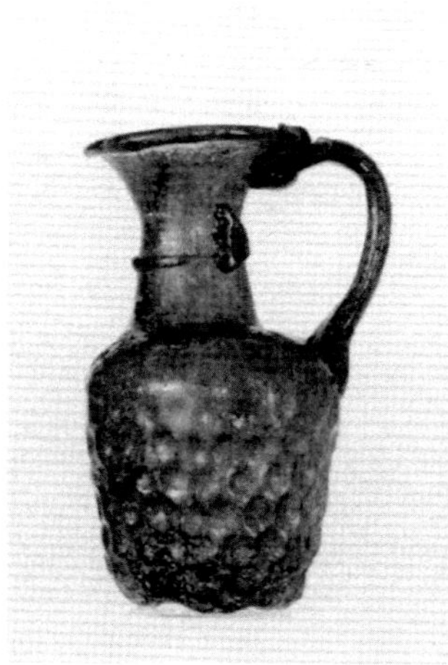
187 (1:2)

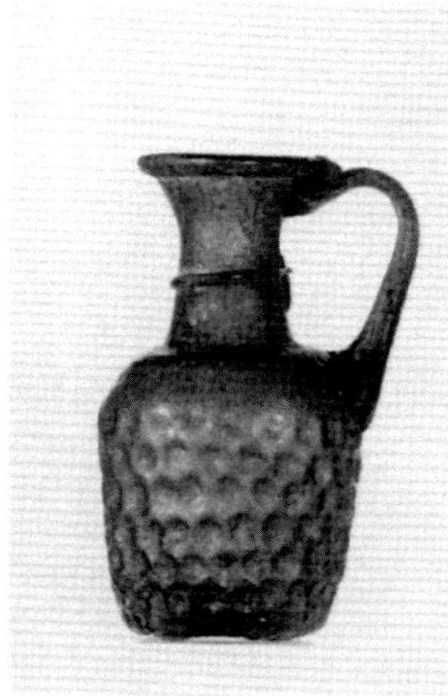
188 (1:2)

to edge of rim where it is folded downward/inward and upward/outward to form a closed loop.
On the body, six decorated panels with designs in low relief. In the panels, from left to right: 1, cross *fourchée* above a rectangular element (column? altar?) decorated with a zigzag and dots in the spaces, below this a thick raised horizontal border (?); 2, stylized palm branch with twelve leaves on the left side and eleven on the right; 3, lattice pattern of seven lozenges with a dot in the center of each; 4, human figure (Stylite saint?) wearing a hood and holding a quartered circle (shield? wheel with four spokes?), the lower part of the figure consisting of three thick parallel vertical lines, possibly representing the column on which the saint perched; 5, same as panel 3; 6, stylized palm branch with ten leaves on one side and twelve on the other. On the handle, a series of horizontal tool marks along length of handle. Underside of base undecorated.

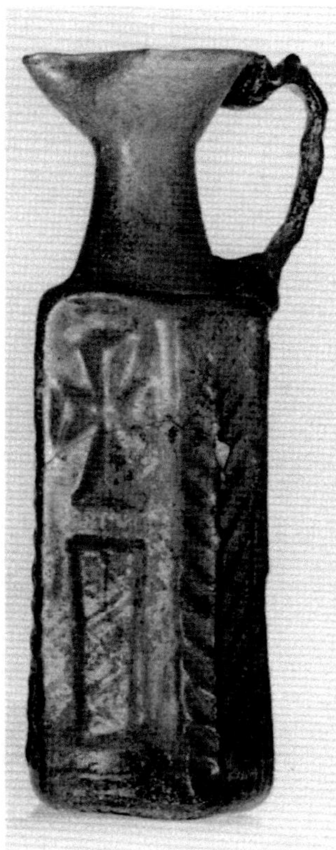
189, panels 1 and 2 (1:2)

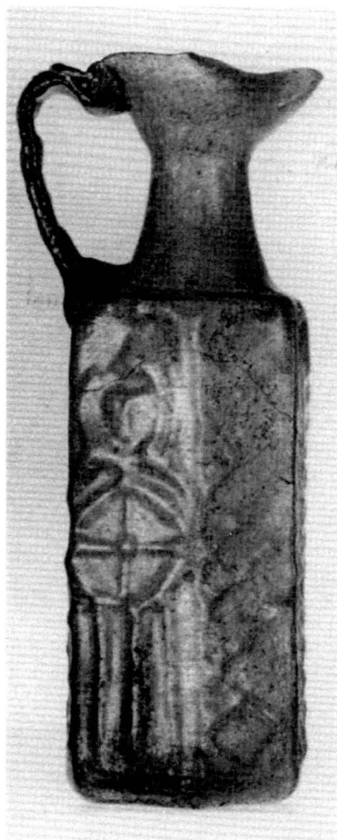
189, panels 4 and 5 (1:2)

*Condition* Upper body broken and repaired. Exterior covered with pink and purple iridescence and severe surface pitting. Interior severely pitted with areas of brown weathering.
*Collection* Thomas E. H. Curtis.
*Published* Joseph Philippe, *Le monde byzantine dans l'histoire de la verrerie* (Bologna 1970) 34, 35, ill.
*Remarks* Panel 4 may represent same symbol as Cat. No. 177, panels 1 and 3: a stylized human (?) figure.[1]
Stylite saints were Christian ascetics who retreated from the world by passing large portions of their lives in meditation on top of giant columns (Greek *stylos*, hence stylite, one who stands on a column). The first and most famous was Saint Symeon Stylites (389-459), whose monastery at Qala'et Sema Can (ancient Tellnesin), near Antioch on the Orontes, remained a site of pilgrimage into the nineteenth century.[2] The saint may be represented twice on this jug, first on panel 1 symbolically as the cross above the column; second on panel 4 more naturalistically wearing a pointed monk's hood and holding a quartered circle (perhaps a simplified version of the victorious cross surrounded by a wreath).[3] Theodoret, the contemporary and biographer of the first Symeon, mentions pilgrims from Spain, Britain, and Gaul. These jugs were possibly made nearby as souvenirs; they have been found as far away as Egypt.

[1] Additional examples (per Matheson 1980, 135, at no. 357):
a) Yale, no. 1955. 6.147, formerly Moore Collection, pale green, H 19.5 cm, Eisen 1927, II, 474, fig. 198; Matheson 1980, 134-135, no. 357;
b) CMG, no. 66.1.230, formerly Sangiorgi Collection, pale yellowish green, H 21.7 cm, *Glass of the Caesars* 1987, 176, no. 97; David Whitehouse, *Glass of the Roman Empire* (Corning, N.Y. 1988) 44-45, color ill.;
c) from Aleppo, Syria, BM, 1911.5-13.1, dull green, H 22.5 cm, *Masterpieces* 1968, 63, no. 80;
d) formerly Constable-Maxwell Collection, *Maxwell* 1979, no handle, greenish, H 21.7 cm, 180, no. 319, ill.

[2] See also Chapter IV: *Jugs with Christian Symbols in Low Relief.*

[3] *Romans and Barbarians*, exh. cat., Museum of Fine Arts (Boston 1977) 184-196, nos. 226 and 227. The terra-cotta pilgrim's tessera or amulet from Beirut, no. 227, honoring Saint Symeon the Younger (521-592), shows the saint as a bearded bust, draped in a cloak and holding a rectangular form (a book or medallion?) to his breast, on top of a large column represented as a basket capital on top of a shaft with a zig-zag lozenge pattern as in panel 1 of this jug. The tessera (D 3.4 cm) is further ornamented with forked crosses, palm branches, seated Mary and Christ Child, pilgrims, and flying angels bearing wreaths.

## 190 Hexagonal Jug with Christian Symbols

Color plate 30, p. 64
Eastern Mediterranean, probably Syrian
Fifth to seventh centuries
H 19.7 cm; D shoulder 5.5-5.3 cm; D rim 4.5 cm; W base 5.0-4.8 cm
1948.13, Purchased with funds from the Libbey Endowment, gift of Edward Drummond Libbey

Transparent decolorized glass with a greenish tinge. Exact color cannot be determined because of weathering. Thin glass. Blowing spirals. A few vertically elongated bubbles in neck and body. Mouth and neck free blown. Body blown into a two-part mold. One continuous mold seam between panels 1 and 6 and between panels 3 and 4 extends across underside of base. Pontil mark ca. 1.2 cm in diameter. Relief crisp. Excess glass at tip of handle continues into neck coil.
Circular mouth, with rim rounded and thickened in flame. Tapering neck. Concave shoulder. Hexagonal body. Kicked base. Curved coil handle applied to shoulder above panel 5 and attached to upper part of neck; pinched thumbrest ca. 1.8 cm from neck.

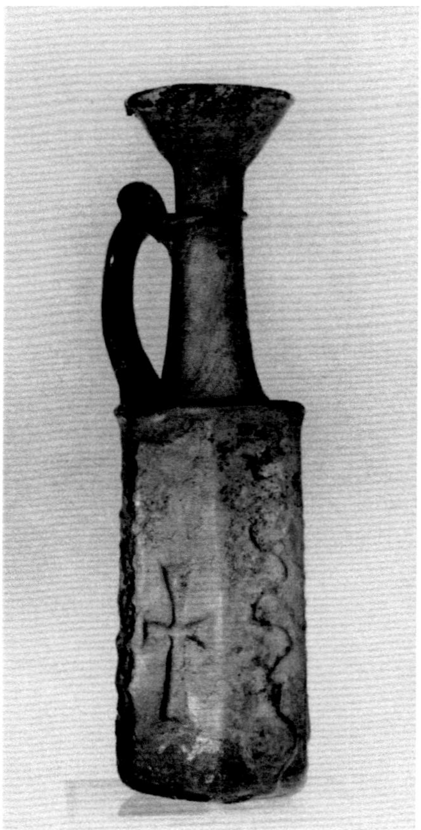
190, panels 1 and 2 (1:2)

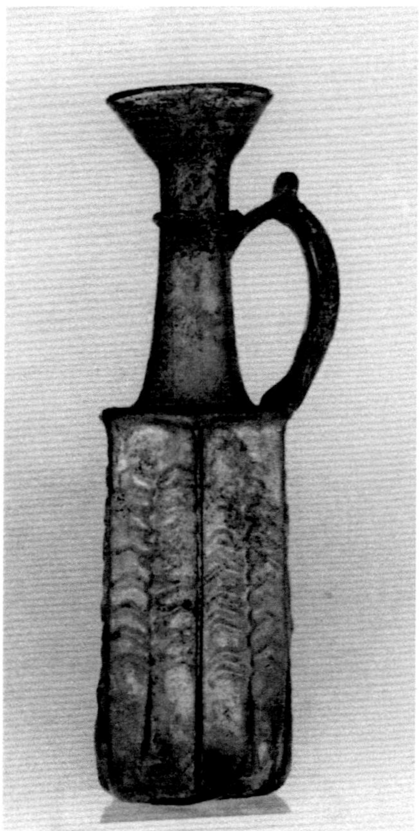
190, panels 3 and 4 (1:2)

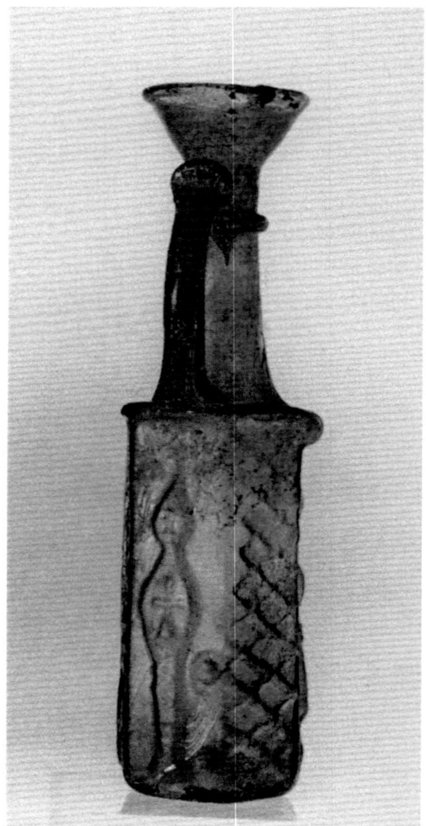
190, panels 5 and 6 (1:2)

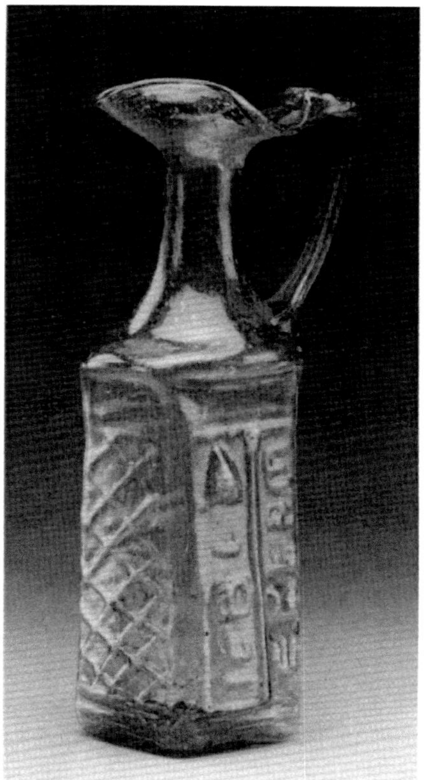

191, panels 2 and 3 (1:2)

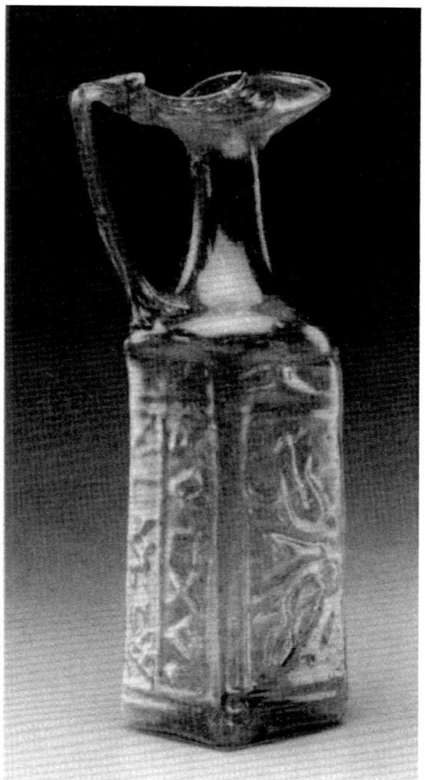

191, panels 4 and 1 (1:2)

Neck coil continuing from handle attachment around upper part of neck from left to right. On the body, six decorated rectangular panels with designs in low relief. In the panels, from left to right: 1, cross *fourchée* below a dot enclosed by a circle; 2, ampulla or jug embossed on the side with a circle out of which a snake or vine curls upward, with several indistinct elements beside to and dangling from it; 3, stylized tree with tall trunk and nine curved branches; 4, stylized tree with ten curved branches on the left side and eleven on the right; 5, figure of a saint in funerary shroud with pointed top of head, with one cross on its torso, a second cross above its head, with two dots on either side of the cross and a zigzag pattern on the lower part of the body (perhaps the wrappings of a mummy for an Egyptian saint); 6, lattice. Underside of base undecorated.

*Condition* Complete, except for small segment of neck coil. Cleaned. Original surface not preserved. Exterior cloudy and severely pitted with orange iridescence and ice floes. Interior pitted with brown and cream-colored weathering.

*Collection* Acquired from Fahim Kouchakiji, New York.

*Published* Riefstahl 1961, 41, ill.; Rudolf M. Riefstahl, "The Complexities of Ancient Glass," *Apollo* 86, no. 70 (December 1967) 434, ill.; Labino 1968, 21, 24, ill.; *TMA, Art in Glass* 1969, 32, ill.; Joseph Philippe, *Le monde byzantin dans l'histoire de la verrerie* (Bologna 1970), 36, ill.; Harden 1971, pl. V, no. b; Charles F. Gunther, "How Glass is Made," *TMA Mus News* 15, no. 1 (1972) 16, ill.; Grose 1978, 90, fig. 33; A. Engle, *Ancient Glass in Its Context* (*Readings in Glass History* 10, Jerusalem 1978) 84, ill.; Vose 1984, 52, color. ill.

*Remarks* See Remarks Cat. No. 189. The figure in panel 5 may be a stylized representation of a Stylite saint, wearing a pointed monk's hood, or possibly of Lazarus, who was frequently represented in Early Christian art in his funeral wrappings.[1]

A jug at Yale offers a close parallel to the six panels on this jug, but it is smaller and arranges the six designs in a different order, suggesting that not only were the motifs copied, but they could also be reduced in scale and the panels composing the mold sections rearranged by the original or other workshops.[2]

For the snake or, more probably, vine rising from the ampulla on panel 2, compare the mosaic floor in the church at ed-Deir Ma'in.[3] If the form represents a snake, it may refer to a legend attached to Saint Symeon.[4]

1 Lazarus, however, is usually shown standing in an aedicula on top of a platform with steps, for example, *Age of Spirituality* 1977, 412, no. 370, marble plaque from a loculus tomb of a child at Rome, dated 290-300, the earliest representation in sculpture of the Raising of Lazarus; see also 444-445, nos. 403 and 404, two cut-glass vessels made in Cologne during the fourth century; 578-580, nos. 518 and 519, two ivory pyxides, Syria-Palestine, fifth to sixth centuries.

2 For example:

a) Yale, no. 1955.6.150, yellowish brown, H 12.0 cm, Eisen 1927, II, 484, fig. 201 pls. 123-124; Matheson 1980, 133, no. 354;
b) *Helbing* 1913, no. 1035, pl. 27;
c) Oppenländer Collection, no. 22322, light green, H 25.7 cm, von Saldern et al. 1974, 181, no. 504.

3 Barag 1985b, 83, fot. 53.

4 As represented more fully on a silver-gilt plaque from a reliquary found at Hama, Syria, Louvre, inv. Bj 2180, *Romans and Barbarians*, exh. cat, Museum of Fine Arts (Boston 1977) 223, no. 226; *Age of Spirituality* 1977, 589-590, no. 529.

## 191 **Rectangular Jug**

Eastern Mediterranean, probably Syrian
Probably fifth to seventh centuries, perhaps modern
H 18.4 cm; max W body 6.0 cm; D rim 5.6 cm; W base 5.0-4.6 cm
1948.17, Purchased with funds from the Libbey Endowment, gift of Edward Drummond Libbey

Transparent pale green. Translucent similarly colored handle. Medium thin bubbly glass. Stone and black specks.

Mouth and neck free blown. Body blown into a multipart mold. Mold seams cannot be detected. Relief crisp. Pontil mark ca. 1.1

cm in diameter. Excess glass at tip of handle folded back against top of handle.
Spouted mouth, set diagonally on body between panels 1 and 2, with rim rounded and thickened in flame. Tapering neck. Strongly sloped shoulder. Rectangular body. Concave base. Angular ribbon handle applied to shoulder between panels 3 and 4 and attached to the edge of the rim where it is folded downward/inward, upward/outward and pinched to form a flat closed loop.
On the body, four decorated rectangular panels with designs in low relief. In the panels, from left to right: 1, two heraldic birds (doves?), with conical object between them; the design is sideways, with the right border at bottom; 2, lattice; 3, second half of Greek inscription in two registers; 4, first half of Greek inscription in two registers.
*Condition* Intact. Scarcely weathered; small areas of brown weathering on interior.
*Collection* Acquired from the Ernest Brummer Gallery, New York.
*Exhibition Early Christian and Byzantine* Art, exh. cat., The Walters Art Gallery (Baltimore, Md. 1947) 124, no. 624 (lent by the Brummer Gallery, erroneously described as having been found with a silver cross and bowl on the banks of the Orontes, near Hama, Syria, per Marvin C. Ross, "A Second Byzantine Silver Treasure from Hamah," *Archaeology* 3 [Autumn 1950] 162).
*Published* Rene Mouterde, *Inscriptions Grecques et Latines de la Syrie* V: *Emesene* (Paris 1959) 197, no. 2468.
*Remarks* The Greek inscription reads ΑΓΙΑϹ ϹΑΒΑϹ ΧΑΙ[ΡΕ], *Hagias Sabas chai(re)*, plus a date according to an eastern calendar: "Father St. Sabas, be favorable, help us the 15th day of Daisios." [1] It has been noted that this vessel may not be ancient, based on observations that it seems to be unique, the glass looks new, with almost no weathering, and the trefoil rim has an edge folded outward, in the opposite manner from similar examples.

[1] Fr. René Mouterde transcribed the complete inscription as follows: "Αγι(ε), ἅγ(ιε) Σά/βας, Χάρ/ιε, β(οήθει). Δ(?)/έσ(ίου) γ′ βώ, translated as "O holy, holy Sabas, be favorable, help us! of the month of Daisios, the 3rd day, (the year) 802" (TMA documentary file correspondence dated December 8, 1948). According to the Sasanian calendar, the year 802 would be A.D. 490.

### 192 **Rectangular Jug**

Probably Syrian
Fourth to sixth centuries
H 11.4 cm; W body 5.4–4.9 cm; D rim 4.4–3.8 cm; W base 4.3–3.9 cm
1923.1916, Gift of Edward Drummond Libbey

Transparent natural dusky yellow green (near 5 GY 5/2). Translucent similarly colored thread and handle. Medium thin glass. Mouth and neck free blown. Body blown into a multipart mold with base plate. Relief moderately indistinct. Pontil mark 1.4 cm in diameter. Excess glass at tip of handle drawn back against top of handle.
Circular mouth, with rim rounded in flame. Tubular neck with constriction at its base. Horizontal shoulder. Rectangular body with downward taper. Kicked base. Curved coil handle applied to shoulder between panels 2 and 3 and attached to edge of rim where it is folded downward and inward against mouth to form a closed loop.
Around mouth, four revolutions of thread trailed on downward left to right from beneath attachment of handle. Neck coil around middle of neck from left to right. On the body, four decorated rectangular panels with designs in low relief. In the panels, from left to right: 1 and 3, perhaps Greek letters in ligature, perhaps forming a monogram; different combination on each panel; 2, stylized tree or branch; 4, unidentified object in raised relief, perhaps two concentric lozenges like Cat. No. 177, panel 4. On underside of base, illegible design, possibly a rosette (or letters?).
*Condition* Complete, except for segments of thread decoration and most of neck coil. Exterior cloudy with ice floes, purple iridescence, and brown weathering. Interior pitted, with cream-colored weathering.
*Collection* Thomas E. H. Curtis.
*Published* Eisen 1927, II, 486, fig. 203.
*Remarks* Eisen noted that several flasks of this type came from tombs in Syria; he proposed an identification as Sasanian.

## *Later Imitation of a Vessel with Christian or Jewish Symbols* (Cat. No. 193)

### 193 **Hexagonal Jug with Unassigned Symbols**

Place of manufacture not determined
Probably early twentieth century (before 1913)
H 13.4 cm; max W body 6.8 cm; D rim 4.7–4.0 cm; W base 6.1–4.8 cm
1923.1353, Gift of Edward Drummond Libbey

Transparent vivid turquoise (not in rock color chart). Similarly colored handle. Medium thin glass. Very bubbly. Black specks and stone inclusions.

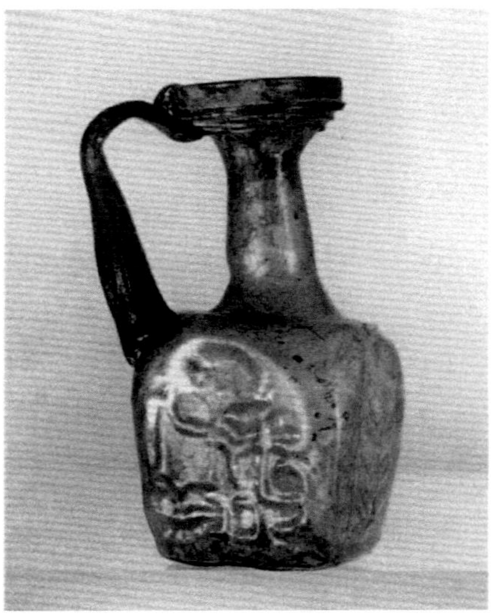
192, panel 3 (1:2)

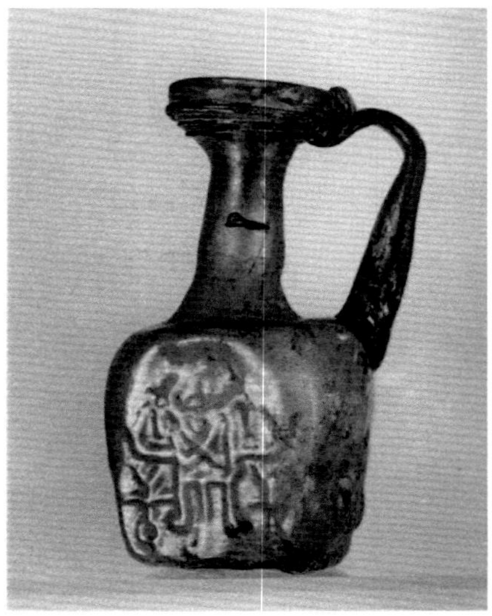
192, panel 1 (1:2)

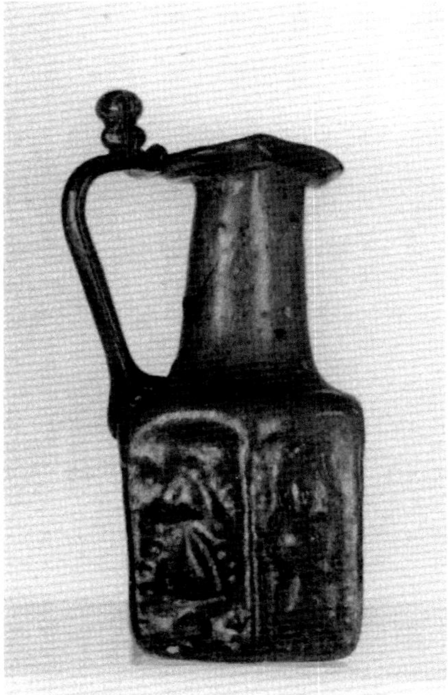

193, panels 2 and 3 (1:2)

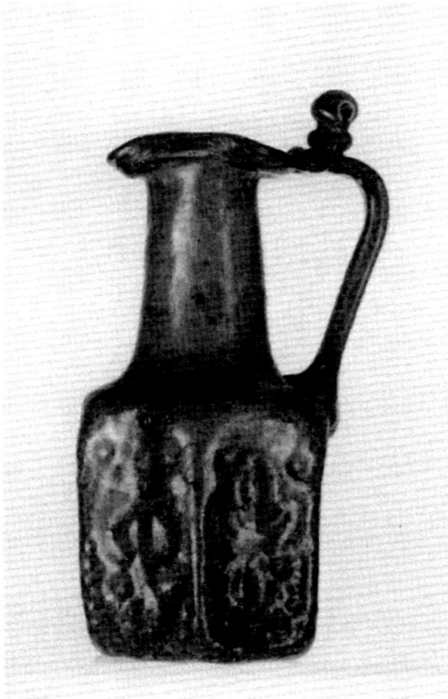

193, panels 5 and 6 (1:2)

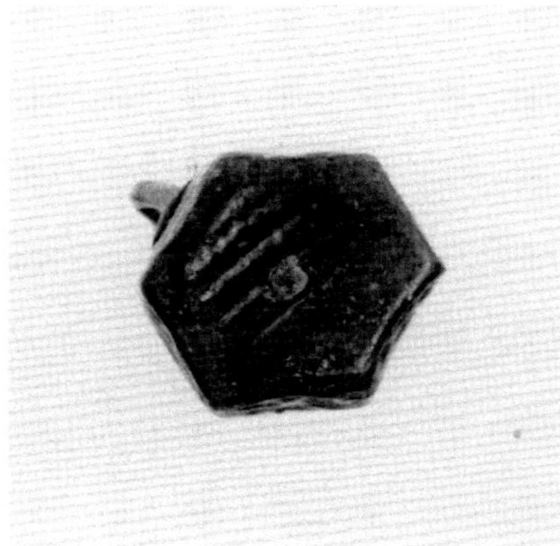

193, underside (1:2)

Mouth and neck free blown. Body probably blown into a multipart mold. Mold seams cannot be detected. Relief moderately crisp. Excess glass at tip of handle folded back against top of handle and pinched against it. Pontil wad of clear colorless glass ca. 0.9 cm in diameter (see detail photograph).

Spouted mouth, with rim folded outward, upward, inward, and downward. Tapering neck. Shoulder slightly sloped. Hexagonal body. Flattened base. Solid coil handle applied to shoulder above center of panel 1 and attached to rim with a double crimped vertical thumbrest near rim.

On the body, six decorated rectangular panels with designs in sunken relief; each panel bordered by recessed dots. In the panels, from left to right: 1, stylized human (?) figure; 2, two inverted V-shaped objects with a triangular element above and a dot below; 3, stylized human (?) figure; 4, two concentric lozenges with dots in each corner of the panel; 5, and 6, stylized human (?) figure. On underside of base, six parallel(?) ribs or indentations.

*Condition* Intact. Very little weathering. Some bubbles broken at surface.

*Collection* Thomas E. H. Curtis.

*Exhibition* Bezalel Narkiss, in *Age of Spirituality* 1977, 386-388, no. 356, 387, ill. (identifies panel 1 as amphora).

*Published* Barag 1971a, 49 (C IV 1), 52, fig. 37.

*Remarks* Probably an imitation of Barag's pilgrim vessels Class C IV with unassigned but doubtless Christian symbols (see Cat. No. 177). The shape of the mouth, the finishing of the base, and the solid coil handle are exceptional for this group of vessels. The extremely seedy glass, vivid color, and colorless pontil wad all suggest that this vessel was made in modern times. On the other hand, Barag mentions one other pilgrim vessel with a solid handle,[1] and fragments of a blue jug from catacomb 20 at Beth Shearim, Israel (the only other blue example noted thus far) that apparently have a similar ribbed pattern on the underside of the base.[2]

1 Tel Aviv, Collection of N. Sadé-Sadowsky, Barag 1970c, 37 n. 11; 56, no. A V 6.

2 Jerusalem, Israel Department of Antiquities and Museums, no. 64-17, Barag 1971a, 49, no. C IV 2; Barag 1970c, 38, n. 14.

# DRAWINGS

# CHAPTER I

## The First Century

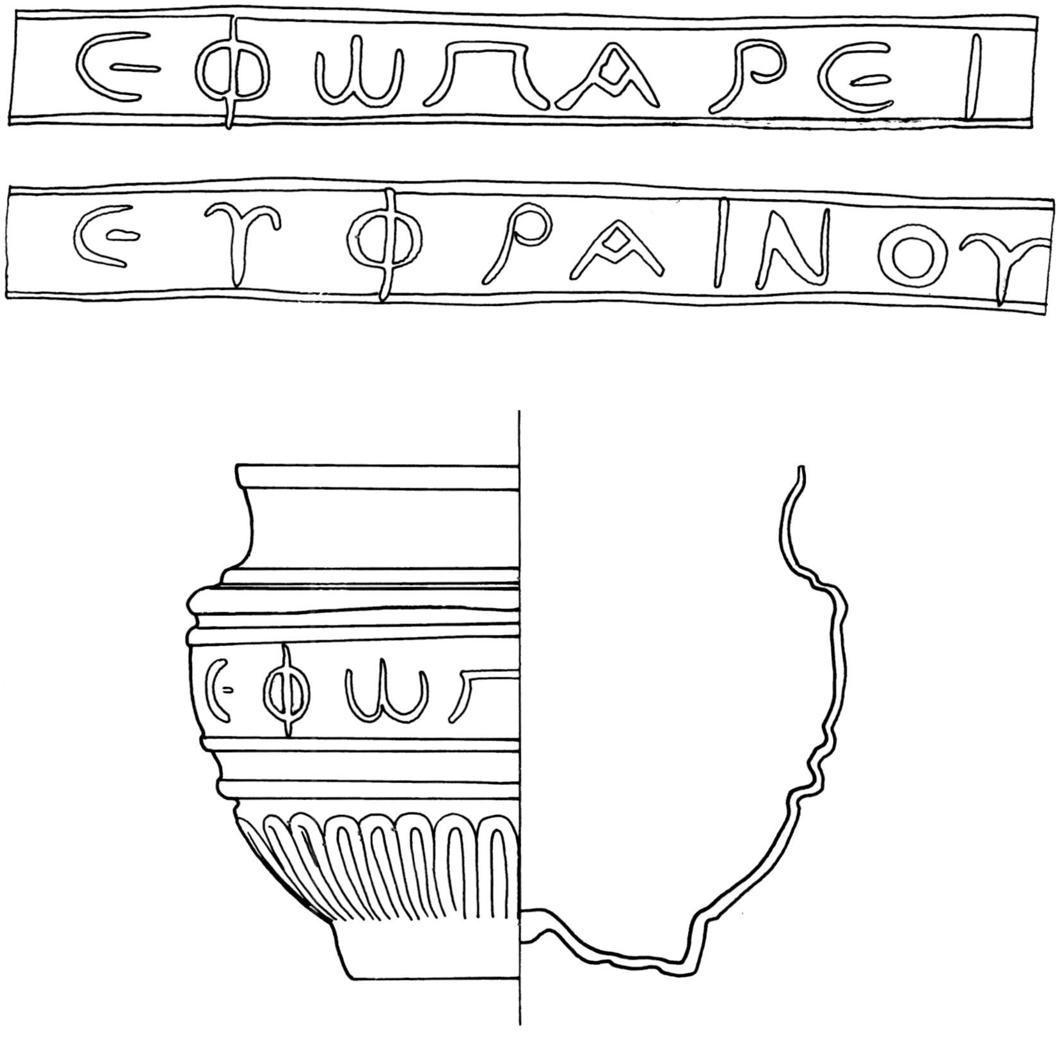


1

*Cups with Mold-blown Inscriptions*

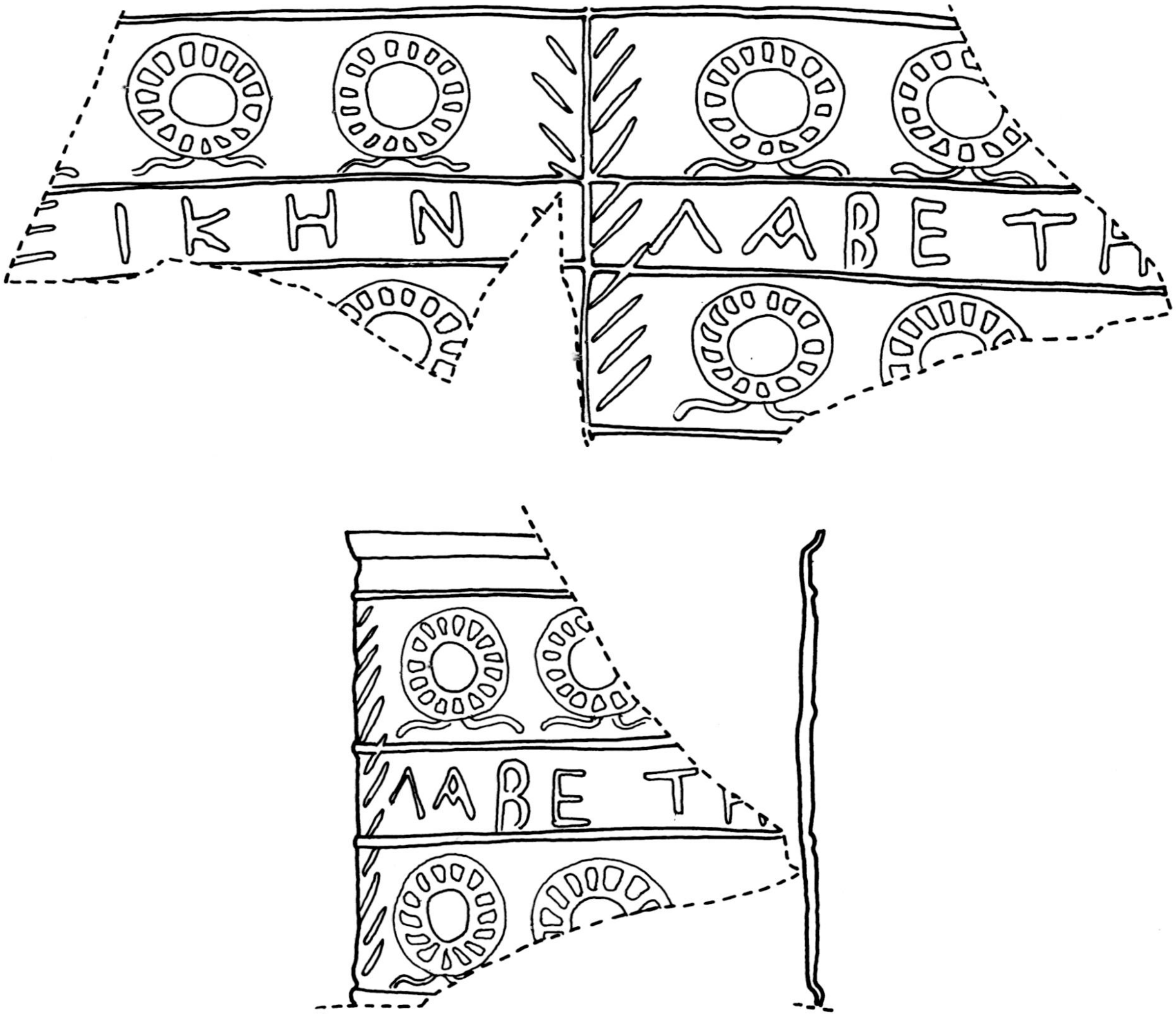


2

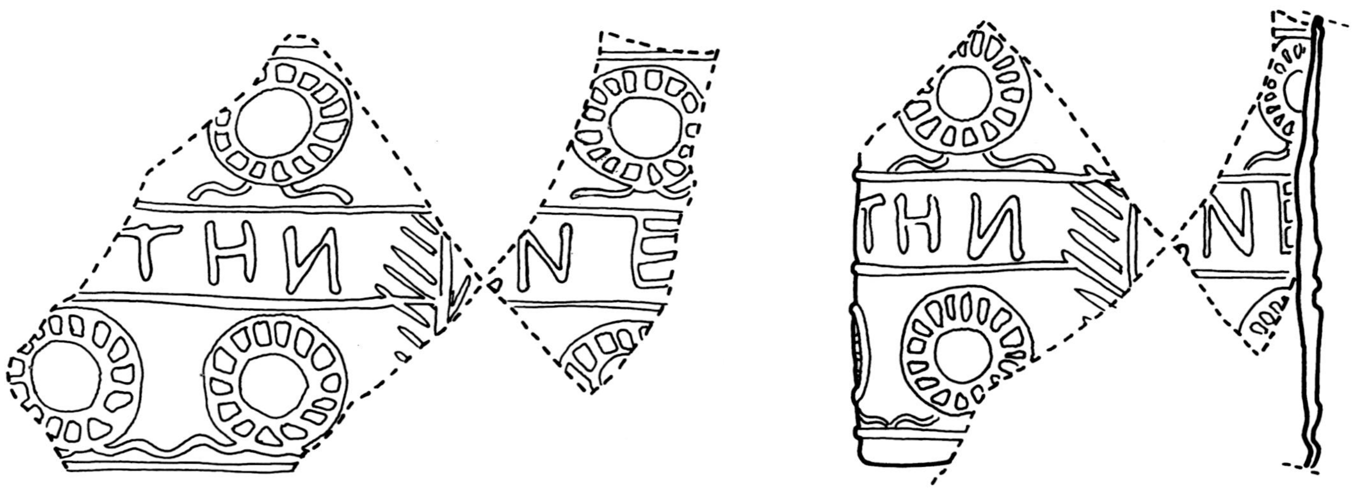


3

*Cups with Mold-blown Inscriptions*

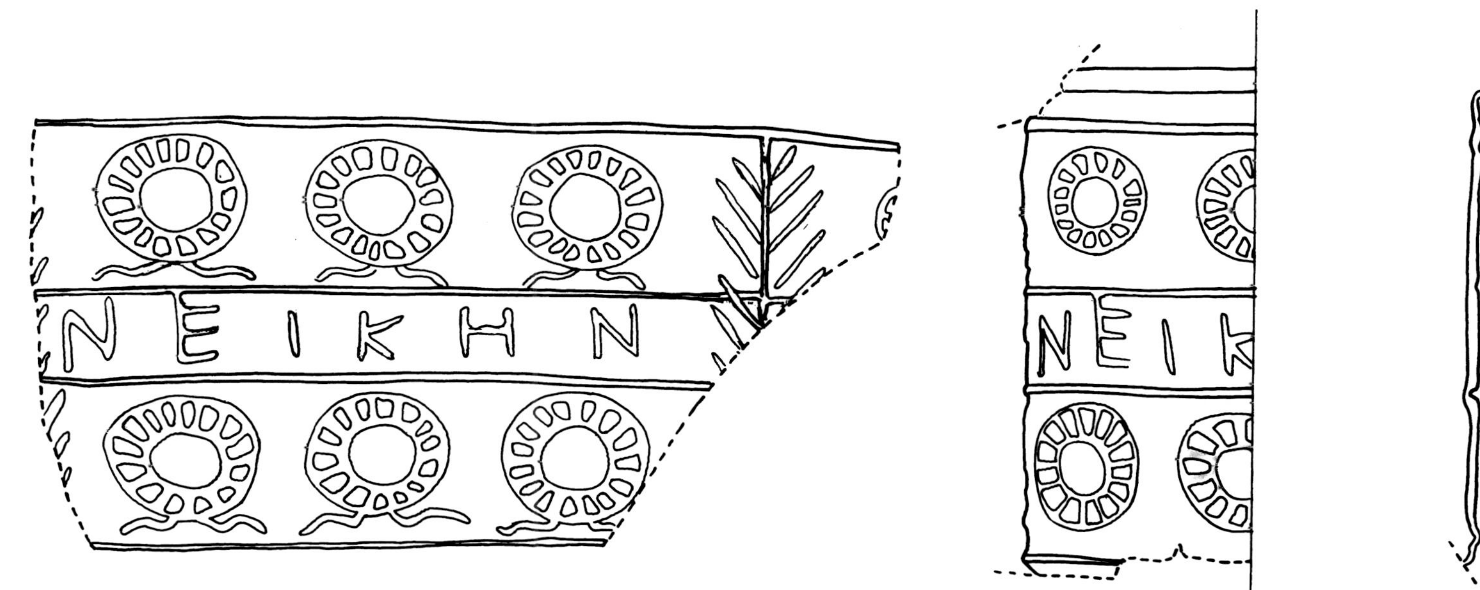


4

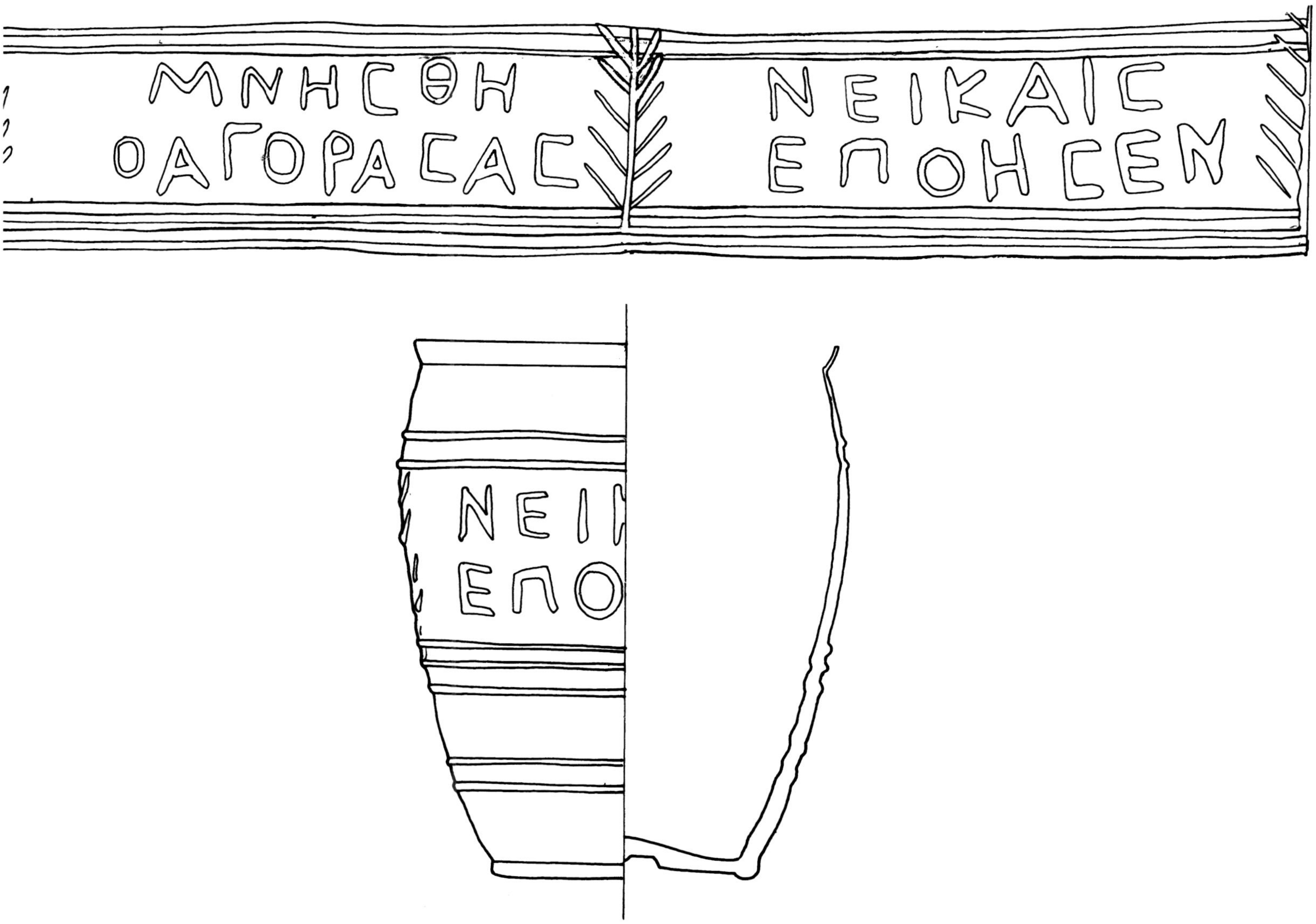


5

*Cups with Mold-blown Inscriptions*

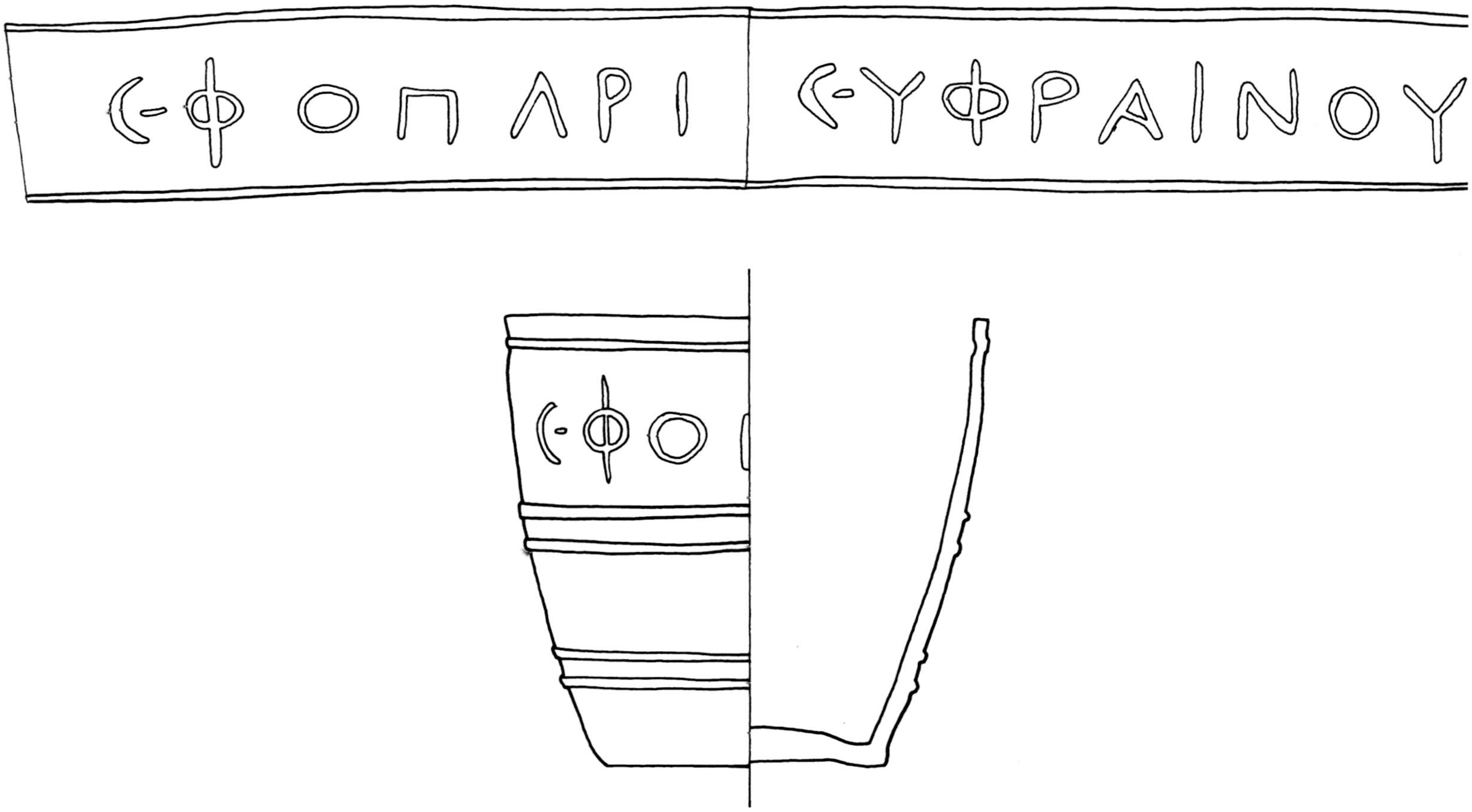


6

*Cups with Mold-blown Inscriptions*

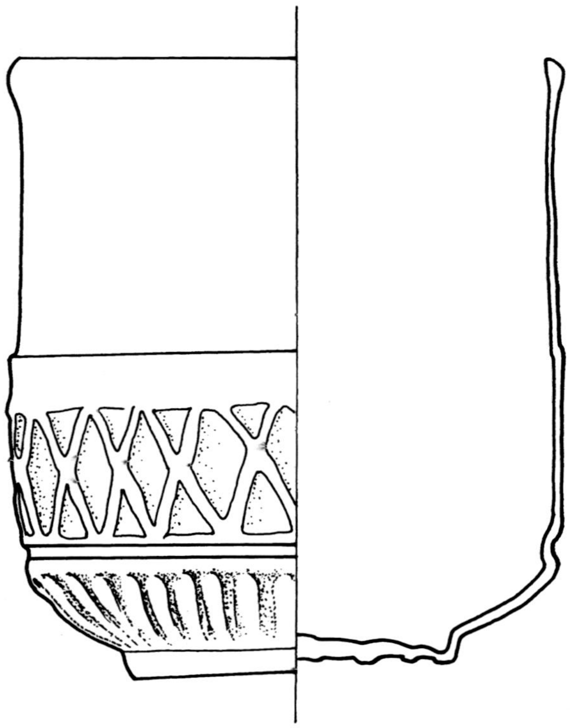

7

*Cup with Non-figural Designs in Low Relief*

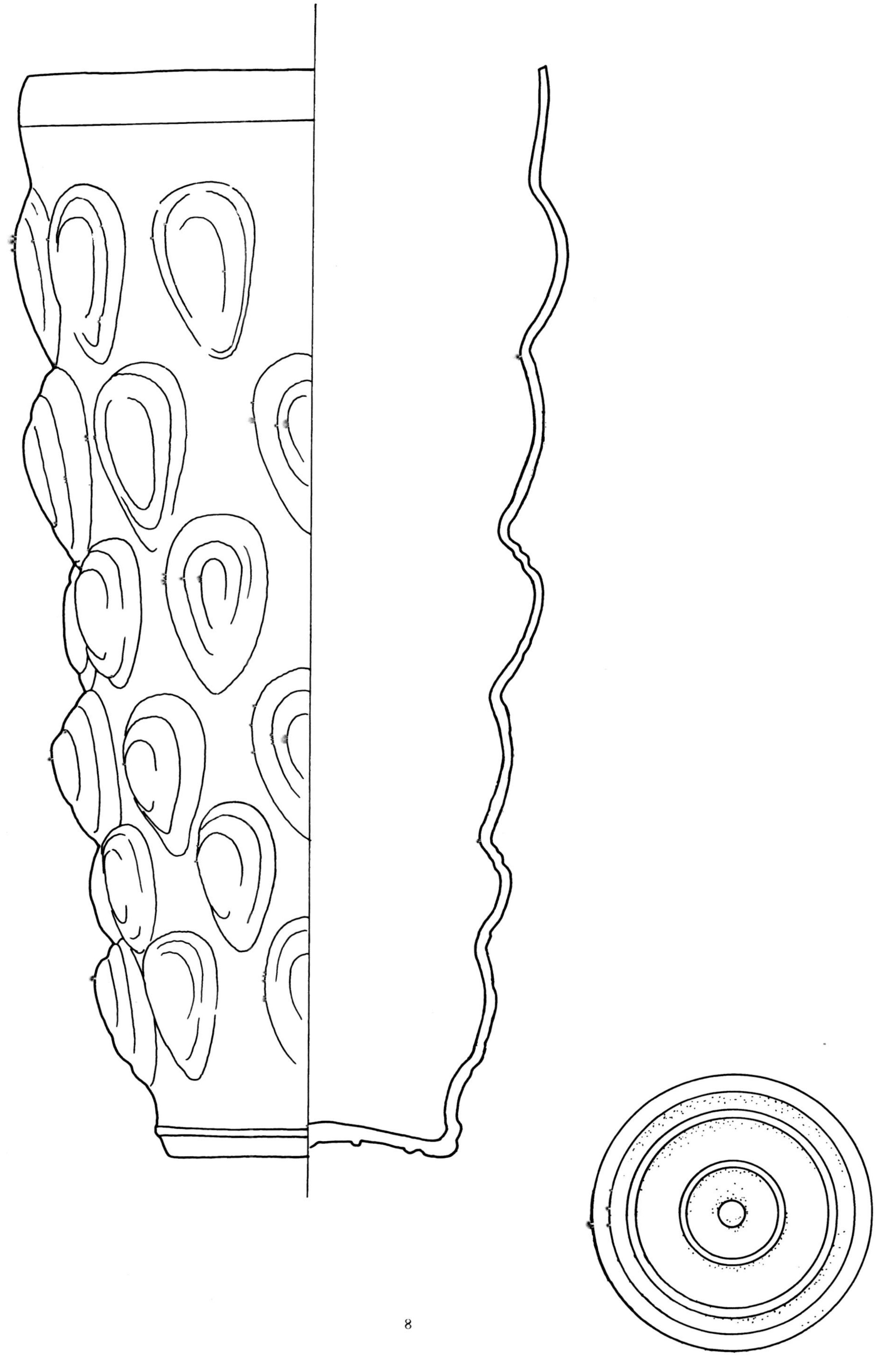

8

*Truncated Conical Beakers with Knot-shaped Knobs*

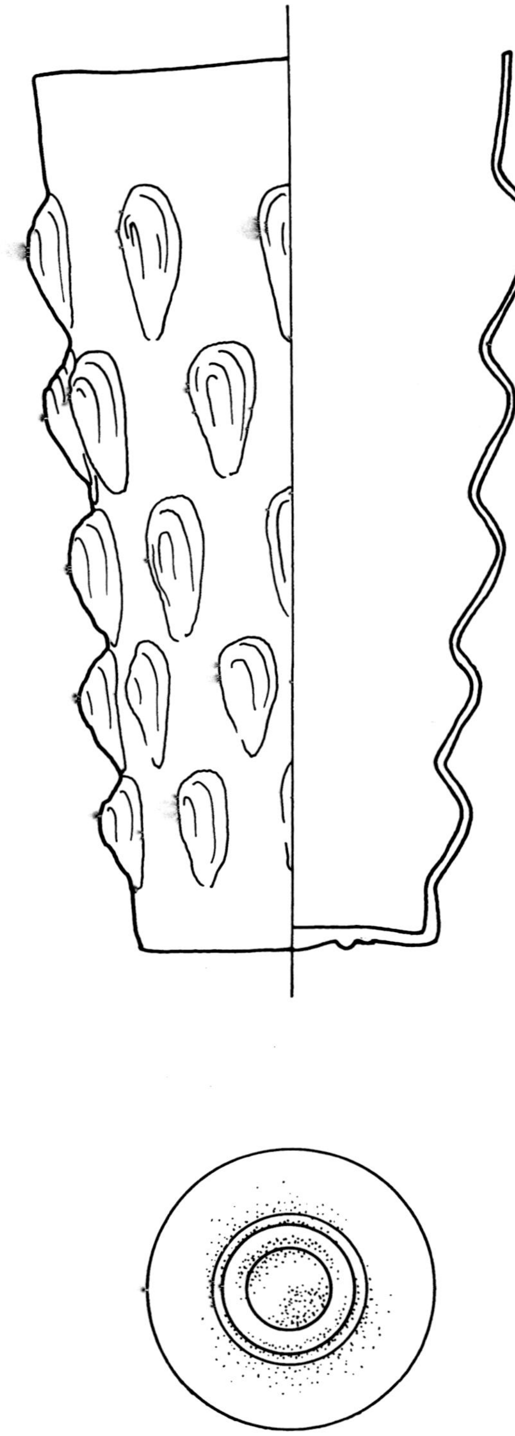

9

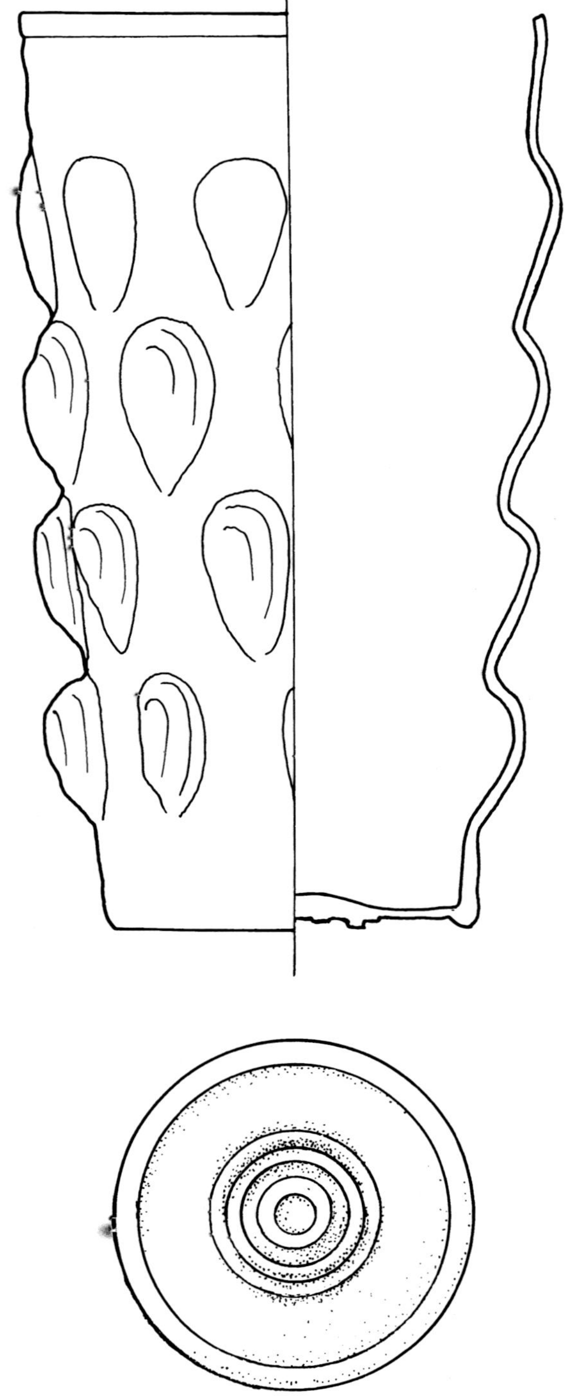

10

*Truncated Conical Beakers with Knot-shaped Knobs*

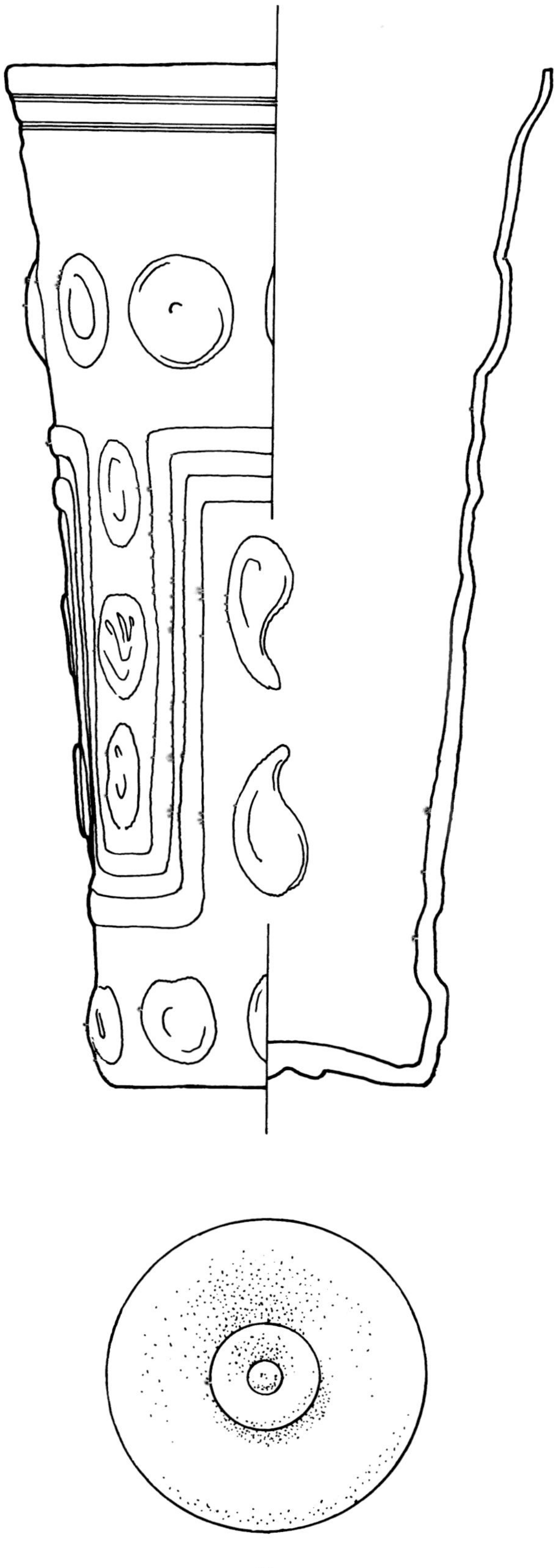

11

Conical Beaker with Isolated Figural Motifs

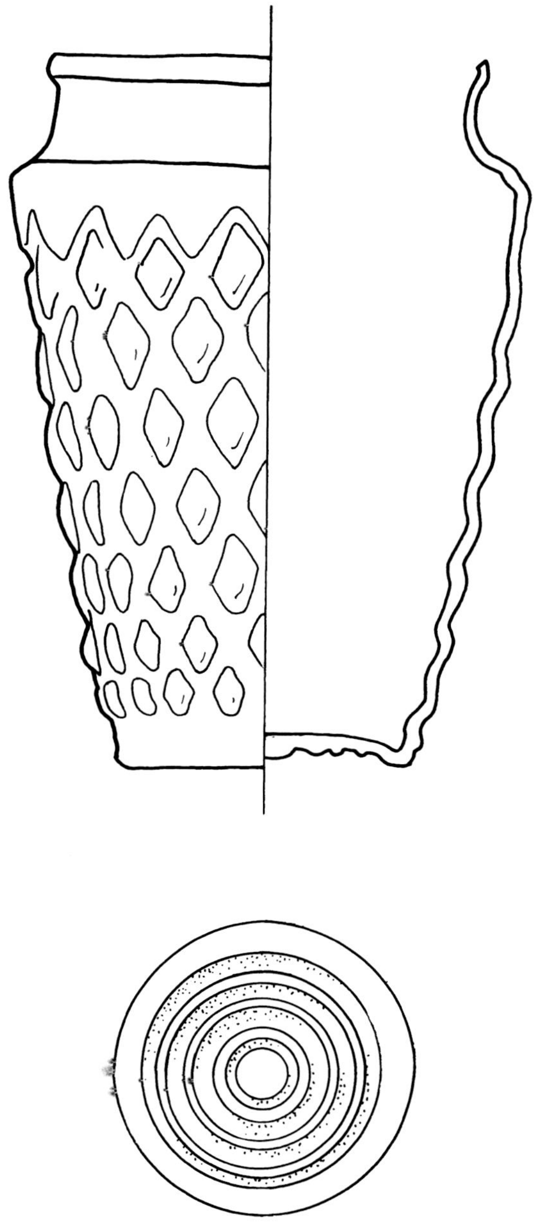

12

Drinking Vessels with a Pronounced Shoulder

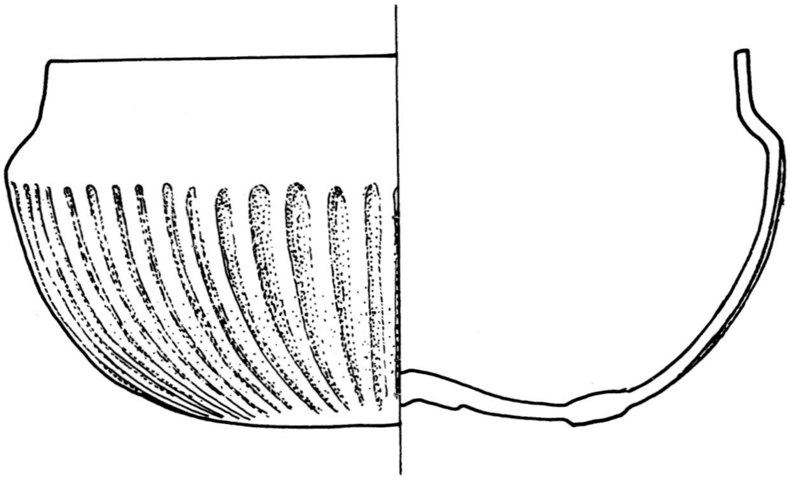

13

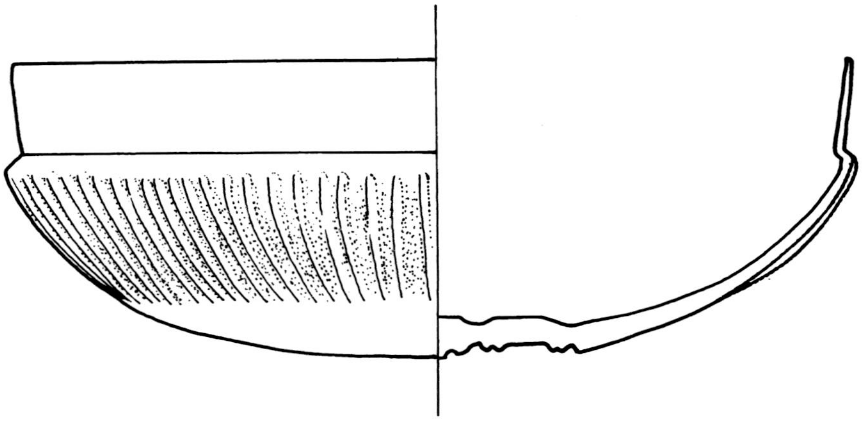

14

*Drinking Vessels with a Pronounced Shoulder*

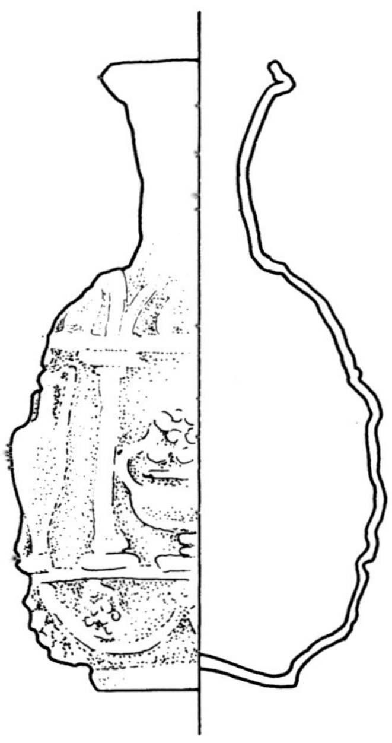

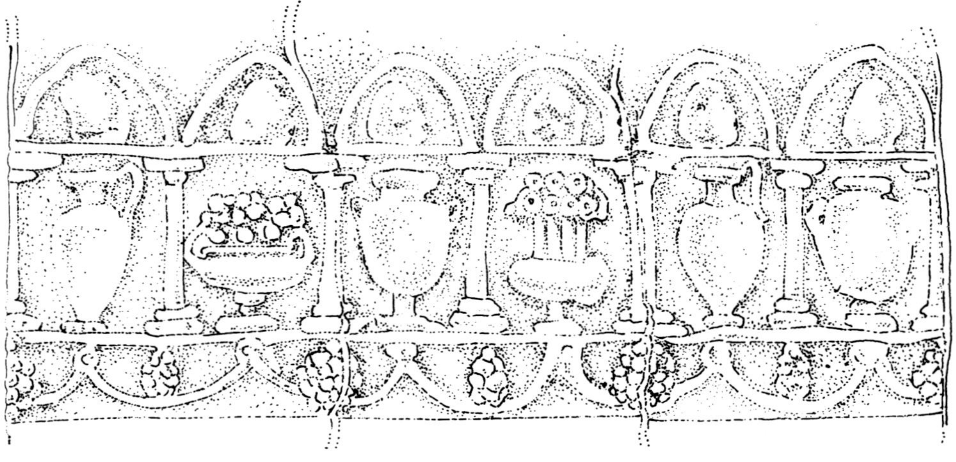

15

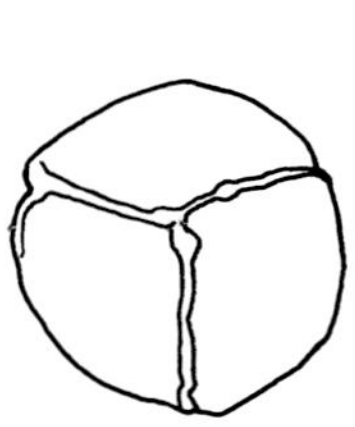

*Hexagonal Bottles with High Relief: Vessels Type, Series A*

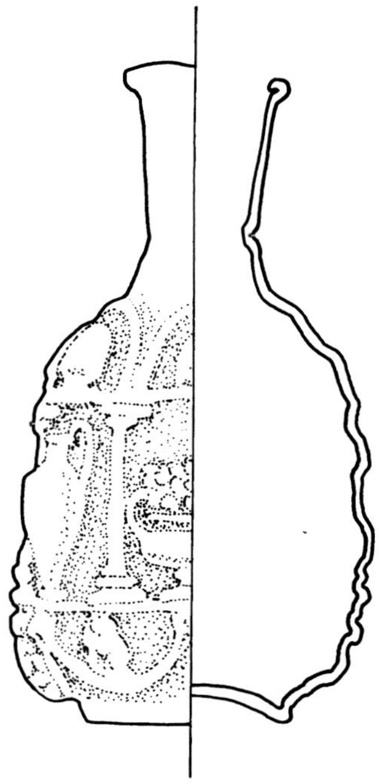

16

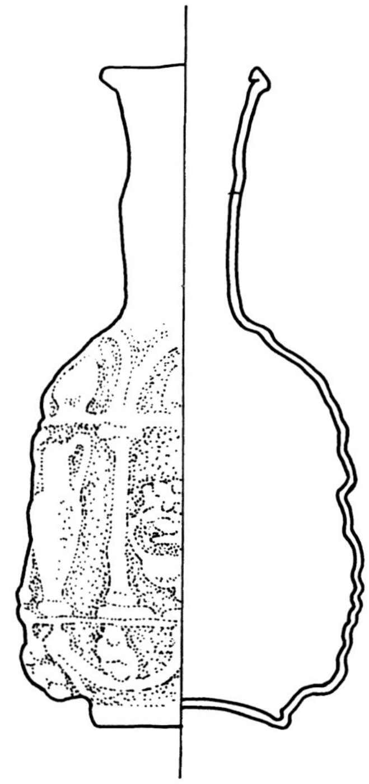

17

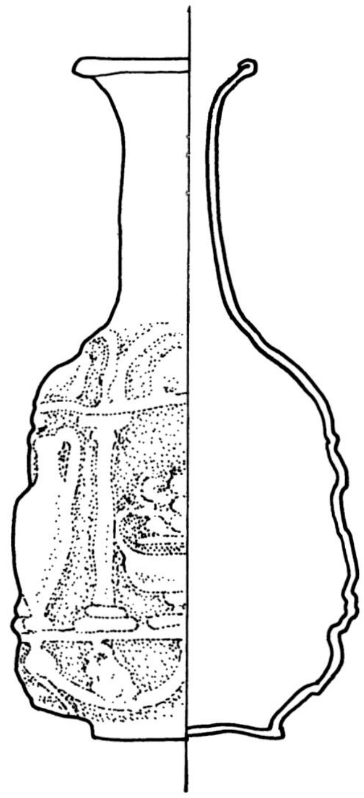

18

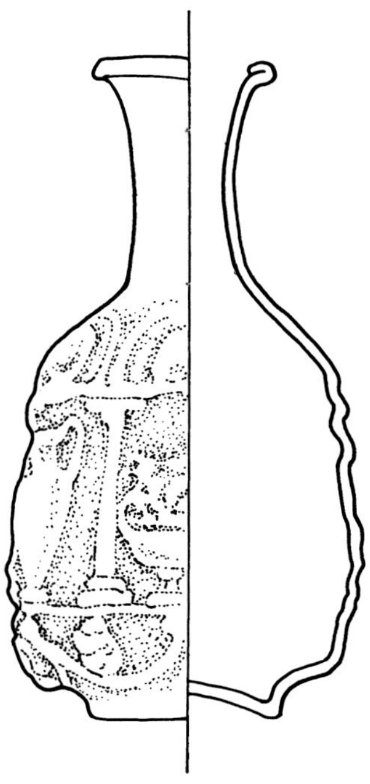

19

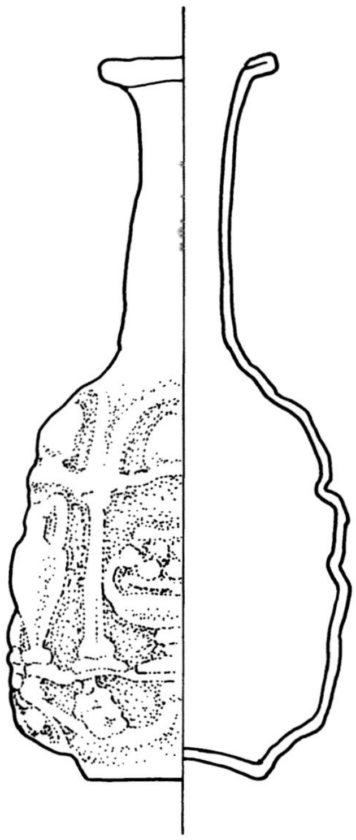

20

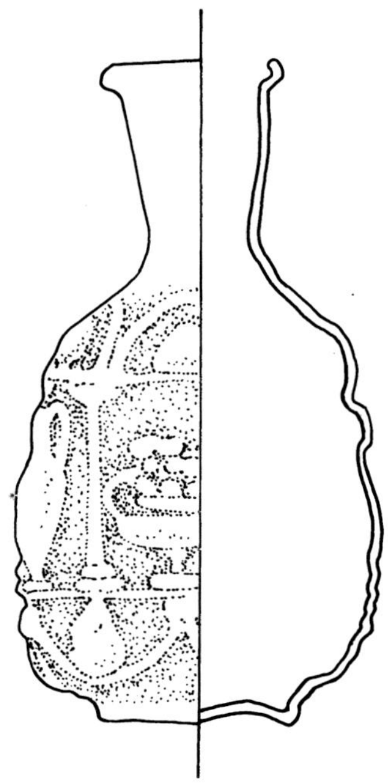

21

*Hexagonal Bottles with High Relief: Vessels Type, Series A*

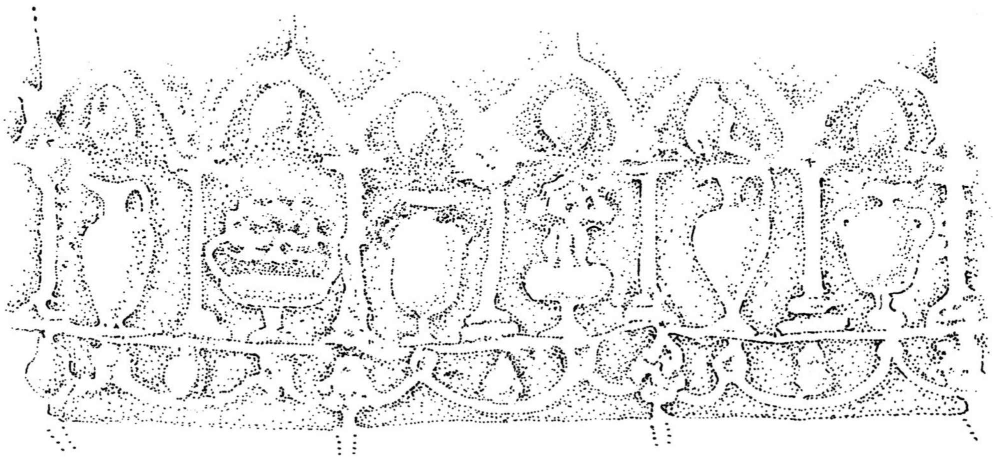
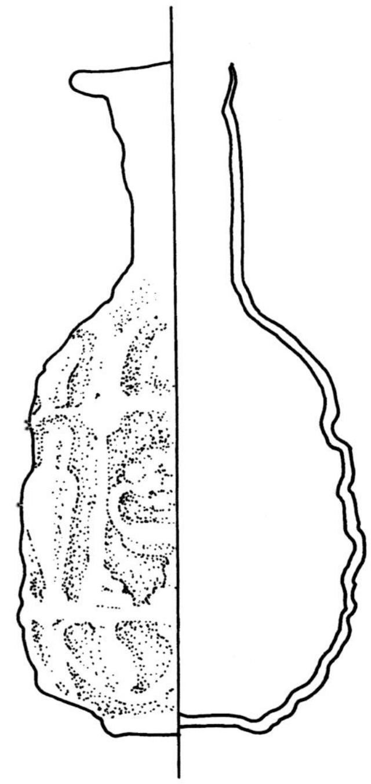

22

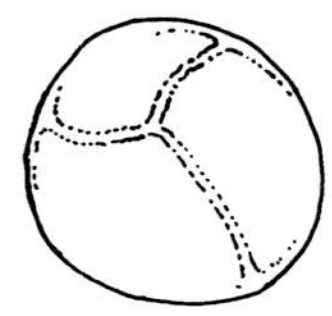

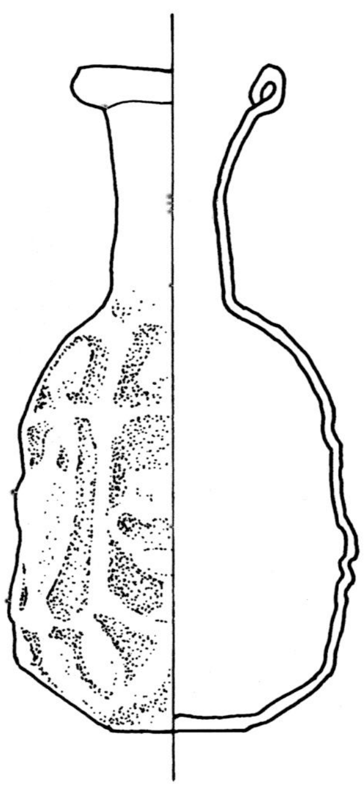
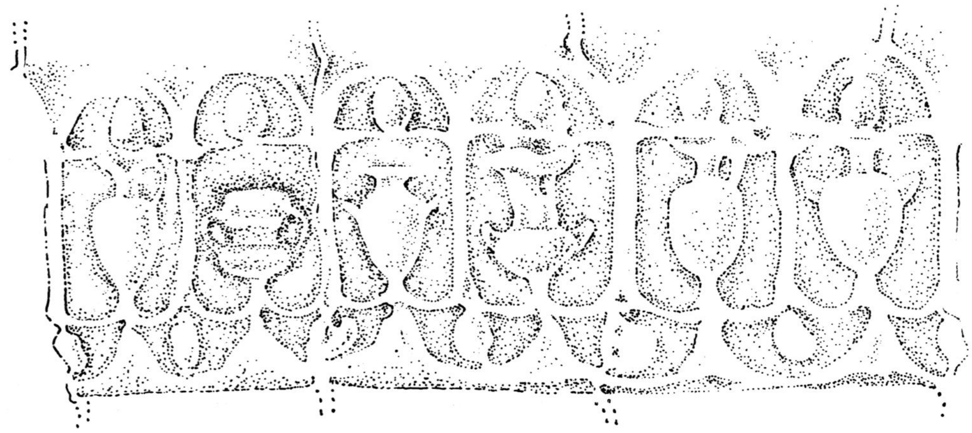

23

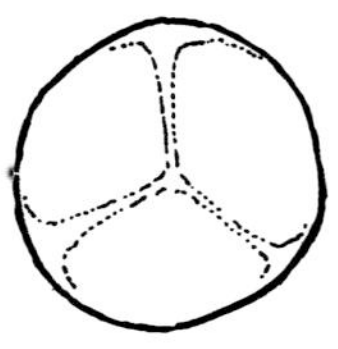

*Hexagonal Bottles with High Relief: Vessels Type, Series A*

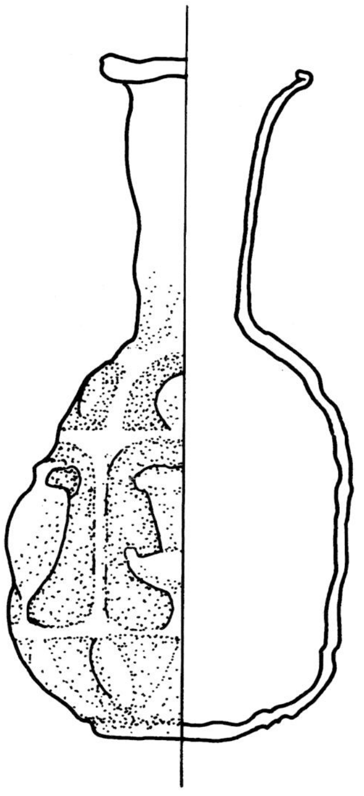

24

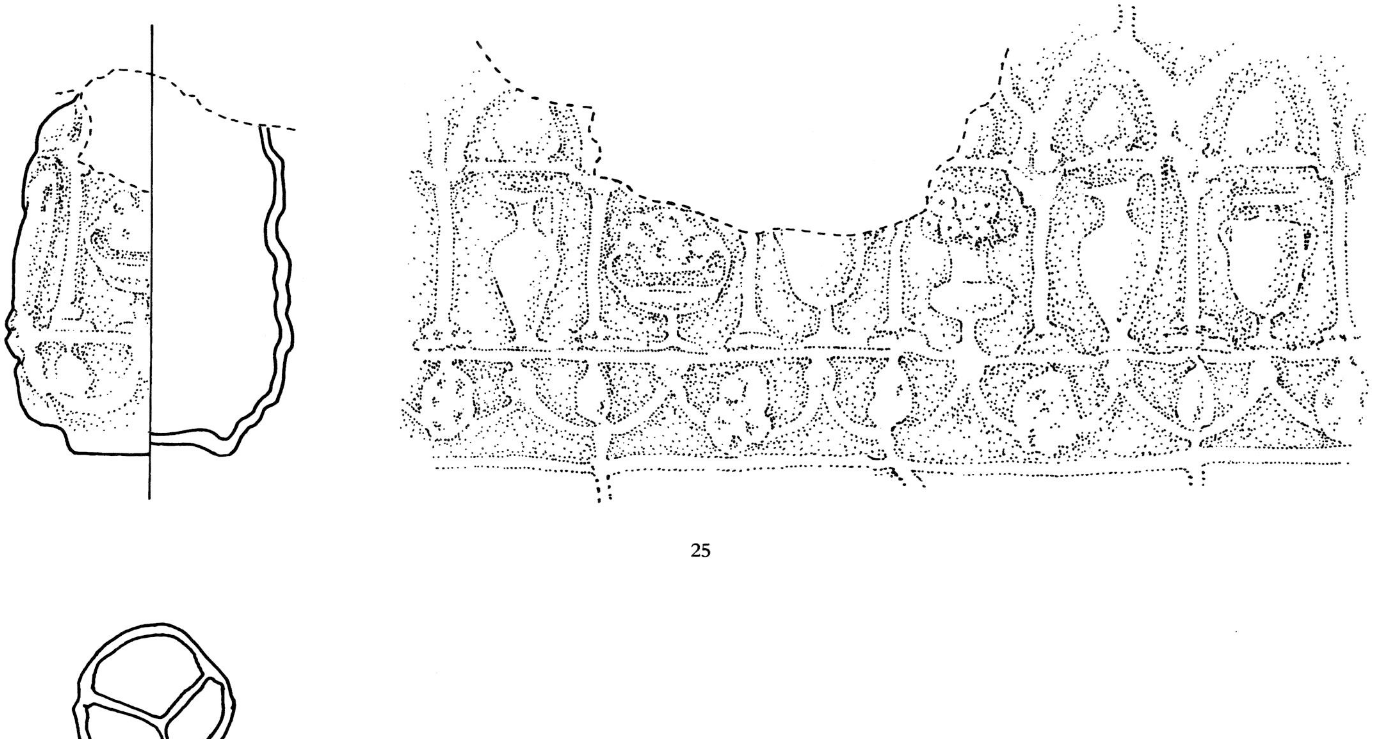

25

*Hexagonal Bottles with High Relief: Vessels Type, Series A*

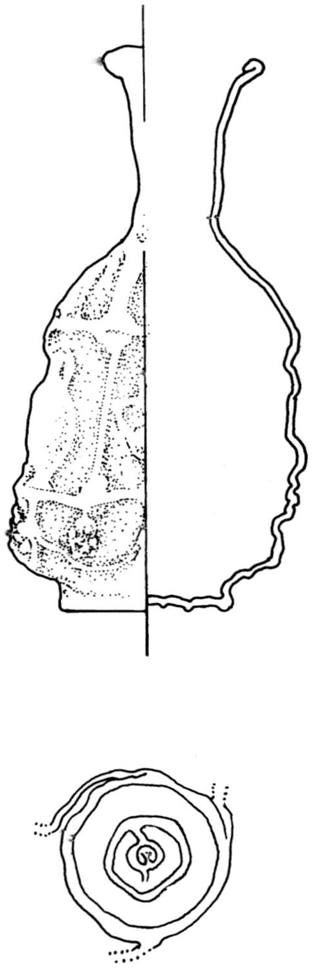

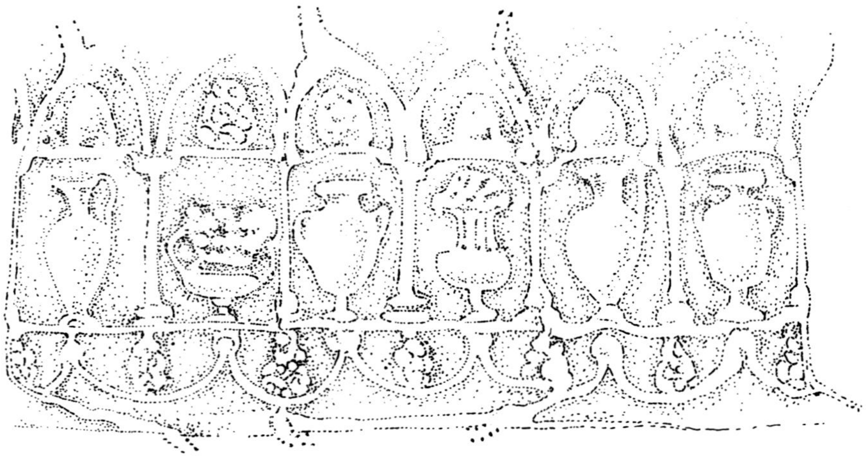

26

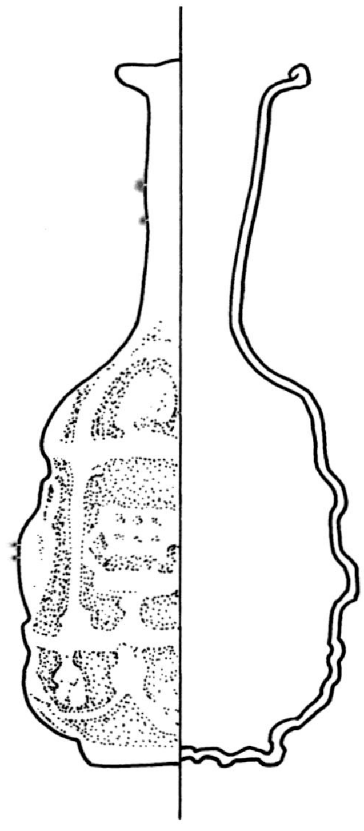

27

*Hexagonal Bottles with High Relief: Vessels Type, Series A*

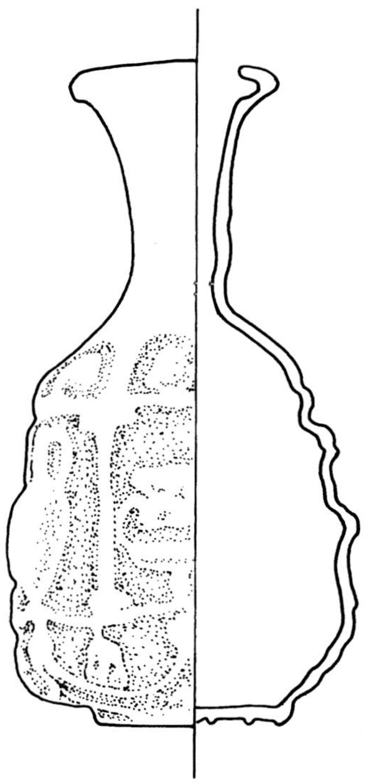

28

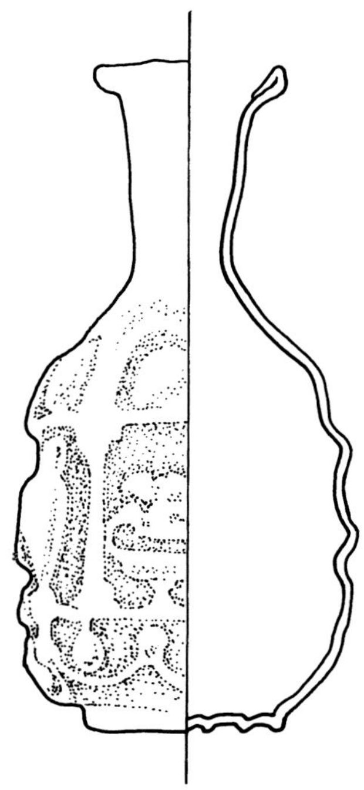

29

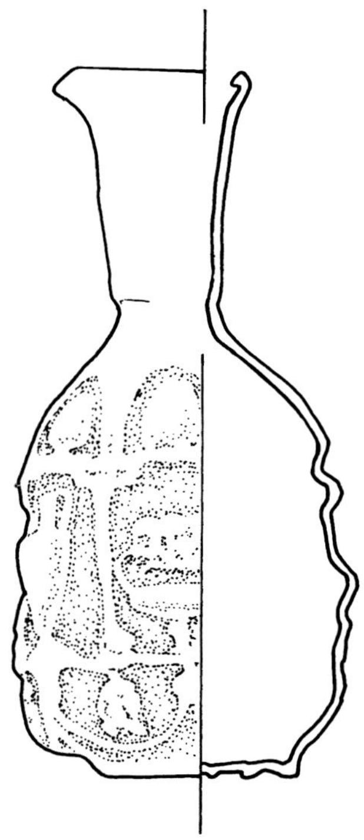

30

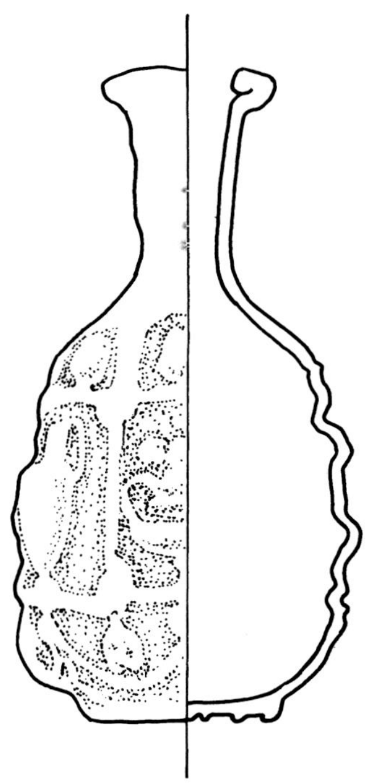

31

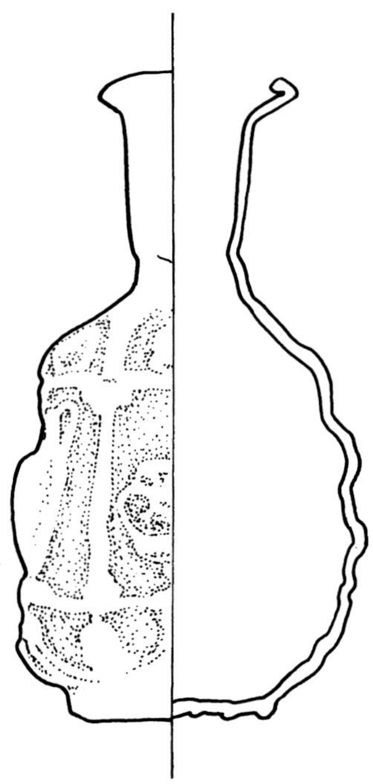

32

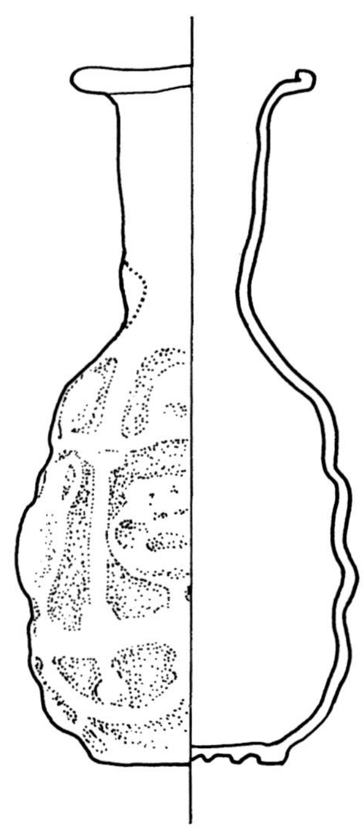

33

*Hexagonal Bottles with High Relief: Vessels Type, Series A*

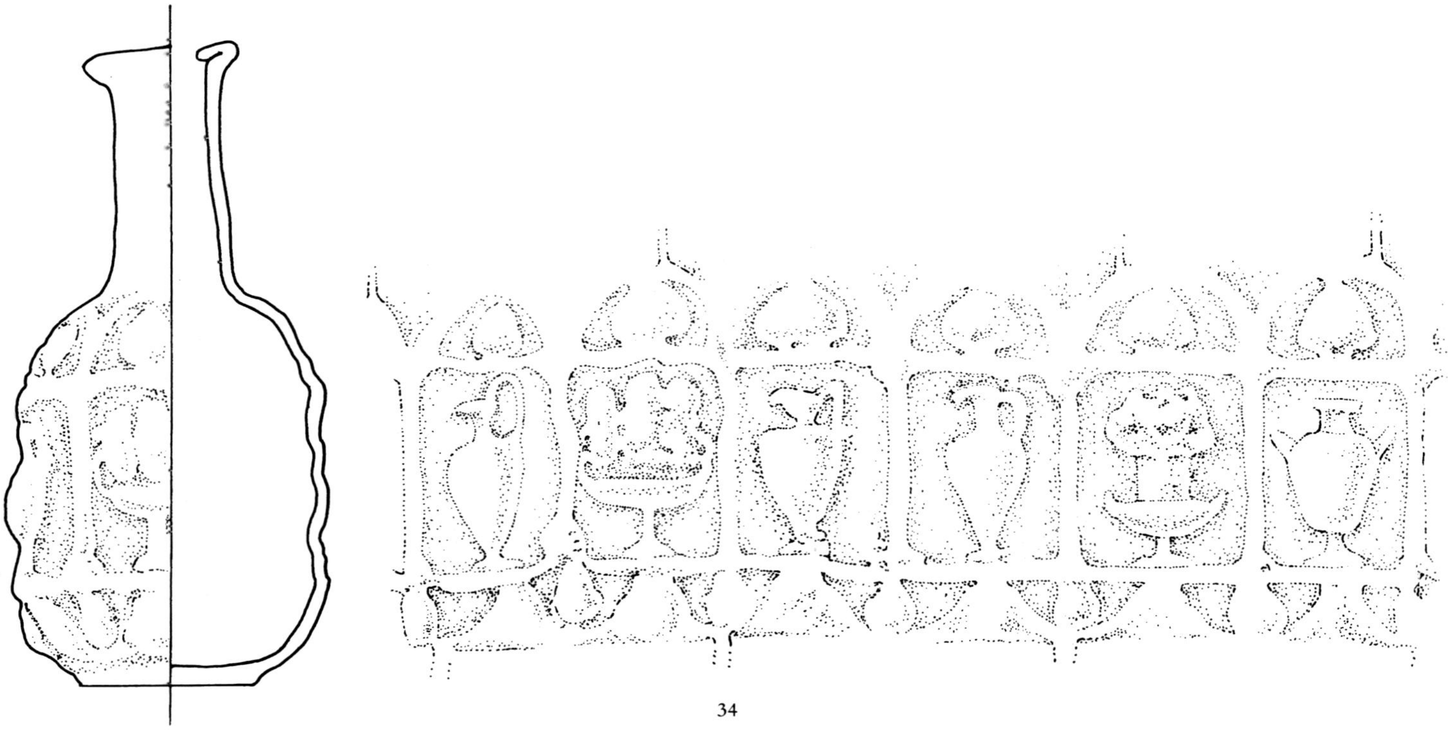

34

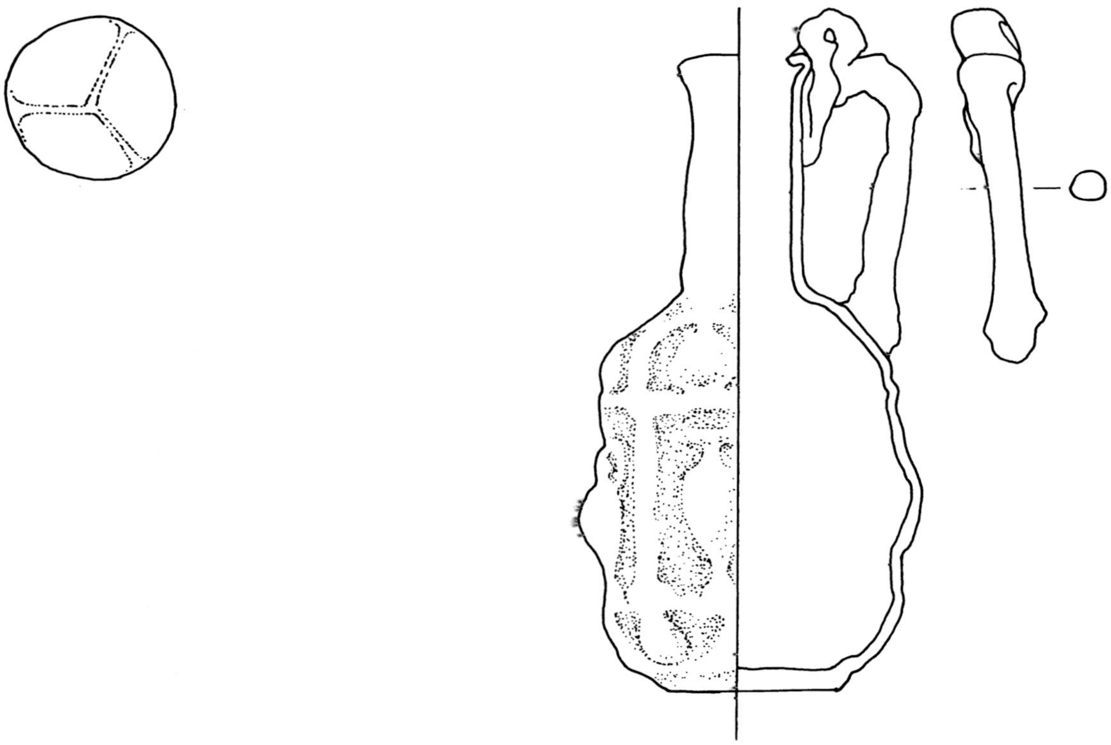

35

*Hexagonal Bottles with High Relief: Vessels Type, Series B*

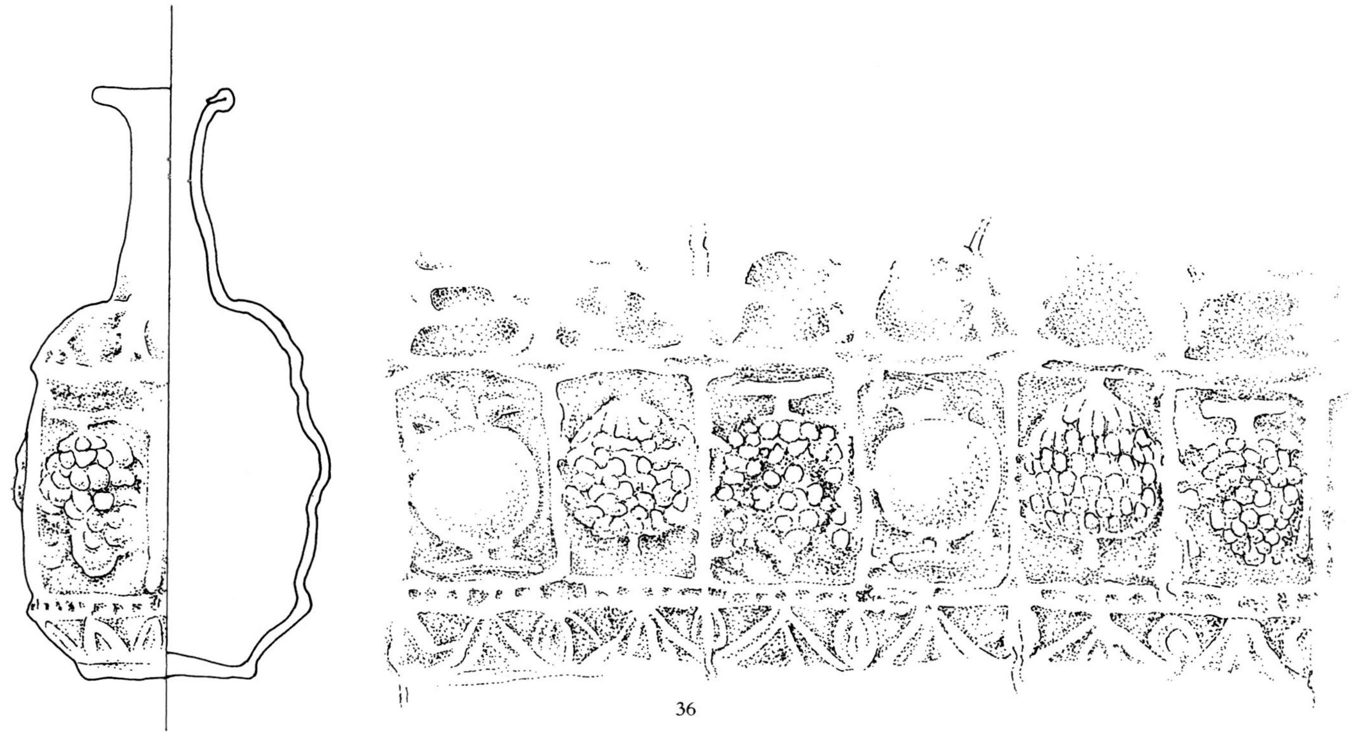

36

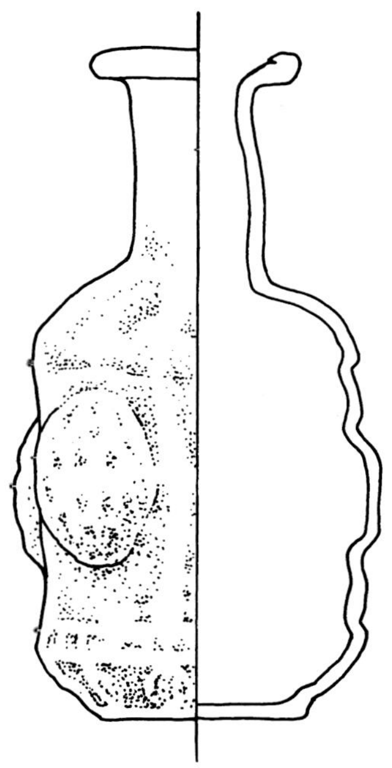

37

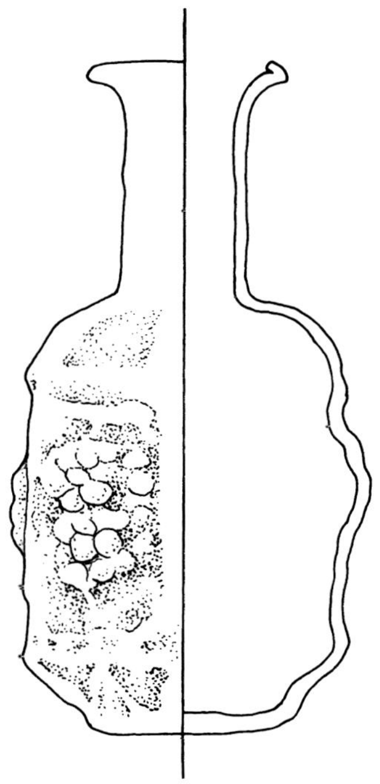

38

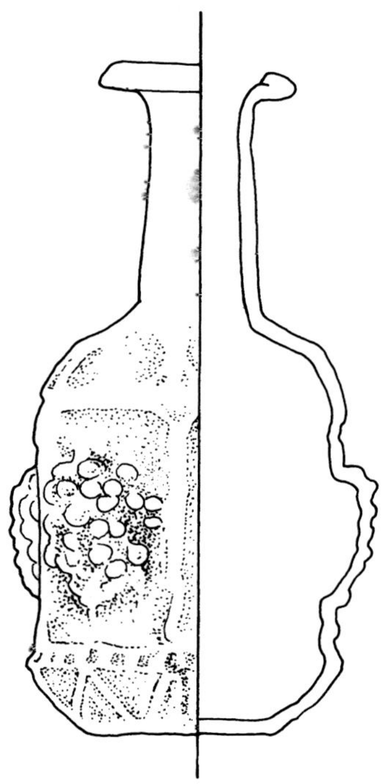

39

*Hexagonal Bottles with High Relief: Fruit Type, Series A*

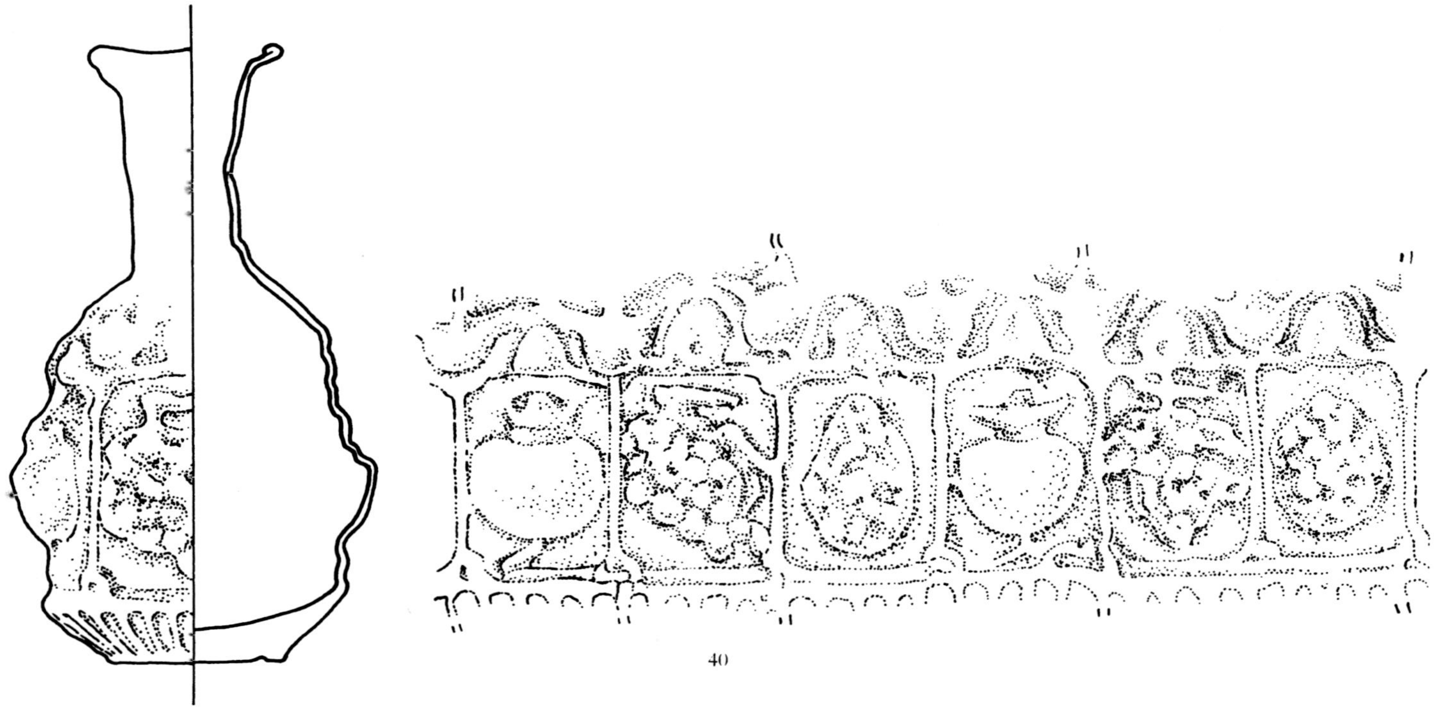

40

*Hexagonal Bottles with High Relief: Fruit Type, Series B*

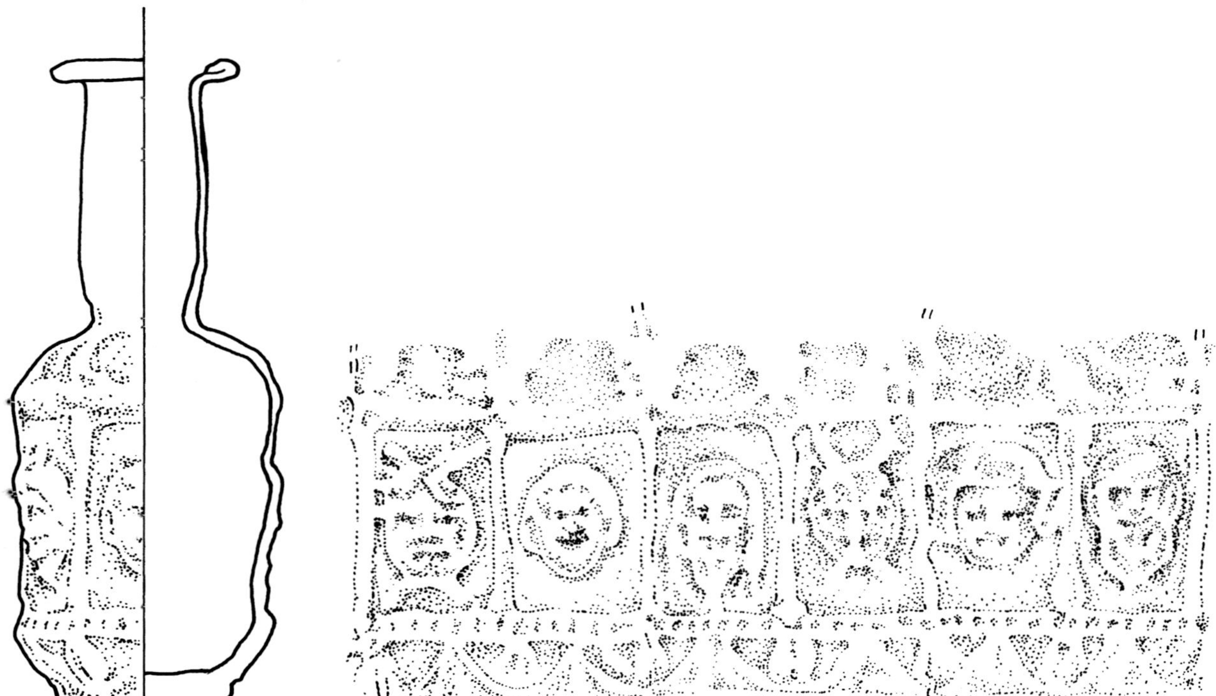

41

*Hexagonal Bottles with High Relief: Masks Type, Series A*

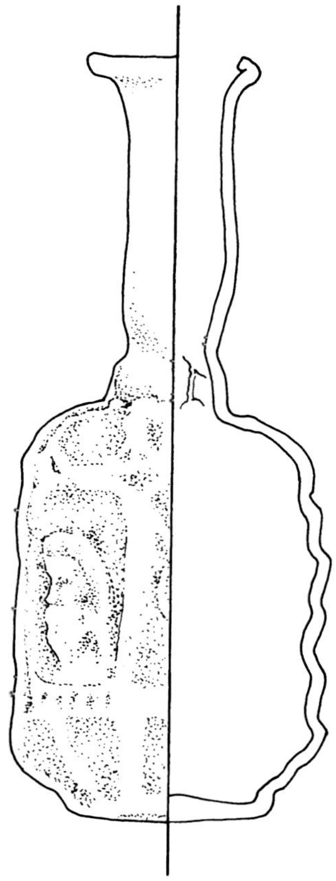

42

*Hexagonal Bottles with High Relief: Masks Type, Series A*

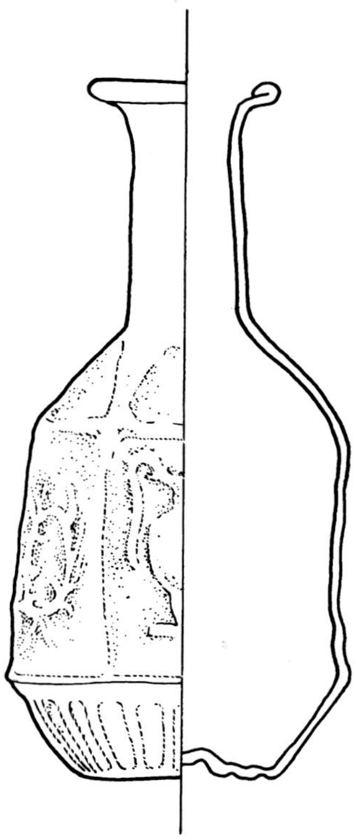

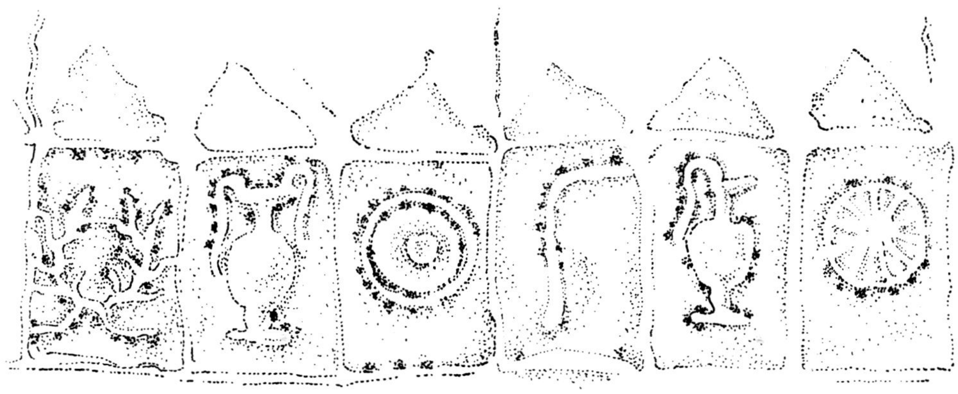

43

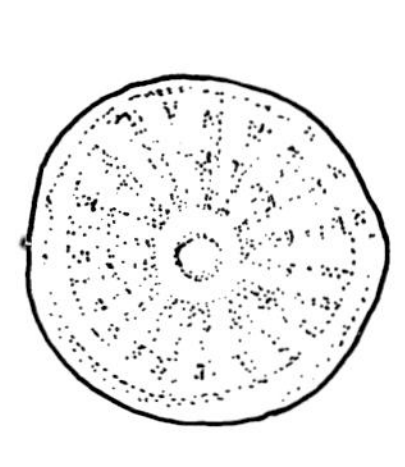

*Hexagonal Bottles with High Relief: Mixed Symbols Type, Series A*

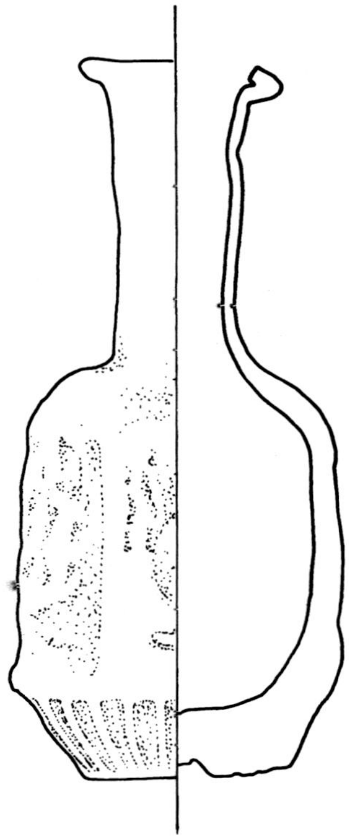

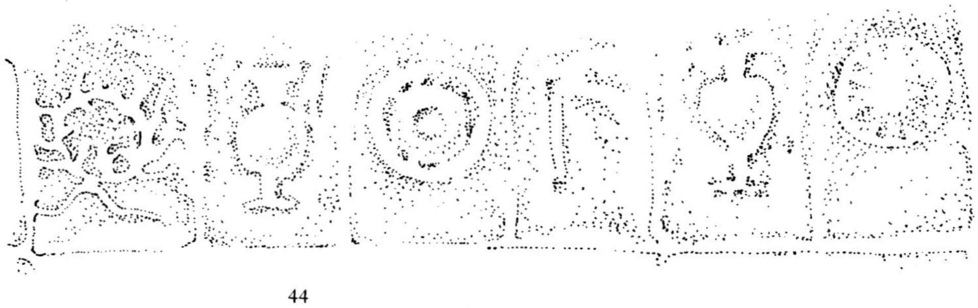

44

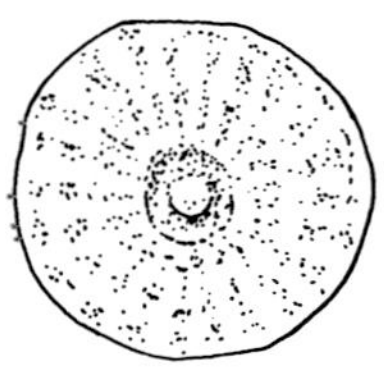

*Hexagonal Bottles with High Relief: Mixed Symbols Type, Series A*

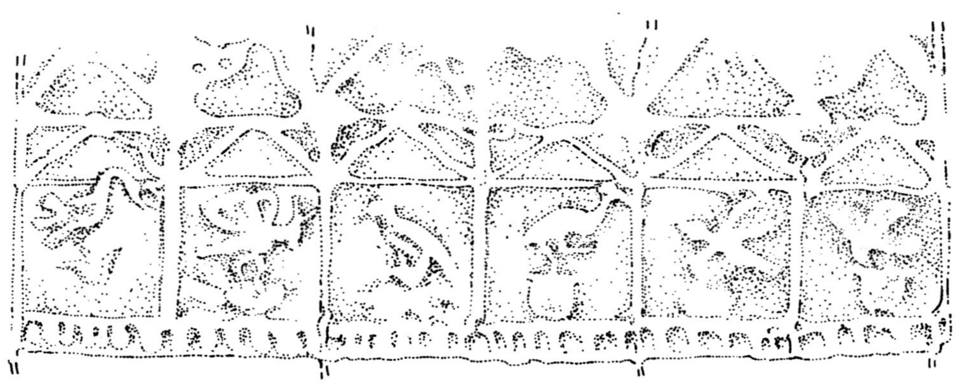

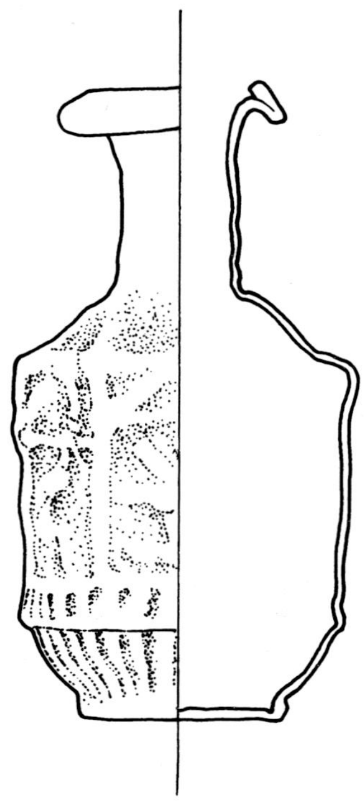

45

*Hexagonal Bottles with High Relief: Birds Type, Series A*

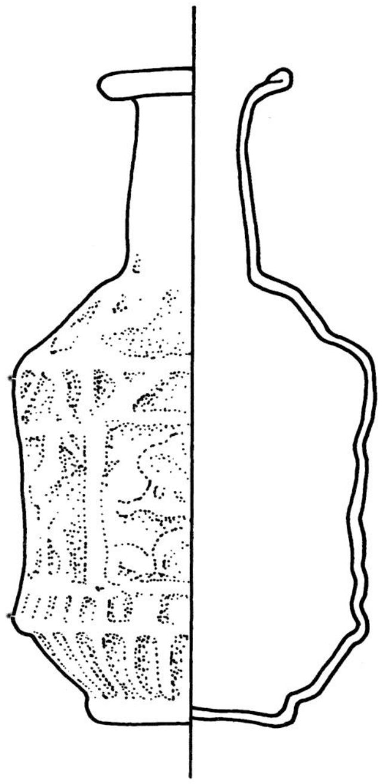
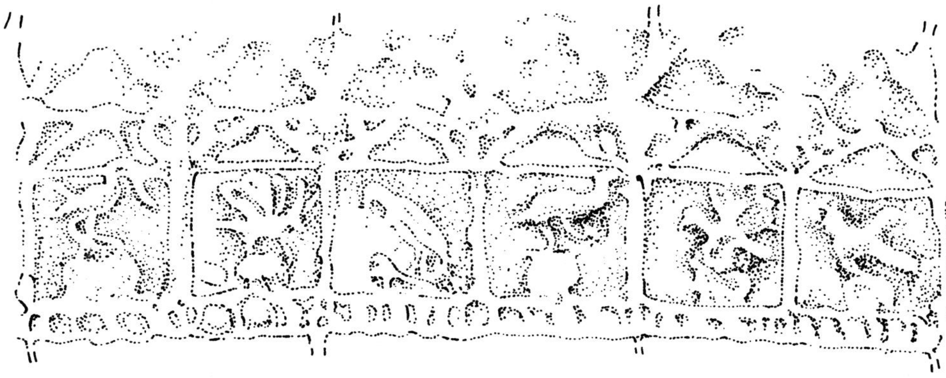

46

*Hexagonal Bottles with High Relief: Birds Type, Series A*

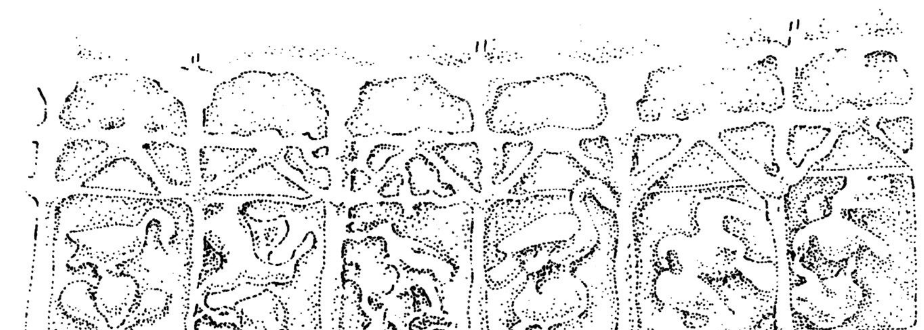

47

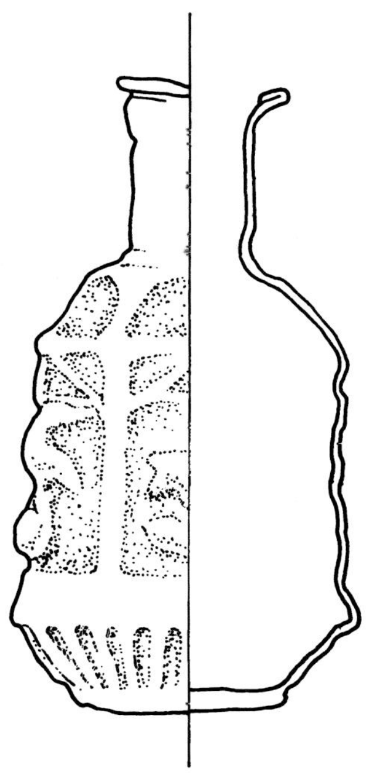

*Hexagonal Bottles with High Relief: Birds Type, Series B*

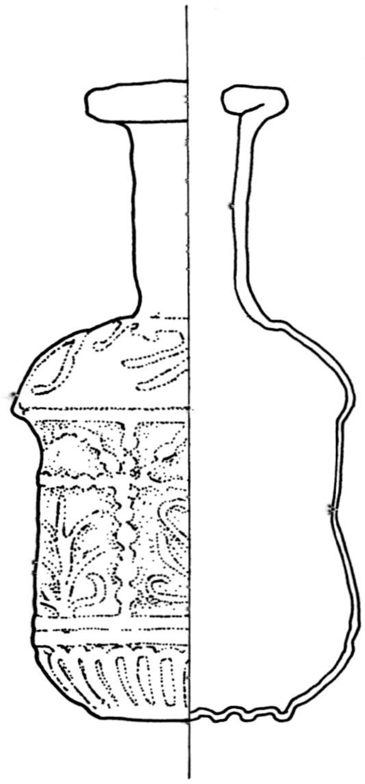

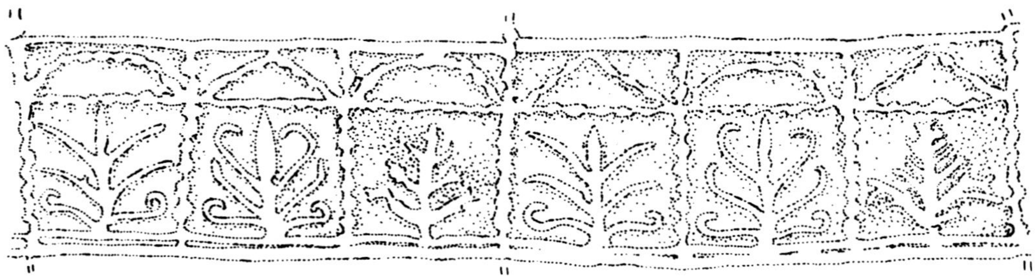

48

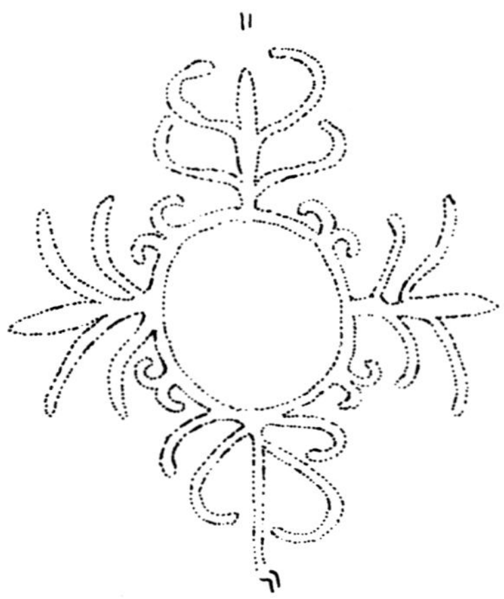

Cylindrical Bottle with Low Relief: Plants Type

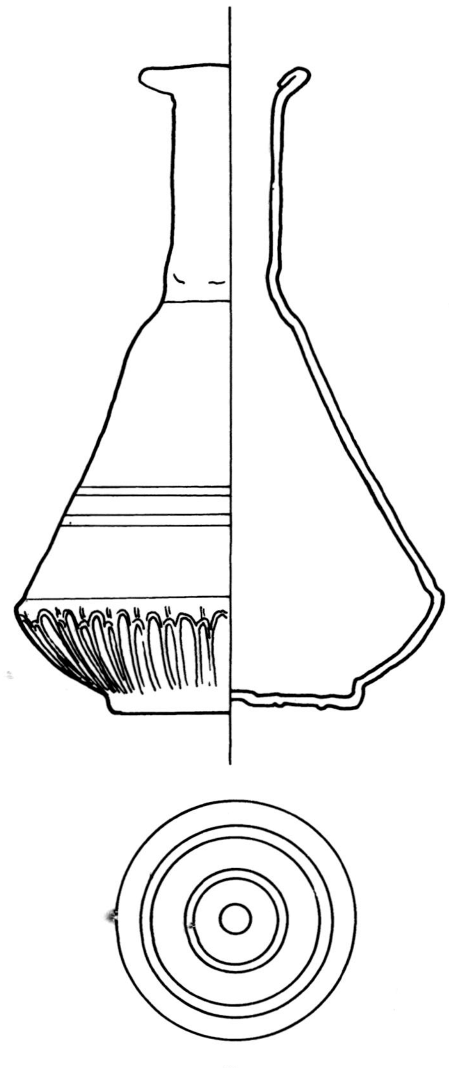

*Carinated Bottle*

49

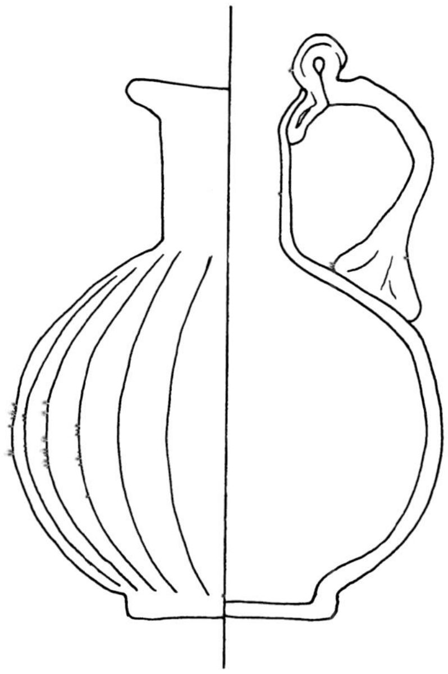

50

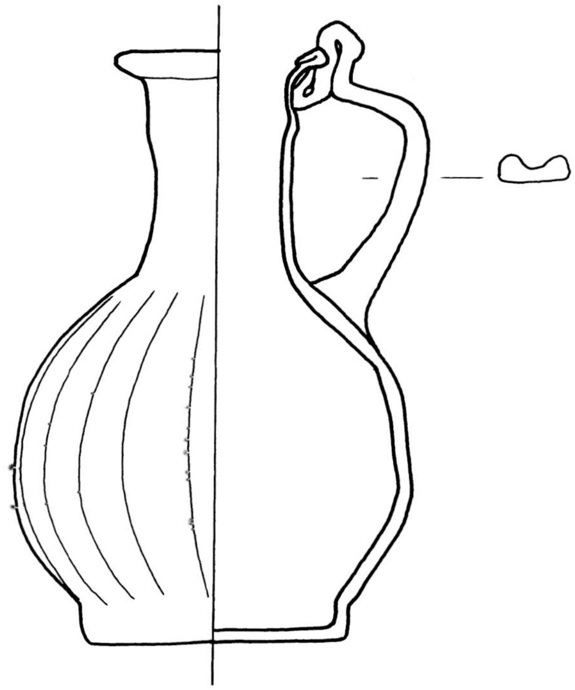

51

*Spherical Bottles with Vertical Ribs*

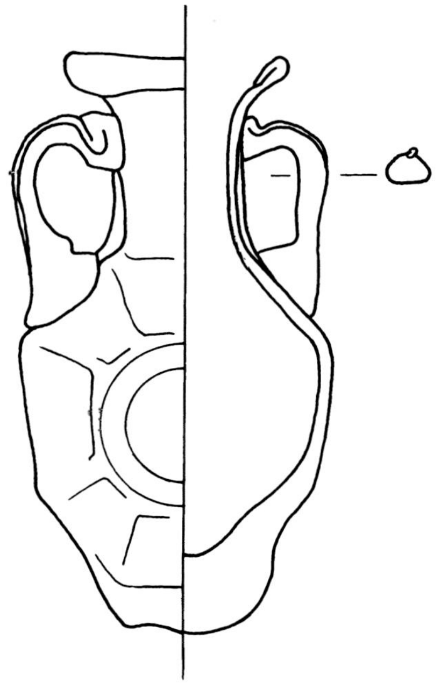

52

*Multisided Bottle*

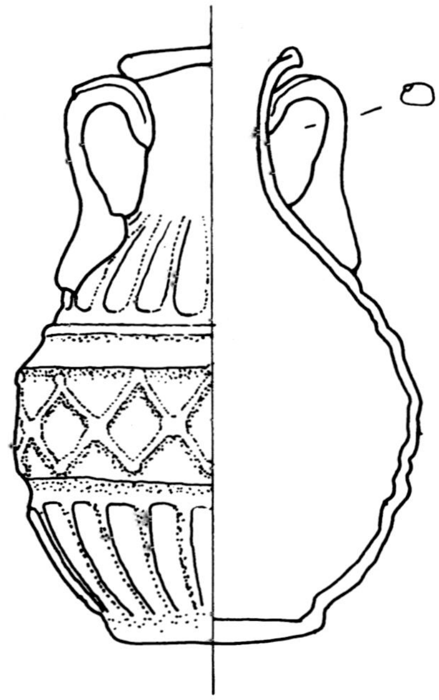

53

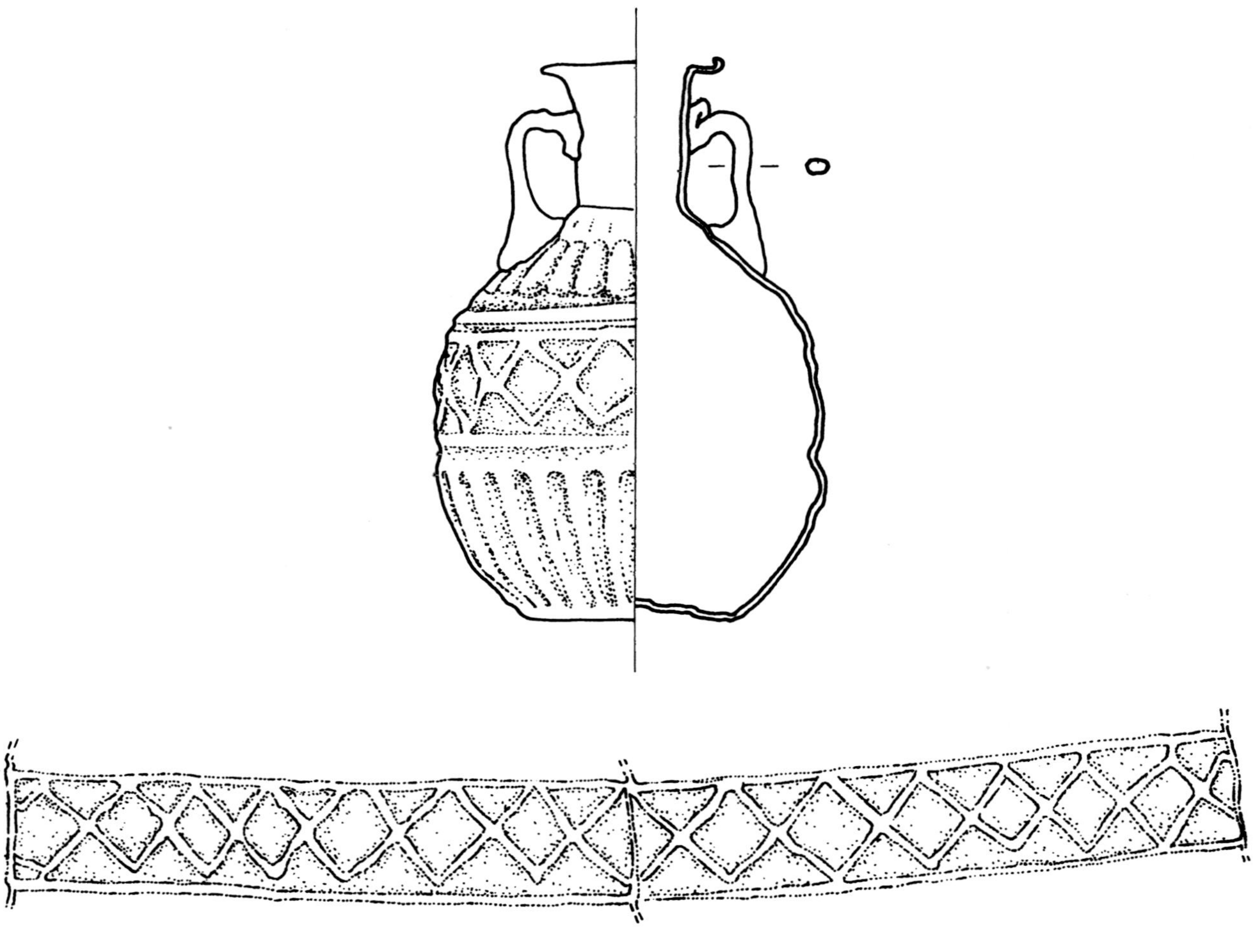

54

*Bulbous Bottles with a Band of Lozenges*

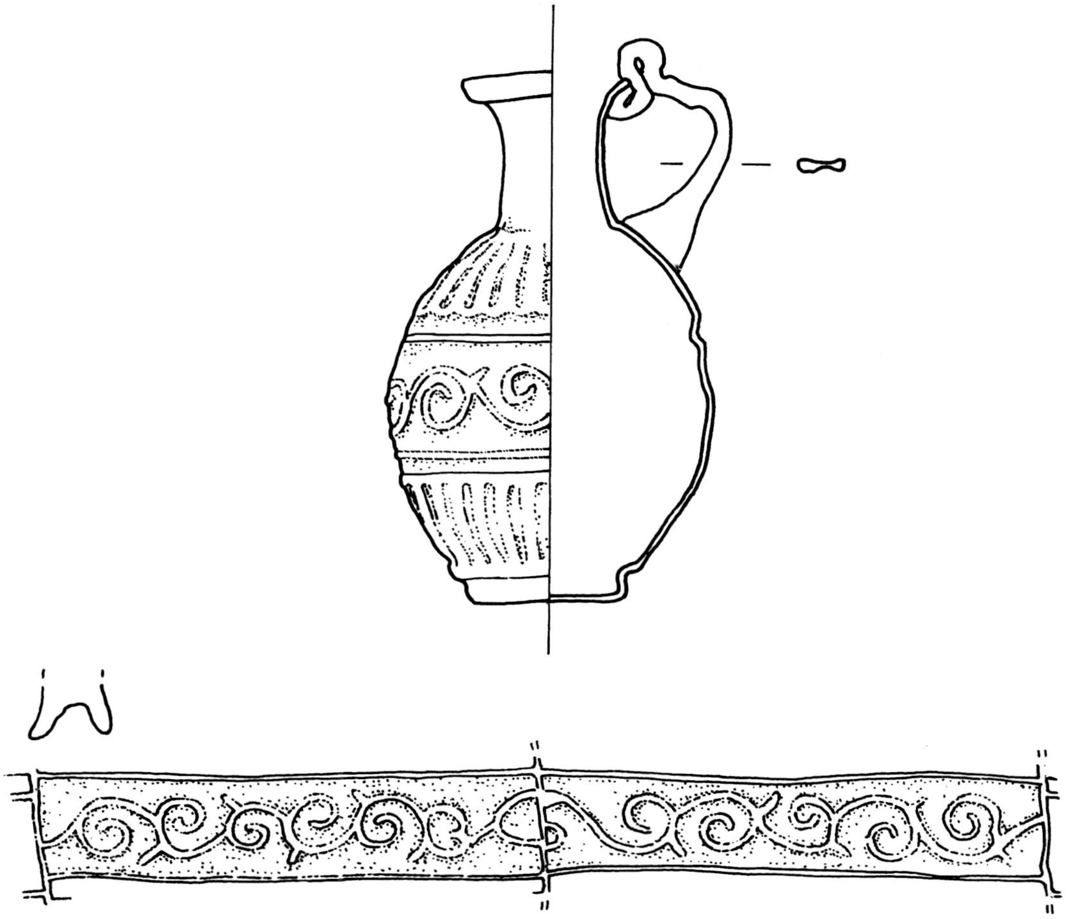

55

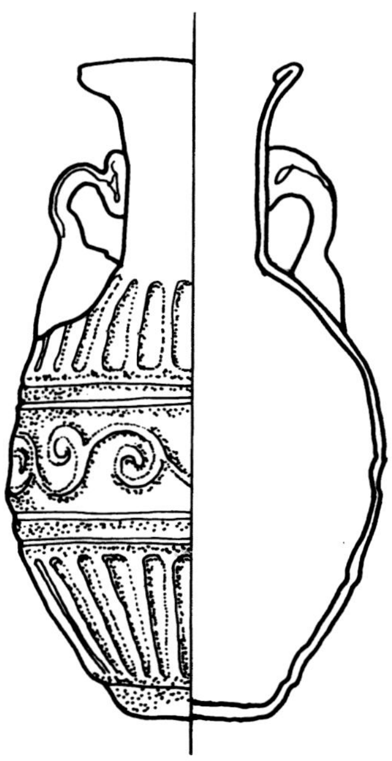

56

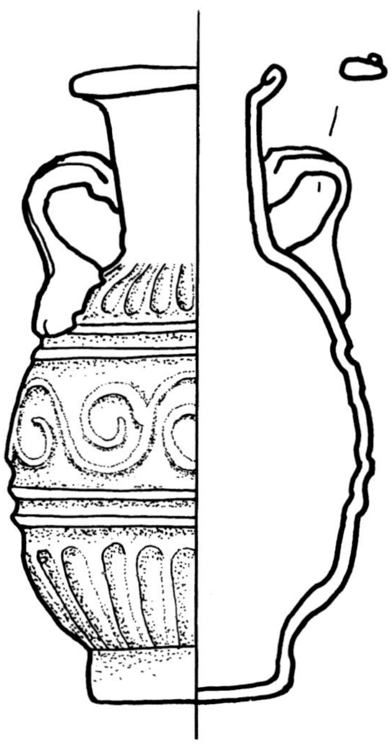

57

*Bulbous Bottles with a Band of Scrolls*

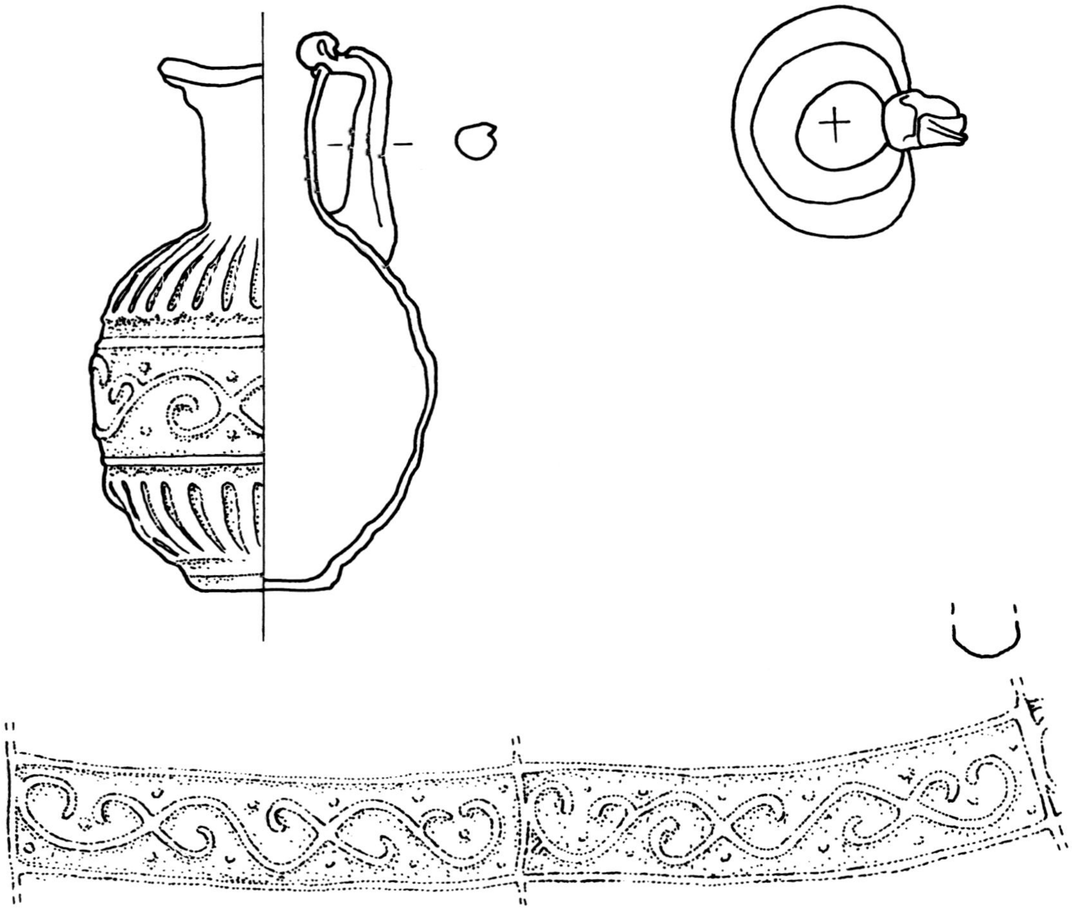

58

*Bulbous Bottles with a Band of Scrolls*

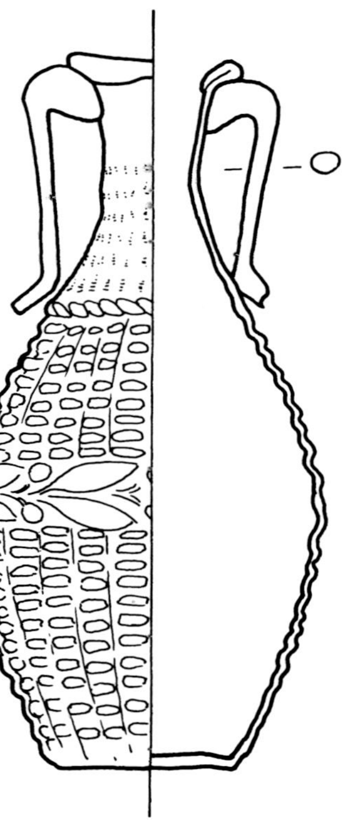

59

*Miniature Amphora with Floating Handles*

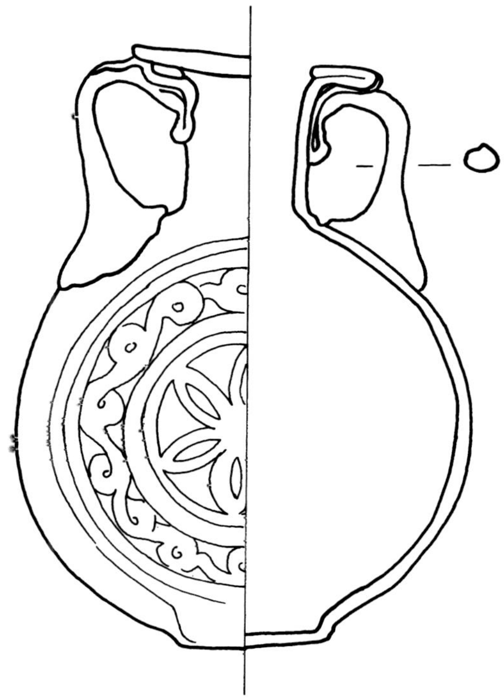

60

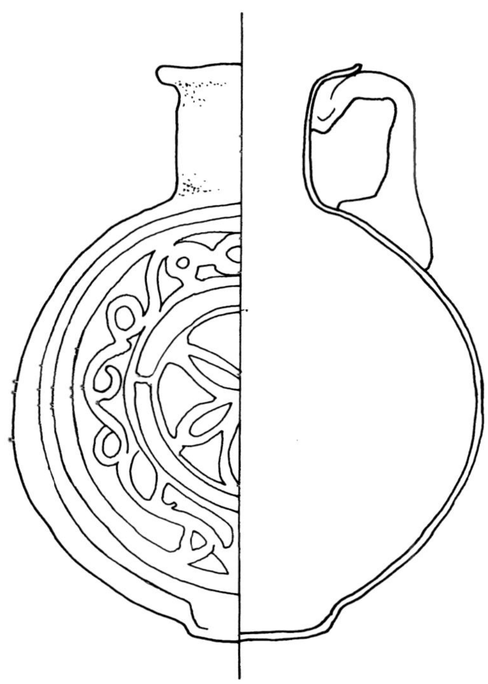

61

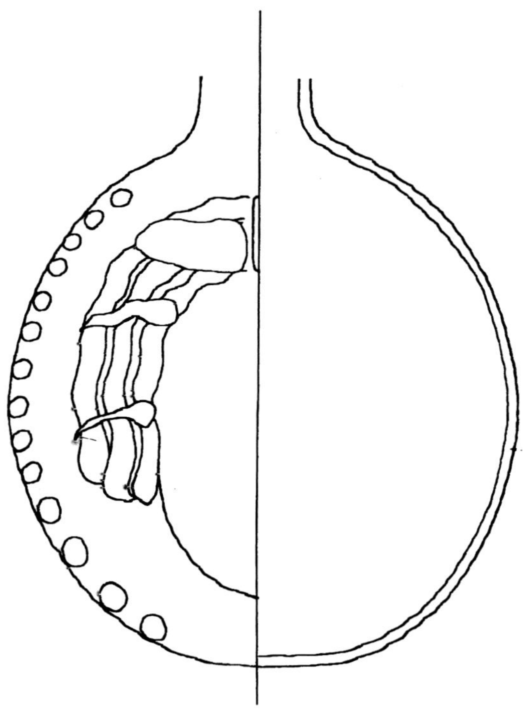

62

*Lenticular Bottles*

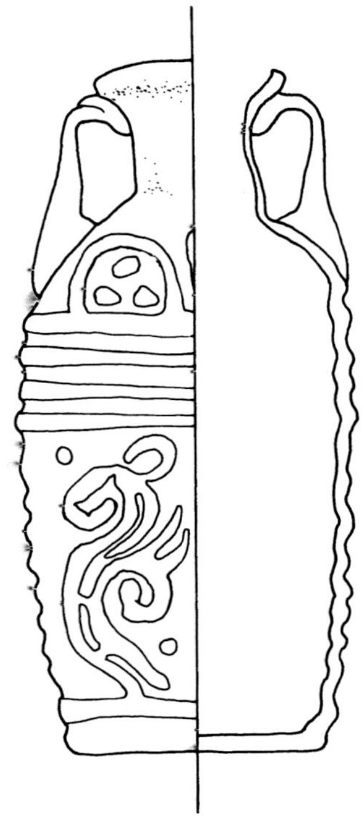

63

*Hexagonal Bottle*

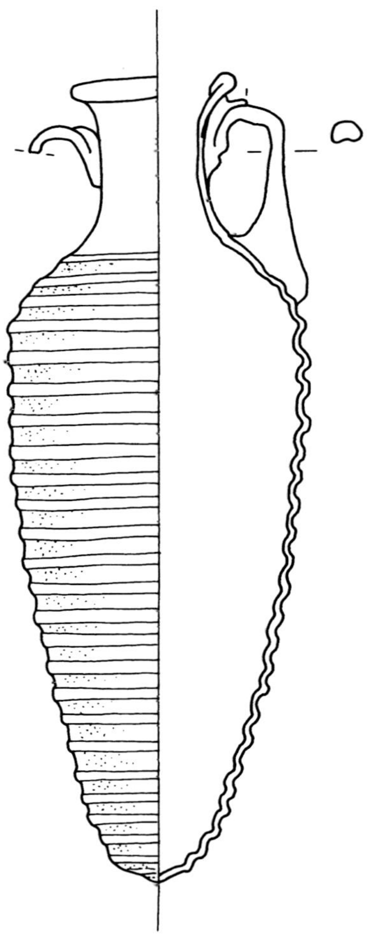

64

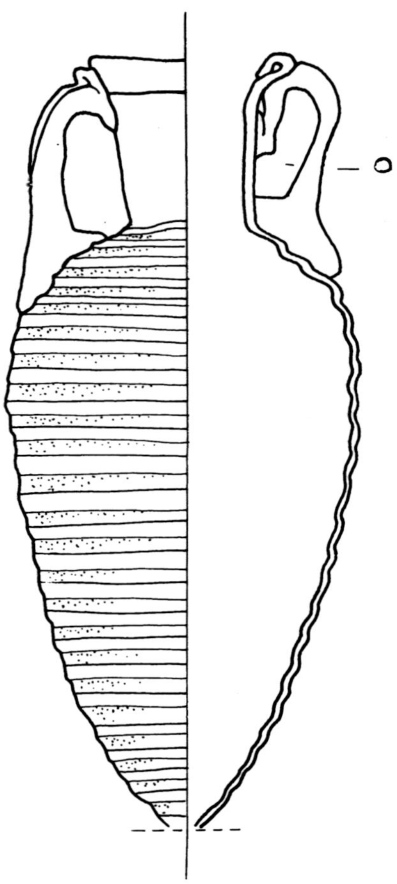

65

*Miniature Transport Amphoras with Horizontal Ribs*

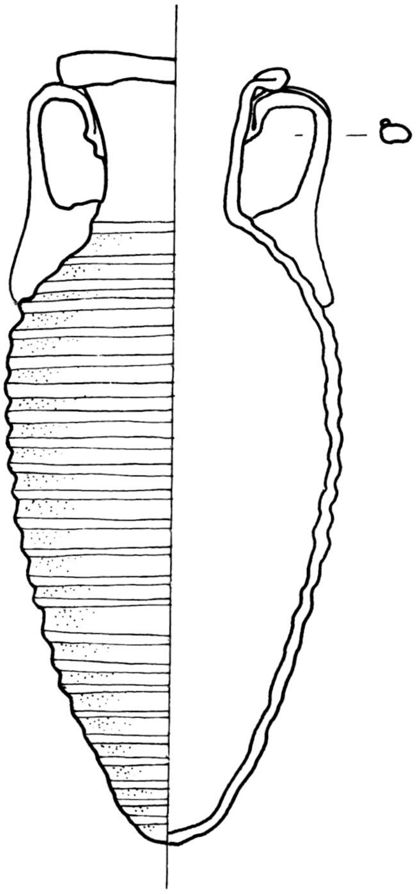

66

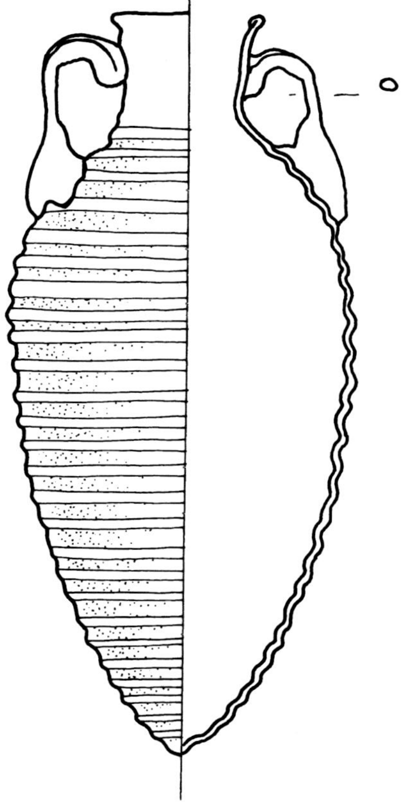

67

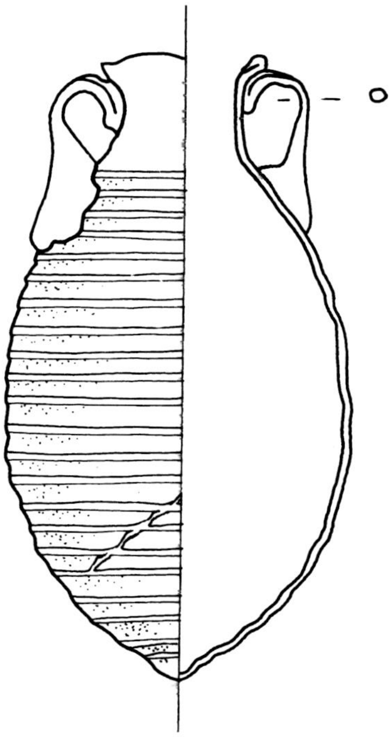

68

*Miniature Transport Amphoras with Horizontal Ribs*

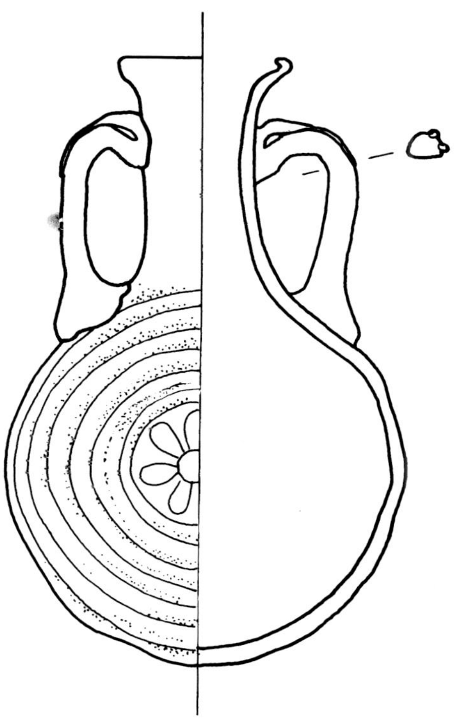

69

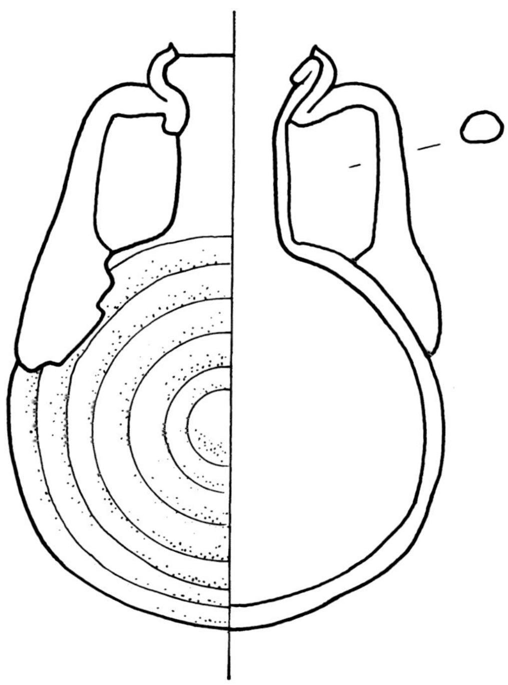

70

*Barrel-shaped Bottles*

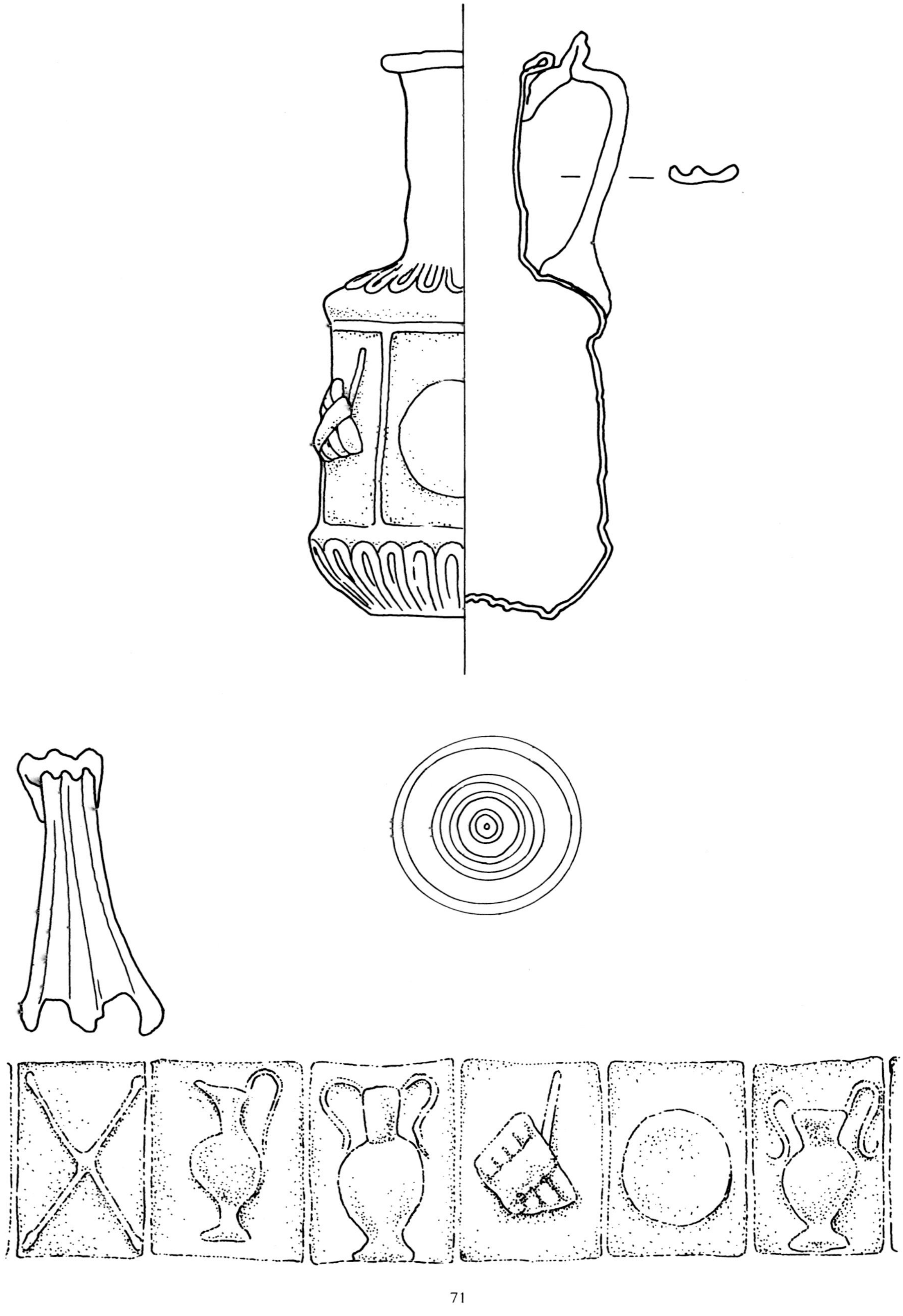

71

*Hexagonal Jugs with Dionysiac Symbols*

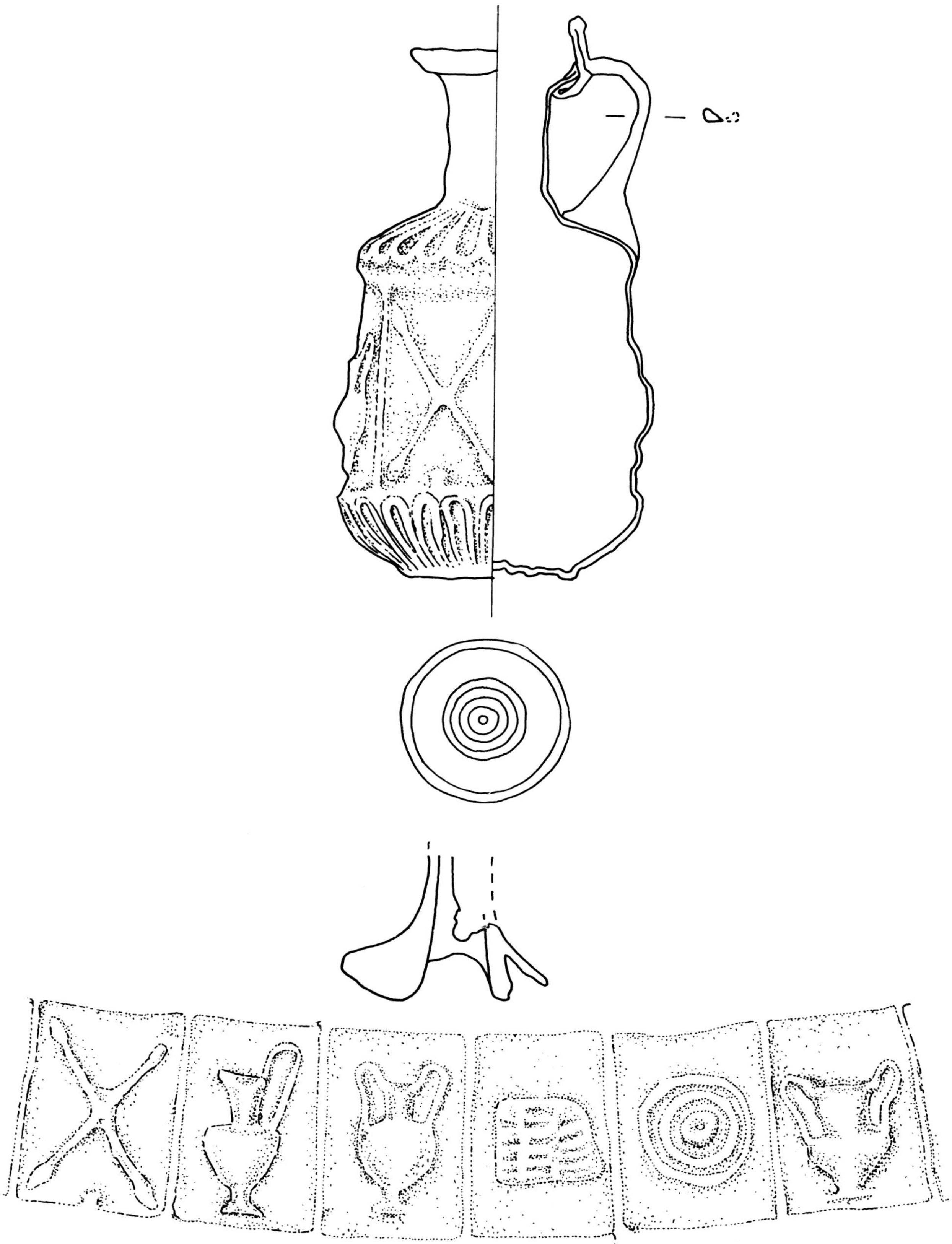

72

*Hexagonal Jugs with Dionysiac Symbols*

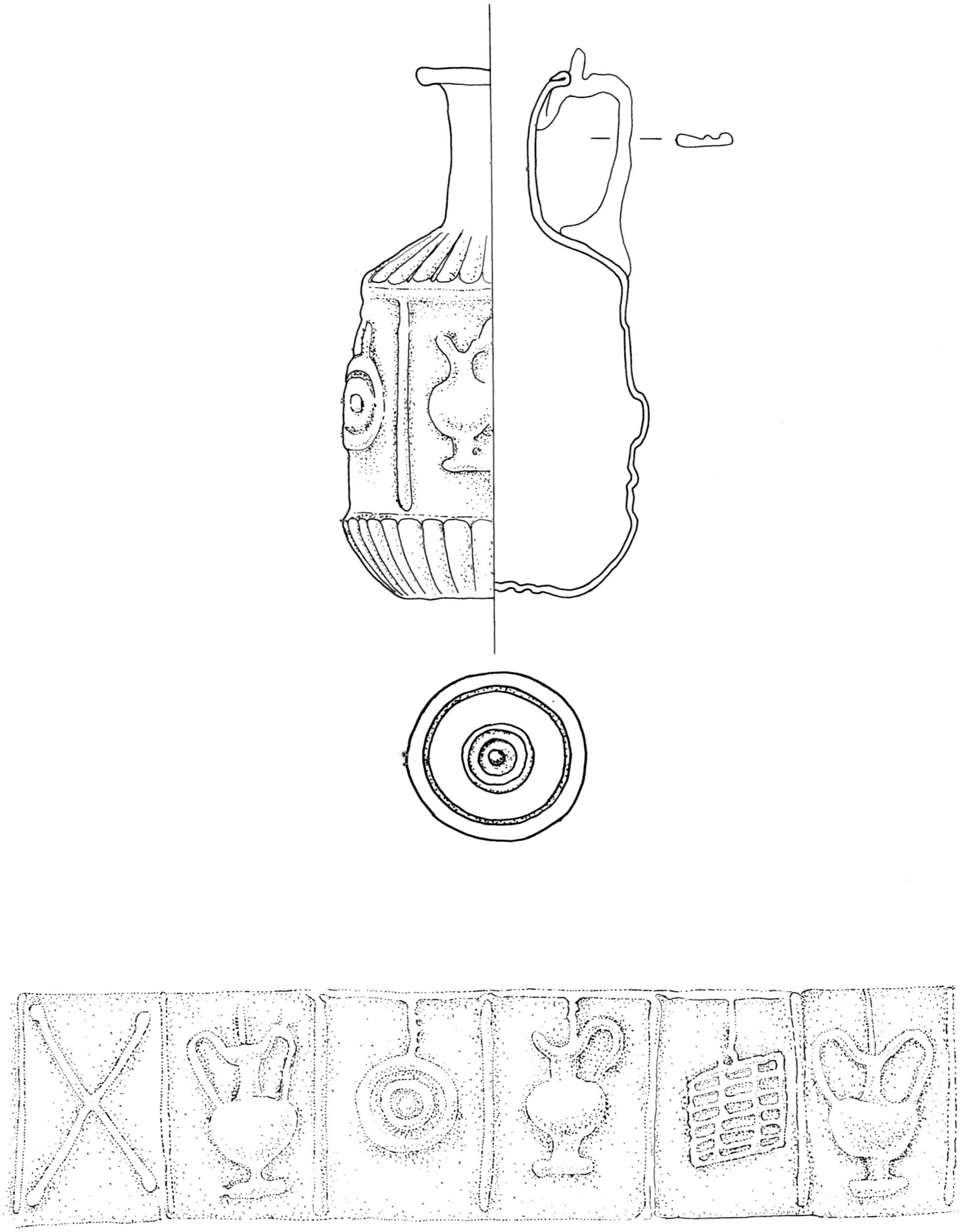

73

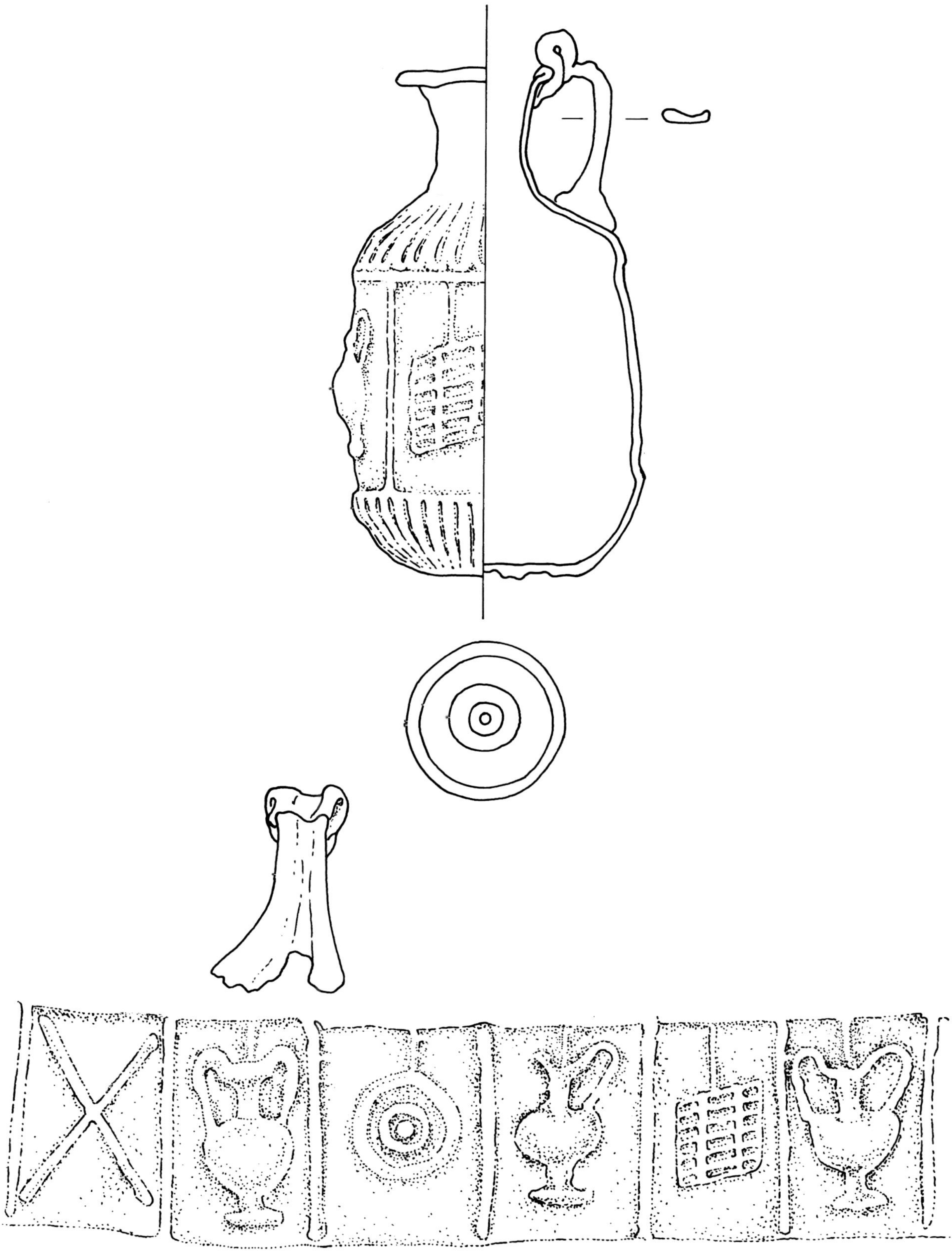

74

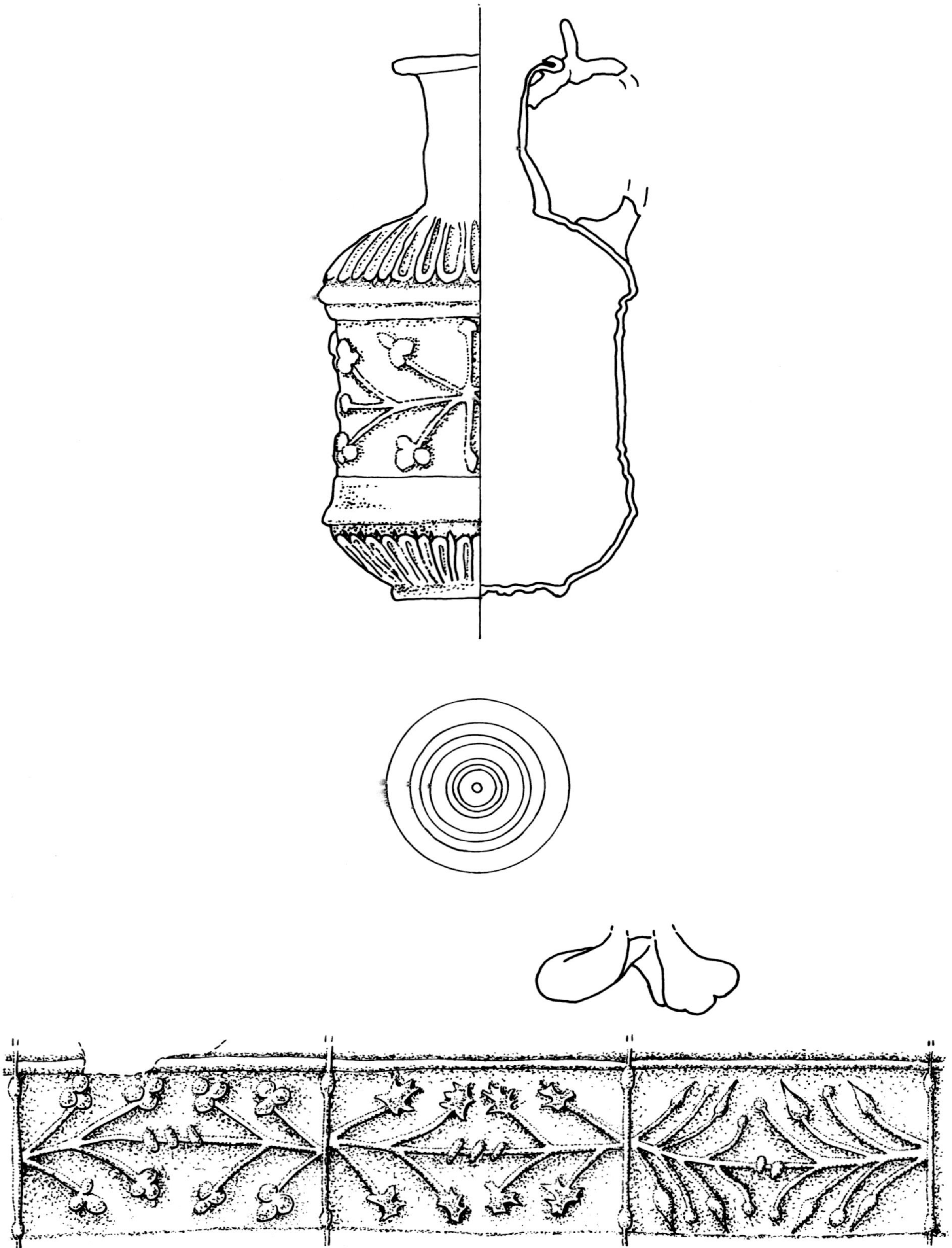

75

*Cylindrical Jugs with Floral Sprays*

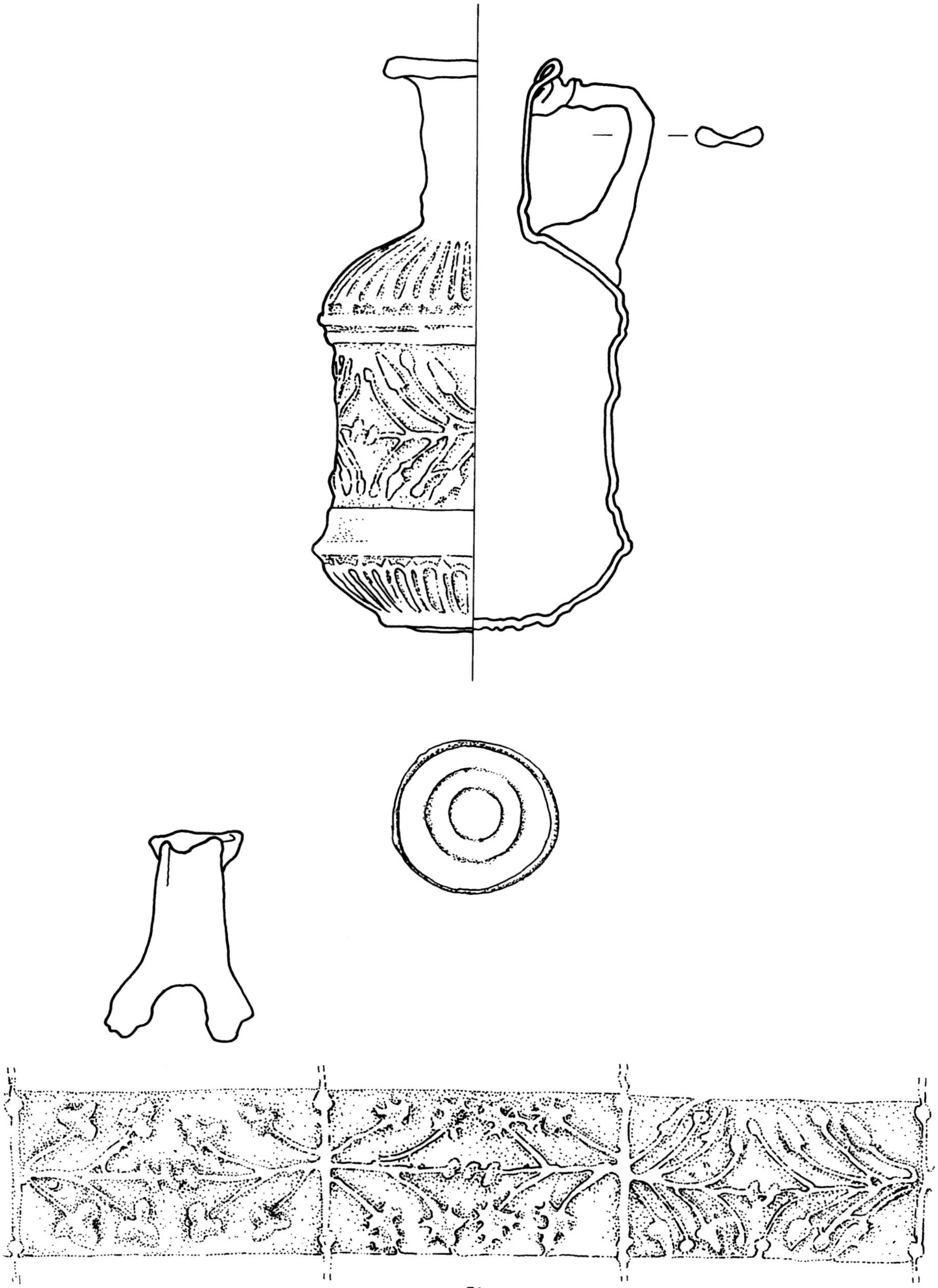

76

*Cylindrical Jugs with Floral Sprays*

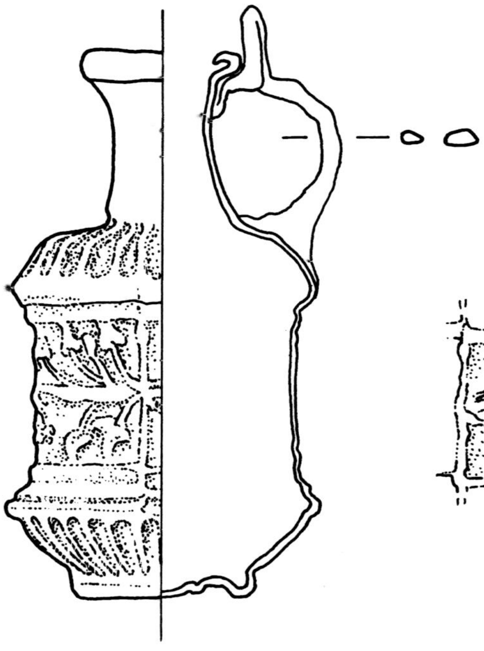
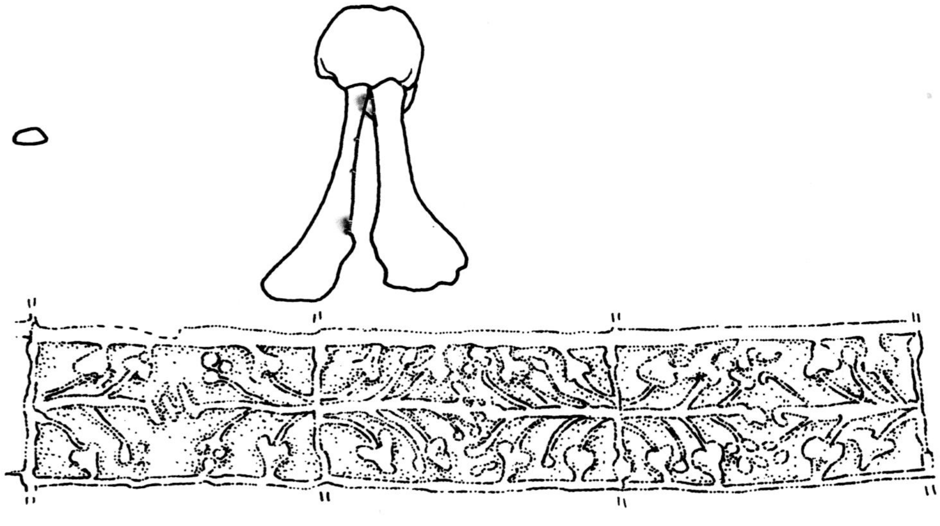
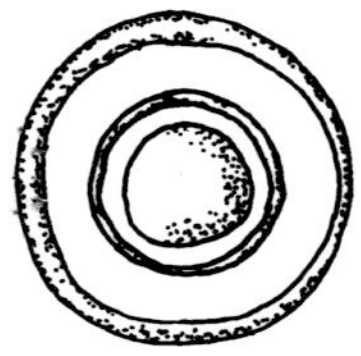

77

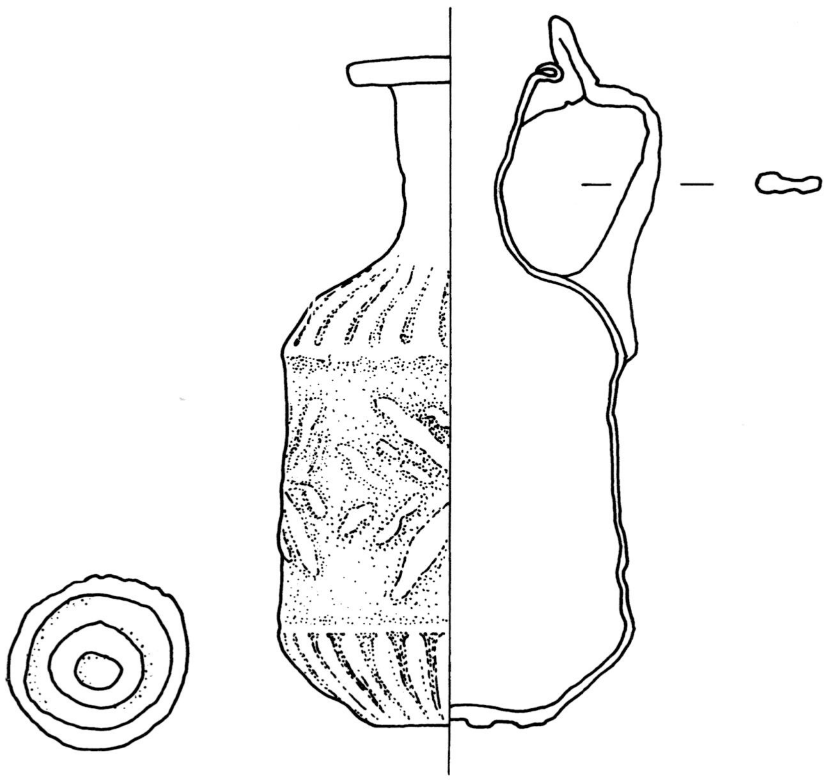
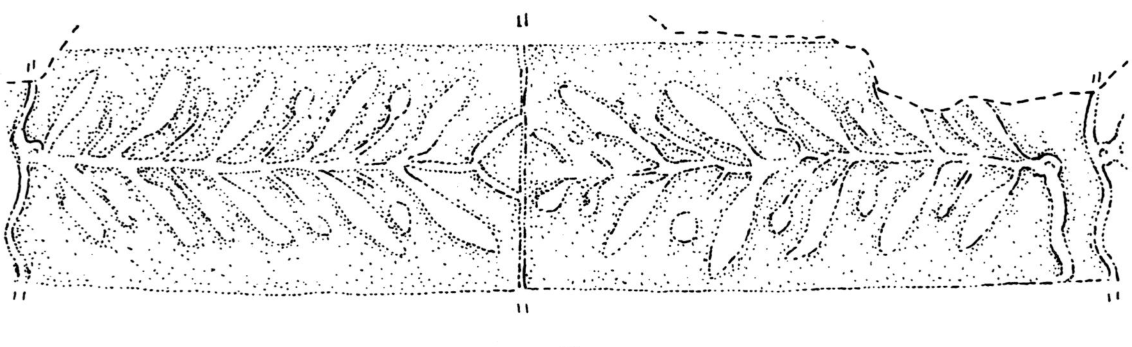

78

*Cylindrical Jugs with Floral Sprays*

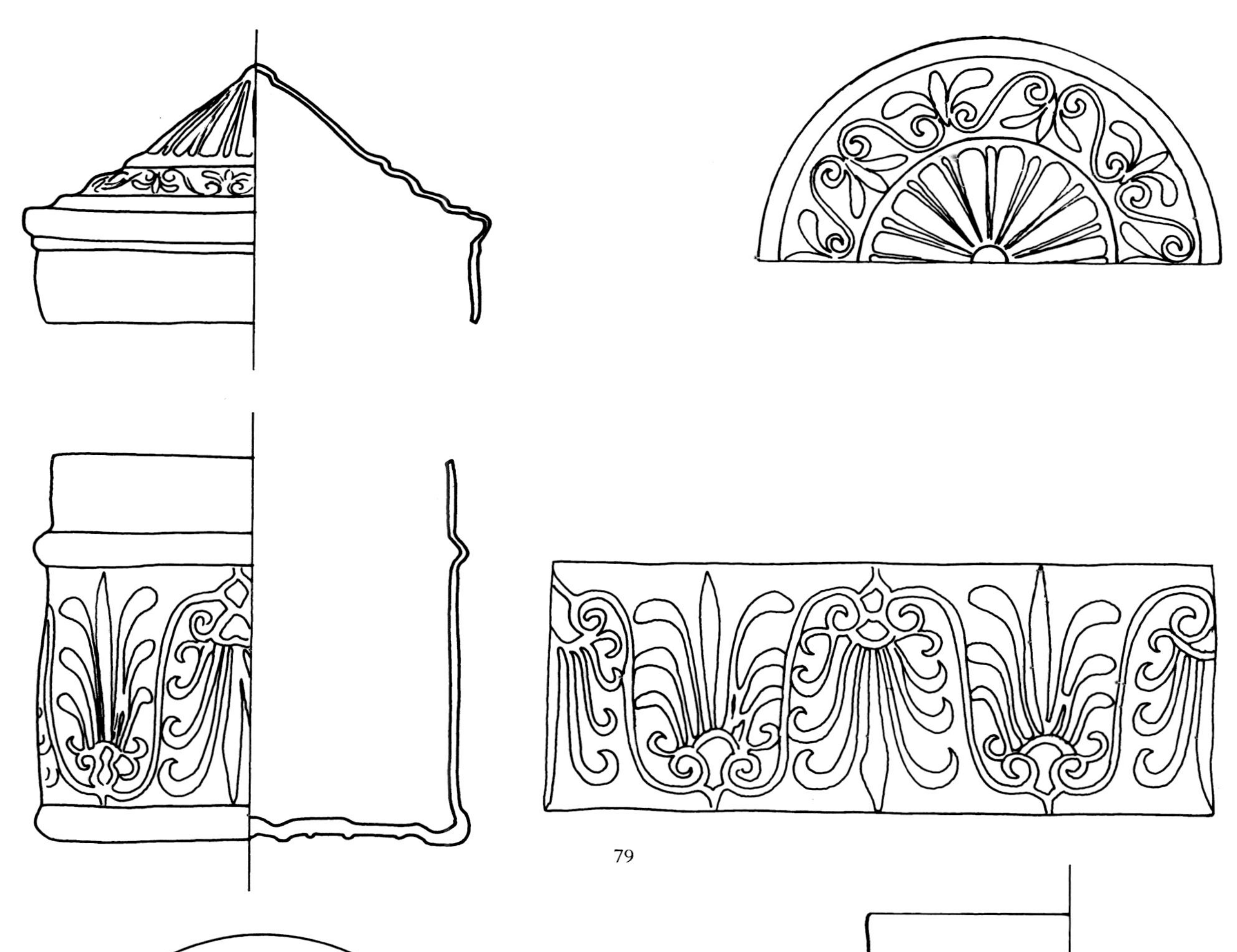

79

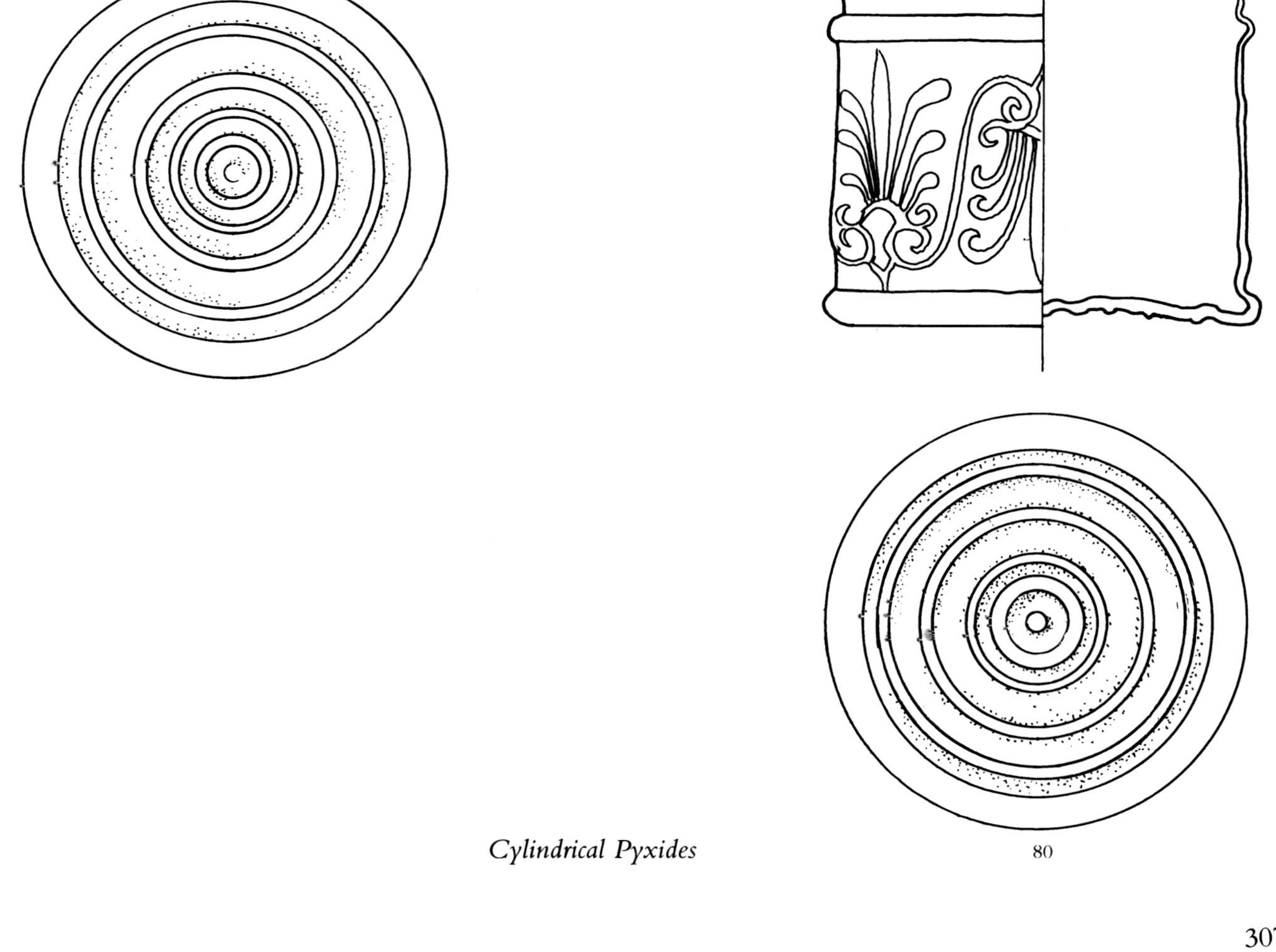

80

*Cylindrical Pyxides*

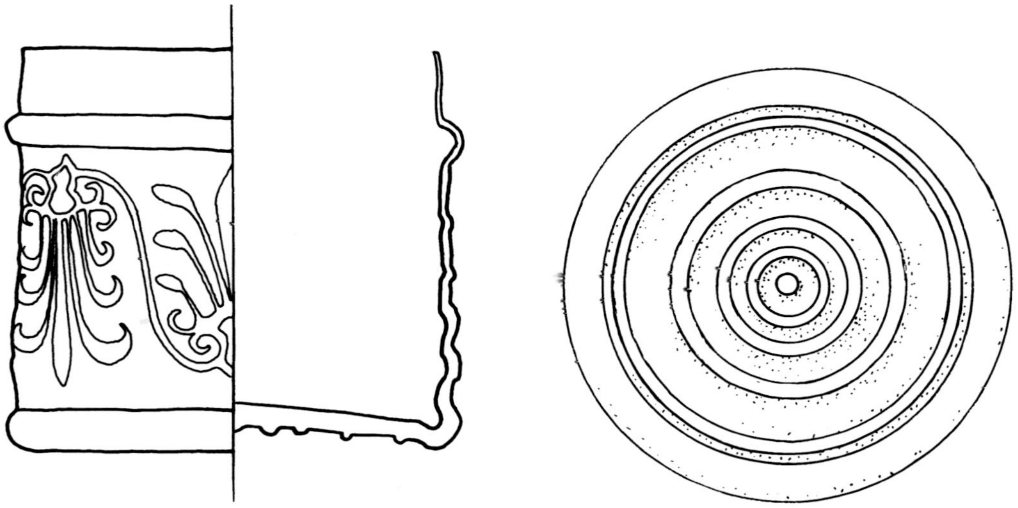

81

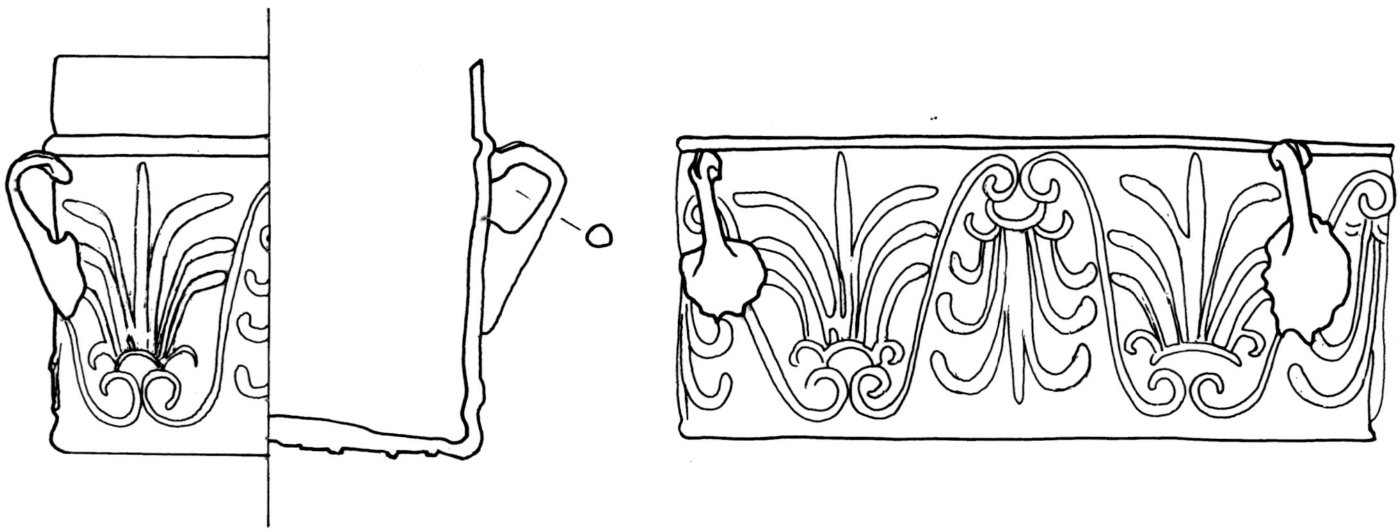

82

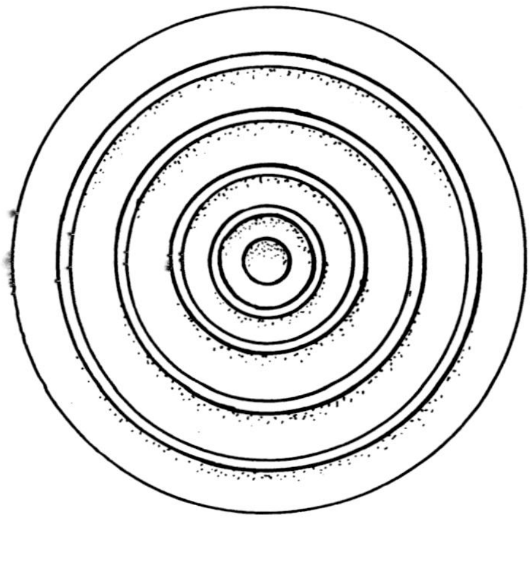

*Cylindrical Pyxides*

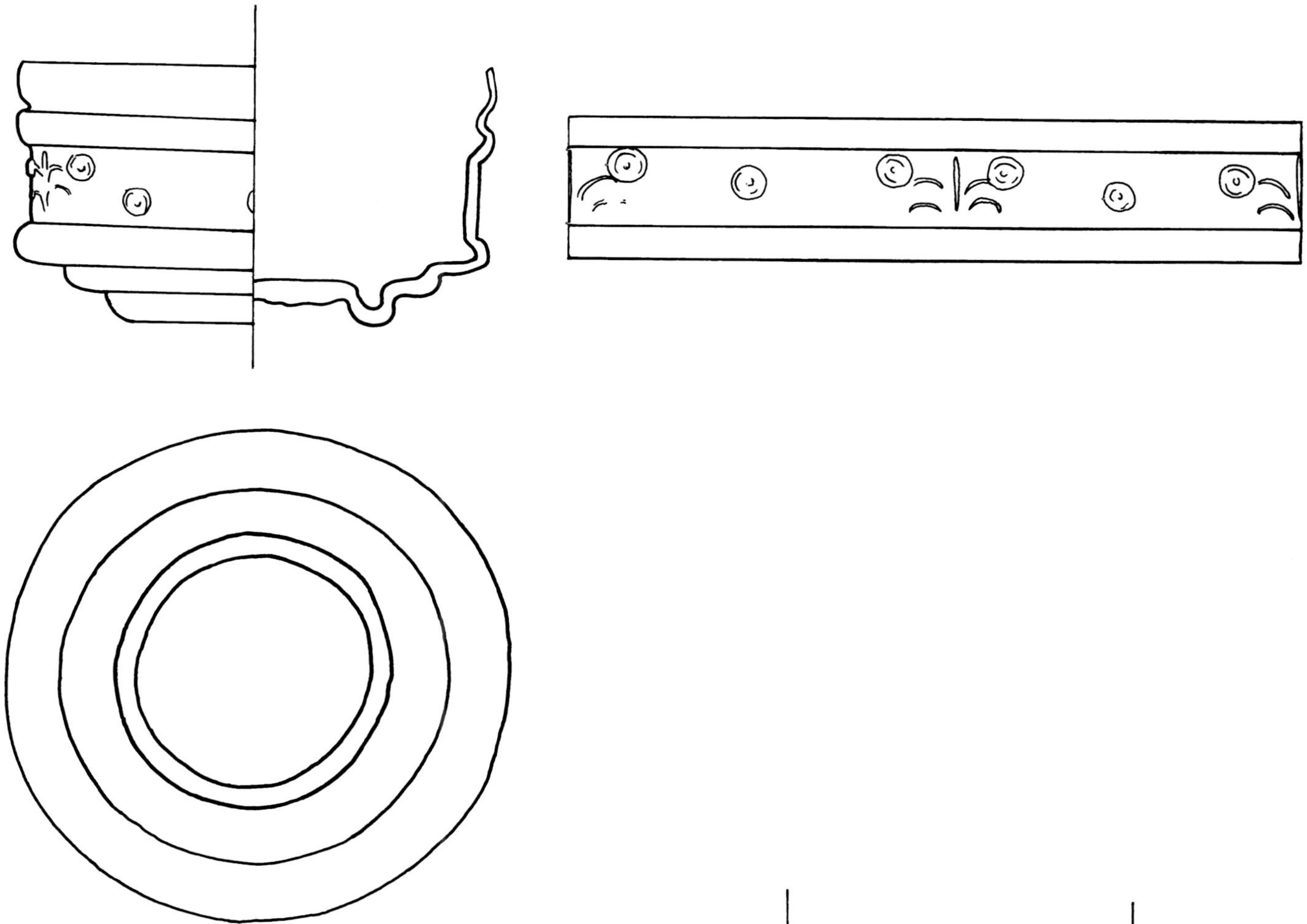

83

*Cylindrical Pyxides*

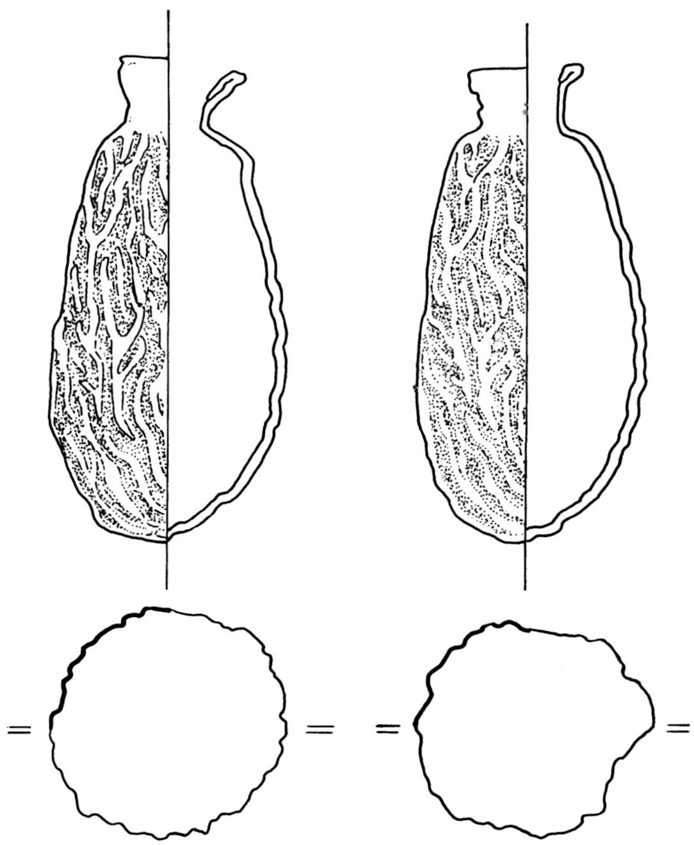

84 85

*Date-shaped Bottles*

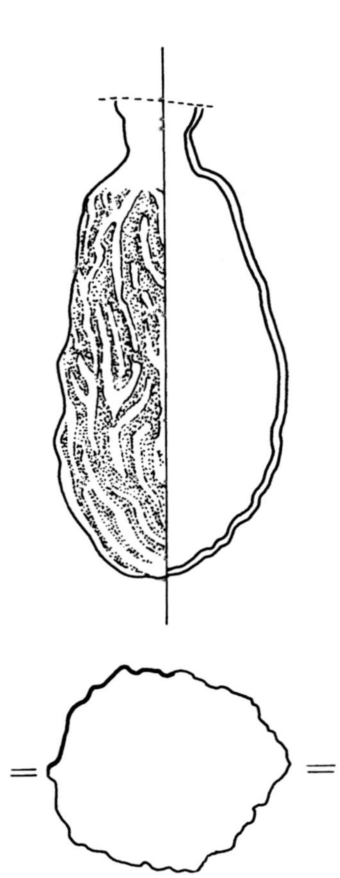
86

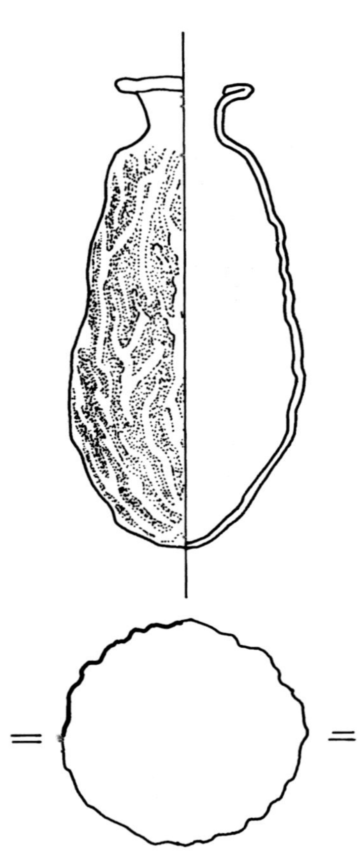
87

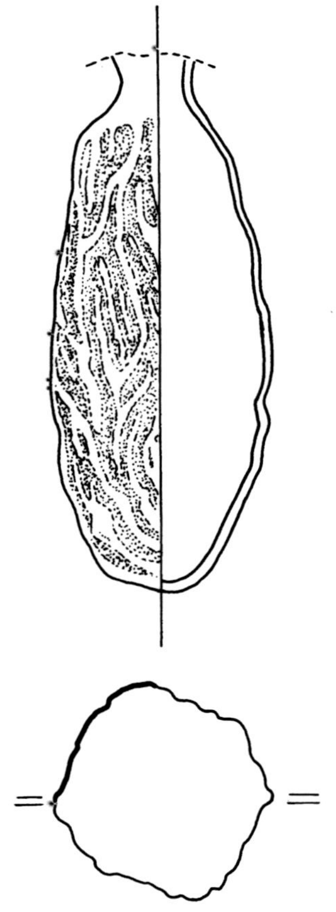
88

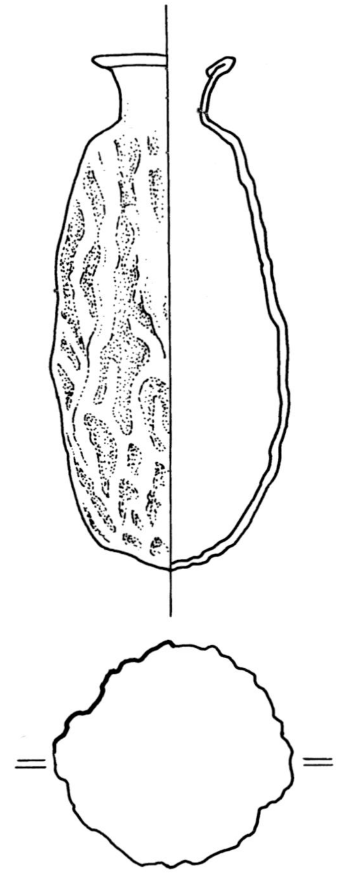
89

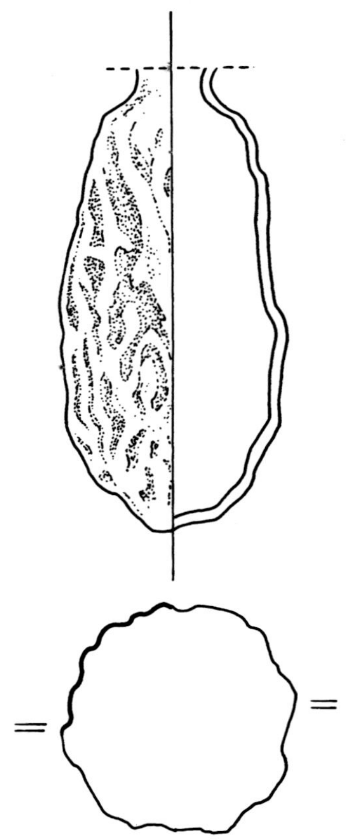
90

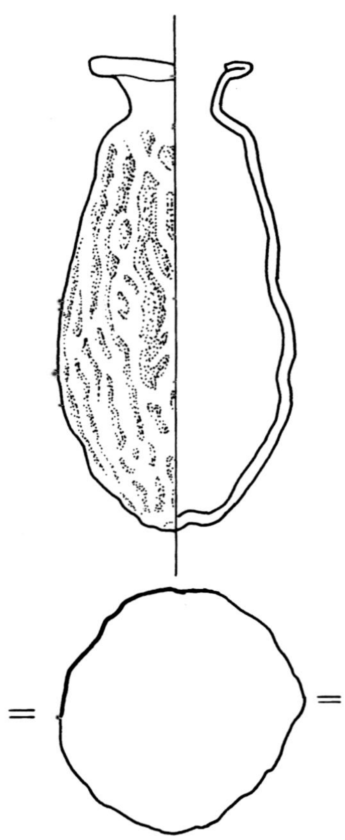
91

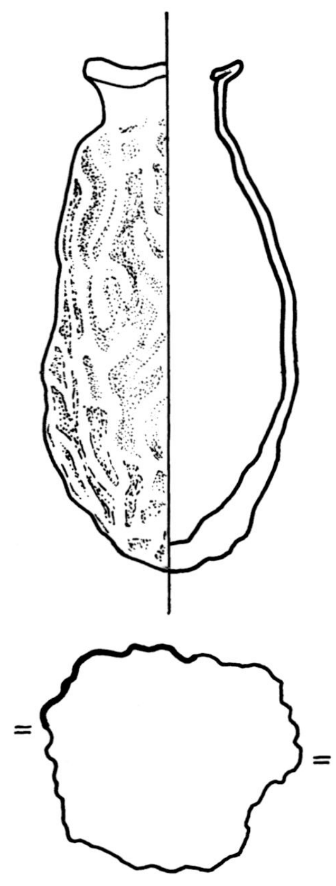
92

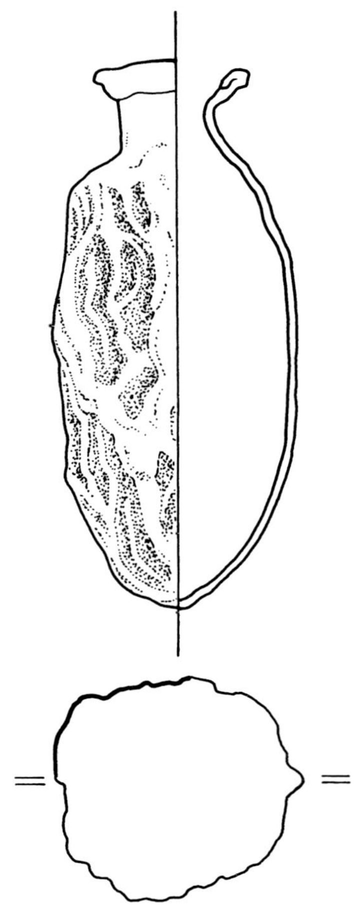
93

*Date-shaped Bottles*

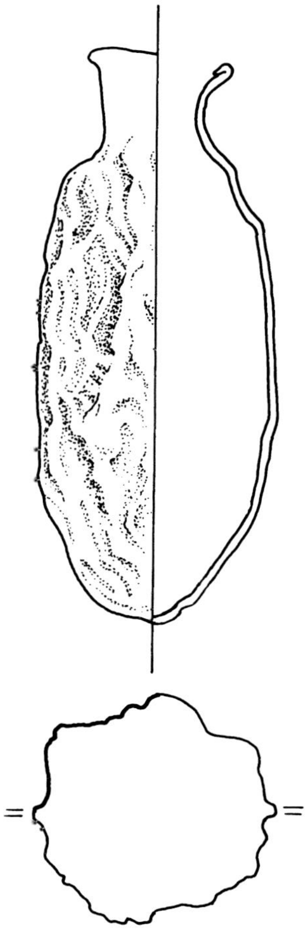
94

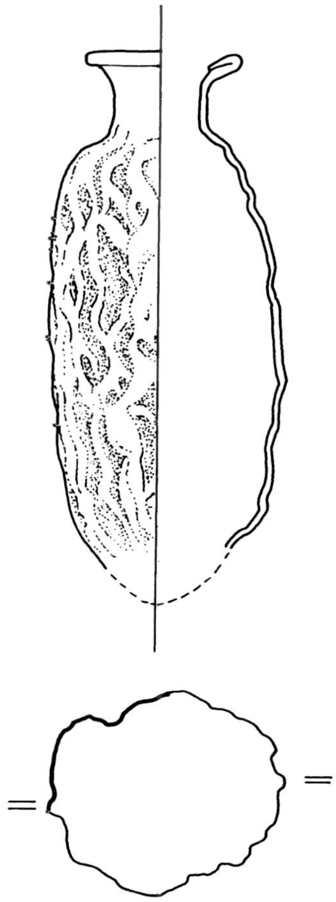
95

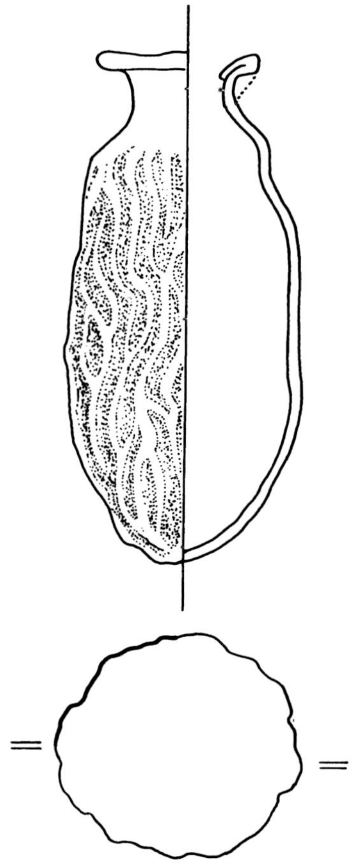
96

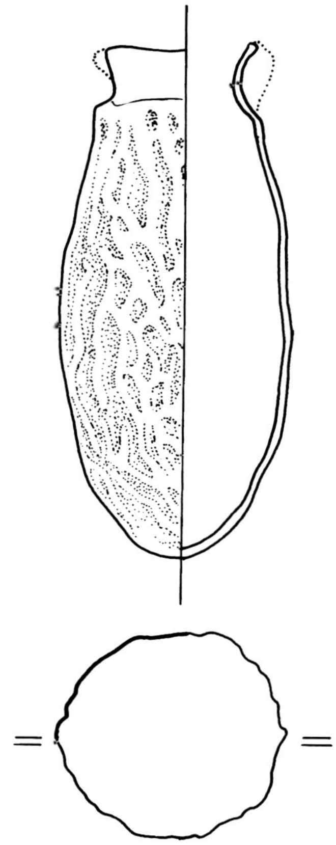
97

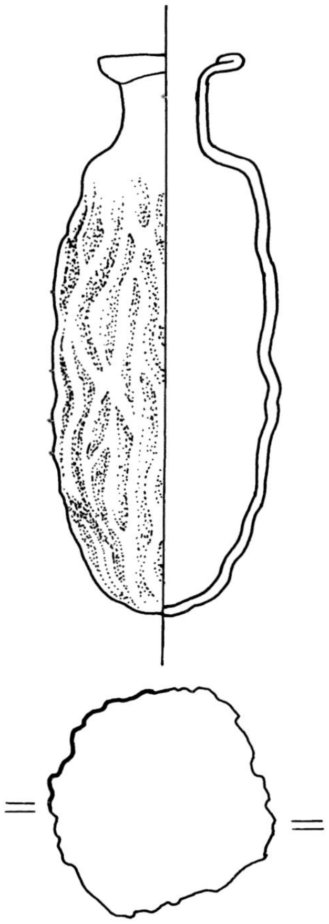
98

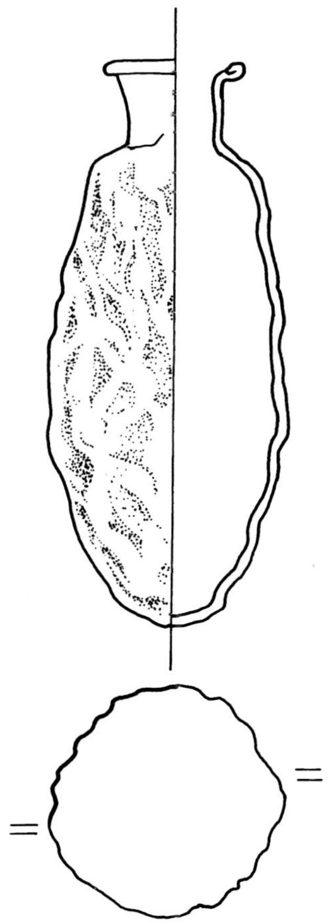
99

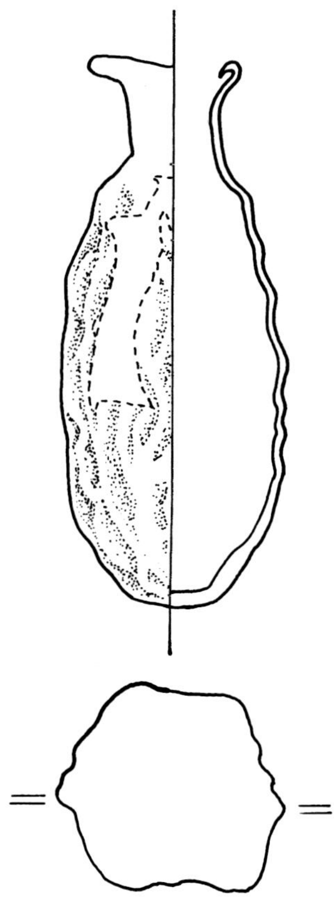
100

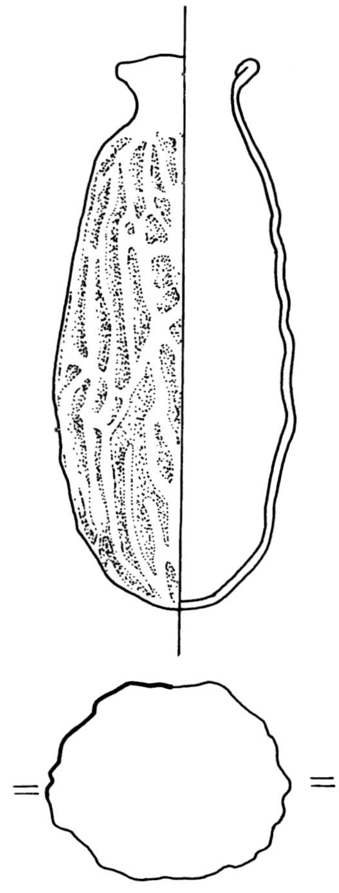
101

*Date-shaped Bottles*

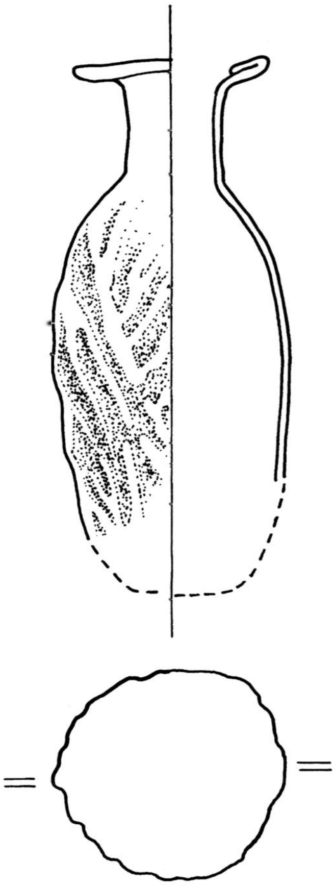

102

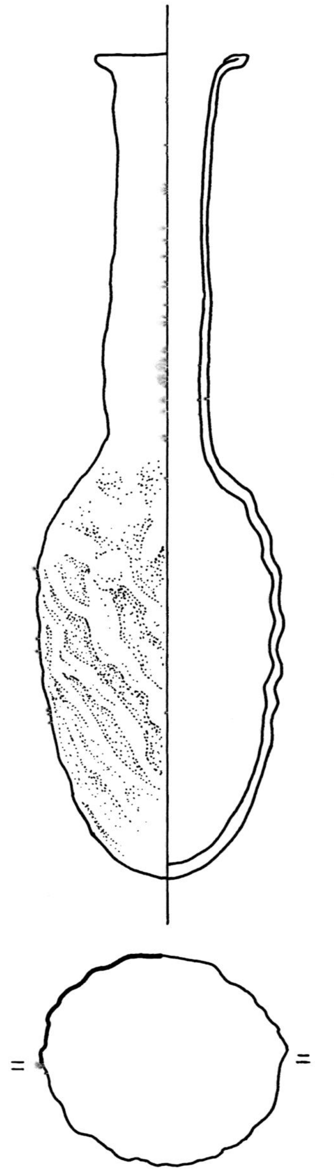

103

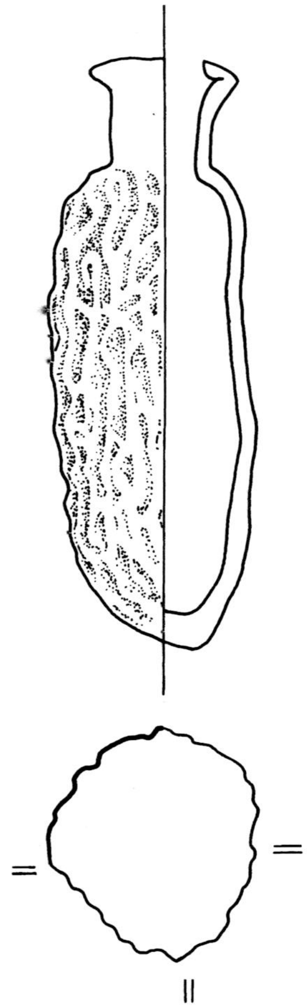

104

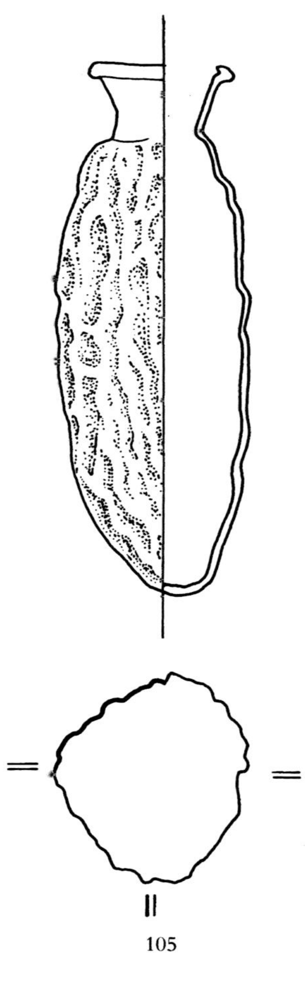

105

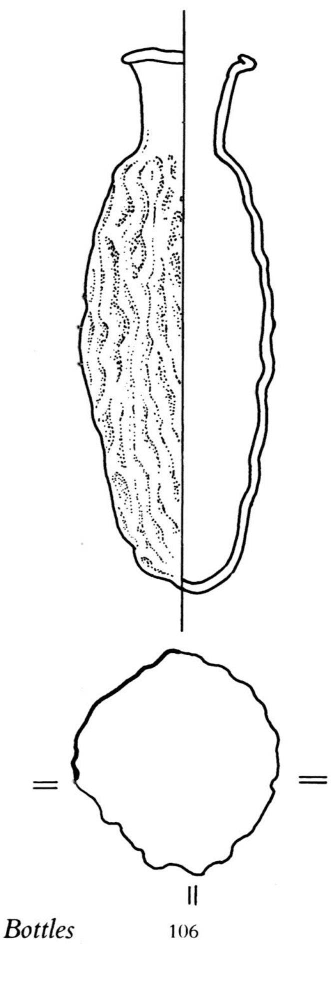

106

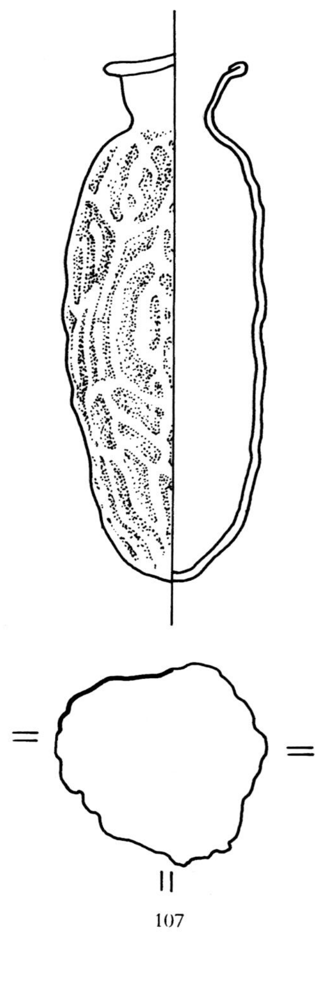

107

*Date-shaped Bottles*

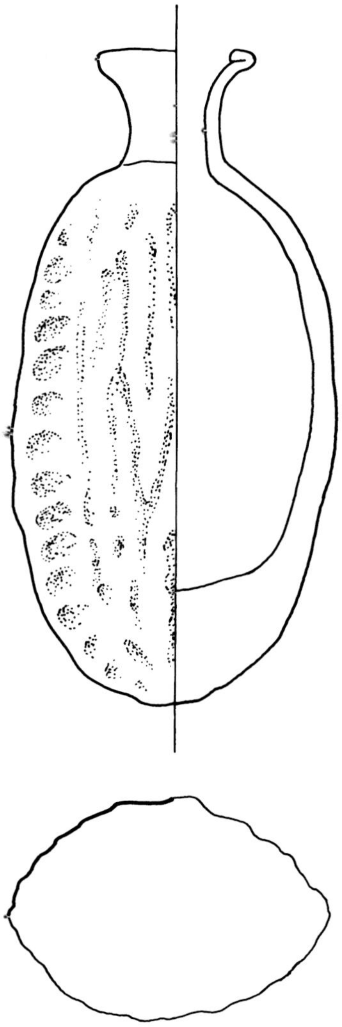

108

*Almond-shaped Bottle*

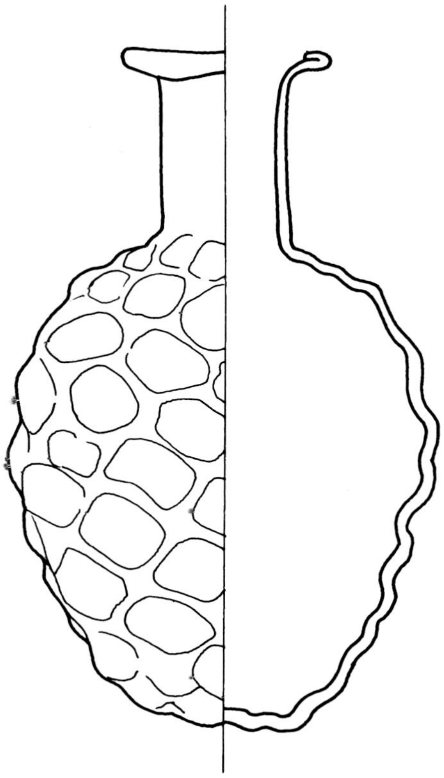

111

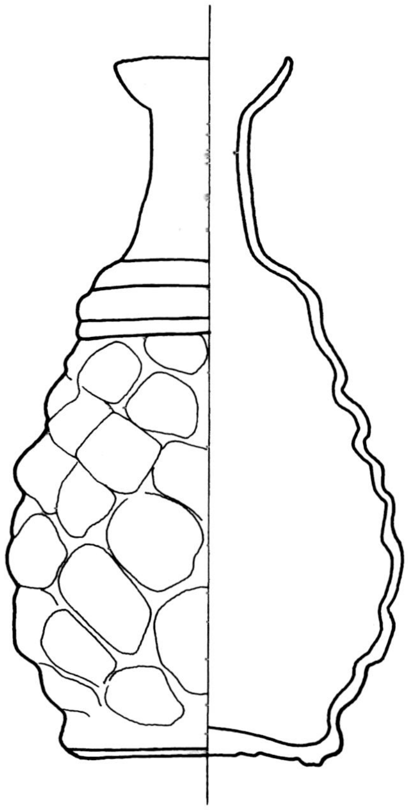

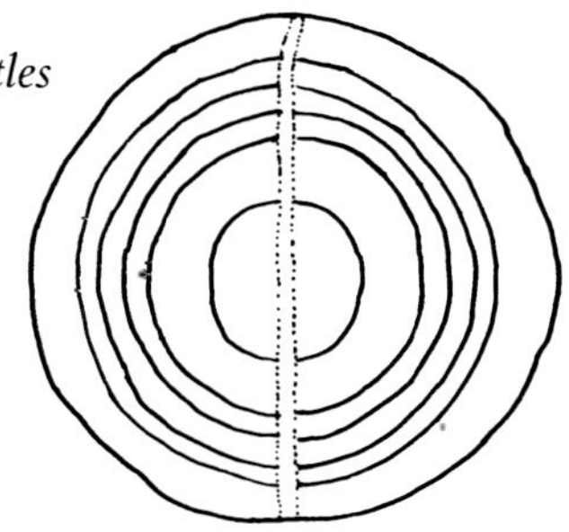

112

*Cone-shaped Bottles*

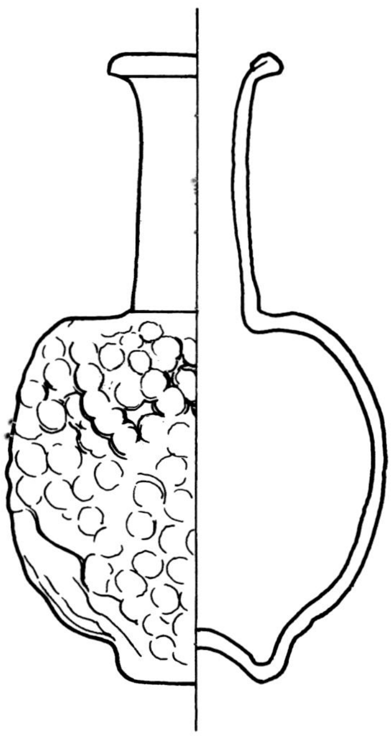

109

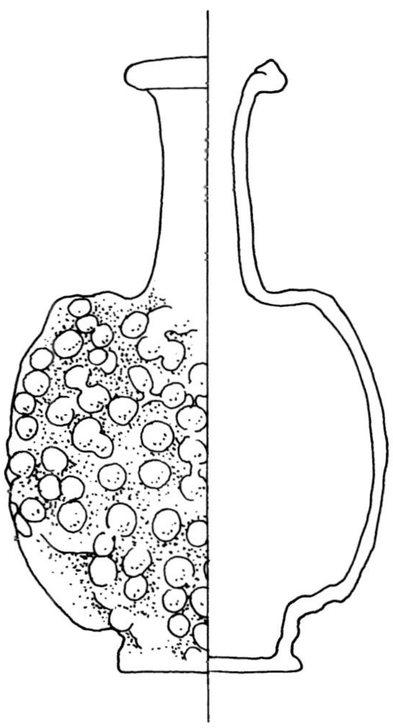

110

*Grape Bunch-shaped Bottles*

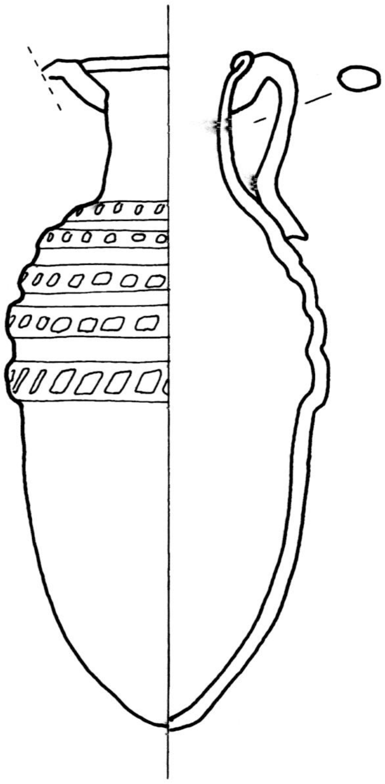

113

*Acorn-shaped Bottle*

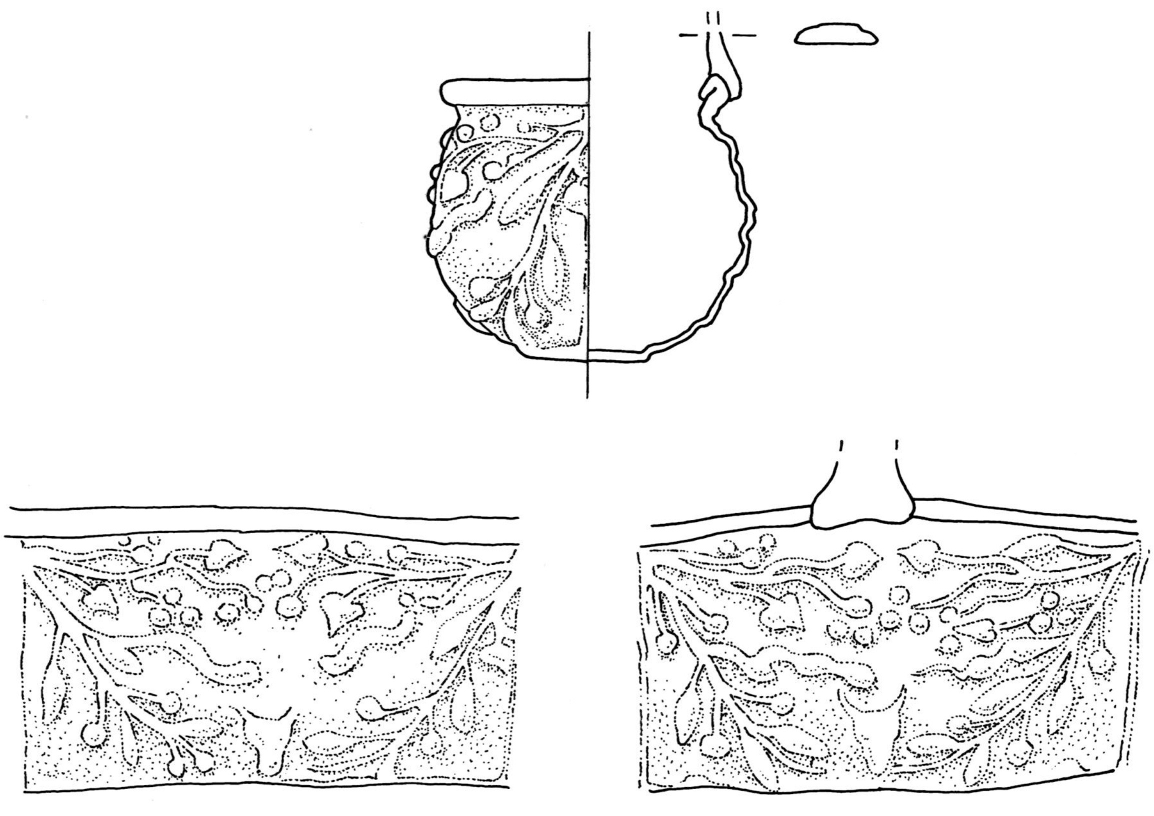

114

*Spherical Jar with Floral Design*

CHAPTER II

# The Second to Fourth Centuries

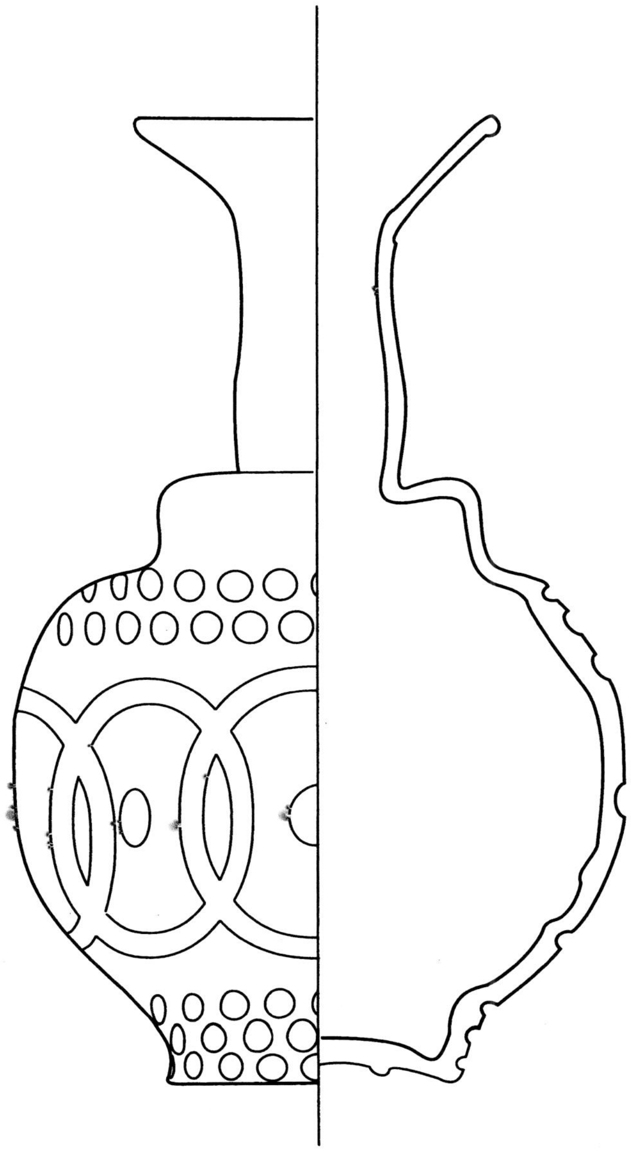

115

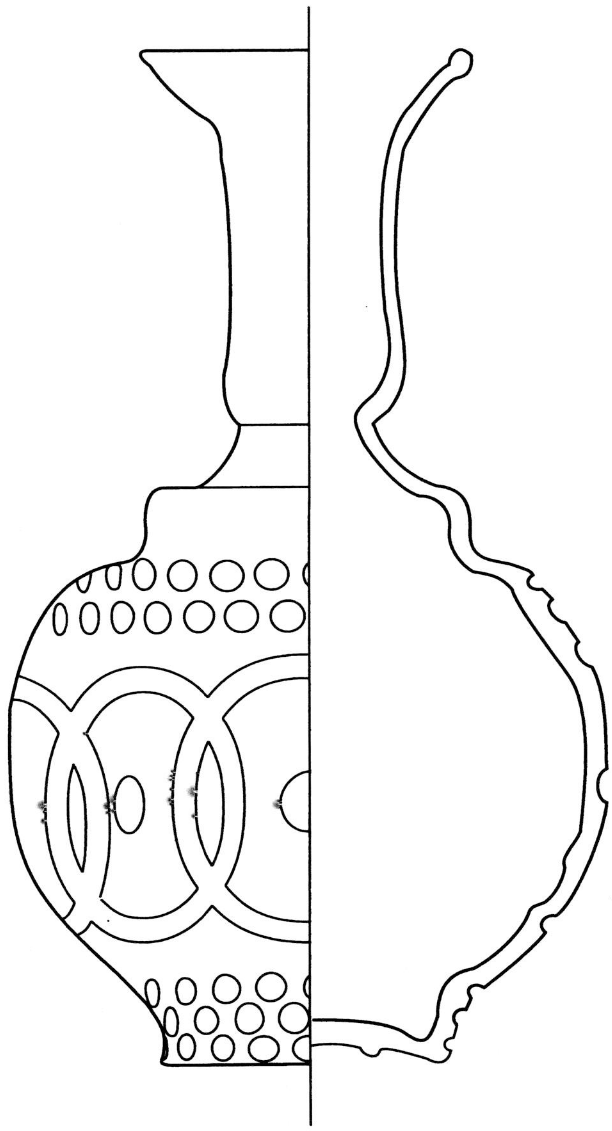

116

*Flasks with Mold-blown Designs in Sunken Relief*

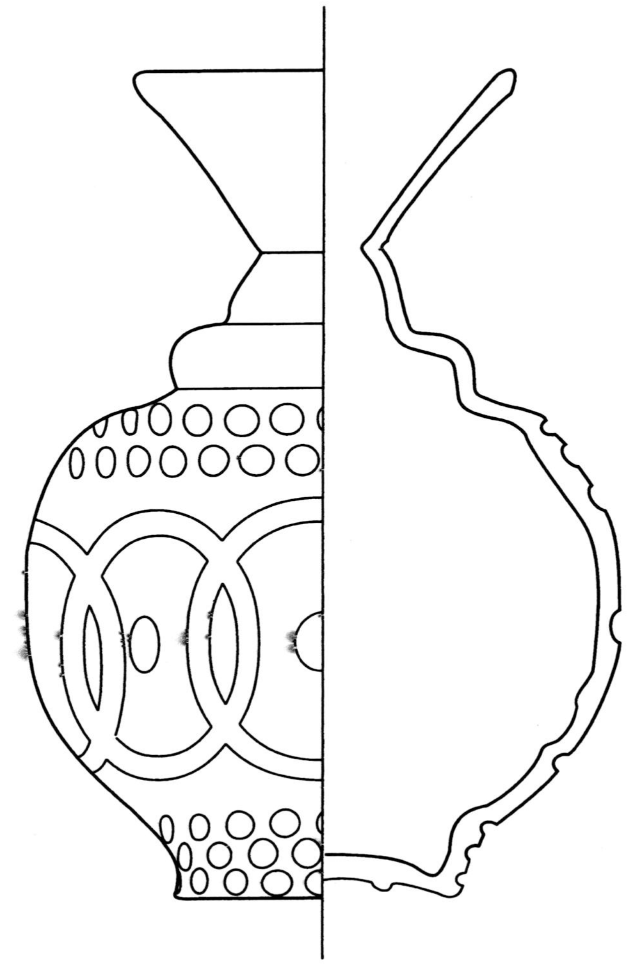

117

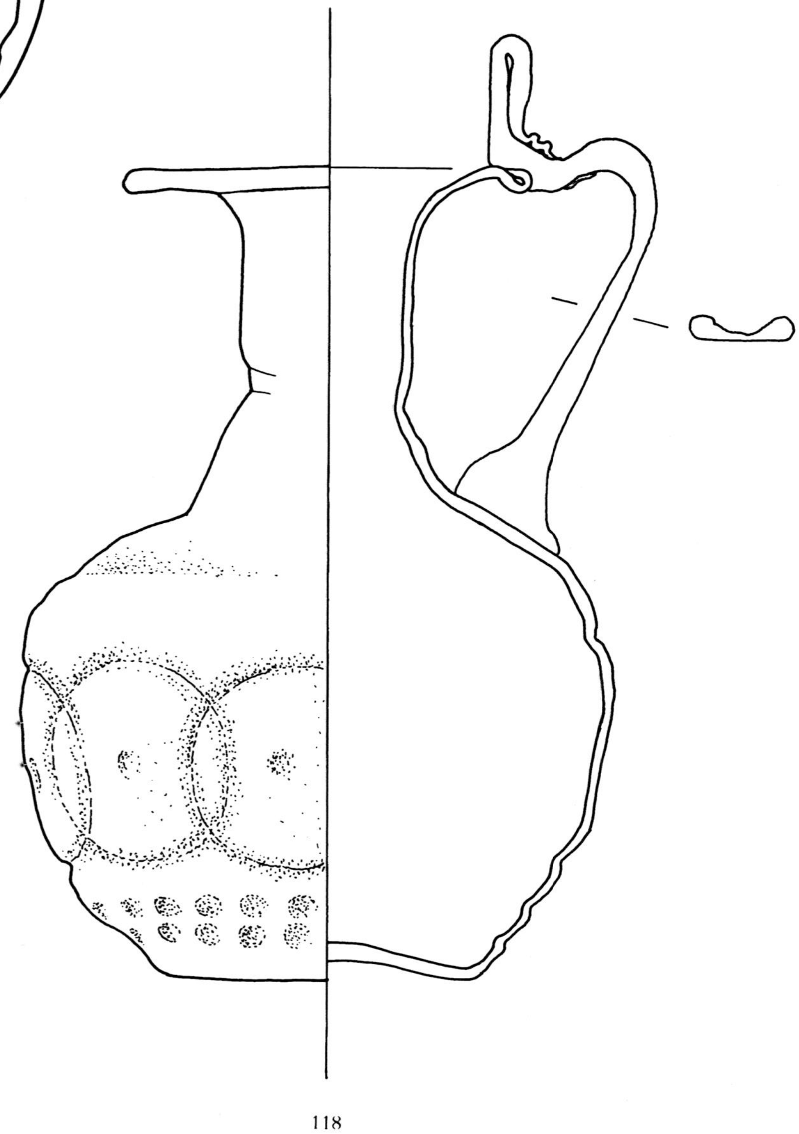

118

*Flasks with Mold-blown Designs in Sunken Relief*

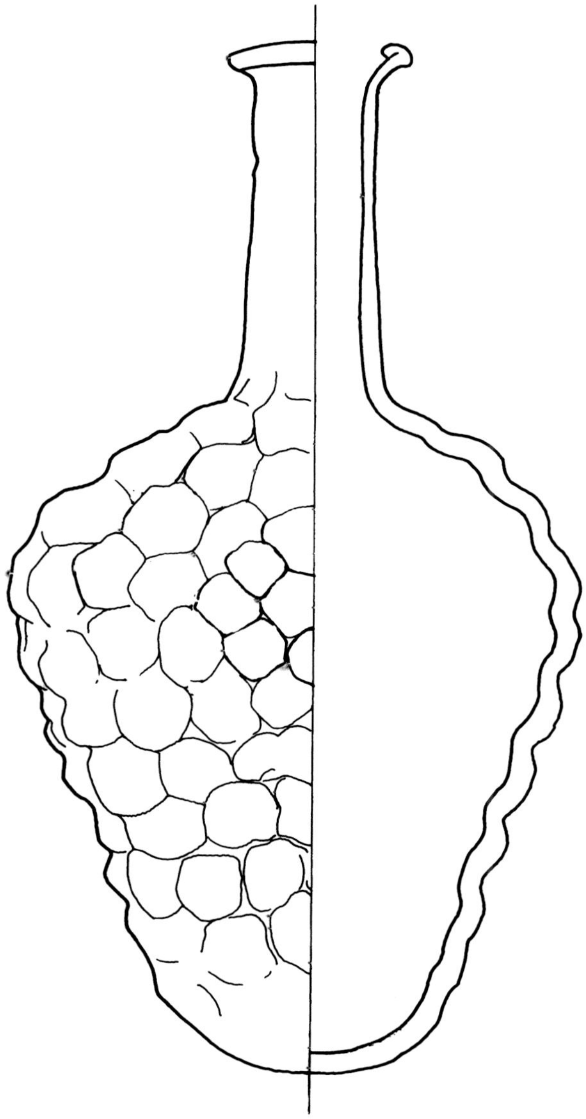

119

*Bottle Shaped like a Naturalistic Cluster of Grapes*

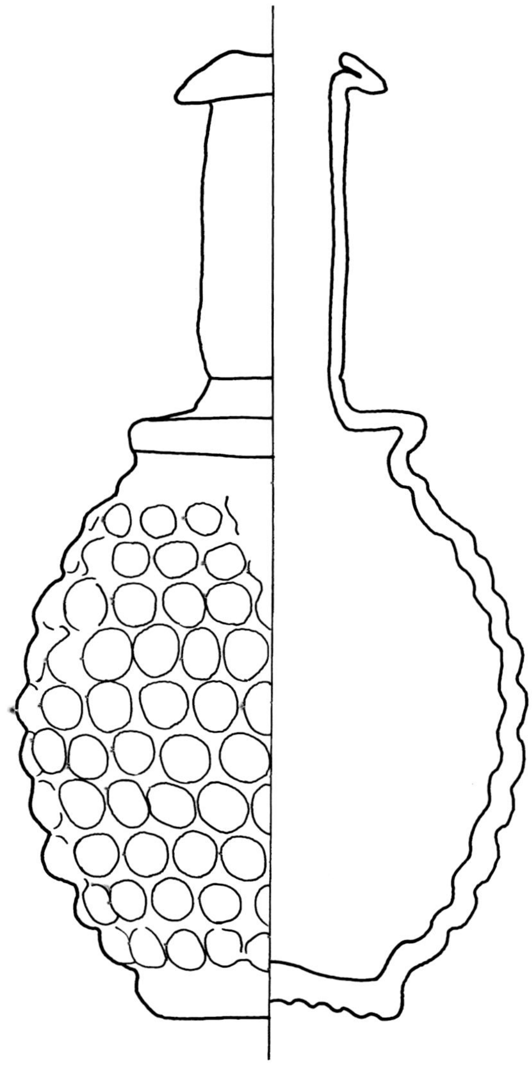

120

*Bulbous Bottles with Stylized Grape Pattern*

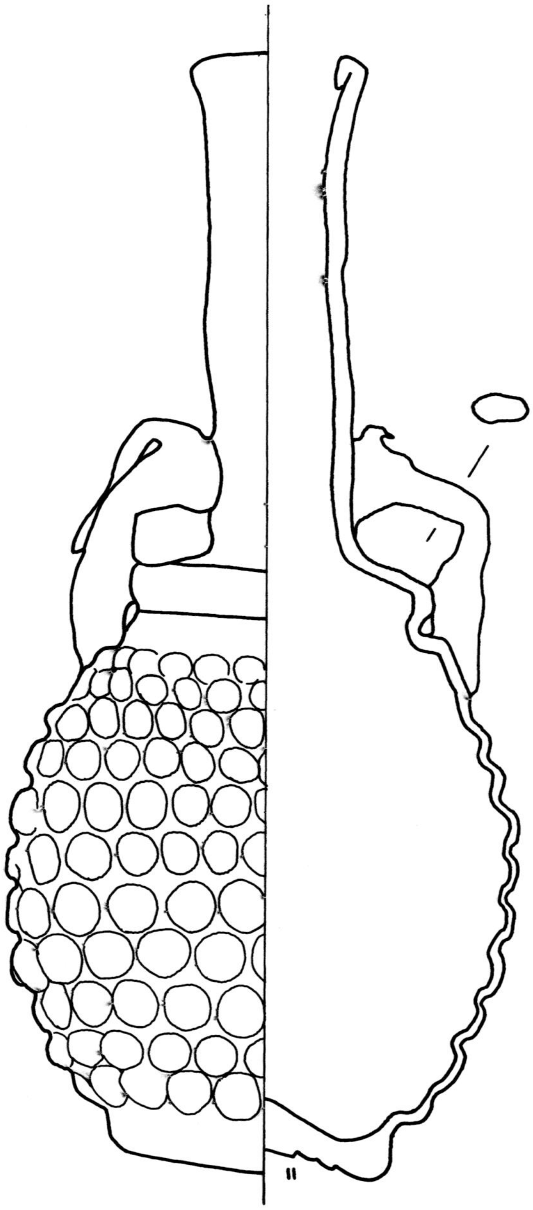

121

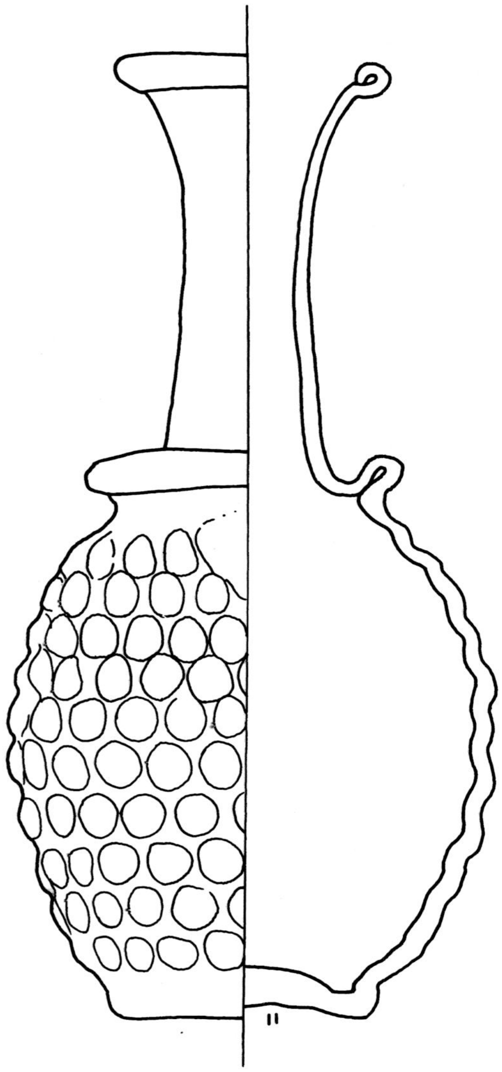

122

*Bulbous Bottles with Stylized Grape Pattern*

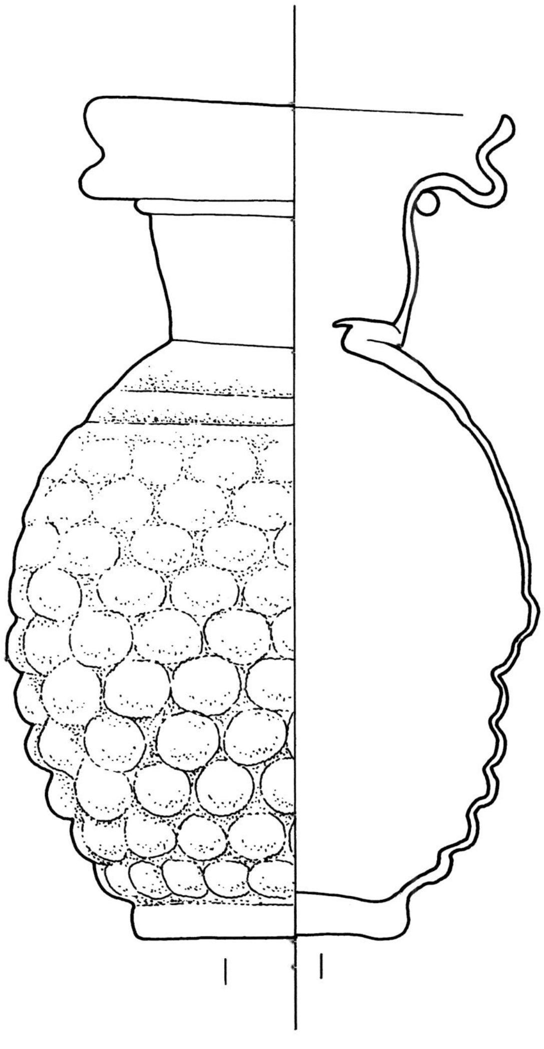

123

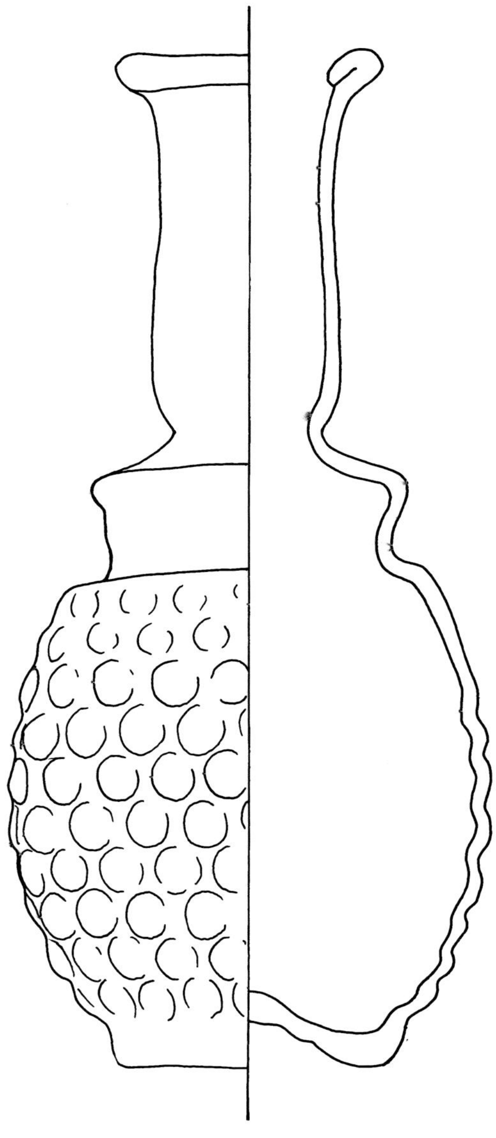

124

*Bulbous Bottles with Stylized Grape Pattern*

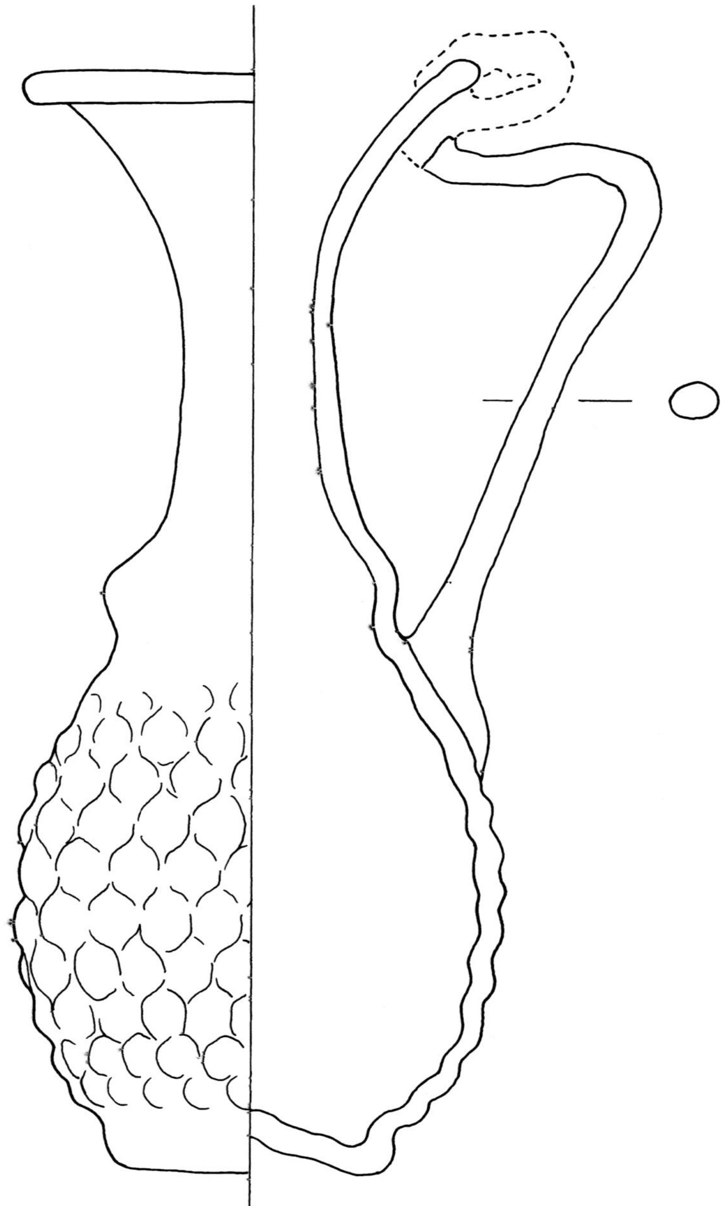

125

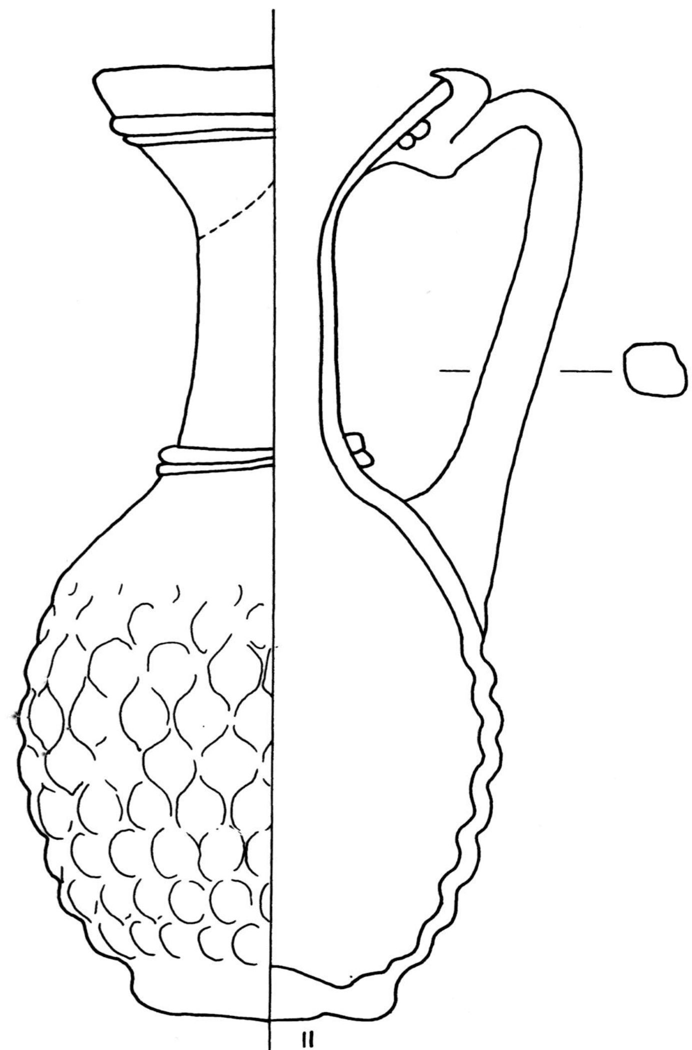

126

*Bulbous Bottles with Stylized Grape Pattern*

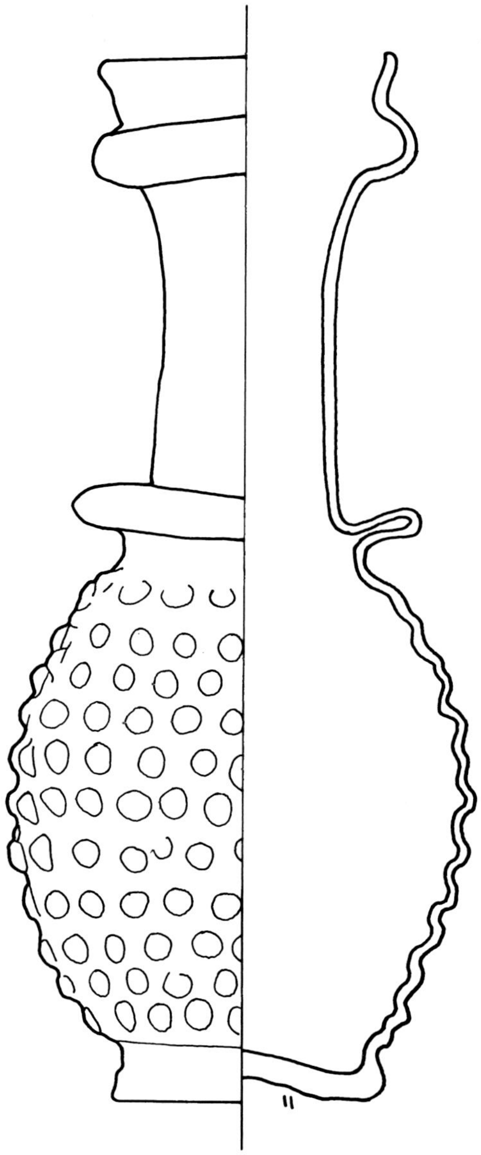

128

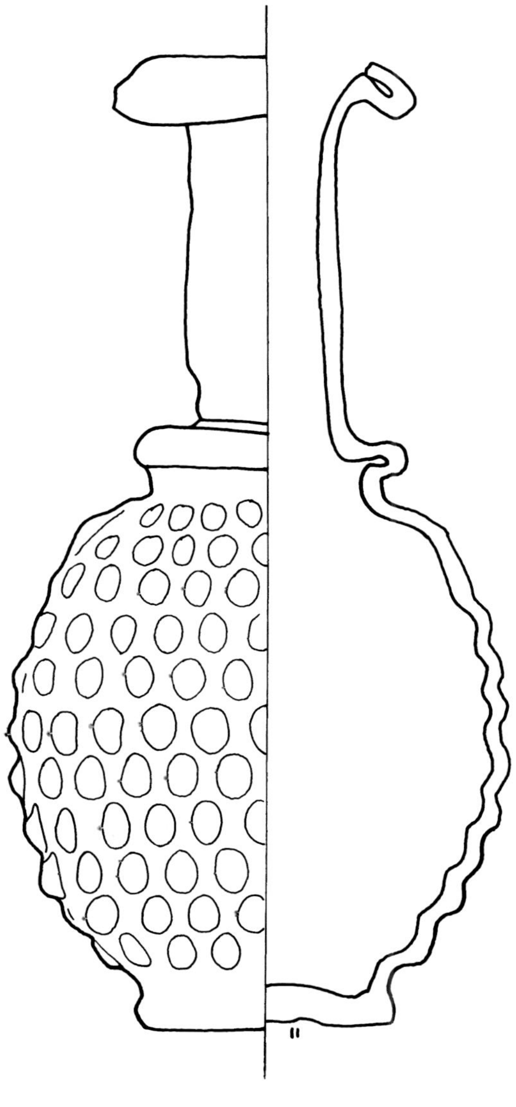

127

*Bulbous Bottles with Stylized Grape Pattern*

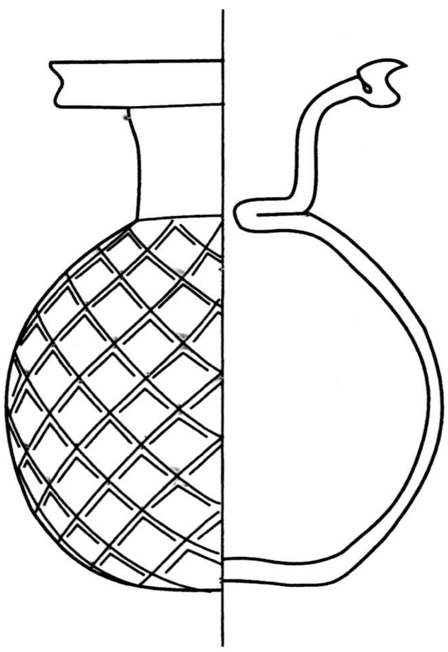

129

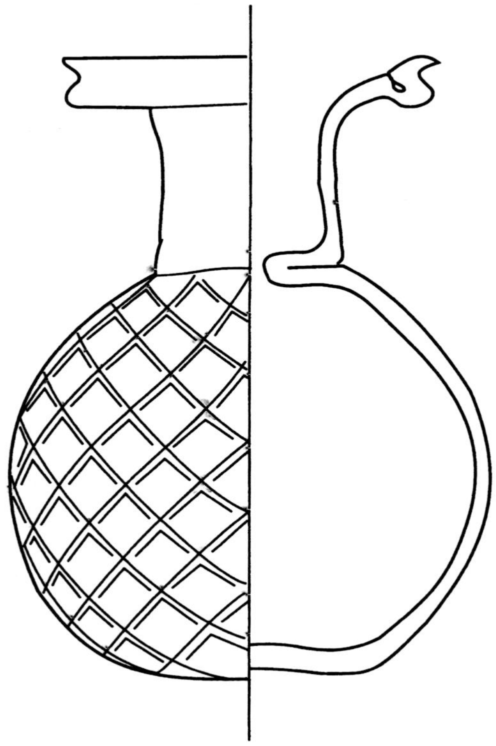

130

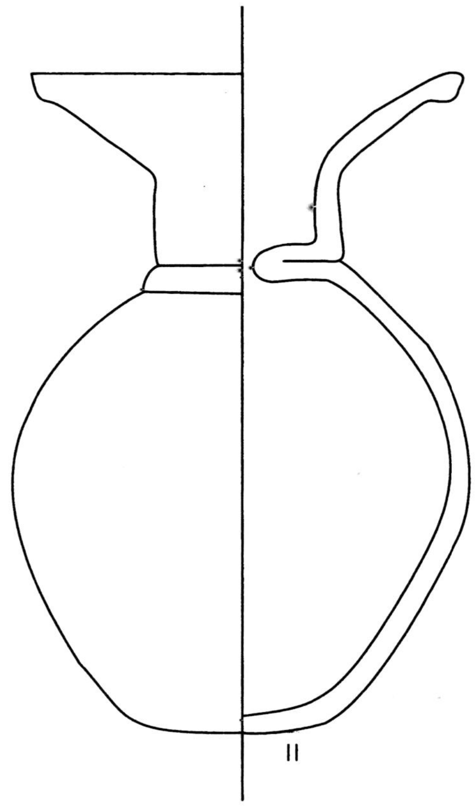

131

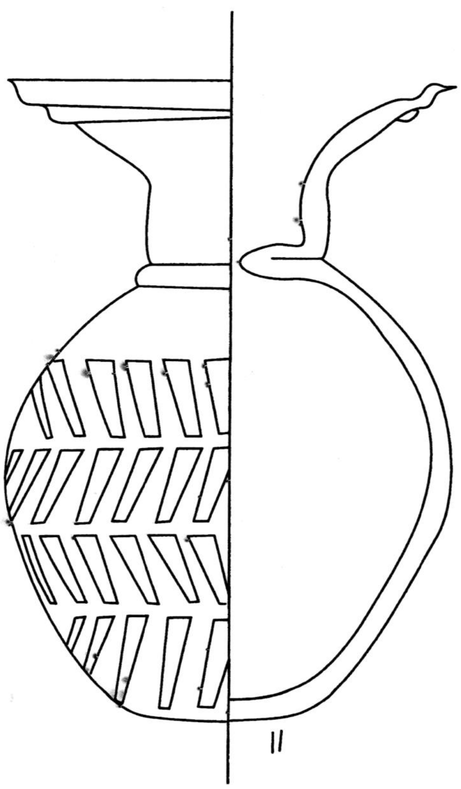

132

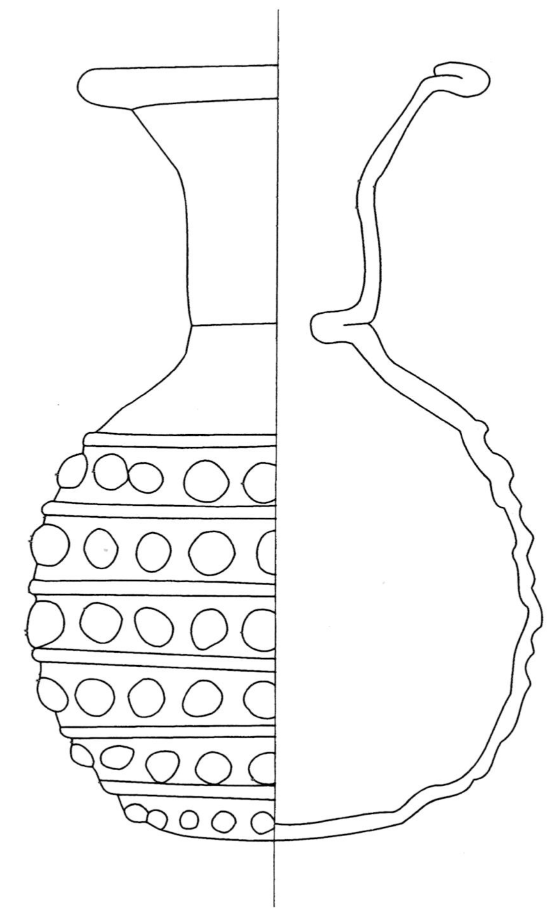

133

*Bulbous Bottles with Geometric Designs*

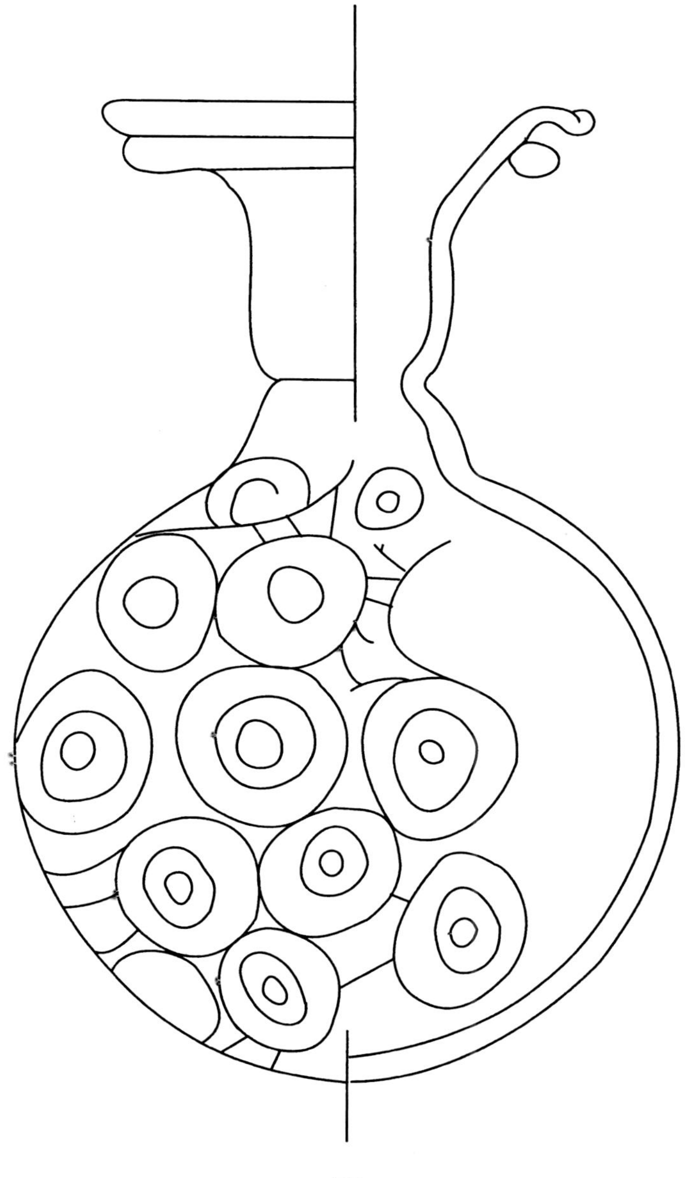

134

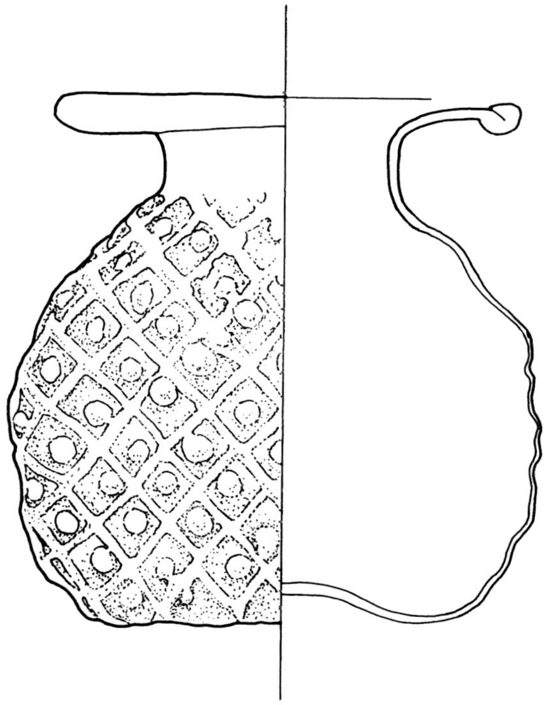

135

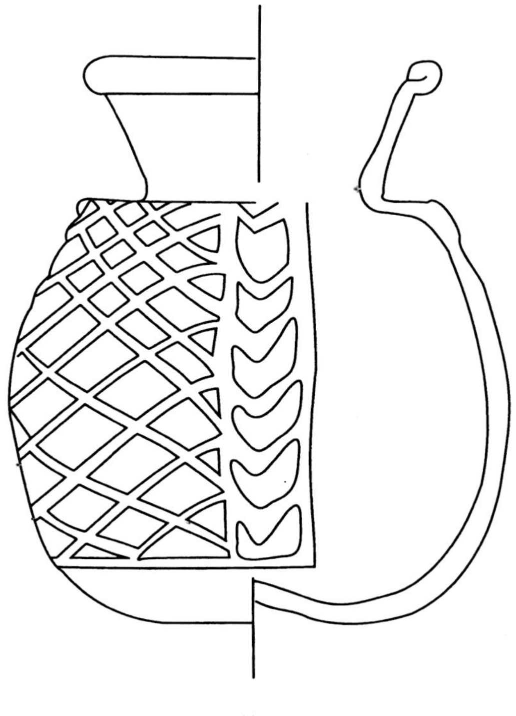

136

*Bulbous Bottles with Geometric Designs*

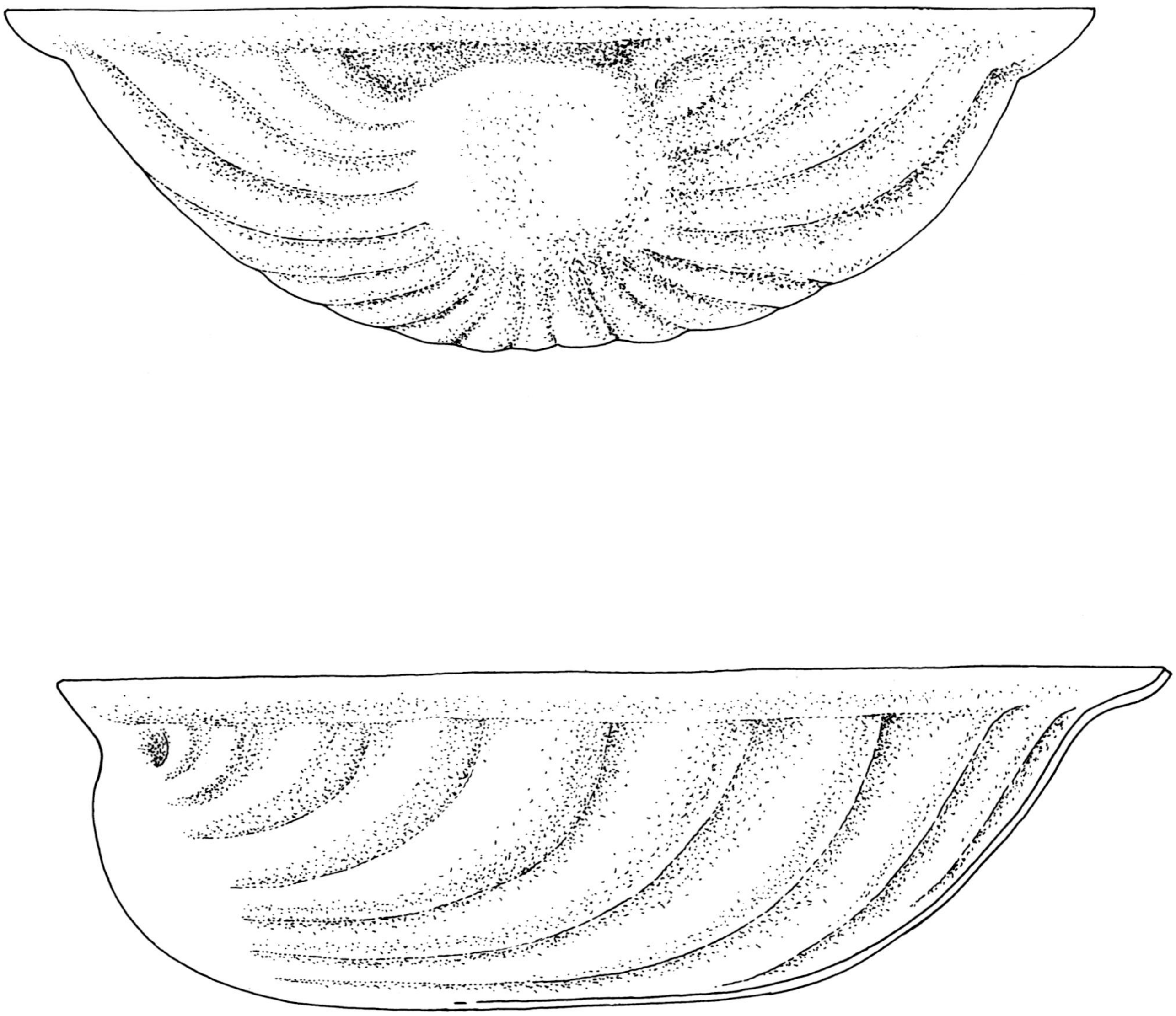

137

*Dish in the Shape of a Shell*

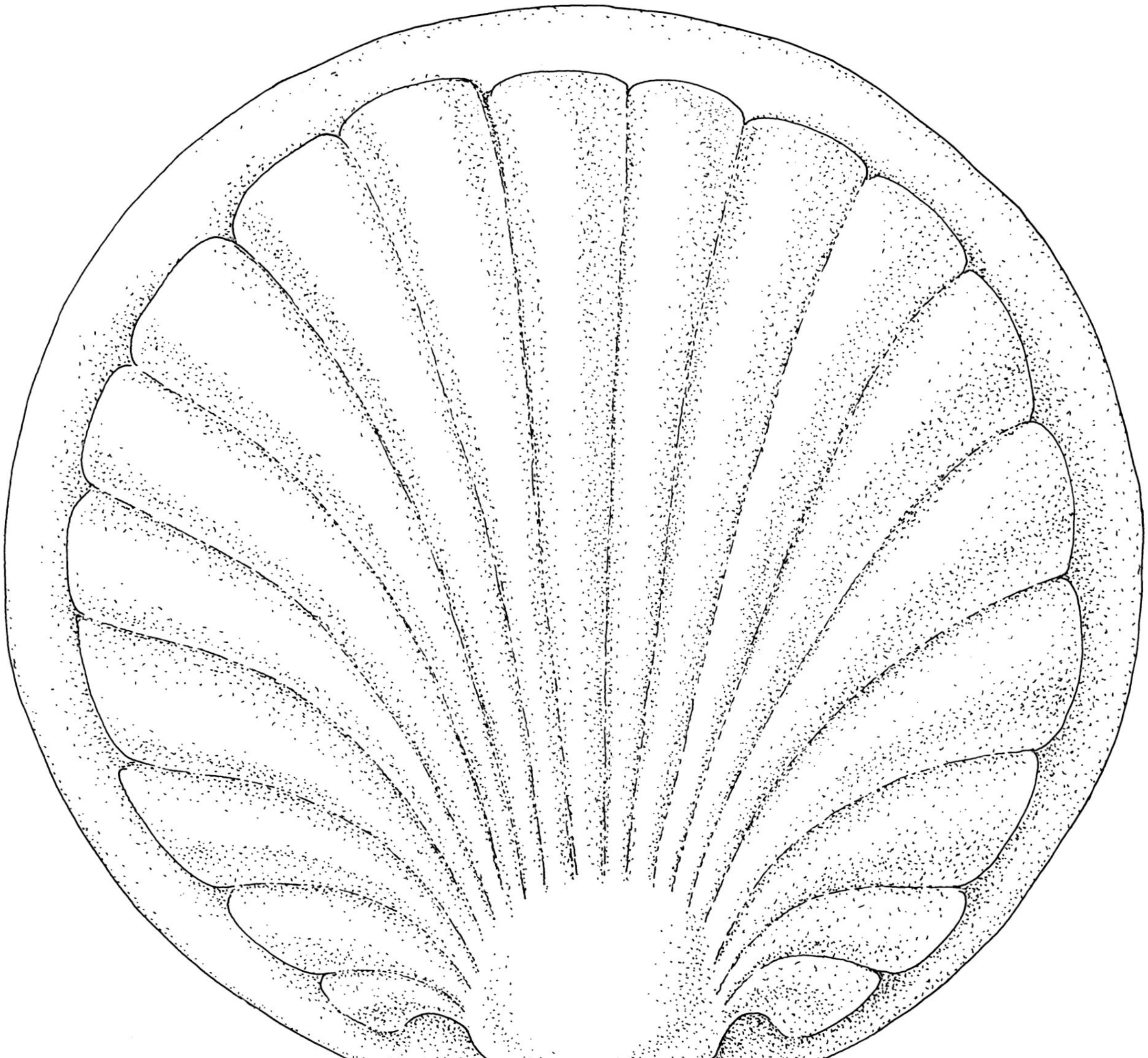

137

*Dish in the Shape of a Shell*

CHAPTER III

# Head-shaped Vessels: The First to Fourth Centuries

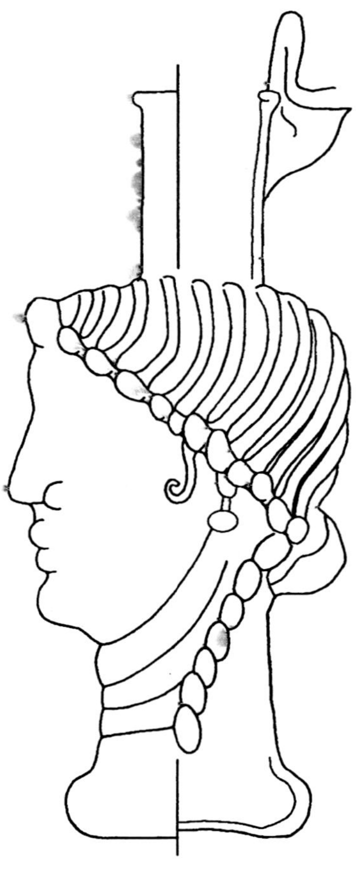

138

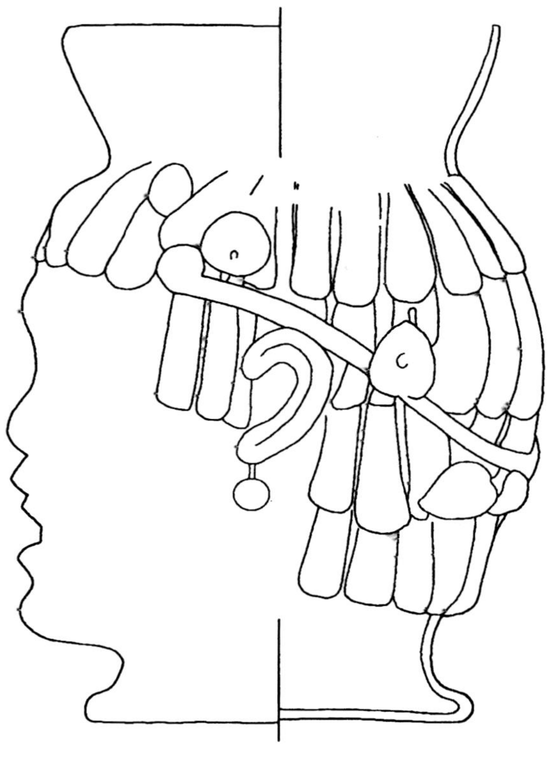

139

*Jug in the Shape of a Female Head*

*Cup in the Shape of a Negroid Head*

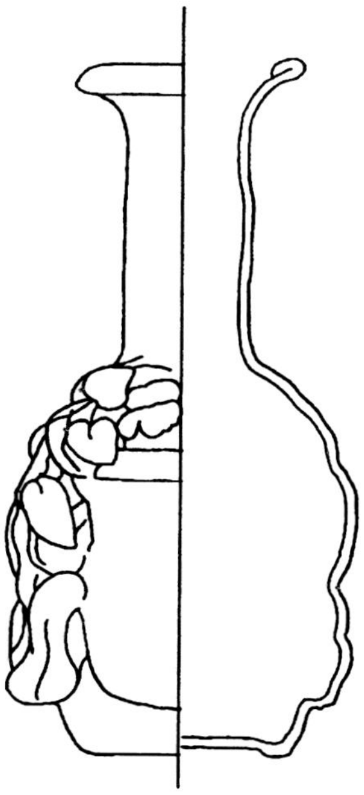

140

*Bottle in the Shape of a Male Head, possibly Dionysos*

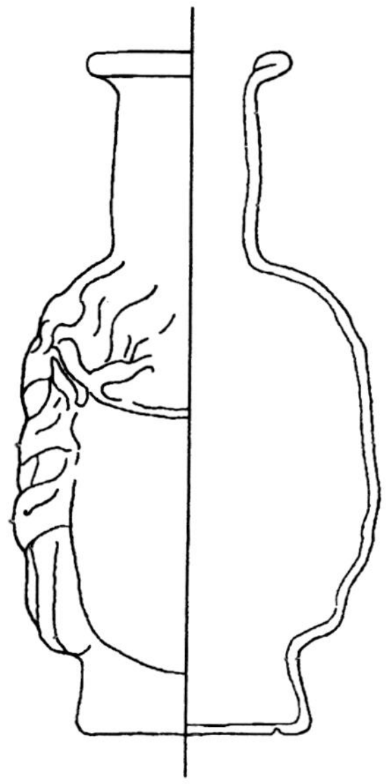

141, side A

*Double Head-shaped Bottle: One Serious Face, One Smiling Face*

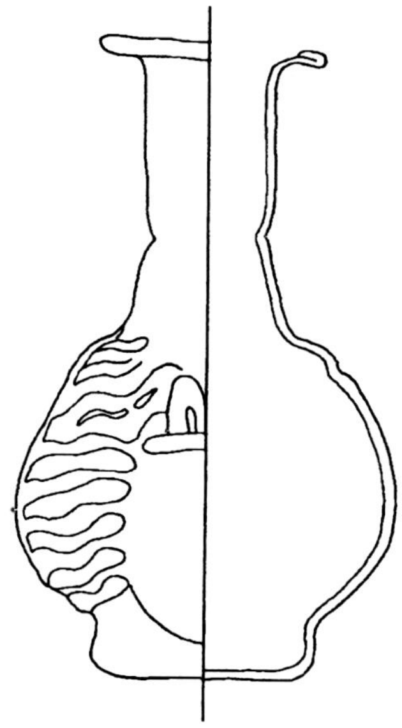

143, side A

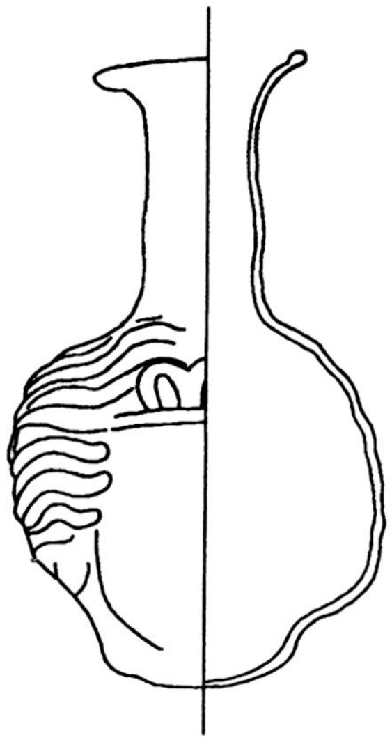

144, side A

*Double Head-shaped Bottles: possibly Dionysos and Satyr*

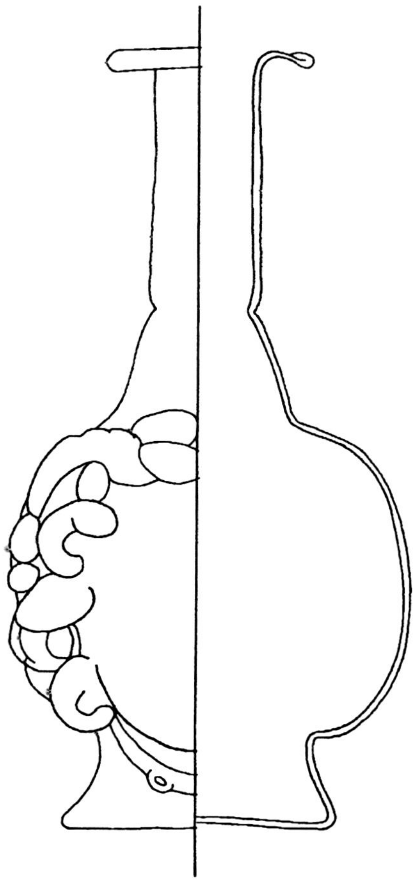

142, side B

*Double Head-shaped Bottle: Two Similar Medusas*

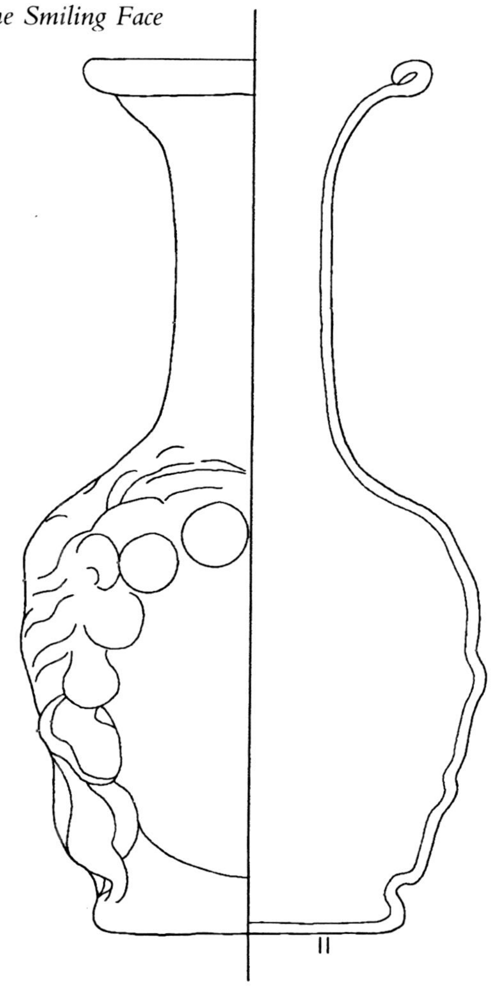

145, side A

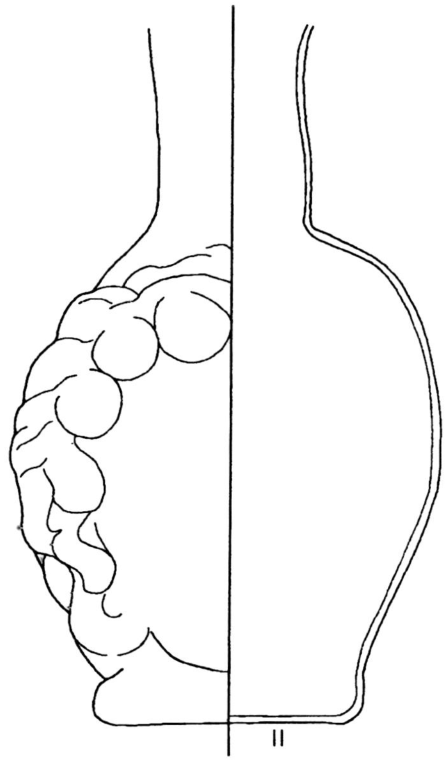

146, side B

*Double Head-shaped Flasks: Two Similar Female Faces*

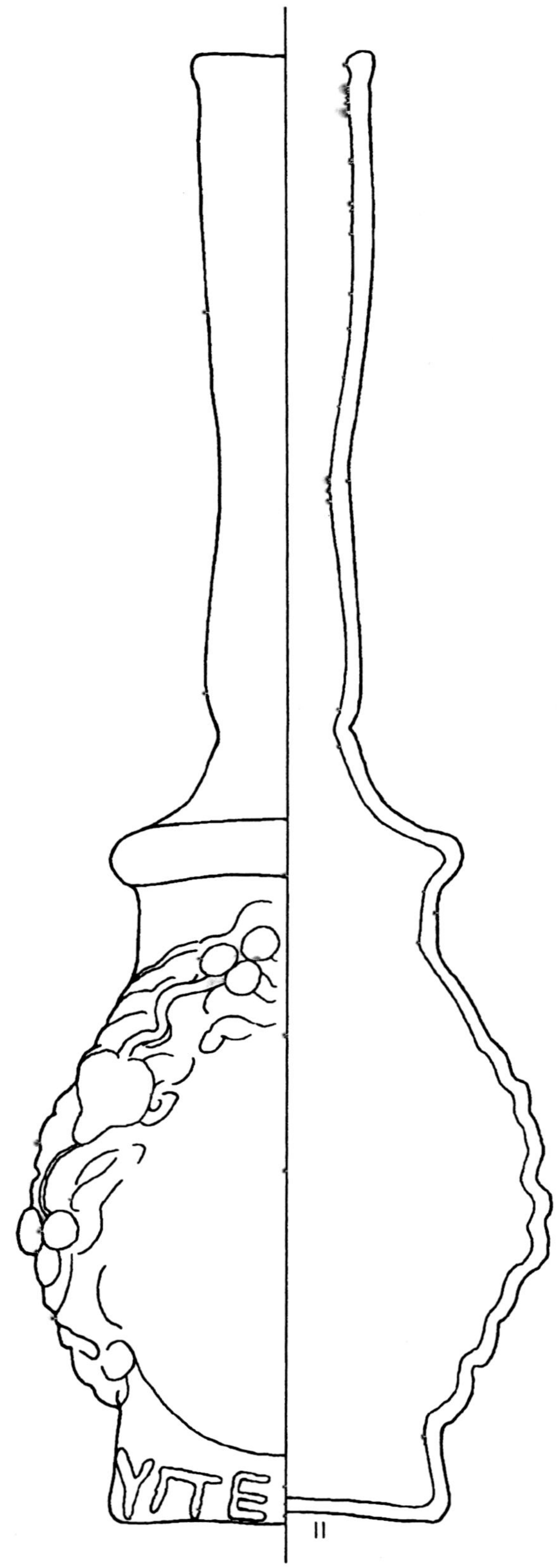


147

*Flask in the Shape of an Idealized Male Head, perhaps Dionysos*

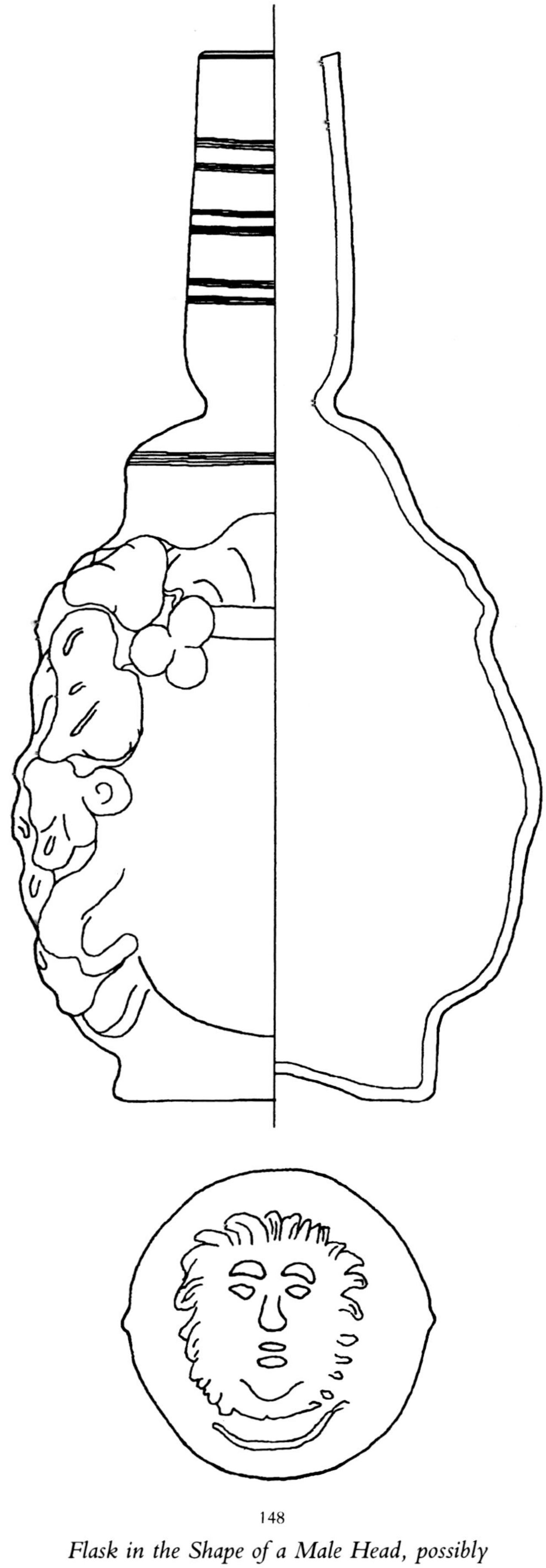

148

*Flask in the Shape of a Male Head, possibly Antinous as Dionysos*

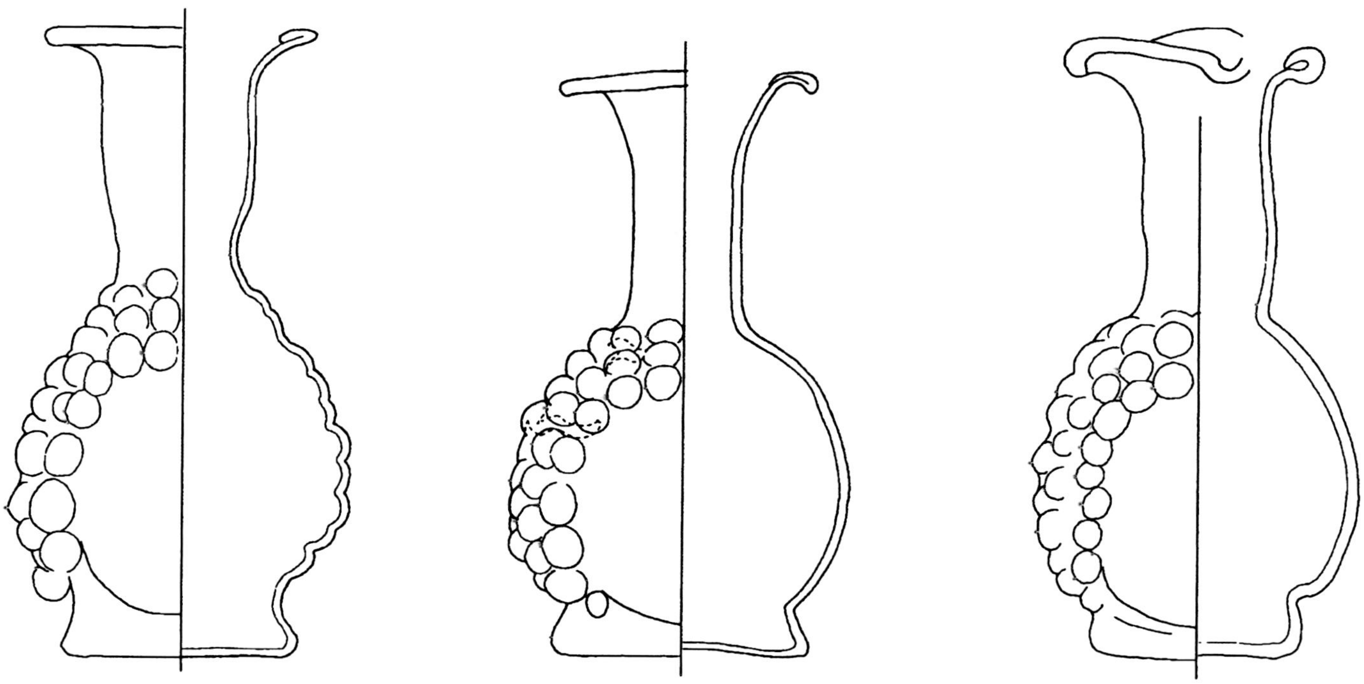

149, side A 150, side B 151, side A

*Double Head-shaped Flasks: Two Similar Chubby Childlike Faces*

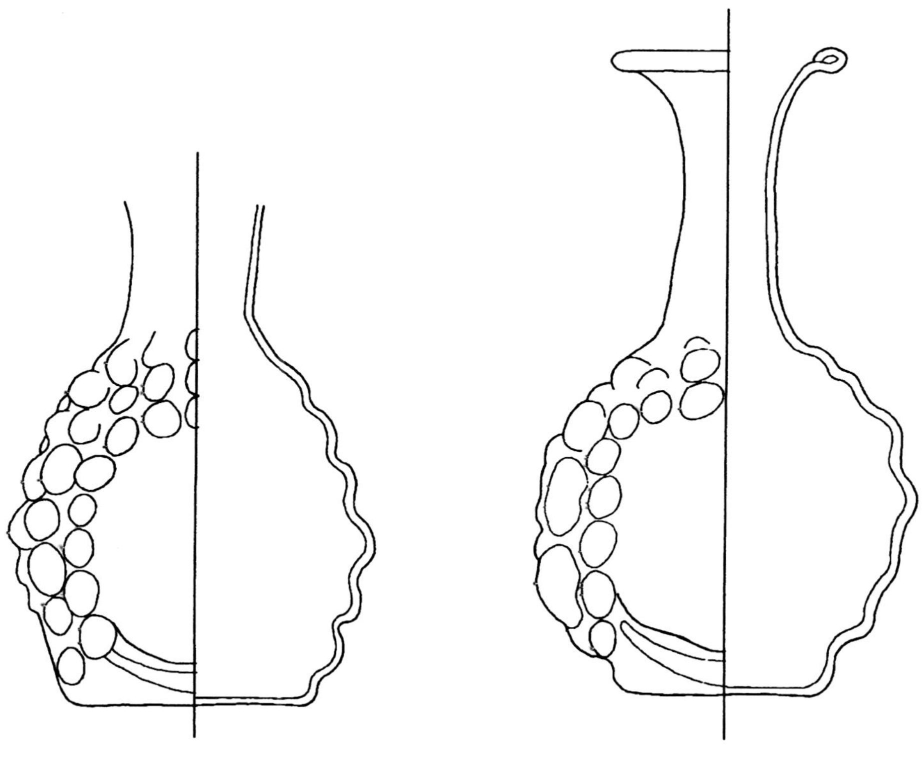

152, side A 153, side A

*Double Head-shaped Flasks: Two Similar Idealized Faces, perhaps Medusas*

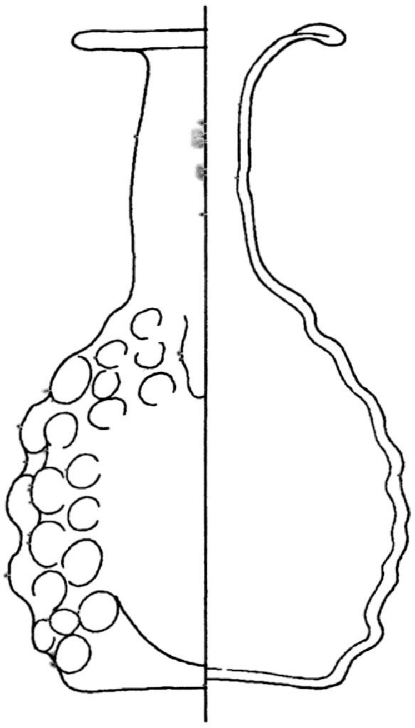

154, side A

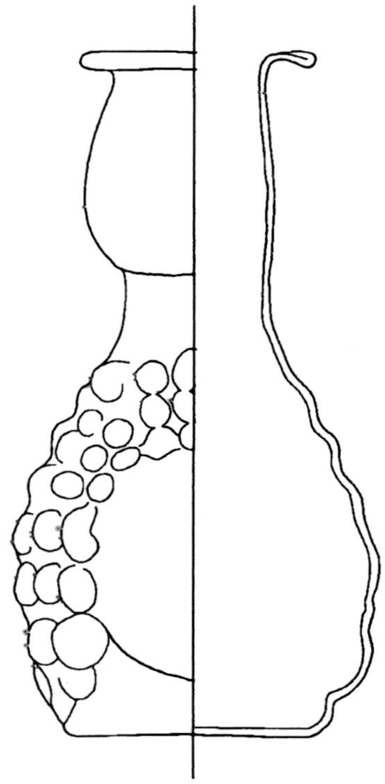

155, side A

*Double Head-shaped Flasks: Two Similar Youthful Faces*

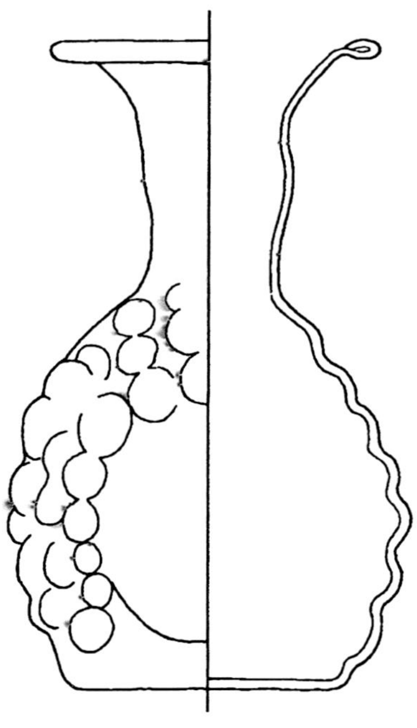

156, side A

*Double Head-shaped Flask: One Smiling Face, One Serious Face*

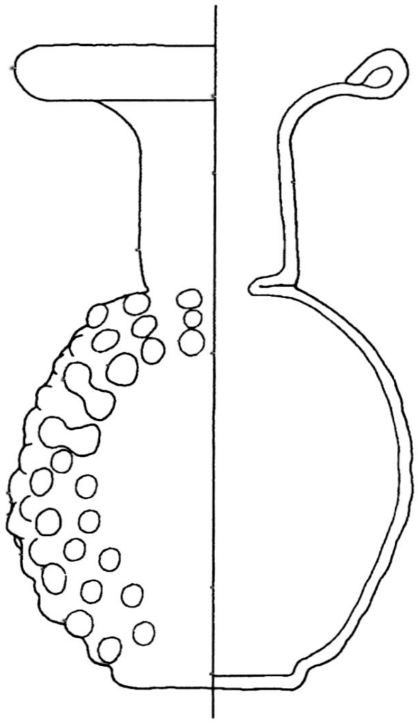

157, side B

*Double Head-shaped Flask with Internal Diaphragm (Sprinkler): Two Similar Stylized Droll Faces*

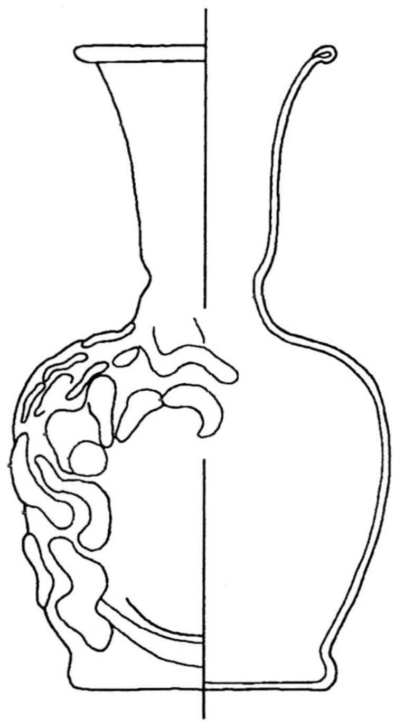

158, side A

*Double Head-shaped Flask: Two Similar Medusas with Wavy Hair*

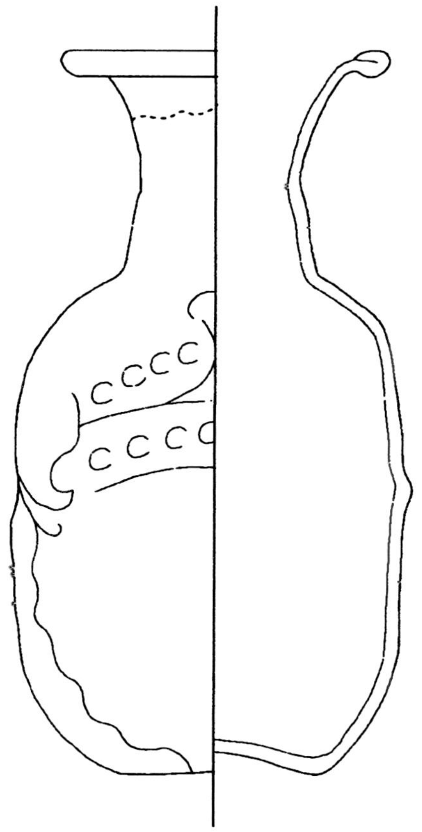

159, side B

*Double Head-shaped Flask: Two Similar Male Heads, probably Hercules wearing a Lion's-Head Helmet*

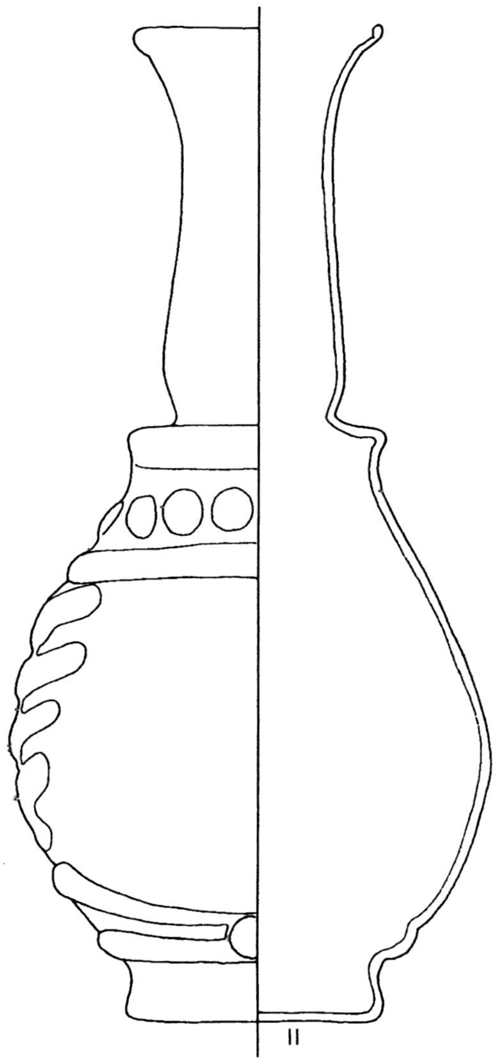

160, side A

*Double Head-shaped Flask: Two Similar Grotesque Faces*

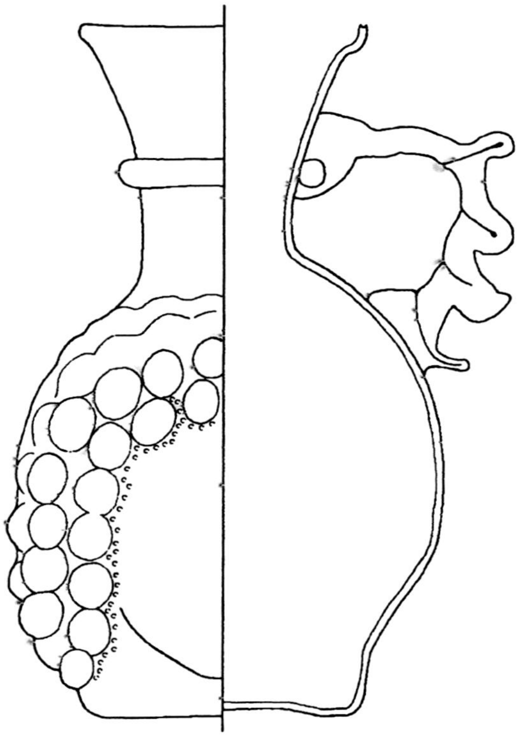

161, side A

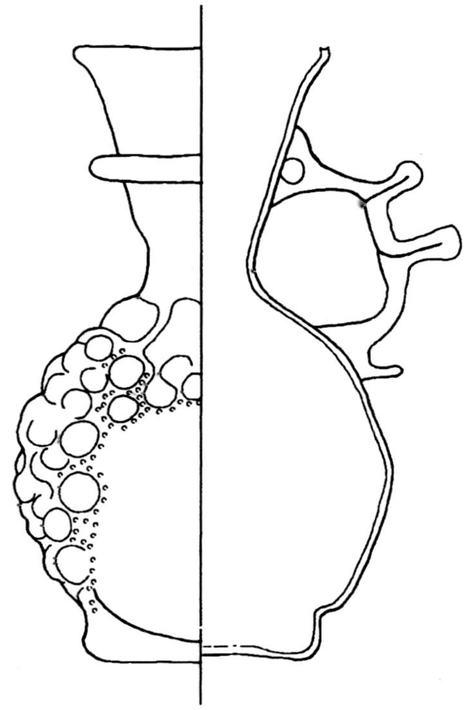

162, side A

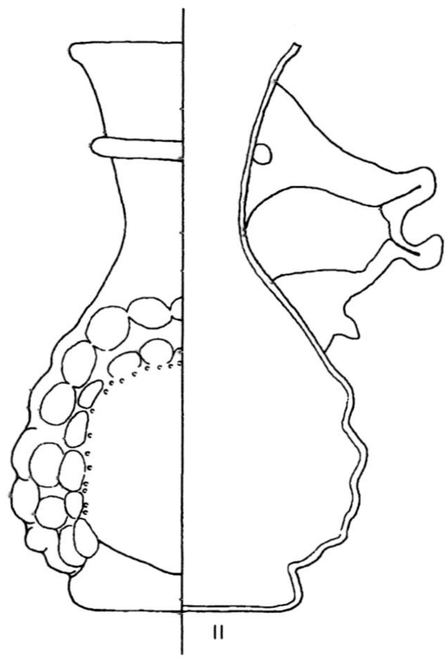

163, side A

*Double Head-shaped Jugs: Two Childlike Faces, One Serious, One Smiling*

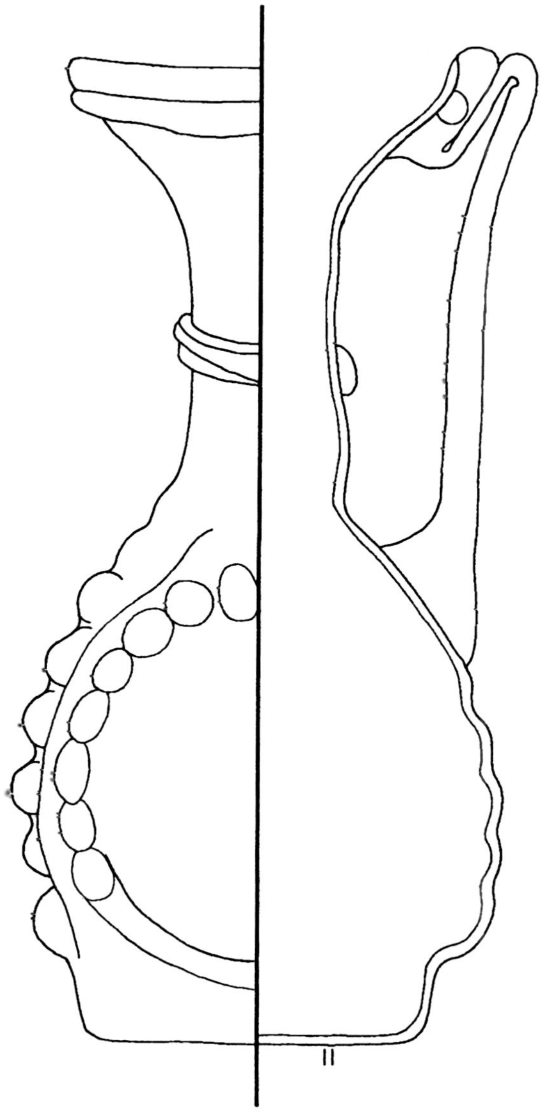

164, side A

*Double Head-shaped Jug: Two Chubby Faces, One Smiling, One Serious*

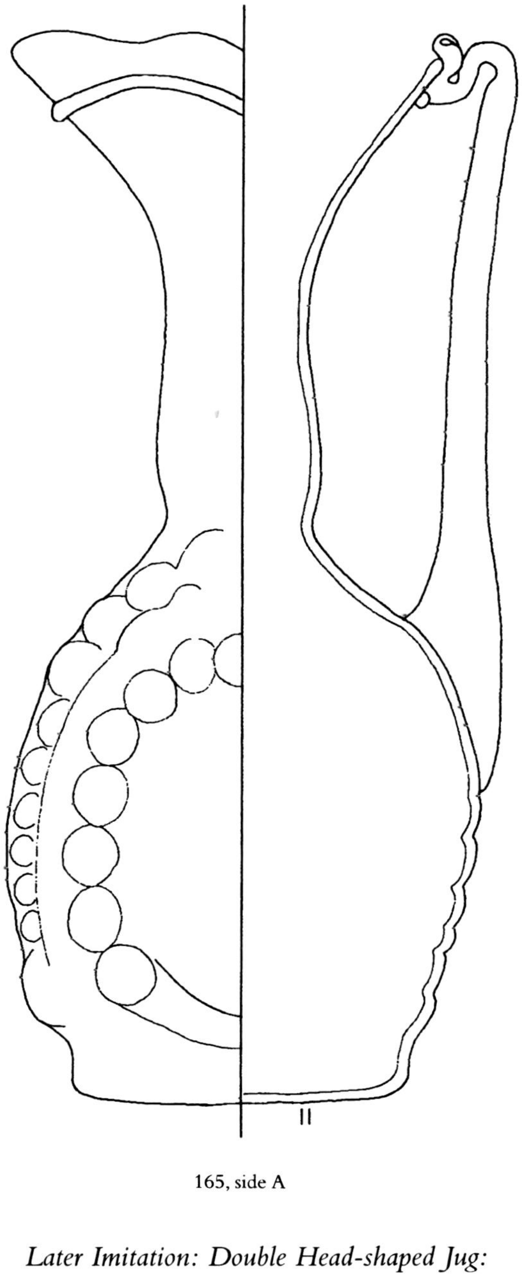

165, side A

*Later Imitation: Double Head-shaped Jug: Two Chubby Faces, One Smiling, One Serious*

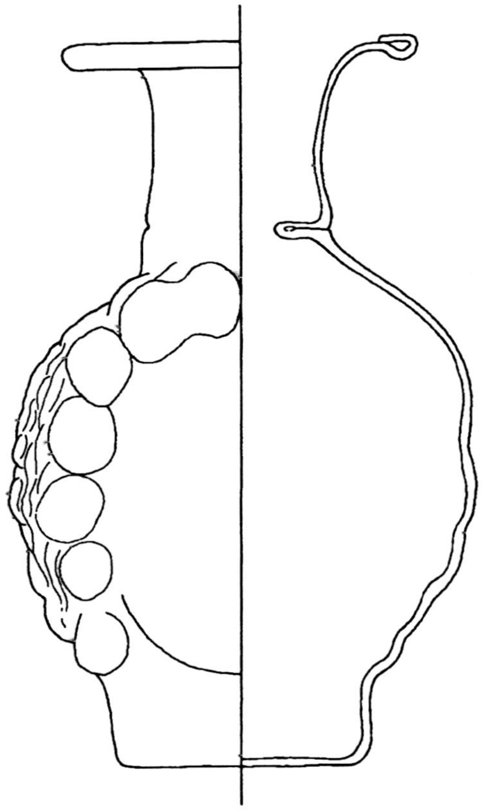

166, side A

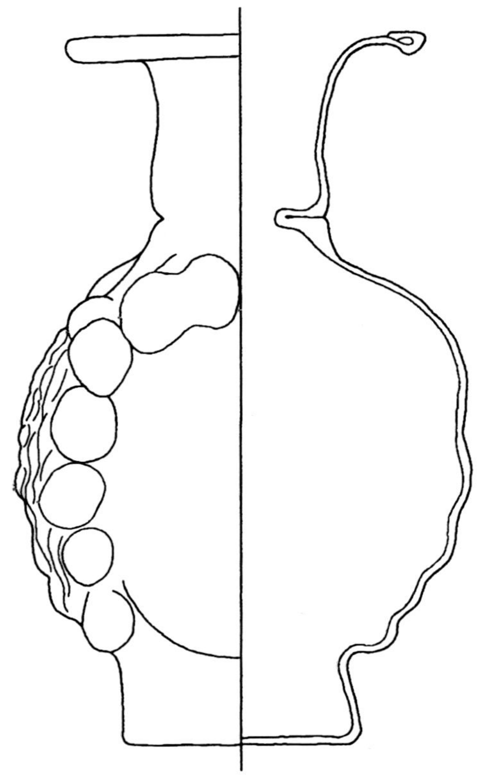

167, side B

*Later Imitation: Double Head-shaped Flasks: Two Similar Smiling Faces*

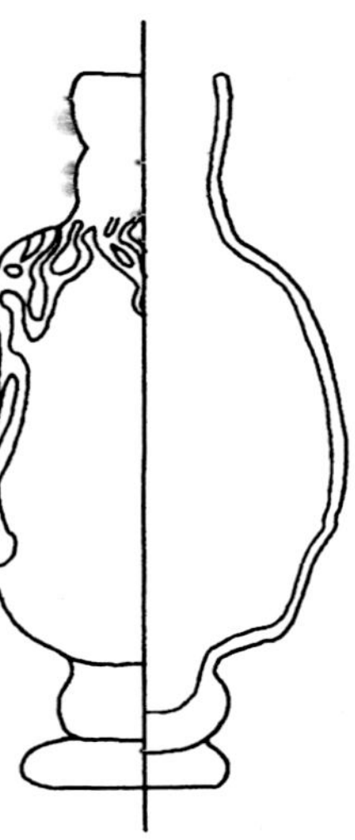

168

*Later Imitation: Head-shaped Bottle on Stemmed Foot*

CHAPTER IV

# The Fourth to Early Seventh Centuries

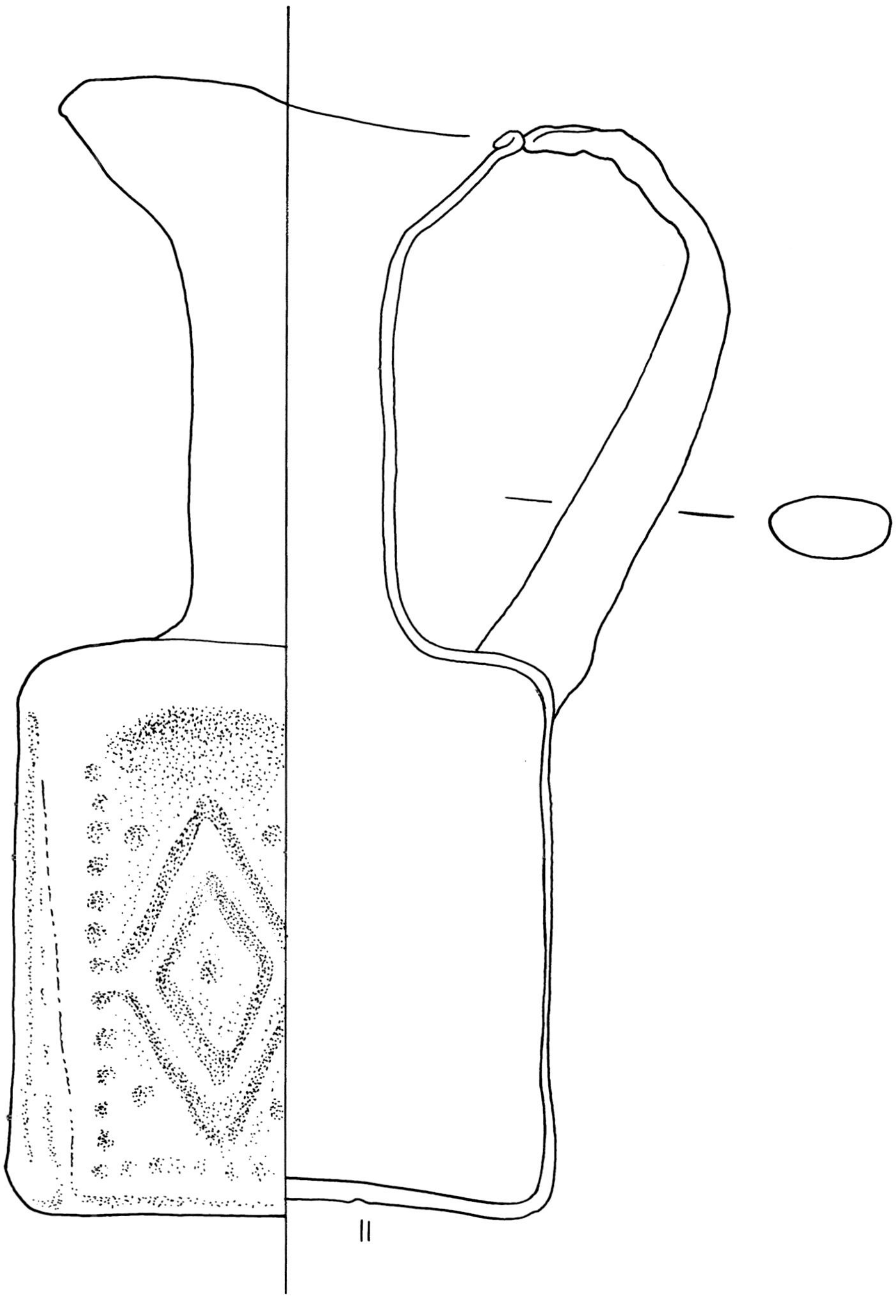

169

*Hexagonal Vessels with Christian Symbols (Barag Class A)*

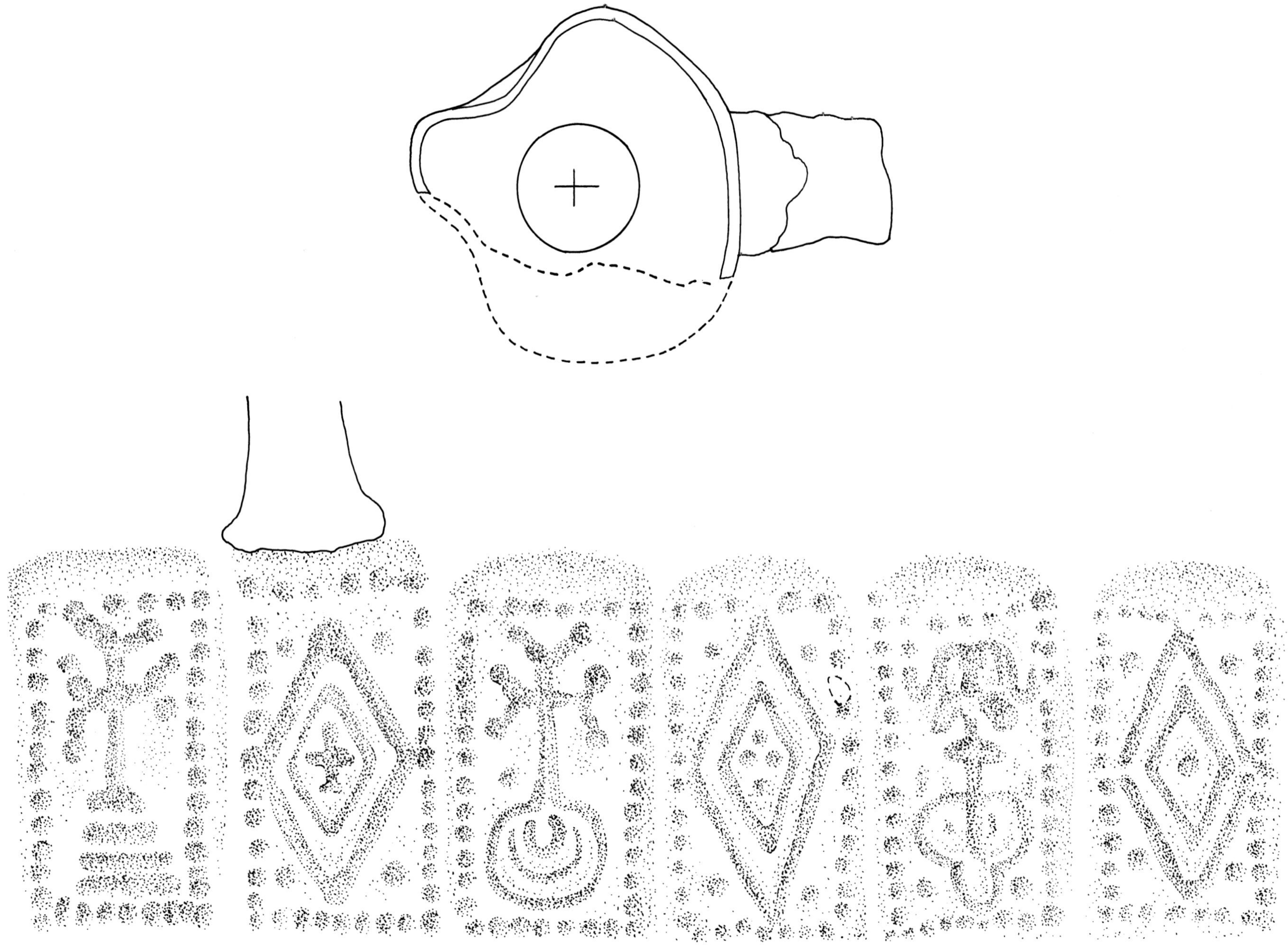

169

*Hexagonal Vessels with Christian Symbols (Barag Class A)*

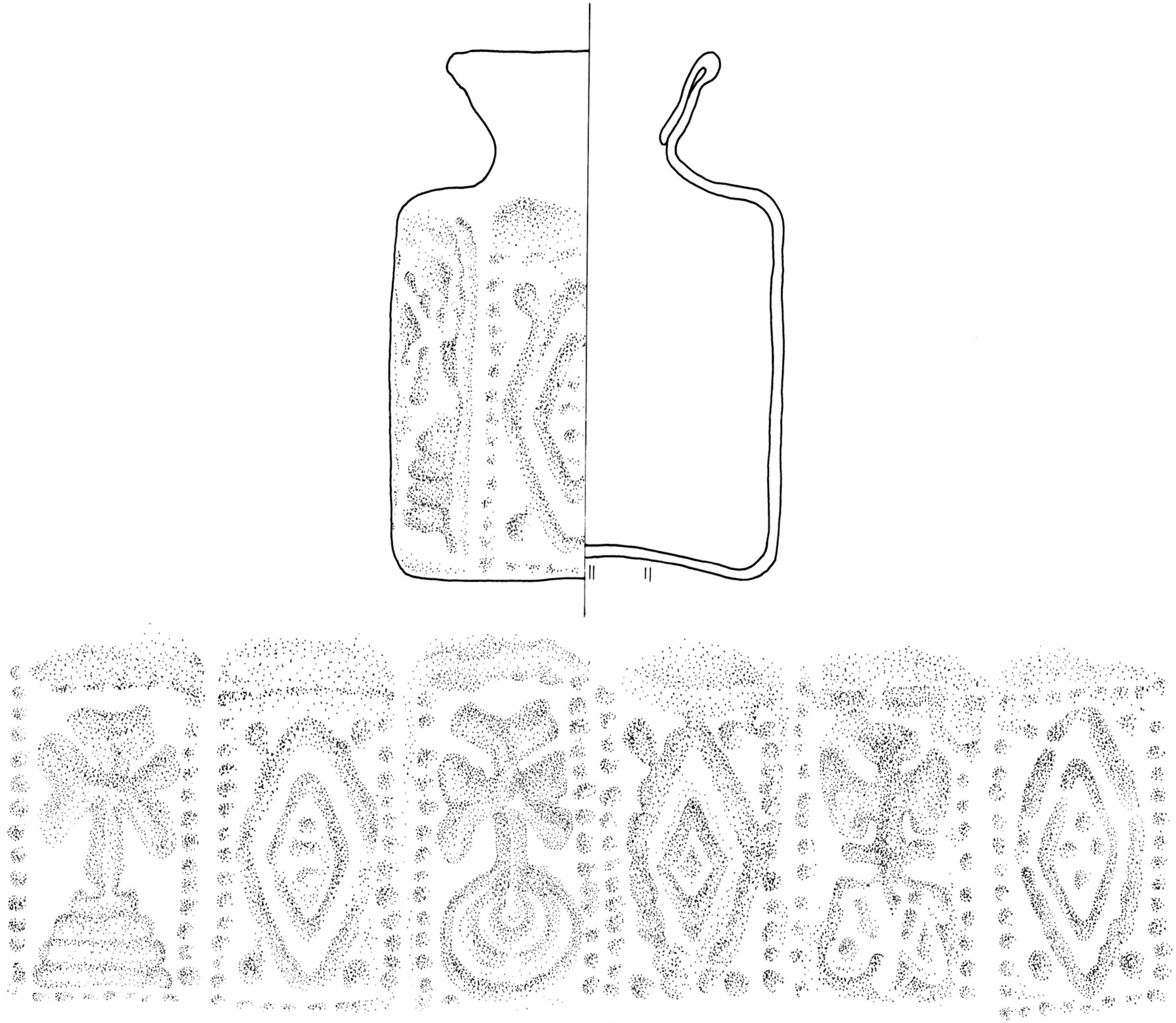

170

*Hexagonal Vessels with Christian Symbols (Barag Class A)*

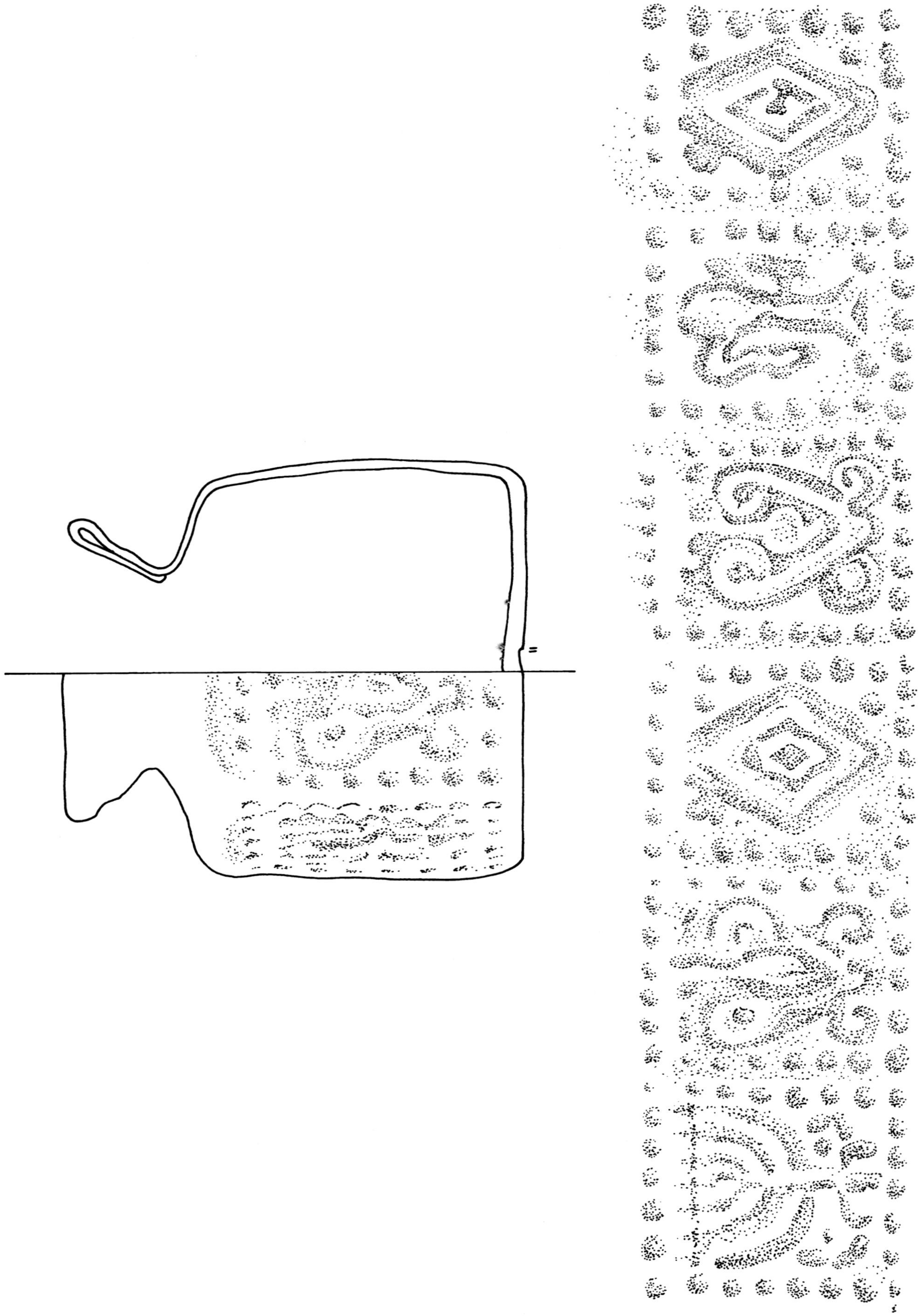

171

*Vessels with Jewish Symbols (Barag Class B)*

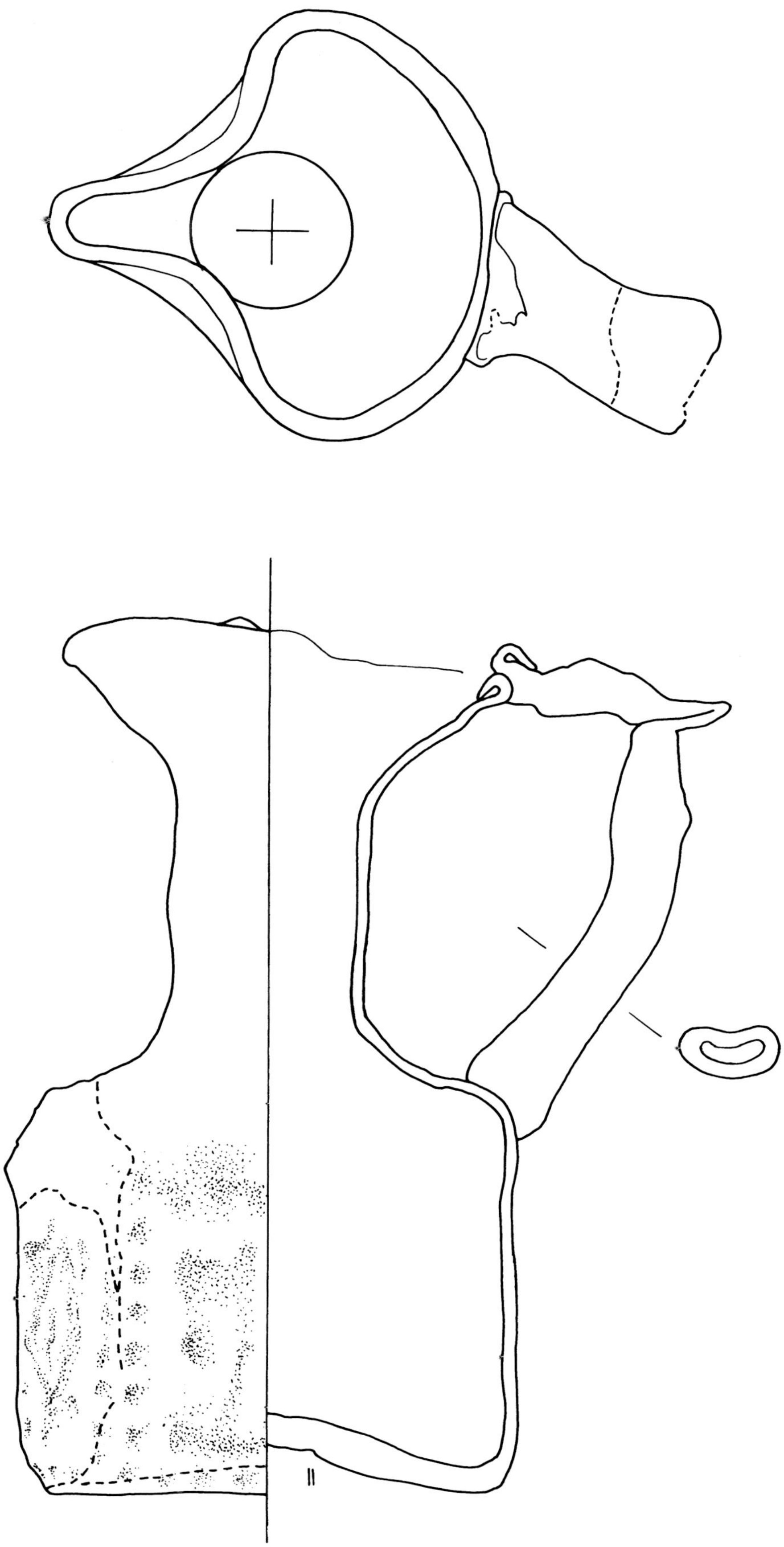

172

*Vessels with Jewish Symbols (Barag Class B)*

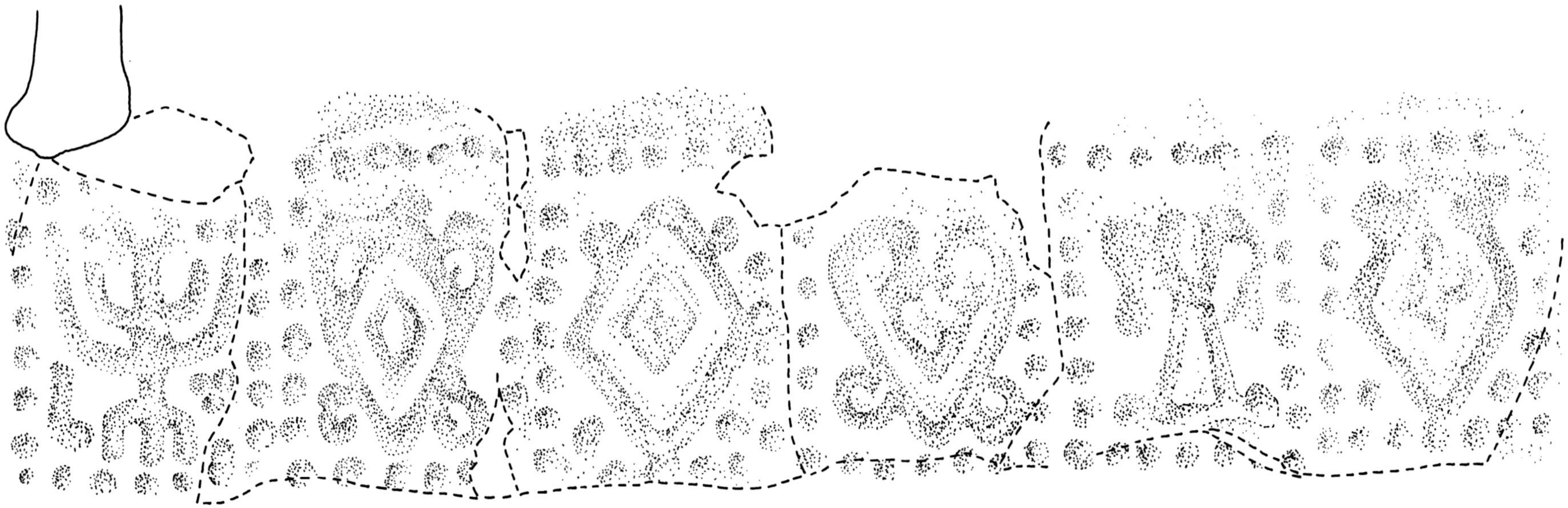

172

Vessels with Jewish Symbols (Barag Class B)

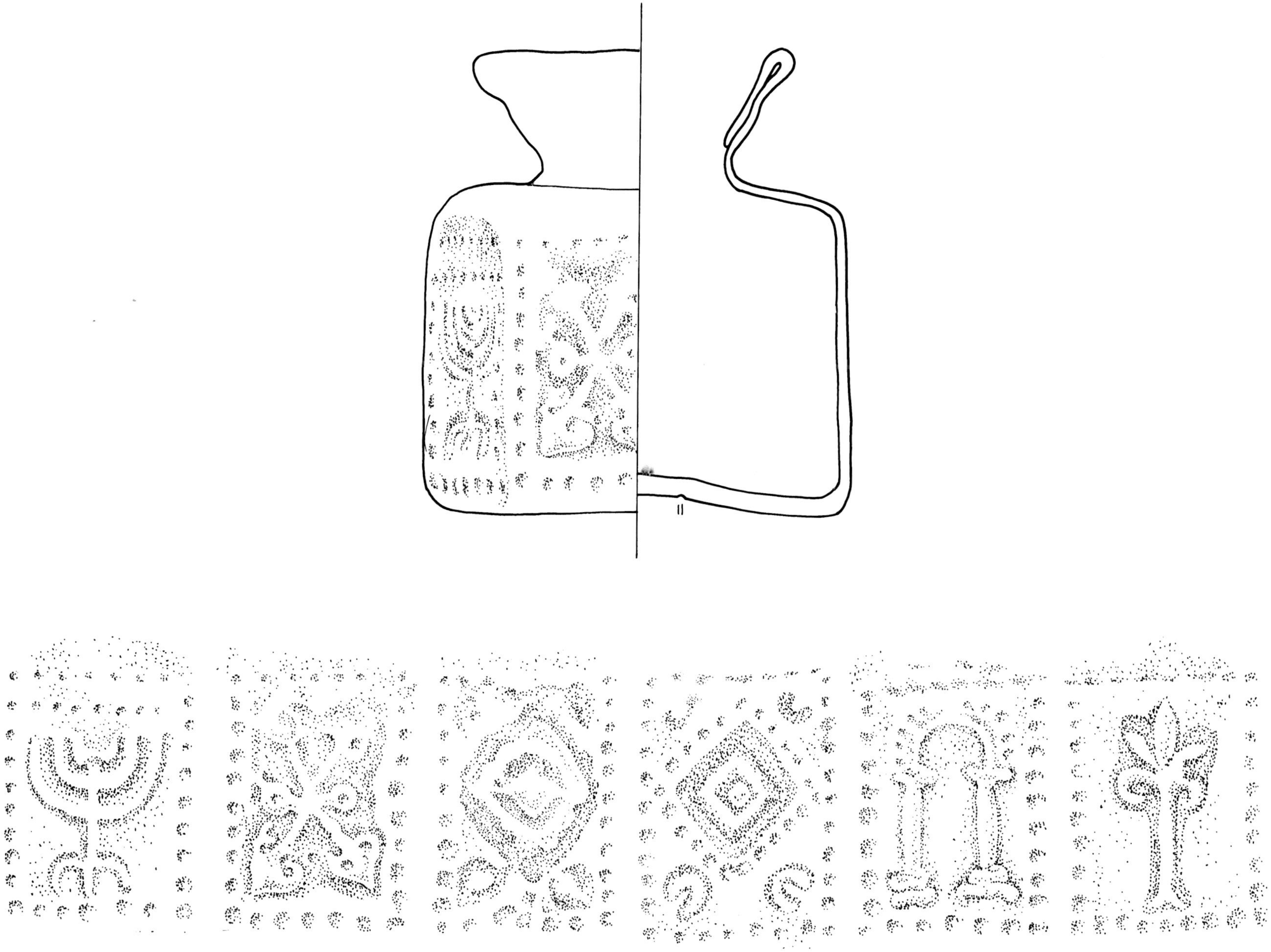

173

*Vessels with Jewish Symbols (Barag Class B)*

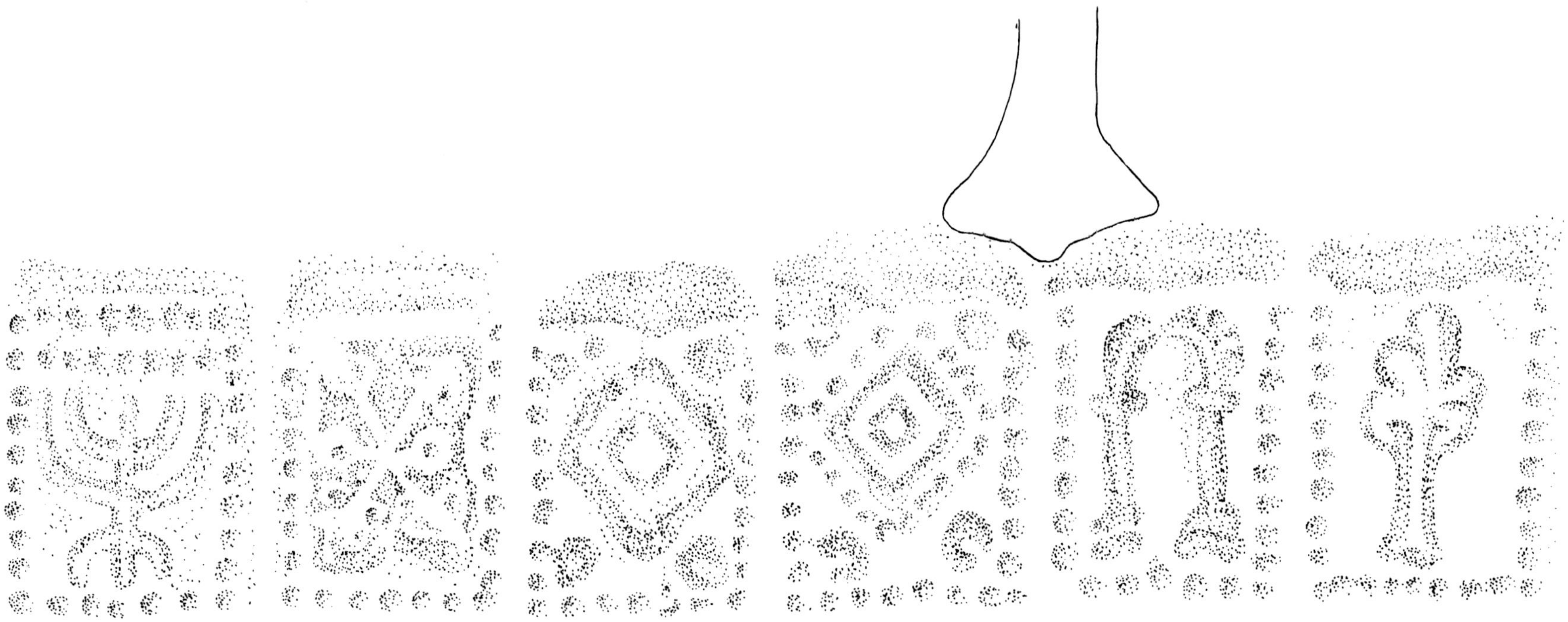

174

*Vessels with Jewish Symbols (Barag Class B)*

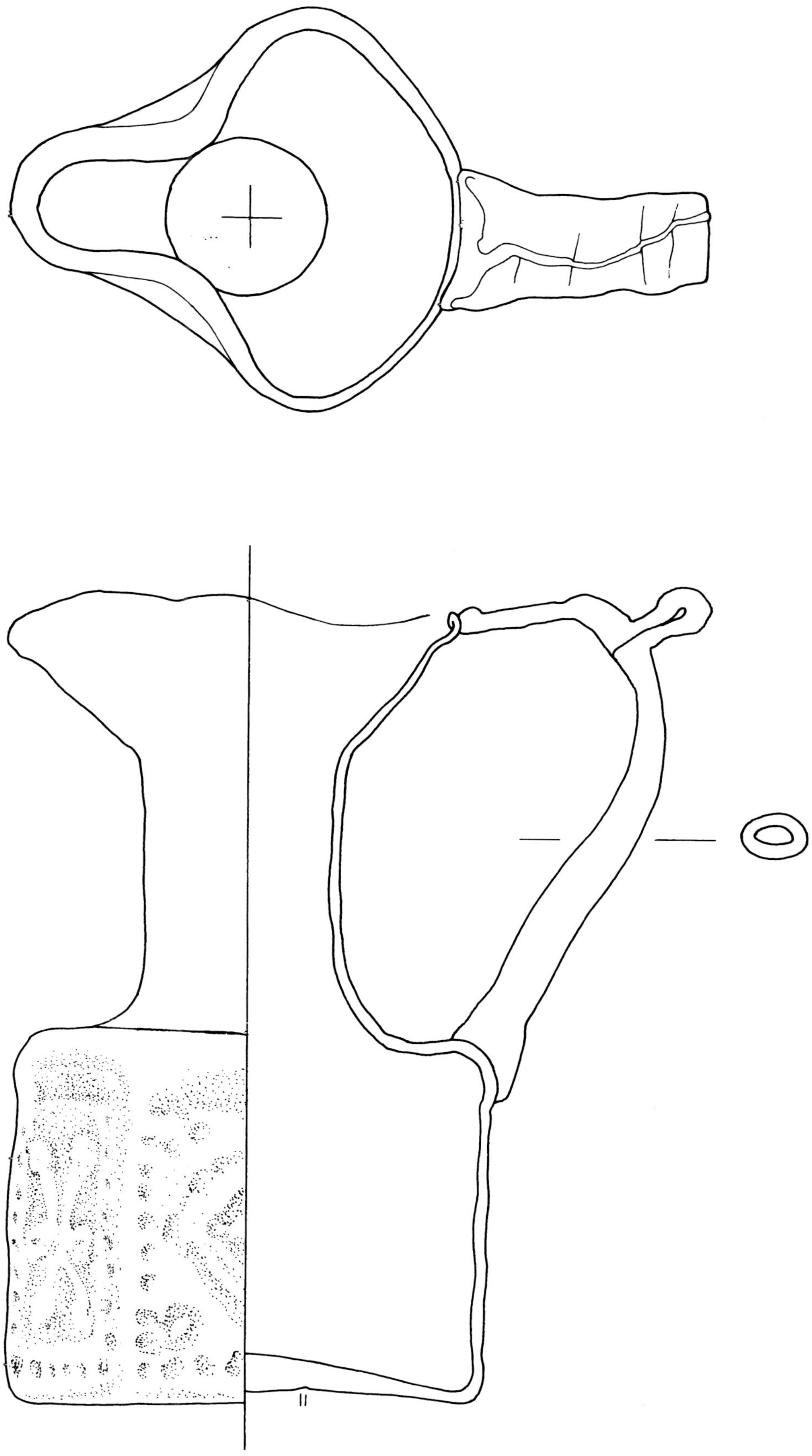

174

*Vessels with Jewish Symbols (Barag Class B)*

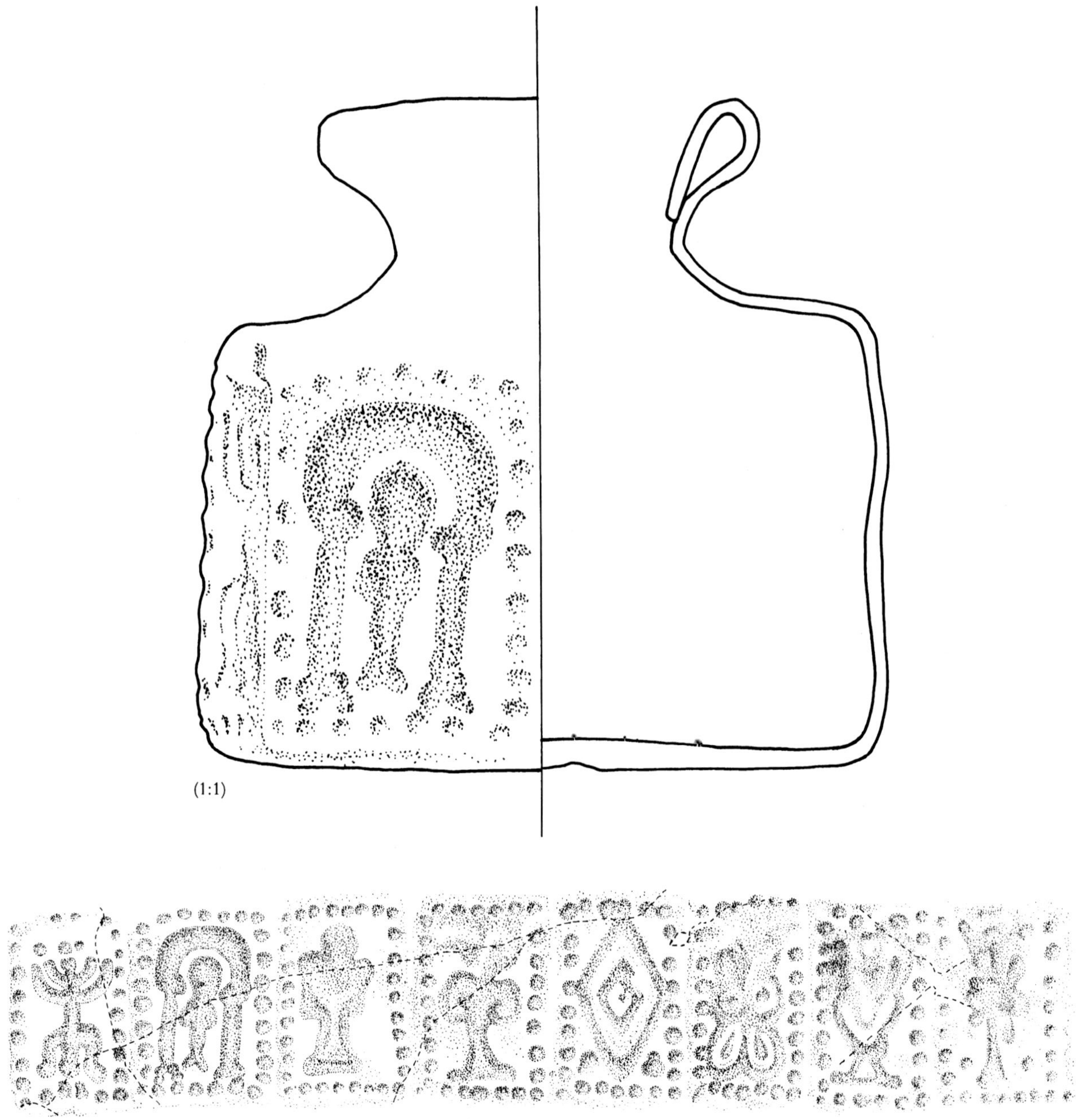

(1:1)

(1:2)

175

*Vessels with Jewish Symbols (Barag Class B)*

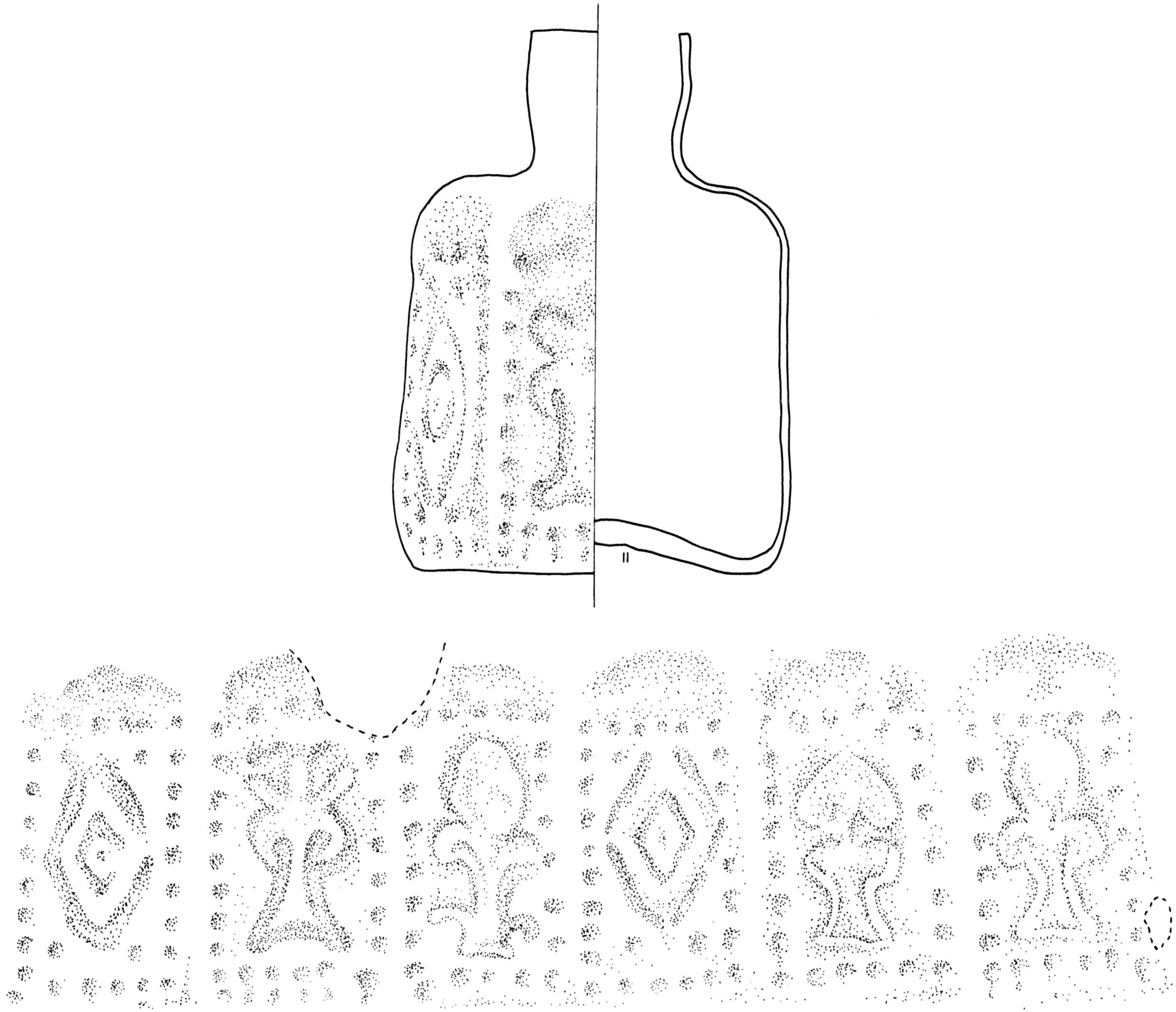

176

*Hexagonal Jug with Unassigned Symbols (Barag Class C)*

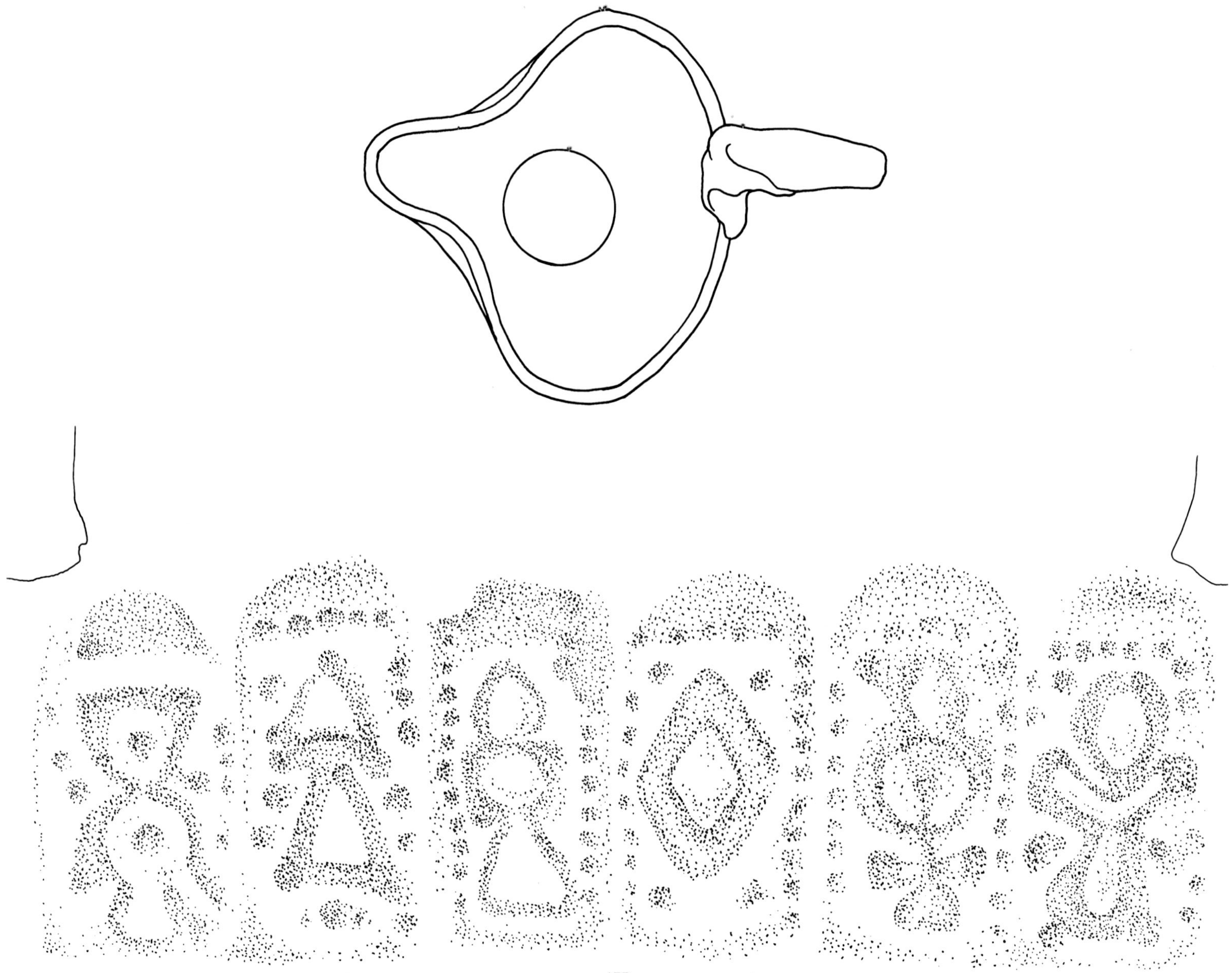

177

*Hexagonal Jug with Unassigned Symbols (Barag Class C)*

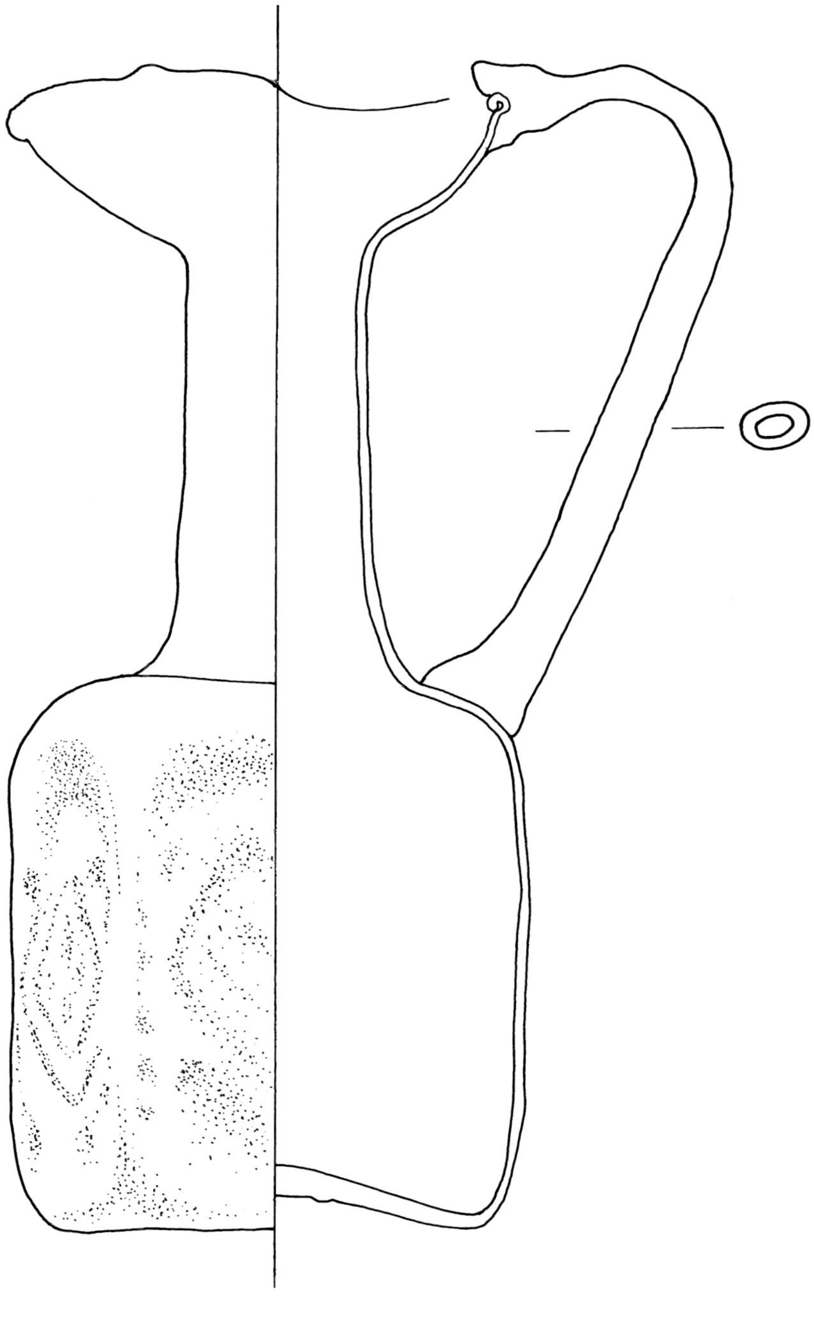

177

*Hexagonal Jug with Unassigned Symbols (Barag Class C)*

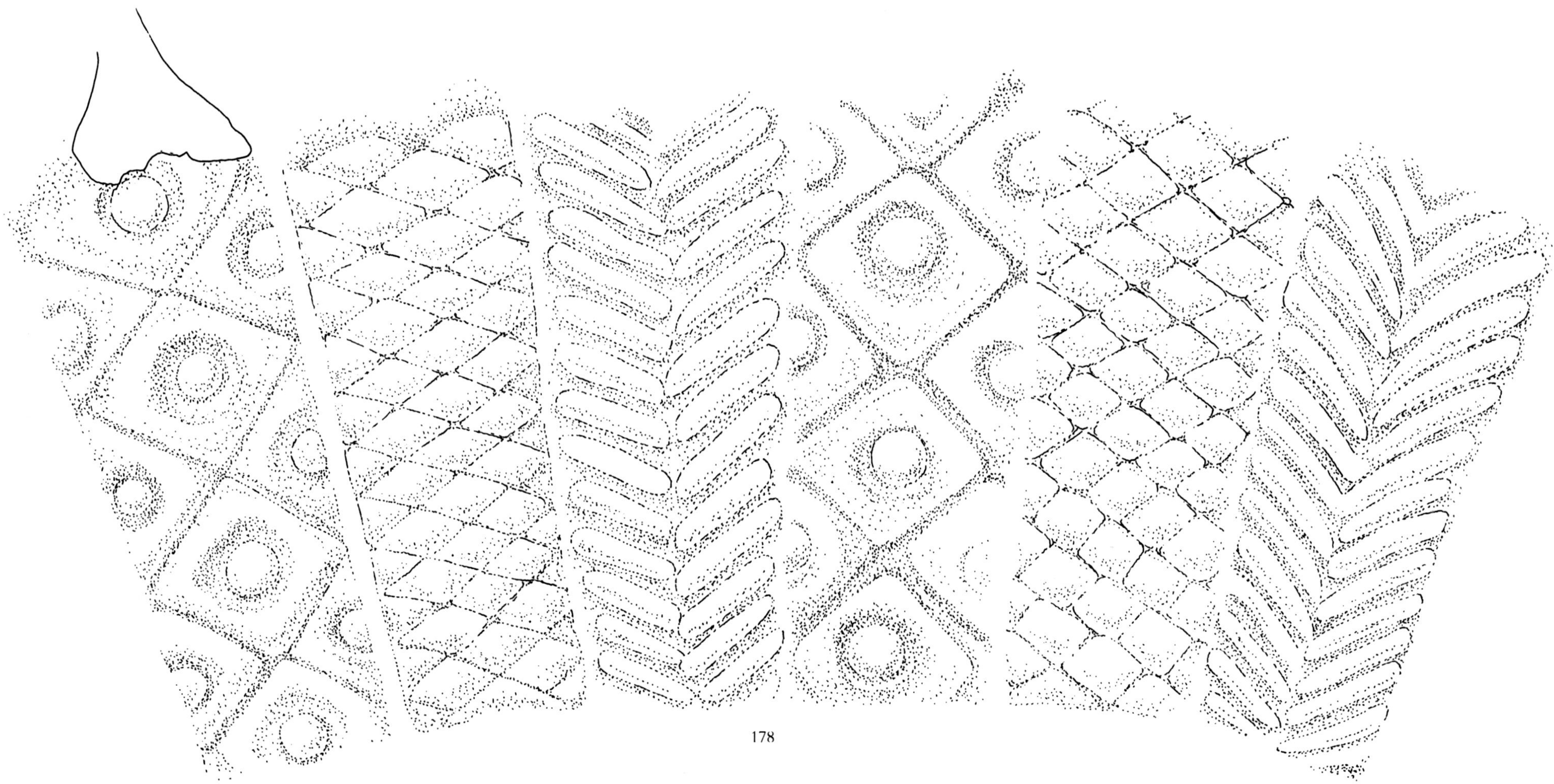

178

*Vessels with Symbols in Sunken Relief, perhaps Christian*

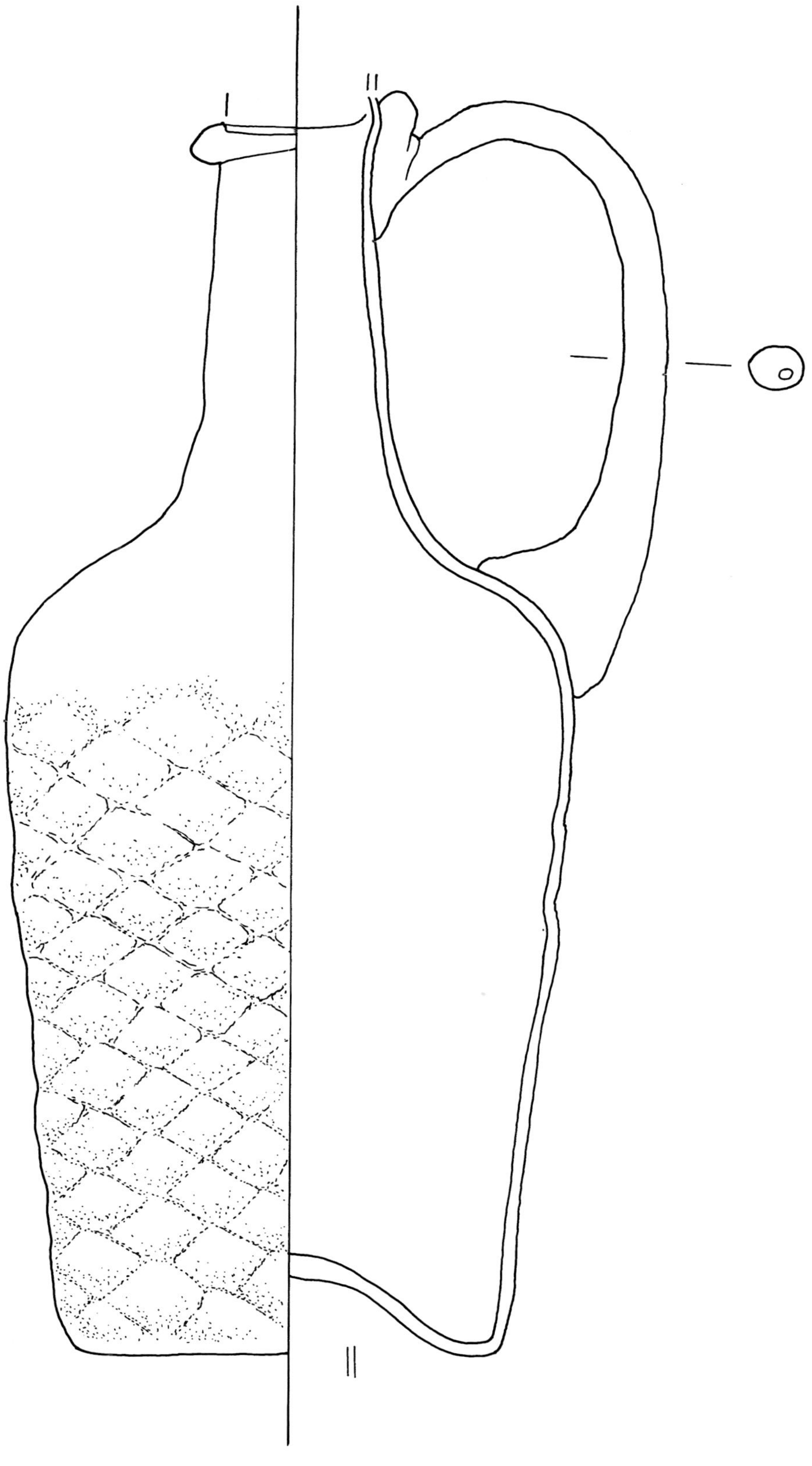

178

*Vessels with Symbols in Sunken Relief, perhaps Christian*

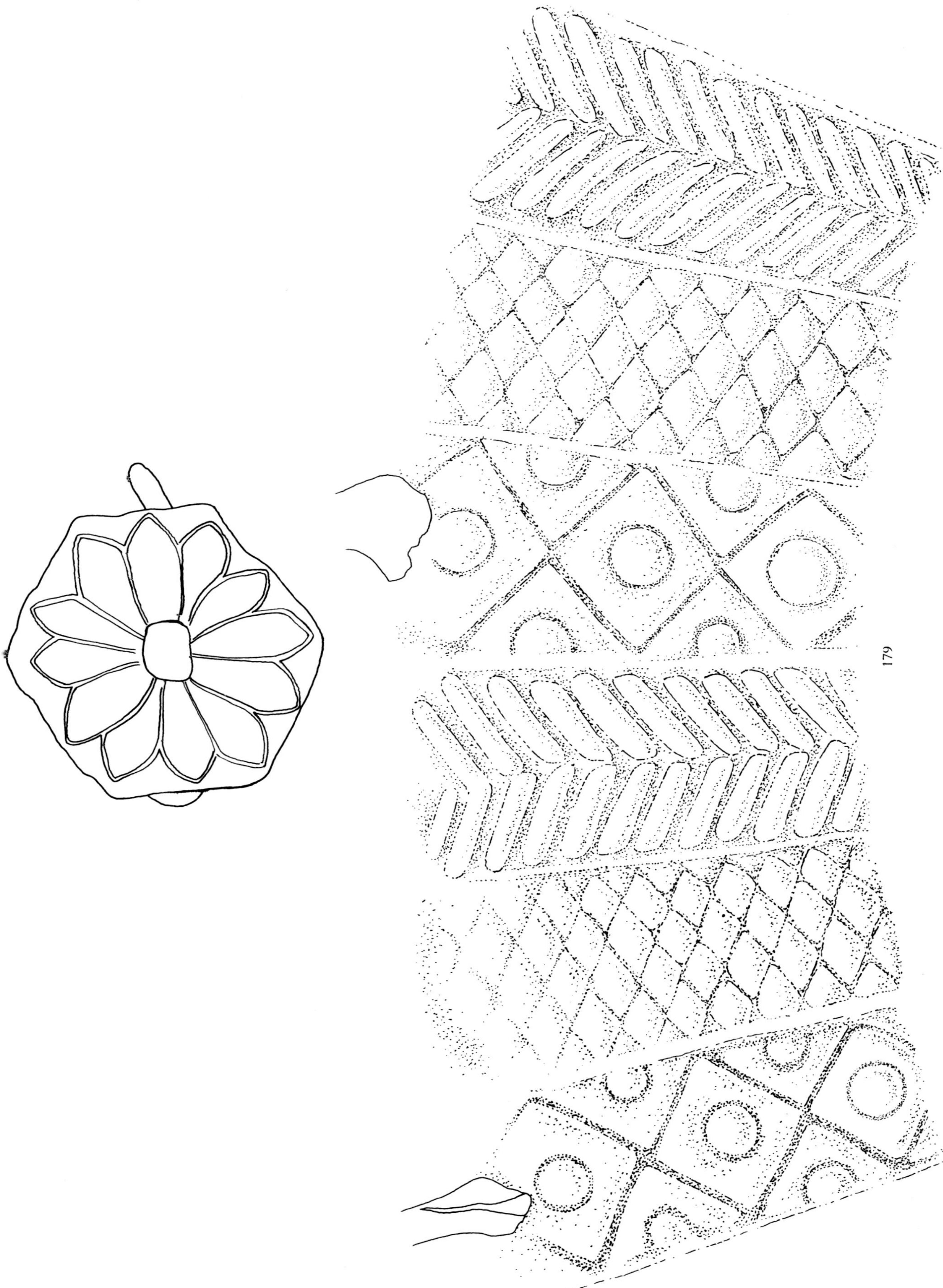

179

*Vessels with Symbols in Sunken Relief, perhaps Christian*

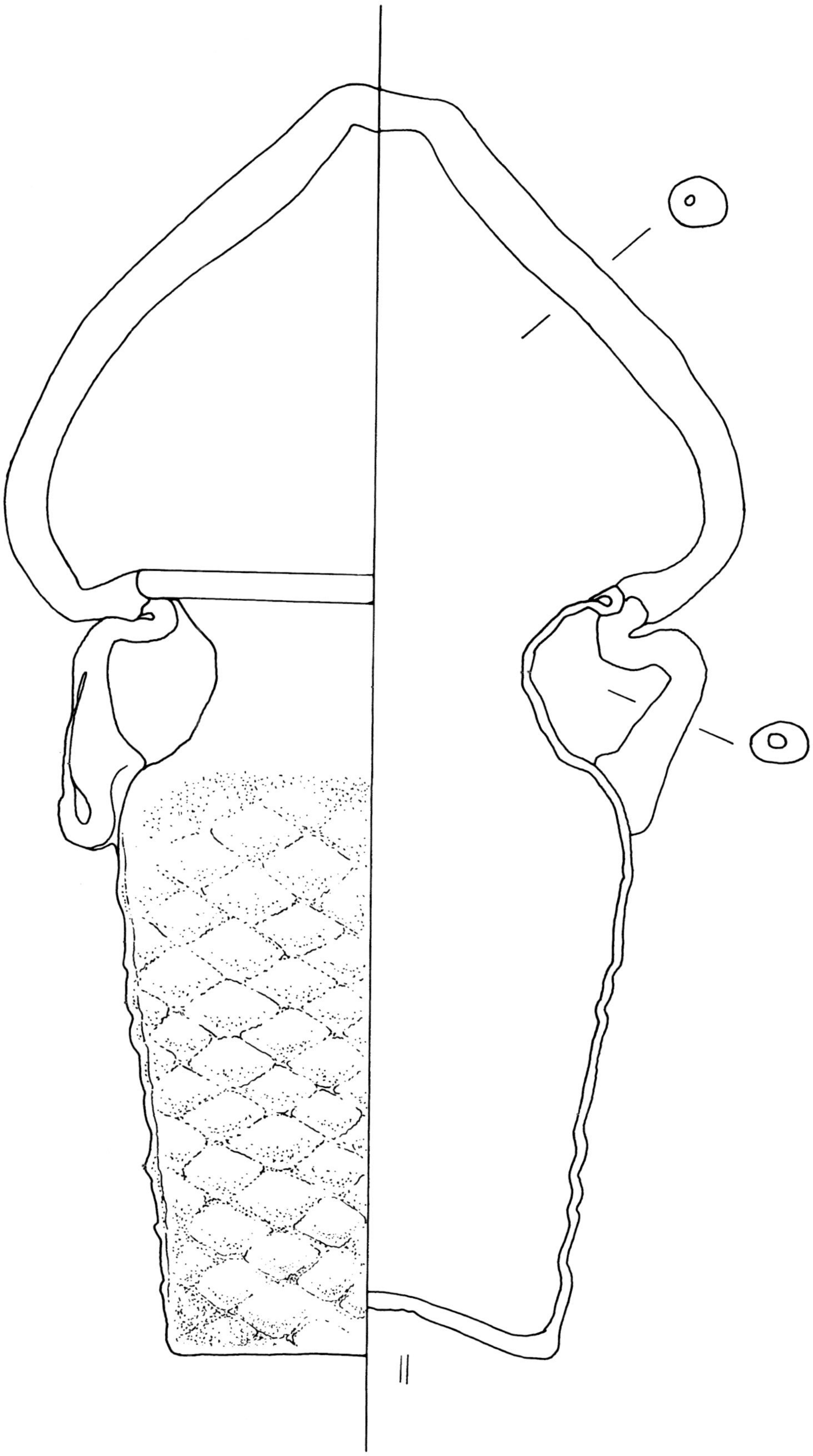

179

*Vessels with Symbols in Sunken Relief, perhaps Christian*

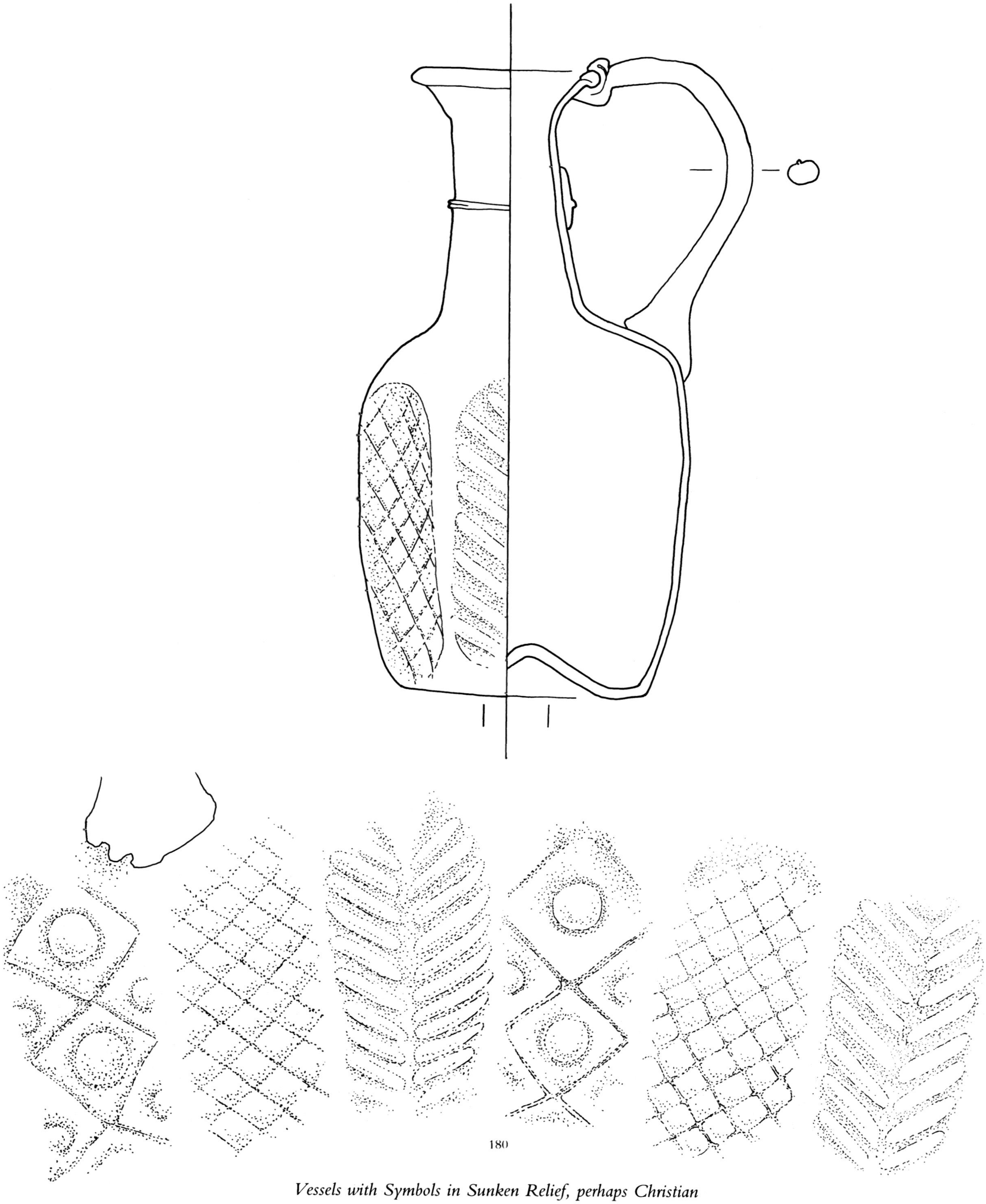

180

*Vessels with Symbols in Sunken Relief, perhaps Christian*

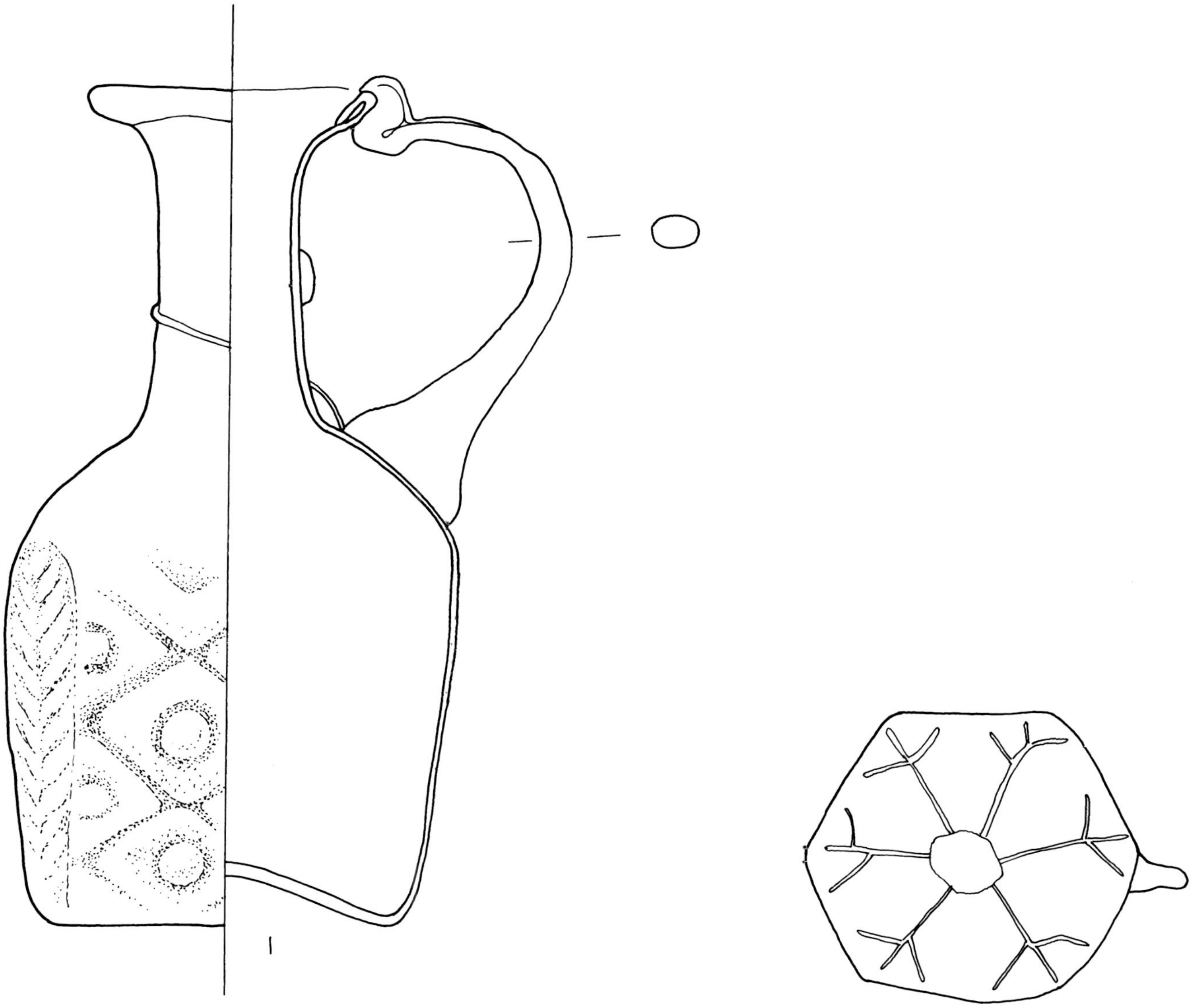

1

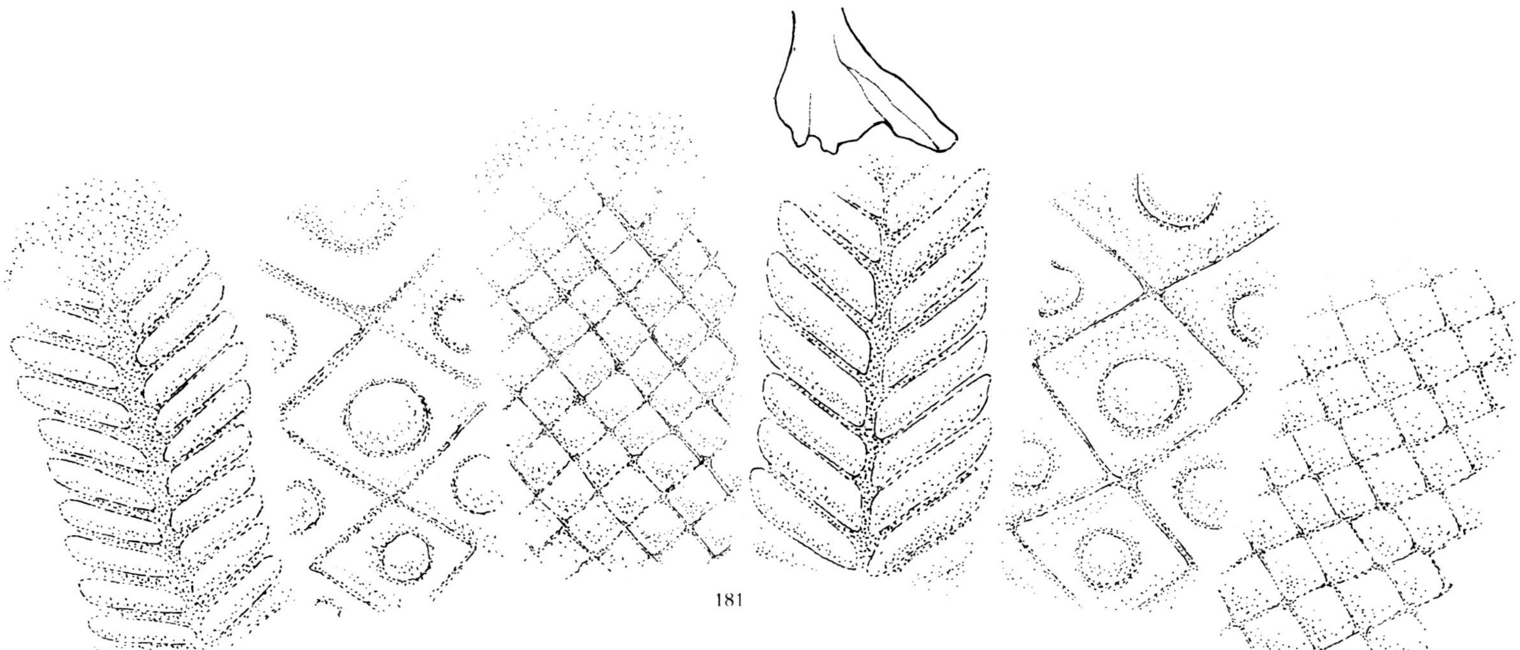

181

*Vessels with Symbols in Sunken Relief, perhaps Christian*

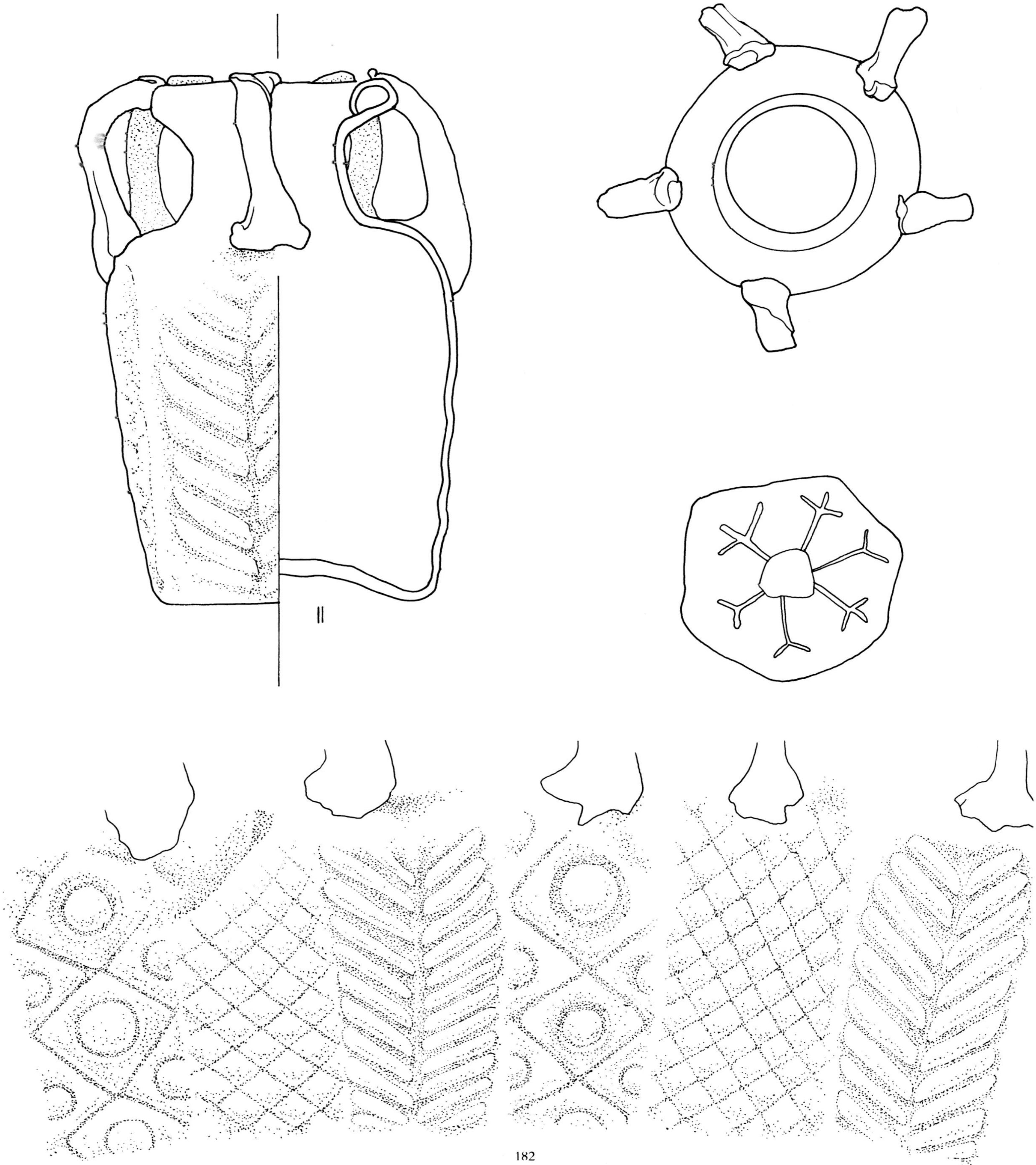

182

*Vessels with Symbols in Sunken Relief, perhaps Christian*

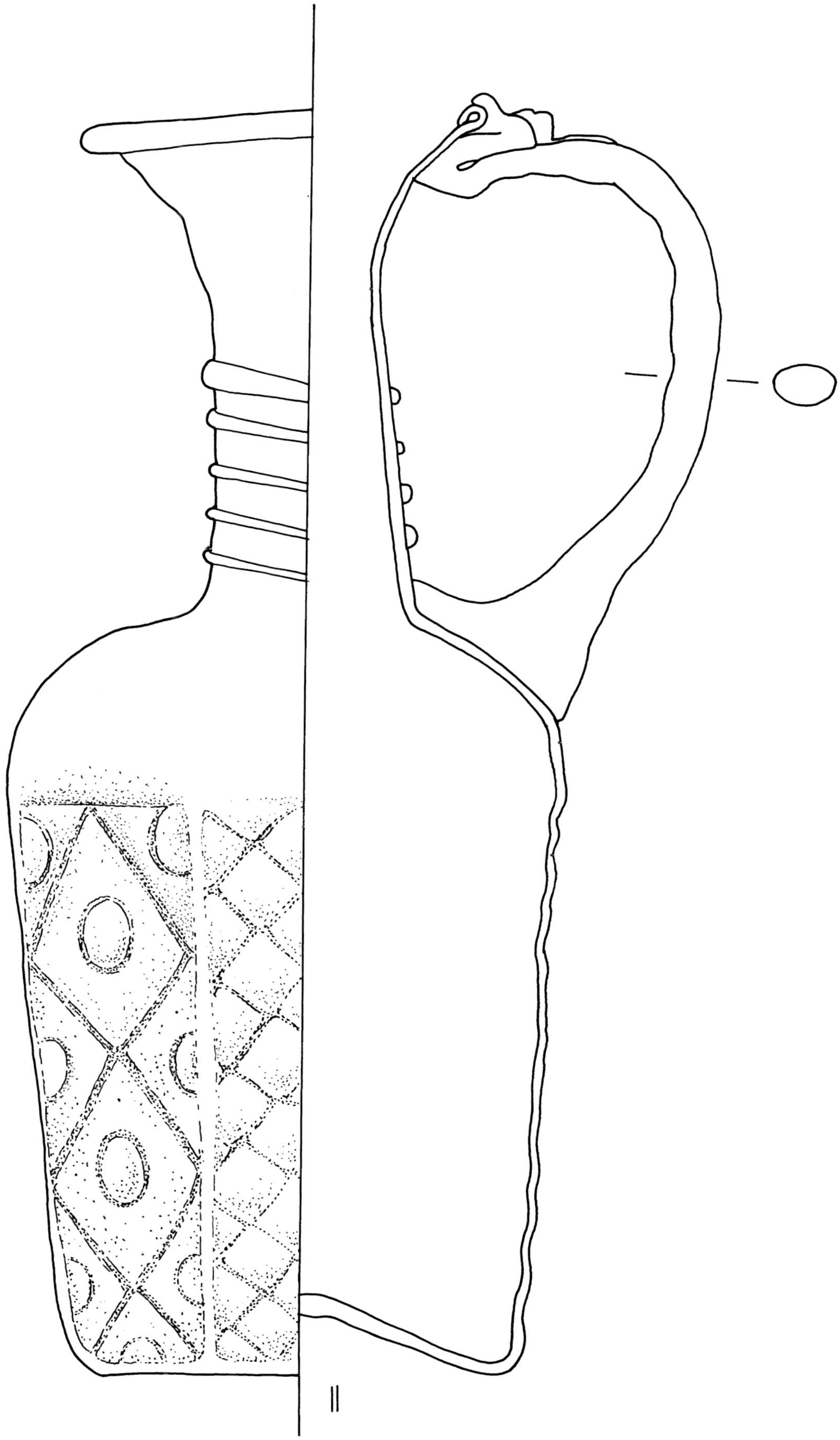

183

*Vessels with Symbols in Sunken Relief, perhaps Christian*

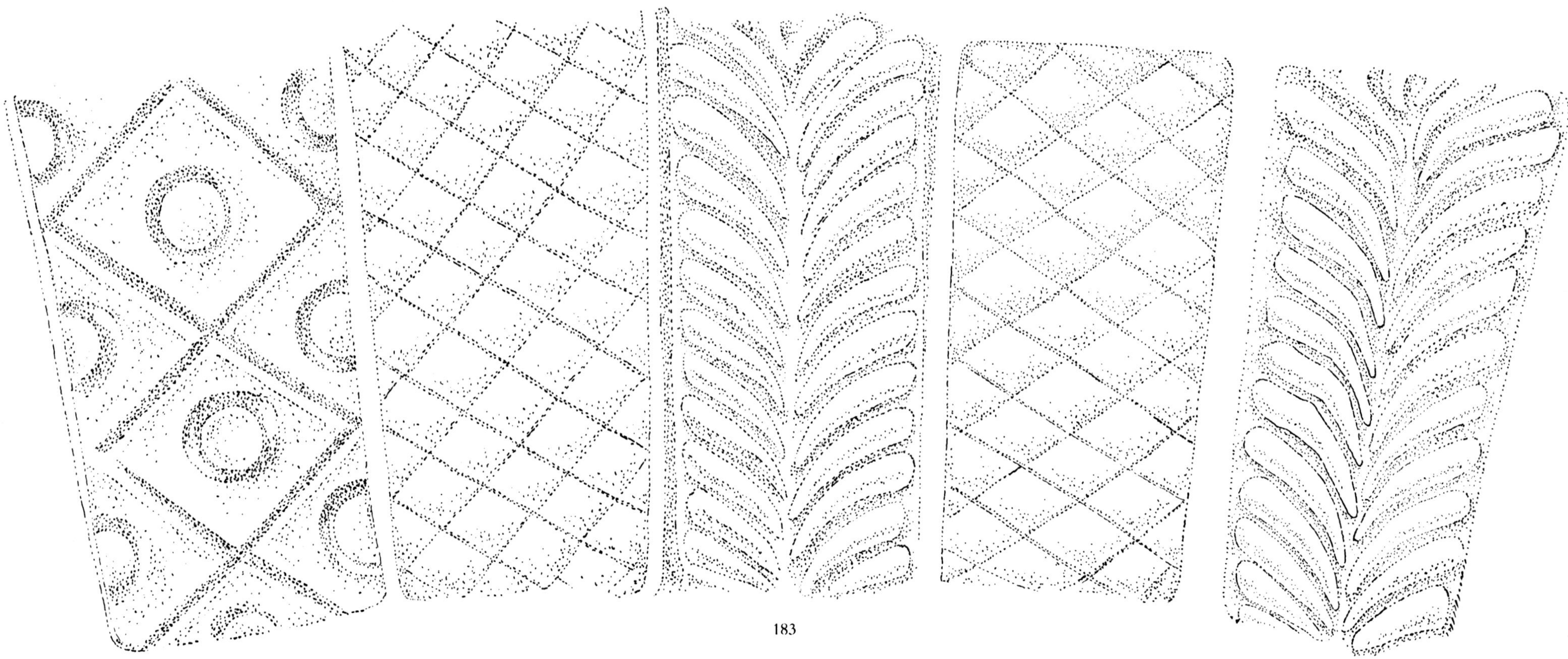

183

*Vessels with Symbols in Sunken Relief, perhaps Christian*

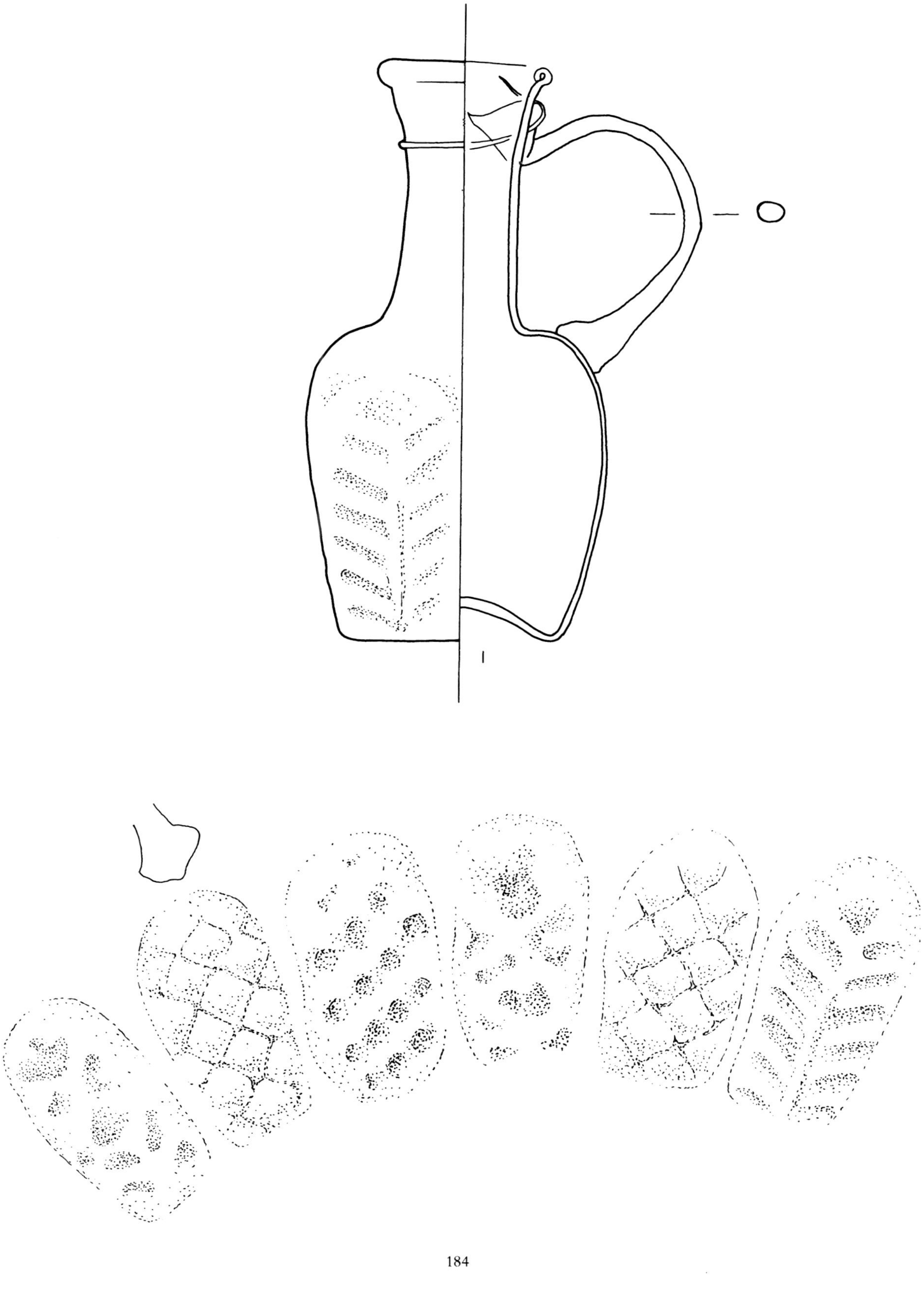

184

*Vessels with Symbols in Sunken Relief, perhaps Christian*

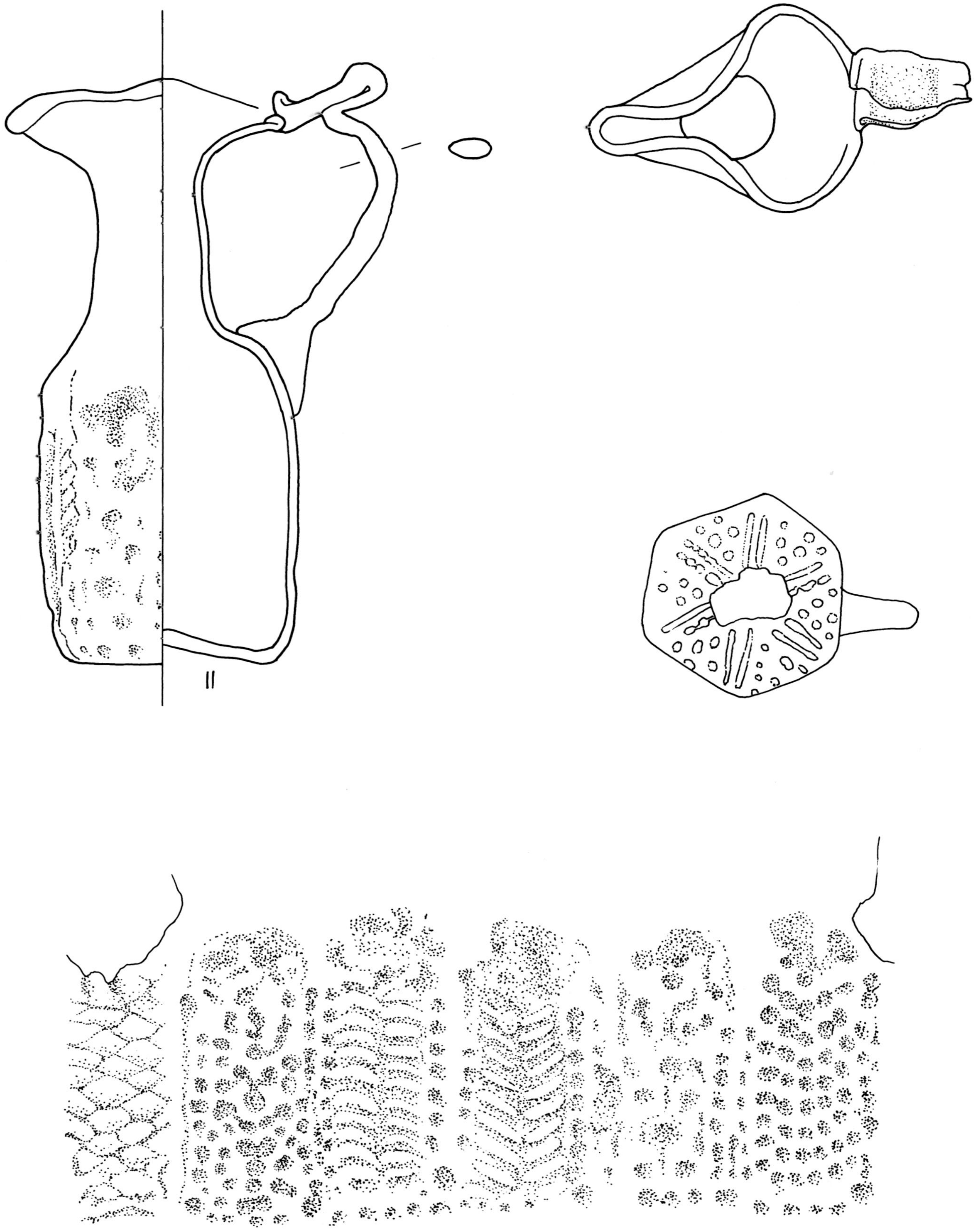

185

*Vessels with Symbols in Sunken Relief, perhaps Christian*

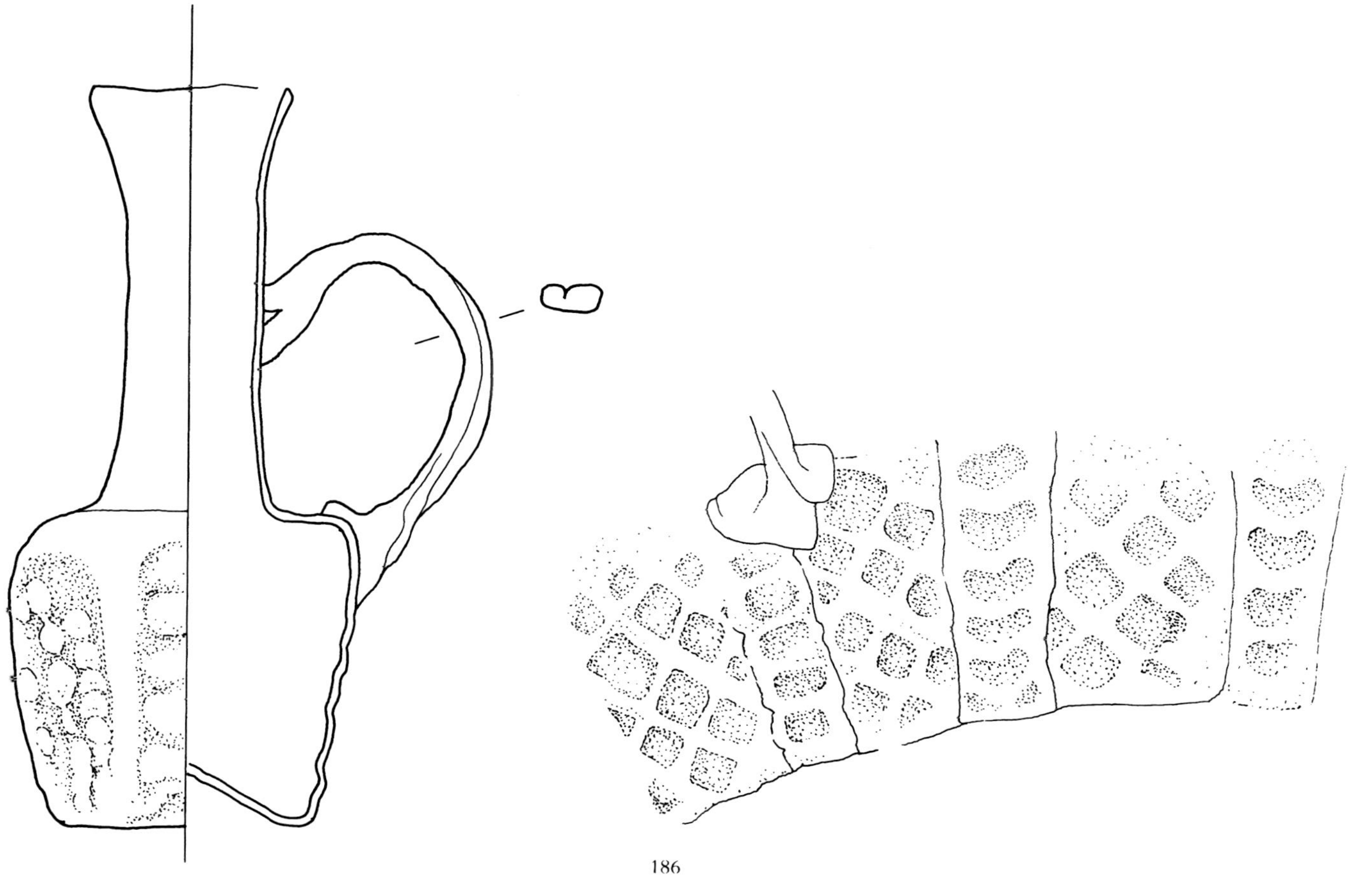

186

*Vessels with Symbols in Sunken Relief, perhaps Christian*

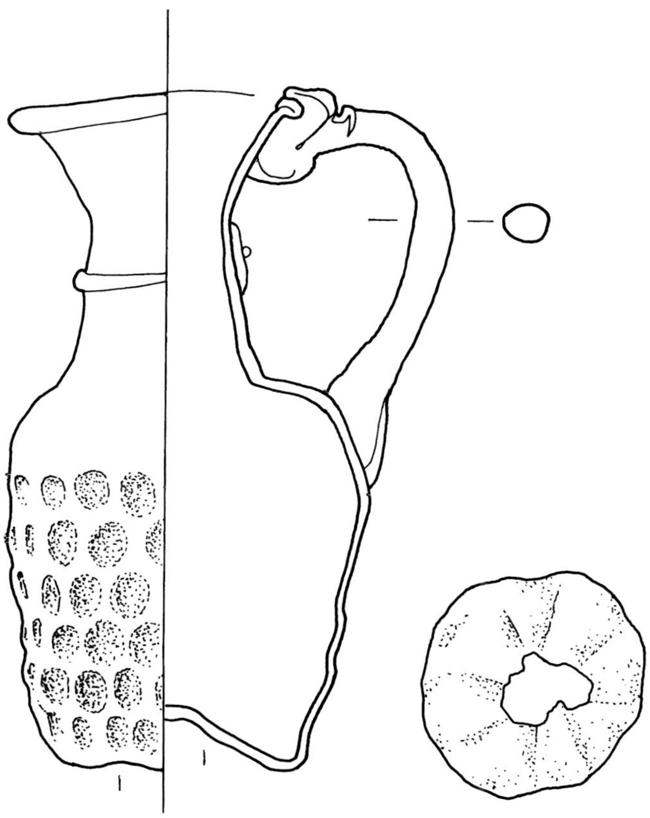

187

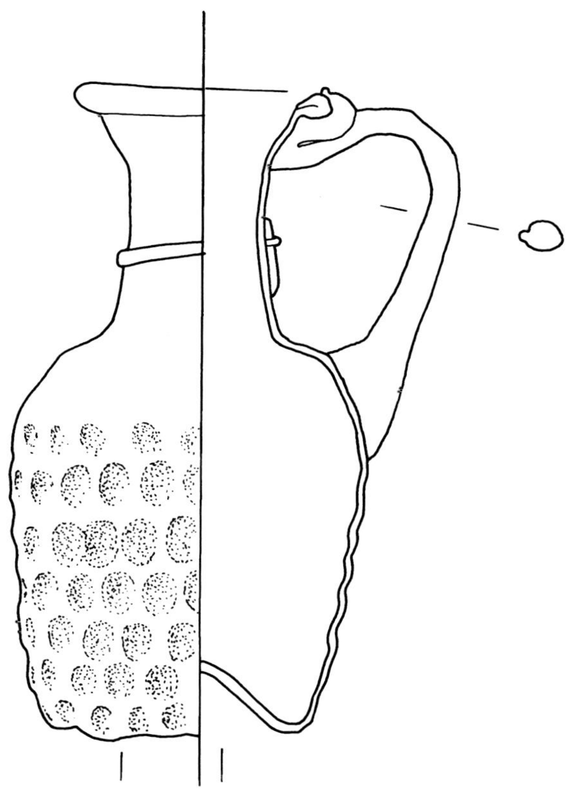

188

*Inverted Conical Jugs with Geometric Designs in Sunken Relief*

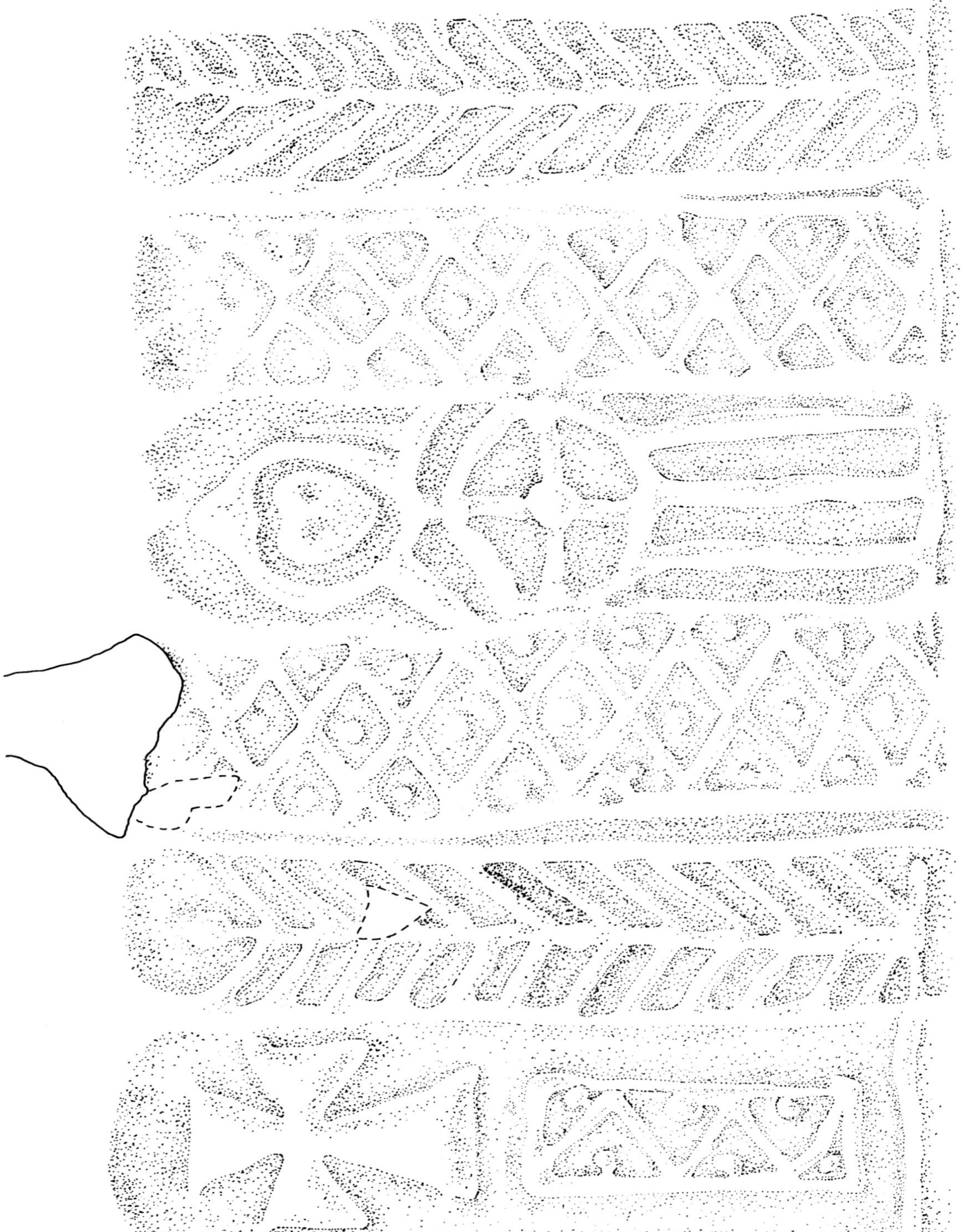

189

*Jugs with Christian Symbols in Low Relief*

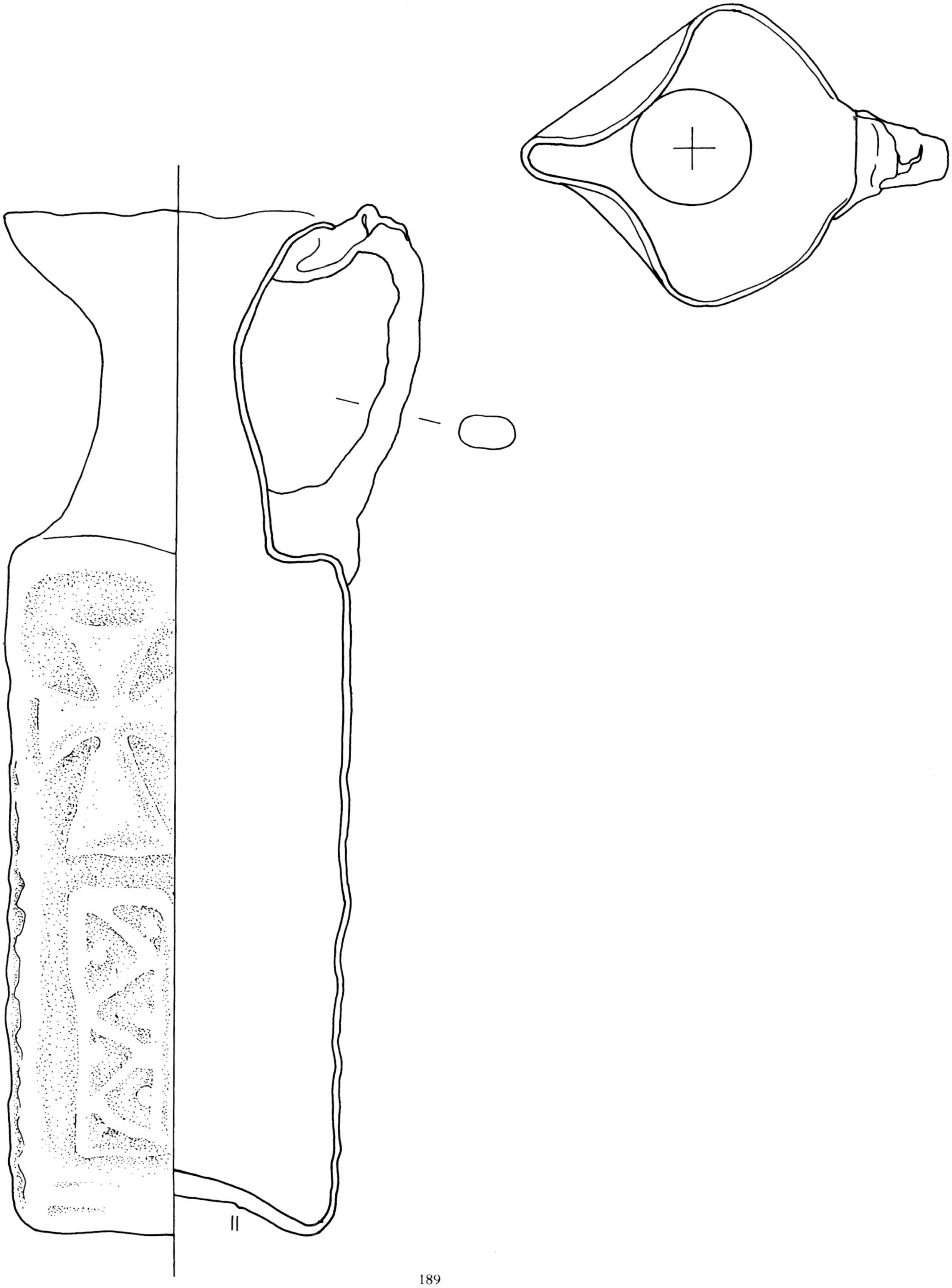

189

*Jugs with Christian Symbols in Low Relief*

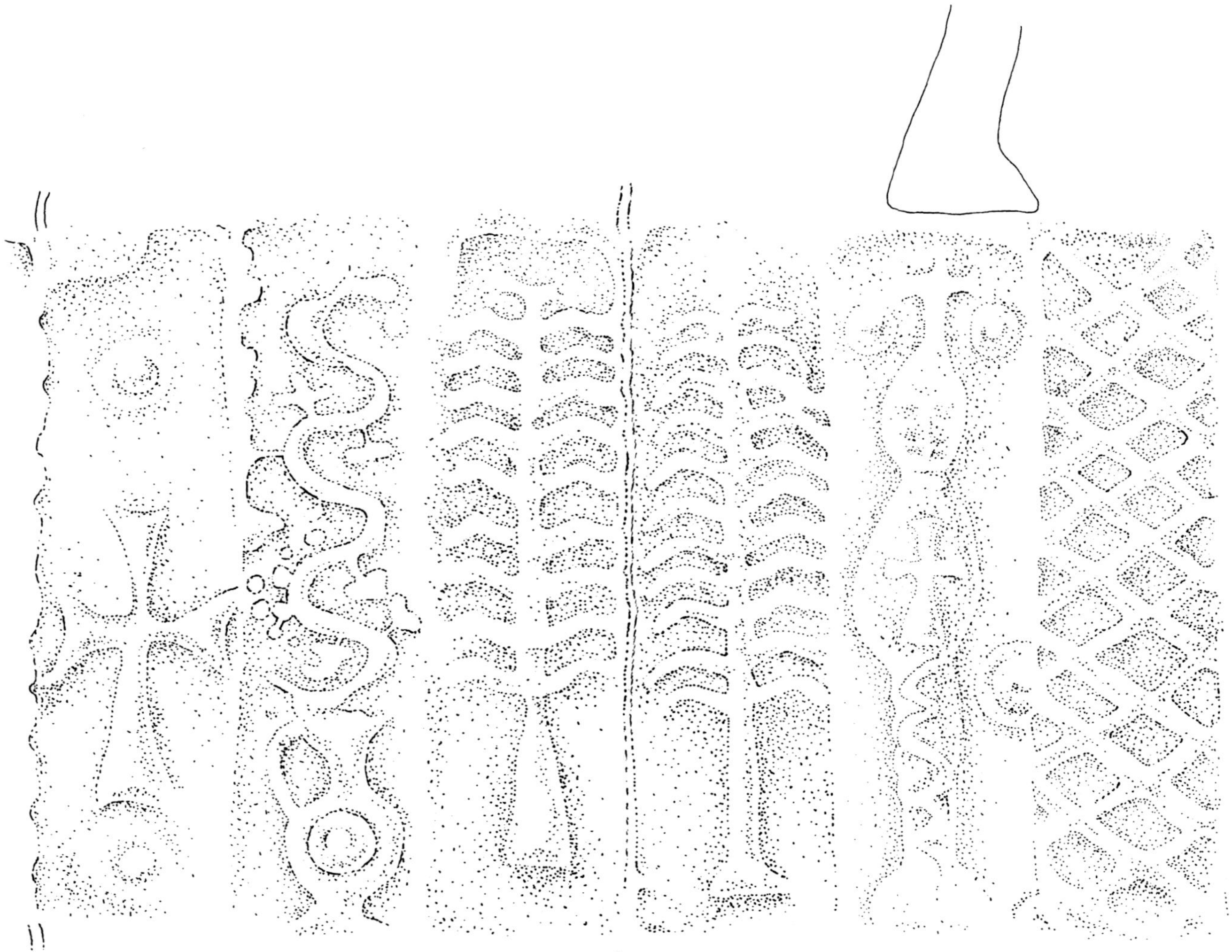

190

*Jugs with Christian Symbols in Low Relief*

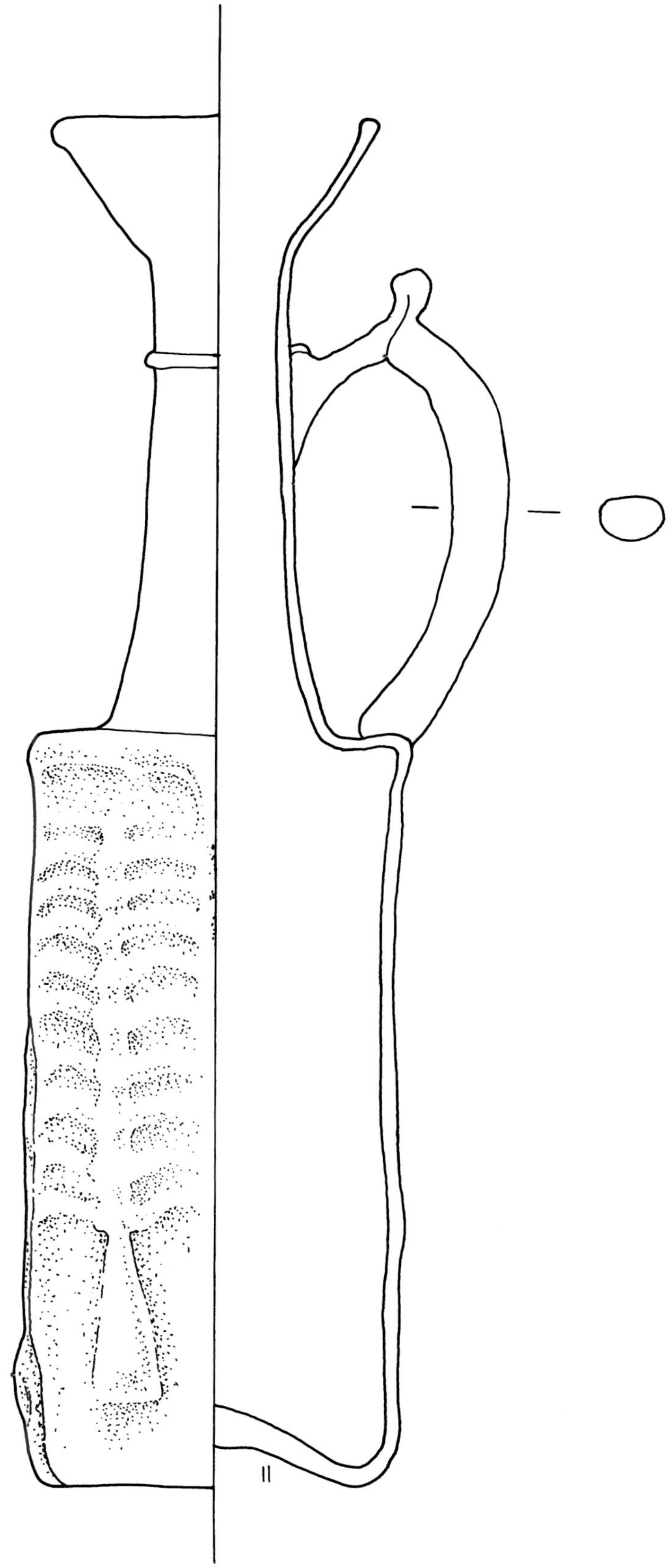

190

*Jugs with Christian Symbols in Low Relief*

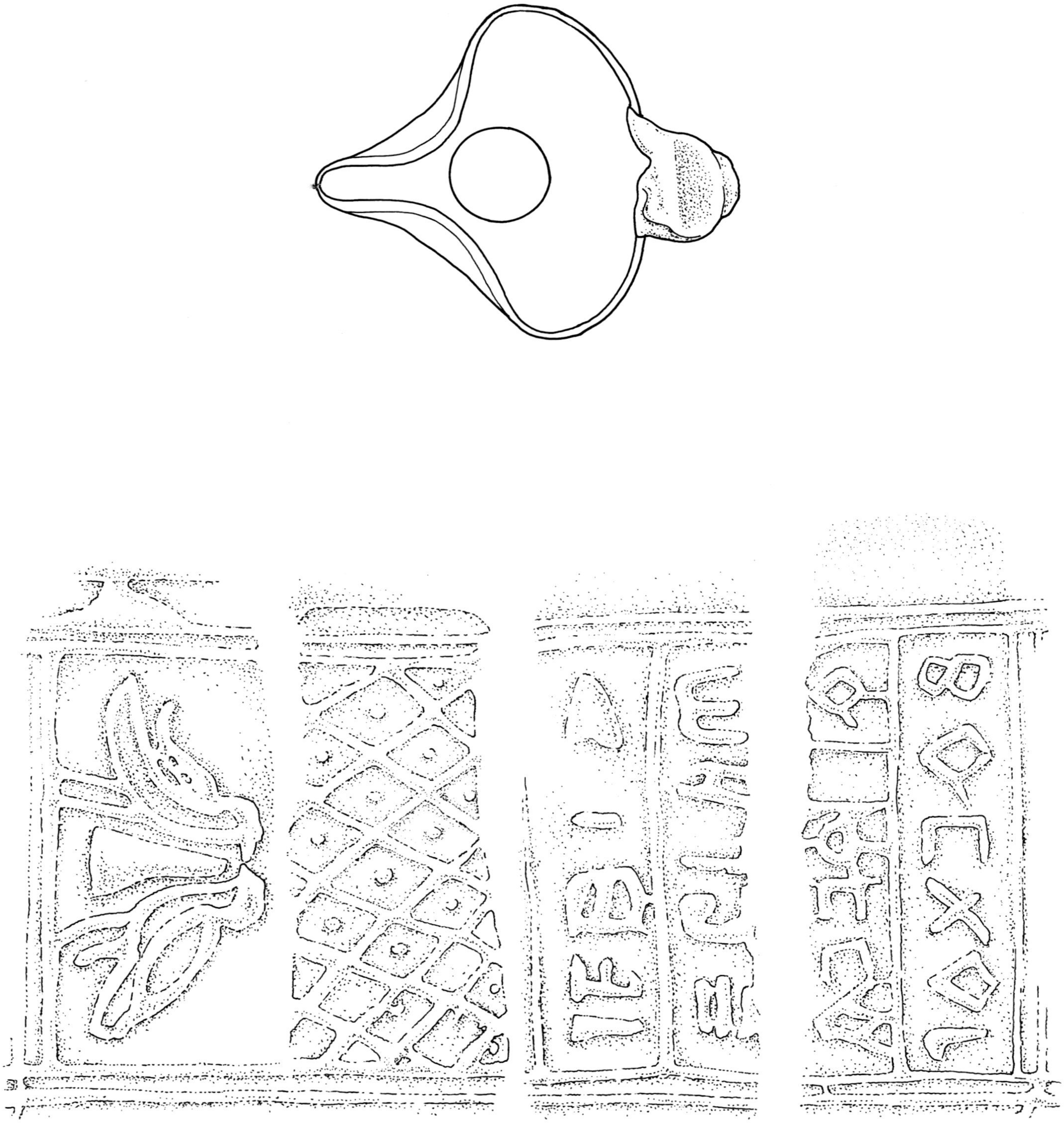

191

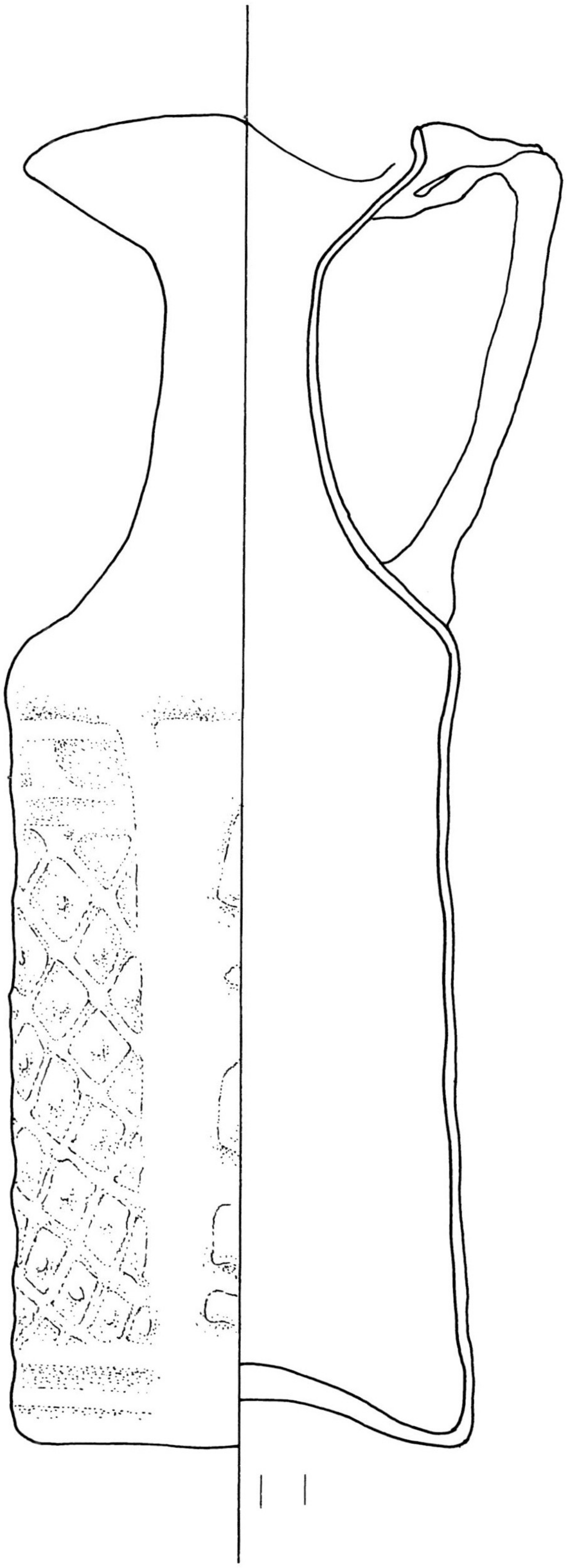

191

*Jugs with Christian Symbols in Low Relief*

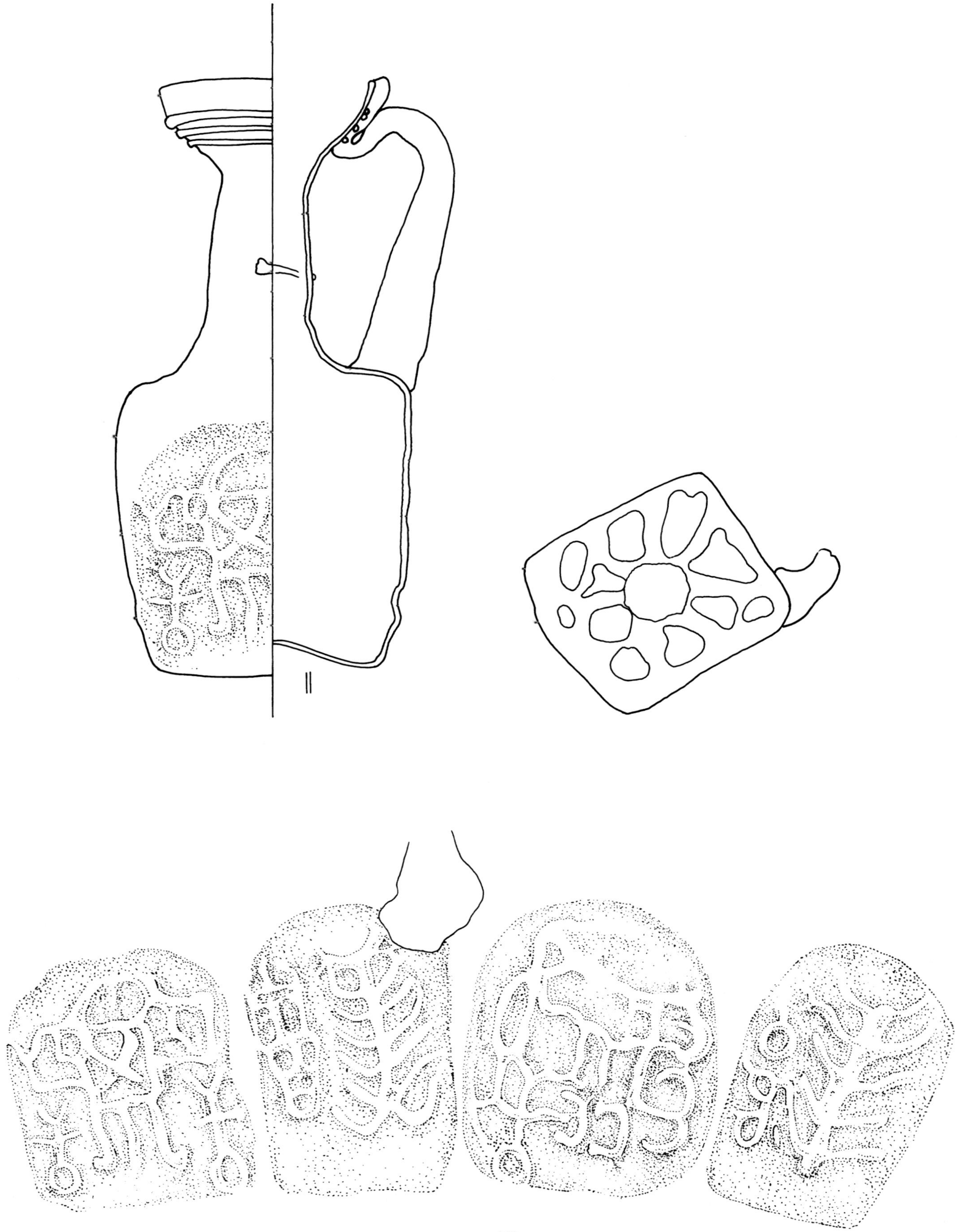

192

*Jugs with Christian Symbols in Low Relief*

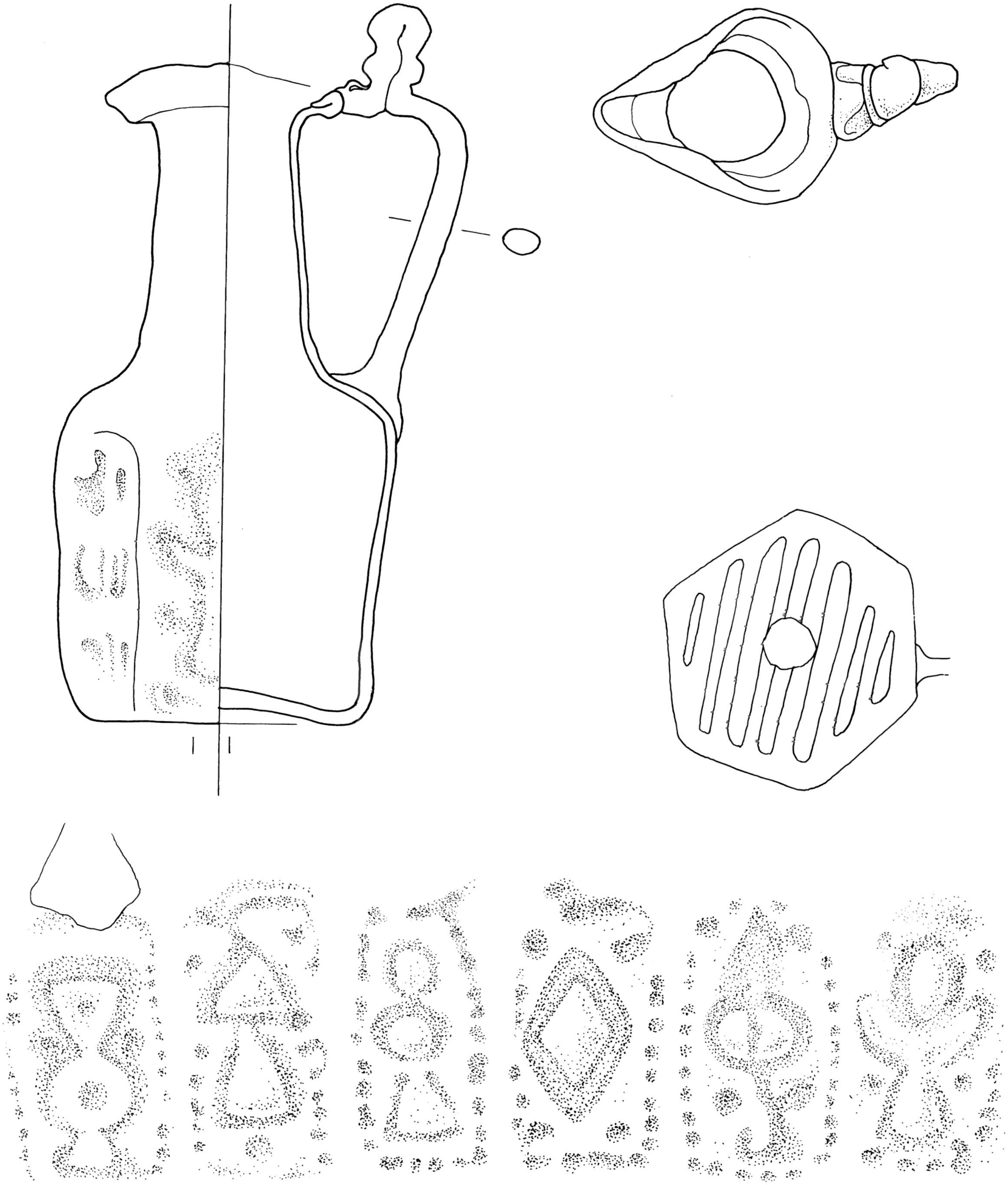

193

*Later Imitation of a Vessel with Christian Symbols (Barag Class C)*

# BIBLIOGRAPHY AND SHORT TITLES

Frequently cited monographs, archaeological and collections catalogues, articles, and exhibition titles have been identified by the short titles listed below. Complete bibliographic information for all other references is given in the footnotes to the text and in the bibliographies within the catalogues entries.


*AA* — *Archäologischer Anzeiger.*

*AAA* — *Archaiologika Analekta ex Athenon.*

*AAS* — *Annales archéologiques arabes syriennes.*

Abdul Hak 1951 — S. Abdul Hak and Andrée Abdul Hak, *Catalogue illustré du département des antiquités gréco-romaines au Museé National de Damas* (Damascus 1951).

Abdul Hak 1965 — S. Abdul Hak, "Contribution d'une découverte archéologique récente à l'étude de verrerie syrienne à l'époque romaine," *JGS* 7 (1965) 26-34.

*AFAV* — *Association française pour l'Archéologie du Verre, Actes des Rencontres.*

*Age of Spirituality* 1977 — *Age of Spirituality: Late Antique and Early Christian Art, Third to Seventh Century*, exh. cat., The Metropolitan Museum of Art (New York 1977, published 1979).

AIHV — Association internationale pour l'histoire du verre.

*AJA* — *American Journal of Archaeology.*

Al-'Ush et al. 1969 — M. Abu -l- Faraj Al-'Ush, Adnan Joundi, and Bachir Zouhdi, translated by G. Saade, *Catalogue du Musée National de Damas* (Damascus 1969).

*AM* — *Mitteilungen des Deutschen Archäologischen Instituts, Athenische Abteilung.*

*AnatSt* — *Anatolian Studies. Journal of the British Institute of Archaeology at Ankara.*

*Ancient Glass in Yugoslavia* — *Ancient Glass in Yugoslavia*, Slovene Archaeological Association, Coloquy [*sic*] on the occasion of 150th anniversary of the National Museum, Ljubljana, 14.X.1971 (*Materijali* 11 [1976], reprinted from: *Arheoloski Vestnik* 25 [1974] 19-210).

*AnnAIHV* — *Annales du congrès de l'Association internationale pour l'histoire du verre.*

*AntJ* — *Antiquaries Journal. Journal of the Society of Antiquaries of London.*

*AntW* — *Antike Welt. Zeitschrift für Archaölogie und Kulturgeschichte.*

Arakelian et al. 1969 — B. N. Arakelian, G. A. Tiratsian, and G. D. Khatchatrian, *Steklo Drebney Armenii (I-IV vv), Arkheologicheskie Pamiatniki Drevney Armenii, 3: Pamiatniki Drevney Epokhi*, 1 (*The Archaeological Monuments and Specimens of Armenia*, 3: *Monuments and Specimens of the Ancient Period*, I: *The Glass of Ancient Armenia, I-IV Centuries*) (Yerevan 1969).

*ArchCl* — *Archeologia classica.*

*ArchDelt* — *'Αρχαιολογικὸν Δελτίον* (*Archaiologikon Deltion*).

*ArchJ* — *Archaeological Journal.*

*'Atiqot* — *'Atiqot. Journal of the Israel Department of Antiquities.*

Auth 1976 — Susan H. Auth, *Ancient Glass at the Newark Museum* (Newark, N. J. 1976).

Avigad 1972 — Nahman Avigad, "Excavations in the Jewish Quarter, Jerusalem," *IEJ* 22 (1972) 193-200.

Avigad 1980 — Nahman Avigad, *Discovering Jerusalem* (Nashville, Camden, and New York 1980).

Avi-Yonah 1975-78 — M. Avi-Yonah, *Encyclopedia of Archaeological Excavations in the Holy Land*, I-IV (Englewood Cliffs, N.J. 1975-78).

Avi-Yonah 1984 — M. Avi-Yonah, *The Jews under Roman and Byzantine Rule: A Political History of Palestine from the Bar Kochba War to the Arab Conquest*, translated and adapted by the author (Jerusalem 1984).

*BABesch* — *Bulletin antieke beschaving. Annual Papers on Classical Archaeology.*

Bagatti 1967a — P. B. Bagatti, "I vetri del Museo Francescano di Nazaret," *Liber Annuus* 17 (1967) 222-240.

Baldoni 1987 — Daniela Baldoni, "Una lucerna romana con raffigurazione di officina vetraria: alcune considerazioni sulla lavorazione del vetro soffiato nell'antichità," *JGS* 29 (1987) 22-29.

*BAR* — *British Archaeological Reports.*

Barag 1963 — Dan P. Barag, "The Glassware," in Y. Yadin, *The Finds from the Bar-Kokhba Period in the Cave of Letters* (Jerusalem 1963) 101-110.

Barag 1970a — Dan P. Barag, *Glass Vessels of the Roman and Byzantine Periods in Palestine*, 2 vols. (Jerusalem 1970).

Barag 1970c — Dan P. Barag, "Glass Pilgrim Vessels from Jerusalem, Pt. I," *JGS* 12 (1970) 35-63.

Barag 1971a — Dan P. Barag, "Glass Pilgrim Vessels from Jerusalem, Pts. II and III," *JGS* 13 (1971) 45-63.

Barag 1971c — Dan P. Barag, "Glass," in *Encyclopedia Judaica* 7 (Jerusalem 1971) cols. 604-611.

Barag 1985a — Dan P. Barag, *Catalogue of Western Asiatic Glass in the British Museum*, I (London/Jerusalem 1985).

Barag 1985b — Dan P. Barag, "Finds from a Tomb of the Byzantine Period at Ma'in," *Liber Annuus* 25 (1985) 365-374.

Barag 1991 — Dan P. Barag, "The Contribution of Masada to the History of Early Roman Glass," in Newby and Painter 1991, 137-140.

Barag transl. — Dan P. Barag, "Glass Vessels from Tomb-Groups in Palestine, Trans-Jordan and Syria," translation by D. P. Barag of Barag 1970a, vol. I, chapter I (manuscript, n.d.).

Barkóczi 1988 — L. Barkóczi, *Pannonische Glasfunde in Ungarn*, Akadémiai Kiadó (*Studia Archaelogica* 9, Budapest 1988).

*BCH* — *Bulletin de correspondance hellénique.*

Behn 1913-14 — Friedrich Behn, "Ausgewählte Neuerwerbungen des Röm.-Germ. Zentral-Museums an Original-Altertümern," *MZ* 8/9 (1913-1914) 5-17.

*Belleten* — *Belleten. Türk tarih Kurumu.*

Berger 1960 — L. Berger, *Römische Gläser aus Vindonissa* (*Veröffentlichungen der Gesellschaft Pro Vindonissa* 4, Basel 1960).

Berger 1981 — L. Berger, "Neufund eines Glasbechers der Form Isings 31," *GPVJ* (1981) 23-28.

Bergman 1974 — Sidney M. Bergman, "Azeez Khayat (1875-1943) — A Noted Collector of Ancient Glass," *Carnegie Magazine* 48 (1974) 238-244.

*BerlMus* — *Berliner Museen. Berichte aus den preussischen Kunstsammlungen.*

*BerROB* — *Berichten van de Rijksdienst voor het Oudheidkundig Bodemondersoek.*

Biaggio Simona 1989 — Simonetta Biaggio Simona, "Glasgefässe ausgewählter Grabgruppen im Tessin," *KölnJb* 22 (1989) 171-176.

Biaggio Simona 1991 — Simonetta Biaggio Simona, *I vetri romani provenienti dalle terre dell'attuale cantone ticino*, 2 vols. (Locarno 1991).

*BICS* — *Bulletin of the Institute of Classical Studies at the University of London.*

*BJb* — *Bonner Jahrbücher des Rheinischen Landesmuseums in Bonn und des Vereins von Altertumsfreunden im Rheinlande.*

*BMMA* — *Bulletin of the Metropolitan Museum of Art, New York.*

Boon 1967 — G. C. Boon, "Roman Glass in Wales," *AnnAIHV* 4 (1967, published 1969) 93-102.

Bremen 1964 — W. Bremen, *Die alten Glasgemälde und Hohlgläser der Sammlung Bremen in Krefeld* (*BJb* Beiheft 13, Bonn 1964).

Brill 1988 — Robert H. Brill, "Scientific Investigations of the Jalame Glass and Related Finds," in Weinberg 1988, 257-294.

Brown 1968 — P. D. C. Brown, "Recent Museum Acquisitions: The Newport Pagnell Beaker," *Burlington Magazine* 110 (1968) 405.

Brusin 1937 — G. Brusin, "Ritrovamenti occasionali," *NSc* 1937, 190.

*BSA* — *Annual of the British School at Athens.*

Bucovala 1968 — M. Bucovala, *Vase antice de sticla la Tomis* (Constanta 1968).

Bucovala 1979 — M. Bucovala, "Pièces romaines d'importation sur le littoral ouest-pontique," *AnnAIHV* 8 (1979, published 1981) 89-95.

*BullAIHV* — *Bulletin de l'Association internationale pour l'histoire du verre.*

Burford 1972 — Alison Burford, *Craftsmen in Greek and Roman Society* (London 1972).

Calvi 1959 — Maria Carina Calvi, "Vetri romani da una tomba di Vigorovea," *Bollettino del Museo Civico di Padova* 48 (1959) 3-9.

Calvi 1965a — Maria Carina Calvi, "La coppa vitrea di Aristeas nella collezione Strada," *JGS* 7 (1965) 9-16.

Calvi 1968 — Maria Carina Calvi, *I vetri romani del Museo di Aquileia* (Aquileia 1968).

Calvi 1975 — Maria Carina Calvi, "I vetri di Brescia romana," in *Atti del convegno internazionale per il XIX centenario della dedicazione del "Capitolium" e per il 150° anniversario della sua scoperta* (Brescia 1975) 205-216.

Carazzetti and Biaggio Simona 1988 — Ricardo Carazzetti and Simonetta Biaggio Simona, *Vetri romani del cantone ticino*, exh. cat., Museo Civico e Archeologico (Locarno 1988).

Carington Smith 1982 — J. Carington Smith, "A Roman Chamber Tomb on the South-East Slopes of Monasteriaki Kephala, Knossos," *BSA* 77 (1982) 255-293.

Cermanović-Kuzmanović 1974 — Aleksandra Cermanović-Kuzmanović, "Ein Diatretglas aus Komini (Municipium S...)," *Archaeologia Iugoslavia* 15 (1974) 34-37.

Cesnola 1903 — Luigi Palma di Cesnola, *A Descriptive Atlas of the Cesnola Collection of Cypriote Antiquities in the Metropolitan Museum of Art*, III (New York 1903).

Chambon 1955 — R. Chambon, *L'Histoire de la verrerie en Belgique* (Brussels 1955).

Charleston 1963 — R. J. Charleston, "Glass 'Cakes' as Raw Material and Articles of Commerce," *JGS* 5 (1963) 54-68.

Charlesworth 1972 — D. Charlesworth, "*The Glass*," in Sheppard Frere, *Verulamium Excavations*, I: (*Reports of the Research Committee of the Society of Antiquaries of London* 28, London 1972) 196-215.

Clairmont 1963 — Christoph W. Clairmont, *The Glass Vessels*, in A. Perkins ed., *The Excavations at Dura Europos, Final Report* 4, Pt 5 (New Haven, Conn. 1963).

Clairmont 1975 — C. W. Clairmont, *Excavations at Salona, Yugoslavia (1969-1972)* (Park Ridge, N. J. 1975).

Clairmont 1977 — C. W. Clairmont, *Catalogue of Ancient and Islamic Glass, Benaki Museum* (Athens 1977).

*Collezioni Napoli* 1989 — Maria Rosaria Borriello et al., *Le Collezioni del Museo Nazionale di Napoli* (Rome 1989).

*Corinth* XII — Gladys R. Davidson, *Corinth* XII: *The Minor Objects* (Princeton 1952).

Crowfoot 1957 — G. M. Crowfoot, "The Glass," in J. W. Crowfoot, G. M. Crowfoot, and K. M. Kenyon, *Samaria — Sebaste* III: *The Objects* (London 1957) 403-422.

Czurda-Ruth 1979 — Barbara Czurda-Ruth, *Die römischen Gläser vom Magdalensberg* (*Kärtner Museumsschriften* 65, *Archäologische Forschungen zu den Grabungen auf dem Magdalensberg* 6, Klagenfurt 1979).

Damevski 1974 — Valerija Damevski, "Pregled tipova staklenog posuda iz italskih, galskih, mediteranskih porajnskih radionica na podrucju Hrvatske u doba rimskog carstva," in *Ancient Glass in Yugoslavia*, 62-87 (English summary).

Daremberg and Saglio — C. Daremberg and E. Saglio, *Dictionnaire des antiquités grecques et romaines* (Paris 1875).

Donati et al. 1987 — Pierangelo Donati, Fulvia Ronchetti-Butti, and Simonetta Biaggio Simona, *Ascona: La Necropoli romana* (*Quaderni d'informazione* 12, Bellinzona 1987).

Durighello 1911 — J. A. Durighello, *Verres et bronzes antiques, verrerie arabe. Ancienne Collection Durighello appartenant à M. Henri Buisset, de Londres. Vente Hotel Drouot* (Paris May 17-19, 1911).

Dusenbery 1959 — Elsbeth B. Dusenbery, "A Samothracian Necropolis," *Archaeology* 12 (1959) 163-170.

Dusenbery 1964 — Elsbeth B. Dusenbery, "The South Necropolis of Samothrace," *Archaeology* 17 (1964) 185-192.

Dusenbery 1967 — Elsbeth B. Dusenbery, "Ancient Glass from the Cemeteries of Samothrace," *JGS* 9 (1967) 34-49.

Dusenbery 1971 — Elsbeth B. Dusenbery, "Ancient Glass in the Collections of Wheaton College," *JGS* 13 (1971) 9-33.

*EAA* — *Enciclopedia dell'arte antica, classica e orientale.*

Edgar 1905 — C. C. Edgar, *Graeco-Egyptian Glass: Catalogue général des antiquités égyptiennes du musée du Caire* (Cairo 1905, repr. 1974).

Eisen 1927 — Gustavus A. Eisen, with Fahim Kouchakji, *Glass: Its Origin, History, Chronology, Technic and Classification to the Sixteenth Century*, 2 vols. (New York 1927).

*ErIsr* — *Eretz-Israel.*

Fasold 1985 — Peter Fasold, "Die früh- und mittelrömischen Gläser von Kempten-Cambodunum," *Schwäbische Geschichtsquellen und Forschung* 14 (1985) 197-230.

*Fitzwilliam* 1978 — *Glass at the Fitzwilliam Museum* (Cambridge 1978).

Foerster and Tsafrir 1988 — "The Beth Shean Project," *Excavations and Surveys in Israel* 6 (1987/88) 7-45.

Foy 1977-80 — Danièle Foy, "Notes sur quelques découvertes de verres dans le Sud Est de la France; verres médiévaux," *BullAIHV* 8 (1977-1980) 214-216.

Foy et al. 1989 — Danièle Foy et al., "Ateliers de verriers de l'Antiquité et du haut Moyen Age en France: méthodologie et résultats, un état de la question," *AFAV* 4 (1989, published 1991) 55-72.

Fraser and Matthews 1987 — P. M. Fraser and E. Matthews eds., *A Lexicon of Greek Personal Names* I: *The Aegean Islands, Cyprus, Cyrenaica* (Oxford 1987).

Fremersdorf 1938 — Fritz Fremersdorf, "Römische Gläser mit buntgefleckter Oberfläche," in *Festschrift für A. Oxé* (Darmstadt 1938) 116-121.

Fremersdorf VI 1961 — Fritz Fremersdorf, *Römisches geformtes Glas in Köln* (*Die Denkmäler des römischen Köln* 6, Cologne 1961).

Froehner 1879 — W. Froehner, *La verrerie antique. Description de la collection Charvet* (Le Pecq 1879).

Froehner 1903 — W. Froehner, *Collection Julien Gréau: Verrerie antique émaillerie et poterie appartenant à M. John Pierpont Morgan* (Paris 1903).

*FuB* — *Forschungen und Berichte. Staatliche Museen zu Berlin.*

Fünfschilling 1985 — Sylvia Fünfschilling, "Römische Gläser aus Baden-Aquae Helveticae (aus den Grabungen 1892-1911)," *GPVJ* (1985) 81-160.

*Gallia* — *Gallia. Fouilles et monuments archéologiques en France métropolitaine.*

*Getty Mus J* — *The J. Paul Getty Museum Journal.*

*Glass of the Caesars* 1987 — Donald B. Harden, Hansgerd Hellenkemper, Kenneth Painter, and David Whitehouse, *Glass of the Caesars*, exh. cat., The Corning Museum of Glass (Milan 1987).

*GlastechBer* — *Glastechnische Berichte. International Journal of Glass Science and Technology*, Frankfurt am Main.

Goethert-Polaschek 1977 — Karin Goethert-Polaschek, *Katalog der römischen Gläser des Rheinischen Landesmuseums Trier* (*Trier Grabungen und Forschungen* 9, Mainz 1977).

Goldstein 1979 — Sidney M. Goldstein, *Pre-Roman and Early Roman Glass in The Corning Museum of Glass* (Corning, N. Y. 1979).

Goodenough I-IV — Erwin R. Goodenough, *Jewish Symbols in the Greco-Roman Period* I: *The Archaeological Evidence from Palestine*; II: *The Archaeological Evidence from Diaspora*; III: *Illustrations*; IV: *The Problem of Method: Symbols from Jewish Cult* (*Bollingen Series* 37, New York 1953).

*GPVJ* — *Gesellschaft pro Vindonissa. Jahresbericht.*

Grose 1973 — David D. Grose, "Roman Glass of the First Century A.D.: A Dated Deposit of Glassware from Cosa, Italy," *AnnAIHV* 6 (1973, published 1974) 31-52.

Grose 1973-76 — David D. Grose, "Découvertes archéologiques récentes de verre: Italie," *BullAIHV* 7 (1973-76) 175-182.

Grose 1978 — David F. Grose, "Ancient Glass," *TMA Mus News* 20 (1978) 67-90.

Grose 1984b — David F. Grose, "The Origins and Early History of Glass," in D. Klein and W. Lloyd eds., *The History of Glass* (London 1984) 8-37.

Grose 1989 — David F. Grose, *The Toledo Museum of Art. Early Ancient Glass: Core-Formed, Rod-Formed, and Cast Vessels and Objects from the Late Bronze Age to the Early Roman Empire, 1600 B. C.–A. D. 50* (New York 1989).

Grose 1991 — David F. Grose, "Early Imperial Roman Cast Glass: The Translucent Coloured and Colourless Fine Wares," in Newby and Painter 1991, 1-18.

Haevernick 1955 — Thea Elisabeth Haevernick and P. Hahn-Weinheimer, "Untersuchungen römischer Fenstergläser," *SaalJb* 14 (1955) 65-73 (reprinted in Thea Elisabeth Haevernick, *Beiträge zur Glasforschung: Die wichtigsten Aufsätze von 1938 bis 1981* [Mainz 1981] 33-38).

Harden 1935 — Donald B. Harden, "Romano-Syrian Glasses with Mould-blown Inscriptions," *JRS* 25 (1935) 163-186.

Harden 1936 — Donald B. Harden, *Roman Glass from Karanis* (*University of Michigan Studies, Humanistic Series* 41, Ann Arbor 1936).

Harden 1944-45 — Donald B. Harden, "Two Tomb Groups of the First Century A.D. from Yahmour, Syria, and a Supplement to the List of Romano-Syrian Glasses with Mould-blown Inscriptions," *Syria* 24 (1944-45) 81-95; 291-292.

Harden 1947 — Donald B. Harden, "The Glass," in C. F. C. Hawkes and M. R. Hull, *Camulodunum First Report on the Excavations at Colchester 1930-1939* (*Reports of the Research Committee of the Society of Antiquaries of London* 14, London 1947) 287-307.

Harden 1958 — Donald B. Harden, "The Glass," in J. du Plat Taylor, "Roman Tombs at Kambi, Vasa," *RDAC* 1940-48 (Nicosia 1958) 46-60.

Harden 1964b — Donald B. Harden, "Syrian Glass from the Earliest Times to the 8th Century A.D.," *BullAIHV* 3 (1964) 19-24.

Harden 1969 — Donald B. Harden, "Ancient Glass, Pt. II: Roman," *ArchJ* 126 (1969) 44-77.

Harden 1971 — Donald B. Harden, "Ancient Glass, Pt. III: Post Roman," *ArchJ* 128 (1971) 78-117, pls. 5-14.

Harden 1982 — Donald B. Harden, "New Light on Mold-blown Glass Sports Cups of the First Century A.D. Bearing Both Chariot Races in *Bigae* and Gladiatorial Combats," *JGS* 24 (1982) 30-43.

Hausmann 1962 — Ulrich Hausmann, "Hellenistische Neger," *AM* 77 (1962) 255-281.

Hayes 1975 — John W. Hayes, *Roman and Pre-Roman Glass in the Royal Ontario Museum* (Toronto 1975).

Helbig[4] — W. Helbig, *Führer durch die öffentlichen Sammlungen klassischer Altertümer in Rom*, 4th edition, supervised by H. Speier (Tübingen 1963-1972).

Helbing 1913 — *Antike und byzantinische Kleinkunst*, auc. cat., Galerie Helbing (Munich October 28-30, 1913).

Hilgers 1969 — Werner Hilgers, "Lateinische Gefässnamen," *BJb* Beiheft 31 (1969).

*Holy Land* 1986 — *Treasures of the Holy Land: Ancient Art from the Israel Museum*, exh. cat., The Metropolitan Museum of Art (New York 1986).

Honey 1946 — W. B. Honey, *Glass: A Handbook for the Study of Glass Vessels of all Periods and Countries and a Guide to the Museum Collection, Victoria and Albert Museum* (London 1946).

Horsfield 1941 — G. Horsfield and A. Horsfield, "Sela-Petra, The Rock of Edom and Nabatene, IV: The Finds," *QDAP* 9 (1941) 105-204.

*IEJ* — *Israel Exploration Journal.*

*IMN* — *Israel Museum News.*

Isings 1957 — Clasina Isings, *Roman Glass from Dated Finds* (*Archaeologica Traiectina* 2, Groningen 1957).

Isings 1964 — Clasina Isings, "A Fourth Century Glass Jar with Applied Masks," *JGS* 6 (1964) 59-63.

Isings 1975 — Clasina Isings, "Sidonian Flask, Ladle," *BABesch* 50 no. 1 (1975) 38-39.

Isings 1976 — Clasina Isings, "Exchanged for Sulphur," in *Festoen Opgedragen aan A. N. Zadoks-Josephus Jitta bij haar zeventigst vernhaardag* (*Scripta Archaeologica Groningana* 6, Groningen 1976) 353-356.

Isings 1979a — Clasina Isings, "Roman Glass from Nijmegen," *AnnAIHV* 8 (1979, published 1981) 101-104.

Isings 1980 — Clasina Isings, "Glass from the *canabae legionis* at Nijmegen," *BerROB* 30 (1980) 281-346.

Israeli 1964 — Yael Israeli, "Sidonian Mold-blown Glass Vessels in the Museum Haaretz," *JGS* 6 (1964) 34-41.

Israeli 1980 — Yael Israeli, "A Rare Glass Vessel from the Dobkin Collection: The Migration of a Design," *IMN* 16 (1980) 111-114.

Israeli 1983 — Yael Israeli, "Ennion in Jerusalem," *JGS* 25 (1983) 65-69.

Israeli 1991 — Yael Israeli, "The Invention of Blowing," in Newby and Painter 1991, 46-55.

*JARCE* — *Journal of the American Research Center in Egypt.*

*JAS* — *Journal of Archaeological Science.*

*JbHamb-KuSamml* — *Jahrbuch der Hamburger Kunstsammlungen.*

*JBL* — *Journal of Biblical Literature.*

*JdI* — *Jahrbuch des Deutschen Archäologischen Instituts.*

*JGS* — *Journal of Glass Studies.*

*JHS* — *Journal of Hellenic Studies.*

*JRA* — *Journal of Roman Archaeology.*

*JRGZM* — *Jahrbuch des Römisch-Germanischen Zentralmuseums, Mainz.*

*JRS* — *Journal of Roman Studies.*

Karageorghis 1964 — Vassos Karageorghis, "Chronique des fouilles à Chypre en 1963," *BCH* 88 (1964) 310-311.

Kern 1954c — J. H. C. Kern, "A Fragmentary Mould-Blown Glass Pyxis from Pompeii," *OMROL* 35 (1954) 33-37.

Khayat 1906 — Azeez Khayat, *Greek Art: Greek Glass, Scarabs, Coins, Greek Vases, Beads, Intaglios, Bronzes, and Other Antiquities*, auc. cat. (New York February 9-10, 1906).

Khayat 1913 — Azeez Khayat, *Catalogue of very rare Egyptian, Phoenician, Greek and Roman Glass, Rhages, Sultanabad and Rakka Potteries, Beautiful Necklaces of Egyptian Beads, Scarabs, Cylinders, Clay Tablets, Coins and Other Rare Egyptian Antiquities, from Recent Excavations in Syria, Egypt and Persia, January 30th and 31st, and February 1st 1913, Fifth Avenue Art Galleries, Silo Building, James P. Silo auctioneer, New York*, auc. cat. (New York 1913).

Khayat 1919 — *A Remarkable Collection of Ancient Works of Art, Iridescent Greek and Roman Glass, Necklaces of Egyptian Beads, Persian Potteries, Egyptian and*

*Roman Bronzes, Tanagra Figures, Greek Vases. Coins, Babylonian Cylinders, Ancient Rings and Intaglios, Greek Marbles, Rhodian Plates, Babylonian Tablets, and a few Chinese Porcelains, Collected by Azeez Khayat of New York, January 31st, February 1st, The Anderson Galleries, New York, Anderson Sales Cat. No. 1397, New York*, auc. cat. (New York 1919).

*Kirchner-Schwarz* 1914 — *Sammlung von Antiken Gläsern, Marmorskulpturen und Bronzen aus dem Besitze von Fr. D. Kirchner-Schwarz, Beirut*, sales catalogue, Galerie Helbing (Munich 1914).

Kisa 1908 — A. Kisa, *Das Glas im Altertum*, 3 vols. (Leipzig 1908, reprinted Rome 1968).

*Kocabaş* 1984 — Yücel Akat, Nezih Fıratli, and Hüseyin Kocabaş, *Catalogue of Glass in the Hüseyin Kocabaş Collection* (Istanbul 1984).

*Kofler-Truniger* 1985 — *Ancient Glass: Formerly the Kofler-Truniger Collection*, auc. cat., Christie, Manson and Woods (London March 5-6, 1985).

*KölnJb* — *Kölner Jahrbuch für Vor- und Frühgeschichte.*

Kotigoroshko 1989 — V. G. Kotigoroshko, "Remeslennoe proizvodsstvo na Dakiickom gerodishtshe 'Malaya Kopanya'," *SovArch* 1982, no. 2, 182-200.

Kunina 1973 — N. Kunina, "Siriiskie vydutye v forme steklyannye sosudy iz nekropolya Pantikapeya" ("Syrian Mould-Blown Glass Vessels from the Necropolis of Panticapaeum"), in K. S. Gorbunova ed., *Pamiatniki antichnogo prikladnogo iskusstva* ("Monuments of Ancient Applied Art") (Leningrad 1973) 101-150; English summary, 153-154.

Kunina and Sorokina 1972 — N. Z. Kunina and N. P. Sorokina, "Glass Unguentaria from the Bosporus," *Trudi Ermitage* 13 (1972) 146-177.

La Baume 1973 — Peter La Baume, *Glas der antike Welt*, I (*Wissenschaftliche Kataloge des Römisch-Germanischen Museums Köln* 1, Cologne 1973).

La Baume and Salomonson 1976 — Peter La Baume and J. W. Salomonson, *Sammlung Karl Löffler* (*Wissenschaftliche Kataloge des Römisch-Germanischen Museums, Köln* 3, Cologne 1976).

Labino 1968 — Dominick Labino, *Visual Art in Glass* (Dubuque, Iowa 1968).

Lamm 1929 — C. J. Lamm, *Mittelalterliche Gläser und Steinschnittarbeiten aus dem Nahen Osten*, 2 vols., text and plates (Berlin 1929-1930).

Landes 1984 — C. Landes, "Note sur un fragment de gobelet en verre céphalomorphe trouvé à Lattes (Hérault)," *RANarb* 17 (1984) 339-343.

Lantier 1929 — R. Lantier, *La Verrerie: Musée des Antiquités Nationaux Saint-Germain-en-Laye* (Paris 1929).

Lehrer 1979 — Gusta Lehrer, *Ennion: A First Century Glassmaker*, exh. cat., Haaretz Museum (Ramat Aviv 1979).

Lehrer-Jacobson 1989 — Gusta Lehrer-Jacobson, *Fakes and Forgeries from Collections in Israel*, exh. cat., Eretz Israel Museum (Tel Aviv 1989).

Liddell-Scott — *A Greek-English Lexicon*, compiled by Henry George Liddell and Robert Scott (Oxford, 9th ed., reprinted 1958).

Liepmann 1982 — Ursula Liepmann, *Glas der Antike*, Kestner-Museum, Hannover (Hanover 1982).

Lierke 1991 — Rosemarie Lierke, "Glass Bowls Made on the Potter's Wheel: A New Approach to Ancient Glass Technology," *GlastechBer* 64 (1991) 310-317.

Lightfoot 1985 — Christopher S. Lightfoot, "A Roman Glass Flask in the Gaziantep Museum," *AnatSt* 35 (1985) 123-129.

Lightfoot 1987 — Christopher S. Lightfoot, "A Group of Early Roman Mold-Blown Flasks from the West," *JGS* 29 (1987) 11-21.

Lightfoot 1989 — Christopher S. Lightfoot, *A Catalogue of Glass Vessels in the Afyon Museum* (*British Institute of Archaeology at Ankara, Monograph* 10, *BAR International Series*, 530, Oxford 1989).

Lightfoot 1990 — Christopher Lightfoot, "Some Types of Roman Cut-Glass Vessels Found in Turkey," in *I. Uluslararasi Anadolu Cam Sanatı Sempozyumu, 26-27 Nisan 1988/1st International Anatolian Glass Symposium, April 26th-27th 1988* (Istanbul 1990) 7-15.

Lightfoot and Arslan 1992 — Christopher Lightfoot and Melih Arslan, *Ancient Glass of Asia Minor: The Yüksel Erimtan Collection* (Ankara 1992).

*LIMC* — *Lexicon Iconographicum Mythologiae Classicae* (Zurich and Munich 1974- ).

van Lith 1978-79 — Sophia M. E. van Lith, "Römisches Glas aus Valkenburg Z. H.," *OMROL* 59-60 (1978-79) 1-150.

van Lith and Randsborg 1985 — Sophia M. E. van Lith and Klaus Randsborg, "Roman Glass in the West: A Social Study," *Berichten van de Rijksdienst voor het Oudheidkundig Bodemonderzoek* 23 (1985) 413-532.

Luzern 1981 — *3000 Jahre Glaskunst von der Antike bis zum Jugendstil*, exh. cat., Kunstmuseum, Luzern (Lucerne 1981).

Maccabruni 1983 — Claudia Maccabruni, *I vetri romani dei musei civici di Pavia. Lettura di una collezione* (Pavia 1983).

Maioli 1974 — M. G. Maioli, "I vetri della necropoli romana de 'le Palazette' nel territorio di Classe," *Felix Ravenna* 107-108 (1974) 15-36.

Mandel 1988 — Ursula Mandel, *Kleinasiatische Reliefkeramik der mittleren Kaiserzeit, Die Oinophorengruppe und Verwandtes* (*Deutsches Archäologisches Institut, Pergamenische Forschungen* 5, Berlin 1988).

*MAntFr* — *Mémoires de la Société nationale des antiquaires de France.*

Mariacher 1970a — Giovanni Mariacher, *Glass from Antiquity to the Renaissance* (London 1970).

Mariacher 1970b — Giovanni Mariacher, *Il Museo Vetrario di Murano* (Milan 1970).

*Masterpieces* 1968 — D. B. Harden, K. S. Painter, R. H. Pinder-Wilson, and Hugh Tait, *Masterpieces of Glass*, exh. cat., The British Museum (London 1968).

Matheson 1980 — Susan B. Matheson, *Ancient Glass in the Yale University Art Gallery* (New Haven, Conn. 1980)

*Maxwell* 1979 — *The Constable-Maxwell Collection of Ancient Glass*, auc. cat., Sotheby Parke Bernet (London June 4-5, 1979).

McClellan 1983 — Murray C. McClellan, "Recent Finds from Greece of First-Century A.D. Mold-blown Glass," *JGS* 25 (1983) 71-78.

McNicoll and Hennessey 1982 — A. W. McNicoll, J. B. Hennessey, A. G. Walmsley, and T. F. Potts, "A Third Season of Excavations at Pella, 1980-81," *ADAJ* 26 (1982) 343-364.

Meconcelli Notarianni 1979 — Gioia Meconcelli Notarianni, *Vetri antichi nelle collezioni del Museo Civico Archeologico di Bologna* (Bologna 1979).

*MEFRA* — *Mélanges de l'École française de Rome, Antiquité.*

Mercando 1974 — L. Mercando, "La necropoli romana di Portorecanati," *NSc* 1974, 142-430.

Mercando 1982 — L. Mercando, "Urbino (Pesaro) — Necropoli romana: tombe al Bivio della Croce dei Missionari e a San Donato," *NSc* 36 (1982) 109-374.

Mikulčić 1974 — Ivan Mikulčić, "Antiko staklo iz Scupia i ostali makedonski nalazi," in *Ancient Glass in Yugoslavia*, 91-210 (English summary).

Montanari 1967 — G. B. Montanari, "Vetri antichi del Museo Nazionale di Ravenna," *Felix Ravenna*, fasc. 44, vol. 95 (1967) 5-39.

Morel et al. 1992 — Jacques Morel, Heidi Amrein, Marie-France Maylan, and Christian Chevalley, "Un atelier de verrier du milieu de 1er siècle apr. J.-C.," *Archäologie der Schweiz* 15 (1992) 2-17.

Morin-Jean 1913 — Morin-Jean, *La Verrerie en Gaule sous l'empire romain* (Paris 1913).

*MüJb* — *Münchner Jahrbuch der bildenden Kunst.*

Mutz 1972 — A. Mutz, *Die Kunst des Metalldrehens bis den Römern* (Basel and Stuttgart 1972).

*MZ* — *Mainzer Zeitschrift.*

Negro Ponzi 1984 — M. Negro Ponzi Mancini, "Glassware from Coche (central Mesopotamia)," in *Arabie Orientale, Mesopotamie et Iran méridional de l'age de fer au début de la période islamique* (Paris 1984) 33-40.

Nesbitt 1871 — A. Nesbitt, *Catalogue of the Collection of Glass Formed by Felix Slade Esq.* (London 1871).

Neuburg 1962 — Frederic Neuburg, *Ancient Glass* (London 1962).

Newby and Painter 1991 — Martine Newby and Kenneth S. Painter eds., *Roman Glass: Two Centuries of Art and Invention* (*The Society of Antiquaries, Occasional Papers* 13, London 1991).

Newton 1980 — Roy Newton, "Recent Views on Ancient Glasses," *Glass Technology* 21 (1980) 173-183.

Newton and Davison 1989 — Roy Newton and Sandra Davison, *Conservation of Glass* (London and Boston 1989).

*NSc* — *Notizie degli Scavi di Antichità.*

Oliver 1977 — Andrew Oliver, Jr., *Silver for the Gods: 800 Years of Greek and Roman Silver*, exh. cat., The Toledo Museum of Art (Toledo, Ohio 1977).

Oliver 1980 — Andrew Oliver, Jr., *Ancient and Islamic Glass in the Carnegie Museum of Natural History, Pittsburgh* (Pittsburgh 1980).

*OMROL* — *Oudheidkundige Mededelingen uit het Rijksmuseum van Oudheden te Leiden.*

*OpArch* — *Opuscula archaeologica.*

Oppenheim et al. 1970 — A. Leo Oppenheim, Robert H. Brill, Dan Barag, and Axel von Saldern, *Glass and Glassmaking in Ancient Mesopotamia* (Corning, N.Y. 1970).

Özet 1993 — Aynur Özet, "An Amphoriskos in the Bodrum Museum of Underwater Archaeology," *JGS* 35 (1993) 142-145.

*PECS* — Richard Stillwell et al. eds., *Princeton Encyclopedia of Classical Sites* (Princeton 1976).

*PEQ* — *Palestine Exploration Quarterly.*

Perdrizet 1904-5 — P. Perdrizet, "Verres de Sidon donnés en prix dans des concours," *MAntFr* ser. 7, 65 (1904-1905) 291ff.

Petru 1974 — "Rimska steklo Slovenije," in *Ancient Glass in Yugoslavia*, 13-34.

Platz-Horster 1976 — Gertrud Platz-Horster, *Antike Gläser*, exh. cat., Antikenmuseum Berlin (Berlin 1976).

Plesničar-Gec 1967 — L. Plesničar-Gec, *La nécropole romaine à Emona* (*Inventaria Archaeologica, Corpus des ensembles archéologiques, Jugoslavija* fasc. 10, Belgrade 1967).

Plesničar-Gec 1972 — L. Plesničar-Gec, *The Northern Necropolis of Emona* (Ljubljana 1972).

Price 1973 — Jennifer Price, "Some Roman Glass from Spain," *AnnAIHV* 6 (1973, publ. 1974) 65-84.

Price 1976a — Jennifer Price, "Glass," in Strong and Brown 1976, 111-125.

Price 1985a — Jennifer Price, "Late Hellenistic and Early Imperial Vessel Glass at Berenice: A Survey of Imported Tableware Found During Excavations at Sidi Khrebish, Benghazi," in Graeme Barker, John Lloyd and Joyce Reynolds eds., *Cyrenaica in Antiquity*, (*Society for Libyan Studies Occasional Papers* I, *BAR International Series* 236, Oxford 1985) chapter 24, 287-296.

Price 1988a — Jennifer Price, "Exhibits at Ballots: An Egyptian Terracotta Group Showing Eros beside a Glass Furnace," *AntJ* 68 (1988) 317-319.

Price 1988b — Jennifer Price, "A Survey of the Hellenistic and Early Roman Vessel Glass Found on the Unexplored Mansion Site at Knossos in Crete," *AnnAIHV* 11 (1988), publ. 1990) 27-36.

Price 1991 — Jennifer Price, "Decorated Mould-Blown Glass Tablewares in the First Century A.D.," in Newby and Painter 1991, 56-75.

*QDAP* — *Quarterly of the Department of Antiquities in Palestine.*

*RA* — *Revue archéologique.*

*RALouvain* — *Revue des archéologues et historiens d'art de Louvain.*

*RANarb* — *Revue archéologique de Narbonnaise.*

*RDAC* — *Report of the Department of Antiquities, Cyprus.*

Reifenberg 1937 — A. Reifenberg, *Denkmäler der jüdischen Antike* (Berlin 1937).

Reifenberg 1950 — A. Reifenberg, *Ancient Hebrew Arts* (New York 1950).

Rendel Harris 1927 — J. Rendel Harris, "Glass Chalices of the First Century," *Bulletin of the John Rylands Library, Manchester* 2 (1927) 286-295.

Renov 1974 — Israel Renov, *Preliminary Report on the Iconography of Jewish Glass Vessels* (Readings in Glass History 4, Jerusalem 1974).

Richter 1914 — G. M. A. Richter, "The Curtis Collection of Ancient Glass," *Art in America* 2 (1914) 72-87.

Richter 1967 — Ilona Richter, *Das Kopfgefäss* (M. A. Thesis Universität zu Köln, Cologne 1967).

Ridder 1909 — A. de Ridder, *Collection de Clerq*, VI: *Les terres cuites et les verres* (Paris 1909).

Riefstahl 1961 — Rudolf M. Riefstahl, "The Tradition of Glass: Ancient and Near Eastern Glass," *TMA Mus News* 4 (1961) 26-46.

Rnjak 1979 — Susan Rnjak, *Anticki teatar na tlu Jugoslavije* (Novi Sad and Beograd 1979).

Roffia 1979b — Elisabetta Roffia, "Osservazioni su alcuni recenti ritrovamenti di vetri in Lomellina," in *Ritrovamenti archeologici nella provincia di Pavia* (Milan 1979) 109-119.

Rohde 1980 — Elisabeth Rohde, "Antikes Glas. Eine Vorlage aus den Beständen der Berliner Antiken-Sammlung," *FuB* 20/21 (1980) 147-187.

Rutgers 1992 — Leonard Victor Rutgers, "Archaeological Evidence for the Interaction of Jews and Non-Jews in Late Antiquity," *AJA* 96 (1992) 101-118.

Rütti 1988 — Beat Rütti, *Vitudurum* IV: *Die Gläser* (*Bericht der Züricher Denkmalpflege, Monographien* 5, *Beiträge zum römischen Oberwinterthur-Vitudurum* 4, Zürich 1988).

Rütti 1991 — Beat Rütti, *Die römischen Gläser aus Augst und Kaiseraugst*, 2 vols. (*Forschungen in Augst* 13, nos. 1 and 2, Augst 1991).

Rütti et al. 1988 — Beat Rütti, Ludwig Berger, Sylvia Fünfschilling, Willem B. Stern, and Norbert Spichtig, "Die Zirkusbecher der Schweiz: Eine Bestandesaufnahme," *GPVJ* 1987 (publ. 1988) 27-104.

*SaalJb* — *Saalburg Jahrbuch. Bericht des Saalburg-Museums.*

Saginashvili 1970 — M. N. Saginashvili, *Stekljannye Sosudy Urbnisskowo Mogilnika ("Glass Vessels of the Urbinskij Burial Ground")* (Tblisi 1970).

von Saldern 1968 — Axel von Saldern, *Ancient Glass in the Museum of Fine Arts, Boston* (Boston 1968).

von Saldern 1974b — Axel von Saldern, *Glassammlung Hentrich, Antike und Islam*, exh. cat., Kunstmuseum (Düsseldorf 1974).

von Saldern 1980a — Axel von Saldern, *Ancient and Byzantine Glass from Sardis* (*Archaeological Exploration of Sardis, Monograph* 6, Cambridge, Mass. 1980).

von Saldern 1980b — Axel von Saldern, *Glass 500 B. C. to A. D. 1900* (*The Hans Cohn Collection, Los Angeles, Calif.*), exh. cat., University of California, Los Angeles (Mainz 1980).

von Saldern et al. 1974 — Axel von Saldern, Birgit Nolte, Peter La Baume, and Thea Elisabeth Haevernick, *Gläser der Antike, Sammlung Erwin Oppenländer*, exh. cat., Museum für Kunst und Gewerbe (Hamburg 1974).

Salomonson 1975 — J. W. Salomonson, "Ein geformtes Glas mit griechischer Inschrift aus einem nordafrikanischen Grabfund," *BABesch* 50, no. 2 (1975) 237-245.

Sambon 1906 — Arthur Sambon, "Les verres antiques," *Le Musée* 3 (1906) 477-524.

Sambon 1928 — Arthur Sambon, *Les verres antiques* (Paris 1928).

Sangiorgi 1914 — G. Sangiorgi, *Collezione di vetri antichi dalle origini al V secolo d. Chr.* (Milan and Rome 1914).

Scatozza Höricht 1986 — Lucia Amalia Scatozza Höricht, *I vetri romani di Ercolano* I (Rome 1986).

Scatozza Höricht 1991 — Lucia Amalia Scatozza Höricht, "Syrian Elements among the Glass from Pompeii and Herculaneum," in Newby and Painter 1991, 76-85.

Schönberger 1980 — Hans Schönberger, *Kastell Oberstimm: Die Grabungen von 1968 bis 1971* (Berlin 1980).

Scholes and Greene 1975 — Samuel R. Scholes, *Modern Glass Practice*, 7th ed., revised and enlarged by Charles H. Greene (Boston 1975).

Simon 1967 — Erika Simon, *Ara Pacis Augustae* (Tübingen 1967).

Simon 1990 — Erika Simon, *Die Götter der Römer* (Munich 1990).

Smith 1957 — R. W. Smith, *Glass from the Ancient World, The R. W. Smith Collection*, exh. cat., The Corning Museum of Glass (Corning, N. Y. 1957).

Sorokina 1962 — N. P. Sorokina, "Steklo ize rasskopok Pantikapeja 1945-1959," ("Glass Vessels from Panticapaeum 1945-1959") *Materiali i Issledovaniya po Arkheologii SSR* 103 (1962) 211-236.

Sorokina 1967 — N. P. Sorokina, "Das antike Glas der Nordschwarzmeerküste," *AnnAIHV* 4 (1967) 67-79.

Sorokina 1968 — N. P. Sorokina, "Un vase de verre sculpté trouvé à Kepoi," *SovArch* 1968, issue 4, 181-189.

Sorokina 1977a — N. P. Sorokina, "Antichnye Stekljannye sosudy iz Paskopole Nekropolja Bosporskowo goroda kepy na Tamanskom Polyostrove," in *Antichny Mir i Archeologia* no. A72 (1977) 115-144.

Sorokina 1977c — N. P. Sorokina, "Facettenschliffgläser der 2[en] und 3[en] Jhd. u. Z. aus dem Schwarzmeer gebiet," *AnnAIHV* 7 (1977) 111-122.

Sorokina 1984 — N. P. Sorokina, "Steklyannaya Posuda," in *Archeologia SSSR Antichnie gosudarstva Severnovo Prichernomorya* (Moscow 1984) 233-236.

Sorokina 1987 — N. P. Sorokina, "Glass Aryballoi (First-Third Centuries A.D.) from the Northern Black Sea Region," *JGS* 29 (1987) 40-48.

*SovArch* — *Sovetskaja archeologija.*

Spartz 1967 — E. Spartz, *Antike Gläser* (Kassel 1967).

Spinazzola 1928 — V. Spinazzola, *Le arti decorative in Pompei e nel Museo Nazionale di Napoli* (Milan and Rome 1928).

Stern 1977a — E. Marianne Stern, *Ancient Glass at the Fondation Custodia (Collection Frits Lugt) Paris* (*Archaeologica Traiectina* 12, Groningen 1977).

Stern 1984 — E. Marianne Stern, "Antikes Glas in der Südtürkei," *GlastechBer* 57.5 (1984) 132-139.

Stern 1985a — E. Marianne Stern, "Ancient and Medieval Glass from the Necropolis Church at Anemurium," *AnnAIHV* 9 (1983, publ. 1985) 35-64.

Stern 1989a — E. Marianne Stern, "The Production of Glass Vessels in Roman Cilicia," *KölnJb* 22 (1989) 121-128.

Stern 1989b — E. Marianne Stern, "Glass Vessels Exhibited in the Bölge Museum-Adana," *Belleten* 153/207-208 (1989) 583-593.

Stern 1991a — E. Marianne Stern, "The Workshop of the Floating Handles," in M. Gnade ed., *Stips Votiva. Papers Presented to C. M. Stibbe* (Amsterdam 1991) 199-206.

Stern 1991d — E. Marianne Stern, "Glass Working before Glass Blowing," *AnnAIHV* 12 (1991), publ. 1993) 21-31.

| | |
|---|---|
| Stern 1992a | E. Marianne Stern, "A Fourth Century Factory for Gathering and Blowing Chunks of Glass?" (review of Weinberg 1988) *JRA* 5 (1992) 491–494. |
| Stern 1992b | E. Marianne Stern, "A Glass Head featuring Livia as Hera?," *Kotinos. Festschrift für Erika Simon* (Mainz 1992) 394-399. |
| Stern 1993 | E. Marianne Stern, "The Glass from Heis," in Johan Desanges, E. Marianne Stern, and Pascale Ballet, *Sur les routes antiques de l'Azanie et de l'Inde. Les Fonds révoil du Musée de l'Homme (Hëis et Damo, en Somalie) (Mémoires de l'Académie des Inscriptions et Belles-Lettres* n.s. 13, Paris 1993) 21-61. |
| Stern 1994 | E. Marianne Stern and Birgit Schlick-Nolte, *Early Glass of the Ancient World. The Ernesto Wolf Collection* (Stuttgart 1994). |
| Sternini 1989 | Mara Sternini, *Una manifattura vetraria di V secolo a Roma* (*Quaderni del Dipartimento di Archeologia e Storia delle Arti, Sezione Archeologica — Università di Siena*, Florence 1989). |
| Strong 1966 | D. E. Strong, *Greek and Roman Gold and Silver Plate* (Ithaca, N.Y. 1966). |
| Strong and Brown 1976 | Donald Strong and David Brown eds., *Roman Crafts* (London 1976). |
| Sunkowsky 1956 | R. Sunkowsky, *Antike Gläser in Carnuntum und Wien* (Vienna 1956). |
| Taborelli 1982a | Luigi Taborelli, "Vasi di vetro con bollo monetale," *Opus: Rivista Internazionale per la Storia Economica e Sociale dell'Antichità* 1 (1982) 315-340. |
| Taborelli 1983 | Luigi Taborelli, "Nuovi esemplari di bolli già noti su contenitori vitrei dall'area centro-italica (Regg. IV, V, VI)," *PICVS, Studi e Ricerche sulle Marche nell'Antichità* 3 (1983) 86-132. |
| Taborelli 1986-88 | Luigi Taborelli, review of Scatozza Höricht 1986, *ArchCl* 38-40 (1986-1988) 136-149. |
| Tait 1991 | Hugh Tait ed., *Glass: 5,000 Years* (New York 1991). |
| Tamassia 1965 | Anna Maria Tamassia, "La mostra dei vetri romani in Lombardia: osservazioni e problemi," *Arte Lombarda* 10, no. 1 (1965) 3-14. |
| TMA, *Art in Glass* 1969 | The Toledo Museum of Art, *Art in Glass: A Guide to the Glass Collections* (Toledo 1969). |
| TMA, *Guide* 1966 | The Toledo Museum of Art, *A Guide to the Collections* (Toledo 1966). |
| TMA, *Roman and Early Byzantine Glass* | David F. Grose, E. Marianne Stern, et al., *The Toledo Museum of Art. Roman and Early Byzantine Glass* (forthcoming). |
| TMA *Mus News* | *The Toledo Museum of Art, Museum News.* |
| Trowbridge 1930 | M. L. Trowbridge, *Philological Studies in Ancient Glass* (*University of Illinois Studies in Language and Literature* 13, Urbana 1930). |
| *TrZ* | *Trier Zeitschrift für Geschichte und Kunst des Trierer Landes und seiner Nachbargebiete.* |
| Turner 1956 | W. E. S. Turner, "Studies in Ancient Glasses and Glass Making Processes. Part V: Raw Materials and Melting Processes," *Journal of the Society of Glass Technology* 40, no. 194 (1956) 277T-300T. |
| Vessberg 1952 | Olaf Vessberg, "Roman Glass in Cyprus," *OpArch* (1952) 109-165. |
| Velickovic 1974 | Milivoje Velickovic, "Tipologija i hronologija rimskog stakla iz Budva u zbirci Narodnog muzeja u Beogradu" ("Typologie et chronologie de la verrerie romaine de Budva (Buthoe) dans la collection du Musée National de Belgrade") in *Ancient Glass in Yugoslavia*, 165-174. |
| Vose 1984 | Ruth Hurst Vose, "From the Dark Ages to the Fall of Constantine," in Dan Klein and Ward Lloyd eds., *The History of Glass* (London 1984) 40-65. |
| Weinberg 1959 | Gladys Davidson Weinberg, "Glass Manufacture in Ancient Crete: A Preliminary Study," *JGS* 1 (1959) 11-32. |
| Weinberg 1972 | Gladys Davidson Weinberg, "Mold-Blown Beakers with Mythological Scenes," *JGS* 14 (1972) 26-47. |
| Weinberg 1988 | Gladys Davidson Weinberg ed., *Excavations at Jalame: Site of a Glass Factory in Late Roman Palestine* (Columbia, Mo. 1988). |
| Weinberg 1992 | Gladys Davidson Weinberg, *Glass Vessels in Ancient Greece: Their History Illustrated from the Collection of the National Archaeological Museum, Athens* (*Publications of the Archaeologikon Deltion* no. 47, Athens 1992). |
| Welker 1985 | Edith Welker, *Die römischen Gläser von Nida-Hedderheim* II (*Schriften des Frankfurter Museums für Vor- und Frühgeschichte* 3, Frankfurt 1985). |
| Wight 1988 | Karol Wight, "Mythological Beakers: Questions of Provenance and Production," *AnnAIHV* (1988) 71-76. |
| Wilhelm 1969 | Eugénie Wilhelm, *Verrerie de l'époque romaine: Musée d'histoire et d'art, Luxemburg* (Luxemburg 1969). |
| Zahn 1921 | R. Zahn, *Galerie Bachstitz, 's Gravenhage* II (Berlin 1921). |
| Zampieri 1984 | G. Zampieri, *Le divisioni agrarie romane nel territorio patavino*, exh. cat., Civici Musei di Padova (Padua 1984). |
| Zoudhi 1960 | Bachir Zoudhi, "La verrerie antique et ses chefs-d'oeuvre au musée national de Damas," *AAS* 10 (1960) 103-134. |
| Zouhdi 1964 | Bachir Zouhdi, "Les verres conservés au département des antiquités syriennes des époques grecque, romaine, et byzantine du musée national de Damas," *BullAIHV* 3 (1964) 41-54. |


Note: The glass in the Murano Museum has recently been published, but it has not been possible to insert the appropriate references: Giovanna Luisa Ravaganan, *Vetri antichi del Museo Vetrario di Murano. Collezioni dello Stato*, Corpus delle Collezioni archeologiche del vetro nel Veneto (Venice 1994).

# CONCORDANCES

## Catalogue Number by Accession Number

| ACC. NO. | CAT. NO. | ACC. NO. | CAT. NO. | ACC. NO. | CAT. NO. | ACC. NO. | CAT. NO. | ACC. NO. | CAT. NO. | ACC. NO. | CAT. NO. |
|---|---|---|---|---|---|---|---|---|---|---|---|
| 1923.410 | 4 | 1923.463 | 44 | 1923.513 | 167 | 1923.581 | 98 | 1923.887 | 112 | 1923.1351 | 183 |
| 1923.411 | 2 | 1923.464 | 43 | 1923.514 | 156 | 1923.582 | 96 | 1923.888 | 118 | 1923.1352 | 189 |
| 1923.412 | 3 | 1923.465 | 20 | 1923.515 | 149 | 1923.583 | 88 | 1923.889 | 117 | 1923.1353 | 193 |
| 1923.413 | 6 | 1923.466 | 21 | 1923.516 | 158 | 1923.585 | 85 | 1923.893 | 52 | 1923.1354 | 177 |
| 1923.414 | 147 | 1923.467 | 16 | 1923.517 | 143 | 1923.586 | 94 | 1923.912 | 11 | 1923.1355 | 184 |
| 1923.415 | 62 | 1923.468 | 18 | 1923.518 | 151 | 1923.588 | 99 | 1923.913 | 12 | 1923.1356 | 186 |
| 1923.416 | 114 | 1923.470 | 17 | 1923.519 | 168 | 1923.589 | 95 | 1923.914 | 54 | 1923.1357 | 176 |
| 1923.424 | 14 | 1923.471 | 19 | 1923.520 | 162 | 1923.590 | 87 | 1923.1075 | 126 | 1923.1358 | 173 |
| 1923.425 | 13 | 1923.472 | 15 | 1923.521 | 150 | 1923.591 | 92 | 1923.1076 | 125 | 1923.1359 | 171 |
| 1923.428 | 160 | 1923.473 | 31 | 1923.522 | 163 | 1923.592 | 102 | 1923.1077 | 120 | 1923.1360 | 174 |
| 1923.429 | 38 | 1923.474 | 30 | 1923.523 | 152 | 1923.593 | 89 | 1923.1078 | 123 | 1923.1361 | 169 |
| 1923.430 | 34 | 1923.475 | 22 | 1923.524 | 144 | 1923.594 | 108 | 1923.1079 | 128 | 1923.1362 | 175 |
| 1923.431 | 41 | 1923.476 | 23 | 1923.525 | 145 | 1923.595 | 103 | 1923.1080 | 122 | 1923.1363 | 172 |
| 1923.432 | 42 | 1923.477 | 28 | 1923.526 | 154 | 1923.596 | 63 | 1923.1082 | 127 | 1923.1364 | 179 |
| 1923.433 | 36 | 1923.478 | 24 | 1923.527 | 165 | 1923.677 | 146 | 1923.1083 | 124 | 1923.1365 | 181 |
| 1923.434 | 37 | 1923.479 | 29 | 1923.528 | 166 | 1923.678 | 111 | 1923.1084 | 115 | 1923.1916 | 192 |
| 1923.435 | 39 | 1923.480 | 27 | 1923.529 | 35 | 1923.679 | 55 | 1923.1085 | 116 | 1923.2413 | 91 |
| 1923.436 | 40 | 1923.481 | 32 | 1923.530 | 33 | 1923.680 | 56 | 1923.1224 | 157 | 1923.2418 | 97 |
| 1923.437 | 45 | 1923.482 | 76 | 1923.531 | 26 | 1923.681 | 53 | 1923.1225 | 121 | 1923.570 | 107 |
| 1923.438 | 77 | 1923.483 | 78 | 1923.532 | 60 | 1923.682 | 69 | 1923.1332 | 134 | 1930.5 | 5 |
| 1923.439 | 48 | 1923.484 | 71 | 1923.533 | 61 | 1923.683 | 70 | 1923.1333 | 135 | 1936.12 | 170 |
| 1923.440 | 75 | 1923.485 | 73 | 1923.536 | 141 | 1923.684 | 50 | 1923.1334 | 129 | 1948.13 | 190 |
| 1923.441 | 113 | 1923.486 | 74 | 1923.537 | 140 | 1923.685 | 51 | 1923.1335 | 130 | 1948.17 | 191 |
| 1923.442 | 80 | 1923.487 | 72 | 1923.571 | 104 | 1923.686 | 66 | 1923.1336 | 188 | 1951.373 | 119 |
| 1923.443 | 7 | 1923.488 | 81 | 1923.572 | 90 | 1923.687 | 67 | 1923.1337 | 187 | 1951.375 | 59 |
| 1923.444 | 49 | 1923.489 | 83 | 1923.573 | 106 | 1923.688 | 65 | 1923.1338 | 136 | 1953.138 | 182 |
| 1923.457 | 139 | 1923.490 | 8 | 1923.575 | 86 | 1923.689 | 68 | 1923.1339 | 133 | 1956.3 | 82 |
| 1923.458 | 142 | 1923.491 | 10 | 1923.576 | 100 | 1923.690 | 64 | 1923.1340 | 131 | 1967.5 | 79 |
| 1923.459 | 153 | 1923.492 | 9 | 1923.577 | 93 | 1923.741 | 159 | 1923.1341 | 132 | 1967.6 | 1 |
| 1923.460 | 155 | 1923.493 | 58 | 1923.578 | 84 | 1923.742 | 161 | 1923.1343 | 185 | 1967.8 | 138 |
| 1923.461 | 47 | 1923.494 | 57 | 1923.579 | 105 | 1923.885 | 110 | 1923.1349 | 178 | 1977.26 | 25 |
| 1923.462 | 46 | 1923.512 | 164 | 1923.580 | 101 | 1923.886 | 109 | 1923.1350 | 180 | 1979.53 | 148 |
| | | | | | | | | | | 1981.93 | 137 |

## Accession Number by Catalogue Number

| CAT. NO. | ACC. NO. | CAT. NO. | ACC. NO. | CAT. NO. | ACC. NO. | CAT. NO. | ACC. NO. | CAT. NO. | ACC. NO. | CAT. NO. | ACC. NO. |
|---|---|---|---|---|---|---|---|---|---|---|---|
| 1 | 1967.6 | 33 | 1923.530 | 65 | 1923.688 | 97 | 1923.2418 | 129 | 1923.1334 | 161 | 1923.742 |
| 2 | 1923.411 | 34 | 1923.430 | 66 | 1923.686 | 98 | 1923.581 | 130 | 1923.1335 | 162 | 1923.520 |
| 3 | 1923.412 | 35 | 1923.529 | 67 | 1923.687 | 99 | 1923.588 | 131 | 1923.1340 | 163 | 1923.522 |
| 4 | 1923.410 | 36 | 1923.433 | 68 | 1923.689 | 100 | 1923.576 | 132 | 1923.1341 | 164 | 1923.512 |
| 5 | 1930.5 | 37 | 1923.434 | 69 | 1923.682 | 101 | 1923.580 | 133 | 1923.1339 | 165 | 1923.527 |
| 6 | 1923.413 | 38 | 1923.429 | 70 | 1923.683 | 102 | 1923.592 | 134 | 1923.1332 | 166 | 1923.528 |
| 7 | 1923.443 | 39 | 1923.435 | 71 | 1923.484 | 103 | 1923.595 | 135 | 1923.1333 | 167 | 1923.513 |
| 8 | 1923.490 | 40 | 1923.436 | 72 | 1923.487 | 104 | 1923.571 | 136 | 1923.1338 | 168 | 1923.519 |
| 9 | 1923.492 | 41 | 1923.431 | 73 | 1923.485 | 105 | 1923.579 | 137 | 1981.93 | 169 | 1923.1361 |
| 10 | 1923.491 | 42 | 1923.432 | 74 | 1923.486 | 106 | 1923.573 | 138 | 1967.8 | 170 | 1936.12 |
| 11 | 1923.912 | 43 | 1923.464 | 75 | 1923.440 | 107 | 1923.570 | 139 | 1923.457 | 171 | 1923.1359 |
| 12 | 1923.913 | 44 | 1923.463 | 76 | 1923.482 | 108 | 1923.594 | 140 | 1923.537 | 172 | 1923.1363 |
| 13 | 1923.425 | 45 | 1923.437 | 77 | 1923.438 | 109 | 1923.886 | 141 | 1923.536 | 173 | 1923.1358 |
| 14 | 1923.424 | 46 | 1923.462 | 78 | 1923.483 | 110 | 1923.885 | 142 | 1923.458 | 174 | 1923.1360 |
| 15 | 1923.472 | 47 | 1923.461 | 79 | 1967.5 | 111 | 1923.678 | 143 | 1923.517 | 175 | 1923.1362 |
| 16 | 1923.467 | 48 | 1923.439 | 80 | 1923.442 | 112 | 1923.887 | 144 | 1923.524 | 176 | 1923.1357 |
| 17 | 1923.470 | 49 | 1923.444 | 81 | 1923.488 | 113 | 1923.441 | 145 | 1923.525 | 177 | 1923.1354 |
| 18 | 1923.468 | 50 | 1923.684 | 82 | 1956.3 | 114 | 1923.416 | 146 | 1923.677 | 178 | 1923.1349 |
| 19 | 1923.471 | 51 | 1923.685 | 83 | 1923.489 | 115 | 1923.1084 | 147 | 1923.414 | 179 | 1923.1364 |
| 20 | 1923.465 | 52 | 1923.893 | 84 | 1923.578 | 116 | 1923.1085 | 148 | 1979.53 | 180 | 1923.1350 |
| 21 | 1923.466 | 53 | 1923.681 | 85 | 1923.585 | 117 | 1923.889 | 149 | 1923.515 | 181 | 1923.1365 |
| 22 | 1923.475 | 54 | 1923.914 | 86 | 1923.575 | 118 | 1923.888 | 150 | 1923.521 | 182 | 1953.138 |
| 23 | 1923.476 | 55 | 1923.679 | 87 | 1923.590 | 119 | 1951.373 | 151 | 1923.518 | 183 | 1923.1351 |
| 24 | 1923.478 | 56 | 1923.680 | 88 | 1923.583 | 120 | 1923.1077 | 152 | 1923.523 | 184 | 1923.1355 |
| 25 | 1977.26 | 57 | 1923.494 | 89 | 1923.593 | 121 | 1923.1225 | 153 | 1923.459 | 185 | 1923.1343 |
| 26 | 1923.531 | 58 | 1923.493 | 90 | 1923.572 | 122 | 1923.1080 | 154 | 1923.526 | 186 | 1923.1356 |
| 27 | 1923.480 | 59 | 1951.375 | 91 | 1923.2413 | 123 | 1923.1078 | 155 | 1923.460 | 187 | 1923.1337 |
| 28 | 1923.477 | 60 | 1923.532 | 92 | 1923.591 | 124 | 1923.1083 | 156 | 1923.514 | 188 | 1923.1336 |
| 29 | 1923.479 | 61 | 1923.533 | 93 | 1923.577 | 125 | 1923.1076 | 157 | 1923.1224 | 189 | 1923.1352 |
| 30 | 1923.474 | 62 | 1923.415 | 94 | 1923.586 | 126 | 1923.1075 | 158 | 1923.516 | 190 | 1948.13 |
| 31 | 1923.473 | 63 | 1923.596 | 95 | 1923.589 | 127 | 1923.1082 | 159 | 1923.741 | 191 | 1948.17 |
| 32 | 1923.481 | 64 | 1923.690 | 96 | 1923.582 | 128 | 1923.1079 | 160 | 1923.428 | 192 | 1923.1916 |
| | | | | | | | | | | 193 | 1923.1353 |

# GENERAL INDEX

Museums and other institutions are listed by city; private collections are listed by name. Site names, ancient place names, and modern place names are cross-referenced and further identified with modern country names. When the same word has different meanings depending on whether it is set lowercased or uppercased, e.g. generation and Generation, the term is indexed first as lowercased. For more information about specific technologies, workshops, and types of vessels, please consult the extended Contents on pages 58.

References include (in this order) figure captions, notes (usually to include references to comparanda excavated at known sites), and catalogue entries. Citations are differentiated as follows: 123 indicates a page number; 123 fig. 67 indicates a comparison illustration on the page; 123 n. 45 indicates a reference *only* in a footnote on the page; 123 cat. no. 89 indicates an object in the Toledo Museum's collection described in a catalogue entry that begins on the page.

# MAP

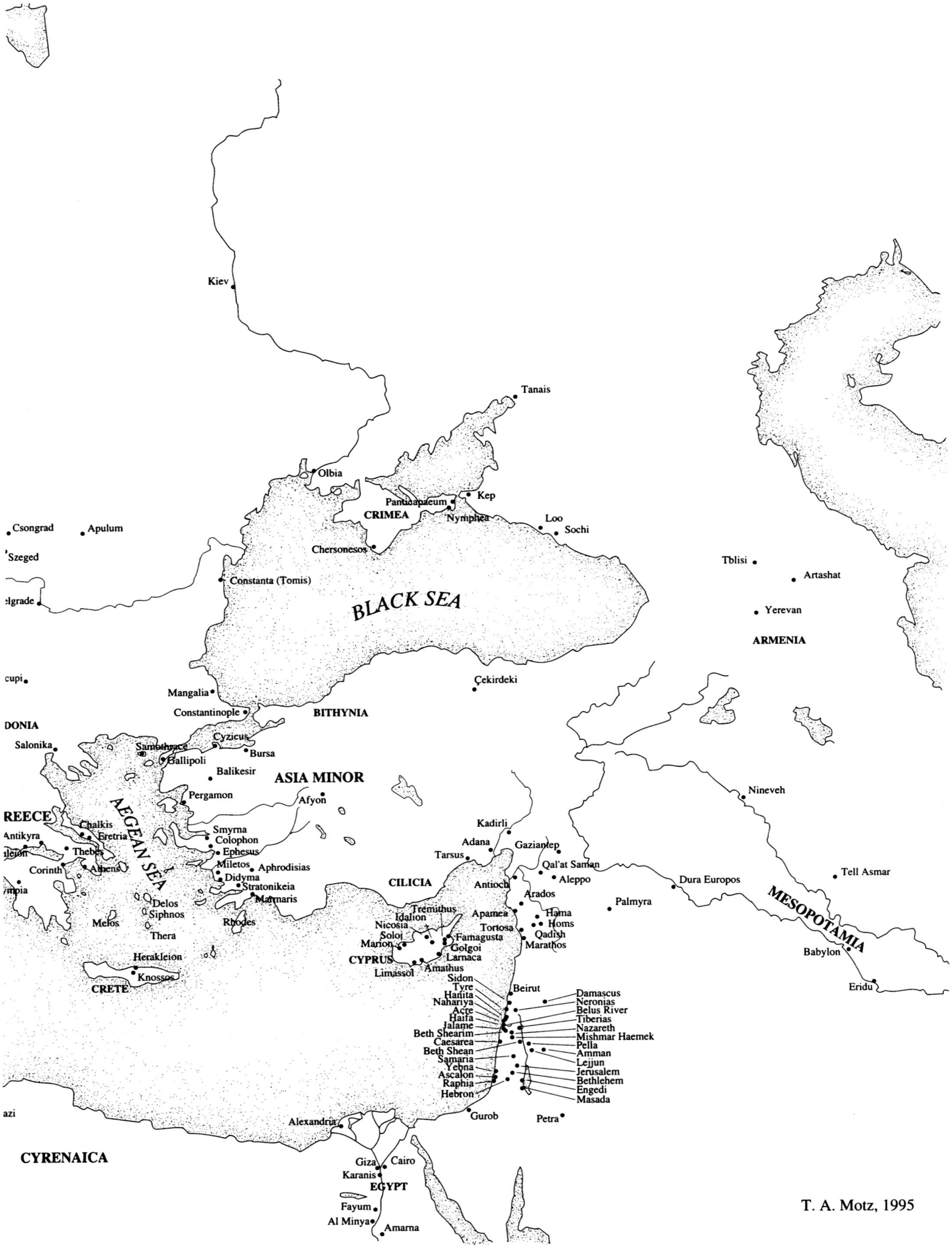

BLACK SEA
AEGEAN SEA
ASIA MINOR
MESOPOTAMIA
ARMENIA
CRIMEA
BITHYNIA
CILICIA
CYPRUS
CRETE
EGYPT
CYRENAICA
Kiev
Tanais
Olbia
Panticapaeum
Kep
Nymphea
Loo
Sochi
Chersonesos
Csongrad
Apulum
Szeged
Constanta (Tomis)
Tblisi
Artashat
Yerevan
Çekirdeki
Mangalia
Constantinople
Salonika
Samothrace
Cyzicus
Gallipoli
Bursa
Balikesir
Pergamon
Afyon
Chalkis
Eretria
Antikyra
Thebes
Athens
Corinth
Smyrna
Colophon
Ephesus
Miletos
Aphrodisias
Didyma
Stratonikeia
Marmaris
Delos
Siphnos
Melos
Thera
Rhodes
Herakleion
Knossos
Kadirli
Adana
Tarsus
Gaziantep
Qal'at Saman
Aleppo
Antioch
Arados
Apamea
Hama
Homs
Tortosa
Qadish
Marathos
Palmyra
Nineveh
Dura Europos
Tell Asmar
Babylon
Eridu
Tremithus
Idalion
Nicosia
Soloi
Marion
Famagusta
Golgoi
Larnaca
Amathus
Limassol
Sidon
Tyre
Hanita
Nahariya
Acre
Haifa
Jalame
Beth Shearim
Caesarea
Beth Shean
Samaria
Yebna
Ascalon
Raphia
Hebron
Beirut
Damascus
Neronias
Belus River
Tiberias
Nazareth
Mishmar Haemek
Pella
Amman
Lejjun
Jerusalem
Bethlehem
Engedi
Masada
Gurob
Petra
Alexandria
Giza
Cairo
Karanis
Fayum
Al Minya
Amarna


T. A. Motz, 1995

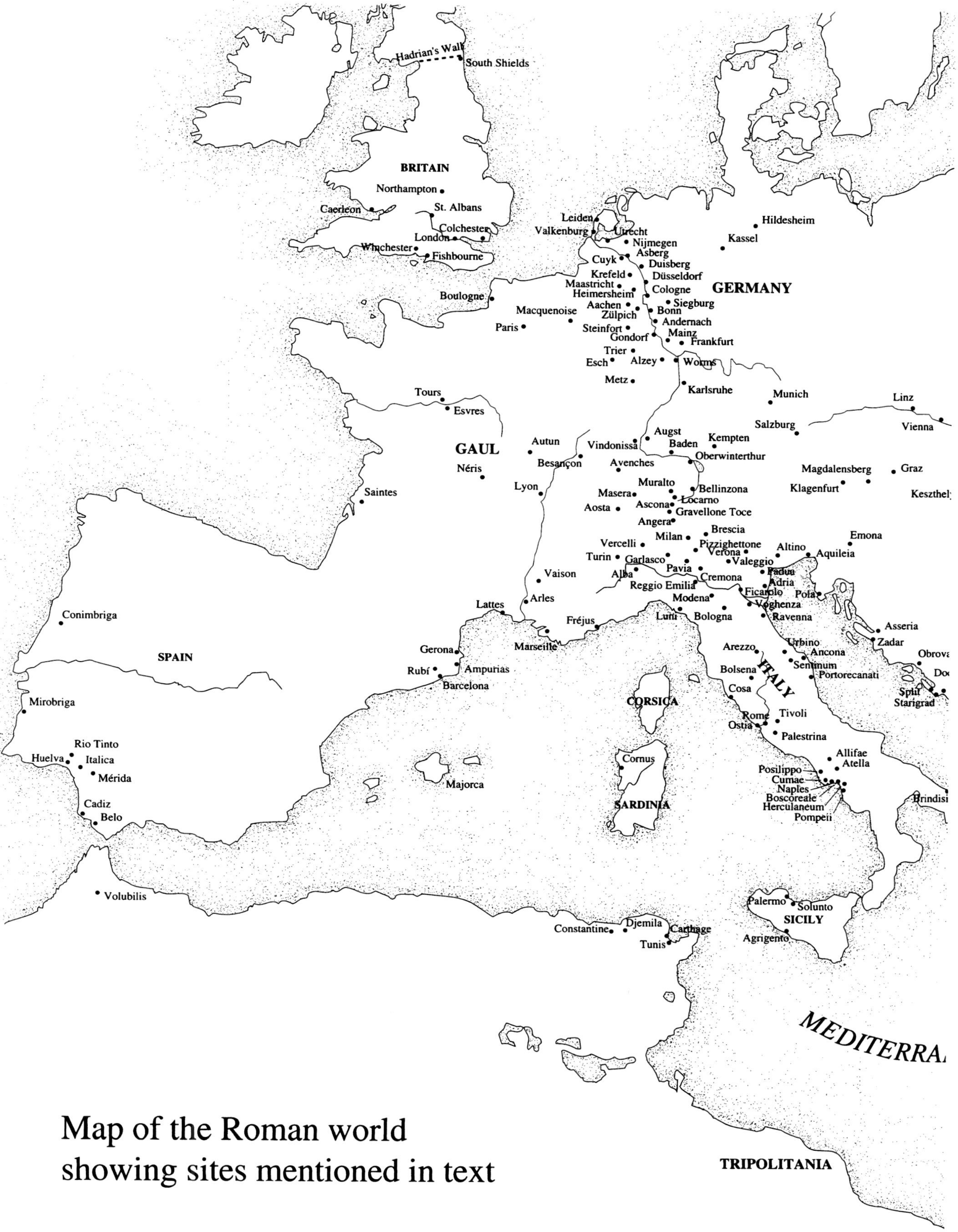


Map of the Roman world
showing sites mentioned in text

Printed in November 2017
on behalf of «L'ERMA» di BRETSCHNEIDER
c/o Services4Media srl - Bari